The Complete Book
of Spaceflight

From **A**pollo 1
to **Z**ero Gravity

David Darling

John Wiley & Sons, Inc.

For general information about our other products and services, please contact
our Customer Care Department within the United States at (800) 762-2974,
outside the United States at (317) 572-3993 or fax (317) 572-4002.

Wiley also publishes its books in a variety of electronic formats. Some content
that appears in print may not be available in electronic books. For more
information about Wiley products, visit our web site at www.wiley.com.

ISBN 0-471-05649-9

Printed in the United States of America

10 9 8 7 6 5 4 3 2 1

Contents

Acknowledgments

A book of this size and scope isn't a one-man enterprise. Dozens of individuals at space agencies, government laboratories, military bases, aerospace companies, and universities generously provided information and illustrations. At John Wiley, I'm particulary grateful to my editor, Jeff Golick, and to Marcia Samuels, senior managing editor, for their excellent suggestions and attention to detail. Any mistakes and inaccuracies that remain are my responsibility alone. As always, my thanks go to my very special agent, Patricia van der Leun, for finding the book a home and providing support along the way. Finally and foremost, my love and gratitude go to my family—my parents, my wife, Jill, and my now-grownup children, Lori-An and Jeff—for making it all possible.

Introduction

It is astonishing to think that there are people alive today from the time when man first flew in an engine-powered, heavier-than-air plane. In the past century, we have learned not only to fly, but to fly to the Moon, to Mars, and to the very outskirts of the Solar System. Look up at the right time and place on a clear night and you can see the International Space Station glide across the sky and know that not all of us are now confined to Earth: always there are a handful of us on the near edge of this new and final frontier of space.

Our first steps beyond our home planet have been hesitant and hazardous. There are some who say, "Why bother?" Why expend effort and money, and risk lives, when there are so many problems to be resolved back on this world? In the end, the answer is simple. We can point to the enormous value of Earth resources satellites in monitoring the environment, or to the benefits of spacecraft that help us communicate among continents or predict the weather or gaze with clear sight across the light-years. We can extol the virtues of mining the Moon or the asteroid belt, or learning about our origins in cometary dust, or the things that can be made or gleaned from a laboratory in zero-*g*. But these reasons are not at the core of why we go—why we must go—on a voyage that will ultimately take us to the stars. Our reason for spaceflight is just this: we are human, and to be human is to be inquisitive. At heart, we are explorers with a universe of billions of new worlds before us.

This book is intended as a companion to the human journey into space. Of course, it has many facts and figures—and acronyms!—as all books on this subject do. But beyond the technical details of rockets and orbits, it tries to capture something of the drama of the quest, the human thread—in a word, the *culture* of space exploration. I hope that many readers will use it to wander from reference to reference and so create their own unique paths through this most unique of adventures. Enjoy the ride!

How to Use This Book

Entries range from simple definitions to lengthy articles on subjects of central importance or unusual interest, and are extensively cross-referenced. Terms that are in **bold** type have their own entries. Numbers that appear as superscripts in the text are references to books, journal articles, and so on, listed alphabetically by author at the back of the book. A list of web sites on subjects dealt with in the text is also provided.

Entries are arranged alphabetically according to the first word of the entry name. So, for example, "anti-*g* suit" precedes "antigravity." Where names are also known by their acronyms or abbreviations, as happens frequently in the language of spaceflight, the definition appears under the form most commonly used. For example, the headwords "NASA" and "TIROS" are preferred to "National Aeronautics and Space Administration" and "Television Infrared Observations System." On the other hand, "Hubble Space Telescope" and "Goddard Space Flight Center" are preferred to "HST" and "GSFC." The alternative form is always given in parentheses afterward. In addition, the Acronyms and Abbreviations section in the back of the book lists all of the alternative forms for easy reference.

Metric units are used throughout, unless it is more appropriate, for historical reasons, to do otherwise. See the "Units" section below for conversion factors.

Exponential Notation

In the interest of brevity, exponential notation is used in this book to represent large and small numbers. For example, 300,000,000 is written as 3×10^8, the power of 10 indicating how many places the decimal point has been moved to the left from the original number (or, more simply, the number of zeroes). Small numbers have negative exponents, indicating how many places the point has been shifted to the left. For example, 0.000049 is written as 4.9×10^{-5}.

Orbits

Orbits of satellites are given in the form:

$$\text{perigee} \times \text{apogee} \times \text{inclination}$$

For example, the Japanese Ohzora satellite is listed as having an orbit of 247×331 km $\times 75°$. This means that the low and high points of the orbit were 247 km and 331 km, respectively, above Earth's surface, and that the orbit was tilted by 75° with respect to Earth's equator.

Units

Distance

1 kilometer (km) = 0.62 mile

1 meter (m) = 3.28 feet (ft) = 39.37 inches (in.)

1 centimeter (cm) = 0.39 in.

1 km = 1,000 m

1 m = 100 cm = 1,000 millimeters (mm)

1 mm = 10^3 microns (μm) = 10^6 nanometers (nm)

1 astronomical unit (AU) = 1.50×10^8 km

1 light-year = 63,240 AU = 9.46×10^{12} km

Area

1 hectare = 2.47 acres

1 square meter (m^2) = 10.76 square feet (ft^2)

Volume

1 cubic meter (m^3) = 35.31 cubic feet (ft^3)

Speed

1 km/s = 2,240 mph

Acceleration

1g (one-gee) = 9.81 m/s^2 = 32.19 ft/s^2

Mass

1 kilogram (kg) = 2.21 pounds (lb)

1 kg = 1,000 grams (g)

1 g = 10^3 milligrams (mg) = 10^9 nanograms (ng)

1 metric ton = 1,000 kg = 2,205 lb = 0.98 long ton

Note: In this book, *tons* refers to metric tons.

Energy

1 joule (J) = 9.48×10^{-4} British thermal unit (Btu)

1 electron-volt (eV) = 1.60×10^{-19} J

1 GeV = 10^3 MeV = 10^6 keV = 10^9 eV

Note: Electron-volts are convenient units for measuring the energies of particles and electromagnetic radiation. In the case of electromagnetic radiation, it is customary to measure longer-wavelength types in terms of their wavelength (in units of cm, μm, etc.) and shorter-wavelength types, especially X-rays and gamma-rays in terms of their energy (in units of keV, MeV, etc.). The wavelength associated with electromagnetic waves of energy 1 keV is 0.124 nm.

Force

1 newton (N) = 0.22 pounds-force (lbf) = 0.102 kilograms-force (kgf)

1 kilonewton (kN) = 1,000 N

Power

1 watt (W) = 0.74 ft-lbf/s = 0.0013 horsepower (hp)

1 kilowatt (kW) = 1,000 W

Temperature

C = ⅝(F − 32)

F = ⅝C + 32

A

"A" series of German rockets

A family of liquid-propellant rockets built by Nazi Germany immediately before and during World War II. With the "A" (Aggregate) rockets came technology that could be used either to bomb cities or to begin the exploration of space. Key to this development was Wernher **von Braun** and his team of scientists and engineers. The series began with the small A-1, which, in common with all of the "A" rockets, used alcohol as a fuel and liquid oxygen as an oxidizer. Built and tested mostly on the ground at **Kummersdorf**, it enabled various design problems to be identified. A reconfigured version, known as the A-2, made two successful flights in December 1934 from the North Sea island of Borkum, reaching a height of about 2 km. The development effort then shifted to **Peenemünde**. In 1937, the new A-3 rocket was launched from an island in the Baltic Sea. Measuring 7.6 m in length and weighing 748 kg, it was powered by an engine that produced 14,700 newtons (N) of thrust. Three flights were made, none completely successful because the A-3's gyroscopic control system was too weak to give adequate steering. Consequently, a new test rocket was developed with the designation A-5–the name A-4 having been reserved for a future military rocket of which the A-5 was a subscale version. The A-5 was built with most of the components from the A-3 but with a larger diameter airframe, a tapered **boat-tail**, and a new steering control system that was incorporated into larger, redesigned fins. Measuring 7.6 m in length and 0.76 m in diameter, it used the same 14,700-N motor as the A-3 and was test-flown from the island of Greifswalder Oie off the Baltic coast. The first flights, conducted in 1938 without gyroscopic control, came close to the speed of sound and reached an altitude of around 8 km. The new guidance system was installed in 1939, enabling the A-5 to maneuver into a ballistic arc, and by the end of its testing the rocket had been launched 25 times, reaching altitudes of nearly 13.5 km. The stage was set for the arrival of the remarkable A-4–better known as the V-2 (see **"V" weapons**).[231]

A.T. (Aerial Target)

Along with the American **Kettering Bug**, one of the earliest experimental **guided missiles**. This British project, begun in 1914 under the direction of Archibald M. Low, was deliberately misnamed so that enemy spies would think the vehicles were simply drones flown to test the effectiveness of antiaircraft weapons. In fact, A.T. concept vehicles were intended to test the feasibility of using radio signals to guide a flying bomb to its target. Radio guidance equipment was developed and installed on small monoplanes, each powered by a 35-horsepower Granville Bradshaw engine. Two A.T. test flights were made in March 1917 at the Royal Flying Corps training school field at Upavon. Although both vehicles crashed due to engine failure, they at least showed that radio guidance was feasible. However, the A.T. program was scrapped because it was thought to have limited military potential.

Abbott, Ira Herbert (1906–)

A prominent aeronautical engineer in the early years of the American space program. After graduating from the Massachusetts Institute of Technology, Abbott joined the **Langley** Aeronautical Laboratory in 1929. The author of many technical reports on aerodynamics, he was instrumental in setting up programs in high-speed research. By 1945, he had risen to be assistant chief of research at Langley. Transferring to **NACA** (National Advisory Committee for Aeronautics) headquarters in 1948 as assistant director of aerodynamics research, he was promoted to director of advanced research programs at NASA in 1959 and to director of advanced research and technology in 1961. In this last capacity, Abbott supervised the **X-15**, supersonic transport, nuclear rocket, and advanced reentry programs. He retired in 1962.

Aberdeen Proving Ground

The U.S. Army's oldest active proving ground. It was established on October 20, 1917, six months after the United States entered World War I, as a facility where ordnance materiel could be designed and tested close to the nation's industrial and shipping centers. Aberdeen Proving Ground occupies more than 29,000 hectares in Harford County, Maryland, and is home to the Ballistic Research Laboratory, where, during the 1950s and early 1960s, important work was done on integrating electronic computers, space studies, and satellite tracking.

ablation

The removal of surface material, such as what occurs in the **combustion chamber** of a rocket, or on the leading surfaces of a spacecraft during atmospheric **reentry** or passage through a dusty medium in space, such as the tail of a comet. An expendable surface made of ablative material may be used as a coating in a combustion chamber or on the heat shield of a reentry vehicle. As the ablative material absorbs heat, it changes chemical or physical state and sheds mass, thereby carrying the heat away from the rest of the structure. See **reentry thermal protection**.

Able

(1) A modified form of the Aerojet AJ-10 second stage of the **Vanguard** rocket used as the second stage of the **Thor**-Able, Thor-Able Star, and **Atlas**-Able launch vehicles. (2) An early, ill-fated American lunar program approved by President Eisenhower on March 27, 1958, and intended to place a satellite in orbit around the Moon. Project Able became the first lunar shot in history, preceding even **Luna** 1, when a Thor-Able took off at 12:18 GMT on August 17, 1958, before a small group of journalists. Unfortunately, only 77 seconds into the flight, the Thor's **turbopump** seized and the missile blew up. Telemetry from the probe was received for a further 123 seconds until the 39-kg spacecraft ended its brief journey by falling into the Atlantic. Although not given an official name, the probe is referred to as Pioneer 0 or Able 1. Before the launch of the second probe, the whole program was transferred to NASA, which renamed it **Pioneer**. (3) A rhesus monkey housed in a biocapsule that was sent on a suborbital flight by a specially configured **Jupiter** missile on May 28, 1959. Able and its companion **Baker**, a female squirrel monkey placed in a second biocapsule, became the first live animals to be recovered after traveling outside Earth's atmosphere. Able died on June 1, 1959, from the effects of anesthesia given to allow the removal of electrodes. An autopsy revealed that Able had suffered no adverse effects from its flight.[236]

abort

The premature and sudden ending of a mission because of a problem that significantly affects the mission's chances of success.

acceleration

The rate at which the **velocity** of an object changes. Acceleration can be linear (in a straight line), angular (due to a change in direction), or negative (when it is known as deceleration). Related terms include: (1) *acceleration stress,* which is the physiological effect of high acceleration or deceleration on the human body; it increases with the magnitude and duration of the acceleration. Longitudinal accelerations cannot be tolerated as well as transverse ones, as the former have a stronger influence on the cardiovascular system, and (2) *acceleration tolerance,* which is the maximum acceleration or **deceleration** that an astronaut can withstand before losing consciousness.

acceleration due to gravity (*g*)

The acceleration that an object experiences when it falls freely close to the surface of a body such as a planet. Its value is given by the formula $g = GM/R^2$, where M is the mass of the gravitating body, R its radius, and G the **gravitational constant**. On Earth, g is about 9.8 m/s^2, although its value varies slightly with latitude.

accelerometer

An instrument that measures **acceleration** or the gravitational force capable of imparting acceleration. It usually employs a concentrated mass that resists movement because of its **inertia**; acceleration is measured in terms of the displacement of this mass relative to its supporting frame or container.

ACCESS (Advanced Cosmic-ray Composition Experiment on the Space Station)

An experiment to study the origin and makeup of **cosmic rays** over a three-year period. ACCESS will be attached to the International Space Station and is due to replace **AMS** (Alpha Magnetic Spectrometer) in about 2007. Its two instruments, the Hadron Calorimeter and the Transition Radiation Detector, will measure the elemental makeup of cosmic rays from lightest nuclei to heaviest and determine if the flux of high-energy electrons in cosmic rays varies with direction, as would be the case if some come from local sources.

ACE (Advanced Composition Explorer)

A NASA satellite designed to measure the elemental and isotopic composition of matter from several different sources, including the solar corona and the interstellar medium. ACE was placed in a **halo orbit** around the first **Lagrangian point** (L1) of the Earth-Sun system, about 1.4 million km from Earth. It carries six high-resolution sensors and three monitoring instruments for sampling low-energy particles of solar origin and high-energy galactic particles with a collecting power 10 to 1,000 times greater than previous experiments. The spacecraft can give about an hour's advance warning of geomagnetic storms that might overload power grids, disrupt communications, and pose a hazard to astronauts.

Launch
    Date: August 25, 1997
    Vehicle: Delta 7920
    Site: Cape Canaveral
    Orbit: halo
    Mass at launch: 785 kg

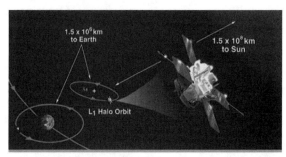

ACE (Advanced Composition Explorer)   ACE and its orbit around the first Lagrangian point. *NASA*

acquisition

(1) The process of locating the orbit of a satellite or the trajectory of a space probe so that tracking or **telemetry** data can be gathered. (2) The process of pointing an antenna or telescope so that it is properly oriented to allow gathering of tracking or telemetry data from a satellite or space probe.

ACRIMSAT (Active Cavity Radiometer Irradiance Monitor Satellite)

A satellite equipped to measure the amount of energy given out by the Sun–the total solar irradiance (TSI)–over a five-year period. ACRIMSAT carries ACRIM-3 (Active Cavity Radiometer Irradiance Monitor 3), the third in a series of long-term solar-monitoring tools built by JPL (Jet Propulsion Laboratory). This instrument extends the database started by ACRIM-1, which was launched on **SMM** (Solar Maximum Mission) in 1980 and continued by ACRIM-2 on **UARS** (Upper Atmosphere Research Satellite) in 1991. ACRIM-1 was the first experiment to show clearly that the TSI varies. The solar variability is so slight, however, that its study calls for continuous state-of-the-art monitoring. Theory suggests that as much as 25% of Earth's global warming may be of solar origin. It also seems that even small (0.5%) changes in the TSI over a century or more may have significant climatic effects. ACRIMSAT is part of NASA's **EOS** (Earth Observing System).

Launch	
Date: December 21, 1999	
Vehicle: Taurus	
Site: Vandenberg Air Force Base	
Orbit: 272 × 683 km × 98.3°	

ACRV (Assured Crew Return Vehicle)

A space lifeboat attached to the **International Space Station** (ISS) so that in an emergency, the crew could quickly evacuate the station and return safely to Earth. This role, currently filled by the Russian **Soyuz** TMA spacecraft, was to have been taken up by the **X-38**, a small winged reentry ferry. However, budget cuts in 2001 forced NASA to shelve further development of the X-38, leaving the future of the ACRV in doubt. Among the possibilities are that the present Soyuz could either be retained for the job or be replaced by a special ACRV Soyuz that has been under development for more than 30 years. Features that distinguish the ACRV Soyuz from the standard model are seats that can accommodate larger crew members and an upgraded onboard computer that assures a more accurate landing.

active satellite

A satellite that carries equipment, including onboard power supplies, for collecting, transmitting, or relaying data. It contrasts with a **passive satellite**.

ACTS (Advanced Communications Technology Satellite)

An experimental NASA satellite that played a central role in the development and flight-testing of technologies now being used on the latest generation of commercial **communications satellite**s. The first all-digital communications satellite, ACTS supported standard fiber-optic data rates, operated in the K- and Ka-**frequency band**s, pioneered dynamic hopping spot beams, and advanced onboard traffic switching and processing. (A hopping spot beam is an antenna beam on the spacecraft that points at one location on the ground for a fraction of a millisecond. It sends/receives voice or data information and then electronically "hops" to a second location, then a third, and so on. At the beginning of the second millisecond, the beam again points at the first location.) ACTS-type onboard processing and Ka-band communications are now used operationally by, among others, the **Iridium** and **Teledesic** systems. ACTS was developed, managed, and operated by the Glenn Research Center. Its mission ended in June 2000.[110]

Shuttle deployment	
Date: September 16, 1993	
Mission: STS-51	
Orbit: geostationary at 100°W	
On-orbit mass: 2,767 kg	

adapter skirt

A flange, or extension of a space vehicle stage or section, that enables the attachment of some object, such as another stage or section.

additive

A substance added to a **propellant** for any of a variety of reasons, including to stabilize or achieve a more even rate of combustion, to make ignition easier, to lower the freezing point of the propellant (to prevent it from freezing in space), or to reduce corrosive effects.

ADE (Air Density Explorer)

A series of balloons, made from alternating layers of aluminum foil and Mylar polyester film, placed in orbit to study the density of the upper atmosphere. Although **Explorer** 9 was the first such balloon launched (as well as being the first satellite placed in orbit by an all-solid-propellant rocket and the first to be successfully launched from **Wallops Island**), only its three identical successors were officially designated "Air Density Explorers." (See table, "Air Density Explorers.") ADE was a subprogram of NASA's **Explorer** series.

ADE (Air Density Explorer) Explorer 24, the second Air Density Explorer, at Langley Research Center. *NASA*

Launch site: Vandenberg Air Force Base
Mass: 7–9 kg
Diameter: 3.7 m

ADEOS (Advanced Earth Observation Satellite)

Japanese Earth resources satellites. ADEOS 1, also known by its national name, Midori ("green"), was the first resources satellite to observe the planet in an integrated way. Developed and managed by Japan's **NASDA** (National Space Development Agency), it carried eight instruments supplied by NASDA, NASA, and CNES (the French space agency) to monitor worldwide environmental changes, including global warming, depletion of the ozone layer, and shrinking of tropical rain forests. Due to structural damage, the satellite went off-line after only nine months in orbit. ADEOS 2, scheduled for launch in November 2002, will continue where its predecessor left off and also study the global circulation of energy and water. Additionally, it will contribute to NASA's **EOS** (Earth Observing System) by carrying NASA's **Seawinds** scatterometer, a microwave radar to measure near-surface wind velocity and oceanic cloud conditions, which scientists hope will improve their ability to forecast and model global weather.

ADEOS 1
Launch
    Date: August 17, 1996
    Vehicle: H-2
    Site: Tanegashima
Orbit (circular): 800 km × 98.6°
Size: 5.0 × 4.0 m
Mass at launch: about 3.5 tons

Advanced Concepts Program

A program managed by NASA's Office of Space Access and Technology to identify and develop new, far-reaching concepts that may later be applied in advanced technology programs. It was set up to help enable unconventional ideas win consideration and possible acceptance within the NASA system. Among the areas that the Advanced Concepts Program is looking into are **fusion**-based space propulsion, optical computing, robotics, interplanetary navigation, materials and structures, ultra-lightweight large aperture optics, and innovative modular spacecraft architectural concepts.

Air Density Explorers

Spacecraft	Launch Date	Vehicle	Orbit
Explorer 19	Dec. 19, 1963	Scout X-3	597 × 2,391 km × 78.6°
Explorer 24	Nov. 21, 1964	Scout X-3	530 × 2,498 km × 81.4°
Explorer 39	Aug. 8, 1968	Scout B	570 × 2,538 km × 80.7°

February 28, 1959, a Thor-Agena placed **Discoverer 1** into the first polar orbit ever achieved by a human-made object. An Agena A carried Discoverer 14 into orbit on August 18, 1960, and sent it back to Earth 27 hours later to become the first satellite recovered in midair after reentry from space. The Agena had primary and secondary propulsion systems. The main engine had a thrust of about 70,000 newtons (N), while the secondary was used for small orbital adjustments. Both engines used liquid propellants and (from the Agena B on) could be restarted in orbit.

aging

The main problem facing future interstellar voyagers is the immense distances involved—and consequently the inordinate lengths of time required to travel—between even neighboring stars at speeds where **relativistic effects** do not come into play. For example, at a steady 16,000 km/s—over 1,000 times faster than any probe launched from Earth has yet achieved—a spacecraft would take about 80 years to cross from the Sun to the next nearest stellar port of call, Proxima Centauri. No astronauts embarking on such a voyage would likely live long enough to see the destination, unless they boarded as children. Volunteers might be hard to find. This problem of limited human life span and extremely long journey times led, earlier this century, to the suggestion of **generation starships** and **suspended animation**.

agravic

A region or a state of weightlessness.

AIM (Aeronomy of Ice in the Mesosphere)

A proposed NASA mission to investigate the causes of the highest altitude clouds in Earth's atmosphere. The number of clouds in the **mesosphere**, or middle atmosphere, over the Poles has been increasing over the past couple of decades, and it has been suggested that this is due to the rising concentration of greenhouse gases at high altitude. AIM would help determine the connection between the clouds and their environment and improve our knowledge of how long-term changes in the upper atmosphere are linked to global climate change. It has been selected for study as an **SMEX** (Small Explorer) mission.

air breakup

The disintegration of a space vehicle caused by aerodynamic forces upon reentry. It may be induced deliberately to cause large parts of a vehicle to break into smaller parts and burn up during reentry, or to reduce the impact speed of test records and instruments that need to be recovered.

Air Force Flight Test Center

A U.S. Air Force facility at **Edwards Air Force Base**, California. The Test Center includes the Air Force Rocket Propulsion Laboratory (formed in 1952 and previously known as the Air Force's Astronautics Laboratory), the Air Force Propulsion Laboratory, and the Air Force Phillips Laboratory, which is the development center for all Air Force rocket propulsion technologies, including solid-propellant motors and liquid-propellant fuel systems and engines.

Air Force Space Command (AFSPC)

A U.S. Air Force facility located at Peterson Air Force Base, Colorado. Among its responsibilities have been or are **BMEWS** (Ballistic Missile Early Warning System), **DSCS** (Defense Satellite Communications System), **FLSATCOM** (Fleet Satellite Communications System), **GPS** (Global Positioning System), and **NATO** satellites.

air-breathing engine

An engine that takes in air from its surroundings in order to burn fuel. Examples include the **ramjet, scramjet, turbojet**, turbofan, and **pulse-jet**. These contrast with a **rocket**, which carries its own **oxidizer** and thus can operate in space. Some vehicles, such as **space plane**s, may be fitted with both air-breathing and rocket engines for efficient operation both within and beyond the atmosphere.

airfoil

A structure shaped so as to produce an aerodynamic reaction (lift) at right angles to its direction of motion. Familiar examples include the wings of an airplane or the Space Shuttle. Elevators, ailerons, tailplanes, and rudders are also airfoils.

airframe

The assembled main structural and aerodynamic components of a vehicle, less propulsion systems, control guidance equipment, and payloads. The airframe includes only the basic structure on which equipment is mounted.

airlock

A chamber that allows astronauts to leave or enter a spacecraft without depressurizing the whole vehicle. The typical sequence of steps for going out of a spacecraft in orbit is: (1) the astronaut, wearing a spacesuit, enters the airlock through its inner door; (2) the airlock is depressurized by transferring its air to the spacecraft; (3) the inner door is closed, which seals the spacecraft's atmosphere; (4) the airlock's outer door is opened into space, and the astronaut exits. The reverse sequence applies when the astronaut returns.

AIRS (Atmospheric Infrared Sounder)

An instrument built by NASA to make extremely accurate measurements of air temperature, humidity, cloud makeup, and surface temperature. The data collected by AIRS will be used by scientists around the world to better understand weather and climate, and by the National Weather Service and NOAA (National Oceanic and Atmospheric Administration) to improve the accuracy of their weather and climate models. AIRS is carried aboard the **Aqua** spacecraft of NASA's EOS (Earth Observing System), which was launched in May 2002.

Ajisai

See **EGS** (Experimental Geodetic Satellite).

Akebono

A satellite launched by Japan's **ISAS** (Institute of Space and Astronautical Science) to make precise measurements of the way charged particles behave and are accelerated within the auroral regions of Earth's magnetosphere. Akebono, whose name means "dawn," was known before launch as **Exos**-D.

Launch
    Date: February 21, 1989
    Vehicle: M-3S
    Site: Kagoshima
    Orbit: 264 × 8,501 km × 75.1°
    Mass at launch: 295 kg

Akiyama, Tokohiro (1944–)

The first Japanese in orbit and the first fee-paying **space passenger**. A reporter for the TBS television station, Akiyama flew to the **Mir** space station in 1992 after his employer stumped up the cost of his ride–$12 million. Alongside him was to have been a TBS colleague, camerawoman Ryoko Kikuchi, but her spaceflight ambitions were dashed when she was rushed to the hospital before the flight for an emergency appendectomy.

Albertus Magnus (1193–1280)

A German philosopher and experimenter who, like his English counterpart Roger **Bacon**, wrote about **black powder** and how to make it. A recipe appears in his *De mirabilis mundi* (On the Wonders of the World): *"Flying fire:* Take one pound of sulfur, two pounds of coals of willow, six pounds of saltpeter; which three may be ground very finely into marble stone; afterwards . . . some may be placed in a skin of paper for flying or for making thunder."

Alcantara

A planned launch complex for Brazil's indigenous **VLS** booster. Located at 2.3° S, 44.4° W, it would be able to launch satellites into orbits with an inclination of 2 to 100 degrees.

Alcubierre Warp Drive

An idea for achieving **faster-than-light travel** suggested by the Mexican theoretical physicist Miguel Alcubierre in 1994.[4] It starts from the notion, implicit in Einstein's **general theory of relativity**, that matter causes the surface of **space-time** around it to curve. Alcubierre was interested in the possibility of whether *Star Trek*'s fictional "warp drive" could ever be realized. This led him to search for a valid mathematical description of the gravitational field that would allow a kind of space-time warp to serve as a means of superluminal propulsion. Alcubierre concluded that a warp drive would be feasible if matter could be arranged so as to expand the space-time behind a starship (thus pushing the departure point many light-years back) and contract the space-time in front (bringing the destination closer), while leaving the starship itself in a locally flat region of space-time bounded by a "warp bubble" that lay between the two distortions. The ship would then surf along in its bubble at an arbitrarily high velocity, pushed forward by the expansion of space at its rear and by the contraction of space in front. It could travel faster than light without breaking any physical law because, with respect to the space-time in its warp bubble, it would be at rest. Also, being locally stationary, the starship and its crew would be immune from any devastatingly high accelerations and decelerations (obviating the need for inertial dampers) and from **relativistic effects** such as time dilation (since the passage of time inside the warp bubble would be the same as that outside).

Could such a warp drive be built? It would require, as Alcubierre pointed out, the manipulation of matter with a *negative* **energy density**. Such matter, known as **exotic matter**, is the same kind of peculiar stuff apparently needed to maintain stable **wormhole**s–another proposed means of circumventing the light barrier. Quantum mechanics allows the existence of regions of negative energy density under special circumstances, such as in the **Casimir effect**.

Further analysis of Alubierre's Warp Drive concept by Chris Van Den Broeck[34] of the Catholic University in Leuven, Belgium, has perhaps brought the construction of the starship *Enterprise* a little closer. Van Den Broeck's calculations put the amount of energy required much lower than that quoted in Alcubierre's paper. But this is not to say we are on the verge of warp capability. As Van Den Broeck concludes: "The first warp drive is still a long way off but maybe it has now become slightly less improbable."[230, 239]

Aldrin, Edwin Eugene "Buzz," Jr. (1930–)

The American astronaut who became the second person to walk on the Moon. Aldrin graduated with honors from West Point in 1951 and subsequently flew jet fighters in the Korean War. Upon returning to academic work, he earned a Ph.D. in astronautics from the Massachusetts Institute of Technology, devising techniques for manned space rendezvous that would be used on future NASA missions, including the **Apollo-Soyuz Test Project**. Aldrin was selected for astronaut duty in October 1963, and in November 1966 he established a new spacewalk duration record on the **Gemini** 9 mission. As backup Command Module pilot for **Apollo** 8, he improved operational techniques for astronautical navigation star display. Then, on July 20, 1969, Aldrin and Neil **Armstrong** made their historic Apollo 11 moonwalk. Since retiring from NASA (in 1971), the Air Force, and his position as commander of the Test Pilot School at Edwards Air Force Base, Aldrin has remained active in efforts to promote American manned space exploration. He has produced a plan for sustained exploration based on a concept known as the **orbital cycler**, involving a spacecraft system that perpetually orbits between the orbits of Earth and Mars. His books include *Return to Earth* (1974),[5] an account of his Moon trip and his views on America's future in space, *Men from Earth* (1989),[6] and a science fiction novel, *Encounter with Tiber* (1996). Aldrin also participates in many space organizations worldwide, including the National Space Society, which he chairs.

Edwin Aldrin   Aldrin in the Lunar Module during the Apollo 11 mission. *NASA*

ALEXIS (Array of Low Energy X-ray Imaging Sensors)

A small U.S. Department of Defense spacecraft that has provided high-resolution maps of astronomical X-ray sources. The mission was also intended to demonstrate the feasibility of quickly building low-cost sensors for arms treaty verification. ALEXIS was equipped with six coffee-can-sized telescopes that worked in pairs to make observations in the soft (longer wavelength) X-ray and extreme ultraviolet (EUV) part of the spectrum. Among its science objectives were to survey and map the diffuse soft X-ray component of the sky, to look at known bright EUV sources, to search for transient (fast-changing) behavior, and to study stellar flares. One of the first of the modern generation of miniature spacecraft, ALEXIS was designed and built over a three-year period by Los Alamos National Laboratory, Sandia National Laboratory, Space Sciences Laboratory at the University of California, Berkeley, and AeroAstro.

Launch
   Date: April 25, 1993
   Vehicle: Pegasus
   Site: Edwards Air Force Base
Orbit: 741 × 746 km × 69.8°
Mass: 115 kg

algae

Simple photosynthetic organisms that use carbon dioxide and release oxygen, thus making them viable for air purification during long voyages in spacecraft. They also offer a source of protein. However, their use is limited at present because they require the Sun's or similar light, and the equipment required to sustain them is bulky.

Almaz

(1) Satellites that carry a **synthetic aperture radar** (SAR) system for high-resolution (10–15 m), all-weather, round-the-clock surveillance of land and ocean surfaces. Developed and operated by the Russian space company NPO Mashinostroyenia, Almaz ("diamond") spacecraft are used for exploration and monitoring in fields such as map-making, geology, forestry, and ecology. The first in the series was placed in orbit by a Proton booster on March 31, 1991. (2) An ambitious, top-secret Soviet project envisioned by Vladimir **Chelomei** as a manned orbiting outpost equipped with powerful spy cameras, radar, and self-defense weapons. The program would also have involved heavy supply ships and multiple reentry capsules. Although Almaz was delayed and eventually canceled after Chelomei fell out of favor with the Soviet government in the late 1960s, its design was used as the basis for **Salyut** 1.

ALOS (Advanced Land Observing Satellite)

A Japanese satellite designed to observe and map Earth's surface, enhance cartography, monitor natural disasters, and survey land use and natural resources to promote sustainable development. ALOS follows **JERS** and **ADEOS** and will extend the database of these earlier satellites using three remote-sensing instruments: the Panchromatic Remote-sensing Instrument for Stereo Mapping (PRISM) for digital elevation mapping, the Advanced Visible and Near Infrared Radiometer type 2 (AVNIR-2) for precise land coverage observation, and the Phased Array type L-band Synthetic Aperture Radar (PALSAR) for day-and-night and all-weather land observation. ALOS is scheduled for launch by Japan's **NASDA** (National Space Development Agency) in 2003.

Alouette

Canadian satellites designed to observe Earth's ionosphere and magnetosphere; "alouette" is French for "lark." Alouette 2 took part in a double launch with **Explorer** 31 and was placed in a similar orbit so that results from the two could be correlated. Alouette 2 was also the first mission in the **ISIS** (International Satellites for Ionospheric Studies) program conducted jointly by NASA and the Canadian Defense Research Board. (See table, "Alouette Missions.")

Launch
    Vehicle: Thor-Agena B
    Site: Vandenberg Air Force Base
Mass: 145 kg

Alouette Missions

Spacecraft	Launch Date	Orbit
Alouette 1	Sep. 29, 1962	987 × 1,022 km × 80.5°
Alouette 2	Nov. 29, 1965	499 × 2,707 km × 79.8°

ALSEP (Apollo Lunar Science Experiment Package)

See **Apollo**.

alternate mission

A secondary flight plan that may be selected when the primary flight plan has been abandoned for any reason other than abort.

altimeter

A device that measures the **altitude** above the surface of a planet or moon. Spacecraft altimeters work by timing the round trip of radio signals bounced off the surface.

Alouette A model of Alouette 1 at a celebration after the launch of the real satellite. *Canadian Space Agency*

altitude

The vertical distance of an object above the observer. The observer may be anywhere on Earth or at any point in the atmosphere. Absolute altitude is the vertical distance to the object from an observation point on Earth's (or some other body's) surface.

aluminum, powdered

The commonest fuel for **solid-propellant rocket motor**s. It consists of round particles, 5 to 60 micrometers in diameter, and is used in a variety of composite propellants. During combustion the aluminum particles are oxidized into aluminum oxide, which tends to stick together to form larger particles. The aluminum increases the propellant density and combustion temperature and thereby the **specific impulse** (a measure of the efficiency of a rocket engine).

American Astronautical Society (AAS)

The foremost independent scientific and technical group in the United States exclusively dedicated to the advancement of space science and exploration. Formed in 1954, the AAS is also committed to strengthening the global space program through cooperation with international space organizations.

American Institute of Aeronautics and Astronautics (AIAA)

A professional society devoted to science and engineering in aviation and space. It was formed in 1963 through a merger of the American Rocket Society (ARS) and the Institute of Aerospace Sciences (IAS). The ARS was founded as the American Interplanetary Society in New

York City in 1930 by David Lasser, G. Edward Pendray, Fletcher Pratt, and others, and it changed its name four years later. The IAS started in 1932 as the Institute of Aeronautical Science, with Orville Wright as its first honorary member, and substituted "Aerospace" in its title in 1960. AIAA and its founding societies have been at the forefront of the aerospace profession from the outset, beginning with the launch of a series of small experimental rockets before World War II based on designs used by the **Verein für Raumschiffahrt** (German Society for Space Travel).

American Rocket Society
See **American Institute of Aeronautics and Astronautics**.

Ames, Milton B., Jr. (1913–)
A leading aerodynamicist in the early days of the American space program. Ames earned a B.S. in aeronautical engineering from Georgia Tech in 1936 and joined the **Langley** Aeronautical Laboratory that same year. In 1941, he transferred to the headquarters of **NACA** (National Advisory Committee for Aeronautics), where he served on the technical staff, becoming chief of the aerodynamics division in 1946. Following the creation of NASA, Ames was appointed chief of the aerodynamics and flight mechanics research division. In 1960, he became deputy director of the office of advanced research programs at NASA Headquarters and then director of space vehicles in 1961. He retired from the space program in 1972.

Ames Research Center (ARC)
A major NASA facility located at Moffett Field, California, in the heart of Silicon Valley. Ames was founded on December 20, 1939, by **NACA** (National Advisory Committee for Aeronautics) as an aircraft research laboratory,

Ames Research Center   An aerial view of Ames Research Center. The large flared rectangular structure to the left of center of the photo is the 80 × 120 ft. Full Scale Wind Tunnel. Adjacent to it is the 40 × 80 ft. Full Scale Wind Tunnel, which has been designated a National Historic Landmark. *NASA*

and it became part of NASA when that agency was formed in 1958. Ames has some of the largest **wind tunnels** in the world. In addition to aerospace research, Ames specializes in space life research—being home to NASA's Exobiology Branch and the recently formed Astrobiology Institute—and the exploration of the Solar System. Among the missions it has been closely involved with are Pioneer, Voyager, Mars Pathfinder, Mars Global Surveyor, Ulysses, SOFIA, Galileo, and Cassini. The center is named after Joseph Ames, a former president of NACA.[212]

ammonium perchlorate (NH_4ClO_4)

The **oxidizer** used in most composite rocket motors. It makes up 68% of the Space Shuttle's Solid Rocket Booster propellant, the rest being powdered **aluminum** and a combustible binding compound.

AMPTE (Active Magnetosphere Particle Tracer Explorer)

An international mission to create an artificial comet and to observe its interaction with the solar wind. It involved the simultaneous launch of three cooperating spacecraft into highly elliptical orbits. The German component (IRM, or Ion Release Module) released a cloud of barium and lithium ions to produce the comet, the American component (CCE, or Charge Composition Explorer) studied its resultant behavior, and the British component (UKS, or United Kingdom Satellite) measured the effects of the cloud on natural plasma in space. (See table, "AMPTE Component Spacecraft.")

Launch
   Date: August 16, 1984
   Vehicle: Delta 3925
   Site: Cape Canaveral

AMS (Alpha Magnetic Spectrometer)

An experiment flown on the Space Shuttle and the International Space Station (ISS) to search for dark matter, missing matter, and antimatter in space. It uses a variety of instruments to detect particles and to measure their electric charge, velocity, momentum, and total energy.

Particle physicists hope that its results will shed light on such topics as the Big Bang, the future of the universe, and the nature of unseen (dark) matter, which makes up most of the mass of the cosmos. AMS1 flew on Shuttle mission STS-91 in May 1998. AMS2 will be one of the first experiments to be fixed to the outside of the ISS and is scheduled for launch in October 2003.

anacoustic zone

The region of Earth's atmosphere where distances between rarefied air molecules are so great that sound waves can no longer propagate. Also known as the zone of silence.

Anders, William Alison (1933–)

An American astronaut, selected with the third group of astronauts in 1963, who served as backup pilot for **Gemini** 11 and Lunar Module pilot for **Apollo** 8. Although a graduate of the U.S. Naval Academy, Anders was a career Air Force officer. He resigned from NASA and active duty in the Air Force in September 1969 to become Executive Secretary of the **National Aeronautics and Space Council**. He joined the Atomic Energy Commission in 1973, was appointed chairman of the Nuclear Regulatory Commission in 1974, and was named U.S. ambassador to Norway in 1976. Later he worked in senior positions for General Electric, Textron, and General Dynamics.

Andøya Rocket Range

A launch facility established in the early 1960s in northern Norway at 69.3° N, 16.0° E and used initially for launching small American sounding rockets. The first launches of Nike Cajun rockets took place in 1962, and until 1965 the range was occupied only at the time of the launching campaigns. In late 1962, **ESRO** (European Space Research Organisation), aware that the rocket range it had planned to build at **Esrange**, Sweden, would not be ready before autumn 1965, reached an agreement with Norway to use Andøya. The first six ESRO rockets were launched from there in the first quarter of 1966, and four were launched on behalf of CNES (the French space agency) the same year. In late 1966, Esrange opened and ESRO shifted its launches to this new location; how-

AMPTE Component Spacecraft

Spacecraft	Nation	Orbit	Mass (kg)
AMPTE-1 (CCE)	United States	1,121 × 49,671 km × 4.8°	242
AMPTE-2 (UKS)	United Kingdom	402 × 113,818 km × 27.0°	605
AMPTE-3 (IRM)	West Germany	1,002 × 114,417 km × 26.9°	77

ever, Andøya continued to be used regularly for bilateral and international sounding rocket programs. Since 1972, the range has been supported through a Special Project Agreement under which it is maintained by and made available to some ESA (European Space Agency) states, and it has been operated for commercial and bilateral programs. Now managed by the Norwegian Space Center, the Andøya range comprises eight launch pads, including a universal ramp able to launch rockets weighing up to 20 tons.

anergolic propellant

A propellant in which, in contrast to a **hypergolic propellant**, the liquid fuel and liquid oxidizer do not burn spontaneously when they come into contact.

Angara

A new series of Russian launch vehicles intended to complement and eventually replace the existing line of **Rokot** and **Proton** boosters. It was conceived in 1992 in order to give the Russian Federation a launch capability independent of the hardware and launch sites in the newly independent republics of the former Soviet Union. Angara (named after a Siberian river) is being developed by the Moscow-based **Khrunichev State Research and Production Space Center** as a family of rockets capable of delivering payloads of 2 to 25 tons into **LEO** (low Earth orbit). The first stage uses a common core module with a single-chamber version of the **Zenit** RD-170 LOX/kerosene engine (known as the RD-191M) plus up to five identical strap-on boosters. Except in the case of the Angara 1.1, the second stage is a newly developed cryogenic, LOX/liquid hydrogen engine (the KVD-1M). Upper stages utilize the Breeze-KM/-M and, in heavy-lift models, the new cryogenic KVRB. In cooperation with KB Salyut, the developer of the **Buran** orbiter, Khrunichev has also designed a reusable flyback booster, the Baikal, to serve as an alternative first stage. Delays in

developing launch facilities for Anagara at the **Plesetsk** cosmodrome have pushed back the initial launch to at least 2003. (See table, "The Angara Family.")

angle of attack

In the theory of airplane wings, the acute angle between the wing profile (roughly, measured along its bottom) and the wing's motion relative to the surrounding air. In the case of a rocket rising through the atmosphere, it is the angle between the long axis of the rocket and the direction of the air flowing past it.

angular momentum

The **momentum** an object has because of its rotation, including spin about its own axis and orbital motion. A spacecraft's spin can be controlled or stopped by firing small rockets or by transferring angular momentum to one or more flywheels. Orbital angular momentum is given by multiplying together the object's mass, **angular velocity**, and distance from the gravitating body. According to the law of conservation of angular momentum, the angular momentum of an object in orbit must remain constant at all points in the orbit.

angular velocity

The rate of rotation of an object, either about its own axis or in its orbit about another body.

anhydrous

Without water. For example, an anhydrous propellant works in the absence of water.

Anik

A Canadian domestic satellite system that supports TV transmissions and carries long-distance voice and data services throughout Canada as well as some transborder service to the United States and Mexico; "anik" is Inuit for "brother." See **Nimiq**.

The Angara Family

	Stage			Payload (tons)	
	1	2	3	LEO	GEO
Lightweight					
Angara 1.1	1 × common core	Breeze-KM	—	2	—
Angara 1.2	1 × common core	KVD-1M	Breeze-KM	3.7	—
Intermediate					
Angara 3	3 × common core	KVD-1M	Breeze-M	14.1	1.1
Heavy-lift					
Angara 5	5 × common core	KVD-1M	Breeze-M + KVRB	24.5	4.0

aniline (C₆H₅NH₂)

A colorless, oily liquid that served as a propellant for some early rockets, such as the American **Corporal**. It is highly toxic, however, and no longer used as a rocket fuel.

animals in space

The menagerie of animals (not to mention plants, fungi, and microorganisms) that have made orbital and suborbital trips includes rats, mice, frogs, turtles, crickets, swordtail fish, rabbits, dogs, cats, and chimpanzees. Spaceflights involving animals began just after World War II and continue today with biological experiments on the International Space Station (ISS). The first primates sent on rocket journeys above most of the atmosphere were the monkeys Albert 1 and Albert 2 aboard nosecones of captured German V-2 (see **"V" weapons**) rockets during American tests in the 1940s. They died, however, as did a monkey and several mice in 1951 when their parachute failed to open after an **Aerobee** launch. But on September 20 of the same year, a monkey and 11 mice survived a trip aboard an Aerobee to become the first passengers to be recovered alive from an altitude of tens of kilometers. On May 28, 1959, monkeys **Able** and **Baker** reached the edge of space and came back unharmed. From 1959 to 1961 a number of primates, including **Ham**, went on test flights of the **Mercury** capsule. During this same period, the Soviet Union launched 13 dogs toward orbit, 5 of which perished, including the first animal space farer—**Laika**. In the pre-Shuttle era, spacecraft carrying a wide variety of different species included the **Bion**, **Biosatellite**, and **Korabl-Sputnik** series.

annihilation

The process in which the entire mass of two colliding particles, one of matter and one of **antimatter**, is converted into radiant energy in the form of **gamma rays**.

ANS (Astronomische Nederlandse Satelliet)

A Dutch X-ray and ultraviolet astronomy satellite notable for its discovery of X-ray bursts and of the first X-rays from the corona of a star beyond the Sun (Capella); it was the first satellite for the Netherlands. The universities of Groningen and Utrecht provided the ultraviolet and soft (longer wavelength) X-ray experiments, while NASA furnished a hard (shorter wavelength) X-ray experiment built by American Science and Engineering of Cambridge, Massachusetts. ANS operated until 1976.

Launch
  Date: August 30, 1974
  Vehicle: Scout D
  Site: Vandenberg Air Force Base
  Orbit: 258 × 1,173 km × 98.0°
  Mass: 130 kg

antenna

A device for collecting or transmitting radio signals, the design of which depends on the wavelength and amplitude of the signals.

anti-*g* suit

A tight-fitting suit that covers parts of the body below the heart and is designed to retard the flow of blood to the lower body in reaction to acceleration or deceleration; sometimes referred to as a *g*-suit. Bladders or other devices are used to inflate and to increase body constriction as *g*-force increases.

The circulatory effects of high acceleration first became apparent less than two decades after the Wright brothers' seminal powered flight. During the Schneider Trophy Races in the 1920s, in which military and specialized aircraft made steep turns, pilots would occasionally experience "grayouts." An early documented case of *g*-induced loss of consciousness, or *g*-LOC, occurred in the pilot of a Sopwith Triplane as long ago as 1917. But the problem only became significant with the dawn of higher performance planes in World War II. In the quarter century between global conflicts, the maximum acceleration of aircraft had doubled from 4.5*g* to 9*g*.

Two medical researchers played key roles in the evolution of the anti-*g* suit during the 1930s and 1940s. In 1931, physiologist Frank Cotton at the University of Sydney, Australia, devised a way of determining the center of gravity of a human body, which made possible graphic recordings of the displacement of mass within the body under varying conditions of rest, respiration, posture, and exercise. He later used his technique to pioneer suits that were inflated by air pressure and regulated by *g*-sensitive valves. At the University of Toronto, Wilbur R. Franks did similar work that eventually led to the Mark III Franks Flying Suit—the first anti-*g* suit ever used in combat. His invention gave Allied pilots a major tactical advantage that contributed to maintaining Allied air superiority throughout World War II, and after 1942 the Mark III was used exclusively by American fighter pilots in the Pacific.

At the same time the anti-*g* suit was being perfected, it was realized that pilots who were able to tolerate the greatest *g*-forces could outmaneuver their opponents. This led to the rapid development of **centrifuge**s.

antigravity

A hypothetical force that acts in the direction opposite to that of normal gravity. In Einstein's **general theory of relativity**, a gravitational field is equivalent to a curvature of space-time, so an antigravity device could work only by locally rebuilding the basic framework of the Universe. This would require negative mass.[31, 237] The theme of antigravity appeared early in science fiction–a typical nineteenth-century example being "apergy," an antigravity principle used to propel a spacecraft from Earth to Mars in Percy **Greg**'s *Across the Zodiac* (1880) and borrowed for the same purpose by John Jacob Astor in *A Journey in Other Worlds* (1894). More famously, in *The First Men in the Moon* (1901),[312] H. G. Wells used moveable shutters made of "Cavorite," a metal that shields against gravity, to navigate a spacecraft to the Moon.[233]

antimatter

Matter composed of **antiparticles**. An atom of antihydrogen, for example, consists of a positron (an antielectron) in orbit around an antiproton. Antimatter appears to be rare in our universe, and it is certainly rare in our galaxy. When matter and antimatter meet, they undergo a mutually destructive process known as **annihilation**, which in the future could form the basis of **antimatter propulsion**.

antimatter propulsion

Devotees of *Star Trek* will need no reminding that the starships *Enterprise* and *Voyager* are powered by engines that utilize **antimatter**. Far from being fictional, the idea of propelling spacecraft by the **annihilation** of matter and antimatter is being actively investigated at NASA's Marshall Space Flight Center, Pennsylvania State University, and elsewhere. The principle is simple: an equal mixture of matter and antimatter provides the highest energy density of any known propellant. Whereas the most efficient chemical reactions produce about 1×10^7 joules (J)/kg, nuclear **fission** 8×10^{13} J/kg, and nuclear **fusion** 3×10^{14} J/kg, the complete annihilation of matter and antimatter, according to Einstein's **mass-energy relationship**, yields 9×10^{16} J/kg. In other words, kilogram for kilogram, matter-antimatter annihilation releases about 10 billion times more energy than the hydrogen/oxygen mixture that powers the Space Shuttle Main Engines and 300 times more than the fusion reactions at the Sun's core.

However, there are several (major!) technical hurdles to be overcome before an antimatter rocket can be built. The first is that antimatter does not exist in significant amounts in nature–at least, not anywhere near the Solar System. It has to be manufactured. Currently the only way to do this is by energetic collisions in giant particle accelerators, such as those at FermiLab, near Chicago, and at CERN in Switzerland. The process typically involves accelerating protons to almost the speed of light and then slamming them into a target made of a metal such as tungsten. The fast-moving protons are slowed or stopped by collisions with the nuclei of the target atoms, and the protons' kinetic energy is converted into matter in the form of various subatomic particles, some of which are antiprotons–the simplest form of antimatter. So efficient is matter-antimatter annihilation that 71 milligrams of antimatter would produce as much energy as that stored by all the fuel in the Space Shuttle External Tank. Unfortunately, the annual amount of antimatter (in the form of antiprotons) presently produced at FermiLab and CERN is only 1 to 10 nanograms (a nanogram is a million times smaller than a milligram).[263] On top of this production shortfall, there is the problem of storage. Antimatter cannot be kept in a normal container because it will annihilate instantly on coming into contact with the container's walls. One solution is the Penning Trap–a supercold, evacuated electromagnetic bottle in which charged particles of antimatter can be suspended (see illustration). Antielectrons, or positrons, are difficult to store in this way, so antiprotons are stored instead. Penn State and NASA scientists have already built such a device capable of holding 10 million antiprotons for a week. Now they are developing a Penning Trap with a capacity 100 times greater.[275] At the same time, FermiLab is installing new equipment that will boost its production of antimatter by a factor of 10 to 100.

A spacecraft propulsion system that works by expelling the products of direct one-to-one annihilation of protons and antiprotons–a so-called beamed core engine–would need 1 to 1,000 g of antimatter for a manned interplanetary or an unmanned interstellar journey.[97] Even with the improved antiproton production and storage capacities expected soon, this amount of antimatter is beyond our reach. However, the antimatter group at Penn State has proposed a highly

antimatter propulsion An antimatter trap at Pennsylvania State University. *Pennsylvania State University*

efficient space propulsion system that would need only a tiny fraction of the antimatter consumed by a beamed core engine. It would work by a process called antiproton-catalyzed microfission (ACMF).[274] Whereas conventional nuclear fission can only transfer heat energy from a uranium core to surrounding chemical propellant, ACMF permits all energy from fission reactions to be used for propulsion. The result is a more efficient engine that could be used for interplanetary manned missions. The ICAN-II (Ion Compressed Antimatter Nuclear II) spacecraft designed at Penn State would use the ACMF engine and only 140 ng of antimatter for a manned 30-day crossing to Mars.

A follow-up to ACMF and ICAN is a spacecraft propelled by AIM (antiproton initiated microfission/fusion), in which a small concentration of antimatter and fissionable material would be used to spark a microfusion reaction with nearby material. Using 30 to 130 micrograms of antimatter, an unmanned AIM-powered probe—AIM-Star—would be able to travel to the Oort Cloud in 50 years, while a greater supply of antiprotons might bring Alpha Centauri within reach.[190]

antiparticle
A counterpart of an ordinary subatomic particle, which has the same mass and spin but opposite charge. Certain other properties are also reversed, including the magnetic moment. Antiparticles are the basis of **antimatter**. The antiparticles of the electron, proton, and neutron are the positron, antiproton, and antineutron, respectively. An encounter between an electron and a positron results in the instantaneous total conversion of the mass of both into energy in the form of gamma rays. When a proton and an antiproton meet, however, the outcome is more complicated. Pions are produced, some of which decay to produce gamma radiation and others of which decay to produce muons and neutrinos plus electrons and positrons, which make more gamma rays.

aphelion
The point in a heliocentric orbit that is farthest from the Sun.

apoapsis
The point in an orbit that is farthest from the body being orbited. Special names, such as **apogee** and **aphelion**, are given to this point for familiar systems.

apogee
The point in a geocentric orbit that is farthest from Earth's surface.

apogee kick motor
A solid rocket motor, usually permanently attached to a spacecraft, that circularizes an elliptical transfer orbit by igniting at **apogee** (leading to the colloquial phrase "a kick in the apogee"). It was first used on the early **Syncom** satellites in 1963 and 1964 to "kick" the satellite from a **geostationary transfer orbit** to a **geostationary orbit**. Also known simply as an *apogee motor.*

Apollo
See article, pages 23–33.

Apollo-Soyuz Test Project (ASTP)

Apollo spacecraft
Launch date: July 15, 1975
Launch vehicle: Saturn IB
Crew
Commander: Thomas **Stafford**
Command Module pilot: Vance **Brand**
Docking Module pilot: Donald **Slayton**
Mission duration: 9 days 1 hr
Splashdown: July 24, 1975

Soyuz 19 spacecraft
Crew
Commander: Aleskei **Leonov**
Flight engineer: Valeriy Kubasov
Mission duration: 5 days 23 hr
Landing: July 21, 1975

The first international manned spaceflight and a symbolic end to the nearly 20-year-long Space Race between the United States and the Soviet Union. Setting political differences aside, the two superpowers successfully carried out the first joint on-orbit manned space mission. ASTP negotiations, begun in 1970, culminated in an agreement for ASTP flight operations being signed at the superpower summit in May 1972.

The project was designed mainly to develop and validate space-based rescue techniques needed by both the American and the Soviet manned space programs. Science experiments would be conducted, and the logistics involved in carrying out joint space operations between the two nations would be tried and tested, paving the way for future joint ventures with the **Space Shuttle, Mir**, and the **International Space Station (ISS)**. As the American and Soviet space capsules were incompatible, a new docking module had to be built with a Soviet port on one side and an American port on the other. This module also served as an airlock and a transfer facility, allowing astronauts and cosmonauts to acclimatize to the atmospheres

(continued on page 34)

Apollo

An American-manned space program that built on the achievements of **Mercury** and **Gemini** and eventually landed 12 astronauts on the Moon. Undertaken at a time of intense military rivalry with the Soviet Union, it demanded rapid progress in all aspects of spaceflight. Apollo hardware was also used for other missions, including **Skylab** and the **Apollo-Soyuz Test Project.**

Apollo history

On July 29, 1960, NASA unveiled a plan to develop a three-man spacecraft, called Apollo, capable of operating in low Earth or circumlunar orbit. President **Eisenhower** initially opposed this development beyond the Mercury Project, but Apollo was given the green light by his successor on May 21, 1961, when President **Kennedy** declared America's goal of placing humans on the Moon by the end of the decade. As a NASA historian observed,[141] the decision "owed nothing to any scientific interest in the Moon. The primary dividend was to be national prestige." Not surprisingly, given that beating the Soviets was the main objective, public and government interest in Apollo rapidly waned after the Space Race was won.

NASA leaders had to choose between three ways of getting astronauts to and from the lunar surface: direct ascent, Earth orbit rendezvous, and lunar orbit rendezvous. Direct ascent meant sending a single spacecraft on a straight shot from Earth's surface to the Moon's surface with enough propellant for the return journey, and could only be done with the development of a huge new rocket known as the **Nova**. Earth orbit rendezvous involved first placing the moonship in low Earth orbit, then fueling its booster–a scheme that called for the launch of two Saturn Vs. Only the lunar orbit rendezvous approach would enable the mission to be accomplished with the launch of a single Saturn V, and mainly for this reason it was the one selected by NASA in June 1962.

North American Aviation was chosen to develop the main part of the craft, the so-called Command and Service Module, which could operate together with or independently of a special Lunar Module. To speed development, Apollo vehicles were built in two configurations: Block 1 and Block 2. The former was intended only for test missions in Earth orbit; the latter was more sophisticated and reserved for the Moon shots themselves. However, following a fire, which cost the lives of three astronauts (see below), the Apollo design was overhauled and no Block 1s were launched with a crew aboard. The rest of the Apollo program proved to be a triumph of technology and human endeavor.[8, 22, 59]

Apollo spacecraft

Carried into space atop a **Saturn** launch vehicle were the Command Module (CM), the Service Module (SM), and the Lunar Module (LM). Unmanned test flights and one manned test flight, Apollo 7, were launched by Saturn IBs; all other manned Apollos were launched by Saturn Vs.

Command Module (CM)

A conical three-man capsule that served as the control center and main living area; the CM was the only part of Apollo built to withstand the heat of **reentry**. The forward section contained a pair of thrusters for attitude control during reentry, parachutes for landing, and a tunnel for entering the LM. At the end of the tunnel was an airtight hatch and a removable docking probe used for linking the CM and the LM.

Crewmen spent much of the time on their couches but could leave them and move around. With the seat portion of the center couch folded, two astronauts could stand at the same time. The astronauts took turns sleeping in two sleeping bags mounted behind the left and right couches. Food, water, clothing, waste management, and other equipment were packed into bays that lined the walls of the craft. The pressurization (about one third of sea-level pressure), temperature (about 24°C), and controlled atmosphere afforded a shirtsleeve environment. Spacesuits were worn only during critical phases of a mission, such as launch, reentry, docking, and crew transfer. The left-hand couch was occupied by the commander, who in addition to assuming the duties of command normally worked the spacecraft's flight controls. The center couch was for the CM pilot, whose main task was guidance and navigation, although he also sometimes flew the craft. On a lunar mission, the CM pilot remained in the CM while his two companions descended to the Moon's surface. In the right-hand couch was the LM pilot, who was mainly responsible for managing the spacecraft's subsystems.

The CM had five windows: two forward-facing for use during docking with the LM and three others for general observation. A hatch opposite the center couch was used to enter and leave the CM on the ground. The aft section contained 10 reentry thrusters, their fuel tanks, and the heat-shield.

COMMAND MODULE FACTS
Height: 3.2 m
Diameter (at base): 3.9 m
Weight
    At launch, including crew: 5,900 kg
    At splashdown: 5,300 kg

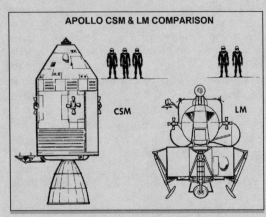

APOLLO CSM & LM COMPARISON

CSM   LM

Apollo Command Service Module and Lunar Module   A comparison of the Apollo CSM and the LM; schematic diagram. *NASA*

Service Module (SM)

An aluminum alloy cylinder at the end of which was the main engine used to place Apollo into lunar orbit and begin the return to Earth. The SM carried the hypergolic (self-igniting) propellants for the main engine, the systems (including fuel cells) used to generate electrical power, and some of the life-support equipment. At four locations on the SM's exterior were clusters of attitude control jets. On the Apollo 15, 16, and 17 missions, the SM also contained a Scientific Instrument Module (SIM) with cameras and other sensors for studying the Moon from orbit.

SERVICE MODULE FACTS
Diameter: 3.9 m
Length: 7.5 m
Engine thrust: 91,000 N
Propellants: hydrazine, UDMH, nitrogen tetroxide

Command and Service Module (CSM)

The combined CM and SM. The CSM orbited the Moon, while the LM conveyed two astronauts to and from the lunar surface and subsequently provided the means of returning to Earth.

Lunar Module (LM)

The part of the Apollo spacecraft in which two astronauts could travel to and from the Moon's surface; it was the first manned spacecraft designed for use exclusively outside Earth's atmosphere. Built by Grumman Aircraft, the LM was a two-stage vehicle consisting of an ascent stage and a descent stage. During descent to the lunar surface and while the astronauts were on the Moon, these stages acted as a single unit. The descent stage contained the components used to de-orbit and land the LM, including the main engine, propellants, and landing gear. Its engine was the first in the American space program that could be throttled, providing a thrust range of 4,890 to 43,900 N, and could swing through 6 degrees from vertical in two planes to give the vehicle maneuverability in landing. The ascent stage, equipped with its own engine of 15,600-N thrust, separated at the start of the climb back to lunar orbit and used the descent stage as a launch platform. Given that every kilogram of the LM had to be paid for with 70 kg of launch vehicle and fuel from Earth, the LM was made as light as possible. Its main cladding was a paper-thin skin of aluminum alloy fixed to aluminum alloy stringers. The ascent stage also had several skins of Mylar to serve as heat and micrometeoroid shields. The ladder enabling the astronauts to climb to the lunar surface was so flimsy that it could only support a man's weight in the one-sixth gravity of the Moon. Weight limita-

Apollo Lunar Module Alan Bean, Lunar Module pilot for the Apollo 12 mission, starts down the ladder of *Intrepid* to join Charles Conrad, mission commander, on the lunar surface. *NASA*

tions also meant the absence of bunks, so that the astronauts could only rest on the floor, and the absence of an airlock, so that the module had to be depressurized and repressurized before and after every excursion. For this reason, and to avoid being locked out of their vehicle so far from home, the astronauts left the hatch slightly ajar during their moonwalks.[166]

LUNAR MODULE FACTS

Height: 6.7 m

Width

    Shortest distance across descent stage: 4.3 m

    Across landing gear, diagonally: 9.4 m

Mass, fully loaded

    Earlier missions: 14,500 kg

    Later missions: 15,900 kg

Launch Escape System (LES)

A 4,000-kg towerlike structure, carrying four solid-propellant motors, mounted on top of the CM at take-off and later jettisoned. In the event of a booster failure or some other imminent danger, the LES could be fired to lift the CM clear of the Saturn V. The LES engines, with a thrust of 654,000 N, were more powerful than the entire Redstone launcher that put the first American into suborbital space, and could provide an acceleration of about $6.5g$. As the maximum acceleration during ascent for Apollo-Saturn V was about $4g$, the CM/LES combination could still be separated, even if the Saturn V engines were running at full thrust.

Lunar Roving Vehicle (LRV)

A two-person open automobile, powered by electric batteries and used on the Moon's surface during the last three Apollo missions. The collapsible LRV was fixed to the side of the LM; it was released and unfolded by pulling a cord. It enabled several trips by the astronauts of Apollo 15, 16, and 17, covering total distances of 28, 27, and 35 km, respectively. The TV camera on the second LRV was used to televise live the launch of the Apollo 16 LM ascent stage. Left on the Moon, the LRVs will be available if needed by future lunar missions.

LUNAR ROVER FACTS

Mass

    Maximum, with payload: 700 kg

    Vehicle only: 210 kg

Length: 3.48 m

Width: 1.83 m

Wheelbase: 2.29 m

Power supply: two 36-V batteries (with one backup)

Maximum speed: 14 km/hr (design), 17.1 km/hr (actual)

Cumulative endurance: 7.8 hours

Design range: 92 km

ALSEP (Apollo Lunar Science Experiment Package)

A set of experiments deployed by astronauts on the Moon. An ALSEP was left by every mission except the first. Each was powered by a small nuclear generator and included a seismometer to measure moonquakes, a solar wind detector, and instruments to measure any trace atmosphere and heat flow from the Moon's interior. All the experiments were turned off in 1978.[20]

Apollo missions

The manned Apollo flights were preceded by a number of unmanned test flights of the Saturn IB and Saturn V launch vehicles. Data from the **Lunar Orbiter** and **Surveyor** missions were used in selecting sites for the Apollo lunar landings.

Disaster struck the program on January 27, 1967. "Gus" **Grissom**, Roger **Chaffee**, and Ed **White**, assigned to the first manned Apollo test flight, had entered the CM for a countdown rehearsal. Just after 6:30 P.M. with the capsule sealed and the countdown at T − 10 minutes, Grissom cried out, "Fire in the spacecraft!" In moments the interior was ablaze in the capsule's pure oxygen atmosphere and the exterior became so hot that technicians were unable to make a speedy rescue. A postmortem revealed that the astronauts had died within seconds, principally from smoke inhalation. Ironically, while everything else but metal inside the capsule was badly burned, a portion of the flight plan survived with only a few pages singed. The postaccident inquiry laid most of the blame on a poorly designed hatch that was impossible to open in under 1½ minutes and on the use of pure oxygen, which had allowed a small spark (possibly from poorly insulated wires under Grissom's seat) to become a conflagration. Flash fires had previously broken out in two boilerplate cabin mockups in September and November 1963. Also, Soviet cosmonaut Valentin Bondarenko was killed in a pure oxygen flash fire in a training simulator in March 1961, although this was only revealed in the late 1980s. A new hatch had been under development at the time of the Apollo tragedy, but the inquiry revealed a catalogue of bad design and shoddy workmanship throughout the Apollo spacecraft. Although the program was delayed by 18 months following the fatal ground test—named Apollo 1 in retrospect to honor the three astronauts who died—the result was a safer vehicle for those destined to fly to the Moon. (See table, "Apollo Test Missions.")

Apollo Test Missions

Mission	Date	Notes
AS-201	Feb. 26, 1966	First unmanned test flight of Saturn IB-Apollo
AS-203	Jul. 5, 1966	Second unmanned test flight of Saturn IB-Apollo
AS-202	Aug. 25, 1966	Third unmanned test flight of Saturn IB-Apollo
Apollo 1 (AS-204)	Jan. 27, 1967	Fatal fire during countdown test
Apollo 4	Nov. 9, 1967	First unmanned test flight of Saturn V
Apollo 5	Jan. 22–24, 1968	Unmanned LM test in Earth orbit (Saturn IB)
Apollo 6	Apr. 4, 1968	Second unmanned test flight of Saturn V

Apollo 7 and 9 were Earth-orbiting missions to test the CM and LM. Apollo 8 and 10 tested various components while orbiting the Moon. Apollo 13 did not land on the Moon due to a major malfunction en route. The six missions that did land (Apollo 11, 12, 14, 15, 16, and 17) brought back a wealth of scientific data and nearly 400 kg of lunar samples. Experiments provided information on soil mechanics, meteoroid impacts, seismic activity, heat flow through the soil, magnetic fields, the solar wind, and the precise distance to the Moon. (See table, "Manned Apollo Flights.")

Apollo 7

Crew
   Commander: Walter **Schirra**
   LM pilot: Walter **Cunningham**
   CM pilot: Donn **Eisele**

The first manned Apollo flight. It was launched by a Saturn IB (unlike all subsequent missions, which used the Saturn V), conducted in Earth orbit, and devoted to testing guidance and control systems, spacesuit design, and work routines. During rendezvous and station-keeping operations, the CSM approached to within 21 m of the spent Saturn IVB stage that had boosted the spacecraft into orbit.[124]

Apollo 8

Crew
   Commander: Frank **Borman**
   LM pilot: William **Anders**
   CM pilot: James **Lovell** Jr.

The first manned flight to and orbit of the Moon, and the first manned launch of the Saturn V. Originally

Manned Apollo Flights

Mission	Launch	Lunar Landing	Recovery	Duration	Crew
Apollo 7	Oct. 11, 1968	—	Oct. 22, 1968	10 days 20 hr	Schirra, Eisele, Cunningham
Apollo 8	Dec. 21, 1968	—	Dec. 27, 1968	6 days 3 hr	Borman, Lovell, Anders
Apollo 9	Mar. 3, 1969	—	Mar. 13, 1969	10 days 1 hr	McDivitt, Scott, Schweickart
Apollo 10	May 18, 1969	—	May 26, 1969	8 days 3 hr	Stafford, Young, Cernan
Apollo 11	Jul. 16, 1969	Jul. 20, 1969	Jul. 24, 1969	8 days 3 hr	Armstrong, Aldrin, Collins
Apollo 12	Nov. 14, 1969	Nov. 19, 1969	Nov. 24, 1969	10 days 4 hr	Conrad, Gordon, Bean
Apollo 13	Apr. 11, 1970	—	Apr. 17, 1970	5 days 23 hr	Lovell, Swigert, Haise
Apollo 14	Jan. 31, 1971	Feb. 5, 1971	Feb. 9, 1971	9 days 0 hr	Shepard, Roosa, Mitchell
Apollo 15	Jul. 26, 1971	Jul. 30, 1971	Aug. 7, 1971	12 days 17 hr	Scott, Worden, Irwin
Apollo 16	Apr. 16, 1972	Apr. 29, 1972	Apr. 27, 1972	11 days 1 hr	Young, Duke, Mattingly
Apollo 17	Dec. 7, 1972	Dec. 11, 1972	Dec. 19, 1972	12 days 14 hr	Cernan, Evans, Schmidt

intended as simply an Earth-orbit test mission, Apollo 8 evolved into an ambitious circumlunar flight at a time when rumors suggested a possible Soviet attempt at a manned orbit of the Moon (see **Russian manned lunar programs**). Its three astronauts became the first human beings to achieve Earth escape velocity and the first to see in person both the farside of the Moon and the whole of our planet from space. During 10 lunar orbits, the crew took star sightings to pinpoint landmarks, surveyed landing sites, took still and motion pictures, and made two TV transmissions. During one transmission on Christmas Eve, they read passages from the Book of Genesis. At 1:10 A.M. EST on Christmas Day, 1968, while on the Moon's farside, the SM's main engine was fired to take the spacecraft out of lunar orbit. As the crew began its return to Earth, Lovell remarked, "Please be informed . . . there is a Santa Claus." Apollo 8 achieved another first when it splashed down in darkness.[125, 319]

Apollo 9

Crew
Commander: James **McDivitt**
LM pilot: Russell **Schweickart**
CM pilot: David **Scott**
Call signs
CM: *Gumdrop*
LM: *Spider*

The first flight of all three main Apollo vehicle elements—the Saturn V, the CSM, and the LM. Following insertion into LEO (low Earth orbit), the Apollo 9 CSM separated from the S-IVB, turned around, docked with the LM, and removed it from the spacecraft lunar adapter. Then McDivitt and Schweickart boarded the LM, undocked it, and flew it independently for over six hours at distances up to 160 km from the CSM. The LM descent stage was jettisoned and left behind in LEO, eventually to burn up in the atmosphere. After the CSM and LM ascent stage redocked, two spacewalks were carried out. McDivitt and Schweickart entered the LM while Scott remained aboard the CSM, and the CSM and LM were both depressurized. The LM hatch was opened, and Schweickart exited, remaining attached to the spacecraft by a foot restraint dubbed the

"golden slipper" because of its gold exterior, and retrieved two experiments from outside the LM. The main purpose of the walk was to test the special spacesuit, known as the EMU (extravehicular mobility unit), and backpack, or PLSS (Portable Life Support System), to be used during the Moon landings. This was the first time an astronaut had ventured into the vacuum of space free of capsule-based life-support equipment. During the second spacewalk, Scott opened the CSM hatch while attached to a life-support umbilical line to demonstrate the ability to prepare for emergency transfer of astronauts between the LM and the CSM should a docking prove impossible once the LM had left the Moon. Although the spacewalks were scheduled to take up to two hours, they were shortened to just 46 minutes because all three astronauts had suffered space sickness earlier in the mission. The LM ascent stage was jettisoned and its engine fired by remote control, placing the craft in a high elliptical orbit. Following separation of the CM and the SM, the CM reentered and splashed down.[117]

Apollo 10

Crew
Commander: Thomas **Stafford**
LM pilot: Eugene **Cernan**
CM pilot: John **Young**
Call signs
CM: *Charlie Brown*
LM: *Snoopy*

The final rehearsal for the first manned lunar landing. Apollo 10's main purpose was to test rendezvous and docking operations between the CSM and the LM in lunar orbit. Having entered orbit around the Moon, Stafford and Cernan transferred to the LM, undocked it, and flew within 15,200 m of the lunar surface. After the LM descent stage had been jettisoned prior to re-docking, the orientation of the ascent stage began to change unexpectedly due, it turned out, to an incorrectly placed switch. The astronauts took manual control of the LM and were able successfully to rendezvous and re-dock with the CSM. The Apollo 10 crew achieved the highest speed ever attained by human beings— 39,896 km/hr.[118]

Apollo 11

Crew
  Commander: Neil **Armstrong**
  LM pilot: Edwin "Buzz" **Aldrin** Jr.
  CM pilot: Michael **Collins**
Call signs
  CM: *Columbia*
  LM: *Eagle*
Duration of moonwalk: 2 hr 33 min
Time spent on Moon: 21 hr 36 min
Samples collected: 21.0 kg

The mission that climaxed with the first manned landing on the Moon. During the final stages of the LM's 12.5-minute descent to the Moon's surface, Armstrong took manual control of the spacecraft and piloted it to a suitable landing site. A low-fuel warning gave Armstrong just 94 seconds to land prior to an abort and return to the CSM. As the LM came down, its descent engine kicked up dust and reduced Armstrong's visibility to a few meters. At 10 m above the surface, the LM lurched dangerously, but Armstrong continued to guide the spacecraft toward a successful touchdown in the Sea of Tranquility at 20:17:40 GMT on July 20, 1969, about 6.5 km from the designated target. The astronauts donned spacesuits and were ready to step onto the Moon about 6.5 hours after their arrival. Armstrong placed a TV camera on the LM ladder then set foot on the Moon. He was watched live on television by an estimated 500 million people. (The only two countries that declined to telecast the moonwalk were the Soviet Union and China.) Aldrin followed about an hour later. The two men set up a flag; deployed a number of experiments including a seismometer, a laser reflector, and a solar wind detector; gathered samples of lunar rock and soil; and took the longest distance phone call in history, from President Nixon. Upon returning to Earth the astronauts were quarantined, initially in a mobile quarantine facility aboard the recovery ship and then for about three weeks in the specially built Lunar Receiving Laboratory at the Johnson Space Center.[10, 11, 100, 121]

Apollo 12

Crew
  Commander: Charles **Conrad** Jr.
  LM pilot: Alan **Bean**
  CM pilot: Richard **Gordon**
Call signs
  CM: *Yankee Clipper*
  LM: *Intrepid*
Duration of moonwalks
  First: 3 hr 56 min
  Second: 3 hr 49 min
Time spent on Moon: 31 hr 31 min
Samples collected: 34.3 kg

A mission planned to build on the success of Apollo 11, with the added goals of making a precision touchdown and of sampling lunar rocks within 0.5 km of the landing site. Apollo 12 began dramatically. Gordon was so convinced that electrical storms would lead to the launch being scrubbed that he fell asleep during the countdown. In fact, Apollo 12 took off on schedule, but, at T + 36 seconds, as the Saturn V passed through low clouds, a lightning bolt discharged through it to the ground. Sixteen seconds later, the rocket was struck again, causing safety mechanisms to disconnect primary power to the CSM and forcing the crew to restore it manually. Once in lunar orbit, the LM separated from the CSM and descended to a pinpoint landing on the Ocean of Storms near a ray of the crater Copernicus and less than 180 m from Surveyor 3, which had soft-landed on the Moon in April 1967. Conrad and Bean went on two moonwalks. During the first, they set up the ALSEP and positioned a color TV camera to provide the first color transmissions from the lunar surface. However, Bean allowed direct sunlight to enter the camera's lens, which damaged its vidicon tube and rendered it useless; television viewers on Earth were able to see the astronauts step onto the Moon but little else. During a second moonwalk, the astronauts walked about 1.5 km, collecting lunar samples and removing parts of Surveyor 3 for return to Earth. For the first time, the astronauts documented each sample they took, including the first double-core tube sample of lunar soil. Later laboratory examination revealed that the Surveyor 3 parts harbored bacteria that had survived 19 months of extreme temperatures, dryness, and the near-vacuum of the lunar environment. Conrad inadvertently carried a *Playboy* photo to the Moon; it had been planted by a NASA employee and Conrad came across it unexpectedly on the lunar surface while flipping through his mission checklist. For the first time, lunar dust tracked into the LM proved to be a problem. Since the dust became weightless after liftoff from the Moon, the astronauts had trouble breathing without their helmets. For the first time, the LM was fired back toward the Moon after its occupants transferred to the CSM. *Intrepid* slammed into the Moon at more than 8,000 km/hr with a force equivalent to an explosion of 9,000 kg of TNT. The resulting artificial moonquake registered on the seismometer that the astronauts had left on the surface, providing valuable data on the Moon's internal makeup. Some lunar dust found its way into the CSM, requiring the astronauts to clean air filter screens every few hours. The splashdown, at 15g, was the hardest ocean landing ever recorded–enough to jar a 16-mm camera from its mounting and hit Al Bean on the head.[122]

Apollo 13

Crew
    Commander: James **Lovell** Jr.
    LM pilot: Fred **Haise**
    CM pilot: John **Swigert**
Call signs
    CM: *Odyssey*
    LM: *Aquarius*

For the superstitious: Apollo 13 was launched on schedule at 13:13 CST (Houston time), April 11, 1970. On April 13, while en route to the moon, an oxygen tank in the SM exploded. The crew got home safely thanks to the consumables and propulsion system of the LM and the ingenuity of ground controllers in improvising LM lifeboat procedures. The S-IVB stage that boosted the mission into translunar trajectory was delivered to the Kennedy Space Center on June 13, 1969–a Friday. Swigert replaced Thomas Mattingly as CM pilot after Mattingly contracted measles (the only preflight substitution of this kind in the history of the American space program). Following liftoff, the second stage S-II booster's center engine cut off 132 seconds early. To compensate, the four remaining S-II engines burned an extra 34 seconds, and the S-IVB third stage burned an extra 9 sec-

onds. The flight continued according to plan. For the first time, the S-IVB third stage was fired on a lunar trajectory following spacecraft separation and struck the Moon so that the resulting moonquake could be measured, at a point about 137 km from the seismometer planted by the Apollo 12 astronauts. Unfortunately, the S-IVB would be the only part of Apollo 13 to reach the lunar surface. About 56 hours after liftoff and more than halfway to the Moon, a spark and resulting fire ruptured the Number Two Oxygen Tank in the SM, causing a violent explosion. This resulted in the loss of all fuel-cell-generated electricity and led to many other complications, including a complete loss of oxygen and water supply from the CSM. The mission was immediately aborted and all efforts shifted to the safe return of the crew. The CSM was powered down, and the crew moved to the LM for the bulk of the return flight. Not wishing to risk complicated maneuvers to turn the spacecraft around, NASA directed Apollo 13 to proceed around the Moon. Virtually all spacecraft systems were shut down to conserve power. The crew squeezed into the LM, which was designed to support two astronauts for about 50 hours but now needed to support all three astronauts for four days. The crew endured temperatures at or below freezing for the bulk of the return flight as well as other hardships, including water rationed at 170 g per astronaut per day. After circling the Moon once, the LM descent engine was fired twice to establish a fast return path. Nearing Earth, Swigert returned to the CSM to power up the craft using onboard batteries. Engineers were not certain that power could be restored due to low temperatures during the flight; however, sufficient power was restored without difficulty. Swigert jettisoned the SM while Lovell and Haise remained aboard the LM. Following jettison, the crew viewed and took dramatic pictures of the explosion's aftermath: an entire side of the SM had been blown out. Eventually, Lovell and Haise joined Swigert on the CM. The LM, which had successfully served as a lifeboat, was jettisoned, and the CM reentered Earth's atmosphere. Under such circumstances, no one knew if the CM would come in at the proper angle to avoid burning up in or skipping off the atmosphere. As in all previous American manned spaceflights, there was a communications blackout of several minutes during reentry. Then, to the cheers of an anxious world, Apollo 13 splashed down within sight of the recovery team and the crew were rescued about one hour later.[61, 123, 168]

Apollo 14

Crew
  Commander: Alan **Shepard**
  LM pilot: Edgar **Mitchell**
  CM pilot: Stuart **Roosa**
Call signs
  CM: *Kitty Hawk*
  LM: *Antares*
Duration of moonwalks
  First: 4 hr 48 min
  Second: 4 hr 35 min
Time spent on Moon: 33 hr 31 min
  Samples collected: 44.8 kg

The launch of Apollo 14 was put back about three months to allow changes to the flight plan and hardware following the experience of Apollo 13. The outbound flight went on schedule, although it took six attempts to successfully dock the CSM and the LM. *Antares* landed on the Moon just 27 m from its target point in the Fra Mauro highlands—the site selected for the aborted Apollo 13 mission. During two moonwalks, Shepard and Mitchell collected rock and soil samples and deployed the ALSEP, a communications antenna, and a color TV camera. For the first time, an astronaut wore a spacesuit that was color-coded. The Apollo 12 astronauts had trouble telling who was who when they reviewed photos taken on the Moon. NASA subsequently decided to place distinguishing marks on one of the spacesuits; Shepard wore red stripes at the knees and shoulders and on the helmet. During the second moonwalk, the astronauts covered about 3 km traveling to and from the rim of Cone Crater. For the first time, a MET (Modularized Equipment Transporter), nicknamed the "rickshaw," was deployed. Resembling a wheelbarrow, it was used mainly to carry tools, photographic equipment, and rock and soil samples. However, as it filled up it tended to tip over, so the astronauts resorted to carrying instead of pushing it. This was the first moonwalk during which astronauts used Buddy Life Support Systems so that they could share life support from one pack in an emergency. Shepard played the first golf shots on the Moon: with a six iron head fixed to a metal rod (the handle of his lunar sample collector), he struck one ball about 180 m and another about twice as far. While Shepard and Mitchell were on the surface, Roosa became the first CSM pilot to carry out extensive onboard experiments from lunar orbit. Concurrent with Apollo

14, the Russian **Lunokhod** 1 probe, operated remotely from ground control, was exploring another part of the lunar surface. The return to Earth went smoothly, and the CM splashed down just 1.5 km from its intended recovery point.[127]

Apollo 15

Crew
  Commander: David **Scott**
  LM pilot: James **Irwin**
  CM pilot: Alfred **Worden**
Call signs
  CM: *Endeavor*
  LM: *Falcon*
Duration of moonwalks
  First: 6 hr 33 min
  Second: 7 hr 12 min
  Third: 4 hr 50 min
Time spent on Moon: 66 hr 55 min
  Samples collected: 78 kg

The first extended-duration manned lunar mission. *Falcon* landed on the Moon in Hadley Rille near the base of the Apennines. Shortly after, Scott stood in the LM upper hatch to photograph the landing area—a scheduled "standup spacewalk" to allow more detailed analysis of the surrounding terrain. For the first time, the Lunar Roving Vehicle (LRV) was taken to the Moon and, following initial difficulties with deployment and steering, used for an excursion to St. George Crater. Scott and Irwin drove the LRV a total of 10 km before returning to set up the ALSEP. During their second outing, the astronauts made a 12-km round-trip to Mount Hadley Delta and found a green crystalline rock, later called the "Genesis Rock" because of its presumed great age. On their third excursion, Scott and Irwin drove to Scarp Crater and Hadley Rille and became the first astronauts to venture beyond the LM's field of view. A feather was dropped during the mission alongside a hammer to illustrate in dramatic style one of Galileo's most significant findings. Sure enough, the feather and hammer hit the Moon's surface simultaneously. For the first time, the liftoff of the LM was photographed by a remote-operated TV camera on the surface. The empty LM was again crashed into the Moon following undocking to measure the impact with seismometers. Also for the first time, a scientific subsatellite was released into lunar orbit from the

CSM; it transmitted data back to Earth for the next year. On the return journey, while about 275,000 km from Earth, Worden went on a 41-minute spacewalk during which he was attached to the CSM by a tether—the most distant EVA up to that time. During it, Worden used handrails and foot restraints to complete three trips to and from the Scientific Instrument Module (SIM) bay on the side of the SM.[265]

Apollo 16

Crew
    Commander: John **Young**
    LM pilot: Charles **Duke** Jr.
    CM pilot: Thomas **Mattingly** II
Call signs
    CM: *Casper*
    LM: *Orion*
Duration of moonwalks
    First: 7 hr 11 min
    Second: 7 hr 23 min
    Third: 5 hr 40 min
Time spent on Moon: 71 hr 2 min
Samples collected: 95 kg

The fifth successful manned lunar mission and the first to visit a highland region of the Moon. Apollo 16's flight went according to plan until the CSM and the LM undocked in lunar orbit. Shortly after, the CSM began to move strangely due to an apparent problem in the craft's thruster controls. This required the CSM and the LM to remain close together until the problem was fixed, delaying the LM's descent by almost six hours. Eventually, Orion touched down on April 20, 1972, in the Descartes highlands, 230 m from the targeted landing area and 5,500 m above lunar "sea level"—the highest manned lunar landing. During the first moonwalk, the astronauts set up the ALSEP and drove the LRV to Flag Crater. Unfortunately, Young tripped and fell over one of the leads attached to the ALSEP, rendering the experiment package useless. However, the day ended well for Young because during his excursion he learned that Congress had approved fiscal year 1973 funding for the Space Shuttle development, without which the program could have been canceled. Young, who later commanded the first Shuttle mission, jumped in the air—or, rather, the vacuum—when he heard the news. During the second moonwalk, the astronauts drove to Stone Mountain, where they made observations and collected rock and soil samples. The third drive, to

Smoky Mountain, was cut short because the water supply for cooling the LM's instrumentation was running low—more water than expected having been used during the delay before landing. In fact, the coolant ran out just moments after the LM and the CSM re-docked. Several records were broken during the mission, including the highest speed by a vehicle on the lunar surface (21 km/hr) and the largest crater yet visited by man—North Ray Crater, about 200 m deep and 1.5 km wide.

Apollo 17

Crew
    Commander: Eugene **Cernan**
    CM pilot: Ronald **Evans**
    LM pilot: Harrison **Schmitt**
Call signs
    CM: *America*
    LM: *Challenger*
Duration of moonwalks
    First: 7 hr 12 min
    Second: 7 hr 37 min
    Third: 7 hr 15 min
Time spent on Moon: 75 hr 0 min
Samples collected: 110 kg

The final Apollo mission to the Moon and the first American manned launch in darkness. *Challenger* landed in the Taurus-Littrow Valley of the Sea of Serenity, a site chosen because a landslide had recently (in geological terms) taken place there, bringing down material from the nearby Taurus Mountains. As with Apollo 16, the first steps onto the Moon were not televised; however, in this case the blackout was planned—the camera gear for recording the first lunar steps having been dispensed with to save weight. During the first of three moonwalks, Cernan and Schmitt planted an American flag that had hung in Mission Control since Apollo 11. They also set up the most advanced ALSEP of the Apollo program and drove the LRV to Steno Crater. On their second excursion—the longest on the Moon to date—the astronauts drove a round-trip of 19 km to South Massif. The final outing, and the last by an Apollo crew, took them to North Massif. Numerous records were set on the mission, including the first flight of a scientist-astronaut—geologist Schmitt—who had been selected by NASA with no prior piloting skills. The Apollo 17 LM and crew logged the longest stay on the Moon; the Apollo 17 CSM completed the most lunar orbits at 75, setting a record manned lunar orbit stay of 147 hours 48 minutes; and Cernan and Schmitt logged the longest total excursion time on the Moon at 22 hours 5 minutes. The Apollo 17 LRV also logged the greatest distance driven on the lunar surface (a total of 35 km), and a record amount of lunar rock and soil samples was collected and returned to Earth.

The last human lunar explorers—to date—left the Moon at 22:45 GMT on December 14, 1971. An economic recession and waning public interest in the Moon led to the cancellation of Apollo 18, 19, and 20, although Apollo hardware did fly again during the Apollo-Soyuz Test Project and the Skylab missions.

Apollo 17 Eugene Cernan, Apollo 17 mission commander, checks out the Lunar Roving Vehicle during the early part of the first Apollo 17 EVA. The mountain in the background is the east end of South Massif. *NASA*

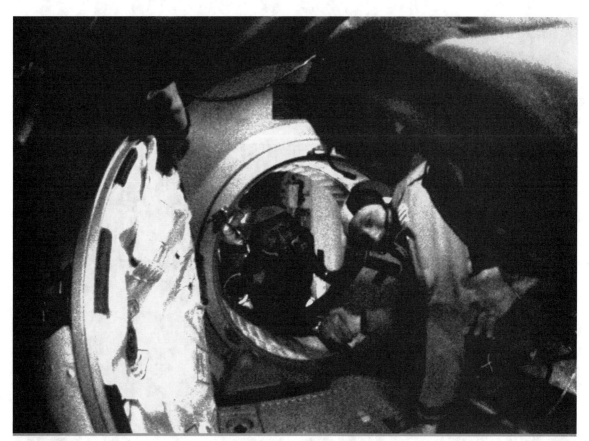

Apollo-Soyuz Test Project   Apollo Commander Thomas Stafford (in foreground) and Soyuz Commander Alexei Leonov make their historic handshake in space during the Apollo-Soyuz Test Project. The handshake took place after the hatch to the Universal Docking Adapter was opened. *NASA*

Apollo-Soyuz Test Project (ASTP)

(continued from page 22)

of each other's vehicles. If the cosmonauts had transferred directly to Apollo, they would have suffered from the bends. Other differences, such as language, were not so easily resolved. The cosmonauts and astronauts agreed to talk with their respective mission controllers in their native tongues, while in-flight communication between the crews would rely mostly on gestures and sign language. National pride also played its part: Americans referred to the mission as the Apollo-Soyuz Test Project, Soviets as the Soyuz-Apollo Test Project.

Soyuz 19 was launched about seven hours ahead of its Apollo counterpart. Once in orbit, the Apollo craft separated from its spent S-IVB booster, turned around, docked with the ASTP docking module, then chased Soyuz 19 to a rendezvous, completing a docking at 12:10 P.M. EDT on

July 17, 1975. Stafford and Slayton entered the docking module and adjusted the air pressure inside, and, finally, in an event broadcast live on global television, the two cosmonauts entered through their side of the docking module and shook hands with the waiting astronauts. The two crews conducted experiments together, shared each other's accommodations and meals, and took part in a variety of press conferences and other live broadcasts. Messages were relayed from the crews directly to President Ford and Premier Brezhnev. The two spacecraft remained docked for two days, then undocked and re-docked for practice purposes, before returning to Earth. Soyuz 19 landed in Russia on July 21, while the Apollo craft remained in space another three days to conduct more on-orbit experiments.

At splashdown a tragedy was only narrowly averted. Difficulties with communications following reentry had distracted Brand so that he forgot to operate the two Earth landing system switches that would deploy the parachutes and deactivate the thrusters. When the drogue chute failed to come out, Brand manually commanded it to deploy, but the swinging of the spacecraft triggered the still-armed thrusters to fire to correct the oscillations. Stafford noticed this and shut them down, but by then the thrusters' nitrogen tetroxide propellant was boiling off and entering the cabin via a pressure-relief valve. So much of the highly toxic gas was drawn into the capsule that the astronauts started to choke. Then the Command Module (CM) hit the water "like a ton of bricks," Stafford said, and turned upside-down. Stafford grabbed the oxygen masks from a locker, but by the time he reached Brand, the CM pilot was unconscious. Later examination showed that the fast-acting gas had blistered the astronauts' lungs and turned them white. Doctors also discovered a shadow on an X-ray of one of Slayton's lungs and, fearing cancer, decided to operate. Fortunately, it proved to be a benign tumor; but had the shadow been found before the flight, Slayton, who had been grounded during the Mercury Project with a heart problem, would probably have been prevented from going into space at all. This was the last manned spaceflight by the United States using a traditional rocket booster, and the last American manned spaceflight prior to the start of the Space Shuttle program.

apolune
The point in an elliptical lunar orbit that is farthest from the Moon.

Aqua
Also known as EOS PM, the second satellite in NASA's **EOS** (Earth Observing System) and a sister craft to **Terra**. Flying three hours behind Terra in the same 705-km-high polar orbit, Aqua maps Earth's entire surface every 16 days and will provide a six-year chronology of the planet and its processes. Its main job is to help investigate the link between water vapor, the most active greenhouse gas, and climate. Three of its instruments—the Advanced Microwave Scanning Radiometer, the Moderate Resolution Imaging Spectroradiometer, and the Clouds' and the Earth's Radiant Energy System Detector—measure cloud cover and surface vegetation, temperatures across Earth's surface and in the atmosphere, humidity, and the flow of energy through the global system. A second package of three instruments—the Atmospheric Infrared Sounder, the Advanced Micro-wave Sounding Unit, and the Humidity Sounder for Brazil—track water as it cycles from Earth's surface

through the atmosphere and back. The first benefit of the mission may be an improvement in daily weather forecasts: Aqua's data are at least a factor of two better than those used in current forecasts.

Launch
    Date: May 4, 2002
    Vehicle: Delta 7920
    Site: Vandenberg
Orbit (circular): 705 km × 98°
Mass: 2,934 kg
Size: 6.6 × 2.6 m

Archytas (c. 428–c. 350 B.C.)
An ancient Greek who figures semimythically in the annals of rocket history. According to Aulus Gellius, a Roman writer, Archytas lived in the city of Tarentum in what is now southern Italy. Around 400 B.C., Gellius relates, Archytas mystified and amused the citizens of Tarentum by flying a pigeon made of wood. Apparently, the bird was suspended on wires and propelled by escaping steam—one of the earliest references to the practical application of the principle on which rocket flight is based. See **Hero of Alexandria**.

arcjet
A simple, reliable form of **electrothermal propulsion** used to provide brief, low-power bursts of thrust, such as a satellite needs for station keeping. A nonflammable propellant is heated, typically changing state from liquid to gas, by an electric arc in a chamber. It then goes out of the nozzle throat and is accelerated and expelled at reasonably high speed to create thrust. Arcjets can use electrical power from solar cells or batteries and any of a variety of propellants. Hydrazine is the most popular propellant, however, because it can also be used in a chemical engine on the same spacecraft to provide high thrust capability or to act as a backup to the arcjet.

ARGOS (Advanced Research and Global Observation Satellite)
The most advanced research and development satellite ever launched by the U.S. Air Force. It carries an ion propulsion experiment, ionospheric instruments, a space dust experiment, a high-temperature semiconductor experiment, and the Naval Research Laboratory's hard X-ray astronomy detectors for X-ray binary star timing observations. Much delayed, it was finally placed into

orbit on the 11th launch attempt along with the secondary payloads **Ørsted** and Sunsat.

Launch
Date: February 23, 1999
Vehicle: Delta 7925
Site: Vandenberg Air Force Base
Orbit: 825 × 839 km × 98.8°

Ariane

A series of launch vehicles developed by **ESA** (European Space Agency). The decision to develop an autonomous European access to space was taken at the same meeting, in Brussels in July 1973, at which ministers also agreed to set up the European Space Agency. The maiden flight of Ariane 1 on December 24, 1979, marked Europe's arrival on the international satellite launch market. Changing payload requirements prompted the development of the Ariane 2 and 3 in the early 1980s, then the Ariane 4, and most recently the Ariane 5. In March 1980, **Arianespace** was formed to handle Ariane production and commercialization. Launches take place from the **Guiana Space Centre**. (See table, "Comparison of the First Three Arianes.")

Ariane 1

Europe's first successful commercial launch vehicle, designed to carry two communications satellites at once. Its development took eight years and was based on the design of a replacement rocket for the **ELDO** (European Launcher Development Organisation) Europa. Ariane 1 flew 11 times from 1979 to 1986, failed twice, and launched a number of communications and other satellites, and the **Giotto** probe to Halley's Comet.

Ariane 2/Ariane 3

Upgrades that provided more lift capability. The first and third stages of the Ariane 2 and 3 were lengthened from those of their predecessor to enable a longer burn time, and the engines of stages one, two, and three were increased in thrust. Ariane 3 had strap-on solid- or liquid-propellant boosters for additional power and flexibility. Out of a total of 17 launches, Ariane 2 flew successfully 5

times between 1987 and 1989, and Ariane 3 flew 11 times from 1984 to 1989.

Ariane 4

A family of six medium- to heavy-lift launch vehicles. The Ariane 4 family builds upon a three-stage liquid-propellant core vehicle, the Ariane 40. The Ariane 42P adds two solid strap-on motors, the Ariane 42L two liquid strap-ons. The Ariane 44P and Ariane 44L use four solid and four liquid boosters, respectively. The Ariane 44LP uses two solid and two liquid strap-on boosters. The Ariane 44L, the most powerful Ariane 4 variant with four liquid boosters, can place more than 4,500 kg into geostationary transfer orbit (GTO). Since its inaugural flight in June 1988, the Ariane 4 has flown more than 100 times and captured almost half the commercial GTO market. Its deployment of two communications satellites in March 2002 marked its 68th successful launch, dating back to 1995. In 2003, Ariane 4 will be retired in favor of the next-generation Ariane 5.

Ariane 5

Originally designed as both a commercial satellite and a manned spacecraft launch vehicle with an LEO (low Earth orbit) payload capacity sufficient to orbit the **Hermes** space plane. After Hermes was cancelled, Ariane 5 was converted to a strictly commercial heavy-lift launcher. Built around a central core with a single liquid hydrogen/liquid oxygen Vulcan engine, it uses two large solid rocket boosters to provide a major fraction of its initial thrust. A small storable-propellant second stage is used for final velocity and insertion maneuvers. After a failure on its first launch attempt in 1996 in which the original **Cluster** mission was lost, Ariane 5 came through a subsequent test launch successfully in October 1997 and made its first successful operational flight in December 1999, carrying ESA's **XMM** (X-ray Multi-Mirror) observatory. Five more successful commercial launches followed through March 2001, but an upper-stage failure during a July 2001 launch left two valuable communications satellites, including ESA's own **ARTEMIS**, in incorrect orbits. Ariane 5's first two launches of 2002—its 11th and 12th in all—successfully orbited the European **Envisat** 1 and the Stellat 5 and N-star C communications satellites. Future developments,

Comparison of the First Three Arianes

	Ariane 1	Ariane 2	Ariane 3
Height	47.4 m	49 m	49 m
Diameter	3.8 m	3.8 m	3.8 m
Liftoff mass	210 tons	219 tons	237 tons
Maximum payload	1,830 kg	2,270 kg	2,650 kg

Ariane An Ariane 5 takes off from Guiana Space Centre. *European Space Agency*

principally to the second stage, will double Ariane 5's GTO payload capacity to about 12 tons by 2006.

ARIANE 5 FACTS
Height: up to 52 m
Core diameter: 5.4 m
Liftoff mass: 710 tons
Maximum payload
    LEO: 16 tons
    GTO: 6.2 tons

Arianespace

A private company incorporated in 1980 with headquarters in Paris to produce, market, and launch **Ariane** launch vehicles and to operate and maintain the **Guiana Space Centre**. Among the shareholders of Arianespace are 36 leading European space and electronics corporations, 13 major banks, and the French space agency, **CNES**. Twelve countries are represented in Arianespace: Belgium, Denmark, Germany, France, Ireland, Italy, the Netherlands, Norway, Spain, Sweden, Switzerland, and the United Kingdom.

Ariel

A series of six British satellites launched by NASA. The first four were devoted to studying the ionosphere, with the remaining two devoted to X-ray astronomy and cosmic-ray studies. Ariel 5 was one of the earliest X-ray astronomy satellites and involved a British-American collaboration. The Science Research Council managed the project for the United Kingdom and the Goddard Space Flight Center for the United States. Several catalogs of X-ray sources stemmed from its observations, which continued until the spring of 1980. (See table, "Ariel Missions.")

ARISE (Advanced Radio Interferometry between Space and Earth)

A mission consisting of one or possibly two 25-m radio telescopes in highly elliptical Earth orbit. The telescope(s) would make observations in conjunction with a large number of radio telescopes on the ground, using very long baseline interferometry to obtain high resolution (10-microarcsecond) images of the most energetic astronomical phenomena in the Universe. ARISE's top science goals would be to image active galactic nuclei (AGN) to learn more about the supermassive black holes at their hearts and to study the formation of energetic jets on scales as small as 100 times the size of the black hole event horizons. Particular targets would be gamma-ray blazars—extremely active AGN that the **Compton Gamma Ray Observatory** has shown emit gamma rays from the very region that can be imaged at radio wavelengths by ARISE. The telescope would be pointed at so-called water megamaser disks in nearby AGN to measure the motions of molecular material within one light-year of the central black hole, thus "weighing" the black hole with unprecedented accuracy. ARISE has been selected for study under the New Mission Concepts for Astrophysics NASA Research Announcement.

Armstrong, Neil Alden (1930–)

A veteran American astronaut and the first human to set foot upon the Moon. Born in Wapakoneta, Ohio, Armstrong received a B.S. in aeronautical engineering from Purdue University and an M.S. from the University of Southern California. He entered the Navy and flew as a naval aviator from 1949 to 1952. In 1955, he joined NACA's (National Advisory Committee for Aeronautic's) **Lewis** Flight Propulsion Laboratory and later transferred to the High Speed Flight Station at Edwards Air Force Base as a civilian aeronautical test research pilot for NACA and NASA. Among the aircraft he tested was the **X-15** rocket plane.[285] He became an astronaut in 1962 and subsequently commanded the **Gemini** 8 and **Apollo** 11 missions. On May 6, 1968, he had a narrow escape when the **Lunar Landing Research Vehicle** he was flying went out of control and he was forced to eject; he landed by parachute and walked away uninjured. Upon returning from the Moon, Armstrong served as deputy associate administrator for the office of Advanced Research and Technology at NASA Headquarters. In 1971, he left NASA to become a professor of aeronautical engineering at the University of Cincinnati, where he taught until 1981. Since then he has been in the business world and is currently chairman of EDO Corp.[44]

Army Ballistic Missile Agency (ABMA)

An organization formed by the U.S. Army on February 1, 1956, at Redstone Arsenal, Huntsville, Alabama, taking over what was previously the Guided Missile Development Division, to develop the **Redstone** and **Jupiter** ballistic missiles. Its first commander was Major General John Mendaris. In July 1960, ABMA's buildings and staff, including Wernher **von Braun**, were transferred to NASA's **Marshall Space Flight Center**, which remains in the midst of the Redstone Arsenal.

Ariel Missions

Spacecraft	Date	Launch Vehicle	Launch Site	Orbit	Mass (kg)
Ariel 1	Apr. 26, 1962	Delta	Cape Canaveral	398 × 1,203 km × 53.8°	60
Ariel 2	Mar. 27, 1964	Scout X-3	Wallops Island	287 × 1,349 km × 51.7°	68
Ariel 3	May 5, 1967	Scout A	Vandenberg	499 × 604 km × 80.6°	90
Ariel 4	Dec. 11, 1971	Scout B	Vandenberg	476 × 592 km × 82.0°	100
Ariel 5	Oct. 15, 1974	Scout B	San Marco	504 × 549 km × 2.9°	129
Ariel 6	Jun. 2, 1979	Scout D	Wallops Island	372 × 383 km × 55.0°	154

Neil Armstrong Armstrong in front of the X-15 that he piloted. *NASA*

Arnold, Henry H. "Hap" (1886–1950)

The commander of the U.S. Army Air Forces in World War II and the only air commander ever to attain the five-star rank of general of the armies. Arnold was especially interested in the development of sophisticated aerospace technology to give America an edge in air superiority, and consequently he helped foster the development of such innovations as jet aircraft, rocketry, rocket-assisted takeoff, and supersonic flight. After a lengthy career as an Army aviator and commander spanning the two world wars, he retired from active service in 1945 but continued to urge that the United States should develop a postwar deterrent force that included long-range ballistic missiles, developed from the V-2 rather than short-range guided missiles.[12, 57, 82] See **guided missiles, postwar development**.

Around the Moon

See *From the Earth to the Moon*.

ARTEMIS (Advanced Relay Technology Mission)

The most advanced—and, at $850 million, most expensive—communications satellite ever developed by ESA (European Space Agency). ARTEMIS supports laser transmissions of voice and data as well as traditional radio frequency (RF) links and is the first non-American satellite to use ion propulsion for station keeping. Positioned on the equator over central Africa, ARTEMIS is supposed to become a key element of Europe's **EGNOS** satellite navigation system, broadcasting GPS-like navigation signals—a role that seemed threatened when a faulty launch in July 2001 placed the spacecraft in a much lower orbit than intended. However, following the partial launch failure,

ESA devised a four-step recovery strategy that should allow ARTEMIS eventually to reach its correct geostationary position and function as originally planned for at least five years. The first two steps of this strategy involved several firings of the satellite's solid-propellant apogee kick motor to raise the apogee (highest point of the orbit) and then circularize the orbit at about 31,000 km, while the second two involved an unforeseen use of the ion engine for maneuvering into geostationary orbit.

Launch
   Vehicle: Ariane 5
   Date: July 12, 2001
   Site: Kourou
Initial orbit
   Intended: 858 × 35,853 km × 2°
   Actual: 590 × 17,487 km × 2.9°

Artemis Project, the
A multi-industry program to establish a commercial lunar base, led by the Lunar Resources Company of Houston. It is supported by the Artemis Society, which publishes *Pleiades,* a monthly newsletter, and *Artemis,* a bimonthly commercial magazine.

artificial gravity
The simulation of the pull of gravity aboard a **space station**, **space colony**, or manned spacecraft by the steady rotation, at an appropriate angular speed, of all or part of the vessel. Such a technique may be essential for long-duration missions to avoid adverse physiological (and possibly psychological) reactions to **weightlessness**.

The idea of a rotating wheel-like space station goes back as far as 1928 in the writings of Herman **Noordung** and was developed further by Werhner **von Braun**. Its most famous fictional representation is in the film *2001: A Space Odyssey,* which depicts spin-generated artificial gravity aboard a spaceship bound for Jupiter. The **O'Neill-type space colony** provides another classic illustration of this technique.

However, there are several reasons why large-scale rotation is unlikely to be used to simulate gravity in the near future. In the case of a manned Mars spacecraft, for example, the structure required would be prohibitively big, massive, and costly in terms of energy to run (except possibly in the case of an **orbital cycler**). A better approach for such a mission, and one being explored, is to provide astronauts with a small spinning bed on which they can lie for an hour or so each day, head at the center and feet pointing out, so that their bodies can be loaded in approximately the same way they would be under Earth-normal gravity.

In the case of space stations, one of the objects is to carry out experiments in zero-g, or, more precisely, **microgravity**. In a rotating structure, the only gravity-free place is along the axis of rotation. At right-angles to this axis, the pull of simulated gravity varies as the square of the tangential speed.

Another way to achieve Earth-normal gravity is not by constant rotation, which produces the required force through angular acceleration, but by steadily increasing straight-line speed at just the right rate. This is the method used in the hypothetical **1g spacecraft**.[279]

ARTV (Advanced Reentry Test Vehicle)
Early American suborbital reentry tests involving mice named Mia, Mia II, and Wickie. (See table, "ARTV Missions.")

Launch
   Vehicle: Thor-Able
   Site: Cape Canaveral

Aryabhata
India's first satellite, named for the Indian mathematician (c. A.D. 450). The Soviet Union assisted India in developing Aryabhata, which carried out satellite technology tests and made observations of the upper atmosphere.

Launch
   Date: April 19, 1975
   Vehicle: Cosmos-3M
   Site: Kapustin Yar
Orbit: 398 × 409 km × 50.7°
Mass: 360 kg

ARTV Missions

Mission	Launch Date	Notes
ARTV 1	Apr. 24, 1958	Failed due to Thor turbopump problem; Mouse Mia not recovered
ARTV 2	Jul. 10, 1958	Mouse Mia II reached 1,600 km altitude, flew 9,600 km range, but reentry vehicle not recovered
ARTV 3	Jul. 23, 1958	Mouse Wickie not recovered; nosecone lost

ASAT (antisatellite)

A satellite or other device whose purpose is to disable an enemy satellite. The method used could involve either the physical destruction of the satellite or interference with its communications or power systems.

ASCA (Advanced Satellite for Cosmology and Astrophysics)

Japan's fourth X-ray astronomy mission, launched by ISAS (Institute of Space and Astronautical Science), and the second for which the United States provided part of the scientific payload. This included four grazing-incidence X-ray telescopes developed at the Goddard Space Flight Center, each of which worked in the energy range of 0.7 to 10 keV. After eight months of instrument validation, ASCA became a guest-observer project open to astronomers in Japan, America, and the ESA (European Space Agency) member states. Among its targets for study were the cosmic X-ray background, active galactic nuclei, galaxy clusters, and supernovae and their remnants. ASCA was the first satellite to use CCDs (charge-coupled devices) for X-ray astronomy. It is also known by its national name, Asuka ("flying bird"), and prior to launch was called Astro-D.

Launch
   Date: February 20, 1993
   Vehicle: M-3S
   Site: Kagoshima
   Orbit: 538 × 645 km × 31°
   Size: 4.0 × 1.2 m
   Mass: 417 kg

ASCE (Advanced Spectroscopic and Coronographic Explorer)

A proposed solar observation satellite to study the physical processes in the outer atmosphere of the Sun that lead to the solar wind and explosive coronal mass ejections. It would carry three solar experiments–the Large Aperture Spectroscopic and Polarimetric Coronograph, the Extreme Ultraviolet Imager, and a deployable mast supporting a remote external occulter (for blocking the Sun's bright disk)–more advanced than their counterparts on **SOHO** (Solar and Heliospheric Observatory). ASCE, led by John L. Kohl of the Smithsonian Astrophysical Observatory, is one of four **MIDEX** (Medium-class Explorer) missions selected by NASA in April 2002 for further development, two of which will be launched in 2007 and 2008.

ASI (Agenzia Spaziale Italiano)

The Italian Space Agency; a government agency, formed in 1988, that identifies, coordinates, and manages Italian space programs and Italy's involvement with ESA.

ASLV (Advanced Satellite Launch Vehicle)

An Indian launch vehicle derived from the **SLV-3**. The all-solid-propellant ASLV was created by adding two additional boosters modified from the SLV-3's first stage and by making other general improvements to the basic SLV-3 four-stage stack. In fact, it is a five-stage vehicle, since the core first stage does not ignite until just before the booster rockets burn out. The first launch of the ASLV on March 24, 1987, failed when the bottom stage of the core vehicle did not ignite after booster burnout. The second attempt, on July 13, 1988, ended with the **Rohini** payload falling into the Bay of Bengal when the vehicle became unstable after release of the boosters. Finally, on May 20, 1992, **SROSS**-3 was inserted into LEO (low Earth orbit) by the third ASLV. However, instead of entering a circular orbit near 400 km, the ASLV achieved only a short-lived orbit of 256 km by 435 km. The fourth ASLV mission in May 1994 successfully reached its programmed orbit of 434 km by 921 km with the SROSS-C2 payload. The vehicle is likely to be phased out shortly in favor of the more powerful **PSLV** (Polar Satellite Launch Vehicle).

ASLV STATISTICS
Thrust at liftoff: 910,200 N
Mass, fully fueled: 41,000 kg
Payload to LEO: 150 kg
Diameter: 1.0 m
Length: 23.5 m

ASSET (Aerothermodynamic Elastic Structural Systems Environmental Tests)

The first part of the U.S. Air Force **START** (Spacecraft Technology and Advanced Reentry Test) project on lifting bodies. ASSET test flights took place from 1963 to 1965 and used surplus **Thor** missiles, returned from the United Kingdom, for 4,000 m/s flights and Thor-Delta for 6,000 m/s flights. A spacecraft known as an aerothermodynamic structural test vehicle (ASV) was flown on a suborbital trajectory to a recovery zone near Ascension Island in the Atlantic to carry out heat-shield experiments.

Asterix

The first French satellite, also known as A-1, which served as a test payload for the **Diamant** rocket. With its launch, France became the first nation other than the two superpowers to place its own satellite in orbit.

Launch
   Date: November 26, 1965
   Vehicle: Diamant
   Site: Hammaguira
   Orbit: 527 × 1,697 km × 34.3°
   Mass: 42 kg

asteroid missions
See **comet and asteroid missions**.

Astra
A **communications satellite** system supported by a fleet of (at mid-2002 count) 12 communications satellites for direct-to-home reception of TV, radio, and computer data. The system is owned and operated by SES Astra (headquartered in Betzdorf, Luxembourg), a subsidiary of SES Global. Seven of the Astra satellites are co-located at 19.2° E, three at 28.2° E, and one each at 5.2° E and 24.2° E. Their **footprints** provide coverage across the whole of Europe.

Astro (shuttle science payload)
A 12-ton Space Shuttle–borne ultraviolet observatory that consists of three ultraviolet instruments: the Ultraviolet Imaging Telescope for imaging in the spectral range of 120 to 310 nm, the Hopkins Ultraviolet Telescope for spectrophotometry in the range of 43 to 185 nm, and the Wisconsin Ultraviolet Photopolarimetry Experiment for spectropolarimetry in the range of 125 to 320 nm. Astro is attached to and controlled from the Shuttle throughout its mission. Astro-1 also gathered X-ray data in the 0.3- to 12-keV range using the **BBXRT** (Broad Band X-ray Telescope). Astro-2 featured a Guest Observer Program but did not carry the BBXRT. (See table, "Astro Flights.")

Astro Flights

Spacecraft	Shuttle Deployment Date	Mission	Orbit
Astro-1	Dec. 2, 1990	STS-35	190 km × 28°
Astro-2	Mar. 2, 1995	STS-67	187 km × 28°

astro-
A prefix (from the Greek *astron*) meaning "star" or "stars" and, by extension, sometimes used as the equivalent of "celestial," as in astronautics.

Astro-
The prelaunch designation of a series of Japanese astronomical satellites built by ISAS (Institute of Space and Astronautical Science) to carry out observations of cosmic sources at X-ray and other wavelengths. This series has included Astro-A (**Hinotori**), Astro-B (**Tenma**), Astro-C (**Ginga**), and Astro-D (**ASCA**). **Astro-E** suffered a launch failure, while Astro-F, in a break with tradition, has already been named **IRIS** (Infrared Imaging Surveyor).

Astrobiology Explorer (ABE)
A proposed orbiting infrared telescope, with a mirror about 40 cm in diameter, that would search for the spectral signatures of complex interstellar organic compounds in the range of 2.5 to 20 microns (millionths of a meter). This data would enable astrobiologists to learn more about the galactic abundance, distribution, and identities of molecules that may play a role in the origin of life. ABE, led by Scott Sandford of the **Ames Research Center**, is one of four **MIDEX** (Medium-class Explorer) missions selected by NASA in April 2002 for further development, two of which will be launched in 2007 and 2008.

astrodynamics
The science concerned with all aspects of the motion of satellites, rockets, and spacecraft. It involves the practical application of celestial mechanics, astroballistics, propulsion theory, and allied fields.

Astro-E
The fifth in a series of Japanese astronomy satellites designed to observe celestial X-ray sources. Launched on February 10, 2000, it failed to reach orbit after the first stage of its M-5 launch vehicle malfunctioned. Astro-E carried three main science instruments developed in partnership with the Goddard Space Flight Center and the Massachusetts Institute of Technology: an X-ray spectrometer to provide high-resolution spectroscopy in the 0.4- to 10-keV range, four X-ray imaging spectrometers to take X-ray images of objects in the 0.4- to 12-keV range, and hard X-ray imaging detectors to measure high-energy X-rays above 10 keV. It was intended to complement the **Chandra X-ray Observatory** and **XMM-Newton** (X-ray Multi-Mirror).

Astro-F
See **IRIS** (Infrared Imaging Surveyor).

Astron
A Soviet astrophysics satellite that carried science instruments from the Soviet Union and France and made observations of cosmic ultraviolet and X-ray sources from a highly elliptical orbit.

Launch
   Date: March 23, 1983
   Vehicle: Proton-K
   Site: Baikonur
Orbit: 28,386 × 175,948 km × 34.7°
Mass: 3,250 kg

astronaut

A person who flies in space as a crew member or a passenger.[87] More specifically, someone who flies aboard an American spacecraft. See **cosmonaut**.

astronautics

The science of spaceflight, including the building and operation of space vehicles.

astronomical unit (AU)

The mean distance between Earth and the Sun. One AU = 149,597,870 km = 92,955,806 miles = 499 light-seconds. One **light-year** = 63,240 AU.

astronomy satellites

Spacecraft dedicated to making observations of natural objects in space, including the Sun and other stars, nebulae, galaxies, and quasars.[188]

Asuka

See **ASCA**.

Athena (launch vehicle)

A family of solid-propellant launch vehicles built by Lockheed-Martin for delivering small payloads into LEO (low Earth orbit), geostationary transfer orbit (GTO), or interplanetary orbits. After a string of seven commercial launch successes, the prospects for Athena I are uncertain with the shrinkage in the market for small commercial satellite launches. However, the future for Athena II brightened in 2001 when NASA added the rocket to its launch services contract along with the **Delta** and the **Atlas**.

The Athena program was begun in January 1993 by Lockheed to apply its expertise in solid-propellant missile technology, developed as a result of the Polaris, Poseidon, and Trident programs, to launching lightweight payloads into space. What was initially called the Lockheed Launch Vehicle (LLV) became the Lockheed Martin Launch Vehicle (LMLV), following Lockheed's merger with Martin Marietta to form Lockheed Martin. The core launch vehicle was named LMLV-1, with a larger version named LMLV-2. Later, LMLV-1 was renamed Athena I, while LMLV-2 became the Athena II. The first launch of an Athena rocket took place in August 1995 and ended in failure. But more recently, Athena I and Athena II vehicles have been used to launch NASA's **Lewis** and **Lunar Prospector** missions, **ROCSAT**-1 for the People's Republic of China, and four small satellites in a single launch from **Kodiak Launch Complex**—a first for the new Alaskan rocket site.

The Athena launch system integrates several different solid motor stages and common equipment to create the two-stage Athena I and the three-stage Athena II. Common to the two vehicles are the avionics package, the separation system, the destruct system, and the launch equipment. The Athena I has a first stage with a Thiokol Castor 120 solid motor, a second stage with a Pratt & Whitney Orbus 21D solid motor, and a hydrazine-fueled Primex Technologies Orbit Adjust Module (OAM). The Athena II uses the same stack but adds another stage with a Castor 120. (See table, "Athena Launch Vehicles.")

Athena Launch Vehicles

	Athena I	Athena II
Overall length	19.8 m	28.2 m
Core diameter	2.4 m	2.4 m
Payload to LEO	794 kg	1,896 kg
Thrust		
First stage	1,900,000 N	1,900,000 N
Second stage	194,000 N	1,900,000 N
Third stage	–	194,000 N

Athena Science Payload

See **Mars Exploration Rovers**.

Atlantic Missile Range (AMR)

An instrumented missile test range extending 8,000 to 9,700 km from Cape Canaveral to a point beyond Ascension Auxiliary Air Force Base. It includes a series of island-based tracking stations and ocean-range vessels to gather performance data.

Atlantis

A Space Shuttle Orbiter, also designated OV-104. *Atlantis* was named in honor of a two-masted ketch that supported oceanographic research for the Woods Hole Oceanographic Institute in Massachusetts between 1930 and 1966. It first flew on October 3, 1985, as mission STS-51J. Other *Atlantis* milestones have included the deployment of **Magellan** (STS-30) and **Galileo** (STS-34), and the first docking of a Space Shuttle to the **Mir** space station (STS-71).

Atlas

See article, pages 44–46.

atmosphere

A layer of gases surrounding Earth or some other planet. Above the **troposphere** (which extends from the ground

(continued on page 46)

Atlas

America's first intercontinental ballistic missile (ICBM), from which evolved a hugely successful family of space launch vehicles. A modified Atlas was used to launch the orbital flights in the **Mercury** Project, and modern versions of the Atlas continue to play a central role in the U.S. space program. (See table, "Current Atlas Family," on page 46.)

History

Atlas has its roots in October 1945, when the U.S. Army Air Corps sought proposals for new missile systems. A contract was awarded to the San Diego–based Consolidated Vultee Aircraft (Convair) Corporation to develop the so-called MX-774 "Hiroc" missile with a range of about 11,000 km and some unusually advanced features for its time. Among these was a single-wall construction of stainless steel so thin that the missile was kept from collapsing only by the internal pressure of its fuel tanks—a design that remains unique to the Atlas family. The weight saved meant increased range. Other new features included a detachable payload section and gimbaled rocket engines for more precise steering instead of exhaust deflector vanes, which were then common.

Despite these innovations, the MX-774 project was canceled in 1947 and priority given to developing the **Navaho**, the **Snark**, and the **Matador**. However, Convair built and test-launched three of the missiles originally authorized. None was a total success, but Convair continued research and conceived another revolutionary idea that would eventually find its way into the Atlas and become its most defining feature. This is the "stage-and-a-half" propulsion system in which three engines—two boosters and a **sustainer engine**—are fed by the same liquid oxygen/RP-1 (kerosene mixture) propellant tanks and all ignited at liftoff. During the first few minutes of flight, the boosters shut down and fall away (to save weight), while the sustainer continues burning.

In 1951, with the outbreak of the Korean War and rising Cold War tensions, Convair received a new Air Force contract to develop a long-range nuclear ballistic missile incorporating the main features of the MX-774. It was called "Atlas," a name proposed by Convair lead engineer Karel **Bossart** and approved by the Air Force. In September 1955, Atlas was given the highest national development priority and by 1959 was being deployed as an ICBM. But operational versions of the Atlas missile, known as the Atlas D, E, and F, were destined to be not only weapons. Having proved themselves reliable and versatile, they became the core boosters for a range of space launch vehicles, including, in chronological order, the Atlas-Able, Atlas-Mercury, Atlas-Agena, and Atlas-Centaur.[47, 200, 228]

Atlas-Able

A four-stage rocket with an Atlas D first stage and **Able** upper stages. After being used in three unsuccessful attempts to send early **Pioneer** probes to the Moon, the short-lived Atlas-Able was retired in 1960.

Atlas-Mercury

See **Mercury-Atlas**.

Atlas-Agena

A series of rockets based on Atlas first stages and **Agena** second stages. Two were important in the space program. The Atlas-Agena B used an Atlas E or F first stage and an Agena B—the first Agena to have multiple restart capability—as the second stage. Among the spacecraft launched by Atlas-Agena Bs were the **Ranger** lunar probes, **Mariner** 1 and 2, **OGO**-1, and the **MIDAS** and **Samos** military satellites. The Atlas-Agena D featured an improved and lightened Agena second stage and, in one of its configurations, solid Burner third and Star 17 fourth stages for geosynchronous launches. Atlas-Agena Ds were responsible for many launches including those of the **Lunar Orbiters**; Mariner 3, 4, and 5; **OAO**-1; and the **Vela** and other reconnaissance satellites. Atlas-launched modified Agena Ds were used as target vehicles in **Gemini** rendezvous and docking missions.

Atlas-Centaur

A family of Atlas-based first- and upper-stage combinations that evolved from the Atlas D and remains in use. The original Atlas-**Centaur**, introduced in 1962, used the D as first stage and the powerful liquid oxygen/liquid hydrogen–propelled Centaur as second stage. An improved version, incorporating both upgraded Atlas and Centaur stages, debuted in 1966. By this time, the Atlas ICBM was nearing the end of its operational life and about to be eclipsed by more advanced missiles such as the **Titan** and the Minuteman. But as a space launch vehicle the Atlas-Centaur had taken on a life of its own, no longer directly tied to its military ancestor.

Atlas II-Centaur

Introduced in the mid-1980s, the II and IIA models incorporated powerful Centaur stages capable of delivering bigger payloads into geosynchronous and other orbits. A third and still more powerful model, the Atlas IIAS—which, together with the IIA, remains in service—uses strap-on solid rocket boosters for the

Atlas Lockheed Martin's Atlas IIA launch of General Electric Americom's GE-1 from Cape Canaveral. *Lockheed Martin*

first time in the Atlas family. Each of the four Castor 4A strap-ons has a thrust of 440,000 N; two are ignited at liftoff and two others about 70 seconds into flight after the first pair burn out. (See table, "Atlas IIA-Centaur and Atlas IIAS-Centaur.")

Atlas IIA-Centaur and Atlas IIAS-Centaur

Total length: 47.5 m; core diameter: 3.1 m

	Maximum Payload (kg)	
	Atlas IIA-Centaur	Atlas IIAS-Centaur
LEO	7,100	8,600
GTO	2,900	3,700
Earth-escape	2,100	2,670

Atlas III

Intended as a gradual replacement for the Atlas II series and as an evolutionary step to the Atlas V series. The Atlas III marks the first break from the traditional Atlas stage-and-a-half combination. Replacing the old two-booster-plus-sustainer configuration is a single Russian liquid-fueled NPO Energomash RD-180—a two-chamber version of the RD-170 that powers the Zenit and the first Russian propulsion system to be used by an American-designed launch vehicle. Atlas manufacturer Lockheed Martin currently offers two versions: the Atlas IIIA with a single Pratt & Whitney RL-10A-4-2 engine on the Centaur upper stage, and the Atlas IIIB with two RL-10A-4-2 engines on a stretched Centaur upper stage to increase geostationary transfer orbit (GTO) performance to about 4,500 kg. The first launches of the Atlas IIIA, carrying Eutelsat W4, on May 25, 2000, and of the Atlas IIIB,

carrying EchoStar 7, on February 21, 2002, were completely successful.

Atlas V

The latest incarnation of the Atlas, developed by Lockheed Martin under the Air Force's EELV (Evolved Expendable Launch Vehicle) contract and introduced in 1999. The Atlas V comes in various configurations based on a Common Core Booster with a RD-180 first-stage engine, a reinforced first-stage structure, and an increased first-stage propellant load. These modifications, combined with the stretched Atlas IIIB-Centaur upper stage, give the Atlas V a minimum GTO payload capacity of over 4,500 kg. Larger variants that use a combination of several common core boosters and solid strap-ons will extend this GTO capacity to over 8,600 kg. Large military payloads and commercial communications satellites will make up the bulk of the Atlas V's workload. Its maiden launch took place successfully on August 21, 2002.

Current Atlas Family

Version	Liftoff Mass (tons)	Payload (tons)	
		LEO	GTO
II A	184.7	7.10	2.90
II AS	237.1	8.62	3.72
III A	214.3	8.64	4.06
III B	225.4	10.76	4.50
V			
401/402	337.4	12.50	4.95
431	474.5	19.30	7.64
501/502	309.0	10.30	3.97
551/552	540.3	20.50	8.67

atmosphere

(continued from page 43)

up to about 20 km) lies the upper atmosphere, which consists of the **stratosphere**, the **mesosphere**, and the **thermosphere**.

atmospheric braking

See **aerobraking**.

atmospheric trajectory

The portion of a return mission from orbit that takes place within the atmosphere.

ATS (Applications Technology Satellite)

A series of NASA satellites designed to explore and flight-test new technologies and techniques for communications, navigation, and weather satellites. Among the areas investigated during the program were spin stabilization, gravity gradient stabilization, complex synchronous maneuvering, communications experiments, and the geostationary orbit (GSO) environment. Although the ATS flights were intended mainly as test-beds, they also collected and transmitted weather data and worked at times as communications satellites. In addition, ATS-6, as well as carrying out technology experiments, became the world's first educa-

ATS Spacecraft

	Launch			
	— Launch site: Cape Canaveral			
Spacecraft	Date	Vehicle	GSO Location	Mass (kg)
ATS 1	Dec. 7, 1966	Atlas-Agena D	10° W	352
ATS 2	Apr. 6, 1967	Atlas-Agena D	Failed to reach correct orbit	370
ATS 3	Nov. 5, 1967	Atlas-Agena D	105° W	365
ATS 4	Aug. 10, 1968	Atlas LV-3C	Failed to reach correct orbit	391
ATS 5	Aug. 12, 1969	Atlas LV-3C	108° W	821
ATS 6	May 30, 1974	Titan IIIC	1.6° W	930

tion satellite, transmitting educational programs to India, the United States, and other countries. It also played a major role in docking during the **Apollo-Soyuz Test Project**. (See table, "ATS Spacecraft.")

Atterley, Joseph
Pseudonym of George **Tucker**.

attitude
The orientation of a spacecraft—its **yaw**, **roll**, and **pitch**—with respect to a particular frame of reference. In the case of spacecraft in Earth orbit, this frame of reference is usually fixed relative to Earth. However, interplanetary space probes often use bright stars as a frame of reference.

attitude control
The process of achieving and maintaining a particular orientation in space in order to satisfy mission requirements. Attitude control is usually accomplished using small thrusters in conjunction with a measuring instrument, such as a star sensor. It is maintained by a stabilizing device, such as a gyroscope, or by **spin stabilization**.

attitude jets
Fixed or movable gas nozzles or small rocket motors that adjust or change a spacecraft's attitude.

augmentor
A duct usually enclosing the exhaust jet behind the **nozzle** exit section of a rocket to provide increased **thrust**.

Augustine, Norman R. (1935–)
A central figure in the American aerospace industry who has played an important role in shaping U.S. space policy. Augustine served as under secretary of the army, assistant secretary of the army for research and development, and assistant director of defense research and engi-neering in the Office of the Secretary of Defense before becoming chairman and chief executive officer of the Martin Marietta Corporation in the 1980s. In 1990, he was appointed head of an advisory committee for the Bush (senior) administration that produced the *Report of the Advisory Committee on the Future of the U.S. Space Program*[245]–a pivotal study in charting the course of the space program in the first half of the 1990s.[15]

Aura
The third spacecraft in NASA's **EOS** (Earth Observing System) following **Terra** and **Aqua**. It carries four instruments that will provide global surveys of atmospheric trace gases, both natural and human-made, and their transformations. Temperature, geopotential heights, and aerosol fields will also be mapped from the ground up through the **mesosphere**. Aura is scheduled for launch in June 2003 and is expected to operate for at least five years. It was formerly known as EOS Chemistry 1.

Australian Space Research Institute
A nonprofit organization that coordinates, promotes, and conducts space research and development projects in Australia involving both Australian and international (mainly university) collaborators. Among its specific activities are the development of sounding rockets, small satellites (especially microsatellites), high-altitude research balloons, and appropriate payloads. It publishes a monthly newsletter and a quarterly technical journal.

auto-igniting propellant
A liquid propellant that ignites, with a small delay, at room temperature.

auto-ignition temperature
The temperature at which combustible materials ignite spontaneously in air.

Avdeyev, Sergei (1956–)
Cosmonaut, selected in 1987, who holds the record for the longest total time spent in space–747 days. This was accumulated during three stays aboard **Mir**: July 27, 1992, to February 1, 1993 (188.9 days); September 3, 1995, to February 29, 1996 (179.1 days); and August 13, 1998, to August 28, 1999 (379.6 days). On these missions, Avdeyev carried out 10 spacewalks lasting a total of 42 hours.

average thrust
The **total impulse** divided by the **burn time**.

avionics
The electronics and instrumentation that help in controlling a flight.

AXAF (Advanced X-ray Astrophysics Facility)
See **Chandra X-ray Observatory**.

axis
One of three lines of reference around which a spacecraft can rotate. These lines are the longitudinal (front-to-back), horizontal, and vertical axes. Rotation along them is known as **roll**, **pitch**, and **yaw**, respectively.

Ayama
See **ECS (Experimental Communications Satellite)**.

Azur
Germany's first scientific satellite; it studied the Van Allen belts, solar particles, and aurora. Its name is German for "blue."

Launch
  Date: November 8, 1969
  Vehicle: Scout B
  Site: Vandenberg Air Force Base
Orbit: 373 × 2,127 km × 102.7°
Mass: 71 kg

B

Bacon, Roger (c. 1214–c. 1292)

An English philosopher and experimenter who described the preparation of **black powder** sometime in the late 1240s. He also wrote in his *Epistola Fratris Rog. Baconis, de secretis operibus artis et naturae et nullitate magiae* (Epistle of Roger Bacon on the Secret Works of Art and of Nature and Also on the Nullity of Magic) about devices that sound like rockets:

> We can, with saltpeter and other substances, compose artificially a fire that can be launched over long distances. . . . By only using a very small quantity of this material much light can be created accompanied by a horrible fracas. It is possible with it to destroy a town or an army. . . . In order to produce this artificial lightning and thunder it is necessary to take saltpeter, sulfur, and *Luru Vopo Vir Can Utriet*.

The last five mysterious words make an anagram that conceals the proportion of powdered charcoal needed to make the explosive.

Baikonur Cosmodrome

One of three major **Russian launch sites**; it is located in a region of flat, semi-arid grasslands, roamed by herds of wild horses and camels, in the former Soviet republic of Kazakhstan, northeast of the Aral Sea at 45.6° N, 63.4° E. Although typically not as active as **Plesetsk**, this site is used for all Russian manned and planetary missions. Construction of the secret missile complex, originally known as Test Area No. 5, began in 1955. It was from here that the first artificial satellite, Sputnik 1, and the first manned spacecraft, Vostok 1, were launched. All subsequent Russian-manned missions—as well as geostationary, lunar, planetary, and many ocean surveillance missions—have lifted off from Baikonur. It is the only site equipped to launch the giant **Proton**. Until recently, the cosmodrome's designation was intentionally misleading. The former Soviet Union used the name and coordinates of a small mining town, Baikonur, to describe its secret rocket complex with the aim of concealing its true location. In fact, the launch complex is about 370 km southwest of Baikonur, near the railway station and village of Tyuratam and close to the city of Leninsk. However, in 1998, Leninsk was renamed Baikonur City.

Baker

A female squirrel monkey who was the traveling companion of **Able** aboard a U.S. **Jupiter** missile, on the first suborbital flight from which live animals were recovered on May 28, 1959. The two animals' survival of up to 38*g* during reentry and a weightless period of about nine minutes was an early step toward demonstrating that humans could safely be sent into space. Baker was displayed alive years later at the United States Space and Rocket Center in Huntsville, Alabama. She died in 1984.[236]

Baikonur The twin launch towers at Baikonur Cosmodrome from which the Buran space shuttle took off. *Stefan Wotzlaw*

ballistic missile

A missile that has a **ballistic trajectory** over most of its flight path, regardless of whether or not it carries a weapon. Ballistic missiles are categorized according to their range—the maximum distance measured along Earth's surface from the point of launch to the point of impact of the last element of the payload. Various schemes are used by different countries to categorize the ranges of ballistic missiles. The U.S. Department of Defense divides missiles into four range classes, as shown in the table ("U.S. Ballistic Missile Categories").

U.S. Ballistic Missile Categories

Type of Missile	Range
Intercontinental ballistic missile (ICBM)	Over 5,500 km
Intermediate-range ballistic missile (IRBM)	2,750–5,500 km
Medium-range ballistic missile (MRBM)	1,100–2,750 km
Short-range ballistic missile (SRBM)	Up to 1,100 km

Ballistic Missile Defense Organization (BMDO)

An American Department of Defense agency formed in 1994 under the **Reagan** administration to develop an antiballistic missile defense, a major part of which would involve Star Wars weaponry based both on the ground and in space. Originally intended to counter the threat of Soviet intercontinental ballistic missiles, BMDO has shifted its emphasis under the Bush (junior) presidency to address similar threats from rogue nations with ballistic and tactical missile technology. Former Strategic Defense Initiative (SDI) programs to develop Star Wars weapons, including **ASAT** (antisatellite) technology, are being emphasized by the Bush (junior) administration.

Ballistic Research Laboratory

See **Aberdeen Proving Ground**.

ballistic trajectory

The path followed by an object that is being acted upon only by gravitational forces and the resistance of the medium through which it passes.

ballistics

The study of the dynamics of an object moving solely under the influence of a **gravitational field**.

ballute

An inflatable balloon-parachute made from a thin, flexible, inelastic membrane that provides a means of **aerobraking** or **aerocapture**. Typically, ballutes are pumpkin-shaped and wrinkled in appearance with longitudinal seams. To reduce heating on the ballute, one option is to have the main spacecraft engine run at low idle thrust during aerocapture to provide an aerodynamic spike, although the ballute would be replaced after each flight. Stability and heat transfer considerations are critical, as ballutes are aerodynamically unstable at high Mach numbers, and radiative heat transfer is significant. A dramatic example of aerocapture into Jupiter orbit using a ballute system was depicted in the movie *2010: The Year We Make Contact* (1984).

bandwidth

The range of wavelengths or frequencies to which an **antenna** is sensitive.

Bantam-X

A NASA project, managed by the Marshall Space Flight Center, to develop a reusable small satellite launcher.

barycenter

The center of mass of a system of objects, for example, the Earth and the Moon, moving under the influence of their mutual gravity.

BBXRT (Broad-Band X-Ray Telescope)

An instrument that formed part of the **Astro-1** payload. Its flight gave the first opportunity for making X-ray observations over a broad energy range (0.3–12 keV) with a moderate energy resolution (typically 90 eV and 150 eV at 1 and 6 keV, respectively). This energy resolution, coupled with an extremely low detector background, made BBXRT a powerful tool for the study of continuum and line emission from cosmic sources. In spite of some technical hitches with the instrument's pointing system during the mission in 1990, BBXRT successfully performed around 160 observations of some 80 celestial sources including clusters of galaxies, active galaxies, supernova remnants, X-ray binaries, cataclysmic variables, and the X-ray background.

BE (Beacon Explorer)

Early satellites within NASA's **Explorer** series that measured the total electron count between the spacecraft and Earth by means of a radio **beacon** and carried out other basic studies of the ionosphere. They were known as Explorers 22 and 27, as well as by their BE designation.

beacon

A low-power **carrier** transmitted by a satellite that supplies controlling engineers on the ground with a means of monitoring **telemetry** data, tracking the satellite, or carrying out experiments on how well the signals travel through the atmosphere. The beacon is usually sent out by a horn or an omnidirectional antenna.

Beagle 2

See **Mars Express**.

Beal Aerospace Technologies

A Dallas-based company, formed in 1997 by industrialist Andrew Beal (1953–), to develop a completely private medium- to heavy-lift rocket to launch commercial satellites. The company went out of business in October 2000, but not before it had tested its BA-810 engine–the second largest liquid-fueled rocket engine ever built in the United States, surpassed only by the **F-1** used on the Saturn V. The BA-810 would have boosted the second stage of Beal's planned three-stage BA-2 launch vehicle, capable of placing satellites of up to 6,000 kg into geostationary transfer orbit (GTO). The BA-2 would have been 64.6 m tall and 6.2 m in diameter, and it would have weighed about 970 tons at liftoff. Andrew Beal claimed that private companies such as his could not compete with Lockheed Martin's Atlas V and Boeing's Delta IV, which received government funding from programs such as the Air Force's **EELV** (Evolved Expendable Launch Vehicle) program and NASA's **Space Launch Initiative** (SLI).

beamed-energy propulsion

A method of propulsion that uses a remote energy source, such as a ground- or space-based laser or microwave transmitter, to send power to a space vehicle via a beam of electromagnetic radiation. If the energy to propel a spacecraft does not have to be carried onboard the vehicle, significant weight reductions and performance improvements can be achieved. At present, beamed-energy is one of the most promising technologies to lower dramatically the cost of space transportation. See **laser propulsion**.

Bean, Alan LaVern (1932–)

An American astronaut who became the fourth human to walk on the Moon. Bean obtained a B.S. from the University of Texas, Austin, entered the U.S. Navy, and even-

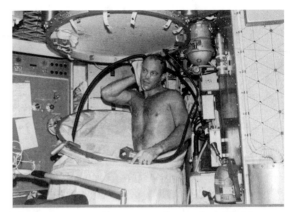

Alan Bean   Bean taking a zero-*g* shower aboard Skylab. *NASA*

tually became a Navy test pilot. He also developed an interest in art and, while serving as a test pilot, enrolled in night classes at a nearby college. In 1963, he was chosen by NASA to train as an astronaut and was later appointed Lunar Module pilot aboard **Apollo** 12. In 1973, Bean again flew in space as commander of **Skylab** 2. This was to be his last space mission, although he was appointed as backup commander for the **Apollo-Soyuz Test Project**. Subsequently, he served as chief of operations and training and acting chief astronaut until the first flight of the Space Shuttle. In 1981, Bean resigned from NASA to devote all his time to painting and to public speaking. His artwork of the Moon is unique in that it is the first to draw upon firsthand experience of another world.[19]

BECO (booster engine cutoff)

The point at which the booster engine stops firing.

Beggs, James Montgomery (1926–)

NASA's sixth administrator. Beggs served in this post from July 10, 1981, to December 4, 1985, when he took an indefinite leave of absence pending disposition of an indictment from the Justice Department for activities that took place prior to his tenure at NASA. This indictment was later dismissed and the attorney general apologized to Beggs for any embarrassment. His resignation from NASA was effective on February 25, 1986. Before his appointment at NASA, Beggs had been executive vice president and a director of General Dynamics, having previously served with NASA in 1968 and 1969 as associate administrator in the Office of Advanced Research and Technology. A 1947 graduate of the U.S. Naval Academy, Beggs served with the Navy until 1954. He currently works as a consultant from his offices in Bethesda, Maryland.

Belyayev, Pavel Ivanovich (1925–1970)

A Soviet cosmonaut and air force and naval pilot who flew aboard **Voskhod** 2 in March 1965. He succumbed to a long stomach illness on January 10, 1970, becoming the first spaceman to die of natural causes.

Bepi Colombo

A planned ESA (European Space Agency) mission to orbit and to observe Mercury. It is named after the late Giuseppe **Colombo**, of the University of Padua, who first suggested how **Mariner** 10 could be placed in an orbit that would bring it back repeatedly to Mercury. Bepi Colombo is in an early design stage but may consist of three elements: a planetary orbiter (about 450 kg), a magnetospheric orbiter (about 60 kg), and a lander (about 30 kg). No date has yet been set for its launch.

BeppoSAX (Satellite per Astronomia a raggi X)

A major project of **ASI** (the Italian Space Agency) with participation of the Netherlands Agency for Aerospace Programs. It is the first X-ray mission with a scientific payload spanning more than three orders of magnitude in energy—0.1–300 keV—and has a relatively large effective area and medium-energy resolution and imaging capabilities in the range of 0.1 to 10 keV. It is named for the Italian physicist Giuseppe "Beppo" Occhialini (1907–1993).

Launch
    Date: April 30, 1996
    Vehicle: Atlas I
    Site: Cape Canaveral
    Orbit: 584 × 601 km × 4.0°

Berezovoi, Anatoly Nikolayevich (1942–)

A Soviet cosmonaut who, in 1982 along with Valentin **Lebedev**, spent a then-record 211 days aboard **Salyut** 7.

Berkner, Lloyd V. (1905–1967)

An influential figure in shaping American space policy in the 1950s and 1960s. Berkner received a B.S. in electrical engineering from the University of Minnesota and later attended George Washington University. Although initially he carried out research on the atmospheric propagation of radio waves, he rose to prominence as a scientific administrator following World War II. Berkner played a central role in the exchange of scientific information dur-

ing the **International Geophysical Year** and subsequently helped formulate America's response to the launch of **Sputnik** 1. From 1951 to 1960, he served as the head of Associated Universities, charged with running the Brookhaven Laboratories for the Atomic Energy Commission. In 1961, he became president of the Graduate Research Center of the Southwest and four years later was named its director. He died less than a year after receiving the NASA Distinguished Public Service Medal for his pioneering work in the advancement of space science.[214]

Bernal, J(ames) D(esmond) (1901–1971)

An Irish physicist who was among the first to discuss the possibility of space colonies (see **Bernal sphere**). Bernal graduated from Emmanuel College, Cambridge, and began his research career in crystallography under William Bragg at the Davy-Faraday Laboratory in London before returning to Cambridge. In 1937, he was appointed to the chair of physics at Birkbeck College, London, where his studies eventually led him to consider the processes leading to the origin of life. In his remarkably perceptive book *The World, the Flesh, and the Devil*,[25] Bernal discussed the possible future evolution of mankind and its spread into the universe.

Bernal sphere

A spherical **space colony** of the type first described in the 1920s by J. D. **Bernal**. As the material and energy needs of the human race grew, Bernal surmised, it would be natural that some day orbiting colonies would be built to harness the Sun's energy and to provide extra living space for a burgeoning population. He conceived of self-sufficient globes, 16 km in diameter, that would each be home to 20,000 to 30,000 inhabitants. Almost half a century later, Gerard K. **O'Neill** based the scheme for his Island One space colony on a smaller Bernal sphere, some 500 m in diameter. Rotating twice a minute, this sphere would generate an Earth-normal **artificial gravity** at its equator. An advantage of the sphere is that it has the smallest surface area for a given internal volume, which minimizes the amount of radiation shielding required. See **Tsiolkovsky, Konstantin Eduardovich**.

BI-1 (Bereznyak-Isayev 1)

The Soviet Union's first high-speed rocket plane. Developed during World War II, it used a liquid-fueled engine built by Isayev with a thrust of 1.5 tons. Its maiden flight, following accidents in ground runs of the rocket engine, came on May 15, 1942, lasted three minutes and reached a speed of 400 km/hr. Problems with corrosion by the acid fuels slowed testing. On its seventh flight, in March

1943, the aircraft reached 800 km/hr (unofficially break-ing the world speed record) but then experienced a previ-ously unencountered tendency to pitch down and crashed, killing the pilot. Plans to put the plane into pro-duction were abandoned, and rocket plane development in the Soviet Union resumed only with the testing of German designs after the war.

Bickerton, Alexander William (1842–1929)

A maverick New Zealand astronomer whose failure to grasp the fundamentals of rocket theory led him to draw this spectacularly wrong conclusion during a speech to the British Association of Science in 1926:

> This foolish idea of shooting at the Moon is an example of the absurd length to which vicious spe-cialization will carry scientists working in thought-tight compartments. . . . For a projectile entirely to escape the gravitation of the Earth, it needs a veloc-ity of 7 miles a second. The thermal energy of a gram at this speed is 15,180 calories. . . . The energy of our most violent explosive–nitroglycerin–is less than 1,500 calories per gram. Consequently, even had the explosive nothing to carry, it has only one-tenth of the energy necessary to escape the Earth. . . . Hence the proposition appears basically impossible.

Big Gemini

A 1967 proposal by McDonnell Douglas to the U.S. Air Force and NASA for a manned orbital logistics spacecraft to provide economical resupply of planned military and civilian space stations. By the end of 1966, the **Gemini** program was nearing an end and the design phase of the Air Force's **Manned Orbiting Laboratory** (MOL) proj-ect was almost complete. McDonnell Douglas had a large manned spacecraft engineering team, built up over eight years on the Mercury, Gemini, and MOL programs, that was facing dissolution. At the same time, both the Air Force and NASA had funded space station projects. The Air Force's MOL and NASA's Apollo Applications Pro-gram Orbital Workshop (later Skylab) were supposed to fly between 1969 and 1974, and both the Air Force and NASA were planning even larger follow-on stations–the LORL and MORL, respectively. The capability of exist-ing spacecraft (the Apollo Command Module and Gem-ini) for such missions was severely limited. From 1970 to 1980, it appeared that at least a dozen launches would be needed for logistics purposes for the Apollo Applications Program (AAP) alone. The MOL and the AAP Workshop would demand 3 to 6 flights a year, each delivering a crew of 2 or 3, with 1 to 7 tons of cargo being sent up and up to 0.6 cubic meter of cargo being returned. Planned late-

1970s stations would have crews of 6 to 24, requiring a resupply craft that could deliver up to 12 passengers and 12 tons of payload 6 to 14 times a year and return up to 7 cubic meters of cargo each time. Big Gemini was intended to provide such a capability by 1971, using Gemini technology applied to Gemini and Apollo hard-ware. However, even at the time the concept was born, both NASA and Air Force manned space projects were being cut back. Within 18 months, the MOL would be canceled and the AAP would be limited to using only spacecraft and boosters surplus to the Moon landing pro-gram. Soon after, the Space Shuttle became the only funded manned space project for the 1970s, and all space station work was abandoned.

big LEO

Orbits, typically a few thousand kilometers high, that are intermediate in size between **little LEO** (low Earth orbit) and **geosynchronous orbit**s. The term is applied espe-cially to some of the latest generation of comsats that support communications using small handheld sets.

bioastronautics

The study of the effects of spaceflight upon living things.

biodynamics

The study of forces acting upon bodies in motion or in the process of changing motion, as they affect living beings.

bioinstrumentation

Equipment used to measure physical, physiological, and biological factors in man or in other living organisms.

Bion

A series of large Soviet spacecraft based on the **Zenit** reconnaissance satellite and designed to study the bio-logical effects of weightlessness and radiation in space. Bion missions were typically given Cosmos designations. The first Bion launch–Cosmos 605 on October 31, 1973–took tortoises, rats, insects, and fungi on a 22-day mission. Other flights have carried mold, quail eggs, fish, newts, frogs, protozoans, and seeds. Starting with Bion 6 (Cosmos 1514), launched on December 14, 1983, flights also carried pairs of monkeys. Experiments were supplied by scientists from various countries, including the United States, France, Germany, China, and Eastern bloc nations. An onboard centrifuge simulated Earth-normal gravity and enabled postmission comparisons to be made between specimens that had floated freely in zero-*g* and their companions who had experienced artificial gravity. Eleven Bions were launched from 1973 to 1996.

biopak
A container for housing a biological organism in a habitable environment.

biosatellite
A spacecraft designed to carry living organisms into space and to conduct experiments on them.

Biosatellite
A series of three NASA satellites, launched from Cape Canaveral, to assess the effects of spaceflight, especially radiation and weightlessness, on living organisms. Each was designed to reenter and to be recovered at the end of its mission. The first two Biosatellites carried specimens of fruit flies, frog eggs, bacteria, and wheat seedlings; the third carried a monkey. Biosatellite 1 was not recovered because of the failure of a retrorocket to ignite. However, Biosatellite-2 successfully de-orbited and was recovered in midair by the U.S. Air Force. Its 13 experiments, exposed to microgravity during a 45-hour orbital flight, provided the first data about basic biological processes in space. Biosatellite 3 carried a 6-kg male pigtailed monkey, called Bonnie, with the object of investigating the effect of spaceflight on brain states, behavioral performance, cardiovascular status, fluid and electrolyte balance, and metabolic state. Scheduled to remain in orbit for 30 days, the mission was terminated after only 8.8 days because of the subject's deteriorating health. Despite the seeming failure of the mission's scientific agenda, Biosatellite 3 was influential in shaping the life sciences flight experiment program, highlighting the need for centralized management, realistic goals, and adequate preflight experiment verification. (See table, "Biosatellites.")

biosensor
A sensor used to detect or to obtain information about a life process.

biotelemetry
The electrical measurement, transmission, and recording of qualities, properties, and actions of organisms and substances, usually by radio from a remote site.

bipropellant
A rocket **propellant** consisting of two substances, the **fuel** and the **oxidizer**, that are held separate prior to combustion. Bipropellants are commonly used in liquid-propellant rocket engines. There are many examples, including **RP-1** (a kerosene-containing mixture) and **liquid oxygen** (used by the Atlas family), and **liquid hydrogen** and liquid oxygen (used by the Space Shuttle).

bird
Slang for a communications satellite located in geosynchronous orbit.

BIRD (Bi-spectral Infrared Detection satellite)
A small (80-kg) German satellite designed to test new-generation, cooled infrared sensors. Built into two cameras, these sensors can detect fires and volcanic activity. BIRD also carries two visible-spectrum cameras for stereo-imaging to gather information about vegetation condition and variations, and performs onboard processing with neural networks. It was successfully launched on October 22, 2001, as a secondary payload aboard an Indian **PSLV** rocket into a circular orbit with an altitude of 572 km. In January 2002, it provided valuable images of forest fires burning around Sydney, Australia.

Biringuccio, Vannoccio (1480–1537)
An early Italian chemist whose reputation stems from a single work, *De La Pirotechnia,* published posthumously in 1540. It was the first printed comprehensive account of the fire-using arts, was a prime source on many practical aspects of inorganic chemistry, and contained a great deal about artillery and the manufacture of gunpowder. *Pirotechnia* offers one of the first written attempts to

Biosatellites

| Spacecraft | Launch | | Orbit | Mass (kg) |
	Date	Vehicle		
Biosatellite 1	Dec. 14, 1966	Delta G	295 × 309 km × 33.5°	425
Biosatellite 2	Sep. 7, 1967	Delta G	297 × 318 km × 33.5°	507
Biosatellite 3	Jun. 29, 1969	Delta N	363 × 374 km × 33.5°	507

explain what causes a rocket to move. Biringuccio attributed the propulsive force to a "strong wind":

> One part of fire takes up as much space as ten parts of air, and one part of air takes up the space of ten parts of water, and one part of water as much as ten parts of earth. Now sulfur is earth, consisting of the four elementary principles, and when the sulfur conducts the fire into the driest part of the powder, fire and air increase. . . . [T]he other elements also gird themselves for battle with each other and the rage of battle is changed by their heat and moisture into a strong wind.

Biringuccio's description of the burning, gas-exhausting phenomenon was correct enough, despite its nontechnical language. Yet it failed to explain why a strong wind blowing downward should cause a rocket to rise upward. It was nearly a century and a half later that Isaac **Newton** provided the key with his third law of motion.

bit

A binary digit, 0 or 1; the smallest unit of information.

Black Arrow

A British three-stage rocket whose development began in 1964 under direction of the Royal Aircraft Establishment. It stood 13 m high, weighed 18.1 tons at takeoff, and could launch nearly 100 kg into LEO (low Earth orbit). Following a failed first attempt at suborbital flight

Black Arrow A Black Arrow launches Britain's first satellite in 1971. *Royal Aircraft Establishment*

in June 1969, a second launch in March 1970 was successful. In September 1970, Black Arrow's initial attempt to reach orbit failed following a premature second-stage shutdown, and in July 1971 the British government decided to drop the program. However, one further launch took place in October 1971, which successfully placed the **Prospero** satellite in orbit.

Black Brant

A Canadian **sounding rocket**, more than 800 of which have been launched since 1962 with a success rate of 98%. The current model, produced by Bristol Aerospace in single or multistage configurations, can carry payloads of 70 to 850 kg to altitudes of 150 to 1,500 km.

black hole

A region of space where the pull of gravity is so strong that nothing can escape from it. The term was coined in 1968 by the physicist John Wheeler (1911–). However, the possibility that a lump of matter could be com-

Black Brant The launch of a Black Brant XII sounding rocket from Wallops Island. *NASA*

pressed to the point at which its surface gravity would prevent even the escape of light was first suggested in the late eighteenth century by the English physicist John Michell (c. 1724–1793), and then by Pierre Simon, marquis de Laplace (1749–1827). Black holes began to take on their modern form in 1916 when the German astronomer Karl Schwarzschild (1873–1916) used Einstein's **general theory of relativity** to find out what would happen if all the mass of an object were squeezed down to a dimensionless point–a singularity. He discovered that around the infinitely compressed matter would appear a spherical region of space out of which nothing could return to the normal universe. This boundary is known as the **event horizon**, since no event that occurs inside it can ever be observed from the outside. Although Schwarzschild's calculations caused little stir at the time, interest was rekindled in them when, in 1939, J. Robert Oppenheimer, of atomic bomb fame, and Hartland Snyder, a graduate student, described a mechanism by which black holes might be created in the real universe. A star that has exhausted all its useful nuclear fuel can no longer support itself against the inward pull of its own gravity. The stellar remains begin to shrink rapidly. If the collapsing star manages to hold on to more than about three times the mass of the Sun, no force in the universe can halt its contraction, and in a fraction of a second the material of the star is squeezed down into the singularity of a black hole. Strong observational evidence now supports the existence not only of stellar black holes but also of supermassive black holes at the centers of galaxies, including our own.

The equations of general relativity also allow for the possibility of space-time tunnels, or **wormhole**s, connected to the mouths of black holes. These could act as short-cuts linking remote points of the universe. Unfortunately, they appear to be useless for travel or even for sending messages, since any matter or energy attempting to pass through them would immediately cause their gravitational collapse. Yet not all is lost. Wormholes leading to remote regions in space might be traversable if some means can be found to hold them open long enough for a signal, or a spacecraft, to pass through.[7, 287]

Black Knight

An experimental rocket built by Britain to support the development of its **Blue Streak** ballistic missile. Conceived in 1955 and first launched in 1958, it enabled studies to be conducted on long-range-missile guidance. As in the case of all large British rockets, except for Blue Streak, it was powered by a mixture of kerosene and **hydrogen peroxide**–the same liquid propellant used by the German World War II Messerschmitt 262 interceptor.

It was also built in sections so that it could be easily shipped to and assembled at its launch site in **Woomera**, Australia. Consisting in its first version of a single stage 10.2 m high, 1.8 m in diameter, and weighing 5.4 tons at takeoff, Black Knight was propelled by an Armstrong-Siddeley Gamma 201 engine equipped with four swiveling nozzles that developed 75,000 newtons (N) of thrust during a burn time of 140 to 145 seconds. Four fins, each with a span of 1.8 m, provided aerodynamic stabilization, and two were equipped with small pods containing tracking equipment. On September 7, 1958, Black Arrow 01 took off from Woomera and reached what was then a new record altitude of 564 km—an impressive achievement, considering that the first Soviet and American satellites, Sputnik 1 and Explorer 1, had already been launched. Following the cancellation of Blue Streak in 1960, Black Knight's future seemed in jeopardy. However, the rocket had proved so successful and cost-effective that it found new life as part of the Gaslight program, carried out in cooperation with the United States and Australia. The main objective of this program was to study the effects of high-speed reentry on dummy nuclear warheads. It required the development of a two-stage version of Black Knight with the second stage, powered by a so-called Cuckoo engine (a **Skylark** booster), pointing downward. This new variant of Black Knight stood 11.6 m high and weighed 6.35 tons. Six Gaslight flights were carried out up to June 1961. In 1962, the tests continued as part of the Dazzle program, in cooperation with the United States, Canada, and Australia. In August 1962, a 95,000-N Gamma 301 engine replaced Gamma 201 on the first stage, and eight more flights were conducted with this more powerful engine. The last of the 22 Black Knight flights took place in November 1965. See **Black Arrow**.

black powder

The oldest of explosives, more commonly known today as gunpowder, though it was first used by the Chinese in firecrackers and to propel black-powder rockets (see **Chinese fire-rockets**) long before guns were invented. Its ingredients are saltpeter, or potassium nitrate (about 75%), charcoal (about 15%), and sulfur (about 10%). Knowledge of the explosive spread from China and other parts of the Far East to the Arab world and then to Western Europe by the mid-thirteenth century. Its preparation was described, for example, by Roger **Bacon** and **Albertus Magnus**.

blackout

(1) A fade-out of radio communications due to environmental factors such as ionospheric disturbances or a plasma sheath surrounding a reentry vehicle. (2) A condition in which vision is temporarily obscured by blackness, accompanied by a dullness of certain other senses, brought on by decreased blood pressure in the head and a consequent lack of oxygen. It may occur, for example, as a result of being exposed to high *g*-forces in an airplane or spacecraft (see **anti-*g* suit**).

Blagonravov, Anatoli Arkadyevich (1895–1975)

A Soviet academician who helped establish cooperation in space between the United States and the Soviet Union. As the Soviet representative to the United Nations' Committee on the Peaceful Uses of Outer Space (COPUOS) in the early 1960s, Blagonravov was a senior negotiator alongside NASA's Hugh **Dryden** for cooperative space projects at the height of the Cold War. During World War II, he specialized in infantry and artillery weapons, and later he worked on the development of rockets.

bleed-cycle operation

A mode of operation in some liquid rocket engines in which the **turbopump** is driven by hot gases bled from the **combustion chamber** assembly during main-stage operation.

blockhouse

A heavily reinforced building, designed to withstand blast and heat, that contains the electronic controls and equipment for preparing and launching a rocket.

blowdown

The draining of a rocket engine's unused liquid propellants. Blowdown may be done before or after firing the engine.

blowoff

The separation of a portion of a rocket or spacecraft by explosive force. The Solid Rocket Boosters of the Space Shuttle, for example, are separated by blowoff.

Blue Scout Junior

A smaller Air Force version of the **Scout** launch vehicle that was used for suborbital military tests.

Blue Streak

A British launch vehicle of the 1950s that started out as an intercontinental **ballistic missile** (ICBM) and became an ancestor of the **Ariane**. Blue Streak was designed to deliver a two-ton nuclear payload over a range of 4,500 km–a capability similar to that of the American **Atlas**. The De Havilland Aircraft Company was given the contract to build the missile; Rolls Royce was to provide the engines. The first tests took place at **Spadeadam Rocket Establishment** in August 1959. By the following February, the engines were being run for their full burn time of three minutes. But then on April 13, 1960, the British government decided to cancel the Blue Streak ICBM project and, subsequently, proposed to other European countries that it be used as the first stage of a space launcher.

boat-tail

The cylindrical section of a ballistic body that gradually decreases in diameter toward the tail to reduce aerodynamic **drag**.

Boeing

A major U.S. aerospace company that, in addition to aircraft and other products, manufactures the modern **Delta** family of space launch vehicles, the **Inertial Upper Stage** (IUS), and a variety of rocket engines through its Rocketdyne subsidiary. It is also the prime contractor for the **International Space Station** (and the builder of the U.S. Destiny laboratory module) and a partner in **Sea Launch**. Boeing was involved in the development of the **Bomarc** missile and of the design of **Dyna-Soar** (X-20), and built the **Lunar Orbiters**, the first stage of the **Saturn V**, the **Apollo** lunar rover, and the **Space Shuttle** main engine. In 1996, it merged with **Rockwell** (which had evolved from North American Aviation and included Rocketdyne) and in 1997 merged with **McDonnell Douglas**.

boilerplate

A piece of test hardware, generally nonfunctioning, that simulates weight, center of gravity, and aerodynamic configuration, and may incorporate interim structural shells or dummy structures. Internal systems may be inert or have some working subsystems to obtain flight data for use in development.

boiloff

The vapor loss from a volatile liquid–for example, liquid oxygen–particularly when stored in a vehicle ready for flight.

Bomarc

The world's first long-range antiaircraft missile. Authorized by the U.S. Air Force in 1949, the Bomarc was developed by Boeing and the University of Michigan Aeronautical Research Center, whose initial letters form its name. The Model B Bomarc had a range of about 650 km and a ceiling of over 24,000 m.

Bond, Alan (1944–)

A British aerospace engineer who began his career with Rolls-Royce Rocket Division, led the design team in Project **Daedalus**, and went on to design the **HOTOL** space plane and its successor, **Skylon**. In 2002, Bond was managing director of Reaction Engines, the developer of the Skylon concept.

bonded grain

Solid rocket propellant cast in a single piece and cemented or bonded to the motor casing.

Bondi, Hermann (1919–)

An Austrian-born British cosmologist and mathematician who was director general of **ESRO** (European Space Research Organisation) from 1967 through the early 1970s and the organization's transformation into ESA (European Space Agency). Bondi later served as science advisor to the British minister for energy.

bone demineralization in space

One of the greatest threats to long-duration manned space missions. Relieved of the normal stresses produced by gravity, bones start to lose calcium and other minerals faster than they can replace them. On short missions, such as those of the Space Shuttle, this is not a problem. However, some of the bones of astronauts and cosmonauts who have spent months aboard space stations have been found to have lost up to one-fifth of their mineral content–significantly increasing the risk of fractures. Not all bones lose minerals at the same rate in space. For example, the bones in the upper body appear not to be affected at all, whereas the weight-bearing bones in the legs and lower back experience a disproportionately high loss.

When bones lose minerals–as they do all the time, even under normal conditions–those minerals are carried by the blood to the kidneys, where they are filtered out for excretion in the urine. However, prolonged heightened levels of calcium in the body, a condition known as hypercalciuria, can lead to the formation of kidney stones, which are potentially disabling. This is one of the reasons that astronauts' diets cannot simply be calcium-enriched. The methods used to remove kidney stones on Earth are simply not available aboard a space station. Furthermore, the drugs that have some effect on bone calcium loss have side effects, and they act on all the bones in the body, not

bone demineralization   American astronaut Shannon Lucid using the treadmill aboard the Russian Mir space station. *NASA*

just those that need the extra calcium. The effect of systemic drugs is a buildup in unwanted calcium in some bones while mineral loss is prevented in others.

Exercise has been suggested and used as a countermeasure to bone demineralization (and also muscle degeneration), but the forces needed to prevent deterioration of the weight-bearing bones seem to be roughly equal to the body weight of the individual (not surprisingly, since most people gain and lose bone minerals at the same rate doing normal activities on Earth). However, it is difficult to design exercise equipment that can produce this level of force in a way convenient for the astronaut to use. Various methods have been proposed and tried, including bungees, springs, bicycles, and treadmills, but attaching the load-bearing portion of the exercise equipment to the body in an acceptable way has been a stumbling block. Straps tend to cut into the shoulders and hips, and the human shoulder is not designed to carry body-weight loads for long durations or intensive exercise.

Researchers at NASA and elsewhere are developing ground-based protocols for understanding the mechanism of bone loss and changes in calcium metabolism that occur during spaceflight. These ground-based research protocols include human and animal studies. The human studies may involve prolonged bed rest as a microgravity analog, and the animal studies may involve limb immobilization and other techniques. Meanwhile, scientists at the National Space Biomedical Research Institute are developing a compact machine to allow precision bone and tissue measurements in space. Known as the advanced multiple projection dual-energy X-ray absorptiometer (AMPDXA), it will give astronauts aboard the International Space Station a readout of their tissue mass, bone density, and bone geometry, enabling them to spot any adverse effects and take steps to counter them before the problems become serious.

Bonestell, Chesley (1888–1987)

An American artist whose imaginative and technically authentic depictions of spacecraft and other worlds had a powerful effect on people in the decade before the start of the Space Age. From 1944 on, he illustrated numerous books and articles, such as Willy **Ley**'s *The Conquest of Space*[191] and Wernher von Braun's pieces for the *Collier's* magazine series on spaceflight in the 1950s (see *Collier's Space Program*). He also provided the backdrops for several well-known science fiction films, including *Destination Moon* (1950), *When Worlds Collide* (1951), and *The Conquest of Space* (1955). Although his artwork appeared at a time when space travel was primarily a subject of fiction, Bonestell used scientific fact and photographic realism to create paintings that offered a believable, plausible, and highly alluring vision of our Solar System. In von Braun's words: "Chesley Bonestell's pictures . . . present the most accurate portrayal of those faraway heavenly bodies that modern science can offer."[48, 205]

boost

The momentum given to a space vehicle during its flight, particularly during liftoff.

Boost Glide Reentry Vehicle (BGRV)

A classified U.S. Air Force program to investigate missile maneuvering at **hypersonic** speeds after reentry into the atmosphere. Upon reentry, flight control was achieved by using aft trim flares and a reaction jet system commanded from an onboard inertial guidance system. A BGRV was launched on February 26, 1968, from Vandenberg Air Force Base by an Atlas F to the area of Wake Island in the Pacific, collecting data that proved valuable in developing later maneuvering reentry vehicles. See **ASSET** and **PRIME**.

boost glider

A winged aircraft that is powered into flight by a rocket motor, then returns to the ground as a glider after ejecting its motor or motor assembly.

boost phase

The part of a rocket's flight in which the **propellant** is generating **thrust**.

booster

Also known as a *booster rocket*. (1) A **stage** of a launch system used to provide an initial **thrust** greater than the total liftoff weight. (2) A rocket engine, either solid- or liquid-fueled, that assists the normal propulsive system or **sustainer engine** of a rocket or aeronautical vehicle in some phase of its flight. (3) A rocket used to set a launch vehicle in motion before another engine takes over. A *booster assembly* is a structure that supports one or more booster rockets.

bootstrap

A self-generating or self-sustaining process. In rocketry, it refers to liquid-fueled rocket engines in which, during main-stage operation, the **gas generator** is fed by the main propellants pumped by the **turbopump**. The turbopump in turn is driven by hot gases from the gas generator system. Such a system must be initiated by a starting system that supplies outside power or propellants. When rocket-engine operation is no longer dependent on outside power or propellants, it is said to be in bootstrap operation.

Borman, Frank (1928–)

An American astronaut, selected by NASA in 1962, who commanded the **Gemini** 7 and **Apollo** 8 missions. He received a B.S. from West Point in 1950 and an M.S. from the California Institute of Technology in 1957. From 1950 to 1970, Borman was a career Air Force officer serving as a fighter pilot in the Philippines, an operational pilot and instructor with various squadrons in the United States, an assistant professor of thermodynamics

Frank Borman   The Apollo 8 crew, left to right: James Lovell Jr., William Anders, and Frank Borman. *NASA*

and fluid dynamics at West Point, and an experimental test pilot at the Air Force Aerospace Pilot School. After leaving NASA in 1969, Borman joined Eastern Airlines, becoming its chairman in 1976, before occupying senior positions at other companies, including Patlex Corporation. Today he lives in New Mexico and restores airplanes.[32]

Bossart, Karel Jan (1904–1975)

A Belgian-born engineer who emigrated to the United States before World War II and became involved in the development of rocket technology at Convair Corporation. In the 1950s, he was largely responsible for designing the **Atlas** with a very thin, internally pressurized fuselage instead of massive struts and a thick metal skin.[200]

braking ellipses

A series of elliptical orbits that skim the atmosphere of a celestial body. Their purpose is to decelerate an orbiting spacecraft by exposing it to the aerodynamic drag of the atmosphere. See **aerobraking**.

braking rocket

See **retrorocket**.

Brand, Vance (1931–)

The Command Module pilot on the **Apollo-Soyuz Test Project** (ASTP) and commander of three **Space Shuttle** missions. Brand received a B.S. in aeronautical engineering from the University of Colorado in 1960 and an M.B.A. from the University of California, Los Angeles, in 1964. A commissioned officer and aviator with the U.S. Marine Corps from 1953 to 1957, he continued serving in Marine Corps Reserve and Air National Guard jet squadrons until 1964, and from 1960 to 1966 was employed with Lockheed Aircraft Corporation. While with Lockheed, he graduated from the Naval Test Pilot School and was assigned as an experimental test pilot on Canadian and German F-104 programs. Following his selection by NASA as an astronaut in April 1966, he served on the ASTP, becoming the first American to fly inside a Russian spacecraft. Brand returned to space on November 11, 1982, as commander of the fifth flight of the Space Shuttle *Columbia* and the first operational mission of the Shuttle fleet. He commanded *Challenger* on the tenth Shuttle flight in February 1984, then returned to *Columbia* for his third Shuttle command in December 1990. In 1992, Brand left the astronaut corps and accepted a NASA assignment as director of plans for the **X-30** National Aerospace Plane Joint Program

Office at Wright-Patterson Air Force Base. He retired from NASA in 2002.

break-off phenomenon

The feeling, which is sometimes reported by pilots during high-altitude flight at night, of being separated and detached from Earth and human society. A similar feeling, which may involve exhilaration or fear, is experienced by sky divers and deep sea divers, and could be a major psychological factor on future, long-duration space missions, for example, to Mars. It is also known as the *breakaway phenomenon*.

Breakthrough Propulsion Physics Program (BPPP)

A small program, managed by the **Glenn Research Center** and sponsored jointly by the Advanced Space Transportation Program and Advanced Concepts Program, to study fundamentally new forms of propulsion that would involve no propellant mass and, if possible, **faster-than-light travel**.

Britain in space

See **Ariel, Black Arrow, Black Knight, Blue Streak, British Interplanetary Society, British National Space Center,** William **Congreve, ELDO, ESA, HOTOL, IRAS, Miranda, MUSTARD, Prospero, ROSAT, Skylark, Skylon, Skynet, Spadeadam Rocket Establishment,** and Frank **Whittle**.

British Interplanetary Society (BIS)

An organization formed in 1933 to promote the exploration and utilization of space. In the 1930s, the BIS put forward plans for a manned lunar spacecraft that bore a remarkable similarity to Apollo. In the late 1970s, it prepared a detailed design for a robot star-probe, Project **Daedalus**, to explore the system of Barnard's Star. It publishes the monthly magazine *Spaceflight* and the technical periodical *Journal of the British Interplanetary Society*.

British National Space Centre (BNSC)

Britain's space agency. BNSC advises and acts on behalf of the British government and provides a focus for British civilian space activities.

broadband

Communications across a wide range of frequencies or multiple channels. In particular, it can refer to evolving digital telephone technologies that offer integrated access to voice, high-speed data services, and interactive information delivery services.

BS- (Broadcasting Satellite)

Experimental Japanese communications satellites, also known by the national name Yuri ("lily"). Launched by **NASDA** (National Space Development Agency), they have been used to test technologies and techniques in satellite TV broadcasting, including transmission to areas with poor reception and of high-definition television signals. (See table, "Broadcasting Satellite Series.")

bubble colony

A proposed colony on the Moon or a planet consisting of individual or group capsules in which an Earth-like living environment is maintained, complete with people, animals, and plants.

Buckley, Edmond C. (1904–1977)

A prominent engineering administrator in the American space program. Buckley joined the **Langley** facility of NACA (National Advisory Committee for Aeronautics) in 1930 after earning his B.S. in electrical engineering from Rensselaer Polytechnic Institute. He became chief of the instrument research division in 1943 and was responsible for instrumentation at Wallops Island and at the Flight Research Center at Edwards, California. In 1959, he became assistant director for spaceflight operations at NASA headquarters. Two years later, his title changed to director for tracking and data acquisition, and from 1962 to 1968 he was associate administrator for tracking and data acquisition. He retired in 1969 as special assistant to administrator James **Webb**.

buffeting

The vibratory motion of a component or the airframe as a whole when subjected to the vibratory impulses contained within an aerodynamic wake.

"Bug," the

Colloquial early name for the **Apollo** Lunar Module.

bulge of the Earth

The extension of the equator caused by the **centrifugal force** of Earth's rotation, which slightly flattens our planet's spherical shape. This bulge causes the planes of satellite orbits inclined to the equator (but not polar) to rotate slowly around Earth's axis. See **station-keeping**.

Bull, Gerald Vincent (1928–1990)

A Canadian aeronautical and artillery engineer at the center of various schemes to develop **space cannon**s. Bull earned a Ph.D. in aeronautical engineering and spent the 1950s researching supersonic aerodynamics in Canada. Inspired in his youth by reading Jules Verne's *From the Earth to the Moon*, Bull's dream was to fire projectiles from Earth's surface directly into space. In 1961, he set up **HARP** (High Altitude Research Project), funded by McGill University in Montreal, where Bull had become a professor in the mechanical engineering department, with support from the Army Ballistic Research Laboratory. Following the collapse of HARP in 1967, Bull set up Space Research Corporation, a private enterprise through which he tried to sell his space cannon ideas to various organizations and nations including the Pentagon, China, Israel, and, finally, Iraq. Bull's ultimate goal was to put a cannon-round into orbit for scientific purposes, but the military potential of his designs led him to become an arms trader. He set up a weapons plant in northern Vermont complete with workshops, artillery range, launch-control buildings, and a radar tracking station. In 1980, he was jailed in the United States for seven months for a customs violation in supplying the South African military. Once out of prison, he abandoned his American enclave to work full-time in Brussels, Belgium. In November 1987, he was contacted by the Iraqi Embassy and was invited to Baghdad. Bull promised the Iraqis a launch system that could place large numbers of

Broadcasting Satellite Series

| Spacecraft | Launch | | | GSO Location | Mass (kg) |
	Date	Vehicle	Site		
BS-1	Apr. 8, 1977	Delta 2914	Cape Canaveral	110° E	350
BS-2A	Jan. 23, 1984	N-2	Tanegashima	110° E	350
BS-2B	Feb. 12, 1986	N-2	Tanegashima	110° E	350
BS-3A	Aug. 28, 1990	H-1	Tanegashima	110° E	1,100
BS-3B	Aug. 25, 1991	H-1	Tanegashima	110° E	1,100

small satellites into orbit for tasks such as surveillance. By 1989, the Iraqis were paying Bull and his company $5 million a year to redesign their field artillery, with the promise of greater sums for Project Babylon—an immense space cannon. The Iraqi space-launcher was to have had a barrel 150 m long and been capable of firing rocket-assisted projectiles the size of a phone booth into orbit. However, it was never built, and Bull soon paid the price for his dangerous liaisons. On March 22, 1990, he was surprised at the door of his Brussels apartment and assassinated.[2, 196]

Bumper WAC

The world's first two-stage liquid-propellant rocket: a 19-m-long, 1.9-m-diameter marriage of the German V-2 (see **"V" weapons**) and the liquid-propellant stage of the U.S. Army's **WAC Corporal**. The WAC Corporal remained atop the nose of the V-2 for the first minute of flight. Then the V-2 shut down, having provided a high-altitude "bump" for the second stage. Eight Bumper WACs were built. The first six were launched from the **White Sands** Proving Ground starting on May 13, 1948, and the last two from **Cape Canaveral** (the first flights from the fledgling missile testing grounds). In 1949, the fifth flight of a Bumper WAC reached an altitude of 390 km, a record that stood until 1957. Although the missile was tracked by radar for most of its flight, more than a year passed before the smashed body section was located. The program was officially concluded in July 1950.

The idea for the vehicle—capable of testing two-stage technology, reaching higher altitudes than ever before, and carrying out new upper atmosphere research—was put forward in July 1946 by Holger **Toftoy**, then colonel, chief of the Research and Development Division, Office of the Chief of Ordnance. On June 20, 1947, the Bumper

Bumper WAC   The first launch of a Bumper WAC from Cape Canaveral on July 24, 1950. *NASA*

Program was inaugurated with overall responsibility being given to the General Electric Company. JPL (Jet Propulsion Laboratory) carried out the necessary theoretical investigations, the design of the second stage, and the basic design of the separation system. The Douglas Aircraft Company was assigned responsibility for detail design, building the second stage, and making the special V-2 parts required. In the final design, the powder rocket booster normally used to launch the WAC Corporal was dispensed with. This was in order to limit the size of the combination missile and allow the smaller rocket to fit as deeply as possible into the V-2, yet retain enough space in the instrument compartment of the V-2 to house the guidance equipment. Also fitted within the instrument section were the guide rails and expulsion cylinders used as a launcher for the WAC Corporal.

Buran

A Soviet reusable space shuttle similar in design to its American counterpart, but with two important differences: it could be flown automatically, and it did not have reusable boosters. Piloted tests of Buran ("snowstorm") fitted with ordinary jet engines were carried out extensively in the atmosphere. However, its first and only orbital flight, launched by a giant **Energia** rocket on November 15, 1988, was unmanned. Two Burans were manufactured, but after the sole unpiloted mission, the program was canceled due to funding problems.

burn

The process in which a rocket engine consumes fuel or other propellant.

burn pond

An artificial pond that contains water a few centimeters above a mechanical burner vent. It serves to dispose of—by burning—dangerous and undesirable gases, such as hydrogen, which are vented, purged, or dispelled from the space vehicle propellant tanks and ground storage tanks.

burn rate

The linear measure of the amount of a solid propellant's grain that is consumed per unit time. Usually expressed in inches or centimeters per second.

burn time

The total operating time of a rocket engine. For solid motors, this is the more-or-less unstoppable period of thrust until all of the propellant is consumed. For liquid engines, it is the maximum rated thrust duration of the engine for a single operation—a quantity generally greater than the time for which the engine thrusts during any given burn in flight.

burnout

The point at which a rocket engine's fuel runs out and no more fuel combustion is possible.

burnout plug

A valve in a rocket engine designed to retain a liquid fuel under pressure until the engine fires and provides the ignition flame.

burnout velocity

The maximum velocity reached by a rocket when all of its **propellant** has been used.

burnout weight

The weight of a spacecraft after burnout occurs. The weight includes any unusable fuel that may be left in the rocket motor's tanks.

bus

(1) The part of a spacecraft's payload that provides a platform for experiments or contains one or more atmospheric entry probes. (2) A main circuit or conductor path for the transfer of electrical power or, in the case of computers, information.

Bush, Vannevar (1890–1974)

One of the most powerful members of the American scientific and technological elite to emerge during World War II. An aeronautical engineer on the faculty at the Massachusetts Institute of Technology, Bush lobbied to create and then headed the National Defense Research Committee in 1940 to oversee science and technology in the federal government. Later, its name was changed to the Office of Science Research and Development, and Bush used it as a means to build a powerful infrastructure for scientific research in support of the federal government. Although he went to the Carnegie Institution after the war, Bush remained a strong force in shaping postwar science and technology by serving on numerous federal advisory committees and preparing several influential reports.[229]

Bussard interstellar ramjet

See **interstellar ramjet**.

buzz bomb

Popular name for V-1 (see **"V" weapons**).

C

Caidin, Martin (1926–1997)

An American writer, pilot, and aerospace specialist. He wrote more than 80 nonfiction books, mostly on aviation and space exploration. His science fiction novel *Marooned* (1964), about an attempt to rescue astronauts trapped in orbit, was turned into a film of the same name in 1969 and has been credited with inspiring the 1975 **Apollo-Soyuz Test Project**. Caidin cofounded the American Astronautical Society in 1953.

CALIPSO (Cloud-Aerosol Lidar and Infrared Pathfinder Observations)

A satellite previously known as PICASSO-CENA that will carry an array of spectral imaging instruments to measure the chemical components of Earth's atmosphere, as well as a **lidar** system to track airborne water vapor particles. It will fly in formation with the **Cloudsat**, **PARASOL**, **Aqua**, and **Aura** satellites, which will also study atmospheric and oceanic processes, thus enabling a pooling of data. CALIPSO is a collaborative project led by the **Langley Research Center** that involves **CNES** (the French space agency), Ball Aerospace & Technologies Corporation, Hampton University (Virginia), and the Institut Pierre Simon Laplace. It is the third selected mission in NASA's **ESSP** (Earth System Science Pathfinder) mission and is scheduled for launch into a 705-km-high orbit, on the same Delta rocket as Cloudsat, in April 2004.

Canada in space

See **Alouette**, **Anik**, **Black Brant**, **Canadian Space Agency**, **Canadian Space Society**, **CloudSat**, **ESA**, Marc **Garneau**, **HARP**, **ISIS**, **MOST**, **Nimiq**, **RADARSAT**, and **Remote Manipulator System**.

Canadarm

See **Remote Manipulator System**.

Canadian Space Agency (CSA)

A federal organization, established in 1989, that contributes to the development of the Canadian civil space industry by procuring contracts from foreign space agencies such as ESA (European Space Agency) and NASA. CSA coordinates major Canadian space program activities including the Canadian Astronaut Office, the David Florida Laboratory (Canada's facility for assembling, integrating, and testing spacecraft and other space hardware), and Canadian contributions to the International Space Station.

Canadian Space Society (CSS)

A federally incorporated nonprofit organization inspired by the old **L5 Society**. Its main aim is to sponsor and promote the involvement of Canadians in the development of space. CSS publishes a newsletter titled *The Canadian Space Gazette,* has organized several space conferences, and has taken part in a number of space design projects, most notably the development of a preliminary design of a solar sail racing spacecraft under the Columbus 500 initiative in 1992.

Canberra Deep Space Communication Complex

One of three complexes around the world that make up NASA's **Deep Space Network** (DSN). It is located at Tidbinbilla, about 30 km southwest of the Australian capital. The Canberra Space Centre at the Complex has a piece of Moon rock weighing 142 g, presented to it by

Canberra Deep Space Communication Complex The Canberra 70-m antenna, with flags from the three Deep Space Networks (in the United States, Australia, and Spain). *NASA*

astronaut John **Young**, which is the largest piece of Moon rock outside the United States.

canted nozzle
A **nozzle** positioned so that its line of thrust is not parallel to the direction of flight. See **vectored thrust**.

Canyon
The first generation of large American **SIGINT** (signals intelligence) satellites. Canyons operated during the 1970s from near-GSO (geostationary orbit) and replaced a series of earlier, low-Earth-orbiting "heavy ferrets." They were themselves superseded by **Chalet/Vortex**.

CAPCOM
Capsule Communicator; the person at mission control who speaks directly to the astronauts. The CAPCOM is also usually an astronaut.

Cape Canaveral
A cape in eastern Florida, located on the Atlantic Ocean at 28.5° N, 80.5° W, about 25 km northeast of Cocoa Beach. It is home to America's largest complex of launch pads and support facilities. The northern part of the complex, including Merritt Island, is operated by NASA and known as **Kennedy Space Center** (KSC). To the south is **Cape Canaveral Air Force Station**, operated by **Patrick Air Force Base**. It is usual to talk about the civilian and military facilities together since NASA and commercial launch companies often use launch pads in the military-run section. In fact, NASA and the Air Force now have a single office at Patrick Air Force Base for dealing with launch customers. They also recently agreed to share police and fire departments, and other functions that in the past were duplicated. Efforts continue to unite the military, civilian, and commercial facilities at Cape Canaveral into a single "spaceport."

Cape Canaveral was chosen as a missile launch site soon after World War II, when it became clear that new rockets were becoming too powerful for inland facilities such as that at **White Sands**, New Mexico. The Cape has good weather, the Atlantic to the east (over which missiles could be fired without risk to human populations), and a string of islands on which tracking stations could be set up. In the late 1940s, Patrick Air Force Base was established as a launch command center and, shortly after, acquired more than 4,500 hectares of land from the state of Florida to use as its new proving ground. To the east extended the **Eastern Test Range** across which missiles, and eventually space vehicles, could be launched safely and tracked for thousands of kilometers. The first launch from Cape Canaveral was of a **Bumper-WAC** in July 1950. This was swiftly followed by many more test

flights of the **Navaho**, **Snark**, and **Matador** guided missiles and of the **Redstone** and **Jupiter** C ballistic missiles. As America's seminal space program began to take shape, Cape Canaveral was the obvious choice as a launch site. Not only were the launch and tracking facilities already in place, but the Cape's location not far from the equator meant that rockets heading eastward for orbit could take advantage of the extra velocity imparted to them by Earth's rotation.

Since 1950, more than 40 launch complexes have been constructed at the Cape, many of them now obsolete or dismantled. At the southern end are pads used to test-fire Trident and Minuteman missiles, and small- and medium-sized unmanned space launchers such as **Scout**, **Delta**, and **Atlas**-Centaur. Toward the center of the site is "Missile Row"—a chain of a dozen or so major pads used for Redstone, Atlas, and **Titan** launches, including those of the manned **Mercury** and **Gemini** flights. At the northern end of Missile Row are Complex 37, used for the **Saturn** Is, and Complex 34, the site of the **Apollo 1** fire in which three astronauts died. These all lie within Cape Canaveral Air Force Station. Still further north is Launch Complex 41, just within the border of the Air Force Station, and Complexes 40 and 39, within Kennedy Space Center. These three Complexes are used to launch the largest American rockets and operate differently than the smaller pads to the south. Instead of each rocket being assembled and tested on the pad, it is put together and checked inside special buildings and then moved to the pad as the launch day approaches. The cores of Titan IIIs and IVs are assembled in the Vertical Integration Building, then moved on railtracks to another building where their solid boosters are attached, and finally transported to either Launch Complex 40 or 41 prior to liftoff.

Launch Complex 39, the most northerly at the Cape, includes two launch pads, 39A and 39B, which were used by Saturn Vs during the Apollo program. Today they are the points of departure for the Space Shuttle, which is put together in the **Vertical Assembly Building** and then moved to its pad by a **crawler-transporter** a few weeks before launch. Other specialized buildings at KSC are used to check the Shuttle after its return from orbit and to prepare its payloads. A 4,500-m runway several kilometers to the west of Complex 39A enables the Shuttle to return directly to the Cape after each mission. (See table, "Principal Active Launch Complexes.")

Cape Canaveral Air Force Station
The part of **Cape Canaveral** from which Air Force–controlled launches take place. In addition to missile and military spacecraft, many unmanned civilian and commercial spacecraft, and all the Mercury and Gemini manned missions, have taken off from here. Cape

Principal Active Launch Complexes

Space Launch Complex	Launch Vehicle
17A, 17B	Delta
36A, 36B	Atlas
39A, 39B	Space Shuttle
40, 41	Titan

Canaveral Air Force Station is run by the adjoining **Patrick Air Force Base**.

Cape Kennedy

The name by which **Cape Canaveral** was known between 1963 and 1973. President Lyndon Johnson authorized the change in November 1963 following the assassination of President Kennedy. At the request of Floridians, the Cape's original name was restored a decade later, although NASA's facility continues to be known as the Kennedy Space Center.

capsule

A sealed, pressurized cabin with a habitable environment, usually for containing a man or an animal for extremely high-altitude flights, orbital spaceflight, or emergency escape.

captive test

A static or hold-down test of a rocket engine, motor, or stage, as distinct from a **flight test**. It is also known as a *captive firing*.

capture

The process in which a spacecraft comes under the influence of a celestial body's gravity.

cardiovascular deconditioning

A state in which the cardiovascular system does not work as efficiently as it can, usually caused by a change of environment, such as long periods of bed rest or space travel.

Carpenter, Malcolm Scott (1925–)

One of the **Mercury Seven** astronauts. Carpenter was the backup pilot for America's first manned orbital spaceflight and became the second American to go into orbit when, on May 24, 1962, he piloted his Aurora 7 capsule three times around the world. Subsequently, while on leave of absence from NASA, he became an aquanaut team leader in the Navy's Man-in-the-Sea Project, spending 30 days on the ocean floor as part of the Sealab II program. Upon returning to his NASA duties as executive assistant to the director of the Manned Space Flight Center, Carpenter participated in the design of the **Apollo** Lunar Module and in underwater extravehicular activity crew training. In 1967, he rejoined the Navy's Deep Sea Submergence Systems Project as director of aquanaut operations during the Sealab III experiment. Carpenter is an active author, public speaker, and consultant on ocean and space technology programs.[41]

Carr, Gerald P. (1932–)

An American astronaut selected by NASA in April 1966. Carr served as a member of the astronaut support crews

Cape Canaveral   An aerial view of "Missile Row." *NASA*

and as CAPCOM (Capsule Communicator) for the **Apollo** 8 and Apollo 12 missions, was involved in the development and testing of the Lunar Roving Vehicle, and was commander of **Skylab** 4. In mid-1974, Carr was appointed head of the design support group within the Astronaut Office; he was responsible for providing crew support for activities such as space transportation system design, simulations, testing, and safety assessment, and he was also responsible for the development of man/machine interface requirements. He retired from the Marine Corps in September 1975 and from NASA in June 1977. Currently, Carr is president of CAMUS, of Little Rock, Arkansas, director of the Arkansas Aerospace Education Center in Little Rock, and consultant on special staff to the president of Applied Research, in Los Angeles. He has also worked with fellow astronaut William **Pogue** on Boeing's contribution to the International Space Station, specializing in assembly extravehicular activity.

carrier

The main frequency of a radio signal generated by a transmitter prior to the application of a modulator.

Casimir effect

A small attractive force that acts between two close parallel *uncharged* conducting plates. Its existence was first predicted by the Dutch physicist Hendrick Casimir in 1948[42] and confirmed experimentally by Steven Lamoreaux, now of Los Alamos National Laboratory, in 1996.[179, 264] The Casimir effect is one of several phenomena that provide convincing evidence for the reality of the **quantum vacuum**–the equivalent in quantum mechanics of what, in classical physics, would be described as empty space. It has been linked to the possibility of **faster-than-light (FTL) travel**.

According to modern physics, a vacuum is full of fluctuating electromagnetic waves of all possible wavelengths, which imbue it with a vast amount of energy normally invisible to us. Casimir realized that between two plates, only those unseen electromagnetic waves whose wavelengths fit into the gap in whole-number increments should be counted when calculating the vacuum energy. As the gap between the plates is narrowed, fewer waves can contribute to the vacuum energy, and so the **energy density** between the plates falls below the energy density of the surrounding space. The result is a tiny force trying to pull the plates together–a force that has been measured and thus provides proof of the existence of the quantum vacuum.

This may be relevant to space travel because the region inside a Casimir cavity has *negative* energy density. Zero energy density, by definition, is the energy density of normal empty space. Since the energy density between the conductors of a Casimir cavity is less than normal, it must be negative. Regions of negative energy density are thought to be essential to a number of hypothetical faster-than-light propulsion schemes, including stable **wormhole**s and the **Alcubierre Warp Drive**.

There is another interesting possibility for breaking the light barrier by an extension of the Casimir effect. Light in normal empty space is slowed by interactions with the unseen waves or particles with which the quantum vacuum seethes. But within the energy-depleted region of a Casimir cavity, light should travel slightly faster because there are fewer obstacles. A few years ago, K. Scharnhorst of the Alexander von Humboldt University in Berlin published calculations showing that under the right conditions light can be induced to break the usual light-speed barrier.[260] Under normal laboratory conditions this increase in speed is incredibly small, but future technology may afford ways of producing a much greater Casimir effect in which light can travel much faster. If so, it might be possible to surround a space vehicle with a "bubble" of highly energy-depleted vacuum in which the spacecraft could travel at FTL velocities, carrying the bubble along with it.

Cassini

A Saturn probe built jointly by NASA and ESA (European Space Agency), and named for the Italian astronomer Gian Domenico Cassini (1625–1712), who observed Saturn's rings and discovered four of its moons. The spacecraft has two main parts: the Cassini orbiter and the Huygens probe, which will be released into the atmosphere of Saturn's largest moon, Titan. Cassini will enter orbit around the sixth planet on July 1, 2004, after a **gravity-assist**ed journey that has taken it twice past Venus and once each past Earth and Jupiter. It will then begin a complex four-year sequence of orbits designed to let it observe Saturn's near-polar atmosphere and magnetic field and carry out several close flybys of the icy satellites–Mimas, Enceladus, Dione, Rhea, and Iapetus–and multiple flybys of Titan. The climax of the mission will be the release of the Huygens probe and its descent into Titan's atmosphere.

Launch
  Date: October 15, 1997
  Vehicle: Titan IVB
  Site: Cape Canaveral
Size: 6.7 × 4.0 m
Total mass: 5,560 kg

Cassini orbiter

Experiments aboard the orbiter include: the Imaging Science Subsystem, the Cassini Radar, the Radio Science Subsystem, the Ion and Neutral Mass Spectrometer, the Visible and Infrared Mapping Spectrometer, the Composite Infrared Spectrometer, the Cosmic Dust Analyzer, the Radio and Plasma Wave Spectrometer, the Cassini Plasma Spectrometer, the Ultraviolet Imaging Spectrograph, the Magnetospheric Imaging Instrument, and the Dual Technique Magnetometer. Cassini is about the size of a 30-passenger school bus and one of the largest interplanetary spacecraft ever launched.

Huygens probe

The ESA component of the Cassini mission, named for the Dutch scientist Christaan Huygens (1629–1695), who discovered Titan. It will be released in December 2004 and enter Titan's atmosphere on January 14, 2005. Initially, the 2.7-m diameter probe will use its heat shield to

Cassini (Huygens probe) An artist's rendering of the Huygens probe descending to the surface of Titan. *European Space Agency*

decelerate. Then, at an altitude of about 175 km, its main parachute will deploy, followed 15 minutes later by a smaller, drogue chute. Throughout the remainder of its 140-km descent, Huygens will take measurements of the temperature, pressure, density, wind speed, and energy balance in Titan's atmosphere and relay this information to the orbiter. It will also send back images, including the first clear views of Titan's surface. After the probe mission is completed, Cassini will turn its high-gain antenna toward Earth and begin transmitting its recorded data.

cast propellant
A solid fuel fabricated by being poured into a mold and solidified to form a hard grain.

Castor and Pollux
A pair of French microsatellites, also known as D5A and D5B, launched unsuccessfully on May 21, 1973, by a Diamant B from Kourou. Castor was to have studied interstellar space and collected aeronomy and geodesic data; Pollux was to have tested a hydrazine propulsion system.

catalyst
A substance that alters the rate of a chemical reaction but is itself unchanged at the end of the reaction.

CATSAT (Cooperative Astrophysics and Technology Satellite)
A small scientific satellite being developed jointly by the University of New Hampshire and the University of Leicester, England, through the USRA (Universities Space Research Association)/NASA STEDI (Student Explorer Demonstration Initiative) program. It will study the origin of gamma-ray bursters from a 550-km circular polar orbit. CATSAT's developers are seeking a launch opportunity for their spacecraft.

Cavendish, Margaret (1623–1673)
The Duchess of Newcastle and the author of *The Blazing World,* in which the heroine makes a round trip of the Moon and planets and thus qualifies as the first fictional female space traveler. Cavendish was a colorful figure and a prolific and popular author. The well-known diarist Samuel Pepys described her less kindly as "mad, conceited, and ridiculous." A tremendous self-publicist, she published under her own name–a radical and deliberate infringement of contemporary proprieties–a huge body of work encompassing historical treatises, essays, poems, plays, and an autobiography. Her personal excess was legendary, and when she made her rare and highly theatrical public appearances, men and women lined the streets of London to catch a glimpse of her.

cavitation
The rapid formation and collapse of vapor pockets in a fluid flowing under very low pressure. It is a frequent cause of structural damage to rocket components.

C-band
See frequency bands.

CCD (charge-coupled device)
An imaging device, often used by Earth resources and astronomy satellites, consisting of a large-scale integrated circuit that has a two-dimensional array of hundreds of thousands of charge-isolated wells, each representing a pixel (picture element).

celestial guidance
A method of spacecraft guidance that uses stars as references for attitude and navigation.

celestial mechanics
The study of how bodies such as planets or spacecraft move in orbit.

Celestis
A Texas company that has made a business out of launching the cremated remains of individuals into space; the company offers to fly 7 grams of ashes for $5,300 (2002 price). In April 1997, Celestis sent some of the ashes of *Star Trek* creator Gene Roddenberry, 1960s icon Timothy Leary, and 22 other individuals into Earth orbit using a Pegasus XL. Since then, there have been several other Celestis funerary missions, typically as secondary payloads. Celestis 04 was launched by a Taurus rocket from Vandenberg Air Force Base on September 21, 2001, alongside the much larger payloads OrbView-4 and Quik-TOMS. Two canisters mounted to the Taurus fourth stage contained lipstick tube–sized capsules of ashes. The canisters were to remain mounted to the orbiting rocket stage until the spent motor naturally reentered about a year after launch. But the Taurus went out of control following second-stage separation and fell back to Earth with its payloads less than two minutes after liftoff.

Centaur
A powerful upper-stage rocket that can be mated to an Atlas or Titan launch vehicle. The Centaur was America's first cryogenic rocket, running on a combination of liquid hydrogen and liquid oxygen, and was developed under the direction of the Lewis Research Center. It first flew in 1962 and became operational in 1966 with the launch of Surveyor 1. Subsequently, it has undergone many improvements. Until early 1974, Centaur was used

exclusively in tandem with Atlas but was then adapted for mating to the Titan III to boost heavier payloads into Earth orbit and interplanetary trajectories. Today, when combined with the Titan IV, it makes up the most powerful expendable launch vehicle in the world. The Centaur is manufactured by Lockheed Martin and powered by two Pratt & Whitney restartable RL10 engines. It interfaces with the payload via a forward adapter, where the avionics, electrical, flight termination, telemetry, and tracking systems are mounted.[200] See **Ehricke, Krafft**. (See also table, "Centaur Rocket Stage.")

center of mass

The point in a body at which its entire mass can be considered to be concentrated.

center of pressure

The point at which the aerodynamic **lift** on a rocket is centered. Half the surface area of a rocket lies on one side of the center of pressure and half on the other side.

centrifugal force

A force that must be included in the calculation of equilibria between forces in a spinning frame of reference, such as that of a rotating space station. In the rotating frame, the forces on a body of mass m are in equilibrium (as evidenced by the body staying at the same place) only if all forces acting on it, plus a "centrifugal force" mv^2/R directed away from the center of rotation at a distance R, add up to zero.

centrifuge

A large motor-driven apparatus with a long arm at the end of which human and animal subjects or equipment can be rotated at various speeds to simulate the prolonged accelerations encountered in high-performance aircraft, rockets, and spacecraft. The earliest centrifuges were set up in the World War II era, at a time when aircraft were coming into use that were capable of such

high-g maneuvers that pilots could easily black out. It was also realized that pilots who were less susceptible to the effects of g-forces would be able to twist and turn their planes more sharply and thereby outperform an enemy. Centrifuges thus served a dual role: to study the medical effects of high acceleration and to select individuals best able to cope with such acceleration. A pioneer in this field was Heinz von Diringshofen of Germany, who, in 1931, began research into the physiology of radial acceleration forces and established the limits of tolerance to g-forces that could be sustained for any length of time without loss of vision or consciousness. Early centrifuges were set up at two laboratories in Germany, at the Mayo Clinic, at the University of Southern California, and at Farnborough in England. Others were built after the war, including the immense **Johnsville Centrifuge**, used to test and train the first American astronauts.

centripetal acceleration

The acceleration associated with motion in a circle and directed to the center of the circle.

centripetal force

The force that acts toward the center of a circle and makes circular motion possible. To enable an object of mass m to move with velocity v around a circle of radius R, a centripetal force of magnitude mv^2/R must be applied.

CERISE (Characterisation de l'Environment Radio-electrique par un Instrument Spatial Embarque)

An experimental **ELINT** (electronic intelligence) microsatellite built for the French arms procurement agency DGA (Delegation Generale pour l'Armement). It is notable for having suffered the first known accidental collision between two independent orbiting objects. On July 21, 1995, a piece of debris from the Ariane 1 third stage that launched SPOT 1 in 1986 collided with CERISE. During the impact, the upper portion of CERISE's gravity-gradient boom was snapped off. The

Centaur Rocket Stage					
Version	**Diameter (m)**	**Length (m)**	**Engine Type**	**Propellant Mass (kg)**	**Thrust (N)**
Centaur I	3.1	9.2	RL-10A-3A (x2)	13,900	147,000
Centaur II	3.1	10.1	RL-10A-3A (x2)	16,800	147,000
Centaur IIA	3.1	10.1	RL-10A-4 (x2)	16,800	185,000
Centaur IIIA (Atlas IIIA)	3.1	10.0	RL-10-4-1 (x1)	16,800	99,000
Centaur IIIB (Atlas IIIB)	3.1	11.7	RL-10-4-1 (x2)	20,800	198,000
Centaur G (Titan IV)	4.3	9.0	RL-10-3A (x2)	20,900	147,000

broken bit was subsequently catalogued as space object 1995-033E.

Launch
  Date: July 7, 1995
  Vehicle: Ariane 5
  Site: Kourou
  Orbit: 666 × 675 km × 98.1°
  Size: 60 × 35 × 35 cm
  Mass: 50 kg

Cernan, Eugene Andrew (1934–)

A veteran American astronaut, one of 14 selected in October 1963, and a career naval aviator. Cernan received a B.S. in electrical engineering from Purdue University and an M.S. in aeronautical engineering from the Naval Postgraduate School in 1964. He was a CAPCOM (Capsule Communicator) during the joint **Gemini** 7/Gemini 6A mission and became the pilot of Gemini 9A following the death of a prime crew member. He was also backup pilot for Gemini 12, backup Lunar Module pilot for **Apollo** 7, Lunar Module pilot for Apollo 10, backup

Eugene Cernan   Cernan, pilot of the Gemini 9A mission, inside the spacecraft during flight. *NASA*

commander for Apollo 14, and commander for Apollo 17, becoming the eleventh and (so far) penultimate person to step onto the Moon's surface and the last to leave his footprints there. Afterward, he served as special assistant to the manager of the **Apollo-Soyuz Test Project** before resigning from NASA and the Navy on July 1, 1976, to become an executive of Coral Petroleum of Houston, Texas. Later he formed the Cernan Corporation, an aerospace consultancy firm, in Houston. In 1999, he published his autobiography, *The Last Man on the Moon*.[44]

Chaffee, Roger Bruce (1935–1967)

An American astronaut selected by NASA to fly on the first **Apollo** manned mission. Unfortunately, Chaffee never got the chance. He died on January 27, 1967, along with crewmates Gus **Grissom** and Edward **White**, during a launch pad test at the Kennedy Space Center. Chaffee received a B.S. in aeronautical engineering from Purdue University in 1957 and immediately joined the Navy. Photographs taken while he piloted a U-2 spy plane in 1963 proved conclusively that the Soviet Union had installed offensive missiles in Cuba and were displayed by President Kennedy during a televised address. That same month, Chaffee was selected by NASA as a member of its third group of astronauts. He was assigned with Grissom and White to fly the first Apollo capsule on an 11-day mission in Earth orbit. However, a month before their scheduled launch, all three were killed aboard their capsule during a countdown rehearsal, when a flash fire raced through their cabin.

Chalet/Vortex

The second generation of U.S. Air Force **SIGINT** (signals intelligence) satellites, launched between 1978 and 1989, that replaced **Canyon**. The program was renamed Vortex after the code name Chalet appeared in *The New York Times*. It was superseded by **Magnum** and **Mercury ELINT**.

Challenger

(1) **Space Shuttle** Orbiter that first flew on April 4, 1983 (STS-6). Among the milestones of the *Challenger* Orbiter were the first spacewalk from a Shuttle (STS-6), the flight of the first American female astronaut (STS-7), the flight of the first African-American astronaut (STS-8), the first use of free-flying **Manned Maneuvering Unit**s during a spacewalk (STS-41B), and the first in-flight repair and redeployment of a satellite (STS-41C). See *Challenger* disaster. (2) Nickname of the **Apollo** 17 Lunar Module. (3) British Navy research vessel that made a prolonged study of the Atlantic and Pacific Oceans between 1872 and 1876, and after which the two spacecraft were named.

Challenger disaster

On January 28, 1986, the 25th **Space Shuttle** mission (STS-51L) ended in tragedy 73 seconds after launch; the *Challenger* Orbiter was destroyed and its seven-member crew killed. A presidential commission set up to investigate the accident concluded that a failure had occurred in the joint between the two lower segments of the right Solid Rocket Booster (SRB). The rubber seal, or O-ring, had hardened overnight in freezing weather and failed when the boosters ignited at launch. The escaping flame breached the External Tank, and the vehicle broke apart as the propellants ignited. The commission also concluded that there were flaws in the Shuttle program. Many changes in the SRB and Orbiter design, as well as in management, were made to improve safety.[298]

chamber pressure

The pressure of gases in the **combustion chamber** of a rocket engine, produced by the burning of fuel. The engine's **thrust** is proportional to the chamber pressure.

CHAMP (Challenging Minisatellite Payload)

A small German satellite mission for geoscientific and atmospheric research and applications, managed by GFZ Potsdam. Using an advanced GPS (Global Positioning System), CHAMP measures how its orbit changes in response to Earth's gravitational pull, enabling researchers to map indirectly the density of Earth's crust. By mapping the density to a high precision over many years, scientists can detect changes in variables such as seawater level, the thickness of the polar ice caps, and even the density of the atmosphere.

Launch
    Date: July 15, 2000
    Vehicle: Cosmos-3M
    Site: Pletesk
Orbit: 421×476 km $\times 87°$
Mass: 522 kg (total), 27 kg (science payload)

Chandra X-ray Observatory

One of NASA's four **Great Observatories**. After being launched by the Space Shuttle, Chandra was boosted into a high elliptical orbit from which it can make long-duration, uninterrupted measurements of X-ray sources in the universe. It uses the most sensitive X-ray telescope ever built, which consists of four pairs of nearly cylindrical mirrors with diameters of 0.68 to 1.4 m, to observe X-rays in the energy range of 0.1 to 10 keV. These mirrors focus X-rays onto two of Chandra's four science instruments: the High Resolution Camera and the CCD (charge-coupled device) Imaging Spectrometer.

***Challenger* disaster**   Following the explosion of *Challenger,* fragments of the Orbiter can be seen tumbling against a background of fire, smoke, and vaporized propellants from the External Tank. The left Solid Rocket Booster flies away, still thrusting. *NASA*

International participants in the mission include Britain, Germany, and the Netherlands. Known before launch as the Advanced X-ray Astrophysics Facility (AXAF), the observatory was renamed in honor of the distinguished Indian-American astrophysicist Subrahmanyan Chandrasekhar.[294]

Shuttle deployment
    Date: July 23, 1999
    Mission: STS-93
    Orbit: 10,157 × 138,672 km × 29.0°
    Length: 12.2 m

Chang Zeng
See **Long March**.

chaotic control
A relatively new spaceflight technique that, like **gravity assist**, enables interplanetary missions to be flown with much smaller amounts of propellant and therefore at lower cost. First used by NASA in the 1980s to maneuver **ICE** (International Cometary Explorer), it involves exploiting the gravitational properties of the unstable **Lagrangian point**s L1, L2, and L3 for navigational purposes. The basic idea is that a tiny nudge to a spacecraft at one of these points, involving the expenditure of very little fuel, can result in a significant change in the spacecraft's trajectory. Following NASA's initial trial-and-error approach with ICE, the mathematics of chaotic control began to be developed properly in 1990, starting with some work by Edward Ott, Celso Gregobi, and Jim Yorke of the University of Maryland. Their method, known after their initials as the OGY technique, involves calculating a sequence of small maneuvers that will produce the desired overall effect. NASA is exploiting this more refined version of chaotic control on the **Genesis** mission to catch specimens of the solar wind. At the end of its two-and-a-half-year sample collection, Genesis will not have enough fuel for a direct return to Earth. Instead, it will first be sent on a long detour to the L2 point, outside Earth's orbit around the Sun, from which it can be brought back very economically to the Earth-Moon L1 point and from there, by way of a few cheap, chaotic orbits of the Moon, into a stable Earth orbit. Finally, its cargo capsule will be released and will parachute down onto the salt flats of Utah in August 2003.[283]

characteristic length
The ratio of the volume of the **combustion chamber** of a rocket engine to the area of the **nozzle**'s throat. Characteristic length gives a measure of the average distance that the products of burned fuel must travel to escape.

characteristic velocity
(1) The change of velocity needed to carry out a space maneuver, such as an orbital transfer. (2) The sum of all the velocity changes that a spacecraft must achieve in the course of a mission—a quantity related to the amount of fuel that must be carried.

Chelomei, Vladimir Nikolaevich (1914–1984)
One of the Soviet Union's three leading rocket designers in the early years of the Space Age, along with Sergei **Korolev** and Mikhail **Yangel**. A specialist in applied mechanics, Chelmei headed the OKB-52 design bureau that produced a line of long-range cruise and ballistic missiles, radar surveillance satellites, the **Proton** launch vehicle, the IS antisatellite system, and the manned **Almaz** orbital complex. Chelomei's career peaked during Khrushchev's reign, when he tactfully hired the Soviet premier's son Sergei. With Khrushchev's patronage,

Chandra X-ray Observatory Chandra in Space Shuttle *Columbia*'s payload bay. *NASA*

Chelomei turned his small organization into a major space enterprise competing even with that of Korolev.

Born into a teacher's family in the small town of Sedletse, Chelomei moved to Kiev at the age of 12 and later attended the Kiev Aviation College, graduating in 1937. He earned the equivalent of a master's degree in 1939 and was awarded the prestigious Stalin Doctoral Scholarship in 1940. In 1941, he joined the Baranov Central Institute of Aviation Motor Building (TsIAM) in Moscow, where his main interest was **pulse-jet** engines. In 1944, the remains of a German pulse-jet-powered V-1 shot down in London were delivered to Moscow by British allies. In June, Chelomei was invited to the Kremlin, where Stalin's deputy responsible for the aviation industry, Giorgiy Malenkov, asked him if the V-1 could be duplicated. Chelomei gave a confident reply and two days later found himself in charge of a new, one-hundred-strong department at TsIAM. By the end of 1944 he had reproduced the German engine and by mid-1945 had built a similar missile of his own design.

After the war, Chelomei proposed to the Soviet air force changes to his cruise missile that would enable it to be air-launched from long-range bombers. But the project ran into technical problems. In 1953, the chief designer of the Soviet MiG fighters, in cooperation with a design bureau led by Sergei Beriya (the son of Stalin's security chief), won approval for their own missile project, and Chelomei's organization was absorbed by this more powerful competitor. But Chelomei's fortunes were set to change. In March 1953, Stalin died and Malenkov became premier. He remembered the young designer and helped him reestablish his collective. Chelomei started promoting his missiles for use aboard submarines, and in 1955 he was made head of the then-new OKB-52 design bureau to pursue his goals. In the face of stiff competition from experienced aviation designers, such as Mikoyan, Iluyshin, and Beriev, Chelomei managed to get his P-5 cruise missile adopted for Navy use. This was the beginning of Chelomei's ascent, which culminated in the Khrushchev era. In the 1960s, he pursued ambitious space projects including the Almaz military space station, as well as continuing to work on long-range cruise missiles. At first, he was able to compete successfully with both Korolev and Yangel. But his poor relationship with D. Ustinov, patron of the rocket industry, led him to fall from favor after Khruschev was deposed. He lost his biggest infrastructural gain made during the Khrushchev years: what is today the **Khrunichev State Research and Production Space Center.**

chemical fuel
A substance that needs an **oxidizer** in order to achieve combustion and develop thrust. Chemical fuels include those used by liquid and solid rockets, jets, and internal-combustion engines. They contrast with **nuclear fuel**.

chemical rocket
A rocket that uses **chemical fuel** to generate **thrust**.

China Great Wall Industry Corporation (CGWIC)
An organization established in 1980 and authorized by the Chinese government to promote and provide **Long March** launch services to the global market. CGWIC is also responsible for trade related to satellite technology and products, acting as the primary foreign trade and marketing channel for China's aerospace industry. Since the Long March launch vehicle was first made available for international use in 1985, CGWIC has set up business relationships with more than 100 companies and organizations throughout the world.

China in space
See **China Great Wall Industry Corporation; China, manned spaceflight; Chinese fire-rockets; Dong Fang Hong; Jiuquan Satellite Launch Center; Long March; Taiyuan Satellite Launch Center; Hsue-Shen Tsien;** and **Xichang Satellite Launch Center.**[160]

China, manned spaceflight
The most populous nation on Earth is swiftly moving toward a manned spaceflight capability that may eventually include an orbiting space station and a manned mission to Mars. As of April 2002, China had conducted three successful test launches of the Shenzhou spacecraft using the **Long March** 2F (also known as the CZ-2F) rocket from **Jiuquan Satellite Launch Center.** Shenzou consists of three modules, including an orbiter module and a return module, and rocket boosters. For the third flight, which began on March 25, 2002, the spacecraft was equipped with dummy astronauts, human physical monitoring sensors, and metabolic simulation apparatus. After almost a week and 108 orbits, the return module landed safely in a designated area in the central part of the Inner Mongolia Autonomous Region, leaving the orbital module to continue a mission expected to last about six months. No date has been announced for China's first attempt at a manned launch, though some observers believe it could be as early as 2003.

Chinese fire-rockets
Although stories of early rocketlike devices appear sporadically in the historical records of various cultures, the Chinese are generally reckoned to have been the first to use rockets in both ceremony and war. By the first century A.D., a simple form of gunpowder appears to have existed in China, used mostly for fireworks in religious

and other festive celebrations. Bamboo tubes were filled with the mixture and tossed into fires to create explosions. Doubtless some of these tubes failed to explode and instead skittered out of the fires, propelled by the gases and sparks produced by the burning gunpowder. In any event, the Chinese began to experiment with the gunpowder-filled tubes and hit upon the idea of attaching them to arrows and launching them with bows. Eventually, it was found that the gunpowder tubes could launch themselves just by the power produced from the escaping gas, and the true rocket was born. The earliest we know that true rockets were used is 1232, a time when the Chinese and the Mongols were at war. During the battle of Kai-Keng, the Chinese repelled their enemy with a barrage of "arrows of flying fire"–the first **solid-propellant rockets**. A tube capped at one end was filled with gunpowder; the other end was left open and the tube attached to a long stick. When the powder was ignited, its rapid burning produced hot gas that escaped out the open end and produced thrust. The stick acted as a simple guidance system that kept the rocket moving in the same general direction throughout its flight. Just how effective fire rockets were as military weapons is not clear, but their psychological effect must have been formidable, because following the battle of Kai-Keng the Mongols began making their own rockets and may have been responsible for the spread of this technology to Europe.

Chinese fire-arrows A set of Chinese fire-arrows being used in battle. *NASA History Office*

Nearly all rockets up to this time were used for warfare or fireworks, but there is an amusing Chinese legend that reports the use of rockets as a means of transportation. With the help of many assistants, a lesser-known Chinese official named Wan-Hu rigged up a rocket-powered chair. Attached to the chair were two large kites, and fixed to the kites were 47 fire rockets. On launch day, Wan-Hu sat himself on the chair and gave the order to light the rockets. Forty-seven aides, each carrying a torch, rushed forward to light the fuses. Moments later, there was a tremendous roar accompanied by billowing clouds of smoke. When the smoke cleared, Wan-Hu and his flying chair were gone. No one knows for sure what happened to Wan-Hu, but it seems likely that if the event really did take place, he and his chair were blown to bits: fire-arrows were as apt to explode as to fly.

CHIPS (Cosmic Hot Interstellar Plasma Spectrometer)

A UNEX (University-class Explorer) mission, funded by NASA, that will carry out an all-sky spectroscopic survey of the extreme ultraviolet background at wavelengths of 9 to 26 nm with a peak resolution of about 0.5 eV. CHIPS data will help scientists determine the electron temperature, ionization conditions, and cooling mechanisms of the million-degree plasma believed to fill the local interstellar bubble. Most of the luminosity from diffuse million-degree plasma is expected to emerge in the poorly explored CHIPS band, making CHIPS data relevant in a wide variety of galactic and extragalactic astrophysical environments. The CHIPS instrument will be carried into space aboard CHIPSAT, a spacecraft scheduled to be launched by a Delta II in December 2002.

chuffing

A characteristic, low-frequency pulsating noise associated with the irregular burning of fuel in a rocket engine.

CINDI (Coupled Ion-Neutral Dynamics Investigations)

A NASA Mission of Opportunity, developed by researchers at the University of Texas, Dallas, to understand better the dynamics of the ionosphere. The two CINDI instruments will fly aboard a satellite in 2003, which begins a U.S. Air Force program called the Communication/Navigation Outage Forecast System. CINDI will study the coupling processes between ionized and neutral gases near an altitude of 400 km.

CIRA (Centro Italiano Richerche Aerospaziali)

The Italian Aerospace Research Center, established in 1984 and located in Capua. It is partly owned by **ASI**, the Italian Space Agency.

circular velocity

The horizontal velocity an object must have in order to remain in a circular orbit at a given altitude above a planet. It is given by

$$V = \sqrt{\frac{GM}{R}}$$

where G is Newton's gravitational constant, M is the mass of the central body, and R is the distance of the orbiter from the center of the gravitating body.

circumlunar trajectory

A trajectory upon which a spacecraft orbits one or more times around the Moon.

circumplanetary space

The space in the immediate neighborhood of a planet, including the upper reaches of the planet's atmosphere.

circumterrestrial space

(1) Inner space, or the atmospheric region that extends from about 100 km to about 80,000 km from Earth's surface. (2) Also sometimes extended to include the Moon's orbit.

cislunar

Pertaining to space between Earth and the orbit of the Moon, or to a sphere of space centered on Earth with a radius equal to the distance between Earth and the Moon.

Clarke, Arthur C. (1917–)

A British author of science fiction novels and popular science books, resident of Sri Lanka since 1956. He worked on radar development during World War II, originated the concept of global **communications satellites** in the October 1945 issue of *Wireless World*,[50] and earned a first-class degree in physics and mathematics from Kings College, London (1948). Clarke's non-fiction books of the early 1950s, including *The Exploration of Space* (1951),[52] brought him fame as an enthusiastic advocate of space travel, although he had begun writing science fiction two decades earlier. His novels include *2001: A Space Odyssey* and *Rendezvous with Rama*.

Clarke belt

The **geostationary orbit** named after Arthur C. **Clarke**, who first described in detail how such an orbit could be used for global communications.

Clauser, Francis H. (1913–)

A leading research aerodynamicist in academe and the aerospace industry from the 1930s through the 1970s. He worked with Douglas Aircraft (1937–1946) and served as chair of aerospace studies at the Johns Hopkins University (1946–1960). Later he held a variety of academic positions, most notably the Clark B. Millikin Chair of Aerospace Engineering at the California Institute of Technology from 1969 until his retirement in 1980.

Cleator, P(hilip) E(llaby) (1908–1994)

A cofounder, with Les Johnson, and the first president of the **British Interplanetary Society** (1933) and author of *Rockets in Space: The Dawn of Interplanetary Travel* (1936)[56] and *Into Space* (1954). Cleator proved to be more prescient than one of his reviewers, the astronomer Richard van der Riet Wooley, who in the March 14, 1936, issue of *Nature* wrote:

> The whole procedure [of shooting rockets into space] . . . presents difficulties of so fundamental a nature, that we are forced to dismiss the notion as essentially impracticable, in spite of the author's insistent appeal to put aside prejudice and to recollect the supposed impossibility of heavier-than-air flight before it was actually accomplished.

Clementine

A small lunar probe built by the U.S. Naval Research Laboratory and jointly sponsored by Ballistic Missile Defense Organization (BMDO) and NASA as the Deep Space Program Science Experiment (DSPSE). Clementine's main goal was to space-qualify lightweight imaging sensors and component technologies for the next generation of Department of Defense spacecraft. Intended targets for these sensors included the Moon and the near-Earth asteroid 1620 Geographos. After two Earth flybys, the probe entered lunar orbit on February 21, 1994. It then mapped the Moon at a number of resolutions and wavelengths, from infrared to ultraviolet, and made laser altimetry and charged-particle measurements over a two-month period in two parts. The first of these consisted of a five-hour elliptical polar orbit with a perilune (nearest point to the Moon) of about 400 km at latitude 28° S. After one month of mapping, the orbit was changed to a perilune at latitude 29° N, where it remained for one more month. This allowed global imaging and altimetry coverage from 60° S to 60° N. Clementine's most important discovery was that of ice under the surface of craters at the lunar poles, later confirmed by **Lunar Prospector**. After leaving lunar orbit, a malfunction in one of the onboard computers on May 7 caused a thruster to fire until it had spent all of its fuel, leaving the spacecraft spinning at about 80 rpm with no attitude control. This ruled out the planned asteroid flyby. The spacecraft remained in Earth orbit and continued testing its components until the mission was ended.

Launch
   Date: January 25, 1994
   Vehicle: Titan II
   Site: Vandenberg Air Force Base
Size: 1.9 × 1.1 m
Mass: 424 kg

closed ecological system

A system that provides for the maintenance of life in an isolated living chamber, such as a spacecraft cabin, by means of a cycle wherein carbon dioxide, urine, and other waste materials are converted chemically or by photosynthesis into oxygen, water, and food. Extended periods of manned travel in space will require such systems unless other means of sustaining life are devised.

CloudSat

Part of a multisatellite experiment designed to measure the vertical structure and chemical composition of clouds from space. CloudSat will carry a millimeter-wave radar capable of seeing practically all clouds and precipitation, from very thin cirrus clouds to thunderstorms producing heavy precipitation. It will fly in a 705-km-high polar orbit in tight formation with **CALIPSO**, which will be launched on the same vehicle and will provide high-resolution vertical profiles of aerosols and clouds, and **PARASOL**. CloudSat, CALIPSO, and PARASOL, in turn, will follow behind **Aqua** and **Aura** in a looser formation. Scheduled for launch in April 2004, CloudSat is a partnership between Colorado State University, JPL (Jet Propulsion Laboratory), the Canadian Space Agency, the U.S. Air Force, and the U.S. Department of Energy. It is the third mission in NASA's **ESSP** (Earth System Science Pathfinder) program.

cluster

Two or more rocket motors that form a single propulsive unit.

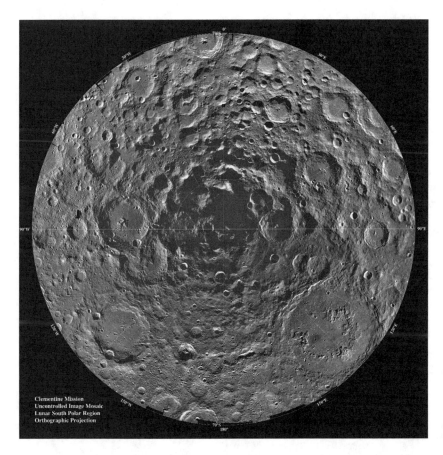

Clementine The Moon's south pole, imaged by Clementine in 1998. *Lawrence Livermore National Laboratory/Department of Defense*

Cluster

An ESA (European Space Agency) mission involving four identical satellites—nicknamed Salsa, Samba, Tango, and Rumba—that fly in formation to study the interaction between the solar wind and Earth's magnetosphere in three dimensions. Each satellite carries 11 instruments to measure electric and magnetic fields, electrons, protons, and ions, and plasma waves. The Cluster I satellites were lost following the explosion of their Ariane 5 launch vehicle on its maiden flight on June 4, 1996. (Reuse of flight software from Ariane 4 was found to be the cause of the accident—faulty program logic that only took effect on Ariane 5's launch trajectory.) Cluster II is a repeat of the original mission and uses some of its spare parts.

Launch
  Dates: July 16, 2000 (Salsa, Samba); August 9, 2000
  (Tango, Rumba)
  Vehicle: Soyuz Fregat
  Site: Baikonur
  Orbit: 25,000 × 125,000 km × 65°
  Mass (each): 1,200 kg

CMBPOL (Cosmic Microwave Background Polarization)

A spacecraft that would provide an important test of the so-called inflation theory in cosmology by searching for evidence of background gravitational waves produced when the universe was less than a second old. CMBPOL was identified in NASA's Office of Space Science Strategic Plan as a potential missions beyond 2007; it remains in the early stages of concept definition.

CNES (Centre National d'Etudes Spatiales)

The French national space agency. It was formed in 1962 to organize French space policy, coordinate major space projects, and promote French space science and technology. CNES has its headquarters in Paris.

coasting flight

(1) A spacecraft's flight between the cutoff of one propulsive stage and the ignition of the next. Between the two stages, the spacecraft coasts along due to the momentum gained from the previous stage. (2) The portion of flight between the end of the final burn and the reaching of peak altitude.

COBE (Cosmic Background Explorer)

A NASA satellite designed to study the residual radiation of the Big Bang. Its historic discovery of cosmic ripples—tiny fluctuations in the temperature of the cosmic microwave background—was announced in 1992. COBE also mapped interstellar and interplanetary dust clouds. Originally planned for launch on the Space Shuttle,

Launch
  Date: November 18, 1989
  Vehicle: Delta 5920
  Site: Vandenberg Air Force Base
  Orbit: 872 × 886 km × 99.0°
  Size: 5.5 × 2.4 m
  Mass: 2,265 kg

COBE was redesigned for launch aboard a Delta following the *Challenger* disaster. Its instruments include the Differential Microwave Radiometer (DMR) to check the thermal and structural uniformity of the early universe, and the Far Infrared Absolute Spectrometer (FIRAS) and Diffuse IR Background Experiment (DIRBE) to search for the remnant radiation emitted from primordial galaxies as they formed. COBE's supply of liquid helium was exhausted in September 1990, causing loss of the FIRAS instrument. The science mission as a whole ended in 1993.

coefficient of drag

The aerodynamic **drag** of a rocket as a function of its speed and cross-section. It provides a measure of a rocket's aerodynamic efficiency.

coherent

A two-way communications mode in which a spacecraft generates its **downlink** frequency based upon the frequency of the received **uplink**.

cold gas attitude control system

See **gaseous-propellant rocket engine**.

cold-flow test

A test of a liquid-propellant rocket engine without firing. The purpose is to check the efficiency of a propulsion subsystem that provides for the conditioning and flow of propellants, including tank pressurization, propellant loading, and propellant feeding.

Collier's Space Program

A series of six lavishly illustrated articles published in *Collier's* magazine over a two-year period, beginning with the March 22, 1952, issue, which sparked widespread public interest in rocketry and space travel.[304] It stemmed from an invitation by *Collier's* managing editor, Gordon Manning, to the pioneering space artist Chesley **Bonestell** to attend a San Antonio conference, "Physics and Medicine of the Upper Atmosphere," at which there would be many leading experts on astronautics and rocketry. The idea was to give the artist—already well-known for his cinematic spacecraft designs and oil paintings of planetary vistas—the biological perspective needed to show readers how human beings might safely travel in space. At the conference, held at the Air Force School of Aviation Medicine, Bonstell was particularly impressed by the enthusiasm and expertise of Wernher **von Braun** and suggested to *Collier's* editors that he was "the man to send our rocket to the Moon." A week later, the two men were at the magazine's New York offices along with some other great names of mid-century space science and illustration: Willy **Ley**, who had already collaborated with both von Braun and Bonestell; the astronomer Fred **Whipple**; the international law expert Oscar Schachter; the artists Fred Freeman and Rolf Klep; and the physicist Joseph **Kaplan**. Together, these spaceflight visionaries set about depicting and explaining for the layperson every element of von Braun's integrated space program, from the first piloted rockets to a mission to Mars.

Collins, Eileen Marie (1956–)

The first woman to command a Space Shuttle mission. A U.S. Air Force colonel, Collins holds various degrees, including a B.A. in mathematics and economics from Syracuse University (1978), an M.S. in operations research from Stanford University (1986), and an M.A. in space systems management from Webster University (1989). Selected by NASA in January 1990, she became an astronaut in July 1991. Collins served as the first female pilot of the Space Shuttle on mission STS-63 (February 2–11, 1995), during which the Shuttle completed its first dock-

Michael Collins   Collins standing between his Apollo 11 crewmates Neil Armstrong (left) and Edwin "Buzz" Aldrin. *NASA*

Eileen Collins   Commander Eileen Collins consults a checklist while seated at the flight deck commander's station aboard *Columbia* during STS-93. *NASA*

ing with **Mir**. She piloted the Shuttle again on the joint American-Russian mission STS-84 (May 15–24, 1997) before being appointed commander of mission STS-93 (July 22–27, 1999), the highlight of which was the successful deployment of the **Chandra X-ray Observatory**.

Collins, Michael (1930–)

An American astronaut, born in Rome, Italy. Collins walked in space during the **Gemini** 10 mission and circled the Moon as the **Apollo** 11 Command Module pilot. He received a B.S. degree from West Point in 1952, then entered the Air Force as an experimental flight test officer at Edwards Air Force Base before being selected as an astronaut in 1963. He retired from the Air Force as a major general and left NASA in 1970. After serving briefly as assistant secretary of state for public affairs, he became the first director of the Smithsonian Institution's National Air and Space Museum (1971–1978), overseeing its construction and development. He has written several books, including *Carrying the Fire, Liftoff,* and *Space Machine,* which blend good humor with incisive journalism. He has recorded, for example, how bad he felt about losing a camera in space on Gemini 10, and how he responded when asked what went through his mind at blastoff: "Well, you think about the fact that you are at the top of six million parts, all made by the lowest bidder!"

colloidal propellant

A solid propellant in which the mixture of fuel and oxidizer is so fine as to form a suspension, or colloid. Alternatively, a colloidal propellant may have the fuel and oxidizer atoms in the same molecule.

Colombo, Giuseppe "Bepi" (1920–1984)

An Italian mathematician and ground-breaking theorist in orbital mechanics at the University of Padua. He proposed **space tethers** for linking satellites together, was one of the initiators of the ESA (European Space Agency) **Giotto** mission to Halley's Comet, and suggested how to put a spacecraft into an orbit that would bring it back repeatedly to the planet Mercury. His calculations were used to determine the course taken by **Mariner** 10 in 1974–1975, which enabled the probe to make three passes of the innermost planet. Colombo also explained, as an unsuspected resonance, Mercury's habit of rotating three times for every two revolutions of the Sun. His name has been given to ESA's own proposed mission to Mercury, **Bepi Colombo**.

Columbia

(1) **Space Shuttle** Orbiter involved in the first orbital Shuttle mission (STS-1) on April 12, 1981. Milestones of the Orbiter *Columbia*, aside from the first launch and test mission of a Shuttle, include the first Department of Defense payload carried aboard a Shuttle (STS-4), the first operational mission of a Shuttle (STS-5), the first satellites deployed from a Shuttle (STS-5), and the first flight of **Spacelab** (STS-9). (2) Nickname of the **Apollo** 11 Command Module. (3) American commercial sloop based at Boston Harbor. On May 11, 1792, Captain Robert Gray and the crew of *Columbia* maneuvered past a dangerous sandbar at the mouth of a river, later named in honor of the sailing vessel, that extends more than 1,600 km through what is today southeastern British Columbia, Washington State, and Oregon. Gray and his crew went on to complete the first American circumnavigation of the globe, carrying a cargo of otter skins to Canton, China, and back to Boston. Other American sailing vessels have also been named *Columbia*, after Christopher Columbus, including a frigate launched in 1836 that became the first U.S. Navy ship to circle the globe.

Columbiad

Jules **Verne**'s Moon gun, as described in his 1865 novel *From the Earth to the Moon*. It consisted of a cannon, 274 m long with a bore of 2.74 m, cast in a vertical well in Florida. The first 61 m of the barrel were filled with 122 tons of guncotton, which when ignited were supposed to propel an aluminum capsule (containing three men and two dogs) at a speed of 16.5 km/s. After deceleration through Earth's atmosphere, the shell would have a residual velocity of 11 km/s—sufficient to reach the Moon. Although Verne made some scientific errors, he used real engineering analysis to arrive at the design of his cannon and lunar projectile.

Columbiad The *Columbiad* fires Jules Verne's three-man spacecraft toward the Moon.

Columbus

The largest single contribution of **ESA** (European Space Agency) to the International Space Station: a laboratory module scheduled for launch in 2004. See **International Space Station** for details.

combustion

A chemical process in which a great deal of heat is produced. Commonly, it involves the chemical reaction of a **fuel** and an **oxidizer**, but it may also involve the decomposition of a **monopropellant** or the burning of solid propellants.

combustion, incomplete

A state in which not all the fuel in the **combustion chamber** burns. Incomplete combustion may result from inadequate chamber design, or it may be deliberately designed into the system so that the unburned fuel acts as a chamber coolant. Generally, incomplete combustion denotes a system not functioning efficiently.

combustion chamber

The chamber of a **liquid-propellant rocket engine** in which the **fuel** and the **oxidizer** burn to produce high pressure gas expelled from the engine **nozzle** to provide **thrust**. To begin with, the fuel and oxidizer pass separately through a complex manifold in which each component is broken down into smaller and smaller flow streams, in the same way that arteries in the body divide into increasingly smaller capillaries. Then the propellants are injected into the combustion chamber via the injector—a plate at the top of the chamber that takes the small flow streams and forces them through an atomizer. The purpose of the injector is to mix the fuel and oxidizer molecules as thoroughly and evenly as possible. Once mixed, the fuel and oxidizer are ignited by the intense heat inside the chamber. To start the combustion, an ignition source (such as an electric spark) may be needed. Alternatively, some propellants are hypergolic—they spontaneously combust on contact—and do not need an ignition source.

combustion efficiency

The ratio of the energy actually released by the fuel during combustion to the energy contained in the fuel. A perfect combustion would release all the energy a fuel contains, in which case combustion efficiency would be 1, or 100%.

combustion instability

Unfavorable, unsteady, or abnormal combustion of fuel in a rocket engine.

combustion limit

In a solid-propellant rocket motor, the lowest pressure at which a given **nozzle** will support the burning of fuel without **chuffing**.

combustion oscillation

High-frequency pressure variations in a **combustion chamber** caused by uneven propellant consumption.

comet and asteroid missions

See, in launch order: **ICE** (August 1978), **Vega** (December 1984), **Sakigake** (January 1985), **Giotto** (July 1985), **Suisei** (August 1985), **Galileo** (October 1989), **NEAR-Shoemaker** (February 1996), **Cassini** (October 1997),

Deep Space 1 (October 1998), **Stardust** (February 1999), **Genesis** (August 2001), **CONTOUR** (July 2002), **MUSES-C** (November 2002), **Rosetta** (January 2003), **Deep Impact** (January 2004), **Dawn** (May 2006), and **Comet Nucleus Sample Return**. Canceled projects include **CRAF** and Deep Space 4 (Champollion).

Comet Nucleus Sample Return (CNSR)

A spacecraft designed to return a pristine sample of material from a comet nucleus to Earth in order to provide clues to the early history of the Solar System. CNSR is identified in NASA's Office of Space Science Strategic Plan but remains at the concept stage.

COMETS (Communications and Broadcasting Experimental Test Satellite)

A Japanese communications satellite launched by **NASDA** (National Space Development Agency); it is also known by the national name Kakehashi ("bridge"). COMETS carries Ka-band communications and inter-satellite data relay payloads (see **frequency bands**). Although premature shutdown 44 seconds into the H-2 second stage burn put the satellite into a much lower orbit than intended, the onboard Unified Propulsion System was able to raise the orbit to a more useful height.

Launch
    Date: February 21, 1998
    Vehicle: H-2
    Site: Tanegashima
    Orbit: 479 × 17,710 km × 30.1°

COMINT (communications intelligence)

A subcategory of **SIGINT** (signals intelligence) that involves messages or voice information derived from the interception of foreign communications.

command destruct

A system that destroys a launch vehicle and is activated on command of the **range safety officer** whenever vehicle performance degrades enough to be a safety hazard. The destruction involves sending out a radio signal that detonates an explosive in the rocket or missile.

Command Module

See **Apollo**.

communications satellite

See article, pages 84–85.

communications satellite

A spacecraft in Earth orbit that supports communication over long distances by relaying or reflecting radio signals sent from the ground to other points on the surface.

The possibility of using artificial satellites for radio communications over global distances had been discussed before World War II, but the modern concept dates from a 1945 *Wireless World* article by Arthur C. **Clarke**.[50] Clarke envisioned three relay stations in **geostationary orbit** by means of which a message could be sent from any point on Earth and relayed from space to any other point on the surface. As an application of such a system, Clarke suggested direct broadcast TV—a remarkably advanced idea, given that television was still in its infancy and it was not yet known whether radio signals could penetrate the **ionosphere**. The concept of the geostationary orbit had been discussed earlier by Herman **Noordung**, but Clarke gave the first detailed technical exposition of satellite communications. His vision was realized through the pioneering efforts of such scientists as John **Pierce** of Bell Labs, the head of the **Telstar** program and a coinventor of the traveling wave tube amplifier, and Harold Rosen of Hughes Aircraft, who was the driving force behind the **Syncom** program.

Most early communications satellites, such as **Echo**, were simply big metal-coated balloons that passively reflected signals from transmitting stations back to the ground. Because the signals bounced off in all directions, they could be picked up by any receiving station within sight of the satellite. But the capacity of such systems was severely limited by the need for powerful transmitters and large ground antennas. **Score**, launched in 1958, was technically an active communications satellite. It had a tape recorder that stored messages received while passing over a transmitting ground station, then replayed them when the satellite came within line of sight of a receiving station. However, real-time active communications came of age in 1962, when **Telstar** 1 enabled the first direct TV transmission between the United States, Europe, and Japan.

In the 1970s, advances in electronics and more powerful launch vehicles made geostationary satellite communications practicable and led to its rapid expansion. The Syncom series demonstrated the viability of the technique, and Early Bird—**Intelsat** 1—

became the first operational geostationary satellite. It was followed over the next three decades by Intelsat series of increasing power and capability.

In 1972, a policy change by the FCC (Federal Communications Commission) cleared the way for domestic satellite services and prompted RCA in 1975 to launch **Satcom** 1. This was immediately used by a group of entrepreneurs to transmit a new type of pay TV known as Home Box Office (HBO) to cable providers throughout the United States.

A number of developing countries, especially those with large areas and diffuse populations, saw satellite communications as an attractive alternative to expensive microwave and coaxial land-based networks. Indonesia led the way with **Palapa** 1 in 1976 and Palapa 2 in 1977.

The late 1970s also saw the beginning of a revolution with the introduction of small, affordable satellite dishes about 3 m across, which consumers could purchase and install on their own property. These units made possible the direct reception of virtually any transmission, including unedited network video feeds and commercial-free shows. A pioneer in this field was Ted Turner, the owner of WTBS, an independent cable channel based in Atlanta, Georgia. In 1976, Turner leased an available transponder and turned WTBS into the first superstation. This led to the rapid expansion of direct satellite TV networks, including Turner's own Cable News Network (CNN), which did not have to rely on the microwave relay network and local affiliate transmitters of the three major American commercial networks. The expansion of community access TV (cable TV) led to greater demand for high-quality satellite feeds. Today, **direct broadcast satellites** (DBS) deliver programs to home antennas measuring less than 50 cm in diameter.

The falling cost of satellites and launches has allowed a number of companies to buy and market satellites, and it has allowed other systems to provide international services in competition with Intelsat. The growth of international systems has been paralleled by domestic and regional systems such as Telstar, **Galaxy**, and Spacenet in the United States; **Eutelsat** and Telecom in Europe; and many single-nation indigenous systems.

The 1990s have seen the development of LEO (low Earth orbit), nongeostationary satellite concepts.

More sensitive radio receivers allow satellites in LEO to receive and relay signals from ground stations without the need for antennas that track the satellite as it moves across the sky. The first commercial service to operate from LEO was **Iridium**, but other similar services are now coming online, including **Globalstar**, **ICO**, and Ellipso. Depending on the height of their orbit, they are characterized as being either **little LEO** or **big LEO** and are used mainly to connect cellular telephones and business facsimile machines. They may also connect computers in areas where landlines and other relays do not exist. New digital coding methods have resulted in a tenfold reduction in the transmission rate needed to carry a voice channel, enhancing the capacity of facilities already in place and reducing the size of ground stations that provide phone service.

Communications satellite systems have entered a period of transition from point-to-point high-capacity trunk communications between large, costly ground terminals to multipoint-to-multipoint communications between small, low-cost stations. The development of multiple access methods has both hastened and facilitated this transition. With time-division multiple access (TDMA), each ground station is assigned a time slot on the same channel for use in transmitting its communications; all other stations monitor these slots and select the communications directed to them. By amplifying a single carrier frequency in each satellite repeater, TDMA ensures the most efficient use of the satellite's onboard power supply.

A technique called **frequency reuse** allows satellites to communicate with a number of ground stations using the same frequency, by transmitting in narrow beams pointed toward each of the stations. Beam widths can be adjusted to cover areas—**footprint**s—as large as the entire United States or as small as a state like Maryland. Two stations far enough apart can receive different messages transmitted on the same frequency. Satellite antennas have been designed to transmit several beams in different directions using the same reflector.

A new method for interconnecting many ground stations spread over great distances was tested in 1993 following the launch of **ACTS** (Advanced Communications Technology Satellite). Known as the hopping spot beam technique, it combines the advantages of frequency reuse, spot beams (tightly focused radio beams, somewhat like spotlights), and TDMA. By concentrating the energy of the satellite's transmitted signal, ACTS can use ground stations with smaller antennas and reduced power requirements.

The concept of multiple spot beam communications was successfully demonstrated in 1991 with the launch of Italsat, developed by the Italian Research Council. With six spot beams operating at 30 GHz on the uplink and 20 GHz on the downlink, the satellite interconnects transmissions between ground stations in all the major economic centers of Italy. It does this by demodulating uplink signals (that is, extracting information superposed on the carrier signal), routing them between uplink and downlink beams, and combining and remodulating them (impressing the information on the new carrier signal) for downlink transmission.

Large reflector antennas have become increasingly important to communications satellites. A large, precisely shaped reflector allows the transmitter to place more of the transmission in a smaller footprint on Earth. It also allows the satellite to collect more of the incoming energy from a ground station. The size of the antenna is limited by the size of the shroud, or nosecone, covering the satellite during launch. Larger rockets and upper stages with wider fairings now permit single-piece antennas up to 4 m in diameter.

Resistance to radiation such as **cosmic rays** is becoming a vital issue to satellite operators, who now rely on advanced electronics to cram more capability into a package. Geostationary orbits lie within the Van Allen belts, where solar flares can easily inject charged particles into sensitive components. Discharges can short-circuit a spacecraft's electronics and render the communications payload or even the spacecraft itself useless. Since the 1970s, several geostationary satellites have been damaged, and a few lost, because of space radiation and sunlight-induced charging.

The application of laser technology to satellite communications continues to be studied. Laser beams can be used to transmit signals between a satellite and Earth, but the rate of transmission is limited because of absorption and scattering by the atmosphere. Lasers operating in the blue-green wavelength, which penetrates water vapor, have been demonstrated for use in communication between satellites and submarines.

Noncommercial satellite systems include **Inmarsat** (a maritime telecommunications network founded in 1979), **TDRSS** (a NASA system set up in the 1980s to link the Space Shuttle, the Hubble Space Telescope, and other high-value satellites to ground controllers), and various military systems including **Milstar** and **Molniya**.

companion body

A portion of a spacecraft or a payload, such as the last stage of a rocket or a discarded part, that orbits unattached to but along with its parent.

complex

The entire area of launch site facilities, including the blockhouse, launch pad, gantry, etc.; also referred to as the launch complex.

composite

Two or more distinct materials, the combination of which produces a new material with more desirable properties such as increased strength, lower density, or resistance to high temperatures. Composites are now used extensively in spacecraft and launch vehicle structures. Carbon-carbon composites, for example, are found in reentry vehicle nose tips, rocket motor nozzles, and leading edges of the Space Shuttle Orbiter.

composite propellant

A solid rocket propellant consisting of an elastomeric (rubbery) fuel binder, a finely ground oxidizer, and various additives.

Compton Gamma Ray Observatory (CGRO)

The second of NASA's **Great Observatories** and the heaviest astrophysical payload ever flown at the time of its launch in 1991. CGRO was named after the American physicist Arthur Holly Compton (1892–1962) and carried four instruments that covered an unprecedented six orders of magnitude in energy, from 30 keV to 30 GeV. In this energy range, CGRO improved sensitivity over previous missions by a full order of magnitude and, during nine years of service, revolutionized our understanding of the gamma-ray sky. Following the failure of one of its three gyroscopes, NASA decided to de-orbit the spacecraft, and it reentered on June 4, 2000.

Shuttle deployment
    Date: April 7, 1991
    Mission: STS-37
Orbit: 448 × 453 km × 28.5°
Size: 9.1 × 4.6 m
Mass: 15,620 kg

comsat

See **communications satellite**.

Comsat (Communications Satellite Corporation)

An organization established by an Act of Congress in 1962 to provide satellite services for the international transmis-

sion of data. Comsat was the driving force behind the creation of **Intelsat** and served as the U.S. signatory to Intelsat. In 2000, Comsat was acquired by Lockheed Martin, which thus became the largest shareholder in Intelsat (privatized in 2001) and also **Inmarsat** (privatized in 1999).

cone stability

The inherent stability of conical shapes to fly without fins, provided the **center of mass** is ahead of the **center of pressure**.

Conestoga

A privately funded commercial launch vehicle built by Houston-based Space Services (SSI), which became the Space Service Division of EER Systems in 1990. Following a 1981 launch failure of Percheron, its first liquid-fueled rocket, SSI successfully tested its solid-propellant Conestoga 1, based on a Minuteman second-stage engine, in 1982. However, the first operational flight of the rocket and the attempted launch of NASA's **Meteor** satellite in 1995 failed when the vehicle was destroyed 45 seconds into its first-stage burn. There have been no further launches.

Congreve, William (1772–1828)

An artillery colonel in the British army whose interest in rocketry was stimulated by the success of Indian rocket barrages against the British in 1792 and again in 1799 at Seringapatam (see **Tipu Sultan**). Congreve's **black powder** rockets proved highly effective in battle. Used by British ships to pound Fort McHenry, they inspired Francis Scott Key to write of "the rockets' red glare" in a poem that later provided the words to *The Star-Spangled Banner*. Congreve's rockets were used in the Napoleonic Wars and in the War of 1812.[314]

conic sections

A family of curves obtained by slicing a cone with planes at various angles. This family includes the circle, the **ellipse**, the **parabola**, and the **hyperbola**. Its members include all the possible orbits an object can follow when under the gravitational influence of a single massive body.

Conquest of Space, The

(1) Book (1949) written by Willy **Ley** and illustrated by Chesley **Bonestell**, based on material published earlier in a series of *Collier's* magazine articles on space travel (see *Collier's* **Space Program**). It is best known for Bonestell's inspirational paintings, including 16 in full color. (2) Film (1955) directed by George Pal and partly inspired by Ley and Bonestell's book. In a thematic sense, *Conquest* was a sequel to Pal's **Destination Moon** (1950), taking

space exploration beyond the Moon to interplanetary space and, in particular, a manned journey to Mars. Although the characterization is poor and the dialogue often inane, there are some memorable scenes, including those of a rocket attempting to outrace a pursuing asteroid and a wheel-like space station designed along the lines proposed by Wernher **von Braun**.

Conrad, Charles "Pete," Jr. (1930–1999)
A veteran American astronaut, selected by NASA in September 1962. He served as pilot on **Gemini** 5 (becoming the first tattooed man in space–he had a blue anchor and stars on his right arm), command pilot on Gemini 11, commander on **Apollo** 12, and commander on **Skylab** 2. In December 1973, after serving for 20 years (including 11 during which he was also an astronaut), Conrad retired from the U.S. Navy to become vice president of operations and chief operating officer with American Television and Communications Corporation in Denver, Colorado. Three years later he joined McDonnell Douglas in St. Louis, Missouri, where one of his projects involved working on the **Delta Clipper**. Conrad died on July 8, 1999, following a motorcycle accident.

conservation of momentum
A fundamental law of motion, equivalent to Newton's first law, which states that the momentum of a system is constant if there are no external forces acting on it.

Constellation-X
A set of powerful X-ray telescopes that will orbit close to each other and work in unison to observe simultaneously the same distant objects, combining their data and becom-

ing one hundred times more powerful than any previous single X-ray telescope. Constellation-X has been designed to perform X-ray spectroscopy with unprecedented sensitivity and spectral resolution. The measurement of large numbers of X-ray spectral lines in hot plasmas will allow astronomers to determine the flow of gas in accretion disks around black holes, in active galactic nuclei, and in binary X-ray sources; to measure the population of newly created elements in supernova remnants; and to detect the influence of dark matter on the hot intergalactic medium in clusters of galaxies. The Constellation-X mission, currently under design, is a key element in NASA's Structure and Evolution of the Universe theme.

contact ion thruster
A form of **electrostatic propulsion** in which **ions** are produced on a heated surface and then accelerated in an electric field to produce a high-speed exhaust. Of the two types of **ion propulsion** that have been studied thoroughly over the past four decades–the other being the **electron bombardment thruster**–it has so far proved much the less useful for practical use in space. The difficulty is that the only propellant that has been shown to work in the contact ion method is cesium, because only cesium atoms have an outer (valence) electron that can be removed when the atoms are adsorbed onto the surface of a suitable metal, such as tungsten. However, cesium is so corrosive that it has been impractical to handle in devices that operate reliably over the long periods required for ion propulsion.

contamination
In spaceflight, the unwanted transfer of microbes by spacecraft from one celestial body to another.

CONTOUR (Comet Nucleus Tour)
A NASA Discovery Program mission that was to have taken images and obtained comparative spectral maps during close flybys of the nuclei of at least three comets, including Comet Encke in November 2003. However, following a successful launch on July 3, 2002, CONTOUR appears to have suffered a catastrophic failure during the rocket burn that took it out of Earth orbit. The probe broke into at least two pieces that are now orbiting the Sun uselessly.

control surface
A surface such as a flap or an elevon used to control the **attitude** of a rocket or aerospace vehicle aerodynamically.

Cooper, Leroy Gordon, Jr. (1927–)
One of the original **Mercury Seven** astronauts and the first man to go into orbit twice. Cooper flew the last and

Charles "Pete" Conrad   Conrad (left) and Gordon Cooper on the deck of the recovery aircraft carrier USS *Lake Champlain* following their Gemini 5 flight. *NASA*

longest **Mercury** orbital mission and spent eight days in space aboard **Gemini** 5. He earned a B.S. in aeronautical engineering from the Air Force Institute of Technology in 1956. Having received an Army commission, he transferred to the Air Force and flew F-84 and F-86 jets in Germany for four years. Back in the United States he received his degree and then attended the Air Force Experimental Flight Test School at Edwards Air Force Base before being assigned as an aeronautical engineer and test pilot in the Performance Engineering Branch of the Flight Test Division at Edwards. In 1959, he was selected by NASA and on May 15–16, 1963, piloted his Faith 7 capsule on the sixth and final Mercury Project mission. Electrical problems near the end of the mission meant he had to manually fire his retrorockets and steer the capsule through reentry. Problems also beset Cooper on his next flight, a then-record eight-day trip aboard Gemini 5 in August 1965. Trouble with fuel supplies, power systems, and a computer-generated command led Gemini 5 to land 166 km short of its designated target. Cooper retired from NASA and the Air Force in 1970, and has since been involved in technical research with several companies.[60]

Copernicus Observatory

See **OAO**-3 (Orbiting Astronomical Observatory 3).

Coriolis

A test mission for the U.S. Air Force that carries two scientific payloads: Wind, and the Solar Mass Ejection Imager (SMEI). Wind is a Navy experiment built by the Naval Research Laboratory to passively measure ocean surface wind directions, while SMEI is an Air Force Research Laboratory experiment to observe coronal mass ejections in visible light. The spacecraft's two payloads will collect data continuously during a three-year mission. Launch was scheduled for January 2003 from Vandenberg Air Force Base by a Titan II rocket into a circular 830 km × 98.7° orbit.

Coriolis effect

The deflection of the flight path of a spacecraft caused by Earth's rotation. Over the northern hemisphere, the deviation is to the right; over the southern hemisphere, it is to the left.

corona

The outermost layer of the Sun's (or some other star's) atmosphere, visible to the eye during a total solar eclipse; it can also be observed through special filters and, best of all, by X-ray cameras aboard satellites. The corona is very hot–up to 1.5 million degrees Celsius–and is the source of the **solar wind**.

Corona

America's first series of photo-reconnaissance or IMINT (imagery intelligence) satellites, involving more than one hundred launches between 1959 and 1972. The program had the cover name Discoverer and was only declassified in 1995. A Discoverer satellite would be placed in a polar orbit by a **Thor**-Agena rocket in order to take photographic swaths as it passed over the Soviet Union. It then collected its exposed film in a heat-resistant "bucket" at the nose, and the bucket would reenter over the Pacific Ocean, deploy two small parachutes, and be recovered in midair by an aircraft towing a trapeze-like snare. Bizarre as this sounds, the program proved successful after a shaky start–the first 12 launches failed, and the thirteenth, though achieving orbit, did not carry a camera. Discoverer 14 marked the program's first triumph. Its returning bucket was caught by a C-119 cargo plane on August 18, 1960, and provided the earliest photos of the Soviet Union's **Plesetsk** rocket base. In that single day, Corona yielded more valuable images of the Soviet Union than did the entire U-2 spy plane program. It proved conclusively that the Soviets' intercontinental ballistic missile (ICBM) arsenal did not number in the hundreds, as was widely feared, but rather amounted to somewhere between 25 and 50. This knowledge, however, was hidden from the American public for years. Corona showed that the supposed missile gap that Kennedy played upon in his presidential campaign was a myth–a fact he would have learned at the time had he taken up Eisenhower's offer of intelligence briefings. Corona allowed the U.S. intelligence community to catalog Soviet air defense and antiballistic missile sites, nuclear weapons–related facilities, and submarine bases, along with military installations in China and Eastern Europe. It also provided pictures of the 1967 Arab-Israeli war and Soviet arms control compliance. Its retrieval system helped NASA develop a safe means of recovering manned spacecraft, and its imaging systems provided the basis for the cameras carried by the Lunar Orbiters in 1966 and 1967.[74]

Coronas

An international project to study the Sun and its interaction with Earth using spacecraft launched by Russia and carrying experiments developed by Russia and Ukraine with involvement from scientists in other European countries and the United States. Coronas satellites are equipped to study solar activity including flares, active regions, and mass ejections, in various regions of the spectrum from radio waves to gamma rays. Coronas-I reentered the atmosphere on March 4, 2001, two days before Mir was de-orbited, leaving Russia temporarily without a single working science payload in orbit. Coronas-F carries three main groups of instruments to

Coronas Missions

Spacecraft	Launch Date	Orbit	Mass (kg)
Coronas-I (Intercosmos 26)	Mar. 2, 1994	$501 \times 504 \times 83°$	2,160
Coronas-F	Jul. 31, 2001	$540 \times 499 \times 82.5°$	2,260

obtain high-resolution X-ray images of solar active regions, measure the strength and polarization of radiation coming from active regions and flares, and measure the flux of solar particles. (See table, "Coronas Missions.")

Launch
    Vehicle: Tsyklon 3
    Site: Pletesk

COROT (Convection, Rotation, and Planetary Transits)

A French-led mission, scheduled for launch in 2005 from Russia, one of the objects of which will be to search for extrasolar planets by photometry (light intensity measurements). It will use a 30-cm telescope equipped with charge-coupled devices to monitor selected stars for tiny, regular changes in brightness that might be due to planets passing in front of the stars as seen from Earth. COROT will also be able to detect the light variations caused by seismic disturbances inside stars and so provide data helpful in determining stellar mass, age, and chemical composition. Techniques developed for COROT will be applied on future European planet-hunting missions, including **Eddington** and **Darwin**.

Corporal

The first American surface-to-surface missile to approach the capability of the German V-2 (see **"V" weapons**). The liquid-fueled Corporal and the **Private** stemmed from Project ORDCIT—a long-range missile program begun by the California Institute of Technology's rocket labora-

Length: 13.8 m
Diameter: 0.76 m
Mass, fully fueled: 5,200 kg
Range: 45–140 km
Altitude (on ballistic trajectory): about 42 km
Top speed: Mach 3.5
Thrust: 20,000 N
Propellants: aniline and red fuming nitric acid

tory at the request of Army Ordnance in 1944. The development began with the Private-A and Private-F and con-

tinued with the **WAC Corporal** and Corporal-E before becoming a separate weapons development program. During the 1950s, Type I and II Corporals were developed and deployed on mobile launchers by the U.S. Army in Europe; they remained active until the mid-1960s, when they were superseded by the **Sergeant**.

Corsa-A

The first Japanese X-ray astronomy satellite. It was lost following a launch failure on February 4, 1976.

Corsa-B

Pre-launch name of **Hakucho**.

Cortright, Edgar M. (1923–)

An influential NASA official during the era of the first planetary probes. Cortright earned an M.S. in aeronautical engineering from Rensselaer Polytechnic Institute in 1949, the year after he joined the staff of **Lewis** Laboratory. His research at Lewis involved the aerodynamics of high-speed air induction systems and jet exit nozzles. In 1958, Cortright joined a small task group to lay the foundation for a national space agency. When NASA came into being, he became chief of advanced technology at NASA headquarters, directing the initial formulation of the agency's meteorological satellite program, including projects TIROS and Nimbus. In 1960, he was appointed assistant director for lunar and planetary programs, supervising the planning and implementation of such projects as Mariner, Ranger, and Surveyor. He became a deputy director and then a deputy associate administrator for space science and applications in the next few years; then, in 1967, he became a deputy associate administrator for manned spaceflight. The following year he took over as director of the **Langley Research Center**, a position he held until 1975, when he went to work in private industry, becoming president of Lockheed-California in 1979.

COS-B

An ESA (European Space Agency) mission, which together with NASA's **SAS**-2 provided the first detailed views of the universe in gamma rays. COS-B carried a single large experiment, the Gamma-Ray Telescope, which was the responsibility of a group of European research laboratories known as the Caravane Collaboration. Although originally projected to last two years, the spacecraft operated successfully for more than six, providing the first complete map of the Milky Way Galaxy in the 2 keV to 5 GeV energy range and identifying about 25 discrete sources.

Launch
Date: April 9, 1975
Vehicle: Delta 2914
Site: Vandenberg Air Force Base
Orbit: 442 × 99,002 km × 90°
Mass: 280 kg

cosmic rays

High-energy subatomic particles from space that include fast-moving protons and small numbers of heavier nuclei. On colliding with particles in Earth's atmosphere, they give rise to many different kinds of secondary cosmic radiation. The lowest energy cosmic rays come from the Sun, while those of higher energy originate in supernovas and other sources both within our galaxy and beyond.

cosmic string

Hypothetical bizarre material that may have formed shortly after the Big Bang. If cosmic string exists, it is predicted to be infinitesimally small in cross-section but enormously long, perhaps forming loops that could encircle an entire galaxy. Theory indicates that strings would be extremely massive (a one-m length weighing perhaps 1.6 times as much as Earth), give rise to peculiar gravitational fields, and be superconducting. It has also been suggested that strings could be used as the basis of an interstellar drive. The idea is that an array of magnet coils attached to a spacecraft would be set up around a string. Magnetic pulses induced in the coils would cause the superconducting string to respond with opposing magnetic fields that would move the spacecraft along. The resulting propulsion would be reactionless in that it would not depend on ejecting mass to gain momentum rocket-style. Instead, the enormous mass of the string would provide a base to push against, effectively storing the energy used to build the spacecraft's momentum. To decelerate, the vessel would simply withdraw the energy from the string, returning it to storage as the destination was reached. Of course, developing the technology for such a propulsion system would be only one aspect of the problem; finding a suitably placed cosmic string would be quite another.[136]

cosmonaut

A Russian/Soviet astronaut. The Russians have a long-standing tradition of calling their people "cosmonauts"

only after they have traveled in space. This differs from the American custom, in which astronauts are so named as soon as they have been accepted into NASA employment after passing the selection tests. See **Avdeyev**, Sergei; **Belyayev**, Pavel; **Berezovoi**, Anatoly; **Dobrovolsky**, Georgy; **Feoktistov**, Konstantin; **Gagarin**, Yuri; **Komarov**, Vladimir; **Lebedev**, Valentin; **Leonov**, Alexei; **Makorov**, Oleg; **Polyakov**, Valery; **Remek**, Vladimir; **Savitskaya**, Svetlana; **Tereshkova**, Valentina; **Titov**, Gherman; **Titov**, Vladimir.

Cosmos (launch vehicle)

A family of small two-stage Russian launch vehicles derived from the R-12 and R-14 ballistic missiles (see **"R" series of Russian missiles**). It stemmed from a government decision in 1960 to develop a means of launching payloads not requiring the more powerful R-7–based (Soyuz-type) boosters. Earlier Cosmos rockets, known as Cosmos and Cosmos-2, consisted of an R-12 first stage and a high-performance second stage that burned a unique liquid oxygen/UDMH propellant combination. These were used to launch Cosmos and Intercosmos satellites until they were phased out in 1977. They were superseded by Cosmos-1, Cosmos-3, and Cosmos-3M, based on the R-14. The Cosmos-3M, with its restartable second stage, was the most prolific and successful of this line, delivering payloads of up to 1,500 kg into LEO (low Earth orbit). Although its production ended in 1995, a number remain in storage.

Cosmos (spacecraft)

A long and ongoing series of Earth satellites launched by Russia for a variety of military and scientific purposes. The first was placed in orbit on March 16, 1962. Subprograms of Cosmos include **Bion**, **EORSAT**, **Geizer**, **Geo-1K**, **Glonass**, **Gonets**, **Luch**, **Parus**, **Rorsat**, **Strela**, and **Tsikada**. The Cosmos label is also used for prototypes of new satellites; for example, Cosmos 122 to 144 were test versions of the weather satellite series known as **Meteor**. Finally, the catchall name was used in the early years of the Space Race to hide the true objectives of missions that failed, including a number of lunar and planetary probes left stranded in low Earth orbit and several vehicles connected with the secret Soviet-manned Moon program.

Cosmos 1

An experimental **space sail**, built for the **Planetary Society** in Russia by the Babakin Space Center and funded by Cosmos Studios, a science-based media and entertainment venture by Ann Druyan (wife of the late Carl Sagan) and A&E Network. The 600 square meter sail, made of 5-micron aluminized mylar in the shape of eight roughly triangular blades, was to be carried into a near-circular 850 km orbit, with an inclination of 78°, aboard a 100-kg spacecraft in late 2002. Imaging systems aboard the spacecraft are designed to show if the sail deploys, using an inflatable tube system to which the sail material will be attached, as planned.

COSPAR (Committee on Space Research)

An organization within the International Council of Scientific Unions, to which the United States belongs through the National Academy of Sciences. COSPAR was formed in 1958 to continue the cooperative research program in space science begun during the **International Geophysical Year**. It has its headquarters in Paris and holds scientific assemblies every two years.

countdown

The step-by-step process leading to a launch. It runs according to a prearranged schedule, measured in terms of T-time (T minus time before the engine start sequence begins, and T plus time thereafter). T − 0:00:00 is not necessarily the liftoff point, however; the Space Shuttle, for example, typically lifts off at T + 0:00:03.

Courier

A U.S. Army experimental communications satellite that followed **SCORE**. (See table, "Courier Satellites.")

course correction

The controlled firing of a rocket engine during the coasting phase of a flight to place a spacecraft on a new heading so that it can arrive successfully at its mission destination.

CRAF (Comet Rendezvous/Asteroid Flyby)

A canceled NASA probe. It was intended to be launched in 1995 and to fly by asteroid Hamburga in June 1998 and comet Kopff in August 2000.

crawler-transporter

An immense vehicle, adapted from earthmoving machinery, that carries the Space Shuttle, and previously the Saturn V, on a **crawlerway** from the **Vertical Assembly Building** to one or the other of launch pads 39A and 39B at **Cape Canaveral**. The crawler-transporter is 40 m long and 35 m wide, weighs 2,721 tons, and contains two diesel generators that provide 5,600 hp for the electric drive motor system. It moves on four double-tracked crawlers with hydraulic jacking pads on 27-m centers. Traveling at a mere 1.5 km/hr, it guzzles fuel at the rate of one liter every 1.6 m (one gallon every 20 feet).

crawlerway

A specially prepared dual roadway providing access for the **crawler-transporter** to the launch pads, arming tower parking areas, and the **Vertical Assembly Building**. The roadway is designed to support 7.9 million kg. The two lanes, each 12 m wide, are spaced on 27-m centers to match the tractor units. The crawlerway has a 5% grade to take the transporter to the raised level of each pad.

crew safety system (CSS)

The necessary sensors, test equipment, and displays aboard a spacecraft that detect and diagnose malfunctions and that allow the crew to make a reasonable assessment of the contingency. For emergency conditions, the CSS is capable of initiating an abort automatically.

Crippen, Robert Laurel (1937–)

An American astronaut and a copilot, alongside John **Young**, on the maiden voyage of the **Space Shuttle** on April 12, 1981. Crippen earned a B.S. in aerospace engineering from the University of Texas at Austin in 1960 and went on to become a Navy captain. In 1966, he entered the **Manned Orbiting Laboratory** program of the Department of Defense, and in 1969 was recruited by NASA. Subsequently, he became commander of the **Skylab** Medical Experiments Altitude Test and was a member of the support crew for the Skylab 2, Skylab 3, Skylab 4, and **Apollo-Soyuz Test Project** missions. After the historic first flight, Crippen flew the Shuttle again, this time as commander, on three more missions, STS-7, STS-41C, and STS-41G, in 1983 and 1984. He then became deputy director of NASA Space Transportation System Operations, before leaving the space agency for the private sector.

Courier Satellites

Launch vehicle: Thor-Able Star; launch site: Cape Canaveral			
Spacecraft	**Launch Date**	**Orbit**	**Mass (kg)**
Courier 1A	Aug. 18, 1960	Exploded 2.5 minutes after launch	225
Courier 1B	Oct. 4, 1960	967 × 1,214 km × 28.3°	230

crawler-transporter The Space Shuttle *Discovery* completes the final Earthbound portion of its journey into space, leaving the Vehicle Assembly Building atop the crawler-transporter on a slow trip to Launch Pad 39B. *NASA*

Crocco, Gaetano Arturo (1877–1968)

A leading Italian aeronautical and aerospace designer, military officer (rising to the rank of general), and founder of the Italian Rocket Society in 1951. Crocco designed a number of airships in the early part of the twentieth century and switched to designing rocket engines in the 1920s. Among his many contributions to the theory of spaceflight was a paper in which he showed that a spacecraft could, in theory, travel from Earth to Mars, perform a reconnaissance Mars flyby (that is, not stop over in Mars orbit), and return to Earth in a total trip time of about one year.[65]

Cromie, Robert (1856–1907)

The Irish author of *A Plunge Through Space* (1890),[66] which passed through many editions and laid the foundation for its author's fame as one of the pioneers of the interplanetary novel. In the book, after a flight to Mars using **antigravity**, the explorers come upon a utopian civilization.

cross range

The distance either side of a nominal reentry track that may be achieved by using the lifting properties of a reentering space vehicle.

Crossfield, Albert Scott (1921–)

A test pilot of the early X-planes and the first human to fly at twice the speed of sound (Mach 2). Crossfield learned to fly with the Navy during World War II and became an aeronautical research pilot with NACA (National Advisory Committee for Aeronautics) in 1950, flying the **X-1** as well as the **Douglas Skyrocket** (D558-II) and other experimental jets. He was the chief engineering test pilot for North American Aviation (1955–1961). In 1953, he achieved Mach 2 in a Skyrocket, and on the first powered flight of the **X-15** in 1959 reached Mach 2.11 and an altitude of 15,953 m.[67]

CRRES (Combined Release and Radiation Effects Satellite)

A NASA and Department of Defense satellite to study the electrical, magnetic, and particle environment of near-Earth space. The Air Force Geophysics Laboratory's Spacerad (Space Radiation Effects) experiment investigated the radiation environment of the inner and outer radiation belts and measured radiation effects on state-of-the-art microelectronics devices. Other magnetospheric, ionospheric, and cosmic ray experiments were supported by NASA and the Office of Naval Research. Marshall Space Flight Center's project involved the release of chemicals from onboard canisters at low altitudes near dawn and dusk perigee (low orbit) times and at high altitudes near local midnight. These releases were monitored with optical and radar instrumentation by ground-based observers to measure the bulk properties and movement of the expanding clouds of photoionized plasma along field lines. Contact with the spacecraft was lost on October 12, 1991, following a battery failure.

Launch
　Date: July 25, 1990
　Vehicle: Atlas I
　Site: Cape Canaveral
　Orbit: 335 × 34,739 km × 18.0°
　Mass: 1,629 kg

CRSP (Commercial Remote Sensing Program)

A program within NASA's Earth Science Enterprise. Based at the **Stennis Space Center**, CRSP was set up to help commercialize remote sensing, geographic information systems, and related imaging technologies. It administers several partnership programs designed to share NASA's remote sensing technology and expertise with American industry.

cruise missile

An unpiloted aircraft with an **air-breathing engine** that uses an onboard automatic navigation system to guide it to its target.

cruise phase

The part of a spacecraft's trajectory during which the vehicle is unpowered except for occasional course corrections.

CRV (Crew Return Vehicle)

See **ACRV** (Assured Crew Return Vehicle).

cryobot

A long, pencil-shaped probe with a heated tip that can melt its way down through a thick layer of ice and deploy equipment for returning data from any watery environment that may lie below. Such a device might eventually be used in the exploration of the subice ocean hypothesized to exist on Europa. Having penetrated to the bottom of the ice crust, the cryobot would release two principal devices for operating at interfaces where life might be expected to occur: an *ice-water interface station* to monitor conditions at the top of the watery sea, and a *sediment exploration station* to descend to the seafloor. The latter might also release a **hydrobot**, which would rise like a bubble to the top of the subice sea, taking measurements and searching for signs of life on the way. A cryobot was tested for the first time in January 2002, successfully melting its way 23 m down into a glacier on the island of Spitsbergen.

cryogenic propellant

A rocket propulsion fluid that is liquid only at very low temperatures. The commonest examples are **liquid hydrogen** and **liquid oxygen**. Cryogenic propellants require special insulated containers and vents to allow gas from the evaporating liquids to escape. The liquid fuel and oxidizer are pumped from the storage tanks to an expansion chamber and injected into the combustion chamber, where they are mixed and ignited by a flame or a spark.

CS- (Communications Satellite)

A Japanese communications satellite program, also known by the national name Sakura ("cherry blossom"). Spacecraft in the CS series, which are launched by

cryobot An artist's rendering of a proposed ice-penetrating cryobot and a submersible hydrobot that could be used to explore the ice-covered ocean on Jupiter's moon Europa. *NASA/JPL*

NASDA (National Space Development Agency), supported domestic telecommunications and also enabled new technologies to be developed and tested. (See table, "NASDA Communications Satellite Series.")

Cunningham, R. Walter (1932–)

An American astronaut who served as Lunar Module pilot on the first manned **Apollo** mission, Apollo 7–an 11-day Earth-orbital flight in October 1968. Cunningham joined the U.S. Navy in 1951 and served with the Marine Corps until 1956 and the Marine Corps Reserve program until 1975. Having received a B.A. in physics in 1960 and an M.A. in physics in 1961 from the University of California, Los Angeles, he worked as a scientist with the RAND Corporation before being selected as an astronaut in October 1963. Cunningham retired from NASA in 1975 and two years later wrote *The All-American Boys* about his experiences as an astronaut. He is now chairman of the Texas Aerospace Commission and serves as a consultant to start-up technology companies and as a lecturer.

cutoff

The termination of burning in a jet or rocket engine brought about by an intentional command; also known as *shutdown*. It is distinct from **burnout**, which signifies the end of burning because of the exhaustion of fuel.

Cyclone

See **Tsyklon**.

NASDA Communications Satellite Series

Spacecraft	Launch			GSO Location	Mass (kg)
	Date	Vehicle	Site		
CS-1	Dec. 15, 1977	Delta 2914	Cape Canaveral	135° E	676
CS-2A	Feb. 4, 1983	N-2	Tanegashima	132° E	772
CS-2B	Aug. 5, 1983	N-2	Tanegashima	136° E	670
CS-3A	Feb. 19, 1988	H-1	Tanegashima	132° E	1,100
CS-3B	Sep. 16, 1988	H-1	Tanegashima	136° E	1,100

Cyrano de Bergerac, Savinien de (1619–1655)

French writer whose works combine political satire and fantasy. He is most familiar in modern times from the 1897 novel *Cyrano de Bergerac* by Edmond Rostand, which describes him as being gallant and brilliant but ugly with a large nose. As a young man, Cyrano joined the company of guards but was wounded at the siege of Arras in 1640 and retired from military life. He then studied under the philosopher and mathematician Pierre Gassendi, who had a significant influence on him. In his *Histoire des Etats et Empires de la Lune* (History of the States and Empires of the Moon), published posthumously in 1657, followed in 1662 by *Histoire des Etats et Empires du Soleil* (History of the States and Empires of the Sun), both eventually collected as *L'Autre Monde* (Other Worlds),[69] Cyrano developed the concepts of rocket power and two-stage rockets. His account is the first description of a manned rocket flight in literature.[70]

CZ (Chang Zeng)

See **Long March**.

D

Daedalus, Project

One of the first detailed design studies of an interstellar spacecraft.[30] Conducted from 1973 to 1978 by a group of a dozen scientists and engineers belonging to the **British Interplanetary Society**, led by Alan **Bond**, it demonstrated that high-speed, unmanned travel to the stars is a practical possibility. Certain guidelines were adopted: the Daedalus spacecraft had to use current or near-future technology, be able to reach its destination within a human lifetime, and be flexible enough in design that it could be sent to any of a number of target stars. These guidelines ensured that the spacecraft would be practical, that those who worked on the project might live to see it achieve its goals, and that several stars could be investigated using the same type of vehicle. The selected target was Barnard's Star, a red dwarf 5.9 light-years from the Sun. Although the Alpha Centauri system is closer, evidence available at the time (now considered unreliable) suggested that Barnard's Star might be orbited by at least one planet. To reach Barnard's Star in 50 years, the flight-time allotted in the study, a spacecraft would need to cruise at about 12% of the **speed of light**, or 36,000 km/s. This being far beyond the scope of a chemical rocket, the Daedalus team had to consider less conventional alternatives. The design they chose was a form of **nuclear-pulse rocket**, a propulsion system that had already been investigated during Project **Orion**. However, whereas Orion would have employed nuclear **fission**, the Daedalus engineers opted to power their starship by nuclear **fusion**–in particular, by a highly efficient technique known as internal confinement fusion. Small pellets containing a mixture of **deuterium** and helium-3 would be bombarded, one at a time, in the spacecraft's combustion chamber by electron beams and thereby caused to explode like miniature thermonuclear bombs. A powerful magnetic field would both confine the explosions and channel the resulting high-speed **plasma** out of the rear of the spacecraft to provide thrust. By detonating 250 pellets a second, and utilizing a two-stage approach, the desired cruising speed could be reached during an acceleration phase lasting four years.

Daedalus would be constructed in Earth orbit and have an initial mass of 54,000 tons, including 50,000 tons of fuel and 500 tons of scientific payload. The first stage would be fired for 2 years, taking the spacecraft to 7.1% of light-speed, before being shut down and jettisoned.

Then the second stage would be fired for 1.8 years before being shut down to begin the 46-year cruise to Barnard's Star. Since the design made no provision for deceleration upon arrival, Daedalus would carry 18 autonomous probes, equipped with artificial intelligence, to investigate the star and its environs. The 40-m diameter engine of the second stage would double as a communications dish. On top of the second stage would be a payload bay containing the probes, two 5-m optical telescopes, and two 20-m radio telescopes. Robot "wardens" would be able to make in-flight repairs. A 50-ton disc of beryllium, 7 mm thick, would protect the payload bay from collisions with dust and meteoroids on the interstellar phase of the flight, while an artificially generated cloud of particles some 200 km ahead of the vehicle would help disperse larger particles as the probe plunged into the presumed planetary system of the target star.

En route, Daedalus would make measurements of the interstellar medium. Some 25 years after launch, its onboard telescopes would begin examining the area around Barnard's Star to learn more about any accompanying planets. The information gathered would be fed to the computers of the probes, which would be deployed 1.8 to 7.2 years before the main craft entered the target system. Powered by nuclear-ion drives and carrying cameras, spectrometers, and other sensory equipment, the probes would fly quickly past any planets looking especially for any signs of life or conditions favorable for biology.

Damblanc, Louis (1889–1969)

A civil engineer at the Institut des Arts et Métiers in Paris. Between 1932 and 1935, he performed experiments with black-powder rockets in order to develop better signaling devices for the Coast Guard and others. He was among the first to carry out methodical tests of the principle of **staging**.[71]

DARA (Deutsche Agentur für Raumfahrtangelegenheiten)

The present German space agency, formed through consolidation of the old West and East German space agencies. While DARA is instrumental in establishing space policy and goals and is the interface with ESA (European Space Agency), DLR (the German Center for Aerospace Research) conducts the technical and scientific research and provides the operational support to implement that policy.

Darwin

A future ESA (European Space Agency) mission, under study, that will look for signs of life on extrasolar planets using a flotilla of six orbiting telescopes. Each telescope will be at least 1.5 m in diameter and will operate at infrared wavelengths in order to pick out planets more clearly from the glare of their parent stars. (At optical wavelengths, a star outshines an Earth-like planet by a billion to one; in the mid-infrared, the star-planet con-

trast drops to a mere million to one!) Another reason for observing in the infrared is that gases, such as water vapor, associated with life as we know it, absorb especially strongly at certain infrared wavelengths, leaving clear spectral fingerprints in this region. Working together, Darwin's multiple telescopes will be about as sensitive as a single instrument 30 m in diameter. A second advantage of the array is that it enables a technique called *nulling interferometry* to be used to cancel out most of the light from the central star. In about 2014, Darwin will be launched into solar orbit at the second **Lagrangian point**, well away from terrestrial interference. It will be

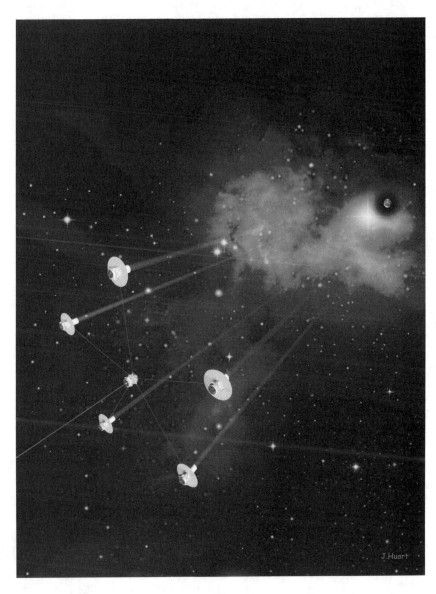

Darwin An artist's rendering of Darwin's flotilla of spacecraft observing an extrasolar planet.
European Space Agency

preceded, in 2006, by **SMART**-2–a two-spacecraft array to demonstrate the type of formation flying that is essential to Darwin's success. NASA is developing a similar mission, the **Terrestrial Planet Finder**, and it is possible that NASA and ESA will collaborate on a single mission that they will launch and operate together.

Dawn

A future NASA mission to two of the largest asteroids in the Solar System, Ceres and Vesta. Dawn, a **Discovery Program** mission, is scheduled to launch in May 2006, arrive at Vesta in 2010, orbit it for nine months, then move on to Ceres in 2014 for a further nine-month orbital stint. Dawn will be NASA's first purely scientific mission to use ion propulsion of the type known as **XIPS** (xenon-ion propulsion system) that was flight-validated by **Deep Space 1**. The probe's suite of science instruments will measure the exact mass, shape, and spin of Vesta and Ceres from orbits 100 to 800 km high; record their magnetization and composition; photograph their surfaces; use gravity and magnetic data to determine the size of any metallic core; and use infrared and gamma-ray spectrometry to search for water-bearing minerals. Flybys of more than a dozen other asteroids are planned along the way. The overall goal of the mission is to learn more about the early history of the Solar System and the mechanisms by which the planets formed.

dawn rocket

A rocket launched into orbit in an easterly direction so that Earth's rotation augments the rocket's velocity. Most launches are of this type.

DE (Dynamics Explorer)

A NASA mission launched on August 3, 1981, involving two spacecraft to study the interaction between the hot, thin, convecting plasma of Earth's magnetosphere and the cooler, denser plasmas and gases in the ionosphere and upper atmosphere. DE 1 and DE 2 were launched together and placed in polar coplanar orbits to allow simultaneous measurements at high and low altitudes in the same field-line region. The spacecraft approximated a short polygon 137 cm in diameter and 115 cm high. The antennas in the x-y plane measured 200 m tip-to-tip, and on the z-axis 9 m tip-to-tip. Two 6-m booms were provided for remote measurements. Science operations continued successfully over a nine-year period until they were terminated on October 22, 1990. The spacecraft were also known as **Explorer** 62 and 63.

De Laval, Gustav (1845–1913)

A Swedish engineer of French descent who, in trying to develop a more efficient steam engine, designed a turbine that was turned by jets of steam. The critical component–the one in which heat energy of the hot high-pressure steam from the boiler was converted into kinetic energy–was the nozzle from which the jet blew onto the wheel. De Laval found that the most efficient conversion occurred when the nozzle first narrowed, increasing the speed of the jet to the speed of sound, and then expanded again. Above the speed of sound (but not below it), this expansion caused a further increase in the speed of the jet and led to a very efficient conversion of heat energy to motion. Nowadays, steam turbines are the preferred power source of electric power stations and large ships, although they usually have a different design–to make best use of the fast steam jet, De Laval's turbine had to run at an impractically high speed. But for rockets the **De Laval nozzle** was just what was needed.

De Laval nozzle

A device for efficiently converting the energy of a hot gas to **kinetic energy** of motion, originally used in some steam turbines and now used in practically all rockets. By constricting the outflow of the gas until it reaches the velocity of sound and then letting it expand again, an extremely fast jet is produced.

Debus, Kurt H. (1908–1983)

An important member of Wernher **von Braun**'s V-2 (see "V" weapons) development team who subsequently supervised rocket launchings in the United States. Debus earned a B.S. in mechanical engineering (1933), and an M.S. (1935) and a Ph.D. (1939) in electrical engineering, all from the Technical University of Darmstadt, before being appointed an assistant professor there. During World War II he became an experimental engineer at the V-2 test stand at **Peenemünde**, rising to become superintendent of the test stand and the test firing stand for the rocket. In 1945, he came to the United States with a group of engineers and scientists headed by von Braun. From 1945 to 1950, the group worked at Fort Bliss, Texas, and then moved to the **Redstone Arsenal** in Huntsville, Alabama. From 1952 to 1960, Debus was chief of the missile firing laboratory of the **Army Ballistic Missile Agency** (ABMA). In this position, he was located at **Cape Canaveral**, where he supervised the launching of the first ballistic missile fired from there–an Army **Redstone**. When ABMA became part of NASA, Debus continued to supervise missile and space vehicle launchings, first as director of the Launch Operations Center and then of the **Kennedy Space Center**, as it was renamed in December 1963. He retired from that position in 1974.

decay

See **orbit decay**.

deceleration
Slowing down, or negative **acceleration**.

decompression sickness
A disorder caused by reduced barometric pressure and evolved or trapped gas bubbles in the body. It is marked by pain in the extremities, occasionally leading to severe central nervous system and neurocirculatory collapse.

Deep Impact
NASA **Discovery Program** mission to collide with a comet and study material thrown out by the impact from beneath the comet's surface. If launched as planned on January 6, 2004, Deep Impact will encounter comet Tempel 1 on July 4, 2005. The mission hardware consists of a flyby spacecraft and a smart impactor that will separate from the flyby probe 24 hours before collision. The 500-kg cylindrical copper impactor has an active guidance system to steer it to its impact on the sunlit side of the comet's surface at a relative velocity of 10 km/s. Prior to collision, the impactor will send back close-up images of the comet. The impact itself will create a fresh crater, larger than a football field and deeper than a seven-story building. Two visible imaging systems on the flyby craft will record the impact events and the subsurface cometary structure, while two near-infrared imaging spectrometers will determine the composition of the cometary material. This is the first attempt to peer beneath the surface of a comet to its freshly exposed interior for clues to the early formation of the Solar System. Images from the cameras on both the impactor and the flyby craft will be sent to Earth in near real-time and be made available on the Internet. The mission is under the direction of principal investigator Michael A'Hearn of the University of Maryland and is managed by JPL.

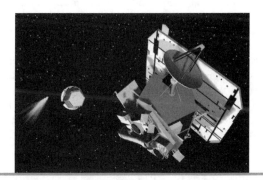

Deep Impact   An artist's rendering of the Deep Impact flyby craft releasing the impactor, 24 hours before the impact event. Pictured from left to right are comet Tempel 1, the impactor, and the flyby craft. *NASA*

deep space
An imprecise term that may be used to refer to all space beyond the Earth-Moon system or, alternatively, all space beyond the orbit of Mars.

Deep Space 1 (DS1)
An experimental probe designed to test 12 advanced spacecraft and science-instrument technologies that may be used on future interplanetary missions. DS1 was powered by a type of ion engine known as **XIPS** (xenon-ion propulsion system). Such systems have been used for station keeping by some satellites for a number of years, but the DS1 ion drive was larger, more efficient, and worked longer than any previously flown. In fact, DS1's engine accumulated more operating time in space than any other propulsion system in the history of spaceflight. Among the other devices and techniques successfully tested during the probe's primary mission were an autonomous navigation system, a miniature camera and spectrometer, an ion and electron spectrometer, a solar-energy concentrator array, and experiments in low-power electronics. DS1's primary mission lasted two years and included a flyby of the 3-km-wide asteroid Braille on July 29, 1999. An extended mission culminating in an encounter with comet Borrelly began in September 1999 but was soon threatened by the failure of the craft's most important navigational instrument, its star tracker, which enabled DS1 to orient itself relative to stellar patterns. Rather than abandon the project, NASA engineers uploaded new software to turn an onboard camera into a replacement star tracker, despite major differences between the two devices. It proved a valuable fix: on September 22, 2001, DS1 flew past Borrelly's nucleus at a distance of only 2,171 km, snapping 30 or so superb black-and-white photos and collecting data on gases and dust around the comet. The spacecraft's ion engine was turned off on December 18, 2001, but its radio receiver remains active in case future generations want to contact the probe.

Launch
   Date: October 24, 1998
   Vehicle: Delta 7325
   Site: Cape Canaveral
Mass, including propellant: 486 kg

Deep Space 2 (DS2)
See **Mars Microprobe Mission**.

Deep Space Communications Complex (DSCC)
Any one of the three **Deep Space Network** tracking sites at Goldstone, California; near Madrid, Spain; and near Canberra, Australia. These sites are spaced about equally

Deep Space 1 An artist's rendering of Deep Space 1 with ion engine thrusting. *NASA/JPL*

around the Earth for continuous tracking of deep-space vehicles.

Deep Space Network (DSN)
An international network of antennas that supports interplanetary missions, some selected Earth-orbiting missions, and radio astronomy. Its three **Deep Space Communications Complex**es combine to make the largest and most sensitive telecommunications system in the world. To give some idea of DSN's sensitivity, the antennas are still able to capture science information from the **Voyager** probes even though the downlink signal reaching a DSN antenna is 20 billion times weaker than the power level at which a digital watch functions. The DSN is a NASA facility managed and operated by JPL (Jet Propulsion Laboratory).

Deep Space Station (DSS)
The antenna front-end equipment at each tracking site of the **Deep Space Communications Complex**.

Defoe, Daniel (1660–1731)
A journalist and the chronicler of the adventures of Robinson Crusoe, who, in *The Consolidator: Or Memoirs of Sundry Transactions from the World of the Moon* (1705),[75] described the discovery of an eponymous spaceship invented 2,000 years before the Flood by a Chinese scientist named Mira-cho-cho-lasmo. The *Consolidator* is a flying machine powered by an internal combustion

engine that also featured hibernation capsules to ease the tedium of long spaceflights.

DeFrance, Smith J. (1896–1985)
An aeronautical engineer who played a major role in wind tunnel design and experimentation before and during the birth of the Space Age. DeFrance was a military aviator with the Army's 139th Aero Squadron during World War I, then earned a B.S. in aeronautical engineering from the University of Michigan in 1922 before beginning a career with **NACA** (National Advisory Committee for Aeronautics). He worked in the flight research section at Langley Aeronautical Laboratory and designed its 9×18 m wind tunnel, completed in 1931 and the largest in existence at that time. He directed the research carried out in that tunnel and designed others before becoming director of the new **Ames** Aeronautical Laboratory in 1940, a position in which he remained until his retirement in 1965. During his time at Ames, the center built 19 major wind tunnels and conducted extensive flight research, including the blunt-body research necessary for returning spacecraft from orbit without burning up.[134, 144]

Delta
See article, pages 102–105.

Delta Clipper
Prototype for a **single-stage-to-orbit** launch vehicle, designed by McDonnell Douglas. The Delta Clipper

Deep Space Network The Deep Space Operations Center, known as the Dark Room, at JPL, which is the nerve center of the Deep Space Network.
NASA/JPL

Experimental, or DC-X, made several successful flights at the **White Sands Missile Range** in the mid-1990s.

delta V

The change in **velocity** needed by a spacecraft to switch from one trajectory to another.

deluge collection pond

A pond at a launch site that collects water used to cool the **flame deflector**.

Demosat (Demonstration Satellite)

The first satellite to be placed in orbit from **Sea Launch** Odyssey. It successfully tested the facility's effectiveness.

Launch
    Date: March 28, 1999
    Vehicle: Zenit
    Site: Sea Launch Odyssey
Orbit: 638 × 36,064 km × 1.2°
Mass (at launch): 4,500 kg

Denpa

An early Japanese satellite launched by **ISAS** (Institute of Space and Astronautical Science) to carry out measurements of the ionosphere; also known as REXS.

Launch
    Date: August 19, 1972
    Vehicle: M-3S
    Site: Kagoshima
Orbit: 238 × 6,322 km × 31.0°
Mass: 75 kg

density

The ratio of the mass of a substance or an object to its volume.

de-orbit burn

The firing of a spacecraft's engine against the direction of motion to cut the spacecraft's orbital speed. The speed reduction places the spacecraft in a lower orbit. If this *(continued on page 106)*

Delta

A family of launch vehicles derived from the **Thor** ballistic missile for the purpose of placing intermediate-mass civilian payloads into orbit. First introduced in 1960 and manufactured by Douglas Aircraft, the Delta was conceived as a short-term solution to NASA's launch needs of that era, but it continued to evolve and remains in heavy use today.

The Thor had already been adapted for satellite launches using a variety of upper stages, giving rise to the Thor-Able, Thor-Able Star, and Thor-Agena A. The intended name, Thor-Delta, reflected the fact that this was the fourth upper-stage configuration of the Thor-based space launch vehicle. However, since the vehicle was designed mainly as a civilian satellite launcher, the military name Thor was dropped. The same naming principle was applied in the late 1950s as the **Redstone**-based and **Jupiter**-based rockets were renamed **Juno** I and Juno II, respectively, for civilian satellite launching. But as far as extant missile-descended rocket families are concerned, the Delta is unique: the **Atlas** and **Titan** retained the names of their military ancestors.

The original Delta was a three-stage launch vehicle consisting of a Thor first stage mated to improved **Vanguard** upper stages. The first Delta, launched on May 13, 1960, and carrying **Echo** 1, failed due to a second-stage attitude-control problem. However, the Echo 1 reflight three months later was a total success. In 1962, Douglas Aircraft began a series of upgrades and modifications that would increase the Delta's capacity tenfold over the next nine years.

The Delta A was almost identical to the original Delta but had an upgraded Rocketdyne first-stage engine. The Delta B, with an improved guidance and electronics system, longer second-stage propellant tanks (to increase fuel capacity), and an upgraded Aerojet second-stage engine, launched the world's first geosynchronous satellite, **Syncom** 2, in 1963. The Delta C was similar to the B but with a bulbous fairing to house bigger payloads and a new, more powerful third-stage solid-rocket motor developed for the Scout rocket.

With the Delta D—originally known as the Thrust Augmented Delta (TAD)—came a major improvement that became a Delta trademark. Three Castor 1 Solid Rocket Boosters (SRBs) were added to augment the thrust of the first-stage engine (which was itself upgraded) and thus the overall payload capacity. This strap-on configuration had already been applied successfully to the Thrust Augmented Thor-Agena D. Ignited at liftoff and jettisoned during flight, the SRBs gave the Delta the extra thrust needed to place operational Syncom satellites into geostationary transfer orbit (GTO).

The Delta E—originally called the Thrust Augmented Improved Delta (TAID)—came with a further upgraded first-stage engine and more powerful Castor 2 SRBs. The second stage was made restartable and its propellant tanks widened. The third-stage motor was from either the Delta D or a more powerful Air Force–developed motor, and the payload space was made still larger using an **Agena** fairing. A two-stage version of the E, known as the Delta G, equipped with a reentry vehicle, was specially made to carry **Biosatellite**-1 and -2 into LEO (low Earth orbit). Delta J featured a new third-stage Star motor.

Referred to as a Long Tank Delta (LTD) or a Long Tank Thrust Augmented Delta (LTTAD), the Delta L used the new Long Tank Thor first stage with its lengthened propellant tanks. The L, M, and N models were identical except for their third-stage configurations. The Delta L used the same third-stage motor as the E. The Delta M used the same third stage as the J, while the M-6 variant was the same as the M but with the addition of three more Castor 2s, for a total of six. The Delta N was a two-stage version of either the L or the M, and the N-6 was a two-stage version of the M-6.

With the 900 series came an important evolutionary step in the Delta program and a vital link to the Deltas that followed. The Delta 900, which came in two-stage (900) and three-stage (904) versions, featured nine strap-on Castor 2s and an improved guidance and electronics package. The second stage used a more powerful Aerojet engine, the AJ-10, previously flown as the Titan Transtage.

All subsequent Deltas used a four-digit numbering system for model identification. The first digit indicates the first stage and SRB type. For example, all 6000-series models employ the Extra Long Extended Tank Thor first stage with a Rocketdyne RS-27A main engine and Thiokol Castor 4A SRBs. The second digit tells how many SRBs are used—3, 4, 6, or 9. The third digit indicates the type of second stage—0 (Aerojet AJ10-118F), 1 (TRW TR-201), 2 (Aerojet AJ10-118K), or 3 (Pratt &

Whitney RL10B-2). The fourth digit tells the third-stage type—0 (no third stage), 2 (FW-4), 3 (Star 37D), 4 (Star 37E), 5 (Star 48B PAM-D derivative), or 6 (Star 37FM).

Introduced in the early 1970s, the 1000 series was the first Delta to be based on the Extended Long Tank Thor configuration that came to be known as the "straight-eight." While previous Deltas were tapered at the top, the Delta 1000 kept the 2.4-m first-stage diameter over its whole length, except for the rounded conical tip of the payload fairing. Since the upper-stage diameters were now the same as that of the first stage, extra room was available at the top of the rocket for larger payloads; otherwise, it was identical to the Delta 900. The 1000 came in eight two- and three-stage variants.

The five models of the 2000 series, flown mainly from 1974 to 1979, incorporated more improvements. Although the same SRBs were used as in the Delta 900 and 1000, the first and second stages were completely upgraded. The first-stage engine was adapted from the **H-1** Saturn I and IB first stages, and the second-stage engine from the **Apollo** Lunar Module main engine. Configuration options included two or three stages, and three, four, six, or nine SRBs.

Born of the need for a rocket capable of carrying payloads too heavy for the Delta 2000 but not heavy enough to require use of an Atlas-Centaur, the 3000 series was introduced in the late 1970s and early 1980s. Since NASA did not envision a need for a new medium-weight launch vehicle so close to the introduction of the Space Shuttle, the agency was reluctant to finance this next generation Delta. However, it did see a use for the 3000 as an interim vehicle to handle medium-weight payloads prior to operational flights of the Shuttle. As a result, McDonnell Douglas was able to secure industrial financing to develop the 3000. NASA financed no R&D but bought the completed vehicles to support civilian and commercial satellite launches. Although similar in design to the Delta 2000, the 3000 used powerful new Castor 4 SRBs. A Payload Assist Module (PAM) was offered as an optional third stage, enabling larger payloads to be carried to geosynchronous transfer orbit (GTO). An improved PAM called PAM-D was later introduced, which further increased the 3000's capacity.

The 4000 was to have been Delta's swan song at a time when virtually all commercial, civilian, and military satellite delivery duties were being switched to the Shuttle. But in the early 1980s, McDonnell Douglas proposed improvements to further boost the Delta's GTO capacity. Although the Delta program was wan-

ing at the time, this goal was met by increasing the first-stage burn time and bringing in other technical innovations. Nine Castor 4s were used as in the Delta 3000, but in the 4000 configuration six of them were typically ignited at launch, with the remaining three ignited following burnout and jettison of the first six. The first-stage fuel tanks were lengthened to give a longer first-stage burn time and improved performance. In the first departure from the straight-eight configuration, the 4000 introduced a payload fairing adapted from the Titan IIIC program to make room for larger payloads. The greater size of the 4000 demanded costly modifications to support structures at Cape Canaveral and Vandenberg Air Force Base. While these expenses may have seemed risky, McDonnell Douglas's investment in the 4000 proved fruitful. Although Delta production officially ceased in 1984 following 24 years of service as a workhorse of the civilian, commercial, and military satellite launch industry, the Delta retirement was short-lived.

Delta II

Like the Atlas family of space launchers, the Delta program faced cancellation in the early 1980s as the Shuttle took on more satellite launching duties. However, the *Challenger* **disaster** graphically showed that the American civilian, commercial, and military satellite industries still needed reliable, expendable launchers. With priorities for the Shuttle fleet dramatically shifted, an improved Delta filled a gap, as had previous Deltas, in providing a rocket capable of launching a wide spectrum of medium-mass payloads. An immediate need for post-*Challenger* expendable vehicles was created when NASA announced that commercial satellites would no longer be carried aboard the Shuttle. However, it was the military that facilitated the official rebirth of the Delta program. The Air Force signed a renewable contract for 20 Deltas to begin launching GPS satellites in 1989. A new and improved generation of Delta launch vehicles was introduced under the name Delta II.

Based largely on the Delta 4000 already in use, the 6000 boosted performance by incorporating nine improved Castor SRBs. Six of these were ignited at liftoff, with the remaining three ignited following burnout and jettison of the first six. The basic nine-booster Delta flight profile has not changed since.

The version of Delta II currently in use, the Delta 7000, which comes in a variety of two- and three-stage forms, uses new SRBs known as graphite epoxy motors

(GEMs) for improved performance. It bears little resemblance to the original Thor-Delta. With a height of 41 m, the Delta 7925 stands over 10 m taller than its ancestor and has a GTO payload capacity of about 1,870 kg–more than 40 times that of the Thor-Delta.

Delta III and IV

Almost four decades of continual development have increased the Delta's payload capacity by steady incre-

ments. The Delta III, however, represents a sudden leap in capacity to over twice that of the largest Delta II. A major part of this improvement is due to the Delta III's second stage, powered by a Pratt & Whitney RL10B engine derived from the RL10 that has been the basis of the **Centaur** for over 30 years. Burning liquid hydrogen and liquid oxygen, the RL10B has a world-record **specific impulse** rating (an efficiency measure) of 462 seconds and represents the first use of a high-energy cryogenic engine in a Delta. Additionally, new SRBs

Delta A Delta II launches an Iridium communications satellite from Vandenberg Air Force Base. *Lockheed Martin*

supply 25% more thrust, and three of the boosters are equipped with thrust vector control for better maneuverability. A new 4-m-diameter payload fairing tops the assembly. The first Delta III was launched in August 1998 and, like the very first Delta, carried a real payload—the Hughes Galaxy-X communications satellite—rather than a test article. Unfortunately, it also suffered the same fate, as a directional control problem doomed the flight. The second Delta III launch, in May 1999, also failed due to a rupture in the thrust chamber of the RL10 engine. The third flight, on August 23, 2000, was successful but carried only a test payload.

Although the Delta III has a backlog of about 20 launches, its manufacturer, Boeing, already has plans to phase it out in favor of the Delta IV. This latest extension of the Delta dynasty, which uses a new liquid oxygen/liquid hydrogen common core booster powered by a single RS68 engine, is being developed under the Air Force EELV contract. Delta IV comes in five variants, ranging from the Delta IVM (medium), which will replace the Delta II, to the top-of-the-line Delta IV "heavy," which combines three common core boosters with a large Delta III-type LOX/LH2 second stage and a new 5-m fairing to place up to 13,130 kg into GTO. Together, the Delta IV variants are capable of replacing not only the current Delta IIs and IIIs but also the more powerful Titan IV. The maiden flight of the Delta IV was scheduled for November 2002. (See table.)

Sample Delta Launch Vehicles

	Length (m)	Diameter Max. (m)	First-Stage Thrust (N)*	Payload (kg) LEO	GTO
Original Delta	27.4	2.4	670,000	270	45
Delta A	27.4	2.4	760,000	320	68
Delta B	28.3	2.4	760,000	375	68
Delta C	28.3	2.4	760,000	410	80
Delta D	28.3	2.4	1,725,000	580	105
Delta E	29.2	2.4	1,644,000	725	150
Delta L	32.4	2.4	1,725,000	n/a	—
Delta M-6	32.4	2.4	2,421,000	1,300	450
Delta 904	32.4	2.4	3,236,000	—	640
Delta 1914	32.4	2.4	3,362,000	—	680
Delta 2914	32.4	2.4	3,362,000	2,000	720
Delta 3000					
Original 3000	32.4	2.4	4,321,000	—	950
With PAM-D	32.4	2.4	4,632,000	—	1,270
Delta 4914	39.0	2.4	5,337,000	3,400	1,360
Delta II 6925	39.0	2.4	5,339,000	3,980	1,450
Delta II 7000					
7320	40.5	3.0	2,387,000	2,730	900
7925H-10L	40.5	3.0	5,382,000	5,090	1,830
Delta III	41.0	4.0	6,540,000	8,290	3,810
Delta IV					
Medium	63.0	5.0	2,891,000	6,760	3,900
Medium-plus (4,2)	63.0	4.0	4,103,000	9,070	5,850
Medium-plus (5,2)	69.0	5.0	4,103,000	7,850	4,640
Medium-plus (5,4)	69.0	5.0	5,315,000	10,300	6,570
Heavy	70.7	5.0	8,673,000	20,500	13,130

*Includes any ground-lit or air-lit strap-ons.

de-orbit burn

(continued from page 101)
lower orbit passes through Earth's atmosphere, the spacecraft reenters.

descent path

The path followed by a spacecraft during its descent to Earth, particularly after reentry through the atmosphere.

de-spun antenna

An **antenna** used in a spin-stabilized spacecraft (see **spin stabilization**) that rotates in the opposite sense and at an equal rate to the body of the spacecraft, thereby continuing to point in the required direction.

Destination Moon

A 1950 film directed by George Pal that depicts the first Moon landing—19 years ahead of the real event. It single-handedly started the boom in science fiction movies in that decade, including Pal's own *When Worlds Collide* (1951), *War of the Worlds* (1953), and *The Conquest of Space* (1955). Although not memorable for its quality of acting or pace of action, *Destination Moon* revealed an extraordinary optimism about the prospects for the Space Age and showed, too, how far scientific and technical knowledge had progressed since the days of Melies's *Le Voyage Dans la Lunes* and Lang's *Die Frau Im Mond* (**Woman in the Moon**). The realism of the film was aided by Chesley **Bonestell**'s stunning artwork.

deuterium

An isotope of hydrogen in which the nucleus contains one proton and one neutron. The abundance of deuterium in natural hydrogen is about 0.015%. It is an important fuel for nuclear **fusion**.

development vehicle

A vehicle whose main purpose is to light-test a rocket under development, rather than to launch satellites or other payloads.

DFVLR

See **DLR**.

Diamant

French three-stage launch vehicles developed by CNES (the French space agency) from a family of sounding rockets and flown from 1965 to 1975. The first stage was liquid-fueled; the upper two stages used solid propellant. Diamant ("diamond") A reached orbit at the first attempt in November 1965 carrying the **Asterix** (A-1) satellite. Three further flights in 1966 and 1967, all from **Ham-**

maguira in the Sahara, launched the D1 geodesy satellites, including the twin Diadémes and Diapason. Diamant B, with its lengthened first stage and, in the case of the later B4 variant, a modified second stage, launched nine times (including two failures) and placed DIAL-WIKA and -MIKA, Peole, Tournesol, and several other satellites in orbit. All Diamant B launches were from the Guiana Space Centre—the site now used by ESA (European Space Agency). After 1975, France stopped its sounding rocket and national launcher programs to concentrate on developing the European **Ariane** launcher. (See table, "Diamant Rockets.")

Diamant Rockets

	Diamant A	Diamant B/B4
Total length	18.9 m	23.5/21.6 m
Maximum diameter	1.4 m	1.4 m
Thrust		
First stage	269,000 N	316,000 N
Second stage	156,000 N	156,000/180,000 N
Third stage	27,000 N	50,000 N

direct ascent

A boost trajectory that goes directly to the final **burnout** and coast trajectory without the need for a **parking orbit** or staging location. Direct ascent of a spacecraft from Earth to the Moon was one of the options considered for the **Apollo** program and was the method depicted for this journey in *Destination Moon*.

direct broadcast satellite (DBS)

A high-powered **communications satellite** that transmits or retransmits signals intended for direct reception by the public, such as television, radio, or telephone signals. The signals are transmitted to small Earth stations or dishes mounted on homes or other buildings.

direct flight

A spaceflight that accomplishes its mission without rendezvousing or combining with another spacecraft after leaving Earth's surface.

directional antenna

An **antenna** used for communication over very long distances, in which the transmitting and/or receiving properties are concentrated along certain directions.

Discoverer

The cover name for **Corona**, America's first series of spy satellites.

Discovery

Space Shuttle Orbiter OV-103; it made its maiden flight on August 30, 1984 (STS-41D). *Discovery* was named after one of two ships captained by the British explorer James Cook, who sailed the South Pacific in the 1770s and discovered the Hawaiian Islands. Cook's other ship, *Endeavour,* also inspired the name of a Space Shuttle. *Discovery's* milestones have included the first flight following the *Challenger* **disaster** (STS-26), the deployment of the **Hubble Space Telescope** (STS-31) and of **Ulysses** (STS-41), the first female Shuttle pilot, Eileen **Collins** (STS-63), and the first Shuttle/**Mir** rendezvous (STS-63).

Discovery Program

A recently implemented and extraordinarily successful NASA program of small probes that operates under the rubric "faster, better, cheaper." The first five missions to be launched were **NEAR-Shoemaker**, **Mars Pathfinder**, **Lunar Prospector**, **Stardust**, and **Genesis**. Other projects in the series include **CONTOUR**, **Deep Impact**, **Dawn**, **Kepler**, and **MESSENGER**. Basic requirements of the Discovery Program are that missions should take less than three years to complete so that they can take advantage of the latest technology.

Disney, Walt (1901–1966)

The creator of Mickey Mouse and other well-known animated characters. In 1955, his weekly television series aired the first of three programs related to spaceflight. The first of these, "Man in Space," premiered on *The Wonderful World of Disney* on March 9, 1955, and attracted an estimated audience of 42 million. The second show, "Man and the Moon," also aired in 1955 and sported the powerful image of a wheel-like space station as a launching point for a mission to the Moon. The final show, "Mars and Beyond," premiered on December 4, 1957, after the launching of Sputnik 1.

DLR (Deutschen Zentrum für Luft und Raumfahrt)

The German Center for Aerospace Research. DLR was formed in 1969 with the merger of national aerospace research and test organizations as DFVLR (Deutsche Forschung und Versuchanstalt für Luft und Raumfahrt), but it was reorganized and renamed in 1989 at the time the German space agency **DARA** was created. One of the consequences of this reorganization was the transfer of major program management functions from DLR to DARA.

DME (Direct Measurement Explorer)

Also known as **Explorer** 31; it was launched together with the Canadian **Alouette** 2. The double-launch project, known as ISIS-X, was the first in a new cooperative

Walt Disney   Disney meets Wernher von Braun, then chief, Guided Missile Development Operation Division at ABMA, Redstone Arsenal, in 1954. A V-2 model is in the background. *NASA*

NASA–Canadian Defense Research Board program for International Satellites for Ionospheric Studies. DME was placed in an orbit with an apogee (highest point) slightly higher than Alouette's and a perigee (lowest point) slightly lower. Eight ionospheric experiments sampled the environment both ahead of and behind the satellite.

Launch
   Date: November 28, 1965
   Vehicle: Thor-Agena B
   Site: Vandenberg Air Force Base
Orbit: 502 × 2,857 km × 79.8°
Mass: 99 kg

DMSP (Defense Meteorological Satellite Program)

A program of U.S. military weather satellites that has been running since the mid-1960s and continuously maintains at least two operational spacecraft in near-polar orbits. More than 30 DMSP satellites have been launched since January 19, 1965, when the first, known as DMSP-Block 4A F1, was lofted from Vandenberg Air

Force base by a Thor Burner. The latest models, designated DMSP-Block 5D-3 and launched by Air Force Titan II rockets, have a mass of about 2.5 tons, orbit at an average altitude of 830 km, and use an array of instruments for monitoring atmospheric and oceanic conditions of interest to military planners. Each satellite crosses any point on Earth up to twice daily and, with an orbital period of about 101 minutes, provides almost complete global coverage of clouds every six hours. Visible and infrared sensors collect images of global cloud distribution across a 3,000-km swath during both daytime and nighttime conditions, while a microwave imager and sounders supply about one-half this coverage. DMSP was previously the exclusive responsibility of the Air Force Space and Missile Systems Center at Los Angeles Air Force Base, but is now run by a tri-agency organization involving the Department of Defense,

Department of Commerce, and **NOAA** (National Oceanic and Atmospheric Administration). The Space and Missile Systems Center remains in charge of developing and acquiring DMSP systems, but as soon as a spacecraft is declared operational it becomes the concern of the National Polar Orbiting Environmental Satellite System (NPOESS) Integrated Program Office at NOAA headquarters in Maryland. See **POES**.

Dnepr

One of the new crop of Russian space launch vehicles based on Cold War hardware. Dnepr is a converted RS-20 (known to NATO as the SS-18 or Satan) three-stage intercontinental ballistic missile. It is marketed by ISC Kosmotras for placing small, typically scientific, satellites into LEO (low Earth orbit), and has a payload capacity of about 4,500 kg.

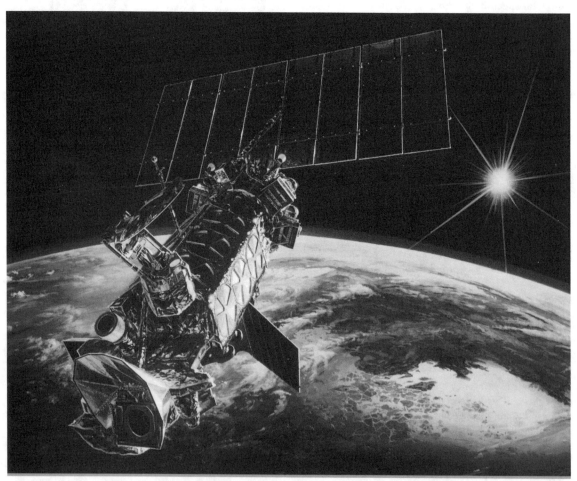

DMSP (Defense Meteorological Satellite Program)   An artist's rendering of a recent DMSP satellite in orbit. *Lockheed Martin*

Dobrovolsky, Georgy Timofeyevich (1928–1971)

A Soviet air force lieutenant colonel and cosmonaut, selected in 1963. In 1971, he commanded Soyuz 11, which docked with **Salyut** 1. The three-man crew remained in space for 24 days, performing meteorological and plant-growth experiments. During reentry of Soyuz 11, a valve in the spacecraft opened accidentally, causing the cabin to depressurize. Dobrovolsky and his crew were found dead upon recovery of the craft.

docking

The mating in space of one vehicle to another. Docking was first achieved on March 16, 1966, when **Gemini** 8 docked with its Agena Target Vehicle.

DODGE (Department of Defense Gravity Experiment)

A U.S. Navy satellite intended mainly to explore **gravity-gradient stabilization** at near-geosynchronous altitude. DODGE carried 10 booms that were radio-commanded to extend or retract along three different axes. Data from in-orbit experiments provided fundamental constants for use in controlling future high-altitude spacecraft. DODGE also carried a number of commandable magnetic-damping devices and two TV cameras to determine satellite alignment. One of those cameras provided the first color pictures of the full Earth.

Launch
   Date: July 1, 1967
   Vehicle: Titan IIIC
   Site: Cape Canaveral
   Orbit: 33,251 × 33,677 km × 11.6°
   Mass: 195 kg

Dong Fang Hong (DFH)

Chinese experimental communications satellites, the first of which was launched on April 24, 1970. The name means "the East is red."

Doolittle, James "Jimmy" Harold (1896–1993)

A longtime aviation promoter, air racer, U.S. Air Force officer, and advocate of aerospace research and development. Doolittle served with the Army Air Corps (1917–1930), then as manager of the aviation section for Shell Oil (1930–1940). In World War II, he won fame for leading the April 1942 bombing of Tokyo, and then as commander of a succession of air units in Africa, the Pacific, and Europe. He was promoted to lieutenant general in 1944. After the war, he was a member of the Air Force's Scientific Advisory Board and the President's Sci-

entific Advisory Committee. At the time of Sputnik, he was chair of NACA (the National Advisory Committee for Aeronautics) and the Air Force Scientific Advisory Board. In 1985 the Senate approved his promotion in retirement to four-star general.[80, 113]

Doppler effect

The phenomenon whereby the wavelengths of light, radio waves, or other kinds of waves coming from an object are changed when the object is in relative motion toward or away from the observer.

Doppler radar

Radar that measures the velocity of a moving object by measuring the shift in carrier frequency of the return signal. The shift is proportional to the velocity with which the object approaches or recedes from the radar station.

Doppler tracking

The most common method of tracking the position of vehicles in space. It involves measuring the Doppler shift of a radio signal sent from a spacecraft to a tracking station on Earth, this signal either coming from an onboard oscillator or being one that the spacecraft has transponded in response to a signal received from the ground station. The second of these modes is more useful for navigation because the returning signal is measured against the same frequency reference as that of the original transmitted signal. The Earth-based frequency reference is also more stable than the oscillator onboard the spacecraft.

Dornberger, Walter (1895–1980)

Wernher **von Braun**'s military superior during the German rocket development program of World War II. He oversaw the effort at **Peenemünde** to build the V-2, fostering internal communication and successfully advocating the program to officials in the German army. He also assembled the team of talented engineers under von Braun's direction and provided the funding and staff organization needed to complete the project. After World War II, Dornberger was brought to the United States as part of Operation Paperclip and helped develop ballistic missiles for the Department of Defense. He also worked for Bell Aircraft for several years, developing hardware for Project BOMI, a rocket-powered space plane.[33, 81]

Douglas Aircraft

See **McDonnell Douglas**.

Douglas Skyrocket

An American experimental rocket-powered plane that became the first aircraft to travel at more than twice the

Douglas Skyrocket   The Skyrocket begins its flight after dropping from its B-29 mothership. *NASA*

speed of sound (Mach 2). On November 30, 1953, pilot A. Scott **Crossfield** nudged the Skyrocket–which had been air-launched from a Boeing B-29–into a shallow dive at 18,900 m and reached 2,136 km/hr (Mach 2.01). The D-558 series, which also included the Skystreak, was developed by Douglas Aircraft under the direction of Edward H. Heinmann of the U.S. Navy. NACA (National Advisory Committee for Aeronautics) used the Skyrocket to explore the flight characteristics of swept-wing aircraft. The Skyrocket set a number of speed and altitude records before the program ended in 1956.

downlink
The radio signal received from a spacecraft, as distinct from the uplink, which is the signal sent to the spacecraft from the ground.

downrange
The area on Earth over which a spacecraft travels after launch and before entering orbit.

drag
Resistance to motion through a fluid. As applied to a spacecraft passing through an atmosphere, it is the component of the resultant force due to relative airflow measured parallel to the direction of motion.

Draper, Charles Stark (1901–1987)
An American engineer and physicist, regarded as the father of inertial navigation. An Institute professor at the Massachusetts Institute of Technology, Draper developed the guidance systems used by aircraft, submarines, and guided missiles that made the Apollo missions possible, and that help steer the Space Shuttle.

drogue
A small parachute used to slow and stabilize a descending spacecraft, usually before the opening of a main landing parachute.

DRTS (Data Relay Test Satellite)
Two Japanese satellites–DRTS-W (West) and DRTS-E (East)–that will carry out experiments in intersatellite communications. Traditionally, keeping in permanent contact with low-Earth-orbiting satellites has involved the use of a number of expensive ground stations. The idea of the DRTS project is to allow constant communication between mission control and an orbiting spacecraft by using the DRTS satellites as relay stations. Launch of the first satellite, DRTS-W, by **NASDA** (National Space Development Agency) was scheduled for September 2002.

dry weight
The weight of a launch vehicle or spacecraft without propellants and pressurizing gases.

Dryden Flight Research Center (DFRC)
NASA's foremost installation for atmospheric flight operations and flight research. Located at **Edwards Air Force Base**, Dryden uses a variety of specialized research aircraft and demonstrators to study high-speed and high-altitude environments. Among the aims of this work are to find ways of cutting long-distance flight travel times and to develop new aircraft configurations and structures to increase agility. DFRC's origins go back to 1946, when a small team of NACA (National Advisory Committee for Aeronautics) engineers came to Muroc Army Airfield (now Edwards Air Force Base) to prepare for **X-1** tests in a joint NACA–Air Force program. The following year NACA's Muroc Flight Test Unit was formed, which in 1949 became the NACA High-Speed Flight Research Station with Walt **Williams** as its chief. The center was eventually named after Hugh **Dryden**. Over the years, all of NASA's high-performance aircraft and flight research vehicles have flown from here, including all of the X-planes, the **Lunar Landing Research Vehicle** or "Flying Bedstead," and the Space Shuttle prototype *Enterprise*. Most recently, Dryden has been involved in test flights of the **X-38** lifeboat for the International Space Station and various solar-powered aircraft.

Dryden, Hugh Latimer (1898–1965)
An aerodynamicist and career civil servant who played a prominent role in American aerospace developments after World War II. Dryden graduated from high school at the remarkably young age of 14 and earned an A.B.

Hugh Dryden   Dryden (left) arrives with John Victory (center), NACA's executive secretary, for a tour of the Langley Memorial Aeronautical Laboratory. Welcoming them is engineer-in-charge Henry Reid. *NASA*

from Johns Hopkins University at age 17. Three years later, in 1918, he was awarded a Ph.D. in physics and mathematics from the same institution even though he had been employed full-time in the National Bureau of Standards since June 1918. His career at the Bureau, which lasted until 1947, was devoted to studying airflow, turbulence, and particularly the problems of boundary layers of air next to moving objects. The work that he carried out in the 1920s measuring turbulence in wind tunnels facilitated research by NACA (National Advisory Committee for Aeronautics) that led to the laminar flow wings used in the P-51 Mustang and other World War II aircraft. From the mid-1920s to 1947, Dryden's publications became essential reading for aerodynamicists around the world. During World War II, his work on a glide bomb named the Bat won him a Presidential Certificate of Merit. He capped his career at the Bureau by becoming assistant director and then associate director during his final two years there. He then served as director of NACA (1947–1958), and later became deputy administrator of NASA under T. Keith **Glennan** and James **Webb**.[276]

DSCS (Defense Satellite Communications System)

The U.S. Department of Defense global network of strategic communications satellites, referred to as "Discus." DSCS provides uninterrupted secure voice and high data rate communications between major military terminals and national command authorities. It was developed in three phases. The first was originally known as **IDCSP** (Initial Defense Communications Satellite

Program) but renamed DSCS 1 upon becoming operational. The second phase involved much larger spacecraft placed in true geosynchronous orbits between 1971 and 1975. Seven of these were still working in 1982 when the first DSCS 3 satellites were launched. A constellation of six DSCS 3 satellites (four operational and two spares) in geosynchronous orbit is now up and running.

DSPS (Defense Support Program Satellite)

See IMEWS.

dual spin

A spacecraft design in which the main body of the satellite is spun to provide altitude stabilization, and the antenna assembly is "de-spun"—in other words, rotated at the same speed but in the opposite direction—by means of a motor and bearing system in order to continually point the antenna earthward. See **spin stabilization**.

dual thrust

Thrust that derives from two propellant grains using the same propulsion section of a missile or space vehicle.

duct

In an **air-breathing engine**, a channel along which a working fluid is forced to travel in order to increase its momentum and thereby produce **thrust**.

duct propulsion

See **air-breathing engine** and **propulsion**.

Duke, Charles Moss, Jr. (1935–)

An American astronaut who served as backup Lunar Module (LM) pilot for **Apollo** 13 and Apollo 17, and LM pilot for Apollo 16. Duke received a B.S. in naval sciences from the U.S. Naval Academy in 1957 and an M.S. in aeronautics from the Massachusetts Institute of Technology in 1964. Following his graduation and commission from the Naval Academy, he entered the Air Force and served three years in Germany before transferring to the Aerospace Research Pilot School. He was among 19 new astronauts selected in April 1969. Duke retired from NASA in 1975 to enter private business.

Dunn, Louis G. (1908–1979)

A South African–born engineer who played a key role in the development of early American missiles and launch vehicles. Dunn earned a B.S. (1936), two M.S.s—in mechanical engineering (1937) and aeronautical engineering (1938)—and a Ph.D. (1940) from the California Institute of Technology, and then joined the faculty there. He became assistant director of JPL (Jet Propulsion Laboratory) in 1945 and then served as its director

(1947–1954), presiding over JPL's early rocketry program, which led to the development of the **Sergeant** missile. Upon leaving JPL, he took charge of the **Atlas** missile project for the recently formed Ramo-Wooldridge Corporation. He remained there through 1957 as associate director and then director and vice president of the guided missile research division, before becoming executive vice president and general manager, then president, and finally chairman of the firm's Space Technology Laboratories. From 1963 on, Dunn assumed various management positions for Aerojet-General Corporation. Besides the Atlas (built by General Dynamics), he played a key role in developing the **Thor** (McDonnell Douglas), **Titan**, and Minuteman missiles (Martin Marietta).[228]

Durant, Frederick Clark, III (1916–)

An American rocket pioneer who was heavily involved in the development of missiles and space launch vehicles between the end of World War II and the mid-1960s. He worked for several aerospace organizations, including Bell Aircraft Corporation, Everett Research Lab, the Naval Air Rocket Test Station, and the Maynard Ordnance Test Station. Later, he became the director of astronautics for the National Air and Space Museum and served as an officer for several spaceflight organizations, including the American Rocket Society (president, 1953), the International Astronautical Federation (president, 1953–1956), and the National Space Club (governor, 1961). Since his retirement, Durant has represented the interests of a number of leading astronomical and space artists and is the author, with Ron Miller, of *The Art of Chesley Bonestell* (2001).

dust detector

A device for measuring the velocity, mass, charge, flight direction, and density of dust particles in space.

DXS (Diffuse X-ray Spectrometer)

An instrument built by the University of Wisconsin–Madison and flown as an attached payload on board the Space Shuttle *Endeavour* (STS-54), January 13 to 19, 1993. Its main goal was to obtain spectra of the diffuse soft X-ray background in the energy range of 0.15 to 0.28 keV.

dynamic behavior

The behavior of a system or component under actual operating conditions, including acceleration and vibration.

dynamic load

A load associated with the elastic deformations of a structure upon which varying external forces are acting. Launch vehicles, for example, experience large dynamic loads because of varying internal pressure, external atmospheric pressure, and vibration.

dynamic pressure

(1) The pressure exerted by a fluid, such as air, by virtue of its motion. (2) The pressure exerted on a body because of its motion through a fluid—for example, the pressure on a rocket moving through the atmosphere.

dynamic response

The time-varying motion of a given structure to a given force input. For example, for a space vehicle that is exposed to side-acting winds during flight, the dynamic response would consist of a rigid body rotation of the vehicle plus a bending deformation, both changing with time.

dynamic stability

The property of a body, such as a rocket or plane, which, when disturbed from an original state of steady flight or motion, dampens the oscillations set up by the restoring movements and thus gradually returns the body to its original state.

dynamics

The study and theory of how and why objects move as they do.

Dyna-Soar (Dynamic Soaring Vehicle)

An early American design for a manned **space plane**. The research program for Dyna-Soar was authorized by the U.S. Air Force a week after the launch of Sputnik 1 in October 1957. It resulted in plans for a military spacecraft that would be launched on a Titan III and then rendezvous with enemy satellites to inspect and possibly destroy them. To emphasize the experimental nature of the program, Dyna-Soar was renamed X-20 in 1962. As it seemed to duplicate plans for the civilian manned spaceflight program, the X-20 project was canceled by the Department of Defense in 1963. Subsequently, NASA continued to study other different **lifting body** designs, including the **M-2**.

dysbarism

A general term that includes a complex variety of symptoms within the body caused by changes in ambient barometric pressure, but not including **hypoxia**. Characteristic symptoms are bends and abdominal gas pains at altitudes above 7,500 to 9,000 m. Also at increased barometric pressure, as in descent from a high altitude, the symptoms are characterized by painful distention of the ear drums and sinuses.

Dyson, Freeman John (1923–)

An English-born theoretical physicist, a president of the Space Sciences Institute, and a professor emeritus at the Institute for Advanced Studies, Princeton, who has had a lifelong interest in space travel and space colonization. In the late 1950s, Dyson became involved with the **Orion** Project to develop a nuclear-powered spacecraft. He is also well known for his audacious 1960 scheme of planetary engineering, which would involve processing the materials of uninhabited planets—including Jupiter—and satellites to fashion many habitats in heliocentric orbits. A shell-like accumulation of such habitats has been called a Dyson sphere.[83] The concept became modified by science fiction writers to include a rigid, monolithic sphere; however, this is probably unrealizable on dynamic and structural grounds.

E

Early Bird (communications satellite)

Also known as **Intelsat 1**, the first operational geostationary communications satellite. It was launched in 1965 and provided 240 telephone channels from its station over the Atlantic at a time when transatlantic cables provided only 412 channels.

Early Bird (Earth resources satellite)

A commercial Earth resources and reconnaissance satellite owned by EarthWatch of Longmont, Colorado. Images from Early Bird–the world's first private "spy" satellite–were to have shown detail as small as three meters across and been available to anyone for a fee of a few hundred dollars. However, contact with the spacecraft was lost shortly after launch.

Launch
   Date: December 24, 1997
   Vehicle: Start 1
   Site: Svobodny, Russia
   Orbit: 479 × 488 km × 97.3°

early warning (EW) satellites

Military spacecraft used to detect the launch of missiles and rockets from Earth's surface. Information from these satellites is used to track the long-term patterns of foreign nations' space programs and would provide first alert of the start of a major missile attack.

Earth Probes

A NASA program that complements **EOS** (Earth Observing System). Earth Probes are specific, highly focused missions that investigate processes requiring special orbits or short development cycles of one to three years. Within this program are the **ESSP** (Earth System Science Pathfinder) missions, the UnESS (University Earth Systems Science) program, and various specialized Earth-observing projects such as **TOMS** (Total Ozone Mapping Spectrometer), **Triana**, and **SRTM** (Shuttle Radar Topography Mission).

Earth Received Time (ERT)

The time at which an event that has occurred on a spacecraft will be seen on Earth. It corresponds to **spacecraft**

event time plus **one-way light time** and does not include any delay in signal processing once the signal has reached the ground.

Earth resources satellites

Spacecraft that collect information about Earth's atmosphere, land, and oceans for scientific research and resource management.

Earth station

Equipment on Earth that can transmit or receive satellite communications, though generally the term refers to receive-only stations. An Earth station usually contains a combination of antenna, low noise amplifier, downconverter, and receiver electronics.

Earth-orbit rendezvous

A means of accomplishing a mission that involves rendezvous and docking, fueling, or transfer in an Earth **parking orbit**. It was one of the approaches considered for the **Apollo** program.[139]

Earth-stabilized satellites

Satellites whose axes maintain a constant relationship to Earth's center, though not to any fixed spot on the surface.

Eastern Space and Missile Center (ESMC)

An American military launch facility located on the east coast of Florida. Operated by the 45th Space Wing of the **Air Force Space Command**, it includes **Cape Canaveral Air Force Station**, where most of the launch pads are located, **Patrick Air Force Base**, where the headquarters is, and the **Eastern Test Range** and other supporting facilities in east central Florida.

Eastern Test Range

The Eastern Test Range extends more than 16,000 km from **Cape Canaveral** across the Atlantic Ocean and Africa into the Indian Ocean and includes tracking stations on Antigua and Ascension Island.

eccentric orbit

A highly elliptical orbit with a high apogee and a low perigee.

eccentricity

A measure of the elongation of an orbit; specifically, the distance between the foci of an ellipse divided by the major axis. The eccentricity of closed orbits varies from 0, for a perfect circle, to almost 1, for an extremely flattened **ellipse**. Open orbits have eccentricites of 1 (parabolic) or greater (hyperbolic).

Echo

The world's first passive **communications satellites.** Each Echo spacecraft was a large aluminized Mylar balloon inflated in orbit, which provided a reflective surface so that two-way voice signals could be bounced from ground stations on the west and east coasts of the United States. Following the failure of the launch vehicle carrying Echo 1, Echo 1A (commonly referred to as Echo 1) was placed in orbit and used to redirect transcontinental and intercontinental telephone, radio, and television signals. Its success proved that microwave transmission to and from satellites in space was possible and demonstrated the promise of communications satellites. Also, because of its large area-to-mass ratio, it provided data for the calculation of atmospheric density and solar pressure. With a diameter of 30.5 m, Echo 1A was visible to the unaided eye over most of the Earth and brighter than most stars. The 41.1-m-diameter Echo 2—the first joint American/Soviet collaboration in space—continued the passive communications experiments, enabled an investigation of the dynamics of large spacecraft, and was used for global geometric geodesy. Although NASA abandoned passive communications systems in favor of active satellites following Echo 2, the Echo program demonstrated several ground station and tracking technologies that would be used by active systems. (See table, "Echo Satellites.")

EchoStar

A constellation of **direct broadcasting satellites** in geosynchronous orbit operated by EchoStar Orbital Corporation. The seventh EchoStar was launched on February 21, 2002; EchoStar 8 was scheduled for liftoff in mid-2002.

ecliptic

The plane in which Earth moves around the Sun.

ecological system

A habitable environment either created artificially, such as in a **space colony,** or occurring naturally, such as the environment on Earth, in which man, animals, and other organisms live in mutual relationship with one another.

ECS (European Space Agency Communications Satellites)

A series of satellites launched in the 1980s to support European telecommunication services, including telephone, telex, data, and television.

ECS (Experimental Communications Satellite)

Two Japanese satellites, also known by the national name of Ayama ("expansion"), launched by **NASDA** (National Space Development Agency) to carry out millimeter-wave communications experiments. Both failed—ECS because the third stage of the N-1 launch vehicle came in contact with the satellite after separation, and ECS-B because of a fault with the **apogee kick motor.**

ECS and ECS-B
Launch
  Date: February 6, 1979 (ECS); February 22, 1980 (ECS-B)
  Vehicle: N-1
  Site: Tanegashima
Size: 1.4-m diameter cylinder
Mass: 130 kg

Eddington

An ESA (European Space Agency) mission to detect seismic vibrations in the surfaces of stars and to search for Earth-like extrasolar planets using precise photometry (light-intensity measurements). Named in honor of the English astrophysicist Arthur Stanley Eddington (1882–1944), it is expected to be launched after 2008, probably by a Soyuz-Fregat rocket from Baikonur. Eddington will

Echo Satellites

Spacecraft	Launch			Orbit	Mass (kg)
	Date	Vehicle	Site		
Echo 1	May 13, 1960	Delta	Cape Canaveral	Failed to reach orbit	56
Echo 1A	Aug. 12, 1960	Delta	Cape Canaveral	966 × 2,157 km × 47.3°	76
Echo 2	Jan. 25, 1964	Thor-Agena B	Vandenberg	1,030 × 1,315 km × 81.5°	256

Echo A test inflation of an Echo satellite in a blimp hangar at Weeksville, North Carolina. *NASA*

orbit at the second **Lagrangian point** beyond the orbit of the Moon on a five-year primary mission. For the first two years, it will look at about 50,000 individual stars for signs of tiny oscillations that betray the stars' internal compositions and allow their ages to be deduced. For the next three years, it will simultaneously watch about 500,000 stars in a single patch of the sky, looking for regular light dips that might indicate the passage across the face of a star or a planet as small as Mars (one-third the size of Earth). Eddington's mission is similar to that of NASA's **Kepler** and will be preceded by three other, smaller photometry missions–**COROT**, **MOST**, and **MONS**.

Edwards Air Force Base

A major American military installation covering 122,000 hectares in southern California, northeast of Lancaster. Established in 1933, it is one of the largest air force bases in the United States and has the world's longest runway. The base is home to the Air Force Flight Test Center, which researches and develops aerospace weapons and rocket-propulsion systems, and NASA's **Dryden Flight Research Center**. It is a proving ground for military aircraft and has been the landing point for several Space Shuttle missions, including the first nighttime Shuttle landing by *Challenger* on September 5, 1983.

EELV (Evolved Expendable Launch Vehicle)

A 1995 U.S. Air Force specification for a new generation of American medium- to heavy-lift launch vehicles intended to reduce the cost of placing large payloads in orbit by at least 25% over existing **Delta, Atlas,** and **Titan** rockets. Contracts for the EELV were eventually awarded to Boeing for the Delta IV and Lockheed Martin for the Atlas V.

effective exhaust velocity

The velocity of an exhaust stream after reduction by effects such as friction, nonaxially directed flow, and pressure differences between the inside of the rocket and its surroundings. The effective exhaust velocity is one of two factors determining the **thrust,** or accelerating force, that a rocket can develop, the other factor being the quantity of reaction mass expelled from the rocket in unit time. In most cases, the effective exhaust velocity is close to the actual exhaust velocity.

As an example, a present-day chemical rocket may achieve an effective exhaust velocity of up to 4 km/s. Although this is not high compared with what may be achieved in the future, a large thrust is nevertheless produced owing to the enormous amount of reaction mass, which is jettisoned every second. Chemical rockets generate high thrust, but only for short periods before their supply of propellant is used up. The final velocity a spacecraft can achieve is fixed by the exhaust velocity of its engines and the spacecraft's mass ratio as shown by the **rocket equation.** Because the exhaust velocity of chemical rockets is so low, they would demand an unachievably high mass ratio in order to propel a spacecraft to the kind of speeds required for practical interstellar flight. Other propulsion strategies must therefore be considered for journeys to the stars.

Efir

Soviet scientific spacecraft based on the Vostok/Zenit design; "efir" is Russian for "ether." These spacecraft were announced under the catchall Cosmos designation. (See table, "Efir Satellites.")

Launch
  Vehicle: Soyuz-U
  Site: Plesetsk
  Mass: 6,300 kg

EGNOS (European Geostationary Navigation Overlay Service)

A European satellite navigation system intended to augment the two operational military satellite navigation systems—the American **GPS** and the Russian **GLONASS**—and make them suitable for safety-critical applications such as flying aircraft or navigating ships through narrow channels. Consisting of three geostationary satellites and a network of ground stations, EGNOS will transmit a signal containing information on the reliability and accuracy of the positioning signals sent out by GPS and GLONASS. It will allow users in Europe and beyond to determine their position to within 5 m, compared with about 20 m at present. EGNOS is a joint venture of **ESA** (European Space Agency), the European Commission, and the European Organisation for the Safety of Air Navigation. It is Europe's contribution to the first stage of the global navigation satellite system (**GNSS**) and is a precursor to the **Galileo satellite navigation system**. EGNOS will become fully operational in 2004; meanwhile, potential users can acquaint themselves with the facility using a test signal broadcast by two **Inmarsat** satellites.

egress

The act of or the mechanism for exit from an enclosure. In a spacecraft this can describe the act of a crew member exiting from the vehicle or the exit chamber, pressure lock, and hatchways themselves.

EGS (Experimental Geodetic Satellite)

A Japanese geodetic satellite used by NASDA (National Space Development Agency) to test a new launch vehicle, the H-1, and determine the accurate location of remote islands; it was also known by its national name, Ajisai ("hydrangea").

Launch
  Date: August 12, 1986
  Vehicle: H-1
  Site: Tanegashima
  Orbit: 1,479 × 1,497 km × 50.0°
  Mass: 685 kg

Ehricke, Krafft Arnold (1917–1984)

A German-born rocket-propulsion engineer who was the chief designer of the **Centaur** and who produced many

Efir Satellites

Spacecraft	Launch Date	Orbit	Duration (days)
Cosmos 1543	Mar. 10, 1984	214 × 401 km × 62.8°	26
Cosmos 1713	Dec. 27, 1985	215 × 397 km × 62.8°	26

other ideas for the development of space, including a space plane design and a strategy for lunar colonization. As a child, he was influenced by Fritz Lang's film *Woman in the Moon* and formed a rocket society at age 12. He studied celestial mechanics and nuclear physics at Berlin Technical University. Injured during World War II, he was transferred to **Peenemünde**, where he served as a propulsion engineer from 1942 to 1945. Upon moving to the United States, he became an American citizen (1954) and during the 1950s, with General Dynamics, helped develop the Atlas missile and then the Centaur upper stage. Later, he carried out advanced studies at Rockwell International while also working independently on schemes for the commercialization and colonization of space. His ashes were placed in orbit aboard the first **Celestis** flight.

Einstein Observatory
See **HEAO**-2.

Eisele, Donn F. (1930–1987)
An American astronaut and Air Force colonel involved with the **Apollo** program. Eisele was selected as an astronaut in October 1963, and he served as Command Module (CM) pilot on Apollo 7 and as backup CM pilot for Apollo 1 and 10. He received a B.S. from the U.S. Naval Academy in 1952 and an M.S. in astronautics from the Air Force Institute of Technology in 1960. After earning his wings in 1954, he flew as an interceptor pilot for four years before attending the Air Force Institute of Technology at Wright-Patterson Air Force Base. After graduating, he worked there as a rocket propulsion and aerospace weapons engineer from 1960 to 1961, attended the Aerospace Research Pilot School, and then worked as a project engineer and experimental test pilot at the Air Force Special Weapons Center at Kirtland Air Force Base. As an astronaut, Eisele was assigned to pilot the first manned Apollo flight but broke his shoulder during training and was replaced by Roger **Chaffee**. The accident saved his life because Chaffee, along with Virgil **Grissom** and Edward **White**, died in a launch-pad fire during a rehearsal for the mission. Eisele resigned from NASA's Astronaut Office in 1970 and became the technical assistant for manned spaceflight at the **Langley Research Center**, a position he occupied until his retirement from both NASA and the Air Force in 1972.

Eisenhower, Dwight D. (1890–1969)
An American president (1953–1961), and the supreme allied commander in Europe during World War II. As president he was deeply interested in the use of space technology for national security purposes and directed that ballistic missiles and reconnaissance satellites be developed on a crash basis. However, he was not an enthusiast of manned spaceflight and in his last budget message advised Congress to consider "whether there are any valid scientific reasons for extending manned spaceflight beyond the Mercury Program."[79, 141]

ejection capsule
(1) A detachable compartment in an aircraft or manned spacecraft that serves as a cockpit or cabin and may be ejected as a unit and parachuted to the ground. (2) A box-like unit in an artificial satellite, probe, or unmanned spacecraft that usually contains recording instruments or records of observed data and may be ejected and returned to Earth by parachute or some other deceleration device.

Ekran
A series of Russian geosynchronous communications satellites. Work on Ekran ("screen") began in the late 1960s, initially using hazardous technologies. The satellite was to have been boosted into orbit by a **Proton** rocket fitted with a new high-performance upper stage using fluorine/amine propellants, and the satellite itself was to have been powered by a 5-kW nuclear reactor. However, by 1973 both these ideas had been abandoned. The first Ekran was launched in October 1976, 27 months after Molniya 1S, the first Soviet geostationary experiment. Early Ekrans were used mainly for test purposes, but they also enabled 18 to 20 million additional Soviet citizens to watch programs of the Central Television. Problems with the Proton booster led to delays in putting the system into operation. The original Ekrans were very short-lived and carried just a single transponder. In the second half of the 1980s, they were replaced by Ekran-Ms, which carried two transponders and generally kept on working well beyond their three-year design life. The final satellite in the Ekran-M series was launched on April 7, 2001, on the maiden flight of Russia's new Proton-M booster.

Ekspress
A series of Russian communications satellites in geostationary orbit. Ekspress will gradually replace the **Gorizont** series.

elasticizer
An elastic substance or fuel used in a solid rocket propellant to prevent cracking of the propellant grain and to bind it to the combustion chamber case.

ELDO (European Launcher Development Organisation)
A multinational consortium formed in the 1960s to build an indigenous European space launch vehicle. It came

about after the cancellation of the **Blue Streak** program by the British government in April 1960. Since the development of this missile was almost complete, it was planned as the first stage of a satellite launcher. Britain then proposed a collaboration with other European countries to build a three-stage launcher capable of placing a one-ton payload into low Earth orbit. ELDO began work in 1962 and was formally signed into existence in 1964, bringing together Germany, France, Belgium, Italy, the Netherlands, and Britain, with Australia as an associate member. Britain was to provide the first stage of the launcher, France the second, and Germany the third stage. Experimental satellites would be developed in Italy and Belgium, telemetry and remote controls in the Netherlands, and launches would take place from **Woomera** in Australia.

ELDO-A, later renamed Europa-1, was to measure 31.7 m in length and weigh more than 110 tons. It was originally planned to put a payload of 1,000 to 1,200 kg into a 500-km circular orbit by 1966. The Blue Streak first stage was to fire for 160 seconds after launch, followed by the French Coralie second stage for the next 103 seconds, and the German third stage for a final 361 seconds, taking the payload into orbit. Testing of the Blue Streak first stage, which began at Woomera in June 1964, went well, but the program nevertheless fell behind schedule. In mid-1966 it was decided to change Europa-1 into a four-stage launcher able to place a satellite in geostationary transfer orbit and to use, for this purpose, the **Guiana Space Centre**. However, a series of subsequent launch failures, and the resignation of Britain and Italy from the project in 1969, led to the cancellation of Europa-1 in 1970. Work began instead on Europa-2, a new launcher 90% funded by France and Germany, which included a **perigee kick stage**. But Europa-2 was launched only once, unsuccessfully, on November 5, 1971, and its failure led to the program's cancellation in favor of a more ambitious project, Europa-3. Although Europa-3 was never realized, its first stage was used as a basis for the **Ariane** launcher.[175]

electric propulsion

A form of advanced rocket propulsion that uses electrical energy for heating and/or directly ejecting propellant. Electric propulsion (EP) provides much lower **thrust** levels than conventional chemical propulsion (CP) does, but much higher **specific impulse**. This means that an EP device must thrust for a longer period to produce a desired change in trajectory or velocity; however, the higher specific impulse enables a spacecraft using EP to carry out a mission with relatively little propellant and, in the case of a deep-space probe, to build up a high final velocity.

The source of the electrical energy for EP is independent of the propellant itself and may be solar (see **solar-electric propulsion**) or nuclear (see **nuclear-electric propulsion**). The main components of an EP system are: an energy source, a conversion device (to turn the source energy into electrical energy at an appropriate voltage, frequency, etc.), a propellant system (to store and deliver the propellant), and one or more thrusters to convert the electrical energy into the kinetic energy of the exhaust material. EP can be subdivided into various types (see diagram on page 120). Of these, several are technologically mature enough to be used on spacecraft, including: the **electron bombardment thruster** (particularly the xenon ion thruster), the **Hall-effect thruster**, the **arcjet**, the **pulsed-plasma thruster**, and the **resistojet**. Other devices such as **magnetoplasmadynamic thrusters**, **contact ion thrusters**, and **microwave plasma thrusters** have not progressed beyond laboratory studies.

electromagnetic field

A region of space near electric currents, magnets, broadcasting antennas, etc., in which electric and magnetic forces operate. In modern physics, an EM field is regarded as a modification of space itself, enabling it to store and transmit energy.

electromagnetic propulsion

A form of **electric propulsion** in which the propellant is accelerated after having been heated to a **plasma** state. There are several subcategories of electromagnetic propulsion, including **magnetoplasmadynamic propulsion**, **pulsed-plasma propulsion**, and **Hall-effect propulsion**.

electromagnetic radiation

Radiation that consists of vibrating electric and magnetic fields. It forms a spectrum from the longest wavelengths (lowest frequencies) of 1,000 m or more to the shortest wavelengths (highest frequencies) of about 10^{-15} m. It includes **radio waves**, **microwave radiation**, **infrared**, **visible light**, **ultraviolet**, **X-rays**, and **gamma rays**.

electron

A negatively charged subatomic particle with a mass 1,837 times less than that of a **proton**. Electrons occur in the outer parts of atoms.

electron bombardment thruster

One of the most promising forms of **electric propulsion** and the *only* form of **ion propulsion** currently employed aboard spacecraft (for example, Deep Space 1). Although various propellants have been tried, the most effective

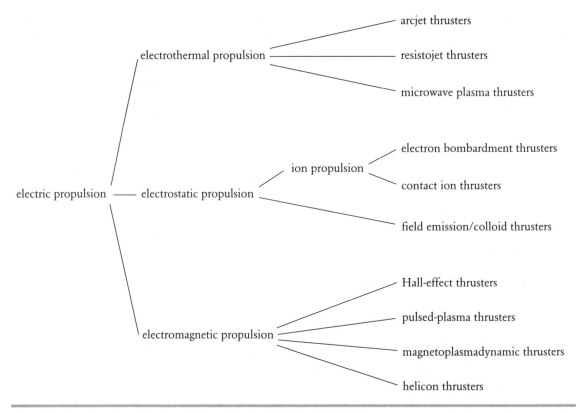

electric propulsion   The various types of electric propulsion.

and popular is the heavy inert gas xenon, which is used in the **XIPS** (xenon-ion propulsion system). The gas propellant enters a discharge chamber at a controlled rate. A hot cathode (negative electrode) at the center of the chamber emits electrons, which are attracted to a cylindrical anode (positive electrode) around the walls of the chamber. Some of the electrons collide with and ionize atoms of the propellant, creating positively-charged ions. These ions are then drawn toward a high-voltage electric field set up between two closely spaced grids at the downstream end of the chamber. These grids contain numerous tiny lined-up holes so that they serve as porous electrodes. The ions are drawn through the first grid (the screen grid), are accelerated in the narrow gap between the first and second grid (the accelerator grid), and then pass through the second grid as a fast-moving ion beam. On the downstream side of the accelerator grid, electrons are injected back into the beam before it is expelled so that the spacecraft remains electrically neutral. If only positively-charged ions were allowed to escape, the vehicle would become more and more negatively-charged until it prevented the thruster working at all.

electrostatic propulsion

A form of **electric propulsion** in which the thrust is produced by accelerating charged particles in an electrostatic field. It includes three types of device: **electron bombardment thrusters**, **ion contact thrusters**, and **field emission/colloid thrusters**. Of these, the first two involve the production and acceleration of separate **ions** and are therefore forms of **ion propulsion**. The third type involves the production and acceleration of charged liquid droplets. Only electron bombardment thrusters have been used operationally aboard spacecraft.

electrothermal propulsion

A form of **electric propulsion** in which electrical energy is used to heat a suitable propellant, causing it to expand through a supersonic nozzle and generate thrust. Two basic types of electrothermal thruster are in use today: the **resistojet** and the **arcjet**. In both, material characteristics limit the effective exhaust velocity to values similar to those of chemical rockets. A third, experimental type is the **microwave plasma thruster**, which potentially could achieve somewhat higher exhaust velocities.

Elektro

See **GOMS**.

Elektron

Two pairs of Soviet spacecraft, launched in 1964, which observed Earth's inner and outer radiation belts, cosmic rays, and the upper atmosphere. One of each pair (Elektrons 1 and 3) were placed in a lower orbit to take measurements of the inner Van Allen belt; the other of each pair (Elektrons 2 and 4) were placed in a higher orbit to study the outer Van Allen belt.

elevon

A control surface on the Space Shuttle that is used after the Shuttle has reentered Earth's atmosphere. The elevon controls roll and pitch, thereby acting as a combination of aileron and elevator.

ELINT (electronic intelligence)

The interception and analysis of electromagnetic signals of other countries, including radar, radio, telephony, and microwave transmissions. It can be carried out effectively using satellites. ELINT is a sub-category of **SIGINT**.

ellipse

The closed path followed by one object that is gravitationally bound to another–for example, by a spacecraft in Earth orbit or a planet going around the Sun. An ellipse is a **conic section** defined by passing a plane through a right circular cylinder at an angle between 0° and 90° or, alternatively, as the locus of a point that moves so that the sum of its distances from two fixed points, known as foci (singular: *focus*), is constant.

elliptic ascent

The profile of the ascent of a spacecraft into Earth orbit.

Endeavour

(1) **Space Shuttle** Orbiter OV-105, which made its first flight on May 7, 1992 (STS-49). (2) The call sign of the **Apollo** 15 Command Module. Both were named after the first ship captained by the explorer James Cook. Cook sailed *Endeavour* on her maiden voyage in August 1768 to the South Pacific to observe and document the rare passage of the planet Venus between Earth and the Sun. He later took her on a voyage that resulted in the discovery of New Zealand, a survey of the eastern coast of Australia, and the navigation of the Great Barrier Reef. A second ship captained by Cook, *Discovery*, also inspired the name of a Shuttle. The selection of the Shuttle name *Endeavour* came from a national competition involving elementary and secondary school students. Shuttle *Endeavour*'s milestones include the first flight of a replacement Shuttle (STS-49), the rescue and redeployment of the Intelsat VI-F3 communications satellite (STS-49), and the first servicing mission to the Hubble Space Telescope (STS-61).

Energia

(1) A Russian aerospace company, known previously as Energia NPO, and now as **Energia Rocket & Space Corporation**. (2) The booster for the Soviet **Buran** space shuttle, mothballed indefinitely in 1988. (3) A heavy launcher, Energia-M, proposed by Energia NPO as a derivative of the Buran launcher; it was capable of delivering 30 to 34 tons to low Earth orbit but was placed on hold through lack of financing and competition from the **Angara** project.

Energia Rocket & Space Corporation (Energia RSC)

A large Russian aerospace company that evolved from Sergei Korolev's design bureau, later known as Energia NPO and now as Energia RSC. It is currently involved in a variety of projects, including development of the Russian modules of the International Space Station, the second stage of the **Angara** rocket (led by the **Khrunichev Center**), and several series of communications satellites.

Energia RSC began as Department No. 3 of Special Design Bureau NII-88, which was given the task, in August 1946, of developing Soviet rockets based on the German V-2. In 1950, the OKB-1 bureau, headed by Sergei **Korolev**, was formed within NII-88 to develop all long-range ballistic missiles until the competing **OKB-586** design bureau was established in 1954 and, later, Chelomei's **OKB-52**. Six years later, OKB-1 was made independent of NII-88. After Korolev's death in 1966, his first deputy, Vasily **Mishin**, was appointed chief designer of the bureau, and from the mid-1960s until 1974, OKB-1 focused on the **Soviet manned lunar program**. When the Moon race was lost to the United States, OKB-1's attention switched to the development of long-term Earth-orbiting stations. In 1974 Mishin was replaced by Valentin **Glushko**, at which point Glushko's old bureau (OKB-456) merged with OKB-1 to form Energia NPO. Under government pressure, Energia NPO was directed to respond to the American Space Shuttle by developing **Buran** and its massive booster. When this project collapsed, Energia NPO proposed evolving the Buran booster rocket into the Energia-M. Currently, this proposal too is in mothballs, having been sidelined by Angara.[129]

energy

(1) A measure of the ability to do work–for example, to lift a body against gravity or drag it against friction, or to

accelerate an object. (2) An intrinsic property of everything in the universe, including **radiation**, matter (see **mass-energy relationship**), and, strangely enough, even empty space (see **vacuum energy**).

energy density

The energy contained per unit volume of space. General relativity provides a way of calculating the energy density associated with a given curvature of **space-time**. Some schemes for **faster-than-light (FTL) travel** call for regions of *negative* energy density–a strange condition, but one that exists in phenomena such as the **Casimir effect**.

Enterprise

The first flight-capable version of the **Space Shuttle** Orbiter but one designed for ground and gliding tests only; its official designation is Orbiter OV-101. NASA

had originally intended to call this vehicle *Constitution* in honor of several U.S. Navy vessels of the same name, including the "Old Ironsides" frigate launched in 1797 and now on display at the Boston Navy Yard. *Constitution* was also appropriate in view of the American Bicentennial, which was being commemorated at the time of factory rollout in 1976. However, *Star Trek* fans petitioned–sending 100,000 letters to the White House–to have the first Shuttle named after the famous fictional starship. NASA relented and changed the name from *Constitution* to *Enterprise* prior to rollout, pointing out that several Navy vessels had also carried the name *Enterprise*, including the first nuclear-powered aircraft carrier. During the summer of 1976, *Enterprise* underwent horizontal vibration tests at the Rockwell plant in Palmdale, California, designed to test the Shuttle structural integrity under simulated launch and landing conditions. Upon completion

Enterprise *Enterprise* flies free after being released from NASA's 747 Shuttle Carrier Aircraft over Rogers Dry Lakebed during the second of five free flights carried out as part of the Shuttle program's Approach and Landing Tests. *NASA*

of these, it was outfitted to perform actual flight tests. But because the vehicle would never leave Earth's atmosphere, it differed markedly from its spacefaring successors. *Enterprise* had no main propulsion system plumbing, fuel lines, or tankage, and its main engines were only mockups. The payload bay contained no mounting hardware for cargo packages, and the payload bay door lacked the hydraulic mechanisms to allow it to open and close. In order to save money, thermal tiles were simulated using black-and-white polyurethane foam. The flight deck controls were much simpler than those required for spaceflight, and the crew compartment was largely empty, since only a pilot and a commander would be aboard for the tests. A feature not included in operational Shuttles, however, was an ejection-seat escape mechanism. *Enterprise* also carried a battery of equipment for making aerodynamic measurements and, unlike future Shuttles, was fitted with a long, pointed air data probe that stuck out from its nose. Following the successful completion of the Approach and Landing Tests, NASA certified the Shuttle as aerodynamically sound and announced that no further flight tests would be necessary. But *Enterprise* was not immediately retired. Beginning in March 1978, a series of mated vertical vibration tests was carried out at **Marshall Space Flight Center** to subject *Enterprise* to the kind of vibrations the Shuttle might experience in flight. Upon their conclusion, NASA had considered returning *Enterprise* to Rockwell to be upgraded into an operational Shuttle. However, as several major design changes had been made to the Shuttle while *Enterprise* was being built and tested, a refit was now considered too expensive. Instead, NASA opted to modify an already existing high-fidelity structural test article (STA-099) into what would become *Challenger*. As a result, *Enterprise* was taken to **Kennedy Space Center** (KSC) to test equipment and procedures that would be necessary to support the first Shuttle space flight. Engineers at KSC had been using a Shuttle mockup nicknamed *Pathfinder*, but *Enterprise* provided a much more realistic tool. Soon after its arrival at KSC, *Enterprise* was transported to the **Vehicle Assembly Building** (VAB), mated to an External Tank and an inert set of Solid Rocket Boosters, and rolled out, in May 1997, to become the first Shuttle on the launch pad. For nearly three months, it supported operational tests there before being returned to the VAB for demating. In early August 1979, it was ferry-flown to Vandenberg Air Force Base, then on to Edwards Air Force Base, and finally transported over land to Rockwell for removal and refurbishment of certain components for use on other Shuttles. *Enterprise* was then taken back to Edwards in September 1981 and put in storage for almost two years. In May 1983, it became the first Shuttle to travel abroad when it was ferry-flown

to France for the Paris Air Show. Eventually, it came back to Vandenberg for use in further validation of Shuttle procedures. In September 1985, it was ferry-flown to KSC to be put on display next to a Saturn V outside the VAB. Finally, on November 18, 1985, *Enterprise* was taken to its permanent home, a facility of the National Air and Space Museum at Dulles Airport in Washington, D.C.

entry
See **reentry**.

environmental engine
A reaction engine that obtains the reactants or propellants it needs from its environment. Examples include the conventional jet engine used by aircraft and the speculative **interstellar ramjet**.

environmental space chamber
A chamber (sometimes a simulated spacecraft) in which variables such as humidity, temperature, pressure, fluid contents, noise, and movement may be controlled so as to simulate different space conditions. It is typically used for astronaut training.

Envisat (Environmental Satellite)
The largest, most expensive, and potentially most important satellite ever launched by the European Space Agency (ESA), and a successor to ESA's **ERS** (Earth Resources Satellite) remote sensing satellites, ERS-1 and ERS-2. Envisat is the size of a double-decker bus and cost $2 billion. Every 35 days throughout its mission of at least five years, it will observe every part of the Earth's surface. Its 10 science instruments will provide vital information on how the planet's land, oceans, ice caps, and atmosphere are changing—data that will be analyzed by scientists and inform European policies on the environment.

Launch
    Date: February 28, 2002
    Vehicle: Ariane 5
    Site: Kourou
Orbit (polar): 800 km × 98.6°
Mass: 8.2 tons
Length: 10 × 4 × 4 m

EO (Earth Observing) satellites
A series of spacecraft in NASA's New Millennium Program Earth Observing series. EO-1 has been launched and has successfully completed its one-year primary mission. EO-2, also known as **SPARCLE**, was cancelled in 1999. EO-3, also known as **GIFTS**, is scheduled to fly in 2004. EO-1's main goal was to flight-validate a set of

Envisat Envisat being prepared for launch. *European Space Agency*

advanced land imaging (ALI) instruments that would reduce the costs of future **Landsat** missions. These instruments range from a hyperspectral imager (a camera that views Earth's surface with unprecedented spectral discrimination) to an X-band phased array antenna that sends back high volumes of data in a unique pencil-beam pattern. The spacecraft was in an orbit that allowed it to fly in formation with Landsat 7 and take a series of the same images. Comparison of these "paired scene" images was one way of evaluating EO-1's instruments. Future use of ALI technology will cut the mass and power consumption by a factor of seven compared to the Landsat 7 imager. EO-1 is also the first spacecraft to use a new generation of **pulsed-plasma thrusters** developed at the Glenn Research Center. EO-1 is a component of NASA's **EOS** (Earth Observing System).

EO-1 FACTS
Launch
    Date: November 21, 2000
    Vehicle: Delta 7320
    Site: Vandenberg Air Force Base
Orbit (circular, sun-synchronous): 705 km × 98.7°

Eole

A French weather microsatellite, also known as CAS (Cooperative Applications Satellite) and, before launch, as FR-2. Eole, named after the god of the wind, interrogated and returned data from over 500 instrumented weather balloons released from Argentina and drifting at heights of 12 km to help study southern-hemisphere winds, temperatures, and pressures.

Launch
    Date: August 16, 1971
    Vehicle: Scout B
    Site: Wallops Island
Orbit: 678 × 903 km × 50.2°
Mass: 84 kg

EORSAT (ELINT Ocean Reconnaissance Satellite)

A system of specialized Russian/Soviet ELINT (electronic intelligence) satellites that has been in operation since 1974. It monitors and locates enemy naval forces by detecting and triangulating on their radio and radar emissions.

EOS (Earth Observing System)

A series of polar-orbiting and low-inclination satellites for conducting long-term global observations of Earth's land surface, atmosphere, oceans, and biosphere. EOS is the centerpiece of NASA's Earth Science Enterprise program (formerly called Mission to Planet Earth), and together with other missions from NOAA (National Oceanic and Atmospheric Administration), Europe, and Japan forms the comprehensive International Earth Observing System (IEOS). Current and future EOS missions include: **SeaStar/SeaWiFS, TRMM, Landsat-7, QuikScat, Terra** (EOS AM-1), **ACRIMSAT, EO-1, SAGE** III, **Jason**-1, **Aqua** (EOS PM-1), **ICESAT** (EOS Laser Alt-1), **ADEOS**-2, **SORCE**, and **Aura** (EOS Chem).

EOS Chem (Earth Observing System Chemistry)

See **Aura**.

EOS PM (Earth Observing System PM)

See **Aqua**.

ephemeris

A set of numbers that specifies the location of a celestial body or satellite in space.

equatorial orbit

An orbit in the same plane as Earth's equator.

Equator-S

A German satellite in a near-equatorial orbit that studies Earth's equatorial magnetosphere out to distances of 67,000 km. It was designed and built by the Max-Planck-Institut fur Extraterrestrische Physik.

Launch
    Date: December 2, 1997
    Vehicle: Ariane 44
    Site: Kourou
    Orbit: 496 × 67,232 km × 4.0°

ERBE (Earth Radiation Budget Experiment)

A NASA project to study the energy exchanged between the Sun, Earth, and space. Absorption and re-radiation of energy from the Sun is one of the main drivers of Earth's weather patterns. ERBE was designed around three Earth-orbiting satellites: **ERBS** and two **NOAA** (National Oceanic and Atmospheric Administration) satellites, NOAA 9 and 10.

ERBS (Earth Radiation Budget Satellite)

Part of NASA's three-satellite **ERBE** (Earth Radiation Budget Experiment). In addition to its ERBE instrumentation, ERBS also carried a Stratospheric Aerosol and Gas Experiment (**SAGE** II), observations of which were used to assess the effects of human activities (such as burning fossil fuels and the use of CFCs) and natural occurrences (such as volcanic eruptions) on Earth's radiation balance. Following deployment from the Space Shuttle, astronaut Sally Ride had to shake the satellite with the remote manipulator arm in order open its solar array.

Shuttle deployment
    Date: October 5, 1984
    Mission: STS-41G
    Orbit: 576 × 589 km × 57.0°
    Size: 4.6 × 3.5 m
    Mass: 226 kg

ERS (Earth Resources Satellite)

The first two **remote sensing satellites** launched by ESA (European Space Agency). Their primary mission was to monitor Earth's oceans, ice caps, and coastal regions. The satellites provided systematic, repetitive global measurements of wind speed and direction, wave height, surface temperature, surface altitude, cloud cover, and atmospheric water vapor level. Data from ERS-1 were shared with NASA under a reciprocal agreement for **Seasat** and **Nimbus** 7 data. ERS-2 carries the same suite of instruments as ERS-1 with the addition of the Global Ozone Measuring Equipment (GOME), which measures ozone distribution in the outer atmosphere. Having performed well for nine years—more than three times its planned lifetime—the ERS-1 mission was ended on March 10, 2000, by a failure in the onboard attitude control system. The length of its operation enabled scientists to track several El Niño episodes through combined observations of surface currents, topography, temperatures, and winds. The measurements of sea surface temperatures by the ERS-1 Along-Track Scanning Radiometer, critical to the understanding of climate change, were the most accurate ever made from space. All these important measurements are being continued by ERS-2 and **Envisat**. (See table, "ERS Missions.")

ERS (Environmental Research Satellite)

Spacecraft designed for piggyback-launching from large primary mission vehicles. With masses of 0.7 to 45 kg and payloads of 1 to 14 experiments, the ERS hitchhikers provided a low-cost, flexible way of making scientific and engineering measurements in space. One of their major roles was to act as a test-bed to determine the reliability of improved components and subsystems destined for use in later generations of spacecraft. A unique feature of the system was its ability to function without a battery by having solar cells fastened to all exterior surfaces. At least 12 satellites were launched from 1962 to 1969 for a variety of missions and sponsors.

ERTS (Earth Resources Technology Satellites)

A series of spacecraft now known as **Landsat**.

ESA (European Space Agency)

A multinational agency formed in 1975 through the merger of **ESRO** (European Space Research Organisation) and **ELDO** (European Launcher Development

ERS Missions

Spacecraft	Launch Date	Orbit	Mass (kg)
Launch vehicle: Ariane 4; launch site: Kourou			
ERS 1	Jul. 17, 1991	774 × 775 km × 98.5°	2,384
ERS 2	Apr. 21, 1995	783 × 784 km × 98.6°	2,516

Organisation). ESA instigates and manages space activities on behalf of its 15 member states: Austria, Belgium, Denmark, Finland, France, Germany, the Republic of Ireland, Italy, the Netherlands, Norway, Portugal, Spain, Sweden, Switzerland, and the United Kingdom. Canada has a cooperative relationship with the agency. While developing an independent capability for Europe in space technology, ESA also works closely with other space agencies, including NASA and RKA (the Russian Space Agency).

ESA has its headquarters in Paris and four major facilities in other countries: **ESTEC** (European Space Research and Technology Centre), at Noordwijk, the Netherlands, the main center for research and management of satellite projects; **ESOC** (European Space Operations Centre), at Darmstadt, Germany, responsible for satellite control, monitoring, and data retrieval; **ESRIN** (European Space Research Institute), at Frascati, Italy, which supports the ESA documentation service and manages the data collected by remote sensing satellites; and **EAC** (European Astronaut Center), at Cologne, Germany, responsible for selecting and training astronauts for space station missions. In addition, ESA operates the **Guiana Space Centre** for launching **Ariane** rockets, sounding rocket launch stations in Norway and Sweden, a meteorological program office at Toulon, France, and satellite tracking stations in Belgium, Germany, Italy, and Spain. Ariane rockets are developed, built, and managed by **Arianespace**, which is a division of ESA. See individual entries

of the major ESA space missions and programs listed in the table ("Major ESA Space Missions and Programs").[176]

escape energy

The energy required per unit mass of a spacecraft for it to escape Earth's gravity.

escape tower

A tower, mounted on top of the Command Module of the **Apollo** spacecraft, that contained a cluster of small rockets to jettison the spacecraft from the booster in the event of an aborted mission.

escape trajectory

The path a body must follow to escape a central force field, such as Earth's gravity.

escape velocity

The minimum velocity that a body, such as a rocket, must have in order to escape completely from the gravitational influence of another, such as a planet or a star, without being given any extra impetus. It is given by:

$$v_{esc} = \sqrt{\frac{2GM}{R}}$$

where G is the **gravitational constant** and R is the distance from the center of the gravitating body of mass M. This is equal to, but in the opposite direction of, the velocity the body would have acquired if it had been

Major ESA Space Missions and Programs

	Past	Present	Under Development/Study
Space science	COS-B	Cluster	Bepi Colombo
	Exosat	COROT	Darwin
	Giotto	Huygens (see **Cassini**)	Eddington
	Hipparcos	SOHO	GAIA
	ISO	Ulysses (with NASA)	Herschel
	IUE	XMM-Newton	INTEGRAL
	Spacelab		LISA
			Mars Express
			Planck
			Rosetta
			SMART
			Solar Orbiter
			XEUS
Earth observation	ERS-1	Envisat	MetOp
		ERS-2	
		Meteosat	
Communication and navigation	MARECS	Artemis	EGNOS
	Olympus		Galileo

Escape Velocities of the Planets

Planet	Escape Velocity (km/s)
Mercury	3.20
Venus	10.08
Earth	11.12
Mars	4.96
Jupiter	59.20
Saturn	35.20
Uranus	20.80
Neptune	24.00
Pluto	1.1

accelerated from rest, starting an infinite distance away. The escape velocity of Earth is about 11 km/s. If the body's velocity is less than the escape velocity, it is said to be gravitationally bound. (See table, "Escape Velocities of the Planets.")

escape-velocity orbit
The orbit achieved around the Sun by a spacecraft escaping from Earth's gravity.

Esnault-Pelterie, Robert A. C. (1881–1957)
A French engineer and aviation pioneer who invented the joystick and built the first monoplane and radial engine. He also tested many liquid-propellant rocket engines, including some that used **cryogenic** fuels, and put forward ideas for long-range ballistic missiles that won him support from the French Army for his rocket experiments. In 1930 he published his fundamental work, *L'Astronautique*,[89] and in 1935 a sequel, *L'Astronautique-Complément*.[90] A crater on the Moon's farside is named after him.

ESOC (European Space Operations Centre)
ESA's center for operating spacecraft, located in Darmstadt, Germany.

Esrange
An international sounding rocket and balloon launch facility located near Kiruna in northern Sweden and operated by the Swedish Space Corporation.[277]

ESRIN (European Space Research Institute)
An ESA (European Space Agency) center in Frascati, Italy, responsible for analyzing and distributing remote sensing data from ESA (**ERS** and **Envisat**) and non-ESA (**Landsat**, NOAA-TIROS, MOS, JERS, SPOT) Earth observation missions.

ESRO (European Space Research Organisation)
An organization formed in 1962 by 10 European nations plus Australia, which made available its rocket-firing range at **Woomera**. Between 1968 and 1972, ESRO launched seven satellites—Iris (ESRO-2B), Aurora (ESRO-1A), HEOS-1, Boreas, HEOS-2, TD-1A, and ESRO-4—using NASA rockets. In 1975, ESRO and its sister organization **ELDO** (European Launcher Development Organisation) merged to form **ESA** (European Space Agency).[175]

ESSA (Environmental Science Services Administration)
An American government agency that was a precursor of **NOAA** (National Oceanic and Atmospheric Administration). ESSA managed the first experimental **TIROS** satellites and **TOS** (TIROS Operational System). TOS satellites were designated ESSA 1, ESSA 2, etc., once they had been successfully launched.

ESSP (Earth System Science Pathfinder)
Small, rapidly developed missions that form part of NASA's **Earth Probes** program. They include **VCL** (Vegetation Canopy Lidar), **GRACE** (Gravity Recovery and Climate Experiment), **CALIPSO**, and **CloudSat**.

ESTEC (European Space Research and Technology Centre)
An ESA (European Space Agency) establishment, at Noordwijk in the Netherlands, responsible for the study, development, and testing of ESA spacecraft. ESTEC is also the home of technological development programs that lay the groundwork for future missions.

Etalon
Soviet passive geodetic satellites; "etalon" is Russian for "standard." Each is spherical and covered with 306 antenna arrays, and each array contains 14 corer cubes for laser reflection. Etalons were designed and launched to enable the complete characterization of Earth's gravitational field at the altitude and inclination planned for the **GLONASS** navigation satellites. They were also used to refine understanding of the Earth-Moon gravitational system, to determine the effect of nongravitational forces on satellites, and for geophysical research. Only two Etalon satellites have been orbited, each accompanied by a pair of GLONASS satellites. (See table, "Etalon Satellites.")

Launch
  Vehicle: Proton
  Site: Baikonur
Size: 1.3 × 1.3 m
Mass: 1,415 kg

Etalon Satellites

Spacecraft	Launch Date	Orbit
Cosmos 1989	Jan. 10, 1989	19,097 × 19,152 km × 64.9°
Cosmos 2024	May 31, 1989	19,095 × 19,146 km × 65.5°

ETS (Engineering Test Satellite)

Satellites launched by Japan's NASDA (National Space Development Agency) to demonstrate new spacecraft techniques and to test new equipment and launch vehicles. They are also known by the national name Kiku ("chrysanthemum"). ETS-7 consisted of two spacecraft—a chase satellite (Hikoboshi) and a target satellite (Orihime)—that carried out experiments in remote-controlled rendezvous-docking and space robotics. ETS-8, scheduled for launch in 2003, will be the world's largest satellite in geostationary orbit. Among other things, it will test multimedia broadcast systems and carry a high-accuracy atomic clock to test its positioning systems using GPS (Global Positioning System) data. (See table, "ETS Missions.")

Eumetsat (European Meteorological Satellite Organisation)

An intergovernmental agency that is responsible for the **Meteosat** weather satellites and provides operational meteorological data for its 17 member states: Austria, Belgium, Britain, Denmark, Finland, France, Germany, Greece, Ireland, Italy, the Netherlands, Norway, Portugal, Spain, Sweden, Switzerland, and Turkey. Eumetsat also has three cooperating members: the Republic of Slovakia, Hungary, and Poland.

EURECA (European Retrievable Carrier)

A reusable free-flying science platform designed to be placed in orbit by the **Space Shuttle** and recovered on a later mission. EURECA was launched by the Shuttle *Atlantis* on July 31, 1992, into a 508-km-high orbit and began experiments in microgravity and materials technology and solar and X-ray observations on August 7. It was retrieved by the Shuttle *Endeavour* on July 1, 1993.

Eurockot

See **Rokot**.

Europa

See **ELDO**.

Europa Lander

A spacecraft that might follow on from the **Europa Orbiter** and sample material on or just below the icy surface of this enigmatic moon of Jupiter. It is identified in NASA's Office of Space Science Strategic Plan as a potential future Outer Planets mission.

Europa Orbiter

A successor to **Galileo** that would probe the surface of Jupiter's moon Europa with a radar sounder in an attempt to determine the thickness of the ice and locate any ice-water interface. Other instruments could include

ETS Missions

Launch site: Tanegashima

Spacecraft	Launch Date	Launch Vehicle	Orbit	Mass (kg)
ETS-1	Sep. 9, 1975	N-1	975 × 1,103 km × 47°	85
ETS-2	Feb. 23, 1977	N-1	35,854 × 35,860 km × 12°	130
ETS-3	Feb. 11, 1981	N-2	240 × 20,680 km × 28°	640
ETS-4	Sep. 3, 1982	N-1	966 × 1,226 km × 45°	385
ETS-5	Aug. 27, 1987	H-1	35,770 × 35,805 km × 2°	550
ETS-6	Aug. 28, 1994	H-2	Failed to reach GSO	3,800
ETS-7	Nov. 28, 1997	H-2	550 km × 35°	2,860

an imaging device capable of resolving surface detail as small as 100 m and an altimeter for accurately measuring the topography and movements of the surface in response to tidal stresses. The Orbiter would serve as a precursor to other projects, such as the **Europa Lander** and the **Europa Subsurface mission**, that would sample the Europan surface and, eventually, penetrate the top layers of ice to discover exactly what lies underneath. Although NASA canceled the Europa Orbiter in 2002, it is a mission likely to be reinstated over the next decade.

Europa Subsurface mission

A mission beyond the **Europa Orbiter** and the **Europa Lander** to carry out the far more technologically difficult task of penetrating the frozen crust of this Jovian moon in order to hunt for life in the putative subsurface ocean. It would probably involve the use of **cryobot**s (melting devices) and **hydrobot**s (minisubs). The Europa Subsurface mission is identified in NASA's Office of Space Science Strategic Plan as a potential mission beyond 2007 but remains in an early concept definition phase.

Eurospace

An organization that represents the interests of the European space industry. Founded in 1961, its membership includes 66 companies from 13 countries in Western Europe that account for about 95% of the total turnover of the European space industry.

Eutelsat (European Telecommunications Satellite Organisation)

An organization that operates about 15 satellites and offers satellite communications services to Europe, Asia, Africa, and America. Founded in 1985 as an intergovernmental entity, it now represents the interests of 47 European countries. In 2001, Eutelsat's structure was streamlined into two tiers: Eutelsat, S.A., a private limited company with headquarters in Paris, to which all assets and activities were transferred; and an intergovernmental organization to ensure that the company observes the principles of pan-European coverage and fair competition.

EUV

Extreme ultraviolet; a portion of the electromagnetic spectrum from approximately 10 to 100 nm.

EUVE (Extreme Ultraviolet Explorer)

A NASA mission to study the universe as it appears in short-wavelength ultraviolet (UV) light. EUVE carried out an all-sky survey in the 7- to 76-nm portion of the spectrum and a "deep survey" of a strip of the sky along the ecliptic (plane of Earth's orbit) with extremely high sensitivity. It performed follow-up spectroscopic observations on bright extreme ultraviolet (EUV) point sources, studying more than 1,000 objects in our galaxy and more than three dozen that lie beyond. Its results have been

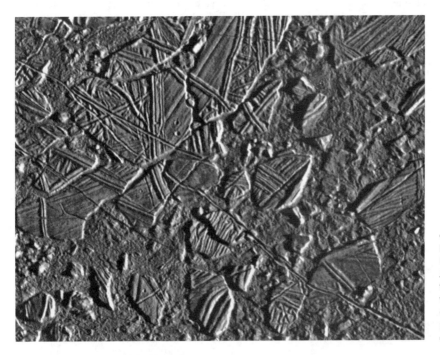

Europa Subsurface mission
Beneath these ice rafts on Jupiter's moon Europa, scientists believe there may be a watery ocean that could be explored by a future robotic probe. *NASA/JPL*

used to study stellar evolution and local stellar populations, energy transport in stellar atmospheres, and the ionization and opacity of the interstellar medium. At the end of the all-sky survey in January 1993, a Guest Observer program was started. EUVE was controlled from the Center for EUV Astrophysics at the University of California, Berkeley, and designed for on-orbit servicing by the Shuttle. Its main instruments were three grazing-incidence UV telescopes and one EUV spectrometer. EUVE's science operations ended in December 2001, and the spacecraft broke up in the atmosphere at the end of January 2002.

Launch
   Date: June 7, 1992
   Vehicle: Delta 6925
   Site: Cape Canaveral
Orbit: 510 × 524 km × 28.4°
Size: 4.5 × 3.0 m
Mass: 3,275 kg

Evans, Ronald E. (1933–1990)

An American astronaut who served as Command Module (CM) pilot on **Apollo** 17. Evans received a B.S. in electrical engineering from the University of Kansas in 1956 and an M.S. in aeronautical engineering from the U.S. Naval Postgraduate School in 1964. He was a combat flight instructor for F-8 jet pilots in 1961–1962, after which he flew carrier-based combat missions for seven months. Having been selected as one of 19 new astronauts in April 1966, Evans served as backup CM pilot on Apollo 14 before taking the CM pilot's seat onboard Apollo 17, in which he circled the Moon for three days while colleagues Eugene **Cernan** and Harrison **Schmitt** explored the lunar surface. He was subsequently the backup CM pilot for the **Apollo-Soyuz Test Project** and then became involved in development of the Space Shuttle. Evans retired from the Navy in 1975 and from NASA in 1977 to become director of Space Systems Marketing for Sperry Flight Systems.

event horizon

The radius to which a spherical mass must be compressed in order to transform it into a **black hole**; or, the radius at which time and space switch properties. Once inside the event horizon, radiation or matter cannot escape to the outside world. Furthermore, nothing can prevent a particle from hitting the singularity in a very short amount of time (as measured by an external observer) once it has entered the horizon. In this sense, the event horizon is a point of no return. If the material of the Earth were squashed hard enough to make a black hole, its event horizon would be about 2 cm in diameter. A star with 10 times the mass of the Sun would form a black hole with an event horizon about 64 km across.

exhaust nozzle

The portion of the **thrust chamber** lying downstream of the nozzle throat.

exhaust stream

The stream of gaseous, atomic, or radiant particles emitted from the nozzle of a reaction engine.

exhaust velocity

The average actual velocity at which exhaust material leaves the nozzle of a rocket engine. In practice, this differs from the **effective exhaust velocity**, which takes into account other factors and is more useful for calculating quantities such as **thrust**.

EXIST (Energetic X-Ray Imaging Survey Telescope)

A proposed NASA mission to carry out the first high sensitivity, all-sky imaging survey at extreme (high energy) X-ray wavelengths using a wide-field aperture telescope array. EXIST is identified in NASA's Office of Space Science Strategic Plan as a potential mission beyond 2007 but remains in the early concept definition phase.

Exos-

The prelaunch designation of a series of satellites built by Japan's **ISAS** (Institute of Space and Astronautical Science) to investigate Earth's upper atmosphere. This series includes Exos-A (**Kyokko**), Exos-B (**Jikiken**), Exos-C (**Ozhora**), and Exos-D (**Akebono**). See **ISAS missions**.

Exosat

An ESA (European Space Agency) satellite for X-ray astronomy that carried out observations of variable X-ray sources in the energy range of 0.05 to 20 keV (corresponding to wavelengths of 0.025 to 1.24 nm). On board were three instruments: an imaging telescope, a proportional counter, and a gas scintillation proportional counter array. Exosat made 1,780 observations of a wide variety of objects, including active galactic nuclei, stellar coronae, cataclysmic variables, white dwarfs, X-ray binaries, clusters of galaxies, and supernova remnants.

Launch
   Date: May 26, 1983
   Vehicle: Delta 3914
   Site: Vandenberg Air Force Base
Orbit: 340 × 191,878 km × 72.5°
End of operation: April 9, 1986
Mass: 510 kg

---

exosphere
The tenuous outermost layers of a planetary atmosphere that merge into the interplanetary medium.

exotic matter
A hypothetical kind of matter that has both a negative **energy density** and a negative pressure or tension that exceeds the energy density. All known forms of matter have positive energy density and pressures or tensions that are always less than the energy density in magnitude. In a stretched rubber band, for example, the energy density is 100 trillion times greater than the tension. A possible source of exotic matter lies in the behavior of certain vacuum states in quantum field theory (see **Casimir effect**). If such matter exists, or could be created, it might make possible schemes for **faster-than-light (FTL) travel** such as stable **wormholes** and the **Alcubierre Warp Drive**.

expansion ratio
The ratio between the area of the combustion chamber exit and the area of the nozzle exit. A large expansion ratio improves the performance of an engine in a vacuum since the exhaust is expanded further, thus converting potential energy into kinetic energy. However, at sea level a high expansion ratio can result in flow separation, which can drastically reduce or eliminate the net thrust of the engine.

expendable launch vehicle (ELV)
A launch system that cannot be recovered for use on subsequent missions.

Experimental Rocket Propulsion Society (ERPS)
A nonprofit liquid fuel rocket engine design and test team based in the San Francisco Bay Area. Its current projects include a **monopropellant** rocket to launch a payload to 100 km; a small two-man suborbital vehicle; a 9-kg to low Earth orbit, **single-stage-to-orbit** demonstrator; a rocket-pack; and a rocket-assist module for high performance gliders.

Exploration of Neighboring Planetary Systems (ExNPS)
One of the key elements of NASA's **Origins Program**. It is intended to be a long-term program of scientific investigation and technological development with the goal of detecting and probing Earth-like planets around nearby stars. The responsibility for identifying a specific plan for ExNPS has been assigned to JPL (Jet Propulsion Laboratory).

Explorer
A long and ongoing series of small American scientific spacecraft. Explorer 1, the first successful American satellite, was launched on January 31, 1958, and discovered the **Van Allen belts**. Explorer 6 took the first photo of Earth from space. Subprograms of the Explorer series have included **ADE** (Air Density Explorer), **AE** (Atmosphere Explorer), **BE** (Beacon Explorer), **DME** (Direct Measurement Explorer), EPE (Energetic Particles Explorer), **GEOS** (Geodetic Earth Orbiting Satellite), **IE** (Ionosphere Explorer), and **IMP** (Interplanetary Monitoring Platform). Among the more recent Explorers are **IUE** (International Ultraviolet Explorer) and **COBE** (Cosmic Background Explorer). Subcategories of the modern Explorer program include **SMEX** (Small Explorer), **MIDEX** (Medium-class Explorer), **UNEX** (University-class Explorer), and **Missions of Opportunity**, while the themes of the program are the Astronomical Search for Origins and Planetary Systems, the Sun-Earth Connection, and the Structure and Evolution of the Universe. (See table, "Explorer Spacecraft.")

Explorer From left to right: William Pickering, director of the Jet Propulsion Laboratory; James Van Allen, from the University of Iowa; and Wernher von Braun hold a full-size model of America's first satellite, Explorer 1, following a successful launch on January 31, 1958. *NASA*

Explorer Spacecraft

Explorer/aka	Launch Date	Vehicle	Site	Orbit	Mass (kg)
1	Feb. 1, 1958	Juno 1	Cape Canaveral	358 × 2,550 × 33°	5
2	Mar. 5, 1958	Juno 1	Cape Canaveral	Failed to reach orbit	5
3	Mar. 26, 1958	Juno 1	Cape Canaveral	186 × 2,799 × 33°	5
4	Jul. 26, 1958	Jupiter	Cape Canaveral	263 × 2,213 × 50°	8
5	Aug. 24, 1958	Juno 1	Cape Canaveral	Failed to reach orbit	17
6	Aug. 7, 1959	Thor-Able	Cape Canaveral	237 × 41,900 × 47°	64
7	Oct. 13, 1959	Juno 2	Cape Canaveral	573 × 1,073 × 50°	42
8	Nov. 3, 1960	Juno 2	Cape Canaveral	417 × 2,288 × 50°	41
9	Feb. 16, 1961	Scout X-1	Wallops Island	545 × 2,225 × 39°	7
10	Mar. 25, 1961	Delta	Cape Canaveral	Highly elliptical	35
11 Met. Sat 1	Apr. 27, 1961	Juno 2	Cape Canaveral	486 × 1,786 × 29°	37
12 EPE-1	Aug. 16, 1961	Delta	Cape Canaveral	790 × 76,620 × 33°	38
13 Met. Sat 2	Aug. 25, 1961	Scout X-1	Wallops Island	Wrong orbit	86
14 EPE-2	Oct. 2, 1962	Delta A	Cape Canaveral	2,601 × 96,189 × 43°	40
15 EPE-3	Oct. 27, 1962	Delta A	Cape Canaveral	300 × 17,438 × 18°	44
16 Met. Sat 3	Dec. 16, 1962	Scout X-3	Wallops Island	744 × 1,159 × 52°	100
17 AE 1	Apr. 5, 1963	Delta B	Cape Canaveral	254 × 891 × 57°	185
18 IMP-1	Nov. 26, 1963	Delta C	Cape Canaveral	192 × 197,616 × 33°	62
19 ADE 1	Dec. 19, 1963	Scout X-4	Vandenberg	597 × 2,391 × 79°	7
20 IE	Aug. 25, 1964	Delta C	Cape Canaveral	855 × 1,001 × 80°	44
21 IMP-2	Oct. 4, 1964	Delta C	Cape Canaveral	191 × 95,590 × 34°	62
22 BE 2	Oct. 9, 1964	Scout X-4	Vandenberg	872 × 1,053 × 80°	52
23 Met. Sat. 4	Nov. 6, 1964	Scout X-4	Wallops Island	463 × 980 × 52°	134
24 ADE 2	Nov. 21, 1964	Scout X-4	Wallops Island	530 × 2,498 × 81°	9
25 Injun 4	Nov. 21, 1964	Scout X-4	Wallops Island	524 × 2,349 × 81°	40
26 EPE-4	Dec. 21, 1964	Delta C	Cape Canaveral	284 × 10,043 × 20°	46
27 BE 3	Apr. 29, 1965	Scout X-4	Wallops Island	932 × 1,309 × 41°	60
28 IMP-3	May 29, 1965	Delta C	Cape Canaveral	229 × 261,206 × 31°	58
29 GEOS 1	Nov. 6, 1965	Delta E	Cape Canaveral	1,114 × 2,273 × 59°	175
30 Solrad 8	Nov. 19, 1965	Scout X-4	Wallops Island	667 × 871 × 60°	57
31 DME	Nov. 29, 1965	Thor Agena A	Vandenberg	502 × 2,857 × 80°	99
32 AE 2	May 25, 1966	Delta C	Cape Canaveral	282 × 2,723 × 65°	225
33 IMP-4	Jul. 1, 1966	Delta E	Cape Canaveral	85,228 × 481,417 × 41°	93
34 IMP-6	May 24, 1967	Delta E	Vandenberg	242 × 214,379 × 67°	75
35 IMP-5	Jul. 19, 1967	Delta E	Cape Canaveral	N/A	104
36 GEOS 2	Jan. 11, 1968	Delta E	Vandenberg	1,079 × 1,572 × 106°	209
37 Solrad 9	Mar. 5, 1968	Scout B	Wallops Island	353 × 433 × 59°	198
38 RAE 1	Jul. 4, 1968	Delta J	Vandenberg	5,835 × 5,861 × 121°	190
39 ADE 3	Aug. 8, 1968	Scout B	Vandenberg	670 × 2,583 × 81°	9
40 Injun 5	Aug. 8, 1968	Scout B	Vandenberg	677 × 2,494 × 81°	70
41 IMP-7	Jun. 21, 1969	Delta E	Vandenberg	80,374 × 98,159 × 86°	174

Explorer Spacecraft *(continued)*

Explorer/aka	Launch Date	Vehicle	Site	Orbit	Mass (kg)
42 SAS 1	Dec. 12, 1970	Scout B	San Marco	$521 \times 570 \times 3°$	143
43 IMP-8	Mar. 13, 1971	Delta M	Cape Canaveral	$1,845 \times 203,130 \times 31°$	288
44 Solrad 10	Jul. 8, 1971	Scout B	Wallops Island	$433 \times 632 \times 51°$	118
45 SSS-1	Nov. 15, 1971	Scout B	San Marco	$272 \times 18,149 \times 3°$	52
46 MTS	Aug. 13, 1972	Scout D	Wallops Island	$492 \times 811 \times 38°$	136
47 IMP-9	Sep. 22, 1972	Delta 1914	Cape Canaveral	$201,100 \times 235,600 \times 17°$	376
48 SAS 2	Nov. 16, 1972	Scout D	San Marco	$526 \times 526 \times 1°$	185
49 RAE 2	Jun. 10, 1973	Delta 1914	Cape Canaveral	$1,053 \times 1,064 \times 56°$	328
50 IMP-11	Oct. 25, 1973	Delta 1914	Cape Canaveral	$141,185 \times 288,857 \times 28°$	371
51 AE 3	Dec. 16, 1973	Delta 1914	Vandenberg	$155 \times 4,306 \times 68°$	658
52 Hawkeye	Jun. 3, 1974	Scout D	Vandenberg	$469 \times 125,569 \times 89°$	27
53 SAS 3	May 7, 1975	Scout F	San Marco	$498 \times 507 \times 3°$	195
54 AE 4	Oct. 6, 1975	Delta 2914	Vandenberg	$151 \times 3,819 \times 90°$	676
55 AE 5	Nov. 20, 1975	Delta 2914	Cape Canaveral	$154 \times 3,002 \times 20°$	721
56 ISEE-1	Oct. 22, 1977	Delta 2914	Cape Canaveral	$436 \times 137,806 \times 13°$	340
57 IUE	Jan. 26, 1978	Delta 2914	Cape Canaveral	$30,285 \times 41,296 \times 34°$	672
58 HCMM	Apr. 26, 1978	Scout D	Vandenberg	$560 \times 641 \times 98°$	134
59 ISEE-3	Aug. 12, 1978	Delta 2914	Cape Canaveral	$181 \times 1,089,200 \times 1°$	479
60 SAGE	Feb. 18, 1979	Scout D	Wallops Island	$456 \times 506 \times 55°$	147
61 Magsat	Oct. 30, 1979	Scout G	Vandenberg	$352 \times 561 \times 97°$	181
62 DE-1	Aug. 3, 1981	Delta 3914	Vandenberg	$568 \times 23,289 \times 90°$	424
63 DE-2	Aug. 3, 1981	Delta 3914	Vandenberg	$309 \times 1,012 \times 80°$	420
64 SME	Oct. 6, 1981	Delta 2914	Cape Canaveral	$335 \times 337 \times 98°$	437
65 AMPTE/CCE	Aug. 16, 1984	Delta 3925	Cape Canaveral	$1,121 \times 49,671 \times 5°$	242
66 COBE	Nov. 18, 1989	Delta 5920	Vandenberg	$872 \times 886 \times 99°$	2,265
67 EUVE	Jun. 7, 1992	Delta 6925	Cape Canaveral	$510 \times 524 \times 28°$	3,275
68 SAMPEX	Jul. 3, 1992	Scout G	Vandenberg	$506 \times 670 \times 82°$	158
69 RXTE	Dec. 30, 1995	Delta 7920	Cape Canaveral	$409 \times 409 \times 29°$	3,200
70 FAST	Aug. 21, 1996	Pegasus XL	Vandenberg	$353 \times 4,163 \times 83°$	187
71 ACE	Aug. 25, 1997	Delta 7920	Cape Canaveral	Halo	785
72 SNOE	Feb. 26, 1998	Pegasus XL	Vandenberg	$529 \times 581 \times 98°$	120
73 TRACE	Apr. 2, 1998	Pegasus XL	Vandenberg	$602 \times 652 \times 98°$	250
74 SWAS	Dec. 6, 1998	Pegasus XL	Vandenberg	$637 \times 651 \times 70°$	288
75 WIRE	Mar. 5, 1999	Pegasus XL	Vandenberg	$540 \times 590 \times 98°$	259
76 TERRIERS	May 18, 1999	Pegasus XL	Vandenberg	$537 \times 552 \times 98°$	120
77 FUSE	Jun. 24, 1999	Delta 7925	Cape Canaveral	$753 \times 769 \times 25°$	1,400
78 IMAGE	Mar. 25, 2000	Delta 7326	Vandenberg	$1,000 \times 45,922 \times 40°$	494
79 HETE-2	Oct. 9, 2000	Pegasus	Kwajalein	$598 \times 641 \times 1.9°$	124
80 MAP	Jun. 30, 2001	Delta 7425	Cape Canaveral	Halo	840
81 RHESSI	Feb. 5, 2002	Pegasus	Cape Canaveral	$600 \times 600 \times 38°$	293

explosive bolt

A bolt incorporating an explosive charge that is detonated on command, thus destroying the bolt and releasing the mated unit. It is used to separate one stage of a launch vehicle from another.

extravehicular activity (EVA)

See spacewalk.

Eyraud, Achille (nineteenth century)

A French writer whose *Voyage à Vénus* (Voyage to Venus),[93] published in 1863, describes the journey of a spacecraft propelled by a "reaction engine"–one of the earliest references to a true interplanetary rocket in science fiction. Unfortunately, Eyraud's grasp of the physics involved is somewhat tenuous, and, in any case, his work was overshadowed by another French space novel that came out in the same year. This was Jules Verne's *From the Earth to the Moon*.

F

F-1

The largest liquid-fueled rocket engine ever built; five propelled the first stage of the **Saturn** V. Each F-1 could be gimbaled (swiveled), was about as big as a two-and-a-half-ton truck, developed 6.7 million N of thrust at sea level (as much as all three Space Shuttle Main Engines combined), and burned three tons of liquid oxygen and RP-1 (kerosene mixture) propellants every second. RP-1 served not only as the fuel for the engine but also as the turbopump lubricant and the control system fluid. A gas generator utilizing the same propellants drove the turbine, which was directly coupled to the **turbopump**.

Faget, Maxime A. (1921–)

A leading aerodynamicist with the American space program. Having received a B.S. in aeronautical engineering from Louisiana State University in 1943, Faget joined the staff at Langley Aeronautical Laboratory in 1946 and soon became head of the performance aerodynamics branch of the pilotless aircraft research division. There, he conducted

F-1 Five F-1 engines being installed in each of three Saturn V first stages in the horizontal assembly area at NASA's Michoud Assembly Facility. *NASA*

research on the heat shield of the Mercury spacecraft. In 1958 he joined the Space Task Group in NASA, the forerunner of the NASA Manned Spacecraft Center—which eventually became the Johnson Space Center—becoming its assistant director for engineering and development in 1962 and later its director. Faget contributed many of the original design concepts for the Mercury Project's manned spacecraft and played a major role in the design of virtually every American-crewed spacecraft since that time, including the Space Shuttle. He retired from NASA in 1981 and became an executive for Eagle Engineering. In 1982, he was one of the founders of Space Industries and became its president and chief executive officer.

FAIR (Filled-Aperture Infrared Telescope)

An orbiting telescope designed to study the evolution of the circumstellar dust disks from which planetary systems may form. Using a sensitive infrared telescope, FAIR would be able to probe deeper than ever before into protostellar disks and jets to investigate the physical processes that govern their formation, development, and dissipation, as well as those that determine their temperature, density, and compositional structure. FAIR is identified in NASA's Office of Space Science Strategic Plan as a potential mission beyond 2007 but remains in the stage of early concept definition.

fairing

The area of a launch vehicle where a payload is attached until it is released into orbit.

fall-away section

A section, such as a lower stage, of a rocket vehicle that separates during flight—especially a section that falls back to Earth.

FAST (Fast Auroral Snapshot Explorer)

A satellite designed to observe and measure rapidly varying electric and magnetic fields and the flow of electrons and ions above the aurora; it is the second **SMEX** (Small Explorer) mission. FAST's data is complemented by

Launch
   Date: August 21, 1996
   Vehicle: Pegasus XL
   Site: Vandenberg Air Force Base
   Orbit: 353 × 4,163 km × 83.0°
   Size: 1.8 × 1.2 m
   Mass: 187 kg

those from other spacecraft, which observe fields and particles and photograph the aurora from higher altitudes.

At the same time, auroral observatories and geomagnetic stations on the ground supply measurements on how the energetic processes FAST observes affect Earth. Although made ready for a mid-1994 launch date, FAST was put in storage for a couple of years until a series of problems with the Pegasus launch vehicle could be corrected.

faster-than-light (FTL) travel

Although Einstein's **special theory of relativity** insists that it is impossible to accelerate an object up to the speed of light and beyond, there has been no shortage of speculation on how ways may be found to circumvent this natural speed restriction. First, Einstein's theory allows in principle the creation of bizarre particles, known as **tachyon**s, that have speeds permanently greater than that of light. Second, an object may be able to travel, apparently superluminally, between two far-flung points in the cosmos, providing it does not make the journey through conventional **space-time**. This opens up the possibility of using space-time gateways and tunnels known as **wormhole**s as a means of circumventing the normal relativistic constraints.[206]

Fastrac Engine

A low-cost rocket engine, with a thrust of 270,000 N, propelled by liquid oxygen and kerosene, that is being developed at the **Marshall Space Flight Center**. It is only the second of about 30 new rocket engines built over the past quarter of a century to originate in the United States. Fastrac is less expensive than similar engines because of an innovative design approach that uses commercial, off-the-shelf parts and fewer of them.

Feoktistov, Konstantin Petrovich (1926–)

A Soviet spacecraft design engineer and cosmonaut. Together with Sergei **Korolev**, Feoktistov began conceptualizing a manned space vehicle that could travel into orbit and return safely to Earth as early as June 1956. He helped design the **Vostok** spacecraft and its subsequent conversion to the multiseater **Voskhod**. For this he was rewarded with the dubious pleasure of flying alongside two crewmates in the extremely risky Voskhod 1 mission in 1964, thus becoming the first scientist-engineer in space. As head of his own bureau, he played a major role in the design of the civilian **Salyut** and **Mir** space stations and served as flight director on the Soyuz 18/Salyut mission in 1975. Although assigned to fly on Soyuz T3, he was grounded a few days before launch because of medical problems. Later he joined Energia NPO before retiring in 1990. Feoktistov graduated from E. N. Bauman Moscow Higher Technical School (MVTU) with a doctorate in physics.[142]

ferret

A satellite whose primary function is to gather electronic intelligence, such as microwave, radar, radio, or voice emissions.

field

A region within which a particular type of force can be observed or experienced; varieties include a **gravitational field**, an electric field, a magnetic field (or when the latter two are linked, an **electromagnetic field**), and a nuclear field. The laws of physics suggest that fields represent more than a *possibility* of force being observed, but that they can also transmit energy and momentum—a light wave, for example, is a phenomenon completely defined by fields.

fin

A surface at the rear of a rocket that serves to stabilize it in flight. Fins are usually planar surfaces placed at right angles to the main body.

Finger, Harold B. (1924–)

An aeronautical and nuclear engineer involved with the early American space program. Finger joined **NACA** (National Advisory Committee for Aeronautics) in 1944 as an aeronautical research scientist at the **Lewis** facility in Cleveland. Having gained experience working with compressors and in nuclear engineering projects, he became head of the nuclear radiation shielding group and the **nuclear propulsion** design analysis group in 1957. The following year, he was appointed chief of the nuclear engine program at NASA headquarters, and by 1962 had become director of nuclear systems. From 1967 to 1969, Finger was NASA's associate administrator for organization and management before taking up a government post in housing and urban development.

firing chamber

See **combustion chamber**.

FIRST (Far Infrared and Submillimetre Space Telescope)

See **Herschel**.

first motion

The first indication of motion of a vehicle from its launcher. Synonymous with **takeoff** for vertically launched rockets and missiles.

firsts in spaceflight

See table.

Firsts in Spaceflight

First	Spacecraft/Individual	Date
Liquid-propelled rocket launch	Robert Goddard	Mar. 16, 1926
Artificial satellite	Sputnik 1	Oct. 4, 1957
Animal in orbit	Dog Laika, Sputnik 2	Nov. 3, 1957
American satellite	Explorer 1	Feb. 1, 1958
Recorded message from space	SCORE	Dec. 18, 1958
Probe to reach escape velocity/lunar flyby	Luna 1	Jan. 2, 1959
Polar orbiting satellite	Discoverer 1	Feb. 28, 1959
Animals recovered alive from space	Monkeys Able & Baker, Jupiter	May 28, 1959
Lunar impact	Luna 2	Sep. 14, 1959
Weather satellite	Explorer 7	Oct. 13, 1959
Active comsat	Courier 1B	Oct. 4, 1960
Human in space	Yuri Gagarin, Vostok 1	Apr. 12, 1961
Day spent in space	Gherman Titov, Vostok 2	Aug. 6, 1961
Satellite telephone call & TV broadcast	Echo 1	Feb. 24, 1962
Real-time active comsat	Telstar 1	Jul. 10, 1962
Successful planetary flyby (Venus)	Mariner 2	Aug. 27, 1962
Woman in space	Valentina Tereshkova, Vostok 6	Jun. 16, 1963
Geosynchronous satellite	Syncom 2	Jul. 26, 1963

(continued)

Firsts in Spaceflight *(continued)*

First	Spacecraft/Individual	Date
Joint US/USSR space project	Echo 2	Jan. 25, 1964
Geostationary satellite	Syncom 3	Aug. 19, 1964
Spacewalk	Alexei Leonov, Voshkod 2	Mar. 18, 1965
Commercial comsat	Early Bird/Intelsat 1	Apr. 6, 1965
Human death in space	Vladimir Komarov, Soyuz 2	Apr. 24, 1967
Manned flight to leave Earth orbit	Apollo 8	Dec. 21, 1968
Human to walk on the Moon	Neil Armstrong, Apollo 11	Jul. 20, 1969
Soft landing on Venus	Venera 7	Dec. 15, 1970
Space station	Salyut 1	Apr. 19, 1971
International docking in space	Apollo-Soyuz Test Mission	Jul. 17, 1971
Flyby of Jupiter	Pioneer 10	Dec. 3, 1973
Gravity-assist to reach other planet	Mariner 10	Feb. 5, 1974
Flyby of Mercury	Mariner 10	Mar. 29, 1974
Soft landing on Mars	Viking 1	Jul. 20, 1976
Flyby of Saturn	Pioneer 11	Sep. 1, 1980
Reusable space vehicle	Space Shuttle *Columbia,* STS-1	Apr. 12, 1981
Probe to go beyond outermost planet	Pioneer 10	Jun. 1983
Untethered spacewalk	Bruce McCandless, Shuttle STS-41B	Feb. 11, 1984
Satellite repaired in orbit	SMM, by Shuttle STS-41C	Apr. 6, 1984
Spacewalk by a woman	Svetlana Savitskaya, Soyuz T-12	Jul. 17, 1984
Flyby of a comet nucleus (Halley)	Giotto	Mar. 13, 1986
Satellites brought back from space	Westar 6 and Palapa B2, Shuttle STS-51A	Nov. 16, 1984
Year in space	Musa Manarov & Vladimir Titov, Mir	Dec. 21, 1988
Flyby of an asteroid (Gaspra)	Galileo	Oct. 29, 1991
Ion-powered deep space probe	Deep Space 1	Oct. 24, 1998
International Space Station mission	Zarya module, Proton	Nov. 20, 1998
Landing on an asteroid (Eros)	NEAR-Shoemaker	Feb. 12, 2001

fission, nuclear

A process in which a heavy atomic nucleus splits into smaller nuclei accompanied by the release of large amounts of energy. A chain reaction results when the neutrons released during fission cause other nearby nuclei to break apart.

flame bucket

A deep cavity built into a launch pad to receive hot gases during rocket thrust buildup. One side is angled to form a blast deflector.

flap

A movable control surface on an aircraft or rocket that deflects air. Ailerons, elevators, and rudders are all flaps. Flaps are also used as air brakes.

flashback

A reversal of flame propagation in a system, counter to the usual flow of the combustible mixture.

Fletcher, James Chipman (1919–1991)

A NASA administrator who oversaw or initiated many of the major American space projects of the 1970s and 1980s during two spells in office: 1971–1976 and 1986–1988. He was in charge at the time of the three **Skylab** missions in 1973–1974 and the **Viking** landings on Mars in 1976. He also approved the **Voyager** probes to the outer planets, the **Hubble Space Telescope** program, and the **Apollo-Soyuz Test Project**. Most significantly, he won approval from the Nixon administration on January 5, 1972, to develop the Space Shuttle as the

James Fletcher NASA administrator Fletcher explains to the U.S. Senate Committee on Aeronautical and Space Science which methods will be used to repair the damage to the Skylab space station. *NASA*

Agency's next endeavor in human spaceflight. During his second spell as administrator, Fletcher was largely involved in efforts to recover from the *Challenger* **disaster**, ensuring that NASA reinvested heavily in the Shuttle program's safety and reliability and making organizational changes to improve efficiency. A critical decision resulting from the accident and its aftermath was to expand greatly the use of expendable launch vehicles. Fletcher received a B.S. in physics from Columbia University and a Ph.D. in physics from the California Institute of Technology. After holding research and teaching positions at Harvard and Princeton Universities, he joined Hughes Aircraft in 1948 and later worked at the Guided Missile Division of the Ramo-Wooldridge Corporation. In 1958, he cofounded the Space Electronics Corporation in Glendale, California, and was later named systems vice president of the Aerojet General Corporation. In 1964, he became president of the University of Utah, a position he held until he was named NASA administrator in 1971. Upon leaving NASA for the first time in 1977, Fletcher served on the faculty of the University of Pittsburgh and as an advisor to key national leaders involved in planning space policy.[180]

flight acceptance test
A test conducted to prove that the operational hardware intended for flight use has been built according to specifications. Flight acceptance tests are carried out under conditions expected to be encountered in actual operations.

flight path
The path followed by the center of gravity of a spacecraft during flight; also known as the spacecraft's trajectory. The path is tracked with reference to any point on Earth, or it may be tracked with reference to a star.

flight path angle
The angle between the velocity vector and the local horizon. In aeronautics this is called the **angle of attack**.

flight profile
A graphic portrayal or plot of a vehicle's flight path in the vertical plane—in other words, the flight path as seen from one side, so that the altitude at any point along the path is apparent.

flight readiness firing
A missile or rocket system test of short duration, conducted with the propulsion system operating while the vehicle is secured to the launcher.

Flight Research Center
A unique and highly specialized facility emphasizing research on manned flight in extreme-performance aircraft and spacecraft. It is located next to **Edwards Air Force Base** in California.

flight simulator
A synthetic flight trainer capable of simulating a complete flight of a specified space vehicle or aircraft.

flight test
A test involving rocket-powered flight in an upward trajectory from Earth's surface. It may involve a suborbital, an orbital, or a translunar flight.

FLTSATCOM (Fleet Satellite Communications System)
A constellation of American military satellites in geostationary orbit which, together with **Leasats**, supported worldwide ultra-high-frequency (UHF) communications among naval aircraft, ships, submarines, and ground stations, and between the Strategic Air Command and the national command authority network. It became fully operational in January 1981 and was gradually replaced during the 1990s by satellites in the **UFO** (UHF Follow-On) series. An outgrowth of the **LES** (Lincoln Experimental Satellite) series and Tacsat 1, FLTSATCOM was developed by the Navy to be the first complete operational system in space to serve the tactical user. However, through special transponders carried aboard the satellites

it also supported **AFSATCOM** (Air Force Satellite Communications System), which is a vital component of the American strategic nuclear capability.

flux

A measure of the energy or number of particles passing through a given area of a surface in unit time. *Luminous flux* is the rate of flow of energy in the form of **photon**s. *Particle flux* is the number of particles (for example, in the solar wind) passing through a unit area per second. *Magnetic flux* is a measure of the strength of a magnetic field perpendicular to a surface.

flux density

The energy in a beam of radiation passing through a unit area at right angles to the beam per unit time.

fly-away disconnects

Umbilical or other connections to the various stages of a space vehicle that disconnect from the vehicle at liftoff, either mechanically or by the actual rising of the vehicle off the launch pad.

flyby

A type of space mission in which the spacecraft passes close to its target but does not enter orbit around it or land on it.

flyby maneuver

See **gravity-assist**.

Fobos

Two nearly identical Soviet spacecraft intended to explore Mars from orbit and land probes on Phobos, the larger (but still very small) Martian moon. One failed completely, the other partially. The mission involved cooperation with 14 other nations including Sweden, Switzerland, Austria, France, West Germany, and the United States (which contributed the use of its Deep Space Network for tracking). As well as onboard instruments, each vehicle carried a lander designed to carry out various measurements on Phobos's surface. Phobos 2 also carried a second, smaller "hopper" lander designed to move about using spring-loaded legs and make chemical, magnetic, and gravimetric observations at different locations. Phobos 2 operated according to plan throughout its cruise and Mars orbital-insertion phases, gathering data on the Sun, interplanetary medium, Mars, and Phobos. However, contact with the probe was lost on March 27, 1989, two months after it had entered Martian orbit and shortly before the final phase of the mission during which the spacecraft was to

approach within 50 m of Phobos's surface and release its landers. The cause of the failure was found to be a malfunction of the onboard computer. Contact with Phobos 1 had been lost much earlier, while the spacecraft was en route to Mars. This failure was traced to an error in the software uploaded to the probe, which had deactivated its attitude thrusters. This caused a loss of lock on the Sun, resulting in the spacecraft orienting its solar arrays away from the Sun, thus depleting the batteries. It remains in solar orbit.

Launch
    Date: July 7, 1988 (Fobos 1); July 12, 1988 (Fobos 2)
    Vehicle: Proton
    Site: Baikonur
Mass (at Mars): 2,600 kg

FOBS (Fractional Orbital Bombardment System)

An orbital variant of the Soviet R-36 intercontinental ballistic missile, known in Russia as the R-360 (see **"R" series of Russian missiles**) and intended to deliver nuclear warheads from space. The Western designation "FOBS" refers to the fact that although the payloads reached orbit, they were intended to reenter over their targets prior to completing a full circuit of the Earth. Four suborbital FOBS tests were carried out between December 1965 and May 1966. These were followed by more tests through 1971, delivering 5-ton payloads into low Earth orbit. FOBS is a direct ancestor of the modern-day commercial **Tsyklon** launch vehicle.

Fontana, Giovanni da (c. 1395–1455)

An Italian author of a sketchbook, *Bellicorum instrumentorum liber* (Book of War Instruments), published in 1420, in which appear various rocket-propelled devices including a cart designed to smash through enemy strongholds and a surface-running torpedo for setting fire to enemy ships.

footprint

(1) In the case of a communications satellite, a map of the signal strength showing the power contours of equal signal strengths as they cover the Earth's surface. (2) In the case of a spacecraft lander, the area within which the vehicle is targeted to touch down.

force

In mechanics, the cause of a change in motion of an object. It is measured by the rate of change of the object's **momentum**.

FORTE (Fast On-orbit Recording of Transient Events)

A U.S. Air Force satellite, built by the Los Alamos National Laboratory to study natural and artificial radio emissions from the ionosphere—information needed to develop technology for monitoring nuclear test ban treaties.

Launch
  Date: August 29, 1997
  Vehicle: Pegasus XL
  Site: Vandenberg Air Force Base
  Orbit: 799 × 833 km × 70.0°

Forward, Robert L(ull) (1932–)

A science consultant, writer, and futurist specializing in studies of exotic physical phenomena and future space exploration with an emphasis on advanced space propulsion. Forward earned a B.S. in physics from the University of Maryland (1954), an M.S. in applied physics from the University of California, Los Angeles (1958), and a Ph.D. in gravitational physics from the University of Maryland (1965). He worked at the Hughes Aircraft Research Labs (1956–1987) before forming his own company, Forward Unlimited. From 1983 to the present, Forward has had a series of contracts from the Department of Defense and NASA to explore new energy sources and propulsion concepts that could produce breakthroughs in space power and propulsion. In 1994, he formed Tethers Unlimited with Robert P. Hoyt to develop and market a new multiline **space tether**.

Foton

A series of recoverable microgravity science satellites based on Zenit-class Soviet spy spacecraft of the type that also served as Gagarin's **Vostok** capsule. The Foton consists of a spherical reentry capsule sandwiched between a battery pack that separates before reentry and an expendable service module with a solid-propellant retro-rocket. The total mass is about 6.5 tons, of which 400 kg is science payload. Foton satellites remain in an approximately 215 × 380 km × 63° orbit for 12 days, carrying out experiments such as materials processing, before deorbiting. The first Foton was launched on April 16, 1985, and the most recent, Foton-12, in 1999.

France 1

A French satellite that investigated the properties of the ionized layers of Earth's atmosphere by observing how very-low-frequency (VLF) waves propagate through the ionosphere.

Launch
  Date: December 6, 1965
  Vehicle: Scout X-4
  Site: Vandenberg Air Force Base
  Orbit: 696 × 707 km × 75.9°
  Mass on-orbit: 60 kg

France in space

See **Asterix, CERISE, CNES, COROT,** Louis **Damblanc, Diamant, ELDO, Eole, ESA,** Robert **Esnault-Pelterie,** Achille **Eyraud, France 1,** Jean **Froissart,** *From the Earth to the Moon,* **Hammaguira, Helios** (reconnaissance satellites), **Hermes, Jason,** Y. P. G. **Le Prieur, NetLander, ONERA, ramjet, Rosetta, SPOT, Stella, STENTOR, TOPEX/Poseidon,** and **Tournesol.**

free fall

The motion of an object under the sole influence of **gravity** and, possibly, drag.

French space agency

See **CNES.**

frequency

The number of back-and-forth cycles per second in a wave or wavelike process. Frequency is measured in hertz (Hz), after the scientist who first produced and observed radio waves in the laboratory, and is often denoted by the Greek letter ν (nu).

frequency bands

Because the complete radio spectrum is so broad, it is considered to be divided into bands ranging from extremely low frequency (longest wavelengths) to extremely high frequency (shortest wavelengths). (See table, "General Radio Band Designations.")

Only a small part of this spectrum, from about 1 GHz to a few tens of gigahertz, is used for practical radio communications. International agreements control the usage and allocation of frequencies within this important region. Communications bands are designated by letters that appear to have been chosen at random—S, C, X, and so on. Indeed, this is exactly how they were chosen. The band letters were originated in World War II, when references to frequency bands needed to be kept secret. Most commercial communications use the C (the first to be used), Ku, K, and Ka bands. The shorter wavelength bands, such as Ka, are used in conjunction with small (including hand-

General Radio Band Designations

Name	Wavelength	Frequency
Extremely low frequency (ELF)	10,000–100 km	30 Hz–3 kHz
Very low frequency (VLF)	100–10 km	3–30 kHz
Low frequency (LF)	10–1 km	30–300 kHz
Medium frequency (MF)	1 km–100 m	300 kHz–3 MHz
High frequency (HF)	100–10 m	3–30 MHz
Very high frequency (VHF)	10–1 m	30–300 MHz
Ultra high frequency (UHF)	1 m–10 cm	300 MHz–3 GHz
Super high frequency (SHF)	10–1 cm	3–30 GHz
Extremely high frequency (EHF)	1 cm–1 mm	30–300 GHz

held) receivers on the ground but suffer severe attenuation from rain. The X-band is used heavily for military applications. (See table, "Some IEEE Radio Band Designations.")

Some IEEE Radio Band Designations

Band	Frequency (GHz)	Wavelength (cm)
L	1–2	30–15
S	2–4	15–7.5
C	4–8	7.5–3.75
X	8–12	3.75–2.5
Ku	12–18	2.5–1.67
K	18–27	1.67–1.11
Ka	27–40	1.11–0.75

frequency coordination

A process to eliminate frequency interference between different satellite systems or between terrestrial microwave systems and satellites.

Froissart, Jean (c. 1337–c. 1410)

A French historian and poet whose *Chronicles*, published in 1410, outlined the design of tube-launched military rockets. He is therefore credited as being the inventor of the bazooka. See Giovanni da **Fontana**.

From the Earth to the Moon

The famous novel by Jules **Verne** in which a capsule containing three men and two dogs is blasted out of an immense cannon, the *Columbiad*, toward the Moon. *From the Earth to the Moon* (1865)[299] and its sequel *Around the Moon* (1870)[300] are packed with technical details, some of which, it was realized at the time of the **Apollo** 8 and 11 missions, were curiously prescient. In the story, as in real-

ity, the United States launched the first manned vehicle to circumnavigate the Moon. Verne gave the cost of his project as $5,446,675—equivalent to $12.1 billion in 1969 and close to Apollo 8's tab of $14.4 billion. Both fictional and real spacecraft had a crew of three: Ardan, Barbicane, and Nicholl in the novel; Anders, Borman, and Lovell on Apollo 8; and Aldrin, Armstrong, and Collins on Apollo 11. Both spacecraft were built mainly of aluminum and had similar dry masses—8,730 kg in the case of Verne's capsule, 11,920 kg in the case of Apollo 8. The cannon used to launch the spacecraft was the *Columbiad;* Apollo 11's Command Module was named *Columbia.* After considering 12 sites in Texas and Florida, Stone Hill (south of Tampa, Florida) is selected in Verne's novel. One hundred years later, NASA considered seven launch sites and chose Merritt Island, Florida. In both cases Brownsville, Texas, was rejected as a site, politics played a major role in the site selection, and site criteria included a latitude below 28° N and good access to the sea. Verne's spacecraft was launched in December from latitude 27°7' N, longitude 82°9' W. After a journey of 242 hours 31 minutes, including 48 hours in lunar orbit, the spacecraft splashed down in the Pacific at 20°7' N, 118°39' W, and was recovered by the U.S. Navy vessel *Susquehanna.* The crew of Apollo 8 was launched in December one hundred years later, from latitude 28°27' N, longitude 80°36' W, 213 km from Verne's site. After a journey of 147 hours 1 minute, including 20 hours 10 minutes in lunar orbit, the spacecraft splashed down in the Pacific (8°10' N, 165°00' W) and was recovered by the U.S. Navy vessel *Hornet.*

Although **space cannon**s are still considered a viable means of launching small satellites, Verne was wildly optimistic in supposing that men and dogs (not to mention some chickens that Arden smuggled aboard with the idea of releasing them on the Moon to astonish his friends) could survive the horrendous *g*-forces associated

From the Earth to the Moon The crew of Jules Verne's moonship float in weightlessness.

with being accelerated almost instantly to escape velocity out of a giant gun. The ingenious system of hydraulic shock absorbers Barbican had devised for the floor of the projectile would have done nothing to save the occupants (only one of which, the dog Satellite, fails to survive the launch). Verne also took a liberty with the crew's means of disposing of the dead animal: by opening a hatch in the capsule "with the utmost care and dispatch, so as to lose as little as possible of the internal air." Aside from these technical implausibilities, Verne's most significant scientific error was his treatment of weightlessness. He believed it occurred only at the neutral point of gravity between Earth and the Moon, and thus allowed his crew only about one hour of it during their flight. But in terms of luxury, Verne's capsule beat Apollo hands down—"even the Pullman cars of the Pacific Railroad could not surpass the projectile vehicle in solid comfort," and at the moment of greatest crisis on the return journey, Arden is able to settle the crew's nerves with some bottles of Tokay Imperial 1863.

frontal area

The surface area of a rocket that faces directly into the air-flow.

Frosch, Robert A. (1928–)

NASA's fifth administrator, serving from 1975 to 1981—throughout the Carter presidency. Frosch was responsible for overseeing the continuation of the development effort on the **Space Shuttle**. During his tenure the *Enterprise* underwent ground testing and the *Columbia* made the first free flight of the Shuttle program. Frosch earned his B.S. and Ph.D. in theoretical physics at Columbia University. Between 1951 and 1963, he worked as a research scientist and director of research programs for Hudson Laboratories of Columbia University, an organization under contract to the Office of Naval Research, before joining ARPA (Advanced Research Projects Agency), where he served as director for nuclear test detection

FUSE FUSE mounted on top of its Delta II launch vehicle at Kennedy Space Center in June 1999. *Johns Hopkins University Applied Physics Laboratory/NASA*

(Project Vela) and then as ARPA's deputy director. In 1966, he became assistant secretary of the Navy for research and development, and then served as assistant executive director of the United Nations Environmental Program (1973–1975). After leaving NASA, he became vice president for research at the General Motors Research Laboratories.

FTL
Faster than light. A general science fiction term for any hypothetical means of circumventing the Einsteinian speed limit. The abbreviation goes back at least as far as the late 1950s. See **faster-than-light (FTL) travel**.

fuel cell
A cell in which a chemical reaction is used directly to produce electricity. The reactants are typically hydrogen and oxygen, which result in water as a by-product. The water can then be used for cooling and for human consumption. Fuel cells are generally used on manned spacecraft.

Fuji
See JAS-1.

FUSE (Far Ultraviolet Spectroscopic Explorer)
A NASA ultraviolet (UV) astronomy satellite. FUSE carries four 0.35-m far UV telescopes, each with an ultraviolet high-resolution spectrograph that covers the ultraviolet band from 91 nm, the hydrogen ionization edge, to 119 nm, just short of the Lyman alpha line. Its observations are used to measure the abundance of deuterium in the universe as well as to study helium absorption in the intergalactic medium, hot gas in the galactic halo, and cold gas in molecular clouds from molecular hydrogen lines. NASA has recommended a two-year extension beyond the three-year primary science mission.

Launch
    Date: June 24, 1999
    Vehicle: Delta 7925
    Site: Cape Canaveral
Mass: 1,400 kg
Orbit: 753 × 769 km × 25.0°

fuselage
The structure or airframe that houses payload, crew, or passengers. Although not inaccurate when applied to rockets, this term is used more commonly with regard to airplanes.

fusion, nuclear
The process by which heavier nuclei are manufactured from lighter ones with (typically) the release of large amounts of energy.

Fuyo
See JERS (Japanese Earth Resources Satellite).

G

g
See **acceleration due to gravity, 1g spacecraft**.

G
See **gravitational constant**.

Gagarin Cosmonaut Training Center
A Russian center for training and preparing crews for manned missions. Located at Zvezdny Gorodok ("Star City") outside Moscow, it includes equipment for simulating missions aboard Soyuz and the Russian modules of the International Space Station. The Center also has a neutral buoyancy facility similar to that at the Marshall Space Flight Center.

Gagarin, Yuri Alexeyevich (1934–1968)
A Soviet cosmonaut who became the first human to travel in space. The son of a carpenter, Gagarin grew up on a collective farm in Saratov (later renamed Gagarin City), west of Moscow. After graduating with honors from the Soviet Air Force in 1957, he was selected as one of 20 fighter pilots to begin cosmonaut training. Immediately before his historic flight, the 27-year-old was promoted from senior lieutenant to major. The flight itself aboard **Vostok** 1 took place on April 12, 1959, lasted 108 minutes, and concluded with Gagarin ejecting from his capsule after reentry and descending by parachute to the ground near the village of Uzmoriye on the Volga. In his orange flight suit, he approached a woman and a little girl with a calf, all of whom began to run away (it had only been a year since U-2 spy plane pilot Gary Powers had been shot down over Russia). Gagarin called out: "Mother, where are you running? I am not a foreigner." Asked then if he had come from space, he replied, "As a matter of fact, I have!"

Arriving on motorcycles, Uzmoriye villagers purloined the cosmonaut's radio and inflatable rubber dinghy and buried it for safekeeping. "The dinghy was a genuine gift for the village fishermen . . . it literally fell down from the sky," *Komsomolskaya Pravda* reported. But then the KGB appeared on the scene and threatened to arrest the entire community if the equipment was not returned. Despite protests from the villagers that the dinghy was torn, the KGB captain put it in his car anyway and drove off.

Gagarin was hailed as a hero and given a luxury apartment in Moscow. However, he found all the attention and publicity hard to deal with and began drinking heavily and sometimes behaving badly. Nevertheless, he was assigned to a second space mission–Soyuz 3, which would involve the first docking between two spacecraft in orbit. On March 27, 1968, he took off on a training flight in a MiG-15 alongside Vladimir Seryogin, a senior test pilot and a decorated military hero. For reasons still not clear, but possibly involving a sudden maneuver to avoid another aircraft, the plane crashed and Gagarin and his copilot were killed. He left a wife, Valentina, and two daughters, and was buried with full honors in the Kremlin wall along with the greatest heroes of the Soviet Union. Statues commemorating Gagarin have been erected in his hometown and in Moscow in the Yuri Gagarin Square. At the spot where he landed after his historic spaceflight, a 40-m-high titanium obelisk has been erected. In addition, a crater on the far side of the Moon has been named after him.[108]

Yuri Gagarin *Joachim Becker*

GAIA (Global Astrometric Interferometer for Astrophysics)

A proposal for an advanced space astrometry (star-distance measurement) mission to build on the work started by **Hipparcos**. It has been recommended within the context of **ESA**'s (European Space Agency's) Horizon 2000 Plus plan for long-term space science. GAIA would measure the distance and velocity of more than a billion stars in the galaxy to an accuracy 100 to 1,000 times higher than that possible with Hipparcos. It would also carry out a special study of some 200,000 stars within 650 light-years of the Sun in a search for Jupiter-mass planets. GAIA is expected to be launched around 2010–2012 and to operate for five years.

Galaxy

A constellation of about 10 geostationary communications satellites owned and operated by PanAmSat. The latest members of the constellation are Galaxy 3C, launched on June 2, 2002, by a Zenit 35L from the **Sea Launch** platform; and Galaxy 13, which was scheduled for launch in the last quarter of 2002.

GALEX (Galaxy Evolution Explorer)

A **SMEX** (Small Explorer) mission designed to make ultraviolet observations that will shed light on the origin and evolution of galaxies. GALEX is equipped with a 50-cm telescope sensitive in the region of 130 to 300 nm and an efficient spectroscope that can obtain the spectra of about 100,000 galaxies per year. Partners in the project include NASA, JPL (Jet Propulsion Laboratory), the Orbital Science Corporation, and various universities in California, France, and Japan. GALEX will probably be launched in 2003.

Galileo

See article, pages 147–148.

Galileo satellite navigation system

A planned European satellite navigation system, similar to **GPS** (the U.S. military's Global Positioning System) but run on a purely civilian basis. It is being developed by **ESA** (European Space Agency) in collaboration with the European Union (the first such joint project) and is expected to be operational by 2008. Its goal is to provide the world in general and Europeans in particular with an accurate, secure, and certified satellite positioning system with applications in road, rail, air, and maritime traffic control, synchronization of data transmission between banks, and so forth. Although Galileo will be in commercial competition with GPS (and so, not surprisingly, has been opposed by the U.S. government), it will also complement GPS and provide redundancy. The Galileo

GALEX   An artist's conception of GALEX in orbit. *NASA/JPL*

system will consist of 30 satellites (27 in operation and 3 in reserve), deployed in three circular medium Earth orbits at an altitude of 23,616 km and an inclination of 56° to the equator. Two control centers will be set up in Europe to monitor the operation of the satellites and manage the navigation system. A precursor of the system is **EGNOS** (European Global Navigation Overlay Service), which refines current GPS data and foreshadows the services Galileo will provide.

Gamma

An orbiting Soviet gamma-ray and X-ray telescope derived from Soyuz-manned spacecraft hardware. It was originally conceived in 1965 as part of a "Cloud Space Station"–a primary space station from which a number of man-tended, free-flying spacecraft would operate. This had evolved by the early 1970s into the MKBS/MOK space station complex. Various spacecraft with specialized laboratories or instrument sets, including Gamma, would fly autonomously away from the **N-1**–launched main station. The Soyuz propulsion system would be used, but the descent and orbital modules were replaced by a large pressurized cylinder containing the scientific instruments. Work on the instrument payload for Gamma began in 1972, and French participation began *(continued on page 148)*

Galileo

NASA spacecraft that carried out the first studies of Jupiter's atmosphere, satellites, and magnetosphere from orbit around the planet. The mission was named in honor of Galileo Galilei (1564–1642), the Italian Renaissance scientist who discovered Jupiter's four biggest moons. JPL (Jet Propulsion Laboratory) designed and developed the Galileo orbiter and operated the mission, Ames Research Center developed the atmospheric probe, and the German government was a partner in the mission through its provision of the propulsion subsystem and two science experiments.

Like Voyager and some other previous interplanetary missions, Galileo used **gravity assists** to reach its target. The spacecraft exploited the gravitational fields of Venus and Earth (twice) to pick up enough velocity to get to Jupiter. This 38-month Venus-Earth–Earth Gravity Assist phase ended with the second Earth flyby on December 8, 1992. En route to Jupiter, Galileo made the first and second flybys of asteroids–Gaspra in October 1991, and Ida (discovering its tiny moon) in August 1993. Galileo was also the only vehicle in a position to obtain images of the far side of Jupiter when more than 20 fragments of Comet Shoemaker-Levy 9 plunged into Jupiter's atmosphere in July 1994.

On December 7, 1995, as Galileo approached the giant planet, it released a small probe that descended into Jupiter's atmosphere (see "Descent Probe," below). After this, the main spacecraft went into orbit around Jupiter to begin its scheduled 23-month, 11-orbit tour, including 10 close satellite encounters. A serious problem had arisen with the spacecraft during its interplanetary cruise: the planned deployment of Galileo's 4.8-m-wide high-gain antenna had failed, forcing the use of a low-gain antenna with a much lower data transmission capacity. However, although this was a blow, it was softened by the skill of technicians on Earth who devised data compression strategies and other ways around the hardware limitations.

The Galileo orbiter carried 10 science instruments: a near-infrared mapping spectrometer, for making multispectral images for atmosphere and surface chemical analysis; an ultraviolet spectrometer, for studying atmospheric gases; a solid-state imager, for visible imaging with an 800 × 800 array CCD (charge-coupled device); a photopolarimeter, for measuring radiant and reflected energy; magnetometers; a dust detection experiment; a plasma investigation; an energetic particle detector; a plasma wave investigation; and a radio.[146]

Deployment
Date: October 18, 1989
Shuttle mission: STS-34
Entered Jupiter orbit: December 7, 1995
Length: 6.2 m
Mass (dry): 2,223 kg (orbiter), 339 kg (descent probe)

Descent Probe

A package that descended through Jupiter's atmosphere by parachute and sent back data until it was destroyed by the crushing pressure. It carried science

Galileo Galileo being prepared for mating to its Inertial Upper Stage booster at Kennedy Space Center in 1989. *NASA*

instruments and the subsystems needed to support them and transmit their data back to the orbiter for storage and later transmission to Earth. The probe consisted of a deceleration module and a descent module. The deceleration module included fore and aft heat shields, the structure that supported the heat shields, and the thermal control hardware for mission phases up to entry. A 2.5-m main parachute was used to separate the descent module from the deceleration module and to control the rate of fall of the descent stage through the atmosphere. The descent module carried six science instruments: an atmosphere structure experiment, to measure temperature, pressure, and deceleration; a mass spectrometer, for atmospheric composition studies; a helium abundance detector; a nephelometer, for cloud location and cloud-particle observations; a net flux radiometer, to measure the difference between upward and downward radiant energy flux at each altitude; and a lightning and energetic particle detector, to measure light and radio emissions associated with lightning and high-energy particles in Jupiter's radiation belts.

Extended Mission

Following funding approval by NASA and the U.S. Congress, Galileo's mission was extended for two more years through the end of 1999. The first part of the extended mission, lasting more than a year, was devoted to searching for further evidence of a subsurface ocean on Europa. During eight orbits of Jupiter, Galileo made close approaches to Europa of 200 to 3,600 km, enabling a camera resolution of down to 6 m. Following the Europa Campaign ("Ice") were scheduled the Perijove Reduction/Jupiter Water Study ("Water") and the **Io** Campaign ("Fire"). The mission was then further extended into 2002 to include additional flybys of moons before a final death plunge into Jupiter's atmosphere.

Gamma

(continued from page 146)

in 1974. However, that same year the N-1 launch vehicle and the MKBS space station were canceled. The Soviet space program was completely reformulated and authorization given to develop the free flyer in conjunction with the DOS-7/DOS-8 space station, which would eventually evolve into **Mir.** The draft project for Gamma was completed in 1978, and production was authorized together with Mir in February 1979. At this point, Gamma included a passive docking port so that the spacecraft could be serviced by Soyuz-manned spacecraft. It was planned that at 6 to 12 months into its one-year mission, Gamma would be visited by a two-crew Soyuz mission that would replace film cassettes and repair or replace instruments. This approach was dropped in 1982 when it became clear that the spacecraft was overweight and that all planned Soyuz spacecraft would be needed to support Mir itself. All film systems were

removed and replaced with purely electronic data return methods. By that time, Gamma was scheduled for launch in 1984, but further technical delays resulted in a 1990 launch, 25 years after the project was first conceived. In the end, the satellite's research in the field of high-energy astrophysics, conducted jointly with France and Poland, did not produce many noteworthy results.

gamma rays

Electromagnetic radiation at the extreme short-wavelength (high-energy) end of the electromagnetic spectrum. Gamma rays lie beyond **X-rays** and have wavelengths less than about 0.1 nm.[3]

gamma-ray astronomy satellites

See, in launch order: **Vela** (1960s), **SAS**-2 (November 1972), **COS-B** (April 1975), **HEAO**-3 (September 1979), **Granat** (December 1989), **Gamma** (July 1990), **Compton Gamma-Ray Observatory** (April 1991), **HETE** (October 2000), **INTEGRAL** (September 2001), **Spectrum-X-Gamma** (2003), **Swift Gamma-Ray Burst Explorer** (2003), **GLAST** (2006), and **EXIST** (2007+).

Ganswindt, Hermann (1856–1934)

A German law student and amateur inventor, born in Voigtshof, East Prussia, who, along with Nikolai **Kibalchich** and Konstantin **Tsiolkovsky**, was one of the first to realize the potential of the rocket for space travel. He came up with a design for an interplanetary spacecraft

---

Launch
    Date: July 11, 1990
    Vehicle: Soyuz
    Site: Baikonur
Orbit: 382 × 387 km × 51.6°
Size: 7.7 × 2.7 m
Mass: 7,350 kg

as early as 1881 that used the reaction principle. In his scheme, steel cartridges charged with dynamite would be placed in a reaction chamber. As a cartridge exploded, half of it would be ejected while the other half struck the top of the chamber to provide the reaction force. Suspended below the chamber on springs was the inhabited part of the ship. Ganswindt even provided his crew with **artificial gravity** by allowing the spaceship to spin.

gantry
The servicing and access tower that stands beside a rocket on its launch pad.

Garneau, Marc (1949–)
The first Canadian in space. Seconded to the Canadian Astronaut Program from the Department of National Defence in February 1984 to begin astronaut training, he flew as a payload specialist on Shuttle Mission 41G in October 1984. He was named deputy director of the Canadian Astronaut Program in 1989. Selected by NASA for astronaut candidate training in July 1992, Garneau worked for the Astronaut Office Robotics Integration Team and served as CAPCOM (Capsule Communicator) during Space Shuttle flights. He subsequently flew as a mission specialist on STS-77 in 1996 and STS-97 in 2000. In November 2001, Garneau was appointed president of the **Canadian Space Agency.** He received a B.S. in engineering physics from the Royal Military College of Kingston (1970) and a Ph.D. in electrical engineering from Imperial College of Science and Technology, London (1973).

GARP (Global Atmospheric Research Project)
An international scientific project to monitor the world's weather and to better understand the physical basis of climate; it spanned the period 1967 to 1982.

Garriott, Owen K. (1930–)
An American astronaut and scientist who flew aboard **Skylab** 3 in 1973 and the Space Shuttle *Columbia* a decade later. Garriott received a B.S. in electrical engineering from the University of Oklahoma (1953) and an M.S. (1957) and a Ph.D. (1960) in electrical engineering from Stanford University. Following graduation from Oklahoma, he served as an electronics officer aboard several destroyers while on active duty with the U.S. Navy (1953–1956). From 1961 to 1965, he taught electronics and ionospheric physics as an associate professor at Stanford before being selected by NASA to be among the first group of scientist-astronauts. Aboard Skylab, Garriott made extensive observations of the Sun during a period of intense solar activity. In November 1983, he returned to orbit aboard *Columbia* and helped conduct medical,

astronomy, Earth survey, atmospheric, and materials processing experiments on the first **Spacelab** mission. Subsequently, Garriott was appointed assistant director for space science at Johnson Space Center before leaving NASA in 1986 to work in the private sector. In 2000, he was appointed interim director of the newly created National Space Science & Technology Center, a research facility associated with the Marshall Space Flight Center.

Garuda
Satellites that provide global support for mobile telephone communications users in Asia; "garuda" is Sanskrit for "eagle." Launched on February 13, 2000, from Baikonur, Garuda-1 is the first satellite of a constellation that comprises the ACeS (Asia System Cellular Satellite) system. The second ACeS satellite, Garuda-2, will first serve as a backup to Garuda-1 and then allow the ACeS system to expand coverage to western and central Asia, the Middle East, Europe, and northern Africa.

gas generator
A combustion chamber used to provide hot gases for a turbine or motor to drive the propellant pumps of a rocket engine, or to provide a source of gas at some predetermined pressure. Gas generators are usually operated fuel-rich to maintain the container temperature at reduced values.

gaseous propellant
A working substance used in a **gaseous-propellant rocket engine**. Nitrogen, argon, krypton, dry air, and Freon 14 have all been employed in spacecraft.

gaseous propellant rocket engine
A small rocket engine that works by expelling a gas, such as nitrogen or helium, stored in (relatively heavy) tanks at high pressure. Cold gas engines were used on many early spacecraft as attitude control systems and are still occasionally used for this purpose. Warm gas engines achieve better performance by heating the propellant, electrically or chemically, before expulsion.

GCOM (Global Change Observing Mission)
Japanese Earth resources satellites planned for launch over the coming decade to improve the accuracy of global observations begun by **ADEOS** (Advanced Earth Observation Satellite) and collect data on worldwide environmental change over a period of up to 15 years. GCOM-A1 will carry instruments to monitor concentrations of ozone, CFCs, and major greenhouse gases such as carbon dioxide and nitrogen oxide. GCOM-B1 will study the large-scale circulation of energy and materials using the Second Generation ion Global Imager (SGLI)

and the follow-on of the Advanced Microwave Scanning Radiometer (AMSR). It may also carry NASA's Alpha SCAT and CNES's (the French space agency's) Polarization and Directionality of Earth's Reflectance (POLDER). GCOM-B1 will be able to study the distribution of aerosols and water vapor and make measurements of ice coverage, phytoplankton concentrations, and sea surface wind directions. Launch is planned for about 2006.

GEC (Global Electrodynamics Connections)

A planned cluster of four NASA satellites in polar orbits, combined with ground-based observations, that will make systematic multipoint measurements to complete our understanding of the role played by the ionosphere and thermosphere in the Sun-Earth connection. GEC is a future Solar Terrestrial Probe mission.

Geizer

A series of satellites that relay telecommunications for the Russian Ministry of Defense.

gelled propellants

Rocket propellants that have additives to make them thixotropic. This means they have the consistency of jelly when at rest but can be made to move as a liquid through pipes, valves, and so on. Experimental rocket engines have shown gelled propellants to be generally safer than liquid propellants yet capable of performing as well.

Gemini

See article, pages 151–158.

Gemini-Titan II

A modified version of the **Titan** II intercontinental ballistic missile used as the launch vehicle for the **Gemini** program. In order for the Titan II to be man-rated, a number of safety systems had to be added, including a malfunction detection system to warn the crew of any failure in the rocket's equipment, backup electrical and hydraulic systems, and monitoring devices to check the rocket before launch and provide telemetry about vehicle performance during flight. In the event of an emergency, the only mode of escape for the crew was by ejection seat. The Titan gave a much smoother ride than the **Mercury-Atlas**—and some astronauts remarked that they were aware of initial motion only through their instrument displays.

general theory of relativity

The geometric theory of **gravitation** developed by Albert Einstein, incorporating and extending the **special theory of relativity** to accelerated frames of reference and introducing the principle that gravitational and inertial forces are equivalent. General relativity treats special relativity as a restricted subtheory that applies locally to any region of space sufficiently small that its curvature can be neglected.[309, 311]

generation ship

An immense, relatively slow-moving spacecraft, also known as an interstellar ark, aboard which many generations would live and die on a voyage between stars. The generation ship has been offered as an alternative to spacecraft that travel at much higher speeds carrying conventional-sized crews. The idea of a vessel carrying a civilization from a dying solar system toward another star for a new beginning was envisioned in 1918 by Robert **Goddard**.[114] But, perhaps concerned about professional criticism, he placed his manuscript in a sealed envelope, and it did not appear in print for over half a century. Konstantin **Tsiolkovsky** and J. D. **Bernal** wrote about artificial planets and self-contained worlds in the 1920s, as did Olaf **Stapledon** in his visionary novels, and by the 1940s the generation ship concept had been fully expanded by science fiction writers in the publications of Hugo Gernsback and others. Robert **Heinlein** (*Orphans of the Sky*, 1958) first raised the possibility that the crew might eventually forget they were aboard a ship and believe instead that they were the inhabitants of a small world (a theme taken up in one of the original *Star Trek* episodes: "For the World Is Hollow and I Have Touched the Sky"). In 1952, L. R. Shepherd examined the idea of the generation ship in more technical detail and described a nuclear-propelled million-ton interstellar vessel shaped as an oblate spheroid, which he called a "Noah's Ark." Such a ship would be a microcosm of human civilization with a substantial and highly varied population, extensive educational, recreational, and medical facilities, food production areas, research laboratories, and so forth—effectively, a miniature, nomadic planet. There is perhaps a strange attraction in the idea, though whether anyone would willingly volunteer to exile themselves to such an environment knowing that they would die some fraction of the way to the ultimate goal is hard to say. Perhaps there would be less of a psychological problem for subsequent generations who were born aboard the ship and therefore never knew what life was like on the surface of a planet under open skies. Then again, how difficult would it be for those who finally reached journey's end to step outside the confines of their artificial world? Like many other old ideas in space travel, including **Lucian**'s waterspout and **Well**'s cavorite, the generation ship now seems quaint and romantic. Almost certainly, interstellar travel will never be accomplished by this means, though it is still possible that the related concept of the **space colony** will come to fruition.

Gemini

A series of 2 unmanned and 10 manned NASA missions conducted between April 1964 and November 1966. Gemini, with its two-seater capsule, 5.8 m long, 3 m in diameter, and weighing 3,800 kg, built on the success of the **Mercury** project and paved the way for the **Apollo** program. Equipped with the Orbital

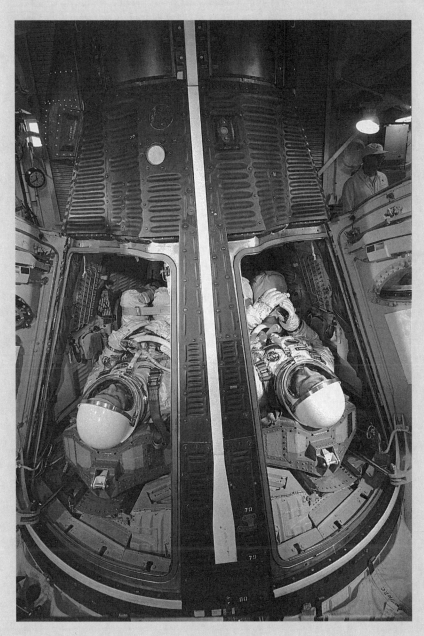

Gemini capsule Astronauts James McDivitt and Ed White inside the Gemini spacecraft for a simulated launch at Cape Canaveral.
NASA

Attitude Maneuvering System (OAMS) to create a controlled orbital and reentry path, the Gemini craft were the first fully maneuverable American manned spacecraft.

The first two Gemini missions were unmanned. Gemini 1 was to check compatibility between the Gemini spacecraft and the Titan II launch vehicle, and no attempt was made to separate the two during their brief flight. Gemini 2 was a suborbital test of the Gemini reentry system. There then followed 10 successful manned missions.[269] (See table, "Gemini Missions.")

Gemini 3

Crew
Command pilot: Virgil **Grissom**
Pilot: John **Young**
Highlight: first American two-manned spaceflight

The first manned Gemini and the first American space mission with two astronauts aboard. During the flight, Gemini 3's orbit was varied using thrusters to rehearse techniques needed for upcoming rendezvous and docking missions. The capsule was unofficially dubbed "Molly Brown" after a survivor of the *Titanic* and in reference to the fact that Grissom's Mercury capsule sank and was lost following his first spaceflight. **Illicit cargo** included a corned beef sandwich reportedly from Wolfie's Restaurant in Cocoa Beach,

which was bought by Grissom, carried aboard by Young, and partly consumed by Grissom during the flight. Young was authorized to eat specially prepared food, and Grissom was supposed to have gone without. The possibility that weightless crumbs from the sandwich might have compromised safety led NASA managers to clamp down on what future astronauts could take into space. Although one of the goals of the program was to achieve pinpoint recoveries, Gemini 3 splashed down about 95 km short of its primary recovery vessel (a result of Molly Brown having less lift than wind tunnel tests had predicted). Unlike the Mercury capsules, which came down upright, the Gemini capsules were designed to splash down on their sides. In the event, the force of the impact hurled the astronauts against their respective windows, causing the faceplate in Grissom's helmet to crack and that of Young's to be scarred.

Gemini 4

Crew
Command pilot: James **McDivitt**
Pilot: Edward **White** II
Highlight: first American spacewalk

Considered the first long-duration American spaceflight, this was also the first mission to be directed from Mission Control at the Johnson Space Center instead of the Mission Control Center at Cape

Gemini Missions

Mission	Launch	Recovery	Duration	Orbits	Crew
Gemini 1	Apr. 8, 1964	—	—	—	Unmanned
Gemini 2	Jan. 19, 1965	—	—	—	Unmanned
Gemini 3	Mar. 23, 1965	Mar. 23, 1965	4 hr 53 min	3	Grissom, Young
Gemini 4	Jun. 3, 1965	Jun. 7, 1965	4 days 2 hr	62	McDivitt, White
Gemini 5	Aug. 21, 1965	Aug. 29, 1965	8 days	120	Conrad, Cooper
Gemini 7	Dec. 4, 1965	Dec. 18, 1965	13 days 19 hr	206	Borman, Lovell
Gemini 6A	Dec. 15, 1965	Dec. 16, 1965	1 day 1 hr 52 min	16	Schirra, Stafford
Gemini 8	Mar. 16, 1966	Mar. 16, 1966	10 hr 41 min	7	Armstrong, Scott
Gemini 9A	Jun. 3, 1966	Jun. 6, 1966	3 days 0 hr 21 min	44	Cernan, Stafford
Gemini 10	Jul. 18, 1966	Jul. 21, 1966	2 days 22 hr 46 min	43	Collins, Young
Gemini 11	Sep. 12, 1966	Sep. 15, 1966	2 days 23 hr 17 min	44	Conrad, Gordon
Gemini 12	Nov. 11, 1966	Nov. 15, 1966	3 days 23 hr	59	Aldrin, Lovell

Gemini 4   Ed White becomes the first American to step outside his spacecraft and drift in the vacuum of space. He is attached to the spacecraft by an umbilical line and a tether line, wrapped together in gold tape to form one cord. *NASA*

Canaveral. Gemini 4's main goals were to evaluate the spacecraft and crew during a lengthy stay in space, rendezvous with the spent Titan II second stage, carry out the first American spacewalk, and continue testing the Orbital Attitude Maneuvering System (OAMS).

The plan to fly in formation with the Titan second stage proved a real learning experience in orbital dynamics. To catch something in Earth's atmosphere involves simply moving as quickly as possible in a straight line to the place where the object will be at the right time. Gemini 4 showed that this does not work in orbit. When the astronauts tried to fly toward the target, they got farther and farther away! The problem is that adding speed also raises altitude, moving the spacecraft into a higher orbit than its target. This paradoxically causes the faster-moving spacecraft to slow down relative to the target because its orbital period increases with increasing distance from the center of gravity (Earth's center, in this case). To catch up with an object ahead of you in orbit, NASA learned, you must drop down and then rise after you catch up, rather than speed up. This was a crucial lesson for future rendezvous maneuvers. But, for Gemini 4, it was learned too late and the rendezvous was called off with half the onboard fuel used up. Gemini 4's tanks were only half the size of later models, and the fuel had to be conserved for essential maneuvering later in the mission.

White also took longer than expected to prepare for his spacewalk because the hatch had jammed. However, just over four hours into the mission, the hatch was opened. White stood upright in his seat, fixed a 16-mm camera to the spacecraft to record his historic feat, and prepared to exit. On his way out, he lost a thermal overglove, which he had elected not to wear and which drifted away to become an unintended new satellite. White then propelled himself away using a handheld maneuvering unit and became the first American to walk in space. For the duration of his spacewalk, White remained tied to the spacecraft by a 7.6-m umbilical line that contained an oxygen supply hose, bioinstrumentation wires, and a support tether. He reported that although the line worked well as a retention device, it was awkward to use and easily got tangled. After 20 minutes, McDivitt called his companion back inside at Mission Control's urgent request, and White remarked, "It's the saddest moment of my life." White again had to struggle with the latch and, during the five minutes he took to deploy it, exceeded the cooling capacity of his spacesuit, which led to severe condensation inside the helmet; his heartbeat peaked at 180. Once again, the intended landing point was missed by a wide margin—Gemini 4 splashed down 724 km east of Cape Canaveral and some 80 km off target.

Gemini 5

Crew
    Command pilot: L. Gordon **Cooper**
    Pilot: Charles **Conrad** Jr. (pilot)
    Highlight: longest American manned spaceflight to date

A mission during which a new space endurance record was set and the United States (temporarily) took the lead in total space man-hours. Gemini 5

showed that astronauts could endure weightlessness for roughly the time needed to fly to the Moon and back. The spacecraft was supposed to meet up with a rendezvous evaluation pod released early in the flight, but the rendezvous had to be canceled because of problems with the Gemini 5 fuel-cell system—the first time such a system had been used for supplying water and electricity in space. Instead, the crew performed a simulated rendezvous, worked out by Buzz **Aldrin**, with a phantom target. A number of other planned on-orbit maneuvering experiments were also called off because of erratic behavior by the OAMS. These cancelations led to plenty of idle time for the astronauts, and Conrad later remarked that he wished he had "brought along a book." Once again, the splashdown was well off the mark, missing the target zone by 143 km—an error blamed on faulty data fed to the spacecraft by ground computers.

Gemini 6A

Crew
    Command pilot: Walter **Schirra** Jr.
    Pilot: Thomas **Stafford**
    Highlight: rendezvous with Gemini 7

A planned rendezvous between Gemini 6 and an unmanned Gemini Agena Target Vehicle (GATV) had to be scrubbed when the GATV exploded 6 minutes after launch. Instead, a new mission, called Gemini 6A, was announced in which the capsule would meet up with the manned Gemini 7 spacecraft. The latter was launched on December 4, 1965, on a two-week mission, but an attempt to launch Gemini 6A on December 12 had to be called off when the Titan II first stage automatically shut down one second after ignition. The shutdown was caused by an electrical umbilical prematurely separating. But it was later found that because of a dust cap left accidentally in place on an engine component, the first stage thrust was in any case in the process of decaying. Without the automatic shutdown caused by the umbilical disconnect, the first-stage decay might have led to a catastrophic explosion. Schirra had the option of manually ejecting the capsule from the launch vehicle by firing the escape tower rockets, but he decided against this, and the astronauts safely left the vehicle after about 90 minutes. Launch eventually took place three days later. A successful rendezvous was completed just under six hours into the flight, the two spacecraft approaching to within about

37 m. Station-keeping maneuvers continued for over three orbits at separation distances of 30 to 90 m—close enough for the crews to see one another. Gemini 6A then maneuvered away and the two spacecraft flew in formation about 48 km apart until Gemini 6A began its reentry sequence. Achieving the most precise Gemini recovery to date, the capsule splashed down just 11 km from its primary target point.[120]

Gemini 7

Crew
   Command pilot: Frank **Borman**
   Pilot: James **Lovell** Jr.
   Highlight: long-duration mission

The longest Gemini flight, rivaling even the Space Shuttle in terms of length of stay in orbit for a single vehicle. Shortly after the capsule separated from the Titan II second stage, the crew moved to within about 18 m of the spent booster and performed a quarter-hour of station-keeping. The crew then prepared to carry out 20 experiments and 5 OAMS tests, including an OAMS burn to place Gemini 7 on the right orbit to serve as a rendezvous target for Gemini 6A. Rendezvous between the two manned spacecraft—the first in spaceflight history—took place on the eleventh day of the mission. (Vostok 3 and 4 passed within 6 km of each other in August 1962, but this was not considered a rendezvous, because the capsules were not maneuverable.) Gemini 7 was also the first American spaceflight during which the crew left off their pressurized suits for much of the time. A new spacesuit, the G5C, was introduced for the mission, which, although lighter, proved uncomfortable when worn for long periods. Both astronauts brought along books to read, heeding the advice given by Conrad after Gemini 5. Gemini 7 splashed down just 10 km off target, barely beating the record set by Gemini 6A two days earlier.

Gemini 8

Crew
   Command pilot: Neil **Armstrong**
   Pilot: David **Scott**
   Highlights: docking and emergency landing

The first successful dual launch and docking with a GATV, and the world's first on-orbit docking of two spacecraft. Originally scheduled for three days, the mission was aborted after just one. During the first six hours of the flight, Gemini 8 maneuvered nine times in order to approach the GATV to within about 45 m. Docking was achieved just over half an hour later; however, a dangerous malfunction soon followed. Just 27 minutes after docking, a short-circuit in the OAMS caused fuel to be lost through one of the capsule's thrusters. Although the astronauts did not know exactly what was wrong, the effects were almost immediate. Seconds after the mishap, the mated spacecraft began to spin rapidly. Armstrong was able temporarily to correct the problem by sequentially firing the OAMS thrusters, but the spin kept restarting. Mission controllers and the astronauts assumed the problem was with the GATV, so promptly undocked; however, Gemini 8 began to spin even more wildly—up to one revolution per second. The motion so disoriented the astronauts that their vision grew blurred and communications with the ground became difficult. Armstrong had to act fast before the astronauts lost consciousness or the spacecraft broke apart. In a rule-breaking move (which may have earned him the commander's spot on Apollo 11), he manually disabled the OAMS thrusters, activated the Reentry Control System (RCS) thrusters, and managed to steady the spacecraft's motion. But the process used up so much RCS fuel that the rest of the mission had to be called off. Gemini 8 returned safely to Earth in a designated emergency area of the Pacific—the only Pacific splashdown of the Gemini program—just 5 km from the recovery ship. Not surprisingly, Armstrong and Scott suffered severe space sickness, which continued after the capsule was recovered.

Gemini 9A

Crew
   Command pilot: Thomas **Stafford**
   Pilot: Eugene **Cernan**
   Highlights: rendezvous and 2-hour EVA

The prime crew for Gemini 9 had been Elliot M. See as command pilot and Charles A. Bassett as pilot. However, these two astronauts were killed in a T-38 jet training crash at St. Louis Municipal Airport, Missouri, on February 28, 1966. Ironically, their plane bounced off the building in which the Gemini 9 spacecraft was being prepared for flight, smashed into a parking lot, and exploded—minutes before Stafford and Cernan touched down at the same airport.

Gemini 9/"Angry Alligator" The Augmented Target Docking Adapter (ATDA) seen from the Gemini 9 spacecraft, 20 m away. Failure of the protective cover to fully separate prevented docking and led to the ATDA being described by the Gemini 9 crew as an "angry alligator." *NASA*

The original mission was to have involved docking with a GATV. However, when the GATV failed to reach orbit, NASA decided to launch an Augmented Target Docking Adapter (ATDA) instead and rename the mission Gemini 9A. The ATDA was launched on June 1, 1966. Two days later, Gemini 9A took off and rendezvoused with the ATDA on the third orbit—only to find that a docking would be impossible. The ATDA's shroud had only partially separated, prompting Stafford and Cernan to nickname their target "the Angry Alligator." Although docking was out of the question, the crew successfully completed several test maneuvers, including a rendezvous using optical equipment only and a rendezvous from above the ATDA rather than below as on previous flights. These were important steps in flight-testing proposed docking techniques between the Apollo Command and Lunar Modules. Later in the mission, Cernan went on a spacewalk—the second of the American space program—leaving the spacecraft manually without a maneuvering unit. He spent 1 hour 46 minutes outside the capsule, tethered by an umbilical line which he disparagingly referred to as "the snake." During the walk, Cernan was supposed to have carried out the first test of a

thruster-powered Astronaut Maneuvering Unit (AMU), stored in the Gemini 9A adapter section and accessible only from outside the spacecraft. The AMU, which had a self-contained life-support unit, would have allowed Cernan to propel himself up to 45 m away from the spacecraft and back. However, he had to work so hard to prepare and don the AMU that his helmet's faceplate fogged up, and, worse, as he struggled with the AMU, Cernan accidentally snagged an antenna on the capsule, which caused several tears in the outer layer of his spacesuit. The Sun beating down on these rips caused hot spots and, together with the other problems, led to the spacewalk being cut short and the AMU test abandoned. Gemini 9A easily took the record for the most accurate splashdown of the Gemini program—just 1.5 km from its prime target.

Gemini 10

Crew
   Command pilot: John **Young**
   Pilot: Michael **Collins**
   Highlights: docking, EVA, and two rendezvous

The second successful dual launch and docking with a GATV. The latter was launched about 100 minutes ahead of Gemini 10. Six hours later, the two vehicles rendezvoused and docked; however, these maneuvers used up more OAMS fuel than expected and forced a revision of the flight plan. Several orbital and docking training maneuvers were canceled, and Gemini 10 remained docked to the GATV for about 39 hours—longer than originally scheduled. During this time, two milestones were passed. At a mission-elapsed time of 23 hours 24 minutes, the hatch was opened and Collins stood up in his seat with his upper body extending outside—the first "standup spacewalk" in American space history. Collins was able to photograph stars in ultraviolet light, only possible outside Earth's atmosphere, and took 22 shots of the southern Milky Way using a 70-mm camera. Also, while the two spacecraft were mated, the GATV's thrusters were used to boost the orbit of the vehicles to a height of 765 km—a record for manned altitude (broken, though, on the next Gemini flight). The thrusters were then fired in a series of six maneuvers to place the mated spacecraft on a path to intercept the GATV used during the Gemini 8 mission. Gemini 10 undocked from its own GATV and about three hours later rendezvoused, but did not dock, with the Gemini 8 target. This was done solely through visual location, because the Gemini 8 GATV no longer had power to run its radar-locating devices. Two days into the mission, Collins went on a second spacewalk, this time tethered to the spacecraft by a life-support umbilical. Using his Personal Propulsion Unit, he jetted himself toward the Gemini 8 GATV, just three meters away, and removed two experiment packages, including a micrometeorite impact detector. While retrieving these packages, he accidentally set the GATV gyrating slightly, causing problems for Young as he wrestled to keep the two spacecraft close together. Using the Gemini 10 thrusters for this purpose caused the fuel to run low, and the spacewalk was cut short. During his EVA, Collins also let go of his camera, which drifted off into space to become another unplanned Earth satellite.

Gemini 11

Crew
Command pilot: Charles **Conrad** Jr.
Pilot: Richard **Gordon**
Highlights: docking, EVA, tether

A mission during which rendezvous and docking was achieved with a previously launched GATV in less than one orbit–a key goal because this procedure might become necessary during the upcoming Apollo program. Other highlights included two spacewalks by Gordon, the setting of a new manned spaceflight altitude record of 1,372 km, and the first fully automated reentry. One day into the mission, Gordon went on his first spacewalk, remaining tied to Gemini 11 by an umbilical while trying to attach a 15-m rope tether from the GATV to a docking bar on Gemini 11. Although the spacewalk had been scheduled to last 105 minutes, Gordon became exhausted while struggling to fix the tether, his helmet fogged up, and consequently he went back inside after just 21 minutes. About 40 hours into the mission, the GATV multiple-start engine thrusters were fired to lift the orbit of the mated spacecraft to an apogee (peak height) of 1,372 km for two orbits, setting a manned altitude record that was not broken until Apollo 8 headed for the Moon. The GATV thrusters were then fired to return the mated spacecraft to their normal orbital altitude. Shortly afterward, the hatch was opened for Gordon to begin a standup spacewalk, during which he took

photos and then managed to fall asleep! After the hatch was closed, Gemini 11 undocked from the GATV and maneuvered to allow it and the tethered GATV to slowly rotate around one another. Although movement of the tethered spacecraft was erratic at first, the motion stabilized after about 20 minutes. Then the rotation rate was increased, and again the motion stabilized. It was a challenge for the astronauts to keep the rope tether tight between the spacecraft– it remained stiff but moved somewhat like a jump rope as the spacecraft executed their pas de deux. This motion between the spacecraft caused artificial gravity to be created in space for the first time–even though it was only 0.0015 that on Earth! Gemini 11's return to Earth marked the first fully automatic splashdown of the American space program. After initial retrorocket firing, computers carried out all of the adjustments needed to bring the capsule down about 5 km from its target point.

Gemini 12

Crew
Command pilot: James **Lovell** Jr.
Pilot: Edwin **Aldrin** Jr.
Highlights: docking, 3 EVAs, tether

The final Gemini mission, during which the crew made the first observation of a solar eclipse from space. After Gemini 12 had docked with its GATV, Aldrin carried out the first of three spacewalks, attaching a 30-m rope from the GATV to a docking bar on the capsule–a tether twice as long as that used on Gemini 11. To combat some of the problems of fatigue experienced on previous spacewalks, a number of astronaut restraints had been fitted to Gemini 12's exterior. These helped Aldrin work at a more methodical, relaxed pace. His spacewalk was also the first to have been rehearsed underwater in NASA's **Neutral Buoyancy Tank**. Gemini 12 splashed down less than 5 km from its primary recovery target area.

Gemini Agena Target Vehicle (GATV)

A modified **Agena** stage launched by an Atlas, which served as a target for rendezvous and docking experiments during the Gemini program. The GATV was an Agena D with a Target Docking Adapter (TDA) on the front. At the start of the Gemini 9 mission, the GATV launch failed and the vehicle fell into the ocean. It was replaced by a backup called the Augmented Target Docking Adapter (ATDA), which was not a modified Agena but simply a TDA with a cylindrical back end to mate it directly to the Atlas. Unfortunately, the ATDA's shroud failed to separate properly, leaving it in "Angry Alligator" mode and unable to support the docking.

Genesis

A NASA mission that will collect 10 to 20 micrograms of particles from the **solar wind** using high purity wafers set in winglike arrays–the agency's first sample-return attempt since Apollo 17 in 1972. For two years, Genesis will orbit around the first **Lagrangian point** (L1) of the Earth-Sun system before returning to enable recovery of its 210-kg sample capsule in September 2004. As the capsule descends by parachute, it will be caught by a helicopter over the Utah desert. Scientists know that the Solar System evolved a little under five billion years ago from an interstellar cloud of gas, dust, and ice, but the exact composition of this cloud remains unknown. As its name implies, Genesis will help unravel this mystery by recovering material that has been shot out of the upper layers of the Sun–material that has not been modified by nuclear reactions in the Sun's core and is thus representative of the composition of the original presolar nebula. At its stable location at the L1 point, Genesis will be well outside Earth's magnetosphere, which deflects the solar wind away from the terrestrial environs.

Launch
    Date: August 8, 2001
    Vehicle: Delta 7326
    Site: Cape Canaveral

GEO

An acronym for either geostationary Earth orbit (see **geostationary orbit**) or geosynchronous Earth orbit (see **geosynchronous orbit**).

Geo-IK

A Soviet geodetic satellite system, also known as Musson ("monsoon"). The first Geo-IK was launched in 1981, to be followed by roughly one more per year until the mid-1990s. Normally one or two satellites were operational at any given time. Each 1,500-kg Geo-IK was launched by a Tsyklon-3 from Plesetsk and placed into an almost circular low Earth orbit with an average altitude of 1,500 km and an inclination of 73.6° or 82.6°. In contrast, the Etalon geodetic satellites were placed in much higher orbits and were completely passive.

geocentric

Relating to or measured from Earth's center.

geodesy

The science that treats the shape and size of Earth through applied mathematics.

geodesy satellites

Spacecraft that are used to measure the location of points on Earth's surface with great accuracy. Their observations help to determine the exact size and shape of Earth, act as references for mapping, and track movements of Earth's crust.

GEOS (Geodetic Earth Orbiting Satellite)

NASA spacecraft flown as part of the National Geodetic Satellite Program (NGSP). Instrumentation varied by mission, with the goals of pinpointing observation points (geodetic control stations) in a three-dimensional Earth center-of-mass coordinate system to within 10 m, determining the structure of Earth's gravity field to five parts in 10 million, defining the structure of Earth's irregular gravitational field and refining the locations and strengths of large gravity anomalies, and comparing the results of the various systems onboard the spacecraft to determine the most accurate and reliable system. GEOS 1 and 2 were part of the **Explorer** series and also designated Explorer 29 and 36. GEOS 3 was designed to be a stepping-stone between the NGSP and the Earth and Ocean Physics Application Program by providing data to refine the geodetic and geophysical results of the NGSP. It proved important in developing gravity models before the launch of **TOPEX/Posiedon** since it was located near that spacecraft's mirror inclination. (See table, "GEOS Missions.")

Geosat

A U.S. Navy satellite designed to measure sea surface heights to within 5 cm. After a year-and-a-half-long classified mission, Geosat's scientific Exact Repeat Mission (ERM) began on November 8, 1986. When the

GEOS Missions

Spacecraft	Launch			Orbit	Mass (kg)
	Date	Vehicle	Site		
GEOS 1	Nov. 6, 1965	Delta E	Cape Canaveral	1,114 × 2,273 km × 59.4°	175
GEOS 2	Jan. 11, 1968	Delta E	Vandenberg	1,079 × 1,572 km × 105.8°	209
GEOS 3	Apr. 9, 1975	Delta 2918	Vandenberg	816 × 850 km × 115.0°	341

ERM ended in January 1990 (due to failures of both onboard tape recorders), more than three years of precise altimetry data were available to the scientific community.

Launch
   Date: March 13, 1985
   Vehicle: Atlas E
   Site: Vandenberg Air Force Base
Orbit: 775 × 779 km × 108.1°
Mass: 635 kg

geospace

The domain of Sun-Earth interactions, also known as the solar-terrestrial environment. It consists of the particles, fields, and radiation environment from the Sun to Earth's space plasma environment and upper atmosphere. Geospace is considered to be the fourth physical geosphere, after solid earth, oceans, and atmosphere.

geostationary orbit (GSO)

A direct, circular **geosynchronous orbit** at an altitude of 35,784 km that lies in the plane of Earth's equator. A satellite in this orbit always appears at the same position in the sky, and its **ground-track** is a point. Such an arrangement is ideal for some **communication satellites** and **meteorological satellites**, since it allows one satellite to provide continuous coverage of a given area of Earth's surface.

geosynchronous orbit

A direct, circular, low-**inclination** orbit around Earth having a period of 23 hours 56 minutes 4 seconds and a corresponding altitude of 35,784 km (22,240 miles, or 6.6 Earth-radii). In such an orbit, a satellite maintains a position above Earth that has the same longitude. However, if the orbit's inclination is not exactly zero, the satellite's **ground-track** describes a figure eight. In most cases, the orbit is chosen to have a zero inclination, and **station-keeping** procedures are carried out so that the spacecraft hangs motionless with respect to a point on the planet below. In this case, the orbit is said to be a **geostationary orbit**.

geosynchronous/geostationary transfer orbit (GTO)

An elliptical orbit, with an **apogee** (high point) of 35,784 km, a **perigee** (low point) of a few hundred kilometers, and an **inclination** roughly equal to the latitude of the launch site, into which a spacecraft is initially placed before being transferred to a **geosynchronous orbit** (GSO). After attaining GTO, the spacecraft's **apogee kick motor** is fired to circularize the orbit and thereby achieve the desired final orbit. Typically, this burn will also reduce the orbital inclination to 0° so that the final orbit is not only geosynchronous but also geostationary. Because the greater the initial inclination, the greater the velocity change (delta v) needed to remove this inclination, it is important that launches of GSO satellites take place as close to the equator as possible. For example, in a Delta or an Atlas launch from Cape Canaveral, the transfer orbit is inclined at 28.5° and the required delta v increment at apogee is 1,831 m/s; for an Ariane launch from Guiana Space Centre, the inclination is 7° and the delta v is 1,502 m/s; while for a Zenit flight from the **Sea Launch** platform on the equator, the delta v is 1,478 m/s. By the **rocket equation**, assuming a (typical) **specific impulse** of 300 seconds, the fraction of the separated mass consumed by the propellant for the apogee maneuver is 46% from Cape Canaveral, 40% from Kourou, and 39% from the equator. As a rough guide, the mass of a geostationary satellite at the start of its operational life (in GSO) is about half its initial on-orbit mass when separated from the launch vehicle (in GTO). Before carrying out the apogee maneuver, the spacecraft must be reoriented in the transfer orbit to face in the proper direction for the thrust. This reorientation is sometimes done by the launch vehicle at spacecraft separation; otherwise, it must be carried out in a separate maneuver by the spacecraft itself. In a launch from Cape Canaveral, the angle through which the satellite must be reoriented is about 132°.

Geotail

A joint **ISAS** (Japanese Institute of Space and Astronautical Science) and NASA mission to investigate Earth's magnetosphere and geomagnetic tail over distances of 8 to 200 Earth-radii.

Launch
>   Date: July 24, 1992
>   Vehicle: Delta 6925
>   Site: Cape Canaveral
>   Orbit: 41,363 × 508,542 km × 22.4°
>   Mass: 1,008 kg

German Rocket Society

See **Verein für Raumschiffahrt.**

Germany in space

See **"A" series of German rockets, AEROS, Albertus Magnus, AMPTE, Azur, BIRD, CHAMP, DARA,** Kurt **Debus, DLR,** Walter **Dornberger,** Krafft **Ehricke, ELDO, Equator-S, ESA, ESOC,** Hermann **Ganswindt, GFZ-1, Helios (solar probes),** Walter **Hohmann,** Johannes **Kepler, Kummersdorf,** and Hermann **Kurzweg.**[102]

Get-Away Special

A small, self-contained experiment carried in the payload bay of the Space Shuttle.

GFO (Geosat Follow-On)

A successor to the **Geosat** program, which flew between 1985 and 1990. GFO provides real-time ocean topography data to 65 U.S. Navy users at sea and onshore. This data on wave heights, currents, and fronts is also archived and made available to scientific and commercial users through NOAA (National Oceanic and Atmospheric Administration). A 13.5-GHz radar altimeter is the primary payload, providing wave-height measurements to an accuracy of 3.5 cm.

Launch
>   Date: February 10, 1998
>   Vehicle: Taurus
>   Site: Vandenberg Air Force Base
>   Orbit: 785 × 788 km × 8.1°
>   Length: 3.0 m
>   Mass: 47 kg (payload)

g-force

The force to which a vehicle or its crew is subjected due to the vehicle's acceleration or deceleration, measured in multiples of normal Earth gravity (*g*).

GFZ-1

A passive geodetic satellite consisting of a 21-cm-sphere covered by 60 retroreflectors; it was designed to study changes in Earth's rotational characteristics and variations in its gravitational field. GFZ-1 was the lowest-flying geodynamic satellite to be ranged by lasers and led to a significant improvement in modeling of Earth's gravity. As it decayed, its orbital motion was also used to calculate atmospheric densities. GFZ-1 was the first satellite designed and funded by the GeoForschungsZentrum of Potsdam, Germany. Built and launched by Russia, it was transported to Mir aboard a Progress supply craft and from there placed into low Earth orbit.

Launch date: April 19, 1995
Orbit: 382 × 395 km × 51.6°
Mass: 20 kg

Gibson, Edward G. (1936–)

An American astronaut who set a joint U.S. space endurance record of 84 days as science pilot on **Skylab** 4. During his 14-year career with NASA, Gibson also served on the support crew of the **Apollo** 12 mission and as the ground communicator with the flight crew while they explored the Moon. He earned Air Force wings, a B.S. in engineering from the University of Rochester, and an M.S. and a Ph.D. in engineering and physics from the California Institute of Technology. After leaving NASA, he entered the private sector and, in 1990, formed his own consulting company, Gibson International Corporation.

GIFTS (Geostationary Imaging Fourier Transform Spectrometer)

An experimental remote-sensing satellite that will use an advanced instrument called an Imaging Fourier Transform Spectrometer to observe atmospheric temperature, water vapor content and distribution, and the concentration of certain other atmospheric gases present at a given altitude over time. By noting changes in temperature, water vapor, and gases that modulate energy in different regions of the electromagnetic spectrum, scientists will get an accurate picture of what is happening in the atmosphere and what weather is likely to develop. Part of NASA's New Millennium Program, GIFTS is mainly intended to space-validate cutting-edge technologies for future use. It is scheduled for launch in 2003.

Gilruth, Robert R. (1913–)

An influential NACA (National Advisory Committee for Aeronautics) engineer who worked at the **Langley** Aeronautical Laboratory (1937–1946), then as chief of the pilotless aircraft research division at **Wallops Island** (1946–1952), and explored the possibility of human spaceflight before the creation of NASA. He served as assistant director at Langley from 1952 to 1959 and as assistant director (manned satellites) and head of the Mercury Project from 1959 to 1961, technically assigned to the Goddard Space Flight Center but physically located at Langley. In early 1961 T. Keith **Glennan** established an independent Space Task Group (already the group's name as an independent subdivision of the Goddard center) under Gilruth at Langley to supervise the Mercury Project. This group moved to the Manned Spacecraft Center, Houston, Texas, in 1962. Gilruth then served as director of the Houston operation from 1962 to 1972.[144]

gimbal

(1) A device with two mutually perpendicular and intersecting axes of rotation on which an engine or other object may be mounted. (2) In a gyro, a support that provides the spin axis with a degree of freedom. (3) The rotation of a rocket's motor nozzle to control the direction of thrust and hence help steer a spacecraft; the nozzles of the Space Shuttle's Solid Rocket Boosters, for example, can gimbal up to 6°.

Ginga

Japanese X-ray satellite, launched by **ISAS** (Institute of Space and Astronautical Science), that measured the time variability of X-rays from active galaxies and quasars in the energy range of 1 to 500 keV. Ginga, whose name means "galaxy," was known before launch as Astro-C. It stopped working in November 1991.

Launch
    Date: February 5, 1987
    Vehicle: M-5
    Site: Kagoshima
    Orbit: 395 × 450 km × 31°
Mass: 420 kg

Giotto

An ESA (European Space Agency) probe—the agency's first interplanetary probe—that encountered Halley's Comet on March 13, 1986; it was named after Giotto di Bondone (1266–1337), who depicted the comet in one of his paintings. Giotto measured Halley's composition and, on March 16, 1986, returned color images of the comet's nucleus. Fourteen seconds before its closest approach to the nucleus (596 km), the spacecraft was struck by a dust particle that knocked its communications antenna out of alignment with Earth. Further impacts

also rendered the probe's camera and some other instruments nonfunctional. However, communications were reestablished after the encounter and Giotto subsequently flew by Comet 26P/Grigg-Skjellerup on July 10, 1992, sending back valuable data. The spacecraft passed close by Earth on July 1, 1999.

Launch
    Date: July 2, 1985
    Vehicle: Ariane 1
    Site: Kourou
Mass: 583 kg

GIRD (Gruppa Isutcheniya Reaktivnovo Dvisheniya)

Group for the Investigation of Reactive Motion. A Moscow-based society, formed by Sergei **Korolev** and Fridrikh **Tsander**, which in 1933 launched the Soviet Union's first liquid-propellant rocket.

GLAST (Gamma-ray Large Area Space Telescope)

A high-energy astrophysics orbiting observatory designed to map the gamma-ray universe with up to 100 times the sensitivity, resolution, and coverage of previous missions.

Giotto Giotto being prepared for tests prior to launch. *European Space Agency*

John Glenn Glenn dons his silver Mercury pressure suit in preparation for launch aboard *Friendship 7* on February 20, 1962. *NASA*

It will carry two instruments: the Large Area Telescope, sensitive to gamma rays in the 200 MeV to 300 GeV range, and the Burst Monitor, for detecting gamma-ray bursts. Following a one-year sky survey, GLAST will be available to scientists for particular research projects. A joint mission involving NASA, the U.S. Department of Energy, and institutions in France, Germany, Sweden, Italy, and Japan, GLAST is scheduled for launch in September 2006.

Glavcosmos

The Soviet Union's contact agency for space affairs, based at the Ministry of Machine Building in Moscow. Following the collapse of the Soviet Union in 1991, the interface between Russia's space activities and the outside world became the newly formed **Russian Space Agency** (RKA).

Glenn, John Herschel, Jr. (1921–)

One of the original **Mercury Seven** astronauts and the first American to orbit the Earth; decades later, he became the oldest human ever to travel in space. Glenn enlisted in the Naval Aviation Cadet Program shortly after Pearl Harbor and was commissioned in the Marine Corps in 1943. He subsequently served in combat in the South Pacific and in the Korean conflict, flying 149 missions in the two wars, and received many honors including the Distinguished Flying Cross (on six occasions) and the Air Medal with 18 clusters. For several years Glenn was a test pilot on Navy and Marine Corps jet fighters and attack aircraft, setting a transcontinental speed record in 1957 for the first flight to average supersonic speeds from Los Angeles to New York. In 1959 he was selected to be one of the first seven astronauts in the American space program. Three years later, on February 20, 1962, he made history aboard *Friendship 7* as the first American to orbit the Earth, completing three orbits in a five-hour flight. He retired from NASA in 1964 and from the Marine Corps in 1965, entering politics and, in 1974, winning a seat in the Senate as a representative of Ohio. To this position he was elected in 1992 for a record consecutive fourth term. Glenn made history again when, at the age of 77, he flew as a payload specialist on the crew of the Space Shuttle mission STS-95. NASA's John H. **Glenn Research Center** at Lewis Field has been named in his honor.[41]

Glenn Research Center (GRC)

NASA's leading center for research and development of aerospace propulsion systems in all flight regimes from subsonic to hypersonic. GRC also carries out research in fluid physics, combustion science, and some materials science, especially with regard to microgravity applications. Many Space Shuttle and International Space Station (ISS) science missions have an experiment managed by Glenn, and the Center has designed power and propulsion systems for spaceflight in support of the ISS, Mars Pathfinder, and Deep Space 1. In addition, Glenn leads NASA's Space Communications Program, including the operation of the ACTS (Advanced Communications Technology Satellite). Established in 1941 by NACA (the National Advisory Committee for Aeronautics), the Center was named after NACA research director George W. Lewis shortly after his death in 1948. Originally known as the Lewis Flight Propulsion Laboratory, the facility was renamed the Lewis Research Center upon becoming part of NASA when the agency was founded in 1958. It was renamed again the John H. Glenn Research Center at Lewis Field in March 1999. Said NASA admin-

Glenn Research Center   The main gate and hangar at Glenn Research Center. *NASA*

istrator Dan Goldin at the time: "I cannot think of a better way to pay tribute to two of Ohio's famous sons—one an aeronautic researcher and the other an astronaut legend and lawmaker—than by naming a NASA research center after them."[73]

Glennan, T(homas) Keith (1905–1995)

The first administrator of **NASA**, serving from 1958 to 1961. Glennan earned a B.S. in electrical engineering from the Sheffield Scientific School of Yale University in 1927 before becoming involved with the newly developed sound motion-picture industry at Paramount Pictures and Samuel Goldwyn Studios. In 1942, he joined the Columbia University Division of War Research, serving first as administrator and then as director of the U.S. Navy's Underwater Sound Laboratories. Shortly after the end of the war, he was appointed president of the Case Institute of Technology, in Cleveland, Ohio, and from October 1950 to November 1952, concurrent with his Case presidency, was a member of the Atomic Energy Commission. During his period as NASA administrator, Glennan presided over the unification of the nation's civilian space projects and the addition to the agency's organization of the **Goddard Space Flight Center**, the **Marshall Space Flight Center**, and JPL (Jet Propulsion Laboratory). Upon leaving NASA in January 1961, he returned to the Case Institute of Technology, where he continued to serve as president until 1966. During this period he helped to negotiate the merger of Case with Western Reserve University, creating Case Western Reserve University. After his retirement in 1966, Glennan spent two years as president of Associated Universities, a Washington-based advocate for institutions of higher learning.[153]

glide

The controlled descent of an air vehicle using control surfaces, not rockets, to maintain aerodynamic stability. The descent is a result of gravity and lifting forces generated by the shape of the air vehicle.

glide angle

The angle, or slope, of glide during descent. The glide angle for the Space Shuttle during the final stage of its descent from orbit is 22°.

glide path

The descent of a gliding air vehicle as viewed from the side.

Globalstar

A consortium of international telecommunications companies formed in 1991 to provide a worldwide satellite-based telephony service. Together with **Iridium, ICO**, and others, Globalstar is one of the new comsat companies supporting global mobile voice and data communications with a constellation of satellites in low Earth orbit. The Globalstar constellation, launched in 1998 and 1999, consists of 48 satellites (with four in-orbit spares) in 1,400-km-high orbits inclined at 52°, allowing a concentration of coverage over the temperate regions of Earth from 70° S to 70° N. The satellites operate in the L-band (see **frequency bands**). Space Systems Loral is the prime contractor for the constellation; in parts of the globe lacking a communications infrastructure, the handheld Globalstar unit, supplied by Qualcomm, sends a signal to a satellite overhead. The satellite then relays this signal to a regional ground station where the signal travels through a terrestrial network to its final destination.

GLOMR (Global Low Orbiting Message Relay)

DARPA (Defense Advanced Research Projects Agency) satellites designed to demonstrate the ability to read out, store, and forward data from remote ground-based sensors. GLOMR was first scheduled for deployment from a modified **Get-Away Special** container on Space Shuttle mission STS-51B, but failed to eject because of a battery problem. It was reflown and deployed successfully from STS-61A and reentered after 14 months. GLOMR-II, also known as USA 55 and SECS (Special Experimental Communications System), was roughly the size of a basketball and had greater data storage, more redundancy, and more space-qualified hardware than its predecessor. (See table, "GLOMR Missions.")

GLONASS (Global Navigation Satellite System)

The Russian counterpart to **GPS** (Global Positioning System); both systems use the same techniques for data transmission and positioning. GLONASS is based on a constellation of active satellites that continuously transmit coded signals in two frequency bands. These signals can be

GLOMR Missions

| Spacecraft | Launch | | | Orbit | Mass (kg) |
	Date	Vehicle	Site		
GLOMR	Nov. 1, 1985	STS-61	Cape Canaveral	304 × 332 km × 57°	52
GLOMR II	Apr. 5, 1990	Pegasus	Edwards AFB	489 × 668 km × 94.1°	68

received by users anywhere on Earth to identify their position and velocity in real time based on ranging measurements. The first GLONASS satellites were placed in orbit in 1982. Each satellite is identified by its "slot" number, which defines the orbital plane (1–8, 9–16, 17–24) and the location within the plane. The three orbital planes are separated by 120°, and the satellites within the same orbital plane by 45°. The GLONASS orbits are roughly circular, with an inclination of about 64.8° and a semimajor axis of 25,440 km. The full constellation is supposed to contain 24 satellites, arranged in three orbital planes; however, Russia's economic troubles prevented replacement satellites from being launched as needed. As of mid-2002, only 9 GLONASS satellites were operational.

Glushko, Valentin Petrovitch (1908–1989)

A Soviet rocket scientist, a pioneer in rocket propulsion systems, and a major contributor to Soviet space and defense technology. He was born in Odessa, Ukraine. After graduating from Leningrad State University (1929), Glushko headed the design bureau of the Gas Dynamics Laboratory in Leningrad and began research on electrothermal, solid-fuel, and liquid-fuel rocket engines. In 1935, he published *Rockets, Their Construction and Utilization*. From 1932 to 1966, Glushko worked closely with renowned rocket designer Sergei **Korolev**. The two achieved their greatest triumphs in 1957 with the launching of the first intercontinental ballistic missile in August and the first successful artificial satellite, **Sputnik** 1, in October. In 1974 Glushko was named chief designer of the Soviet space program, in which he oversaw the development of the **Mir** space station. He received numerous official honors, including the Lenin Prize (1957) and election to the Soviet Academy of Sciences (1958).

GMS (Geostationary Meteorological Satellite)

A series of weather satellites managed by the Japanese Meteorological Agency and **NASDA** (National Space Development Agency); their indigenous name, Himawari, means "sunflower." All have been located in geostationary orbit at 140° E. The first GMS was launched in 1977 and the most recent, GMS-5, on March 18, 1995. GMS-5 is equipped with a VISSR (Visible and Infrared Spin Scan Radiometer), which scans Earth's surface line by line, each line consisting of a series of pixels. For each pixel the radiometer measures the radiative energy at three different wavelength bands—one in the visible region and two in the infrared.

GMT (Greenwich Mean Time)

A worldwide time standard; the time as measured on the Greenwich meridian. Also known as Universal Time.

go, no-go

A term used to indicate a condition of a part, component, system, and so forth, which can have only two states. "Go" indicates "functioning properly"; "no-go" means "not functioning properly."

Goddard Institute for Space Studies

An arm of NASA's **Goddard Space Flight Center** that conducts theoretical and experimental research in global change. It is located next to Columbia University's Morningside Campus in New York City. Work at the Institute focuses on humanity's large-scale impact on the environment and fosters interaction among scientists in atmospheric, geological, and biological sciences. It includes long-range climate modeling, and analysis of biogeochemical cycles, Earth observations, and planetary atmospheres.

Goddard, Robert Hutchings

See article, pages 166–167.

Goddard Space Flight Center (GSFC)

NASA's foremost field center for programs related to space science and observation from space of the Earth and its environment. Founded in 1959, it was named after the rocket pioneer Robert **Goddard** and is located in Greenbelt, Maryland, 16 km northeast of Washington, D.C. Goddard Space Flight Center has designed, built, and operated numerous near-Earth orbiting spacecraft, including members of the **Explorer** series, and controls several astronomy satellites, including the **Hubble Space Telescope**. It manages the National Space Science Data Center and is responsible for the **Wallops Flight Facility**.

Godwin, Francis (1562–1633)

Bishop of Hereford and author of what is arguably the first science fiction story to be written in the English language. In *The Man in the Moone*,[131] published posthumously in 1638, Godwin conveys the astronaut Domingo Gonsales to the Moon in a chariot towed by trained geese. (Gonsales had intended a less ambitious flight but discovered that the geese are in the habit of migrating a little further than ornithologists had supposed!) In keeping with both popular and scientific opinion of his day, Godwin accepted the notion that air filled the space between worlds and that the Moon was inhabited by intelligent human beings. See **Wilkins, John** and **Cyrano de Bergerac, Savinien de**.

GOES (Geostationary Operational Environmental Satellite)

A series of **weather satellites** that superseded the **SMS** (Synchronous Meteorological Satellite) program and,

(continued on page 167)

Robert Hutchings Goddard (1882–1945)

An American physicist and a rocket pioneer. Although historians point to Konstantin **Tsiolkovsky** and Hermann **Oberth** as being the founders of rocket *theory*, Goddard, a native of Worcester, Massachusetts, is generally regarded as the father of the practical modern rocket. Yet it was only after his death that the true value of his work was widely recognized.

Goddard's early attraction to rocketry is clear from an autobiography written in 1927 but not published until 1959, in which he recalls as a 17-year-old climbing into a cherry tree to prune branches and finding himself instead daydreaming of interplanetary travel:

> It was one of the quiet, colorful afternoons of sheer beauty which we have in October in New England, and as I looked toward the fields at the east, I imagined how wonderful it would be to make some device which had even the *possibility* of ascending to Mars, and how it would look on a small scale, if sent up from the meadow at my feet.

From this point on, Goddard's ambition, to develop the practical means of achieving spaceflight, began to take shape. He earned his Ph.D. in physics at Clark University, Worcester, Massachusetts, in 1911, and went on to become head of the Clark physics department and director of its physical laboratories. His first serious work on rocket development began in 1909, although it was not until 1915 that he carried out his first actual experiments, involving solid-fueled rockets, having already been granted patents covering such key components as combustion chambers, nozzles, propellant feed systems, and multistage launchers. In 1916, the Smithsonian Institution awarded him $5,000 to perform high-altitude tests. But a year later, following the United States' entry into World War I, he found himself temporarily in California, working on rockets as weapons, including the forerunner of the bazooka.

In 1919 Goddard published the first of his two important monologues on rocketry, *A Method of Attaining Extreme Altitude*. Based on the report that had earned him the Smithsonian grant, it was written in typically cautious Goddard style. It would probably have attracted little attention but for its final section,

"Calculation of Minimum Mass Required to Raise One Pound to an 'Infinite' Altitude." Despite Goddard's sober analysis of the problems involved in sending a payload from Earth to the Moon and of proving that the target could be reached (by an explosion of flash powder), he was lampooned by the popular press. However, he continued with his research and moved into the field of liquid-fueled rockets. On March 16, 1926, Goddard carried out the world's first launch of such a system, using a strange-looking, 3-m-long rocket powered by a mixture of liquid oxygen and gasoline ignited by a blowtorch. The device rose for two-and-a-half seconds, reaching a height of 67 m and a maximum speed of 96 km/hr before its fuel was exhausted. His work eventually drew the attention of Charles **Lindbergh**, who in 1929 arranged for a $50,000 grant to the rocket pioneer from the Guggenheim Fund for the Promotion of Aeronautics. Goddard's increasingly ambitious tests demanded more

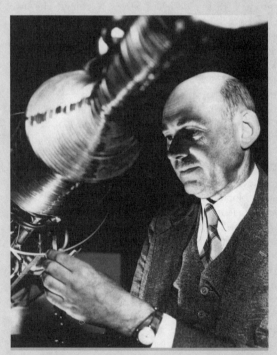

Robert Goddard Goddard examining one of his larger liquid-fueled engines. *NASA*

open space than rural Massachusetts could safely provide (the local fire chief having already banned him from raining down any further missiles on his neighbors' land), so Goddard took his experiments to the more open skies near Roswell, New Mexico. There, he and his wife, together with a few assistants, conducted a remarkable, decade-long program of tests that resulted in the first gyro-controlled rocket guidance system and flights of large, variable-thrust, liquid-fueled rockets to heights of up to 2,300 m and speeds of over 800 km/hr. Some of his results are summarized in his second classic monologue, submitted to the Guggenheim Foundation in 1936, *Liquid-Propellant Rocket Development*.[115] His final major work,

Rocket Development: Liquid-Fuel Rocket Research, 1929–1941,[116] was published posthumously in 1948.

Goddard died in 1945, enthusiastic to the end about the prospects of rocket-propelled spaceflight but disappointed by the failure of officialdom to back his work to a substantial degree—a failure due in part to the secrecy in which Goddard kept many of the technical details of his inventions. Only after his death were his contributions properly recognized. In 1959, one of NASA's major facilities, the **Goddard Space Flight Center**, was named in his honor. The following year, his widow, together with the Guggenheim Foundation, was awarded $1 million in settlement for government use of 214 of the pioneer's patents.[187]

GOES (Geostationary Operational Environmental Satellite)

(continued from page 165)

since the late 1970s, has formed the backbone of short-term weather forecasting in the United States. The real-time data gathered by GOES satellites, combined with measurements from Doppler radars and automated surface observing systems, are disseminated by NOAA (National Oceanic and Atmospheric Administration) and help weather forecasters provide warnings of thunderstorms, winter storms, flash floods, hurricanes, and other severe weather. Two GOES satellites must be active and correctly located in geosynchronous orbit (at 75° W and 135° W) to be able to monitor continuously Earth's full disk about the meridian approximately in the center of the continental United States.

Like their predecessors, the latest in the series, GOES-M (launched in 2001) and GOES-N (due to be launched in early 2004), are equipped with flexible-scan imaging systems and sounders, allowing scientists to collect information about cloud cover, wind speed, temperature, moisture, and ozone levels. But they also carry solar X-ray imagers to provide early detection of especially violent solar activity. Such monitoring systems will give scientists advance warning of possible disruptions in the upper atmosphere—disruptions that in the past have led to spectacular auroras as well as devastating power outages.

Goldin, Daniel Saul (1940–)

NASA's longest-serving administrator. Appointed in 1992, Goldin pursued a (not always successful) policy of "faster, better, cheaper" with respect to robotic missions, reformed the agency's management, and implemented a more balanced aeronautics and space program by reducing the funding of human spaceflight from 48% of the total budget to 38% and increasing funding for science and aero-

space technology from 31% to 43%. Other innovations under Goldin's leadership included new programs of exploration focused on Mars and the Moon, and the Origins program to understand how the universe has evolved and how common life is on other worlds. Goldin played a major role in having the International Space Station redesigned and in bringing Russia into the project. Starting with the Space Shuttle program, he oversaw the transfer of day-to-day space operations to the private sector. Before coming to NASA, Goldin was vice president and general manager of the TRW Space and Technology Group in Redondo Beach, California. He began his career in 1962 at **Lewis Research Center**, where he worked on electric propulsion systems. Goldin was succeeded as NASA administrator in December 2001 by Sean **O'Keefe**.

Goldstone Tracking Facility

A 70-m radio dish located at Camp Irwin, Barstow, California, that in 1966 became the first antenna of the **Deep Space Network**. It is also used for very-long-baseline interferometry in conjunction with other radio telescopes.

GOMS (Geostationary Operational Meteorological Satellite)

A Soviet weather satellite placed in geostationary orbit over the Indian Ocean; also known as Elektro. Originally proposed for a maiden flight in 1978–1979, GOMS has suffered both technical and budgetary problems.

Launch
    Date: October 31, 1994
    Vehicle: Proton
    Site: Baikonur
Orbit: geostationary at 76° E
Mass: 2,400 kg

Gonets

A Russian messaging constellation—"gonets" means "messenger"—with dual military and commercial uses. The current system comprises 18 satellites in two orbital planes, and the system is planned to grow to 36 satellites in six orbital planes once the market is established. One set of 8 satellites entered the incorrect orbit, and its plane will drift with respect to the others. The orbits are 1,400 km high inclined at 82.5°, with individual cell sizes 500 to 800 km in diameter. Only some of the satellites carry digital store and forward payloads. The system is being run by Smolsat, and the satellites will operate as medical and civil data relays.

Gordon, Richard Francis, Jr. (1929–)

An American astronaut who walked in space during the **Gemini** 11 mission and orbited the Moon on **Apollo** 12. Gordon received a B.S. from the University of Washington in 1951 before entering the U.S. Navy and serving as a flight test pilot. In 1960, he joined Fighter Squadron 121 at the Miramar, California, Naval Air Station as a flight instructor and won the Bendix Trophy Race (an annual air race across America, sponsored by the Bendix Aviation Corporation, held from 1931 to 1962) from Los Angeles to New York in May 1961, setting a new speed record of 1,399 km/hr and a transcontinental record of 2 hours 47 minutes. He was selected as an astronaut in 1963 and made his first spaceflight as pilot alongside Charles **Conrad** on the three-day Gemini 11 mission in 1966. Gordon and Conrad served together again in 1969 aboard Apollo 12, with Gordon as Command Module pilot. In 1971, Gordon became chief of advanced programs for the Astronaut Office and worked on the design and testing of the Space Shuttle and development equipment. A year later he retired from NASA and the Navy to become executive vice president of the New Orleans Saints football team. He subsequently held management positions with various energy and aerospace companies.

Gorizont

General-purpose Russian communications satellites owned by Informcosmos and operated by several firms, including Intersputnik; "gorizont" is Russian for "horizon." Thirty Gorizonts have been successfully deployed, the last of them on June 5, 2000; three others were lost in launch mishaps.

GPS (Global Positioning System)

Also known as Navstar-GPS (Navigation Satellite Time and Ranging–GPS), a global system of 24 Department of Defense navigation satellites (21 operational plus 3 spares), completed in 1993 and designed to provide time, position, and velocity data for ships, planes, and land-based vehicles, and for many other purposes. The satellites are arranged in six planes, each in a 12-hour, 20,000-km-high orbit. GPS satellites transmit signals that allow the determination, to great accuracy, of the locations of GPS receivers. These receivers can be fixed on Earth, or in moving vehicles, aircraft, or in satellites in low Earth orbit. GPS is used in navigation, mapping, surveying, and other applications where precise positioning is necessary. Each satellite broadcasts two **L-band** (see **frequency bands**) radio signals containing ranging codes, ephemeris parameters, and Coordinated Universal Time (UTC) synchronization information. Both military and civilian users can use GPS receivers to receive, decode, and process the signals to gain 3-D position, velocity, and time information. Civilian GPS receivers are somewhat less accurate than their military counterparts, owing to their inability to read the coded portions of the satellite transmissions.

GRAB

The first series of American **ELINT** (electronic intelligence) satellites, operated by the U.S. Navy between July 1960 and August 1962. Documents declassified only in 1998 reveal that the project was originally called Tattletale, then renamed GRAB (an acronym of the spurious name "Galactic Radiation and Background"), which was later revised to GREB (Galactic Radiation Experimental Background). Each GRAB carried two payloads—the classified ELINT satellite itself and an unclassified satellite designed to measure solar radiation. The latter, known as **Solrad**, was publicly disclosed by the Department of Defense at the time and used as a cover for the intelligence mission. However, the Solrad experiments were not merely for show: by revealing the effect of solar radiation on the ionosphere and hence on high-frequency radio communications, they supplied data of military value.

Work on GRAB began in 1958, around the time of the first successful Vanguard launch. Reid Meyo of the Naval Research Laboratory (NRL) Countermeasures Branch had developed an electronic intelligence antenna for submarine periscopes. NRL was seeking quick military exploitation of the Vanguard satellite it had developed, and it occurred to Reid that the NRL could simply put his periscope antenna into orbit aboard a Vanguard. The original calculations behind this idea were done, in the best tradition of aerospace engineering, on a restaurant placemat.

From 800 km above the Earth, a GRAB satellite's circular orbit passed it through the beams from Soviet radar, whose pulses traveled beyond the horizon into space. GRAB's task was to receive each radar pulse in a certain bandwidth and to transpond a corresponding signal to collection huts at ground sites within GRAB's field of view. Operators in the huts recorded the transponded data and couriered it to NRL for evaluation.

GRACE (Gravity Recovery and Climate Experiment)

Two identical spacecraft that over a five-year period will measure and map variations in Earth's gravitational field and track ocean circulation and the movement of ocean heat to the poles. Selected as the second mission in NASA's **ESSP** (Earth System Science Pathfinder) program, the dual satellites of GRACE fly in the same orbital plane, one 220 km behind the other, at an altitude of about 550 km. However, the distance between them changes as they are influenced differently by subtle gravitational changes over such terrestrial landforms as ice sheets and mountain ranges. By continuously measuring the distance between the two satellites, scientists will be able to chart Earth's gravitational field to a new degree of accuracy. In addition, GRACE will yield novel information about Earth's atmospheric profile and ocean circulation. The mission is a joint partnership between NASA and the German center for aerospace research, DLR.

Launch
Date: March 17, 2002
Vehicle: Rockot
Site: Plesetsk
Orbit: 472 × 493 km × 89°
Mass: 432 kg

grain

An extruded length into which solid propellants are formed.

Granat

An orbiting Russian X-ray and gamma-ray observatory for studying the sky in the energy range of 2 keV to 100 MeV; "granat" is Russian for "garnet." After an initial period of pointed observations, Granat went into survey mode in September 1994 and continued operating until November 1998. The mission was carried out in collaboration with scientists from France, Denmark, and Bulgaria.

Launch
Date: December 1, 1989
Vehicle: Proton
Site: Baikonur
Orbit: 53,697 × 149,862 km × 86.6°
Mass: 3,200 kg

Grand Tour

A proposal, first put forward in the late 1960s, for a single spacecraft to fly by all four major outer planets—Jupiter, Saturn, Uranus, and Neptune—by taking advantage of a once-in-179-year alignment of these worlds and the multiple **gravity-assist**s this celestial configuration allowed. The Grand Tour, as originally planned, was a budget casualty of 1970. However, shortly after the cancellation, JPL proposed a more modest mission, tentatively called Mariner Jupiter/Saturn. The two-spacecraft project thus developed was eventually approved and renamed **Voyager**.

grapple

A fixture attached to a satellite or payload so that the end effector (the "hand" part) of the Space Shuttle Orbiter's **Remote Manipulator System** can grasp and maneuver it.

graveyard orbit

An orbit in which a satellite, particularly a geosynchronous satellite, may be placed at the end of its operational life and from which it will eventually reenter Earth's atmosphere and burn up. In a graveyard orbit, a spacecraft is not in danger of accidentally colliding with an active satellite.

gravipause

The point or boundary in space at which the gravitational force of one body is exactly balanced by the gravity pull of another. Also known as the neutral point.

gravisphere

The region within which the force of a given celestial body's gravity predominates over that of other celestial bodies.

Gravitational Biology Facility

A suite of habitats aboard the **International Space Station** to support organisms for research in cell, developmental, and plant biology. The habitats will provide food, water, light, air, humidity control, temperature control, and waste management for the organisms. The Gravitational Biology Facility includes a Cell Culture Unit, an Aquatic Habitat, an Advanced Animal Habitat, a Plant Research Unit, an Insect Habitat, and an Egg Incubator. The Cell Culture Unit will be used for research in cell and tissue biology, and it will be able to maintain and monitor microbial, animal, aquatic, and plant cell and tissue cultures for up to 30 days. The Aquatic Habitat will accommodate small freshwater organisms, such as Zebrafish, for up to 90 days to support egg-to-egg generation studies for examination of all life stages. The

Advanced Animal Habitat will house up to six rats or a dozen mice and be compatible with a compartment called the Mouse Development Insert that will accommodate pregnant mice and subsequently their offspring from birth through weaning. The Plant Research Unit will support plant specimens up to 38 cm in height (root + shoot) through all stages of growth and development. The Insect Habitat will house *Drosophila melanagoster* and other insects for multigenerational studies and for radiation biology, and the Egg Incubator will support the incubation and development of small reptilian and avian eggs prior to hatching. All of the Gravitational Biology Facility habitats will have the experimental capability of selectable gravity levels. The Aquatic Habitat, the Cell Culture Unit, the Advanced Animal Habitat, and the Plant Research Unit will be used on the Centrifuge Facility's 2.5-m-diameter centrifuge when selectable gravity levels of up to 2*g* are needed. The Insect Habitat and Egg Incubator will be equipped with internal centrifuges, which will provide selectable gravity levels from zero to 1.5*g*.

gravitational constant (G)

The constant that appears in **Newton's law of gravitation**. It is the attraction between two bodies of unit mass separated by a unit distance, and has the value 6.672×10^{-11} N.m^2/kg^2.

gravitational energy

The energy released by an object falling in a **gravitational field**; a form of potential energy.

gravitational field

The region of space around a body in which that body's gravitational force can be felt.

gravity

The universal force by which every piece of matter in space attracts every other piece of matter. Its effects become obvious only when large masses are involved. So, for instance, although we feel the strong downward pull of Earth's gravity, we feel no (gravitational!) pull at all toward smaller masses such as coffee tables, vending machines, or other people. Yet a mutual attraction does exist between all things that have mass—you pull on Earth, just as Earth pulls on you. Until the beginning of the twentieth century, the only universal law of gravitation was that of Isaac **Newton**, in which gravity was regarded as an invisible force that could act across empty space. The force of gravity between two objects was proportional to the mass of each and inversely proportional to the square of their separation distance. Then, in 1913, Einstein published a revolutionary new theory of gravita-

tion known as the **general theory of relativity**, in which gravity emerges as a consequence of the geometry of **space-time**. In the "rubber sheet" analogy of space-time, masses such as stars and planets can be thought of as lying at the bottom of depressions of their own making. These gravitational wells are the space-time craters into which any objects coming too close may fall (for example, matter plunging into a **black hole**) or out of which an object must climb if it is to escape (for example, a spacecraft leaving Earth for interplanetary space).[209]

gravity gradient

The variation in the force of attraction between two bodies expressed as a function of the separation distance.

gravity gradient stabilization

A useful, passive method to achieve **stabilization of satellites** by using the **gravity gradient** of the primary body. An orbiting spacecraft will tend to align its long axis (more precisely, the axis of minimum moment of inertia) with the local vertical—that is, in a radial direction.

Gravity Probe A

A NASA geodetic satellite designed to test the **general theory of relativity**. Gravity Probe A carried a space-qualified hydrogen maser clock to an altitude of 10,000 km on a two-hour suborbital flight and verified the gravitational red shift predicted by Einstein's theory to a precision of 70 parts per million.

Launch
    Date: June 18, 1976
    Vehicle: Scout D
    Site: Wallops Island

Gravity Probe B

A satellite being developed by NASA and Stanford University to test two extraordinary, unverified predictions of Einstein's **general theory of relativity**. As the probe orbits at an altitude of 640 km directly over Earth's poles, tiny changes in the direction of spin of four onboard gyroscopes will be measured very precisely. So free are the gyroscopes from disturbance that they will provide an almost perfect inertial reference system. They will measure how space and time are warped by the presence of Earth, and, more profoundly, how Earth's rotation drags space-time around with it. These effects, though small for Earth, have far-reaching implications for the nature of matter and the structure of the universe. Gravity Probe B is scheduled for launch in April 2003.

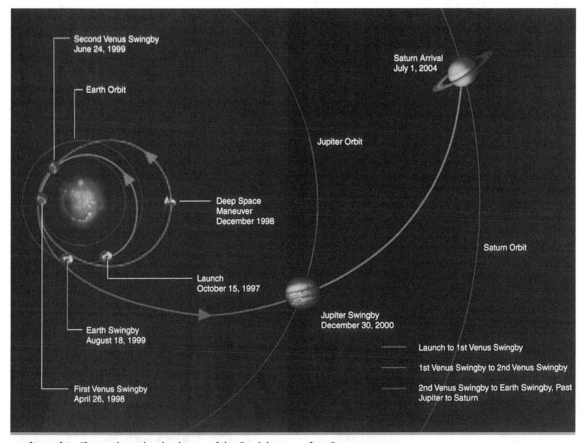

Second Venus Swingby
June 24, 1999

Earth Orbit

Saturn Arrival
July 1, 2004

Jupiter Orbit

Deep Space
Maneuver
December 1998

Saturn Orbit

Launch
October 15, 1997

Earth Swingby
August 18, 1999

Jupiter Swingby
December 30, 2000

Launch to 1st Venus Swingby

1st Venus Swingby to 2nd Venus Swingby

First Venus Swingby
April 26, 1998

2nd Venus Swingby to Earth Swingby, Past
Jupiter to Saturn

gravity-assist The gravity-assisted trajectory of the Cassini spacecraft to Saturn. *NASA/JPL*

gravity-assist

An important fuel-saving technique, also known as the slingshot effect, which has been used on numerous interplanetary missions, including those of **Voyager, Galileo,** and **Cassini**. It involves taking advantage of the gravitational field of a planet to increase the speed of a spacecraft or change its direction, or both. The inbound flight path is carefully chosen so that the spacecraft will be whipped around the assisting body, being both accelerated (or possibly decelerated) and deflected onto a new trajectory. At first sight, it may seem as if something has been gained for nothing. However, the additional speed of the spacecraft is won at the expense of the planet, which, as a result of the encounter, slows imperceptibly in its orbit and therefore moves fractionally closer to the Sun. One of the earliest and most dramatic applications of the technique came in 1970 when the world watched as NASA used a lunar gravity-assist to rescue the Apollo 13 astronauts after an onboard explosion had severely damaged their spacecraft en route to the Moon. By using a relatively small amount of fuel to put the spacecraft onto a suitable trajectory, NASA engineers and the astronauts were able to use the Moon's gravity to turn the ship around and send it back home.

grayout

Temporary impairment of vision, due to reduced flow of blood to the brain, caused by prolonged high acceleration. See **blackout**.

Great Observatories

A series of four major NASA orbiting observatories covering different regions of the electromagnetic spectrum. They are the **Hubble Space Telescope, Compton Gamma Ray Observatory, Chandra X-ray Observatory,** and **Space Infrared Telescope Facility**.

Greg, Percy (1836–1889)

An English poet, novelist, and historian who wrote a number of science fiction tales, including *Across the Zodiac: The Story of a Wrecked Record* (1880).[137] This was possibly the first story to involve a powered spaceship (the craft in Jules **Verne**'s early Moon yarn having been shot out of a huge cannon). Greg describes a propulsion system based on apergy–an **antigravity** force. Arriving on a Lowellian Mars, the protagonist finds a race of technically advanced humanoids, one of whom he marries (qualifying the book perhaps for another SF first), though it all ends unhappily when his Martian bride is killed.

Griffith, George (1857–1906)

A prolific author, traveler, and adventurer, described by Sam Moskowitz as "undeniably the most popular science fiction writer in England between 1893 and 1895 . . . it is entirely conceivable that Griffith's science fiction outsold that of Wells." His work, however, was overshadowed by that of **H. G. Wells** and was never influential in the United States because of the author's anti-American views.

Grissom, Virgil Ivan "Gus" (1926–1967)

One of the original **Mercury Seven** astronauts, the first person to fly in space twice (unless Joseph Walker's suborbital X-15 flights are counted), and the victim of a fire during a prelaunch test before the first manned Apollo flight. Grissom received a B.S. in mechanical engineering from Purdue University before joining the Air Force and flying 100 combat missions in Korea. Upon returning to the United States, he became a jet instructor at Bryan, Texas, studied aeronautical engineering at the Air Force Institute of Technology, and then enrolled at the Air Force Test Pilot School at Edwards Air Force Base. In 1957, he went to Wright-Patterson Air Force Base, Ohio, as a test pilot in the fighter branch. Having been selected as an astronaut in 1959, he flew aboard *Liberty Bell 7* on

Virgil Grissom   Grissom, pilot of the Mercury spacecraft *Liberty Bell 7*, arrives aboard the recovery ship USS *Randolph* following his suborbital space mission. He is flanked by military medical officers. *NASA*

the second American manned spaceflight, commanded **Gemini** 3, the first two-man mission, was backup command pilot for Gemini 4, and was scheduled to command the first Apollo flight when he and his crewmates, Edward **White** II and Roger **Chaffee**, were killed in a launch pad fire on January 27, 1967.[138]

ground-track
An imaginary line on the Earth's surface that traces the course of another imaginary line between Earth's center and an orbiting satellite.

GSLV (Geosynchronous Satellite Launch Vehicle)
India's latest four-stage launch vehicle, capable of placing 2,500-kg **Insat**-type communications satellites into geostationary transfer orbit. The GSLV improves on the performance of the **PSLV** (Polar Satellite Launch Vehicle) by the addition of liquid strap-on boosters and a cryogenic upper stage. The solid first and liquid second stages are carried over from the PSLV, while the cryogenic upper stage is being supplied by Russia until India has developed an indigenous version. Carrying a 1.5-ton experimental satellite, the GSLV was launched successfully for the first time on April 18, 2001—an important first step toward India's independence in space and competing in the lucrative geosynchronous market.

Height: 49 m
Maximum diameter: 2.8 m
Mass at liftoff: 402 tons
Payload to GTO: 2,500 kg

GSO
Abbreviation for either **geosynchronous orbit** or **geostationary orbit**.

g-suit
See **anti-*g* suit**.

GTO
See **geosynchronous/geostationary transfer orbit**.

Guiana Space Centre
Europe's main launch facility, owned by **ESA** (European Space Agency) and operated by **CNES** (the French space agency). Located near Kourou in French Guiana, on the northern coast of South America at 5.2° N, 52.8° W, it is ideal for launching **geosynchronous satellites** and also supports missions into **polar orbit**. It benefits from Earth's rotation near the equator, which gives a free boost of 460 m/s to eastward-launched rockets. Originally selected by CNES, the Kourou site has been vastly expanded with funding from ESA for its **Ariane** launcher programs. The Ariane 5 facilities alone cover 21 square km out of a total of 850 square km occupied by the spaceport.[242]

guidance
The process of directing the movements of an aeronautical or space vehicle, with particular reference to the selection of a flight path or trajectory.

guidance, inertial
See **inertial guidance**.

Guiana Space Centre An Ariane 5 stands on its launch pad at the Guiana Space Centre.
European Space Agency

guidance, midcourse
Guidance of a spacecraft while it is en route to its destination.

guidance system
One of the systems in a space vehicle designed to put the vehicle on a desired trajectory prior to thrust cutoff, or a system in a vehicle that establishes the desired path from launch to target.

guided missile
An unmanned weapon that is controlled by some system from its launch until it hits its target.

guided missiles, postwar development
In the decade after World War II, the U.S. government, persuaded by the conservative arm of the military, decided that the nation's main deterrent force should consist of manned intercontinental bombers supplemented by air-breathing **guided missiles** evolved from the German V-1 (see **"V" weapons**). Despite the early postwar warnings of Gen. Henry H. **Arnold** and others, who argued for the immediate development of long-range ballistic missiles capable of carrying atomic warheads, this approach was deemed too costly and difficult

guided missiles, postwar development The first instrumented Hermes A-1 test rocket fired from White Sands Proving Ground. Hermes, based externally on the V-2, contributed to the development of the Redstone missile. *NASA*

at the time. Just after the war, former NACA (National Advisory Committee for Aeronautics) chairman Vannevar **Bush**, then director of the Office of Scientific Research and Development, expressed the prevailing mood in a testimony before a Congressional committee: "There has been a great deal said about a 3,000-mile high-angle rocket. In my opinion, such a thing is impossible today and will be impossible for many years."

Thus, in the late 1940s and early 1950s, the United States focused on developing relatively small guided missiles for air-to-air, air-to-surface, and surface-to-air interception and as tactical surface-to-surface weapons. The Soviets, by contrast, threw themselves into building huge ballistic missiles capable of carrying the heavy nuclear weapons of that period over thousands of miles—a decision that gave them a long-lasting lead in space exploration.

Interservice rivalry led the Army, the Navy, and the Air Force to pursue essentially separate missile-development programs. The great success of the Air Corps during the war had conditioned it to think in terms of winged vehicles. As a result, the Air Force put more effort at first into two subsonic missile programs, **Snark** and **Matador**, than the supersonic, long-ranged **Atlas**. Later came the **Navaho**; but by this time it was clear that such weapons were not effective compared to longer-range ballistic missiles.

While the Air Force concentrated on relatively small guided missiles, the Army, having inherited a stockpile of V-2s and a team of German rocket specialists, naturally chose to build on this ready-made pool of hardware and expertise. But as the supply of Nazi missiles diminished and it was evident that rocket technology had moved beyond the V-2 stage, the Army initiated the Hermes program and, later, the development of the **Redstone** missile based on the liquid-propellant engine of the Navaho.

By the early 1950s, intelligence revealed that the Soviets were evolving long-range nuclear ballistic missiles that posed a serious threat to the West. In response, the priority in the United States switched from the development of guided missiles to intermediate-range and intercontinental ballistic missile programs, including the **Jupiter**, Atlas, and **Thor**.

Gulfstream
To train Space Shuttle pilots in landing the vehicle prior to making their first spaceflight, NASA uses a modified Gulfstream 2 executive jet. The cockpit has been divided into two—with Shuttle controls on the left and normal Gulfstream controls on the right. The flight dynamics of the Shuttle were calculated before the Shuttle actually flew, and the Gulfstream 2 trainer was modified accordingly. The calculations proved to be so accurate that very

Gurwin Missions

	Launch			
				Mass: 50 kg
Spacecraft	Date	Vehicle	Site	Orbit
Gurwin 1	Mar. 28, 1995	Start	Plesetsk	Failed; fell in Sea of Okhotsk
Gurwin 1B	Jul. 10, 1998	Zenit-2	Baikonur	817 km × 819 km × 98.8°

few changes have been made since the Shuttle entered service.

Gurwin
Satellites built by the Technion Institute of Technology, Israel, and named for the philanthropist Joseph Gurwin (1920–). Onboard instruments include CCD (charge-coupled device) cameras, radiation detectors, ozone monitors, and amateur radio transmitters. Gurwin 1B replaced an earlier satellite that failed to reach orbit. (See table, "Gurwin Missions.")

gyroscope
A device that utilizes the angular momentum of a spinning rotor to sense angular motion of its base about one or two axes at right angles to the spin axis.

H

H series (Japanese launch vehicles)

A family of **Japanese launch vehicles** used by NASDA (National Space Development Agency) that consists, in chronological order of development, of the H-1, H-2, and H-2A and their variants.

H-1

The first stage of the H-1 was essentially the same as that of the N-2 (see **N series [Japanese launch vehicles]**), a license-built **Delta** first stage, with a liquid oxygen (LOX)/kerosene main engine and six to nine small solid-propellant strap-on boosters. The second stage was of Japanese origin, built by Mitsubishi Heavy Industries, and burned LOX/liquid hydrogen. A small solid-propellant third stage designed by Nissan enabled payloads of up to 1,100 kg to be placed in geostationary transfer orbit (GTO). The H-1 program concluded in 1992 with nine successes and no failures.

H-2

To provide greater payload capacity and to permit unencumbered commercial space transportation offerings (the Delta licensing agreement restricted the use of the H-1 for commercial flights), Japan developed the H-2 launch vehicle based on all-Japanese propulsion systems. The H-2 can lift payloads four times heavier than those of the H-1—up to 10 tons into LEO (low Earth orbit) or up to 4 tons into GTO—and opened the door to NASDA spacecraft designed to explore the Moon and the planets. The first mission, on February 3, 1994, deployed one payload into LEO and then carried an experimental package VEP (Vehicle Evaluation Payload) into GTO. Dwarfing its predecessor, the H-2 consisted of a two-stage core vehicle, burning LOX and liquid hydrogen in both stages, with two large solid-propellant strap-on boosters. Nissan produced the 4-segmented strap-on boosters, which are considerably larger than the main stages of ISAS's (Institute of Space and Astronautical Science's) M-3 and M-5 series vehicles. The LE-7 first-stage main engine overcame numerous developmental difficulties, while the LE-5A engine used by the second stage is an upgraded version of the proven LE-5, which had been flown on the second stage of the H-1.

H-2A

An upgraded version of the H-2 currently in service; the first H-2A was launched successfully from Tanegashima on August 29, 2001, although a test satellite failed to separate. A similar problem marred the second test flight on February 3, 2002. The first operational flight, to launch the **DRTS**-W satellite, took place successfully on September 10, 2002. Intended to compete commercially on the world market, the H-2A builds upon its predecessor and incorporates a simplified design and upgraded avionics and engines. Although the core vehicle is similar to the H-2's, the H-2A uses new solid and liquid boosters to improve payload performance. There are five variants. The basic H-2A 202 configuration uses a core vehicle with two solid rocket boosters to place four tons into a 28.5° GTO. The H-2A 2022 and H-2A 2024 configurations add two and four solid strap-on boosters, respectively, to increase GTO capability to 4.5 tons. Adding liquid strap-on boosters (LRBs) to the basic H-2A 202 configuration creates the H-2A 212 (one LRB) and H-2A 222 (two LRBs) configurations, which can place 7.5 to 9.5 tons into GTO.

H-1 (American rocket engine)

A liquid-propellant rocket engine, eight of which were used on the first stage of the **Saturn** I and IB launch vehicles. Employing liquid oxygen and RP-1 (kerosene mixture) as propellants, the gimbaled H-1 developed 830,000 N of thrust at sea level. See **F-1** and **J-2**.

Haas, Conrad (c. 1509–1579)

An Austrian artillery officer who may have been the first to describe the principle of the multistage rocket. The evidence comes from a 450-page manuscript in the national archive of Sibiu, Romania (formerly Hermannstadt), dealing with problems of artillery and ballistics, the third part of which was written by Haas. Born in Dornbach, near Vienna, Haas served as an artillery guard and a commissioned officer of the Imperial court and, in this capacity, it seems, came in 1551 with Imperial troops to Transylvania and became chief of the artillery camp of the arsenal of Hermannstadt. Between 1529 and 1569, he wrote the work mentioned above, which describes and depicts rockets that use two and three stages, stabilizing fins, and liquid fuel.

Hagen, John P. (1908–1990)

Director of the **Vanguard** program during the 1950s. Hagen had been an astronomer at Wesleyan University (1931–1935) before working for the Naval Research Laboratory from 1935 to 1958, during which time he earned his Ph.D. In 1955, he became director of the Vanguard satellite program. When that program became part of NASA on October 1, 1958, he remained chief of the NASA Vanguard division and then became assistant director of spaceflight development (1958–1960). In February 1960, he became director of NASA's office for the United Nations conference and, later, assistant director of NASA's office of plans and program evaluation. In 1962, Hagen returned to higher education as a professor of astronomy at Pennsylvania State University, becoming head of the department in 1975.

Hagoromo

A small (12-kg) lunar orbiter ejected from the Japanese **Hiten** spacecraft on March 19, 1990; its name means "angel's robe." Contact with the probe was lost after its release.

Haise, Fred Wallace, Jr. (1933–)

An American astronaut who served as Lunar Module pilot on the ill-fated **Apollo** 13 mission. Haise received a B.S. in aeronautical engineering from the University of Oklahoma in 1959 and joined the Navy in 1962, subsequently serving as a fighter pilot in the Oklahoma National Guard. He was the Aerospace Research Pilot School's outstanding graduate of Class 64A and served with the Air Force from 1961 to 1962 as a tactical fighter pilot. NASA then tapped him as a research pilot at its Flight Research Center at Edwards, California, and at its Lewis Research Center. Haise was one of 19 selected by NASA in its fifth class of astronauts in April 1966. He was backup Lunar Module pilot for Apollo 8 and Apollo 11 before being named to that slot on Apollo 13. From 1973 to 1976, Haise was technical assistant to the manager of the Space Shuttle Orbiter Project. He was commander of one of the two-man crews that piloted Shuttle approach and landing test flights in 1977. These flights evaluated the Shuttle's capabilities after the test vehicle *Enterprise* was released from the back of a Boeing 747 jet above the California desert. Haise retired from NASA in June 1979 and held several managerial positions with Grumman Aerospace Corporation before retiring in 1996.

Hakucho

The first Japanese X-ray astronomy satellite; it was named "swan" in Japanese because one of the most powerful cosmic X-ray sources is Cygnus X-1. Like many other X-ray satellites deployed around this time, Hakucho was designed to study and monitor transient (quick-changing) phenomena, with special emphasis on X-ray bursts in the energy range of 0.1 to 100 keV. It was known prior to

launch as Corsa-B. Hakucho stopped operating in April 1985.

Launch
Date: February 21, 1979
Vehicle: M-3S
Site: Kagoshima
Orbit: 543 × 566 km × 29.8°
Mass: 96 kg

HALCA (Highly Advanced Laboratory for Communications and Astronomy)

A Japanese satellite that forms the first spaceborne element of the **VSOP** (VLBI Space Observatory Program), led by **ISAS** (Institute of Space and Astronautical Science) in collaboration with the National Astronomical Observatory of Japan. HALCA's main instrument is an 8-m-diameter radio telescope. The satellite's highly elliptical orbit facilitates VLBI (Very Long Baseline Interferometry) observations on baselines up to three times longer than those achievable on Earth. Observations are made at 1.6 GHz, 5 GHz, and 22 GHz. HALCA is also known by the national name Haruka ("far away")—a reference to its great distance from Earth at apogee (the highest point of its orbit). Before launch, the satellite was called MUSES-B.

Launch
Date: February 12, 1997
Vehicle: M-5
Site: Kagoshima
Orbit: 569 × 21,415 km × 31.4°

Hale, Edward Everett (1822–1909)

A prolific American writer, a contributing editor to *The Atlantic Monthly*, and almost certainly the only science fiction writer to serve as chaplain to the U.S. Senate; he appears to have been the first to describe an artificial Earth satellite. His short story "The Brick Moon" (1869) and its sequel, "Life on the Brick Moon,"[140] both published in *The Atlantic Monthly*, tell of a 200-foot diameter sphere (built of bricks to "stand fire very, very well") that is due to be launched into an orbit 4,000 miles high. Since its purpose is to provide a longitude fix for navigators who will see it from the ground as a bright star, Hale reasons correctly that a **polar orbit** is needed. In effect, the Brick Moon will move around a giant Greenwich meridian in the sky, fulfilling the same role for the measurement of longitude that the Pole Star does for latitude. Two huge, spinning flywheels are set up to throw the artificial moon into its correct orbital path—but something goes wrong. The brick sphere rolls prematurely down "upon these angry flywheels,

Edward Hale The Brick Moon's inhabitants signal in Morse code by jumping up and down, while people on Earth throw them books and other provisions.

and in an instant, with all our friends [those building the moon and their visiting families], it had been hurled into the sky!" Later, a German astronomer spots the new moon in orbit, complete with its marooned and unscheduled inhabitants–apparently adapting well to their new life. Hale therefore manages to portray not only the first artificial satellite but also the first **space station**.[152]

Hale, William (1797–1870)
An English inventor who developed the technique of **spin stabilization**. Even following William **Congreve's** work, the accuracy of rockets was not much improved. The devastating nature of war rockets was not their accuracy or power but their numbers. During a typical siege, thousands of them might be fired at the enemy. All over the world, rocket researchers experimented with ways to improve accuracy. It was Hale who thought of the idea of allowing the escaping exhaust gases to strike small vanes at the bottom of the rocket, causing it to spin like a rifled bullet in flight.[314]

Hall-effect thruster
A small rocket engine that uses a powerful magnetic field to accelerate a low density **plasma** and so produce thrust. A radial magnetic field is set up between two concentric annular magnet pole pieces. The interior volume surrounded by the magnet is filled with a low pressure propellant gas through which a continuous electric discharge passes between two electrodes. The positive electrode is upstream of the magnet pole pieces, and the negative electrode is directly downstream. The axial electric field developed between the electrodes interacts with the radial magnetic field to produce, by the so-called Hall effect, a current in the azimuthal direction. This current, in turn, reacts against the magnetic field to generate a force on the propellant in the downstream axial direction. Although conceived in the United States, the Hall thruster was developed into an efficient propulsion device–the SPT (stationary plasma thruster)–in the Soviet Union. See **electromagnetic propulsion**.

halo orbit
An orbit in which a spacecraft will remain in the vicinity of a **Lagrangian point**, following a circular or elliptical loop around that point. The first mission to take advantage of such an orbit was **SOHO** (Solar and Heliospheric Observatory).

Ham
A 44-month-old male chimpanzee launched onboard **Mercury** Capsule No. 5 at 11:55 A.M. on January 31, 1961, for a suborbital flight. He was named for the initials of Holloman Aerospace Medical Center, New Mexico, where the space chimps lived and trained, and also after the commander of Holloman Aeromedical Laboratory, Lt. Col. Hamilton Blackshear. Ham's mission, known as **Mercury-Redstone** 2, was launched from Pad 5/6 at Cape Canaveral. Because of over-acceleration of the launch vehicle plus the added energy of the escape rocket, a speed of 9,426 km/hr was reached instead of the 7,081 km/hr planned, resulting in an apogee of 253 km rather than the intended 185 km. This meant that Ham was weightless for 1.7 minutes longer than the 4.9 minutes scheduled. He landed 679 km downrange after a 16.5-minute flight. His peak reentry *g* was 14.7–almost 3*g* greater than planned. The capsule splashed down about 97 km from the nearest recovery vessel. Tears in the capsule's landing bag capsized the craft, and an open cabin pressure relief valve let still more seawater

Ham Ham shakes hands with the recovery ship commander after his suborbital flight. *NASA*

enter the capsule. When a Navy helicopter finally latched onto and picked up the capsule at 2:52 P.M., there was about 360 kg of seawater aboard. After a dangling flight back to the recovery ship, the spacecraft was lowered to the deck and nine minutes later Ham was out. He appeared to be in good condition and readily accepted an apple and half an orange. Ham's successful flight paved the way for the flight of America's first man in space, Alan Shepard, aboard *Freedom 7* atop Mercury-Redstone 3 on May 5, 1961.

Hammaguira

The launch site, at a military base in the Sahara desert, of four French satellites in the mid-1960s. Following the establishment of Algerian independence in 1967, France was forced to abandon the facility.

HAN (NH₂OH⁺NO₃)

HAN ($NH_2OH^+NO_3$)

Hydroxyl ammonium nitrate; a relatively new synthetic rocket fuel. It has the potential to be used both as a liq-

uid **monopropellant** and as an ingredient in solid propellants.

Hand-held Maneuvering Unit

A small jet held by astronauts to maneuver during space walks; it was used in some flights prior to the Space Shuttle. The unit employed nitrogen as fuel, which was fed to the unit under pressure through the astronaut's umbilical cord from the spacecraft.

hard landing

The deliberate, destructive impact of a space vehicle on a predetermined celestial target. Among the objectives of such a mission may be to test propulsion and guidance to prepare the way for a soft landing, to create a seismic disturbance that can be registered by sensors on the surface (as in the case of spent rocket stages and Lunar Module ascent stages in the **Apollo** program), or to splash material from beneath the surface into space so that it can be collected and/or analyzed by a mother craft (as in the case of **Deep Impact**).

hard mockup

A full-size replica of a spacecraft vehicle or engine equipped with all of the requisite hardware, such as instrumentation, for use in demonstration, training, and testing.

hard vacuum

A **vacuum** that approximates the vacuum of space.

HARP (High Altitude Research Project)

A program to study the upper atmosphere using instrumented projectiles shot from a cannon, conducted in the 1960s by researchers at McGill University, Montreal. The projectiles were cylindrical finned missiles (20 cm wide and 1.7 m long, with masses of 80 to 215 kg) called Martlets, from an old name for the martin bird, which appears on McGill's shield. The cannon that propelled the Martlets was built by the Canadian engineer Gerald **Bull** from two ex-U.S. Navy 16-inch- (41-cm-) caliber cannon connected end to end. Located on the island of Barbados, it fired almost vertically out over the Atlantic. Inside the barrel of the cannon, a Martlet was surrounded by a machined wooden casing known as a sabot, which traveled up the 16-m-long barrel at launch and then split apart as the Martlet headed upward at about 1.5 km/s, having undergone an acceleration of 25,000*g*. Each shot produced a huge explosion that could be heard all over Barbados and a plume of fire rising hundreds of meters into the air. The Martlets carried payloads of metal chaff, chemical smoke, or meteorological balloons, and they were fitted with telemetry antennas for tracking their flight. By the end of 1965, HARP had fired more than a hundred missiles to heights of over 80 km. In November 19, 1966, the Army Ballistics Research Laboratory used a HARP gun to launch an 84-kg Martlet to an altitude of 179 km—a world record for a fired projectile that still stands.[37, 94, 207] See **space cannon**.

Haruka

Alternative name of **HALCA**; "haruka" is Japanese for "far away."

Haughton-Mars Project

An ongoing NASA-led international field program to study the Haughton meteorite crater on Devon Island in the Canadian high arctic and to use the crater's environment to learn how best to explore Mars, by testing robotic and human exploration technologies and strategies.

HCMM (Heat Capacity Mapping Mission)

A NASA satellite that produced thermal maps of the atmosphere and studied dust and liquid droplets in the upper atmosphere. Also known as AEM-1.

Launch
    Date: April 26, 1978
    Vehicle: Scout D
    Site: Vandenberg Air Force Base
Orbit: 560 × 641 km × 97.6°
Mass: 134 kg

HEAO (High Energy Astrophysical Observatory)

A series of three very large scientific satellites launched by NASA, beginning in 1977, to carry out detailed observations of the sky at short wavelengths, from ultraviolet to gamma rays. (See table, "HEAO Missions.")

Haughton-Mars Project   An investigator on a field geology traverse at Haughton Crater, Devon Island, wearing the upper torso of the Hamilton-Sundstrand advanced space exploration concept suit. *NASA Haughton-Mars Project/Pascal Lee*

HEAO Missions

Spacecraft	Launch Date	Orbit	Mass (kg)
HEAO-1	Aug. 12, 1977	$429 \times 447 \times 22.7°$	2,720
HEAO-2	Nov. 12, 1978	$526 \times 548 \times 23.5°$	3,150
HEAO-3	Sep. 20, 1979	$487 \times 503 \times 22.7°$	3,150

HEAO-1

An X-ray astronomy satellite that surveyed the sky in the 0.2 keV to 10 MeV energy band, providing nearly constant monitoring of X-ray sources near the ecliptic poles. More detailed studies of a number of objects were made through pointed observations lasting typically three to six hours. The spacecraft remained active until January 9, 1979.

HEAO-2

The first fully imaging X-ray telescope to be placed in orbit; later renamed the Einstein Observatory to honor the centenary of the great physicist's birth. Its angular resolution of a few arcseconds, field-of-view of tens of arcminutes, and sensitivity several hundred times greater than any previous mission provided, for the first time, the capability to create images of extended objects and diffuse emission and to detect faint sources at such high energies (0.2–4.5 keV). It revolutionized astronomers' view of the X-ray sky. Observations ceased in April 1981.

HEAO-3

Like its predecessor HEAO-1, a survey mission operating in the hard X-ray and gamma-ray (50 keV–10 MeV) band. Its High Resolution Gamma-Ray Spectrometer Experiment, built by JPL (Jet Propulsion Laboratory), was the largest germanium spectrometer placed in orbit at that time. The mission effectively ended when the cryogenic coolant for the germanium detectors ran out in May 1981.

Launch
   Vehicle: Atlas Centaur SLV-3D
   Site: Cape Canaveral

heat barrier

The speed above which friction with an atmosphere may damage a spacecraft. The heat barrier is of particular importance during reentry. Special systems are needed to cool the spacecraft when it goes beyond the heat-barrier speed.

heat-shield

See **reentry thermal protection**.

heavy-lift launch vehicle (HLV)

The most powerful type of space **launch vehicle**, sometimes defined as one having the capacity to place payloads of at least 10,000 kg into low Earth orbit or payloads of 5,000 kg into geostationary transfer orbit. Examples include the **Space Shuttle**, **Atlas** V, and **Ariane** 5.

helicon thruster

A device that works in a similar way to the **pulsed-plasma thruster** but with one important difference: a traveling electromagnetic wave interacts with a current sheet to maintain a strong force on a plasma moving along an axis. This circumvents the pulsed-plasma thruster's problem of the force falling off as the current loop gets larger. The traveling wave can be created in a variety of ways, and a helical coil is often used.

heliocentric

Related to or having the Sun as its center.

heliopause

The boundary of the **heliosphere**, where the pressure at the outer part of the heliosphere equals that of the

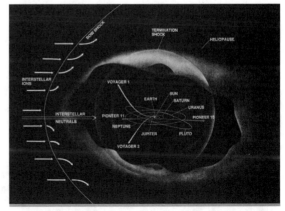

heliopause   The heliopause and other structures shown together with the exit trajectories of the twin Voyager and Pioneer probes. *NASA*

Helios Reconnaissance Missions

Spacecraft	Launch Date	Orbit	Mass (kg)
Helios 1A	Jul. 7, 1995	679 × 682 km × 98.1°	2,537
Helios 1B	Dec. 3, 1999	679 × 681 km × 98.1°	2,537

surrounding interstellar medium (ISM). It is believed to lie 100–150 astronomical units (AU) from the Sun in the direction in which the Sun is moving, but much further out downstream. The heliopause distance probably varies in response to changes in ISM and **solar wind** conditions. Its thickness is unknown but could be tens of astronomical units. Upstream of the heliopause, in the ISM, theory suggests the existence of a bow shock where the incoming interstellar wind first reacts to its impending collision with the Sun's magnetosphere. Downstream from the Sun is thought to be a long, turbulent tail. The extent to which interstellar matter penetrates the heliopause is open to question but may be determined by the first **interstellar precursor missions**.

Helios (reconnaissance satellites)

Advanced French optical military reconnaissance satellites based on **SPOT** 4. The Missile and Space Directorate (DME) of the French armament agency (DGA) heads the overall Helios program. The imaging system uses 4,096 × 2,048 pixel linear CCDs (charge-coupled devices) to provide images with 1- to 5-m resolution. (See table, "Helios Reconnaissance Missions.")

Launch
  Vehicle: Ariane 4
  Site: Kourou

Helios (solar probes)

Two West German space probes launched by NASA and designed to study the Sun and the interplanetary medium

Launch
  Vehicle: Titan IIIE
  Site: Cape Canaveral

from within the orbit of Mercury, as close as 45 million km (0.3 astronomical unit) to the Sun. They operated successfully for more than 10 years. (See table, "Helios Solar Probes.")

heliosphere

The region of space surrounding the Sun that is inflated by the **solar wind** and within which the Sun exerts a magnetic influence. Despite its name, the heliosphere is almost certainly greatly elongated due to the movement of the Sun with respect to the interstellar medium. It is bounded by the **heliopause**. Data returned by **Pioneer** 10, **Voyager** 1, and Voyager 2 (contact with Pioneer 11 has been lost), as these probes head for interstellar space, will help further characterize the heliosphere.

Henry, James Paget (1914–1996)

A pioneer of **aerospace medicine** who developed the first partial pressure suit, a forerunner of today's **spacesuit**s. An American citizen born in Germany, Henry came to the United States as a child but received both his M.A. in medicine (1938) and his doctorate in medicine (1952) from Cambridge University, England. He began his career at the University of Southern California (USC) in 1943, and while there he designed a partial pressure suit for emergency extreme altitude protection that later became standard issue for military pilots. Henry also worked on acceleration physiology with the USC Human Centrifuge. He left USC to continue his work on the pressure suit at Wright-Patterson Air Force Base, Ohio, and to direct the Physiology of Rocket Flight research project for the Air Force, which led to the first successful animal rocket flights. He returned to USC in 1963 as a professor of physiology.

Helios Solar Probes

Spacecraft	Launch Date	Solar Orbit	Mass (kg)
Helios 1	Dec. 10, 1974	0.31 × 0.99 AU × 0°	370
Helios 2	Jan. 15, 1976	0.39 × 1.00 AU × 0°	376

HEO

(1) Highly elliptical orbit. Definitions vary, one being an orbit with a **perigee** (low point) below 3,000 km and an **apogee** (high point) above 30,000 km. (2) High Earth orbit. An orbit whose apogee lies above that of a **geostationary orbit** at 35,800 km.

HEOS (Highly Eccentric Orbiting Satellite)

ESRO (European Space Research Organisation) satellites that traveled in extremely elongated orbits in order to study the interplanetary magnetic field outside Earth's magnetosphere. (See table, "HEOS Missions.")

Hermes

A proposed small, manned space plane for servicing European components of the **International Space Station**. It began as a purely French project, but as development costs increased, France managed to obtain other European funding, primarily from Germany, for development of the craft. Hermes would have provided independent European manned access to space, and the weight growth of the spacecraft during design and development was a major influence in the specification for the **Ariane** 5 launcher. Hermes was intended to take three astronauts to orbits of up to 800-km altitude on missions lasting 30 to 90 days. Payload to be returned to Earth could be contained in a small payload bay, but most experiments were to be carried in an expendable pressurized supply module equipped with a docking port and mounted to the base of the glider. After the *Challenger* **disaster**, many safety features were added to Hermes, leading to further increases in weight and cost. In an emergency the crew cab could have been separated from the remainder of the glider. Later, however, the crew escape capsule idea was abandoned and was replaced by three ejection seats to save weight. Hermes was to have been launched atop an Ariane 5 by 1998, but by 1992, estimated development costs rose to DM 12 billion and the project was abandoned.

Hero of Alexandria (A.D. 1st century)

Greek inventor of the rocketlike device called an **aeolipile**. He lived about three centuries after another ancient rocketeer, **Archytas**.

Herschel Space Observatory

A giant **ESA** (European Space Agency) space telescope with a 3.5-m-diameter main mirror (by comparison, Hubble's is only 2.4 m across) designed to observe the universe in unprecedented detail at far infrared and submillimeter wavelengths from 80 to 670 microns and to carry out sensitive photometry and spectroscopy. Herschel will be launched together with **Planck** in early 2007 by an Ariane 5. Once in space, the two satellites will separate and proceed to different orbits around the L2 **Lagrangian point**, some 1.5 million km from Earth. Herschel, named after the great Hanoverian-born British astronomer William Herschel (1738–1822), was previously known as FIRST (Far Infrared and Submillimetre Space Telescope).

HESSI (High Energy Solar Spectroscopic Imager)

See **RHESSI**.

HETE (High Energy Transient Experiment)

A satellite whose main task is a multiwavelength study of gamma-ray bursts using ultraviolet, X-ray, and gamma-ray instruments. A unique feature of the mission is its ability to localize bursts with an accuracy of several arcseconds in near real-time. These positions are transmitted to the ground and picked up by a global network of primary and secondary ground stations, enabling rapid follow-up studies. The original HETE was lost following its launch on November 4, 1996, when it (and its co-passenger, SAC-B) failed to separate from the third stage of a **Pegasus** launch vehicle. However, the scientific importance and continuing relevance of the mission provided the impetus for the mission collaborators, including NASA, CESR (Centre d'Etude Spatiale des Rayonnements), CNES (the French space agency), and RIKEN (Japan's Institute of Chemical and Physical Research), to fund a replacement satellite. HETE-2 retains nearly all the original HETE design elements but carries a soft (longer wavelength) X-ray camera in place of the original ultraviolet cameras. Its main instrument is FREGATE, the French Gamma Telescope, a hard (shorter wavelength) X-ray spectrometer operating in the 6 to 400 keV energy range. This gamma-ray burst detector, together with a Wide Field X-ray Monitor, is used to trigger searches with the

HEOS Missions

Spacecraft	Date	Launch Vehicle	Site	Orbit	Mass (kg)
HEOS 1	Dec. 5, 1968	Delta E	Cape Canaveral	20,020 × 202,780 km × 60.5°	108
HEOS 2	Jan. 31, 1972	Delta L	Vandenberg	405 × 240,164 km × 89.9°	123

two Soft X-ray Imagers, which have 33-arcsecond resolution. This gives astronomers precise locations of gamma-ray bursts and allows detailed follow-up with optical instruments on the ground. The Massachusetts Institute of Technology operates the satellite, and the program is managed by the Goddard Space Flight Center as an Explorer Mission of Opportunity.

Launch
    Date: October 9, 2000
    Vehicle: Pegasus
    Site: Kwajalein Missile Range, Marshall Islands
    Orbit: 598 × 641 km × 1.9°

high mass fraction

A high value of the ratio of propellant mass to total motor mass. Values of 0.9 and greater characterize high-mass fraction motors.

High Resolution X-ray Spectroscopy Mission

A mission that would provide an exciting new approach to studying the origin of the chemical elements in stellar nucleosynthesis as well as active galaxies and galaxy clusters. It is identified in NASA's Office of Space Science Strategic Plan as a potential mission beyond 2007 but remains in the early concept definition phase.

high-energy radiation

Penetrating particle or electromagnetic radiation with energies of more than a few thousand electron volts. It may include electrons, neutrons, protons, mesons, X-rays, and gamma rays.

high-gain antenna (HGA)

A type of **antenna** used on board spacecraft to provide high amplification of either transmitted or received radio signals. HGAs typically consist of parabolic dishes and must be accurately pointed to be useful.

Highwater

Two suborbital tests of the **Saturn** I launch vehicle that resulted in the creation of massive ice clouds in the iono-sphere. The second and third stages of the Saturn contained 87,000 liters of water to simulate the mass of fully fueled stages. Following the successful launch of the first stage, the rocket was detonated by explosives with a signal from the ground. The released water formed a massive cloud of ice particles several kilometers in diameter. By this experiment, scientists had hoped to obtain data on atmospheric physics, but poor telemetry made the results questionable. Highwater 2 marked the third straight launch success for the Saturn I and the first with maximum mass on board. (See table, "Highwater Missions.")

Himawari

See **GMS (Geostationary Meteorological Satellite)**.

Hinotori

A Japanese satellite launched by **ISAS** (Institute of Space and Astronautical Science) principally to investigate X-rays from the Sun. It carried high-resolution soft (longer wavelength) X-ray spectrometers and hard (shorter wavelength) X-ray instruments that obtained the first images of solar flares at energies above 25 keV and also observed celestial X-ray bursts. Hinotori, whose name means "phoenix," was known before launch as **Astro**-A.

Launch
    Date: February 21, 1981
    Vehicle: M-3S
    Site: Kagoshima
    Orbit: 576 × 644 × 31°
    Mass: 188 kg

Hipparcos

An ESA (European Space Agency) **astrometry** satellite for measuring the position, brightness, and proper motion (movement against the distant stellar background) of relatively nearby stars to an unprecedented degree of accuracy. Its name is an abbreviation of High Precision Parallax Collecting Satellite and was chosen for its (somewhat strained) similarity to that of the Greek astronomer Hipparchus of Rhodes (190–120 B.C.). Although a faulty launch put Hipparcos in a highly elliptical orbit instead of

Highwater Missions

| | | Launch site: Cape Canaveral | | |
| | | **Detonation** | | |
Spacecraft	**Date**	**Altitude (km)**	**Time after Launch**	**Length of Experiment**
Highwater 1	Apr. 25, 1962	105	2.5 min	1.62 days
Highwater 2	Nov. 16, 1962	167	4 min 53 sec	1.46 days

the intended **geosynchronous** one, its mission was ultimately a triumph and resulted in two catalogues: the Hipparcos catalogue of 118,000 stars with positions, parallaxes, and proper motions measured to an unprecedented accuracy of 2 milliarcseconds (see arcseconds), and the Tycho catalogue of over a million stars with measurements of somewhat lower accuracy. Its mission ended in 1993 and its final catalogue was published in 1997.

Launch
    Date: August 8, 1989
    Vehicle: Ariane 44
    Site: Kourou
    Orbit: 542 × 35,836 km × 6.7°
    Mass: 1,130 kg

Hiten

A Japanese lunar probe, launched by **ISAS** (Institute of Space and Astronautical Science), that made multiple flybys of the Moon from an extremely elliptical Earth orbit and released **Hagoromo**, a smaller satellite, into lunar orbit. Hiten itself was put into lunar orbit in February 1992 and crashed into the Moon on April 10, 1993. It was the first lunar probe not launched by the United States or Russia. Hiten, whose name means "celestial maiden," was known before launch as **MUSES**-A.

Launch
    Date: January 24, 1990
    Vehicle: M-3S
    Site: Kagoshima
    Orbit: 262 × 286,183 km × 30.6°
    Mass: 185 kg

HNX (Heavy Nuclei Explorer)

A space laboratory for determining the properties of cosmic rays. HNX would directly sample and measure the properties of these high-speed subatomic particles using silicon and glass detectors. It would be launched by the Space Shuttle and be retrieved three years later. HNX has been selected by NASA for study as a possible **SMEX** (Small Explorer) mission.

Hohmann orbit

An elliptical trajectory, named after Walter **Hohmann**, along which a spacecraft may move from one orbit to another with the minimum expenditure of energy. Such an orbit just touches the original and destination orbits, and may be used for changing the orbit of an Earth satellite or for sending a probe to another planet. It involves two firings of the spacecraft's engine: one to break out of the original orbit and another to enter the destination

orbit. Its chief disadvantage is that it requires relatively long flight times. This can be overcome by judicious use of **gravity-assist**s.

Hohmann, Walter (1880–1945)

A German rocket engineer who was a prominent member of the **Verein für Raumschiffahrt** (Society for Space Travel) in the late 1920s. His book, *Die Erreichbarkeit der Himmelskörper* (The Attainability of Celestial Bodies), published in 1925, was so technically advanced that it was consulted decades later by NASA when planning its first interplanetary probes. In it, he describes his "power tower" spacecraft, a huge cone-shaped rocket with an egg-shaped manned capsule at the top, and, more important, the interplanetary transfer orbits that have been named after him (see **Hohmann orbit**). He also wrote popular works in the field of rocketry, as did his contemporaries Willy **Ley** and Max **Valier**.

hold

A scheduled pause or unscheduled delay in the launching sequence or countdown of a space vehicle or missile.

hold-down test

The testing of a system or subsystem in a launch vehicle while the vehicle is restrained in a stand.

holding fixture

A device or equipment used to support and position the upper launch vehicle stages and the spacecraft modules during test, checkout, and handling operations.

holding pond

Also known as a skimming basin, a human-made basin into which spilled propellants, deluge water, and washdown water are drained from the launch pad, launcher area, and launcher platform. The pond is so constructed that the water can be drained and the propellants skimmed for disposal.

Hoover, George W.

An early advocate of American satellite launches and an instigator of Project **Orbiter**. Hoover entered the Navy in 1944 and became a pilot before moving to the Office of Naval Research to conduct a program in all-weather flight instrumentation. Later he helped develop the idea of high-altitude balloons for use in a variety of projects. These included **Skyhook**, which supported cosmic-ray research and served as a research vehicle for obtaining environmental data relevant to supersonic flight. In 1954, Hoover was project officer in the field of high-speed, high-altitude flight, with involvement in the Douglas D558 project leading to the **X-15**. He was also instrumental in establishing Project Orbiter with **von Braun** and others, resulting in the launch of **Explorer** 1, the first American satellite.

HOPE (H-2 Orbital Plane)

A Japanese unmanned space plane that would be launched by an H-2 booster and used to ferry materials to and from the International Space Station. The project has not yet progressed beyond the stage of flight-testing subscale models.

Length: 16.5 m
Maximum diameter: 5.0 m
Mass
   Total: 13,000 kg
   Payload: 2,000 kg
   Propellant: 1,000 kg

horizontal preflight checkout

A checkout accomplished with the vehicle in a horizontal position, thereby reducing the requirements for gantries, cranes, and similar items. Upon completion, the vehicle is erected into vertical firing position. This kind of testing is generally used with smaller vehicles and spacecraft.

hot test

A propulsion system test that is carried out by actually firing the propellants.

HOTOL (Horizontal Takeoff and Landing vehicle)

A British design, conceived by Alan **Bond**, for a single-stage-to-orbit winged launch vehicle using a unique **air-breathing engine** design. The RB545 air/liquid hydrogen/liquid oxygen (LOX) engine was to be developed by Rolls-Royce. Work on HOTOL began in 1982 by a Rolls-Royce/British Aerospace team and had reached the stage of detailed engine design and mockup when, in the mid-1980s, the British government withdrew further funding. HOTOL would have taken off horizontally with a rocket-propelled trolley, switched to pure rocket propulsion at Mach 5 or 6, ascended to orbit, and glided back to Earth like the Space Shuttle to a runway landing. The HOTOL airframe was derived from conventional vertical takeoff rockets with the engines mounted at the rear of a blunt based fuselage. Since such a vehicle's empty **center of gravity** is dominated by the location of the engine, the wings and the tank for the dense LOX also had to be at the rear. The payload bay and hydrogen tankage were placed in a projecting forebody. The resulting configuration suffered from a severe **center of pressure**/center of gravity

mismatch during the air breathing ascent. The center of pressure shifted 10 m forward due to the wide Mach range, the large fuselage cross-section-to-wing area ratio, and the long overhang of the forward fuselage. Various design changes were made to address these problems, all of which eroded the payload. Conventional landing gear was replaced by a specially designed takeoff trolley to improve the marginal payload fraction. The final design had serious operational disadvantages and a small payload, and the only way the designers could continue to claim to put a reasonable payload into orbit was by specifying untried and speculative structural materials. HOTOL subsequently evolved into **Skylon**, which is being developed by Bond and his colleagues at Reaction Engines.

Length: 75.0 m
Core diameter: 7.0 m
Liftoff thrust (from 3 engines): 3,153,200 N
Mass: 250,000 kg (total), 50,000 kg (empty)

housekeeping

Routine tasks required to maintain spacecraft in habitable and/or operational condition during flight.

Hubble Space Telescope (HST)

An orbiting observatory built and operated jointly by NASA and ESA (European Space Agency), and named for the American astronomer Edwin Hubble (1889–1953). It has a 2.4-m-diameter main mirror and, following upgrades and repairs since its launch, is equipped with the Wide-Field and Planetary Camera 2 (WFPC-2), the Near Infrared Camera and Multi-Object Spectrometer (NICMOS), the Space Telescope Imaging Spectrograph (STIS), and, most recently installed, the Advanced Camera for Surveys (ACS).

During initial on-orbit checkout of Hubble's systems, a flaw in the telescope's main reflective mirror was found that prevented the proper focusing of incoming light–a problem caused by the incorrect adjustment of a testing device used while building the mirror. Fortunately, Hubble was designed for regular on-orbit maintenance by the Space Shuttle. The first servicing mission, STS-61 in December 1993, fully overcame the problem by installing a corrective optics package and upgraded instruments. A second servicing mission, STS-82 in February 1997, installed two new instruments in the observatory, and a third, STS-103 in December 1999, replaced Hubble's six gyros. The most recent Shuttle visit to Hubble, STS-109 in March 2002, was to install the ACS and refurbish the NICMOS.

The program includes significant participation by ESA, which provided one of the science instruments, the solar arrays, and some operational support to the program. The responsibility for conducting and coordinating the science operations of the Hubble Space Telescope rests with the Space Telescope Science Institute (STScI) at Johns Hopkins University, which operates Hubble for NASA as a general observer facility available to astronomers from all countries.

Hubble is the visible/ultraviolet/near-infrared element of the **Great Observatories** program and provides an order of magnitude better resolution than that of ground-based telescopes. Its startling views of the cosmos have captured the public's imagination and revolutionized astronomy. The observatory is expected to remain in service until about 2010, by which time the

Hubble Space Telescope   Attached to the Space Shuttle's robot arm, the Hubble Space Telescope is unberthed and lifted up into the sunlight during the second servicing mission. *NASA*

NGST (Next Generation Space Telescope) should be in use.

Shuttle deployment
Date: April 25, 1990
Mission: STS-31
Orbit: 590 × 596 km × 28.5°
Size: 13.3 × 4.3 m
Mass: 10,863 kg

human factors

A term used broadly to cover all biomedical and psychosocial considerations pertaining to man in the system. It includes principles and applications in the areas of human engineering, personnel selection training, life-support requirements, job performance aids, and human performance evaluation.

Human Research Facility (HRF)

A facility at the **Ames Research Center**, opened in 1971, where ground-based simulation studies of the physiological responses of astronauts during spaceflight can be carried out. It provides an environment with temperature, light intensity, and day-length automatically controlled and is suitable for studies on both ambulatory and bed-rested volunteer subjects. Test equipment and facilities include a lower-body negative-pressure device used in studies of fluid shifts; upright and horizontal bicycles, a treadmill, and other exercise testing devices; a water-immersion tank to simulate the effects of microgravity; a human-rated **centrifuge** (a rotating device used to expose humans to high degrees of gravitational force) and other rotating devices; and a tilt-table for testing the body's ability to respond to an upright position after being weightless or in a head-down position for an extended time. Studies in the HRF have used healthy volunteers from various backgrounds, ranging in age from 21 to 65. For varying periods, volunteers lie in beds tilted head-down at a 6° angle. Continuous head-down bed rest is used to simulate the effects of prolonged microgravity on the human body, such as cardiovascular deconditioning, muscle atrophy, decreased bone strength, and shifts in fluid and electrolyte balance. This method of simulating the effects of weightlessness has enabled extensive study on the ground of the changes responsible for the physiological effects of spaceflight. Scientists also study exercise, diet, fluid-loading, and drugs for their effectiveness in preventing these changes. Physiological responses to a Space Shuttle reentry acceleration profile have been tested on the man-rated centrifuge. The HRF is also equipped for isolation, group interaction, human performance, and physiological rhythm studies.

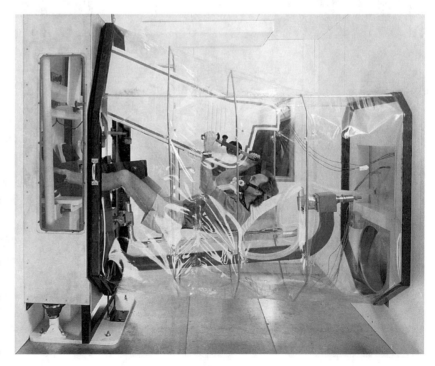

Human Research Facility The EVA (Extra Vehicular Activity) Exercise Device for evaluation of weightlessness on astronauts during long-duration space-flights. *NASA*

Hunsaker, Jerome C. (1886–1984)

A senior aeronautical engineer at the Massachusetts Institute of Technology. He was heavily involved in the development of the science of flight in America for the first three-quarters of the twentieth century.[181]

Huygens probe

See **Cassini**.

hybrid motor

A rocket motor that burns a combination of propellants of different composition and characteristics, such as a liquid oxidizer and a solid fuel, to produce propulsive force.

hybrid propellant rocket engine

An engine that combines some of the advantages of both liquid- and solid-fueled rocket engines. The basic design of a hybrid consists of a combustion chamber tube, similar to that of ordinary solid-propellant rockets, packed with a solid chemical, usually the fuel. Above this tube is a tank containing a complementary reactive liquid chemical, usually the oxidizer. The two chemicals are **hypergolic** (self-igniting), and when the liquid chemical is injected into the combustion chamber containing the solid chemical, ignition occurs and thrust is produced. The engine can be throttled by varying the rate of liquid injected, and stopped and started by cutting off the liquid flow and later restarting it. Other advantages of hybrid engines are that they provide higher energy than standard-solid propellant rockets, can be stored for long periods like solid-propellant rockets, and contain less than half the complex machinery, such as pumps and plumbing, of standard liquid-propellant engines. They are also less sensitive to damage to the solid fuel component than are standard solid-propellant systems. A disadvantage is that they generate less energy per kilogram of propellant than liquid-propellant engines and are more complex than standard solid-fueled engines. Although hybrid propellant rocket engines are still under development, versions with more than 100,000 kg of thrust have been demonstrated.

hydrazine (N_2H_4)

A liquid rocket fuel used both as a **monopropellant**, especially in attitude control thrusters, and a **bipropellant**. It is ignited by passing it over a heated **catalyst** (alumina pellets impregnated with iridium) that decomposes the fuel and produces ammonia, nitrogen, and hydrogen exhaust gases. See **MMH** and **UDMH**.

hydrobot

A remote-controlled submarine of the type expected to be deployed eventually on Jupiter's moon Europa, which is thought to have a subterranean sea. Hydrobots will be released into the subsurface liquid ocean that may exist on this world and will image their surroundings, take a variety of measurements of the physical and chemical environment, and search for possible forms of life.

hydrogen (H)

The lightest chemical element; a flammable, colorless, tasteless, odorless gas in its uncombined state. **Liquid hydrogen** is used as a rocket fuel.

hydrogen peroxide (H_2O_2)

An oxidizer that was used in some early liquid-propellant rocket engines, including those of the German Messerschmitt 163 aircraft and the British **Black Knight** missile. Hydrogen peroxide has not been used subsequently because it is dangerous to handle and easily decomposes, making it difficult to store.

hyperbola

A curve formed by the intersection of a cone with a plane that cuts both branches of the cone (see **conic sections**). Everywhere on a given hyperbola the difference between two fixed points, known as the foci, remains constant.

hyperbolic orbit

An open orbit in the shape of a **hyperbola**. A hyperbolic orbit is followed, for example, by a spacecraft making a flyby of a planet and, in general, is followed by any object that escapes from the gravitational field of a larger body.

hyperbolic velocity

A velocity that exceeds the **escape velocity** of a planet, star, or other massive body. It allows an object, such as a spacecraft, to break free of the gravitational attraction of the larger body by following an open, **hyperbolic orbit**.

hypergolic fuel

A rocket fuel that ignites spontaneously upon contact with an oxidizer, thereby eliminating the need for an ignition system.

hypersonic

Descriptive of something that travels at speeds greater than **Mach** 5 (five times the ambient speed of sound).

hypervelocity gun

Test equipment for accelerating particles to the velocity range of meteoroids in space. The equipment is used in investigations of meteoroid damage to spacecraft and methods of meteoroid protection.

hypoxia

Oxygen deficiency in blood cells or tissues in such a degree as to cause psychological and physiological disturbances. Hypoxia may result from a scarcity of oxygen in the air being breathed or from the inability of the body's tissues to absorb oxygen under conditions of low barometric pressure. It produces a variety of reactions in the body, from mild intoxication and stimulation of the nervous system to progressive loss of attention and judgment, unconsciousness, and eventual brain damage. The continuing physiological requirement for oxygen is a critical consideration in high-altitude flight and in spaceflight.

I

IBSS (Infrared Background Signature Survey)

A test of infrared sensor technology for **SDIO** (Strategic Defense Initiative Organization). The IBSS sensors were mounted on a **SPAS**-II pallet and used to observe the Space Shuttle's engines firing from a distance away. They also observed a variety of common rocket propellants, including nitrogen tetroxide, MMH, and UDMH, released by three Chemical Release Observation subsatellites launched from **Get-Away Special** canisters on the Shuttle.

Shuttle deployment
Date: May 1, 1991
Mission: STS-39
Orbit: 245 × 255 km × 57.0°
Mass: 1,901 kg

ICAN (Ion Compressed Antimatter Nuclear) rocket

See **antimatter propulsion**.

ICE (International Cometary Explorer)

A probe that was originally known as **ISEE**-3 (International Sun-Earth Explorer 3) but was renamed ICE when it was reactivated and diverted, after four years' service, in order to pass through the tail of comet Giacobini-Zinner and to observe Halley's Comet from a distance of 28 million km. ICE became the first spacecraft to be steered by **chaotic control**–a technique that allowed the probe to be directed to its cometary encounters despite having very little remaining fuel. With major rocket maneuvers out of the question, mission controllers decided to see if they could exploit the gravitational instability in the region of two **Lagrangian point**s, the Sun-Earth and Earth-Moon L1 points, to coax ISEE-3 onto its new deep-space trajectory. On June 10, 1982, they began a series of 15 small burns to gradually nudge the satellite onto a path toward the Earth-Moon L1 point. Once they had the satellite in the Earth-Moon system, they flew it past the Earth-Moon L1 point five times, giving it a tiny boost on each lunar flyby, until they had it on a path that would eventually lead to a rendezvous with Giacobini-Zinner. The fifth and final lunar flyby took place on December 22, 1983, when the satellite passed just 119 km above the Apollo 11 landing site. It was at this point that the spacecraft was renamed ICE. On June 5, 1985, ICE passed through Giacobini-Zinner's plasma tail, 26,550 km behind the comet's nucleus, making measurements of particles, fields, and waves. In March 1986, it passed between the Sun and Halley's Comet, adding to the data collected by other spacecraft–Giotto, Planet-A, MS-T5, and Vega–that were in the vicinity of Halley on their rendezvous missions. Thus ICE became the first spacecraft to directly investigate two comets. Its mission was further extended by NASA in 1991 to include studies of coronal mass ejections in coordination with ground-based observations, continued cosmic ray studies, and special periods of observation such as when ICE and Ulysses were on the same solar radial line. ICE was finally shut down in May 1997. The spacecraft will return to the vicinity of Earth in 2014, when it may be possible to capture it and bring it home. NASA has already donated it to the Smithsonian Institute in the event of a successful recovery.

ice frost

Ice on the outside of a launch vehicle over surfaces supercooled by liquid oxygen.

ICESAT (Ice, Cloud, and Land Elevation Satellite)

A NASA satellite that will accurately measure the elevations of Earth's ice sheets, clouds, and land, and will help answer fundamental questions about the growth or retreat of Earth's polar ice sheets and future global sea level changes. ICESAT will also measure the heights of clouds for studies of Earth's temperature balance, and measure land topography for a variety of scientific and potential commercial applications. Its primary instrument is a laser altimeter, developed at the Goddard Space Flight Center, which works by transmitting short pulses of infrared light and visible-green light. The distance from the spacecraft to clouds and to Earth's surface will be determined from measurements of the time taken for the laser pulses to travel to and from these targets. Formerly known as EOS Laser Alt-1, ICESAT is part of NASA's **EOS** (Earth Observing System). It will be placed into a 606-km-high orbit with an inclination of 94°. Launch was scheduled for December 2002.

ICO

A constellation of satellites, owned by ICO Global, which is a component of a new global mobile communications system. The system will support digital data and voice services and the satellite equivalent of third-generation (3G) wireless, including wireless Internet and other packet-data services. Ten satellites plus two spares will make up the constellation, operating at an altitude of 10,390 km and divided equally between two planes, each inclined at 45° to the equator and at right angles to the other, to provide complete, continuous overlapping coverage of the globe. The satellites will communicate with terrestrial networks through ICONet, a high-bandwidth global Internet Protocol network, consisting of 12 Earth stations around the world connected via high-speed links. ICO aims to launch commercial satellite services late in 2003.

IDCSP (Initial Defense Communications Satellite Program)

The American military's first near-geosynchronous satellite communications system. A total of 26 IDCSP satellites were launched in four groups between 1966 and 1968 by Titan IIICs to near-equatorial, 29,500-km-high orbits. Each weighed about 45 kg and drifted from west to east at a rate of up to 30° per day. IDCSP satellites transmitted reconnaissance photos and other data during the Vietnam War. When IDCSP reached initial operational capability, the system was renamed **DSCS** I (Defense Satellite Communications System I). It was succeeded by the **NATO** and DSCS II true geosynchronous satellites in the 1970s.

ideal burning

A term applied to solid propellants. Ideal burning occurs when the propellant burns in such a way that the thrust produced by the rocket motor and the pressure of combustion remain constant throughout the burn period.

ideal rocket

A hypothetical rocket engine that satisfies a number of assumptions that make carrying out calculations on rocket performance much easier. Among these assumptions are that the propellant flow is completely steady, that all the reactants and products are perfectly and evenly mixed, that no heat escapes through the rocket walls, and that all processes are frictionless. Although these are idealizations and simplifications, they are good enough for many purposes. For example, the measured actual performance of a chemical rocket propulsion system is typically only 1% to 6% below that predicted for the ideal case.

IEH (International Extreme Ultraviolet Hitchhiker)

A package of experiments to study sources of high-energy ultraviolet radiation, including the Sun, hot stars, and more remote objects, and the interaction of solar ultraviolet with Earth's upper atmosphere. IEH is carried inside the Space Shuttle's payload bay, deployed in orbit, and

then returned to Earth at the end of the Shuttle's mission. It has been flown several times in a variety of configurations.

IEOS (International Earth Observing System)

A global system of missions made up of **EOS** (Earth Observing System) satellites together with other Earth observation missions from NOAA (National Oceanic and Atmospheric Administration), Europe, and Japan.

ignition

The start of combustion in a rocket engine.

ignition lag

The time interval measured from the instant that electrical contact is made to the first perceptible rise of pressure or thrust, whichever occurs first.

IKI

See **Russian Space Research Institute**.

Ikonos

The first high-resolution commercial imaging satellite; "ikonos" is Greek for "image." It carries a 1-m resolution panchromatic camera and a 4-m resolution color imager, with a 13-km swathe width. The first attempt to launch Ikonos, on April 27, 1999, was unsuccessful following the failure of the payload shroud to separate.

Launch
    Date: September 24, 1999
    Vehicle: Athena 2
    Site: Vandenberg Air Force Base
Orbit: 678 × 679 km × 98.2°
Mass: 726 kg

illicit cargo

Among the many items that have been carried into space without official approval are (1) a small set of handbells that Tom **Stafford** brought for the Gemini 6 mission that took place just before Christmas, 1965; (2) a batch of first-day cover postage stamps, stowed in the personal lockers aboard Apollo 15 and taken to the Moon, which broke NASA regulations and led to the crew being severely reprimanded; and (3) a corned beef sandwich smuggled aboard Gemini 3 and eaten by Gus **Grissom**, who later brought it back up, thus ruining a dietary experiment that had been set for him.

IMAGE (Imager for Magnetopause-to-Aurora Global Exploration)

A NASA satellite that carries some of the most sophisticated imaging instruments ever to be flown in the near-

Earth space environment. From an elliptical orbit about Earth's poles, IMAGE's two-year mission is to obtain the first global images of the major **plasma** regions and boundaries in Earth's inner **magnetosphere**, and to study the dynamic response of these plasma populations to changes in the flow of charged particles from the Sun. It is the first of NASA's **MIDEX** (Medium-class Explorer) missions.

Launch
Date: March 25, 2000
Vehicle: Delta 7326
Site: Vandenberg Air Force Base
Orbit: 1,000 × 45,922 km × 40°
Mass: 494 kg

IMEWS (Integrated Missile Early Warning Satellite)

American military satellites designed to operate from **geosynchronous orbit** and detect launches of enemy intercontinental and submarine-launched ballistic missiles using infrared and optical sensors. Four were launched–the first unsuccessfully–by Titan IIICs in the early 1970s. The IMEWS program was a successor to **MIDAS**.

IMINT (imagery intelligence) satellite

A satellite that uses film and electronic cameras, or radar, to produce high-resolution images of objects on the ground at ranges of up to 1,000 km. Orbiting at altitudes of several hundred kilometers, today's IMINT satellites can identify and distinguish different types of vehicles and equipment with a resolution of up to 10 cm. They are used both for peacetime collection of intelligence, including verification of arms control agreements, and the location of military targets in wartime. The 1991 Gulf War was the first time that data returned from IMINT satellites directly supported combat operations. The first generation of IMINT satellites, known as **Corona**, returned high-resolution photographs to Earth via small reentry capsules. Film return satellites were superseded in the 1980s by electronic imaging satellites, which return pictures to Earth via telemetry.

IMP (Interplanetary Monitoring Platform)

A series of NASA probes, managed by the Goddard Space Flight Center, aimed at investigating plasma (ionized gas) and magnetic fields in interplanetary space. The placement of IMPs in a variety of solar and Earth orbits enabled study of spatial and temporal relationships of geophysical and interplanetary phenomena simultaneously by several spacecraft. The IMP network also provided a crucial early warning of solar flare activity for Apollo manned missions that ventured beyond the Van Allen belts. IMPs were a subprogram of the **Explorer** series and also designated Explorers 18, 21, 28, 33, 34, 35, 41, 43, 47, and 50. The last of them, IMP 8 (Explorer 50), launched in October 1973 into a near-circular orbit of the Earth at just over half the distance to the Moon, was finally switched off in October 2001.

impulse

In general terms, a sudden unidirectional force. More precisely, in physics, impulse is **thrust** multiplied by time. It is measured in newton seconds (N.s).

impulse, total

The integral of the **thrust** (which can vary) over the **burn time**. If the thrust is constant, the total impulse is just the product of thrust and burn time. The total impulse is proportional to the total energy released by the propellant in a propulsion system.

impulse-weight ratio

A quantity used to measure the efficiency of rocket fuels. It is the ratio of the total impulse to the liftoff weight of a space vehicle.

inclination

(1) One of the principal **orbital elements** used to define an orbit; it is the angle between the orbital plane and a reference plane. For the orbits of planets and other bodies around the Sun, the reference plane is the **ecliptic** (the plane in which Earth's orbit lies); for satellite orbits, it is the plane of the equator of the parent planet. (2) The angle between a body's rotation axis and a reference plane, normally taken to be the body's orbital plane.

India in space

See **Aryabhata**, **Indian launch vehicles**, **Indian Space Research Organisation**, **Rohini**, Vikram **Sarabhai**, **Sriharikota High Altitude Range Center**, **SROSS**, **Tipu Sultan**.

Indian launch vehicles

In order of development, these are **SLV-3** (Space Launch Vehicle 3), **ASLV** (Advanced Space Launch Vehicle), **PSLV** (Polar Satellite Launch Vehicle), and **GSLV** (Geosynchronous Satellite Launch Vehicle).

Indian Space Research Organisation (ISRO)

The body that oversees all space activities in India. It operates several facilities, including the Vikram Sarabhai Space Center at Trivandrum and the **Sriharikota High Altitude Range Center**.

inertia

The property of an object to resist changes to its state of motion. Being an inherent property of **mass**, it is present even in the absence of gravity. For example, although a spacecraft may be located well away from any gravitating mass, its inertia must still be overcome in order for it to speed up, slow down, or change direction. Inertia is a familiar concept in physics, but its origin is poorly understood. It is believed to be a property of bodies primarily due to their interaction with masses at the edges of the universe. Momentum (mass times velocity) is a measure of inertia; changes in momentum, resulting from the action of an unbalanced force, permit a measure of inertial mass.

inertia wheel

A rotating mass used to absorb minor torques (twisting forces) created during the stabilization and control of a spacecraft.

inertial force

(1) The force produced by the reaction of a body to an accelerating force, equal in magnitude and opposite in direction to the accelerating force. An inertial force lasts only as long as the accelerating force does. (2) A force that must be added to the equations of motion when Newton's laws are used in a rotating or an otherwise accelerating frame of reference. It is sometimes described as a fictional force because when the same motion is solved in the frame of the external world, the inertial force does not appear.

inertial guidance

A type of guidance for a missile or a space vehicle implemented by mechanisms that automatically adjust the vehicle after launch to follow a given flight path. The mechanisms measure inertial forces during periods of acceleration, integrate the data obtained with previously known position and velocity data, and signal the controls to bring about changes needed in direction, altitude, and so forth.

inertial platform

A gyro-stabilized reference platform that forms part of a guidance system.

inertial reference frame

A frame of reference in which the first of **Newton's laws of motion** is valid. According to the **special theory of relativity**, it is impossible to distinguish between such frames by means of any internal measurement. For example, if someone were to make measurements using a stopwatch or a ruler inside a spaceship traveling at constant high speed (even close to the speed of light) relative to

Earth, they would get exactly the same results as if the ship were at rest relative to Earth.

inertial stabilization

A type of stabilization in which a spacecraft maintains a fixed attitude along one or more axes relative to the stars.

inertial upper stage (IUS)

A two-stage solid-rocket motor used to boost heavy satellites out of low Earth orbit into a higher orbit. It can be used in conjunction with both the **Space Shuttle** and **Titan**.

Inflatable Antenna Experiment (IAE)

A large antenna that was carried aboard the Space Shuttle, deployed on the **Spartan** 207 platform, then inflated in orbit. This experiment laid the groundwork for future inflatable structures in space such as telescopes and satellite antennas. IAE was developed by L'Garde of Tustin, California, a small aerospace firm, under contract to JPL (Jet Propulsion Laboratory). Once in low Earth orbit, the Spartan served as a platform for the antenna, which, when inflated, was about the size of a tennis court. First, the IAE struts were inflated, followed by the reflector canopy. High-resolution video photography recorded the shape and the smoothness of the reflector surface over a single orbit. Finally, the antenna was jettisoned from the Spartan platform and disintegrated as it fell through the atmosphere.

Shuttle deployment
    Date: May 20, 1996
    Mission: STS-77
    Orbit: 180 km × 201 km × 39.1°

infrared

Electromagnetic radiation lying between the regions of visible light and **microwave radiation** with wavelengths of 0.75 to 1,000 microns. Very little of the infrared spectrum from space reaches to sea level, although more of it can be observed by high-altitude aircraft, such as **SOFIA** (Stratospheric Observatory for Infrared Astronomy), or telescopes on high mountaintops, such as the peak of Mauna Kea in Hawaii. Most cosmic infrared is best studied by **infrared astronomy satellites**. The *near infrared* spans the shorter infrared wavelengths from 0.75 (just beyond visible red) to 1.5 microns, the *intermediate infrared* from 1.5 to 20 microns, and the far infrared from 20 to 1,000 microns (the start of the submillimeter band).

infrared astronomy satellites

In launch order, these are: **IRAS** (January 1983), **Hubble Space Telescope** (April 1990), **IRTS** (March 1995), **ISO**

(November 1995), **WIRE** (March 1999), **SIRTF** (December 2001), **IRIS** (2003), **NGSS** (2003–2004), **PRIME** (2006), **FIRST** (2007), **FAIR** (2007+), **SPIRIT** (2007+), **NGST** (2010), and **IRSI** (2014).

infrared radiometer
An instrument that measures the intensity of infrared energy coming from a source.

ingress
The act of, or the mechanism for, entrance to an enclosure. In spacecraft this can relate to the act of a crew member entering the space vehicle, or it can describe the entrance chamber, pressure lock, and hatchways.

inhibitor
A substance bonded, taped, or dip-dried onto a solid propellant to restrict the burning surface and give direction to the burning.

initiator
A unit that receives electrical or detonation energy and produces a chemical deflagration reaction.

injection
(1) The process of putting an artificial satellite into orbit. (2) The time of such action. *Injection altitude* is the height at which a space vehicle is turned from its launch trajectory into an orbital trajectory. *Injection mass* is the mass of a space vehicle at the termination of one phase of a mission that continues into the next phase.

injector
In a liquid-propellant rocket engine, the device that injects fuel and oxidizer into the **combustion chamber**. The injector also mixes and atomizes the propellants for efficient combustion. *Injection pressure* is the pressure at which the fuel is injected into the combustion chamber.

Inmarsat
A network of satellites operated by the International Maritime Satellite Organization (IMSO) to support global mobile transmissions at sea, in the air, and on land.

INPE (Instituto Nacional de Pesquisas Espacias)
Brazil's National Institute for Space Research, a government body that manages and coordinates the country's involvement in space-related activities.

Insat
India's multipurpose satellites for telecommunications, meteorological imaging and data relay, radio and television program distribution, and direct television broadcasting for community reception. The most recent, Insat-3C, was launched on January 31, 2002.

Inspector
A spacecraft intended to serve as a remote inspection device for the **Mir** space station. Built by Daimler-Benz Aerospace, it was delivered to Mir by a Progress-M supply vessel and deployed on December 17, 1997. However, Inspector failed, possibly due to a faulty gyro, and was abandoned by cosmonauts. Mir was boosted out of the way, and the inspection craft became an unintended microsatellite before its orbit decayed on November 2, 1998.

Launch
   Date: October 5, 1997
   Vehicle: Soyuz-U
   Site: Baikonur
Orbit: 377 × 387 km × 51.7°
Mass: 72 kg

INTA (Instituto Nacional de Técnica Aeroespacial)
The Spanish space agency. INTA implements and develops Spain's contributions to international space projects, especially those of **ESA** (European Space Agency).

INTEGRAL (International Gamma-Ray Astrophysics Laboratory)
An ESA (European Space Agency) satellite that will provide the sharpest cosmic gamma-ray pictures to date and identify many new gamma-ray sources. It is designed to carry out fine spectroscopy and imaging of cosmic gamma-ray sources in the 15 keV to 10 MeV energy range, with concurrent source monitoring in the X-ray (3–35 keV) and optical (V-band, around 550 nm) energy ranges. Scheduled to be placed in a highly elliptical orbit in October 2002 by a **Proton** rocket from Baikonur, INTEGRAL has a nominal lifetime of two years with a possible extension of up to five years.

Intelsat (International Telecommunications Satellite Organization)
An organization formed in 1964 that designs, builds, and operates a global system of communications satellites. For most of its life, Intelsat was run by a consortium of government telecommunications authorities with shares in the organization in proportion to their use of it. However, in 2000, the nearly 150 member nations agreed to turn the enterprise from a treaty-based organization into a privately held company. The following year, Bermuda-based Intelsat was formed.

Intelsat   Three crew members of Shuttle mission STS-49 hold onto the 4.5-ton Intelsat 6 after a six-handed "capture" made minutes earlier. *NASA*

Intelsat 1, also known as Early Bird, was launched in April 1965 and became the first comsat to provide regular commercial telecommunications. It could support either one TV channel or 240 voice circuits but not both, a limitation that made it costly to use. In the spring of 1967, Early Bird was joined by two larger companions—Intelsat 2 over the Pacific and Intelsat 3 over the Atlantic. With these three satellites, all the world's TV networks could be linked together, and the first global telecast was broadcast on June 25, 1967, which included the Beatles' live performance of "All You Need Is Love."

During the 1960s and 1970s, message capacity and transmission power of the Intelsat 2, 3, and 4 generations were progressively increased by segmenting the voice circuits into more and more transponder (transmitter-receiver) units, each having a certain bandwidth. The first of the Intelsat 4 series, launched in 1971, provided 4,000 voice circuits. With the Intelsat 5 series (1980), the introduction of multiple beams directed at the Earth resulted in even greater capacity. A satellite's power could now be concentrated on small regions of the planet, making possible lower-cost ground stations with smaller antennas. An Intelsat 5 satellite could typically carry 12,000 voice circuits. The Intelsat 6 satellites, which entered service in 1989, can carry 24,000 circuits and feature dynamic onboard switching of telephone capacity among six beams, using a technique called SS/TDMA (satellite switched/time-division multiple access). Intelsat 7 satellites provide up to 112,500 voice circuits and three TV circuits each, depending on the market needs in the orbital location. Satellites in the most recently completed series, Intelsat 8 and 8A, launched in the late 1990s, can simultaneously handle 112,500 phone calls, or 22,000 phone calls plus three color TV broadcasts. These are now being joined by a 10-strong fleet of the even more powerful Intelsat 9, each carrying 76 C-band and 24-Ku-band transponders (see **frequency bands**). When the new fleet is in place by 2003–2004, Intelsat expects to have a constellation of 24 operational satellites supplemented by over 600 Earth stations. (See table, "Intelsat Series.")

Interbol

Two pairs of spacecraft designed to measure different parts of Earth's magnetic field: one pair with orbits of 500 km by 200,000 km (tail probes), the other with orbits

---

Launch
    Vehicle: Molniya-M
    Site: Plesetsk

---

of 500 km by 20,000 km (auroral probes). Originally an Intercosmos project with a launch planned for the late 1980s, Interbol was delayed until 1995 and 1996 following the breakup of the Soviet Union. Each pair of spacecraft consists of a Russian Prognoz-M (1,250-kg tail probe and 1,400-kg auroral probe) and a Czech Magion (50 kg) satellite. Both Prognoz probes carried a variety of plasma and charged particle detectors, including Swedish, French, and Canadian instruments. The Magion subsatellites can fly in close proximity to the Prognoz or maneuver to as much as 10,000 km from the mother craft. (See table, "Interbol Missions.")

Intelsat Series

| Spacecraft | No. | Launch | | Circuit Capacity | Mass (kg) |
		Dates	Vehicle		
Intelsat 1	1	Apr. 1965	Delta D	240 voice or 1 TV	39
Intelsat 2	3	Jan. 1966–Sep. 1967	Delta E	240 voice or 1 TV	87
Intelsat 3	5	Dec. 1968–Apr. 1970	Delta M	1,500 voice or 4 TV	287
Intelsat 4	7	Jan. 1971–May 1975	Atlas-Centaur	4,000 voice or 2 TV	1,410
Intelsat 4A	6	Sep. 1975–Mar. 1978	Atlas-Centaur	7,250 voice or 2 TV	1,520
Intelsat 5	8	Dec. 1980–Jun. 1984	Atlas-Centaur, Ariane 1, Atlas G	12,000 voice + 2 TV	2,000
Intelsat 5A	6	Mar. 1985–Jan. 1989	Atlas G, Ariane 2/3	15,000 voice + 2 TV	2,013
Intelsat 6	4	Oct. 1989–Oct. 1991	Ariane 4	24,000 voice + 3 TV	4,300
Intelsat K	1	Jun. 10, 1992	Atlas IIA	32 TV	2,930
Intelsat 7	6	Oct. 1993–Jun. 1996	Ariane 4, Atlas IIAS	18,000 voice + 3 TV	4,200
Intelsat 7A	3	May 1995–Feb. 1996	Ariane 4, CZ-3B	18,000 voice + 3 TV	4,500
Intelsat 8	4	Mar. 1997–Dec. 1997	Ariane 4, Atlas IIAS	22,000 voice + 3 TV	3,400
Intelsat 8A	2	Feb.–Jun. 1998	Atlas IIAS	22,000 voice + 3 TV	3,520
Intelsat 9	7	Jun. 2001–	Ariane 4, Proton-K	76 C-band + 24 Ku-band transponders	4,723

Interbol Missions			
Spacecraft	**Launch Date**	**Orbit**	**Mass (kg)**
Interbol 1	Aug. 2, 1995	4,426 × 188,331 km × 68.2°	1,250
Interbol 2	Aug. 29, 1996	239 × 1,093 km × 62.8°	1,400

Intercosmos
International cooperative satellites, launched by Soviet boosters, with a variety of missions.

interference
Unwanted electrical signals or noise causing degradation of reception on a communications circuit.

intergalactic travel
This poses all the difficulties of interstellar travel multiplied a millionfold—the factor by which the typical distances between neighboring galaxies exceed those between stars in the neighborhood of the Sun. If intergalactic travel is ever to be achieved, it will require the development of relativistic starships that can accelerate to well over 99% of the speed of light or, alternatively, some method of bypassing normal **space-time** altogether.

International Astronautical Federation (IAF)
A nongovernmental association, with members drawn from government organizations, industry, professional associations, and learned societies from around the world, that encourages the advancement of knowledge about space and the development and application of space assets for the benefit of humanity. It plays an important role in disseminating information and in providing a worldwide network of experts in space development and utilization of space. Founded in 1951, the IAF today has 152 institutional members from 45 countries. Together with its associates, the International Academy of Astronautics and the International Institute of Space Law, the IAF organizes an International Astronautical Congress, which is held each year in a different country, and other seminars and events.

International Geophysical Year (IGY)
A comprehensive series of global geophysical activities spanning the period from July 1957 to December 1958 and timed to coincide with the high point of the 11-year cycle of sunspot activity. It was proposed by the International Council of Scientific Unions (ICSU) in 1952 following a suggestion by Lloyd **Berkner**. In March 1953, the National Academy of Science appointed a U.S. National Committee to oversee American participation in the IGY. The American program included investigations of aurora and airglow, cosmic rays, geomagnetism, glaciology, gravity, the ionosphere, determinations of longitude and latitude, meteorology, oceanography, seismology, solar activity, and the upper atmosphere. In connection with upper-atmosphere research, the United States undertook to develop an orbiting satellite program. It was from the IGY rocket and satellite research that the United States developed its space program.[164]

International Launch Services (ILS)
An organization formed in June 1995 with the merger of Lockheed Martin Commercial Launch Services, the marketing and mission management arm for the **Atlas** vehicle, and Lockheed-Khrunichev-Energia International, the marketing and mission management arm for commercial **Proton** launches.

International Quiet Sun Years (IQSY)
The period from January 1, 1964, to December 31, 1965, near solar minimum, when solar and geophysical phenomena were studied by observatories around the world and by spacecraft to improve our understanding of solar-terrestrial relations.

International Solar Polar Mission
See **Ulysses**.

International Space Science Institute
A forum in which space scientists, astronomers, ground-based observers, and theorists can work together on the joint analysis and interpretation of their data to achieve a deeper understanding of space mission results. ISSI funding comes from ESA (European Space Agency), the Swiss federal government, the canton of Bern, and the Swiss National Science Foundation.

International Space Station (ISS)
See article, pages 199–204.

interplanetary magnetic field
The weak magnetic field filling interplanetary space, with field lines usually connected to the Sun. The IMF is kept out of Earth's magnetosphere, but the interaction of the

(continued on page 204)

International Space Station (ISS)

The largest structure ever built in space and the most complex international scientific project in history. When complete, the ISS will be over four times larger than **Mir** and longer than a football field (including the end zones) and have a pressurized living and working space equivalent to the volume of a 747 jumbo jet. Its solar panels, spanning more than half an acre, will supply 60 times more electrical power than that available to Mir. Its 400-km-high orbit, inclined at 51.6°, allows it to be reached by the launch vehicles of all the international partners for delivery of crews, components, and supplies. The orbit also enables observations to be made of 85% of the globe and overflight of 95% of Earth's population. When complete, ISS will have six scientific laboratories for research–the U.S. Destiny laboratory, the ESA (European Space Agency) Columbus Orbital Facility, a Japanese experiment module, the U.S. Centrifuge Accommodation Module, and two Russian research modules.

INTERNATIONAL SPACE STATION FACTS

Mass: 470 tons
Size: 109 × 88 m
Pressurized volume: 1,300 cubic m
Solar array area: 2,500 square m
Length of electrical wire: about 13 km
Number of computers: 52

In 1984, President Reagan committed the United States to developing a permanently occupied space station and, along with NASA, invited other countries to join the project. Within little more than a year, 9 of ESA's 13 member countries had signed on, as had Canada and Japan. In 1991, President Bush (senior) and Soviet Premier Gorbachev agreed to joint Space Shuttle–Mir missions that would lay the groundwork for cooperative space station efforts. Since 1993, NASA has had to contend with numerous cost overruns and tight federal space budgets that have eroded the station's capabilities and have delayed its completion. Fresh budget constraints in 2001 put the planned Habitation Module (providing more crew accommodation) and the X-38 **ACRV** on hold, effectively cutting the maximum crew size from seven to three. This seriously threatens the amount of research that can be done on the ISS. Nevertheless, by streamlining the program, simplifying the station's design, and negotiating barter and cost-sharing agreements with other nations, NASA and its international partners have made the ISS a reality. On-orbit assembly of the station began on November 20, 1998, with the launch of the Russian-built Zarya control module, and is due for completion in 2005–2006.

International Contributions

Although the United States, through NASA, leads the ISS project, 15 other countries are involved in building and operating various parts of the station–Russia, Canada, Japan, Brazil, and 11 member nations of ESA (Belgium, Denmark, France, Germany, Italy, the Netherlands, Norway, Spain, Sweden, Switzerland, and the United Kingdom). Contributions include:

United States

- The truss structures that provide the ISS framework
- Four pairs of large solar arrays
- Three nodes (Unity, Node 2, and Node 3, the last of which is in doubt) with ports for spacecraft and for passage to other ISS elements
- An airlock that accommodates American and Russian spacesuits
- The American laboratory
- A Habitation Module (in doubt) and a Centrifuge Accommodation Module
- Power, communications, and data services
- Thermal control, environmental control, and life support

Russia

- Two research modules
- A service module with its own life support and habitation systems
- A science power platform that supplies about 20 kilowatts of electrical power
- Logistics transport using Progress vehicles and Soyuz spacecraft crew rotation

International Space Station   An artist's rendering of the International Space Station after assembly is complete. *NASA*

ESA

• The Columbus Orbital Facility, which includes a pressurized laboratory and external payload accommodations

• Logistics transport vehicles to be launched by the Ariane 5

Canada

• The Mobile Servicing System, which includes a 17-m-long robotic arm along with a smaller manipulator attachment, to be used for assembly and maintenance tasks

• A Mobile Remote Servicer Base to allow the robotic arm to travel along the truss

Japan

• The on-orbit Kibo facility, which includes a pressurized laboratory, a Logistics Module, and an attached facility exposed to the vacuum of space serviced by a robotic arm

• Logistics resupply using the H-2 launch vehicle

Brazil

• A pallet to house external payloads, unpressurized logistics carriers, and an Earth observation facility

Prelude to ISS: The Shuttle-Mir Program

Between 1995 and 1998, nine Shuttle-Mir docking missions were flown, and American astronauts stayed aboard Mir for lengthy periods. Nine Russian cosmonauts rode on the Shuttle, and seven American astronauts spent a total of 32 months aboard Mir, with 28 months of continuous occupancy starting in March 1996. By contrast, it took the Shuttle fleet more than a dozen years and 60 flights to accumulate one year in orbit. Valuable experience was gained in training international crews, running an international space program, and meeting the challenges of long-duration spaceflight for mixed-nation astronauts and ground controllers. Dealing with the real-time challenges of the Shuttle-Mir missions also fostered a new level of cooperation and trust between those working on the American and Russian space programs.

The ISS Takes Shape

Construction of the ISS began in late 1998 and will involve a total of 45 assembly missions, including 36 by the Shuttle, and numerous resupply missions by unmanned Progress craft and rotations of Soyuz crew-return vehicles. American and Russian astronauts will take part in about 850 clock-hours of spacewalks during the five-year building period. (See tables, "Early ISS Missions" and "First ISS Crews.")

Major ISS Components and Servicing Equipment

Zarya ("Dawn") Control Module

A 21-ton, 12.5-m-long, 4.1-m-wide module, equipped with solar arrays and six nickel-cadmium batteries capable of generating an average of 3 kW of power, which provided early propulsion, power, fuel storage, and communication, and served as the rendezvous and docking port for Zvezda. Zarya's construction was funded by NASA and undertaken in Moscow by Boeing and the Khrunichev State Research and Production Space Center. Following its launch, it was put through a series of tests before being commanded to fire its two large engines to climb to a circular orbit 386 km high. The module's engines and 36 steering jets have a six-ton reservoir of propellant to enable altitude and orientation changes. Its side docking ports are used by Russia's Soyuz piloted spacecraft

Early ISS Missions

Launch Date	Description
Nov. 20, 1998	A Proton rocket places the Zarya module in orbit.
Dec. 4, 1998	Shuttle mission STS-88 attaches the Unity module to Zarya.
May 27, 1999	STS-96 delivers tools and cranes to the two modules.
May 19, 2000	STS-101 conducts maintenance and delivers supplies to prepare for arrival of Zvezda and the station's first permanent crew.
Jul. 12, 2000	A Proton rocket delivers Zvezda.
Sep. 8, 2000	STS-106 delivers supplies and outfits Zvezda.
Oct. 11, 2000	STS-92 delivers the Z1 Truss, a pressurized mating adapter for Unity, and four gyros.
Nov. 2, 2000	Arrival of Expedition One crew aboard a Soyuz spacecraft.
Nov. 30, 2000	STS-97 installs the first set of American solar arrays.
Feb. 7, 2001	STS-98 delivers the Destiny laboratory module and relocates a pressurized mating adapter from the end of Unity to the end of Destiny.
Mar. 8, 2001	STS-102 brings Expedition Two crew plus equipment for Destiny and returns on Mar. 17 with Expedition One crew.
Apr. 19, 2001	STS-100 delivers Remote Manipulator System and more laboratory equipment.
Jul. 12, 2001	STS-104 delivers the station's joint airlock.
Aug. 12, 2001	STS-106 brings Expedition Three crew and returns on Aug. 22 with Expedition Two.
Sep. 16, 2001	Delivery of the Russian docking compartment by a Soyuz rocket.
Dec. 7, 2001	STS-108 arrives with Expedition Four crew and departs on Dec. 15 with Expedition Three crew.
Apr. 10, 2002	STS-110 delivers the S0 Truss.
Jun. 7, 2002	STS-111 arrives with Expedition Five crew and brings back Expedition Four crew.

The ISS Takes Shape The STS-88 mission specialist James Newman waves at the camera while holding on to a handrail of the Unity module during early construction work on the ISS. The Shuttle Orbiter can be seen reflected in his visor. *NASA*

and Progress remote-controlled supply vehicles. As assembly progressed, Zarya's roles were assumed by other Station elements, and it is now used primarily as a passageway, docking port, and fuel storage site.

Zvezda Service Module

The first fully Russian contribution to the ISS and the early cornerstone for human habitation of the station. The 19-ton, 13.1-m-long Zvezda provided the first living quarters aboard the station, together with electrical power distribution, data processing, flight control, and propulsion systems. It also has a communications system enabling remote command from ground controllers. Although many of its systems will eventually be supplemented or superseded by American components, Zvezda will remain the

First ISS Crews

| Expedition | Period aboard ISS | Crew Members | |
		Commander	Flight Engineers
One	Nov. 2, 2000–Mar. 14, 2001	Bill Shepherd	Yuri Gidzenko, Sergei Krikalev
Two	Mar. 14–Aug. 9, 2001	Yuri Usachev	Susan Helms, Jim Voss
Three	Aug. 9–Dec. 15, 2001	Frank Culbertson	Vladimir Dezhurov, Mikhail Tyurin
Four	Dec. 15, 2001–Jun. 2002	Yuri Onufrienko	Daniel Bursch, Carl Walz
Five	Jun. 7–Oct. 2002	Valery Korzan	Sergei Treschev, Peggy Whitsun

structural and functional center of the Russian segment of the ISS.

Truss Structure

An American-supplied framework that serves as the backbone of the ISS and the mounting platform for most of the station's solar arrays. The truss also supports a mobile transporter that can be positioned for robotic assembly and maintenance operations and is the site of the Canadian Mobile Servicing System, a 16.8-m-long robot arm with 125-ton payload capability. The truss consists of 9 segments that will be taken up one at a time; four elements will fix to either side of the S0 segment.

Nodes

American-supplied structural building blocks that link the pressurized modules of the ISS together. Unity Node provides six attachment ports, one on each of its sides, to which all other American mod-

ules will join. Node 2, in addition to attachment ports for non-U.S. modules, contains racks of equipment used to convert electrical power for use by the international partners. Node 3, which seems likely to be axed due to budget cuts, was to house life support equipment for the Habitation Module, which may also be scrapped.

Destiny Laboratory Module

America's main workstation for carrying out experiments aboard the ISS. The 16.7-m-long, 4.3-m-wide, 14.5-ton Destiny will support research in life sciences, microgravity, Earth resources, and space science. It consists of three cylindrical sections and two end-cones. Each end-cone contains a hatch through which crew members will enter and exit the lab. There are 24 racks inside the module, 13 dedicated to various experiments, including the **Gravitational Biology Facility**, and 11 used to supply power, cool water, and provide environmental control.

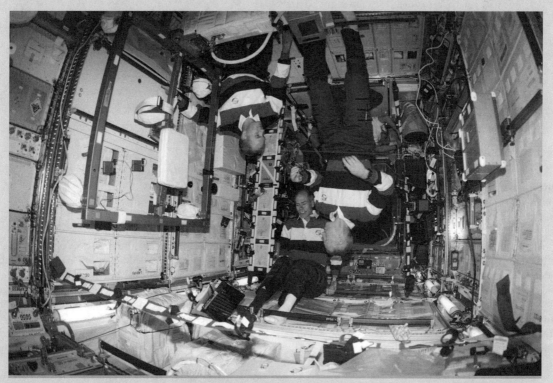

Destiny Laboratory Module   Astronauts Rick Sturckow (left), STS-105 pilot, and Daniel Barry and Patrick Forrester, both mission specialists, work in the ISS Destiny Laboratory in August 2001. *NASA*

Columbus Laboratory

A pressurized research module, supplied by ESA (the European Space Agency) similar in structure to the Multi-Purpose Logistics Module (MPLM). It is due to be attached to Node 2 of the ISS in 2004. Columbus is designed as a habitable, general purpose laboratory capable of supporting research in materials and fluid sciences, life sciences, and technology development.

Kibo Laboratory

Japan's major contribution to the ISS. It consists of a pressurized laboratory module, 11.2 m long and 15.9 tons in mass, to which is attached an exposed facility (for carrying out experiments in the hostile environment of space), a robot arm, and an experiment logistics module (designed to carry pressurized and unpressurized cargo). The first element of Kibo, the logistics module, is scheduled for launch in 2004.

Russian Research Modules

Two pressurized science laboratory modules, built in Ukraine and Russia, scheduled for launch by Soyuz rockets in 2005 and 2006.

Centrifuge Accommodation Module (CAM)

An American pressurized laboratory that will enable long-term study of the effects of varying gravity levels on the structure and function of generations of living organisms. CAM will house the Centrifuge Facility, a 2.5-m-diameter centrifuge rotor, and the **Gravitational Biology Facility**, with racks for holding plants, animals, and habitats.

Science Power Platform (SPP)

A Russian element of the station, scheduled for launch in 2004, that will provide about 25 kW of power and heat rejection for the station's science experiments and general operation.

Canadian Mobile Servicing System

See **Remote Manipulator System**.

Multi-Purpose Logistics Modules

Effectively the ISS's moving vans. Built by ASI (the Italian Space Agency), they allow the Space Shuttle to ferry experiments, supplies, and cargo back and forth during missions to the station. There are three versions of the module, known as Leonardo, Raffaello, and Donnatello. Each can carry up to 16 standard ISS equipment racks or other supplies.

Automated Transfer Vehicle (ATV)

An unmanned servicing and logistics vehicle for the periodic resupply of the ISS. Supplied by ESA, it will be launched by Ariane 5. The ATV can deliver up to 7.5 tons of cargo, including experiments, fuel, food, water, and compressed air, to the Station and reentry (to burn up in the atmosphere) with up to 6.5 tons of waste.

interplanetary magnetic field

(continued from page 198)

two plays a major role in the flow of energy from the solar wind to Earth's environment.

interplanetary space

The space beyond the orbit of the Moon extending to the limits of the Solar System.

intersatellite link

A message transmission circuit between two communication satellites, as opposed to a circuit between a single satellite and Earth.

interstage

A section of a rocket between two stages, such as the interstage spacing ring between the first and second stages of the **Saturn** V.

interstellar ark

See **generation ship**.

interstellar medium (ISM)

Material between the stars. It consists of about 99% gas (mostly hydrogen) and 1% dust, spread incredibly thinly. The average density of the ISM is only about one particle per cubic cm. However, in some regions, such as molecular clouds, the density can be as high as 10^{10} (10 trillion) particles per cubic cm.

interstellar precursor mission

A robot spacecraft, equipped with an advanced propulsion system, that represents an intermediate step between today's deep-space missions, such as **Voyager**, and the first true star probes. One such mission, known as the **Interstellar Probe**, has been proposed by NASA for launch in 2010 or beyond. Other missions that would travel out to distances of 500 to 1,000 astronomical units (AU) are being considered. For example, the astronomy community is beginning to consider using the Sun as a gravity lens, which would require placing a telescope at 550 AU.

interstellar probe

A robot spacecraft, designed to travel at very high speed to a neighboring star system (such as Alpha Centauri or Barnard's Star) and return data to Earth. There have been various designs for such probes, including projects **Daedalus** and **Longshot**.

Interstellar Probe

A proposed NASA **interstellar precursor mission** that could be launched as early as 2010. The Interstellar Probe is designed to travel at least 250 astronomical units (AU)–37 billion km–from the Sun in order to send back data from the solar termination shock and **heliopause** and make a significant penetration into the local **interstellar medium**. Identified in NASA's Office of Space Science Strategic Plan as a potential future mission, it would require the development of a propulsion system capable of speeds of at least 10 AU per year. Solar sails and high-power electric rocket engines are among the options being considered.

interstellar ramjet

An ingenious extension of the **ramjet** concept to provide, potentially, a highly effective form of interstellar propulsion. It was first suggested by the American physicist Robert W. **Bussard** in 1960, and consequently is sometimes referred to as the Bussard ramjet.

Set against the desirability of achieving speeds for star travel that are a significant fraction of the **speed of light** is the perennial problem of rocketry—having to carry, in addition to the **payload**, the **reaction mass** needed for propulsion. The interstellar ramjet neatly avoids this

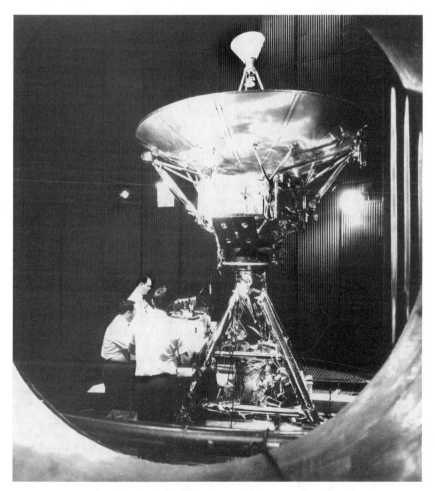

interstellar probe Pioneer 10, the first spacecraft launched from Earth to escape the Sun's gravity and head for interstellar space. *NASA*

problem by harvesting hydrogen for use as a propellant from the **interstellar medium**. The captured hydrogen is fed to a nuclear **fusion** reactor, which supplies the energy for a high-speed exhaust.

Bussard's original design envisaged atomic hydrogen being mechanically scooped up by the spacecraft as it went along. However, his calculations suggested that, in order to achieve the ideal acceleration of 1g (see **1g spacecraft**), a 1,000-ton spacecraft would need a frontal collecting area of nearly 10,000 square km. Even assuming a knowledge of materials science far in advance of our own, it seems inconceivable that such a scoop could be constructed with a mass less than that budgeted for the entire vehicle. A 10,000-square-km structure made of 0.1-cm-thick Mylar, for example, would weigh about 250,000 tons.

A way around this problem is to ionize the hydrogen ahead of the spacecraft using a powerful laser. The hydrogen ions—naked protons—can then be drawn in by a relatively small Bussard collector that generates a powerful magnetic field. Since the harvesting process is electromagnetic rather than mechanical, the scoop does not have to be solid (it can be a mesh), nor does it have to be unrealistically large, because the field can be arranged to extend far beyond the physical structure of the scoop.

However, difficulties remain. One is the enormous power needed to generate the Bussard collector's magnetic field and to operate the ionizing laser. Another problem concerns the way the ram scoop works. As the lines of the magnetic field converge at the inlet funnel, they will tend to bounce away incoming charged particles rather than draw them in. In effect, the scoop will act like a magnetic bottle, trapping material in a wide cone in front of the vehicle and preventing it from being injected as fuel. A solution might be to pulse the magnetic field, but the implementation would not be easy. Yet another problem is that most of the collected matter will be ordinary hydrogen, which is much harder to induce to fuse than either **deuterium** or tritium, hydrogen's heavier isotopes. Finally, the Bussard ramjet will work only when the vehicle is moving fast enough to collect interstellar mass in usable amounts. Therefore a secondary propulsion system is needed to boost the spacecraft up to this critical speed—about 6% of the speed of light.

A modified design known as RAIR (ram-augmented interstellar rocket), proposed by Alan **Bond** in 1974, tackles the fusion-reaction problem by using the scooped-up interstellar hydrogen not as fuel but simply as reaction mass. The incoming proton stream is decelerated to about 1 MeV, then allowed to bombard a target made of lithium-6 or boron-11. Lithium-proton or boron-proton fusion is easy to induce and releases more energy than

any other type of fusion reaction. The energy produced in this way is added to the mass stream, which then exits the reactor. In the exhaust nozzle, the energy created by initially braking the mass stream is added back to it.[29, 156]

The so-called catalyzed-RAIR offers an even more efficient approach. After the incoming mass stream has been compressed, a small amount of antimatter is added. The reaction cross section is not only enormous compared to fusion, it happens at much lower temperatures. According to one estimate, the energy release is such that the drive reactor of a 10,000-ton antimatter catalyzed-RAIR accelerating at 1g and maintaining 10^{18} particles per cubic cm within the reactor only has to be about 3.5 m in diameter. The downside is that large amounts of antimatter would be needed for sustained interstellar flight.[274]

interstellar space

Space starting from the edge of the Solar System and extending to the limit of the Milky Way Galaxy. Beyond this lies intergalactic space.

ion

Usually, an atom from which one or more **electrons** have been stripped away, leaving a positively charged particle. Positive ions are used in **ion propulsion**. Negative ions are atoms that have acquired one or more extra electrons.

ion propulsion

A form of **electric propulsion** in which **ions** are accelerated by an electrostatic field to produce a high-speed (typically about 30 km/s) exhaust. An ion engine has a high **specific impulse** (making it very fuel-efficient) but a very low **thrust**. Therefore, it is useless in the atmosphere or as a launch vehicle but extremely useful in space, where a small amount of thrust over a long period can result in a big difference in velocity. This makes an ion engine particularly useful for two applications: (1) as a final thruster to nudge a satellite into a higher orbit, or for orbital maneuvering or **station-keeping**; and (2) as a means of propelling deep-space probes by thrusting over a period of months to provide a high final velocity. The source of electrical energy for an ion engine can be either solar (see **solar-electric propulsion**) or nuclear (see **nuclear-electric propulsion**). Two types of ion propulsion have been investigated in depth over the past few decades: **electron bombardment thrusters** and **ion contact thrusters**. Of these, the former has already been used on a number of spacecraft. A particular type of electron bombardment thruster, known as **XIPS** (xenon-ion propulsion system), has proved to be particularly effective and will be used increasingly to propel satellites and deep-space probes.

ion propulsion The xenon ion engine being tested at JPL.
NASA/JPL

ION-F (Ionospheric Observation Nanosatellite Formation)

A constellation of three 10-kg nanosatellites to be launched from the Space Shuttle in 2003 into a 380-km-high circular orbit inclined at 51°. The mission is jointly sponsored by the U.S. Department of Defense and NASA, and has been designed by groups at the University of Washington, Utah State University, and Virginia Polytechnic Institute. The goal of ION-F is to investigate satellite coordination and management technologies—tight space formation-flying and intersatellite communication—while at the same time making distributed ionospheric measurements.

ionization

The process by which an atom, or a group of atoms, becomes an **ion**. This may occur, for instance, by absorption of light (photoionization) or by collision with a fast-moving particle (impact ionization). Also, some substances are made up of natural ions (for example, table salt—sodium chloride—contains Na^+ and Cl^- ions), held together by electrical attraction, which separate when dissolved in water, enabling the solution to conduct electricity.

ionized layers

Layers of increased ionization within the **ionosphere**. They are responsible for the absorption and reflection of radio waves and are important in connection with communication and the tracking of satellites and other space vehicles.

ionized plasma sheath

A layer of ionized particles that forms around a spacecraft during reentry and interrupts or interferes with communication with the ground.

ionosphere

A region of Earth's atmosphere that extends from about 80 to 500 km above the surface and is made up of multiple layers dominated by electrically charged, or ionized, atoms.

Ionosphere Mappers

A mission that would address three areas of space weather effects: human radiation exposure, space systems and technology, and global climate change. Ionospheric Mappers is a proposed future mission within NASA's Living with a Star initiative.

IRAS (Infrared Astronomy Satellite)

The first space mission to survey the universe in the thermal **infrared** region of the spectrum. A multinational project involving Britain, the Netherlands, and the United States, it carried a 0.6-m telescope and operated successfully for 300 days, resulting in a catalogue of about 250,000 infrared sources. Among its discoveries were warm dust disks around certain stars, such as Vega and Beta Pictoris, which may indicate systems of planets in the making, "starburst" galaxies in which vast numbers of new stars are forming, and several new comets and asteroids.

Launch
    Date: January 26, 1983
    Vehicle: Delta 3914
    Site: Vandenberg Air Force Base
Orbit: 884 km × 903 km × 99.0°
Mass: 1,073 kg

IRBM (Intermediate Range Ballistic Missile)
A **ballistic missile** with a range of 2,750 to 5,500 km.

IRFNA (inhibited red fuming nitric acid)
See **nitric acid**.

Iridium
A global wireless digital communication system based on a constellation of 66 Lockheed-built **communications satellites** (plus six on-orbit spares) in 780-km-high (**little LEO**) orbits inclined at 86.6° to the equator. Begun in 1997, Iridium was the first commercial satellite system to operate from low Earth orbit. Named because the initial design involved 77 satellites (the number of electrons in an iridium atom), it provides voice, messaging, and data services to mobile subscribers using Motorola-built handheld user terminals. The satellites are arranged in six planes, each 60° out of phase with its neighbors, to provide global coverage. The current operator, Iridium Satellite of Leesburg, Virginia, acquired the system after its previous owner became bankrupt and it appeared that the armada of satellites would be de-orbited.

Iris
A solid-propellant **sounding rocket** for exploring the upper atmosphere.

IRIS (Infrared Imaging Surveyor)
The second infrared astronomy mission of Japan's **ISAS** (Institute of Space and Astronautical Science). IRIS will use a 70-cm telescope, cooled to −267°C using liquid helium, to carry out a highly sensitive survey of the infrared sky and so help astronomers investigate further the formation and evolution of galaxies, stars, and planets. Also known as Astro-F, the mission has been under development since 1997 and is due to be launched into a sun-synchronous polar orbit by an M-5 rocket in 2004.

IRS (Indian Remote Sensing satellite)
A series of Earth resources satellites launched by the Indian Space Research Organisation (ISRO). The first of them, IRS-1A, followed on from the two Bhaskara experimental resources satellites launched in 1979 and 1981. (See table, "IRS Launches.")

IRSI (Infrared Space Interferometer)
See **Darwin**.

IRTS (Infrared Telescope in Space)
A highly successful 15-cm-diameter orbiting infrared telescope developed by Japan's **ISAS** (Institute of Space and Astronautical Science), launched aboard Japan's **SFU** (Space Flyer Unit), and retrieved by the Space Shuttle *Endeavour* on mission STS-72 in December 1995. Attached to the telescope were four instruments: a near-infrared spectrometer (NIRS), a mid-infrared spectrometer (MIRS), a

Launch
    Date: March 18, 1995
    Vehicle: H-2
    Site: Tanegashima
Orbit: 467 × 496 km × 29°

IRS Launches

| Spacecraft | Launch | | | | Mass (kg) |
	Date	Vehicle	Site	Orbit	
IRS-1A	Mar. 17, 1988	Vostok M	Baikonur	894 × 912 × 99°	975
IRS-1B	Aug. 29, 1991	Vostok M	Baikonur	890 × 917 × 99°	980
IRS-1E	Sep. 20, 1993	PSLV	Sriharikota	Launch failure	—
IRS-1C	Dec. 28, 1995	Molniya-M	Baikonur	805 × 817 × 99°	1,250
IRS-P3	Mar. 21, 1996	PSLV	Sriharikota	818 × 821 × 99°	930
IRS-1D	Sep. 29, 1997	PSLV	Sriharikota	737 × 827 × 99°	1,200

far-infrared line mapper (FILM), and a far-infrared photometer (FIRP). MIRS and FIRP were built in collaboration with American researchers. IRTS operated from March 30, 1995, to April 26, 1995, when its supply of cryogenic coolant ran out.

Irwin, James Benson (1930–1991)

An American astronaut and the Lunar Module pilot for **Apollo** 15. Irwin received a B.S. in naval science from the U.S. Naval Academy in 1951 and M.S. degrees in aeronautical engineering and instrumentation engineering from the University of Michigan. In the Air Force, he served as a fighter pilot and a test pilot. His first application to be an astronaut in 1964 failed because he was still recovering from injuries sustained in a plane crash. However, he was among 19 astronauts selected by NASA in April 1966. Although he served on the backup crews for Apollo 12 and Apollo 17, his only flight was on Apollo

James Irwin   Irwin, Lunar Module pilot, uses a scoop in making a trench in the lunar soil during an Apollo 15 moonwalk. Mount Hadley rises approximately 4,500 meters above the plain in the background. *NASA*

15, a mission during which he drove the first lunar rover. Irwin resigned from NASA and the Air Force in July 1972, and set up an evangelical organization. His autobiography, *To Rule the Night,* first published in 1973 and revised in 1982,[154] recounts his astronaut career and the spiritual revelation he experienced while on the lunar surface. Irwin became the first to die of the 12 men who walked on the Moon.

ISAS (Institute of Space and Astronautical Science)

One of Japan's two space agencies, the other being **NASDA** (National Space Development Agency). Based at the University of Tokyo, ISAS is responsible for the development and launch of Japanese astronomy and space science satellites, and lunar and interplanetary probes. Successful past and ongoing **ISAS** space missions include, in chronological order: **Ohsumi, Tansei**-1, **Shinsei, Denpa**, Tansei-2, **Taiyo**, Tansei-3, **Kyokko, Jikiken, Hakucho**, Tansei-4, **Hinotori, Tenma, Ohzora, Sakigake, Suisei, Ginga, Akebono, Hiten, Yohkoh, Geotail, ASCA, HALCA**, and **Nozomi**. Future missions will include **Astro-F, Lunar-A, MUSES-C, SELENE**, and **Solar-B**. ISAS launches take place from the **Kagoshima Space Center** using **M-series** rockets. See **Japan in space** and **Japanese satellite names**.

ISEE (International Sun-Earth Explorer)

A series of three spacecraft built and operated as a fleet by NASA and ESA (European Space Agency) to study the influence of the Sun on Earth's space environment and **magnetosphere**. Specific goals of the mission were to investigate solar-terrestrial relationships at the outermost boundaries of Earth's magnetosphere, the structure of the solar wind near Earth and the shock wave that forms the interface between the solar wind and Earth's magnetosphere, motions of and mechanisms operating in the plasma sheets, and cosmic rays and solar flare emissions in the near interplanetary region. The three spacecraft carried a number of complementary instruments for making measurements of plasmas, energetic particles, waves, and fields and extended the

investigations of previous **IMP** missions. (See table, "ISEE Missions.")

ISEE-3 was initially placed into an elliptical **halo orbit** about the L1 **Lagrangian point**—the first spacecraft to use such an orbit—235 Earth-radii out on the sunward side of Earth, where it continuously monitored changes in the near-Earth interplanetary medium. In conjunction with the ISEE-1 and ISEE-2 mother and daughter spacecraft, which had eccentric geocentric orbits, it explored the coupling and energy transfer processes between the incoming solar wind and Earth's magnetosphere. ISEE-3 also provided a near-Earth baseline for making cosmic-ray and other planetary measurements for comparison with corresponding measurements from deep-space probes. In 1982, it carried out a series of maneuvers based on the revolutionary new technique of **chaotic control**, which resulted in it being ejected out of the Earth-Moon system and into a heliocentric orbit ahead of Earth, on a trajectory intercepting that of comet Giacobini-Zinner. At this time, ISEE-3 was renamed **ICE** (International Cometary Explorer).

Launch
    Vehicle: Delta 2914
    Site: Cape Canaveral

ISIS (International Satellites for Ionospheric Studies)

Joint Canadian/American satellites that carried out measurements over an entire 11-year solar cycle to determine how the ionosphere reacts to changes in the Sun's radiation. The project stemmed from a cooperative agreement between the two countries signed following the success of **Alouette** 1 in 1962. The first spacecraft in the ISIS program was Alouette 2, launched in 1965. Data from this were correlated with results from the more advanced spacecraft, ISIS 1 and 2. (See table, "ISIS Missions.")

Launch
    Vehicle: Delta E
    Site: Vandenberg Air Force Base

ISEE Missions

Spacecraft	Launch Date	Orbit	Mass (kg)
ISEE-1	Oct. 22, 1977	436 × 137,806 km × 12.7°	340
ISEE-2	Oct. 22, 1977	406 × 137,765 km × 13.5°	166
ISEE-3	Aug. 12, 1978	181 × 1,089,200 km × 1.0°	479

ISIS Missions

Spacecraft	Launch Date	Orbit	Mass (kg)
ISIS 1	Jan. 30, 1969	574 × 3,470 km × 88.4°	241
ISIS 2	Apr. 1, 1971	1,353 × 1,423 km × 88.2°	264

Island One

See **O'Neill-type space colony**.

ISO (Infrared Space Observatory)

ESA (European Space Agency) 0.6-m-diameter space-borne infrared telescope. With instrumentation that included a camera, an imaging photopolarimeter, and two spectrometers, ISO was designed to make observations at wavelengths of 2.5 to 200 microns. Among its most important discoveries is that a large fraction of young stars are surrounded by disks of gas and dust out of which planetary systems might form. ISO operated between November 1995 and May 1998.

```
Launch
    Date: November 17, 1995
    Vehicle: Ariane 44P
Orbit: 1,110 × 70,054 km × 5°
Size: 5.3 × 3.6 × 2.8 m
Mass at launch: 2,400 kg
```

Israel in space

Israel's space interests are managed by the Israeli Space Agency (ISA), which is based in Tel Aviv and was established in 1983 as part of the Israel Defense Forces. ISA pursues space programs jointly with Israel Aircraft Industries (IAI), the Interdisciplinary Center for Technological Analysis and Forecasting of Tel Aviv University, and the National Committee for Space Research of the Israel Academy of Sciences and Humanities. IAI produces the **Shavit** launch vehicle, which was used to place Israel's first satellite, **Ofeq**-1, in low Earth orbit on September 19, 1988. Several other Ofeq launches from the **Palmachim** military base–not all successful–have taken place since, the most recent in 2002. Israel also takes part in international space missions, and the first Israeli astronaut, Ilan Ramon, was scheduled to fly aboard the Space Shuttle on mission STS-107 in the second half of 2002.

ISS (Ionospheric Sounding Satellite)

A pair of Japanese satellites, launched by **NASDA** (National Space Development Agency) and also known by their national name Ume ("plum"), that collected data on the ionosphere to aid in short-wave radio communication. Their instruments included an ionospheric sounder, a radio noise receiver, plasma measuring equipment, and an ion mass spectrometer. (See table, "ISS Missions.")

```
Launch
    Vehicle: N-1
    Site: Tanegashima
Size: 3.9 × 0.8 m
```

ITM Wave Imaging

A cluster of satellites equipped with high-resolution visible and infrared sensors to image atmospheric waves that link the troposphere and upper atmosphere and redistribute energy within the ionosphere-thermosphere-mesosphere (ITM) system. The observations of such a system would improve understanding of the generation and loss mechanisms of ITM waves, their interactions, and their role in energy transport within the upper atmosphere. The ITM Wave Imaging mission remains in the early concept definition phase, but it has been identified in NASA's Office of Space Science Strategic Plan as a potential mission beyond 2007.

ITOS (Improved TIROS Operating System)

A series of American weather satellites, launched between 1970 and 1976, that followed on from the **TIROS** series.

ISS Missions

Spacecraft	Launch Date	Orbit	Mass (kg)
ISS 1	Feb. 26, 1976	988 × 1,003 km × 69.7°	139
ISS 2	Feb. 16, 1978	973 × 1,216 km × 69.4°	140

ITOS-1 was also known as TIROS M, but later spacecraft in the series were given **NOAA** (National Oceanic and Atmospheric Administration) designations after NOAA took over the management of the program from **ESSA** (Environmental Science Services Administration).

IUE (International Ultraviolet Explorer)

The first satellite totally dedicated to ultraviolet astronomy. A joint NASA-ESA (European Space Agency) mission, IUE carried a 45-cm-diameter telescope and two spectrographs for use in the range of 115 to 320 nm. It operated successfully for more than 18 years—the longest-lived astronomical satellite—until it was switched off in September 1996.

Launch
  Date: January 26, 1978
  Vehicle: Delta 2914
    Site: Cape Canaveral
  Orbit: 30,285 × 41,296 km × 34.3°
  Mass: 672 kg

Ivan Ivanovich
See **Korabl-Sputnik**.

J

J-1 (Japanese launch vehicle)

A Japanese three-stage solid fuel rocket able to place payloads of about 1,000 kg into low Earth orbit. The J-1 was the first launch vehicle in Japan to be made from a combination of existing indigenous rockets–the solid rocket booster of the H-2 (see **H series**) and the upper stage of the M-3S II (see **M series**). The Japanese space agency **NASDA** had hoped that using existing systems would keep cost and development time down. However, when a report showed that the J-1 program cost more than similar projects in other countries, NASDA switched to developing a more capable rocket, the J-2, using less expensive parts imported from overseas. The J-2 will employ a single Russian NK-33 first-stage engine and fuel tanks derived from the Atlas; it will be able to place approximately 3,500 kg into a 160-km-high orbit and 1,000 kg into a sun-synchronous orbit. The first test flight is planned for 2006.

J-2 (American rocket engine)

An American liquid-propellant rocket engine, manufactured by Rocketdyne, that ranks among the most important in the history of manned spaceflight propulsion. The J-2 was the first manned booster engine that used liquid hydrogen as a fuel and the first large booster engine designed to be restarted multiple times during a mission. It was so versatile that it was used for both the second and third stages of the Saturn V and, in modified form, was also used to demonstrate principles that led to the development of the Space Shuttle Main Engine. Capable of providing about one million N of thrust, the J-2 featured independently driven pumps for liquid oxygen and liquid hydrogen, a gas generator to supply hot gas to two turbines running in series, and pneumatic and electrical control interlocks.

Japan in space

See **Japanese launch vehicles**, **Japanese satellite names**, Tokohiro **Akiyama**, **HOPE**, **ISAS**, **ISAS missions**, **Kagoshima Space Center**, **NASDA**, and **Tanegashima Space Center**.

Japanese launch vehicles

The postwar story of Japanese rocketry began in 1955 with small solid-propellant rockets at the Institute of Industrial Science, Tokyo University, and led to a series of sounding rockets known as Kappa. Following this came the larger-scale Lambda, or **L series**, which launched Japan's first artificial satellite in 1970. More recent developments have included the **M series**, used by **ISAS** (Institute for Science and Astronautical Science), and the **N series**, **H series**, and **J-1**, used by **NASDA** (National Space Development Agency).

Japanese satellite names

Both of Japan's space agencies, **ISAS** (Institute of Space and Astronautical Science) and **NASDA** (National Space Development Agency), tend to give two names to their satellites, one English and one Japanese. However, these two names are completely unrelated, which can be confusing. For example, NASDA's ETS (Engineering Test Satellite) series is also referred to by the Japanese name Kiku, which means "chrysanthemum." Many of NASDA's satellites are similarly named after flowers–Momo ("peach blossom") for MOS (Marine Observation Satellite), Sakura ("cherry blossom") for the CS series of communications satellites, and Ajisai ("hydrangea") for EGS (Experimental Geodetic Satellite). ISAS, on the other hand, uses national names that are more diverse and sometimes loosely linked to the spacecraft's mission, such as Haruka ("far away"), Tenma ("galaxy"), and Hinotori ("phoenix"). A more important distinction between the two agencies' naming schemes is that where NASDA uses English and Japanese names as alternatives for its operational satellites, ISAS uses the English name before launch and then switches to a Japanese name after launch–but only if the launch is successful. The ISAS prelaunch name indicates to which series the craft belongs; for example, Astro-D is the fourth spacecraft in the **Astro-** series devoted to astronomical observations, mainly in the X-ray region of the spectrum. After its successful launch, Astro-D became known as Asuka ("flying bird"). On the other hand, Astro-E never received a Japanese name, because it failed to reach orbit. To add to the confusion, some ISAS spacecraft have an alternative postlaunch name, especially if they have involved collaboration with other nations or institutions. For example, Asuka (Astro-D) is also known as ASCA (Advanced Satellite for Cosmology and Astrophysics). Furthermore, NASDA has recently tended to move away from the pattern of always providing a Japanese name. TRMM (Tropical Rainfall Measuring Mission) and SELENE (Selenological and Engineering Explorer), for example, have no other designations. In this book, all NASDA missions are identified by their English names and all successfully launched ISAS missions by their Japanese names.

JAS (Japanese Amateur Satellite)

Small (50-kg) satellites launched by NASDA (National Space Development Agency) as co-passengers with larger payloads, for use in amateur radio communications. (See table, "JAS Launches.")

JAS Launches

Launch site: Tanegashima; orbit: polar LEO

Spacecraft	Launch Date	Launch Vehicle
JAS-1B (Fuji-2)	Feb. 17, 1970	H-1
JAS-2 (Fuji-3)	Aug. 17, 1996	H-2

Jason

Joint CNES (the French space agency) and JPL (Jet Propulsion Laboratory) oceanography missions that build on the success of **TOPEX/Poseidon** and form part of NASA's **EOS** (Earth Observing System). Jason-1 was launched on December 11, 2001, alongside **TIMED** (which see for launch details). Its instruments map variations in ocean surface topography as small as 2 to 5 vertical cm to monitor world ocean circulation, study interactions of the oceans and atmosphere, help improve climate predictions, and observe events like El Niño. The satellite was named after Jason of Greek mythology—an adventurer fascinated by the sea.

If global warming continues to increase, the polar ice-caps will lose more and more of their mass to the oceans, adding volumes of seawater potentially great enough to swallow islands and permanently flood coastal areas. Jason-2, scheduled for launch in 2004, will monitor this transformation by resolving global sea-level variations as small as one mm per year.

Jastrow, Robert (1925–)

A distinguished American space scientist. Jastrow earned a Ph.D. in theoretical physics from Columbia University in 1948 and carried out further research at the Institute for Advanced Studies, Princeton, and the University of California, Berkeley, before becoming an assistant professor at Yale (1953–1954). He then served on the staff at the Naval Research Laboratory (1954–1958) before being appointed chief of the theoretical division of the **Goddard Space Flight Center** and then director of the

Goddard Institute of Space Studies in 1961, a post he held for 20 years. Subsequently, he was appointed professor of earth sciences at Dartmouth College. Jastrow's work has ranged across nuclear physics, plasma physics, geophysics, and the physics of the Moon and terrestrial planets.

JAWSAT

A payload adapter developed jointly by the U.S. Air Force and Weber State University, Utah. It was first used operationally on January 26, 2000, when it was carried into orbit by a **Minotaur** launcher and then released several microsatellites, including FalconSat (an experimental satellite built by the Air Force Academy), **ASUSat**, and **Opal**.

JERS (Japanese Earth Resources Satellite)

A Japanese Earth observation satellite launched by **NASDA** (National Space Development Agency); also known by the national name Fuyo. Following on the success of **MOS**, JERS tested the performance of optical sensors and a synthetic aperture radar and made observations for use in land survey, agriculture, forestry, fishery, environmental preservation, disaster prevention, and coastal surveillance. Some of its data were shared with the University of Alaska for research purposes.

Launch
   Date: February 11, 1992
   Vehicle: H-1
   Site: Tanegashima
Orbit: 567 × 568 km × 97.7°
Size: 3.1 × 1.8 m
Mass: 1,340 kg

Jet Propulsion Laboratory

See **JPL**.

jet steering

The use of fixed or movable gas jets on a missile to steer it along a desired trajectory during propelled and coasting flight.

jetavator

A control surface that may be moved into or against a rocket's jet stream to change the direction of the jet flow for thrust vector control.

Jikiken

A Japanese satellite, launched by **ISAS** (Institute of Space and Astronautical Science), that made measurements of ionized gases and electric fields in Earth's magnetosphere from a highly elliptical orbit. Before launch it was known as **Exos**-B.

Launch
    Date: September 16, 1978
    Vehicle: M-3H
    Site: Kagoshima
    Orbit: 251 × 21,192 km × 31°
    Mass: 100 kg

Jindai
See **MABES** (Magnetic Bearing Satellite).

Jiuquan Satellite Launch Center
A Chinese launch center, also known as Shuang Cheng Tzu, located in the Gobi desert at 40.6° N, 99.9° E. It is used for launches into low Earth orbit with inclinations of 57° to 70°.

JMEX (Jupiter Magnetospheric Explorer)
A mission to study the magnetosphere of Jupiter. JMEX would be an Earth-orbiting ultraviolet telescope optimized for detecting light from Jupiter's aurora, from Io's atmosphere, and from the plasma ring around Jupiter, which is generated by Io. It was selected for study by NASA as a possible **SMEX** (Small Explorer).

Jodrell Bank
The site in Cheshire, England, of a large radio telescope with a paraboloidal receiver (76 m in diameter, 18 m deep) used to track space probes and conduct radio astronomy. This instrument, completed in 1957 and operated by the radio astronomy department of the University of Manchester, was the first giant steerable dish in the world. It was used by NASA to receive signals from the **Pioneer** series of Moon probes and as a sensitive receiver of signals bounced off the passive **Echo** satellites.

John C. Stennis Space Center
See **Stennis Space Center**.

John F. Kennedy Space Center
See **Kennedy Space Center**.

John H. Glenn Research Center at Lewis Field
See **Glenn Research Center**.

Johnson, Clarence L. "Kelly" (1910–1990)
A prominent American aircraft designer. As the head of the Lockheed Aircraft Corporation's famous "Skunk Works" design center, Johnson led the effort to build the U-2 reconnaissance aircraft in the 1950s. He also worked on the F-80 Shooting Star—the first American jet aircraft—and the SR-71 Blackbird reconnaissance plane that still holds speed records. During World War II, he was responsible for the design of the P-38 Lightning twin-tailed fighter. Johnson worked for Lockheed from 1933 until his retirement as senior vice president in 1975.[159]

Johnson Space Center (JSC)
NASA's primary site, located in Houston, Texas, for the development and operation of manned space missions, including the selection and training of astronauts, JSC was established in 1961 as the Manned Spacecraft Center and, in 1973, renamed in honor of the late president and Texas native Lyndon B. Johnson. Since 1965, it has served as Mission Control for the **Gemini** (beginning with Gemini 4), **Apollo**, and **Skylab** projects, through to today's **Space Shuttle** and **International Space Station** (ISS) programs. It operates separate control centers for Shuttle and ISS missions. JSC also houses the program offices that direct the development, testing, production, and delivery of all U.S. manned spacecraft—including Shuttle Orbiters and Station components—and coordinates Station development efforts among NASA centers and between the United States and its partner nations. Additionally, JSC leads NASA's efforts in space medicine, and houses and investigates the agency's collection of Apollo lunar samples and of meteorites believed to have come from Mars.[77]

Johnsville Centrifuge
A facility at the Aviation Medical Acceleration Laboratory of the Navy Air Development Center, Johnsville, Pennsylvania, used throughout the early American space program; it was the most powerful **centrifuge** then in existence. With a 17-m radius, it had a rate change of $10g/s$ and could reach $40g$. The 10-m-diameter gimbal-mounted chamber was fully air-conditioned.

Johnsville Centrifuge   The centrifuge at the Naval Air Development Center, Johnsville, Pennsylvania. *NASA/U.S. Navy*

Joule

An X-ray observatory that would obtain detailed energy spectra from extreme environments in the universe ranging from the million-degree coronas of nearby stars to the supermassive black holes at the cores of distant galaxies. Joule was selected by NASA for study as an **SMEX** (Small Explorer) mission and remains in the early design stage.

JPL (Jet Propulsion Laboratory)

An establishment in Pasadena, California, operated for NASA by the California Institute of Technology. It plays a major role in the development, operation, and tracking of American interplanetary space probes. JPL is responsible for the **Deep Space Network** and has managed the Viking, Voyager, Magellan, Galileo, Ulysses, and Mars Pathfinder missions. JPL was founded by Theodore **von Karman** and Frank **Malina** in 1944 and was a focal point of American rocket development in World War II. In 1958, it became part of NASA and provided America's first successful Earth satellite, *Explorer* 1.[170, 315]

JPL (Jet Propulsion Laboratory)   At Guggenheim Aeronautical Laboratory, Caltech (CALCIT) students and coworkers fired their first liquid-propellant rocket on October 31, 1936.
NASA/JPL

Jumpseat

American military intelligence satellites launched in the 1970s and early 1980s into Molniya-type orbits. These orbits are very similar to those of SDS (Satellite Data System) classified communications satellites, and identification of a launch in either series as SDS or Jumpseat may differ.

Juno

One of the most significant launch vehicles in the history of spaceflight. Juno formed a crucial link between American Cold War military missiles and the rockets that first launched human beings toward another world.

Juno I

A modified **Redstone** missile specifically designed to carry lightweight payloads into LEO (low Earth orbit). Otherwise identical to the three-stage **Jupiter** C, it had a fourth stage that remained attached to the satellite in orbit. To reflect its nonmilitary role, its name was officially changed from Jupiter to Juno at the request of JPL (Jet Propulsion Laboratory) chief William **Pickering** on November 18, 1957. It later became Juno I to distinguish it from the Juno II. A Juno I successfully launched **Explorer** 1, the first American satellite, on January 31, 1958, and in the same year two more placed Explorer 3 and Explorer 4 into orbit.

Juno II

A marriage of a **Jupiter** first stage and Juno I upper stages capable of delivering a 45-kg payload into LEO. The Juno name was also applied to the next generation of space launch vehicles designed by the Army Ballistic Missile Agency (ABMA), including a proposed Juno V super-booster. Eventually, the Juno program was transferred to NASA and renamed **Saturn**. Thus, the pioneering work of the ABMA, culminating in the successful use of Juno I and Juno II, led directly to the development of the rockets that would carry astronauts to the Moon.

Jupiter

See article, pages 117–118.

Jupiter

An intermediate-range ballistic missile (IRBM) designed by the U.S. **Army Ballistic Missile Agency** (ABMA) as a successor to the **Redstone**. Its development led to America's first successful space launch vehicle.

In late 1955, ABMA was directed to work with the Navy to develop both a land-based and a sea-based version of the Jupiter. One year later, however, in the "Wilson Memorandum," a top-secret document from Secretary of Defense Charles Wilson on human research related to atomic, biological, and chemical warfare, the Army was stripped of all missiles with a range of over 320 km. Subsequently, the Navy abandoned the Jupiter in favor of the solid-fueled, submarine-based Polaris, and the Jupiter was transferred to the Air Force. Although it eventually became the first operational American IRBM, the Jupiter was never given much attention by the Air Force as it competed directly with the Air Force–developed **Thor**. The Army-developed Jupiter had been a source of bitter rivalry between the two branches of service since its conception. (See table, "Comparison of Jupiter and Thor IRBMs.")

Comparison of Jupiter and Thor IRBMs

	Jupiter	Thor
Length	17.7 m	19.8 m
Diameter	2.7 m	2.4 m
Mass at liftoff	50,000 kg	50,000 kg
Propellants	LOX/RP-1	LOX/RP-1
Range	2,600 km	2,600 km

Built by Chrysler, the Jupiter first stage was powered by a Rocketdyne engine that burned liquid oxygen (LOX) and RP-1 (kerosene mixture) to produce a liftoff thrust of 670,000 N. An ablative technology reentry vehicle made by Goodyear, which shrouded a one-megaton warhead, separated from the Jupiter following the detonation of explosive bolts. A Thiokol solid-fueled motor supplied final velocity trimming. Operational Jupiter missiles were delivered to the Air Force beginning in November 1957. Jupiter components were test-flown aboard rockets known as Jupiter A and Jupiter C, which were not really Jupiters at all but rather modified Redstones. An actual Jupiter was first test-flown in April 1957, and the missile was deployed by NATO in Europe before being withdrawn from service in 1964. Although short-lived, the Jupiter program played a key role in the history of the American space program. It led directly to the launch of America's first satellite aboard a Juno I, a rocket adapted from the Jupiter C, and the Jupiter missile acted as the first stage for the Juno II, which carried a number of scientific payloads into space. Inert Jupiter missile shells ballasted with water were used as dummy upper stages for the earliest version of the Apollo-Saturn I.

Jupiter A

The first exploitation of the **Redstone** for testing components that would be incorporated into the Jupiter IRBM. Although the Jupiter A was identical to the Redstone, its name was intended to associate it with the Jupiter missile development program. However, Jupiter A flights really served a dual role: as tests of Jupiter missile hardware and as tests of Redstone missile performance at a time when the Redstone was still under development.

Jupiter C and America's Space Debut

A multistage adaptation of the Redstone and a follow-on from the Jupiter A, designed to test Jupiter missile nosecone technology. In 1955, a rocket nearly identical to the Jupiter C was described in the plans for Project **Orbiter**. When Orbiter was dropped in favor of Project Vanguard, the multistage Redstone faced extinction. However, on September 13, 1955, four days after the official rejection of Orbiter, the Army and Navy were told to work together in developing the Jupiter IRBM. Engineers at the Redstone Arsenal were authorized to develop the Jupiter C for high-altitude reentry tests of Jupiter missile components. Although, like the Jupiter A, the C was really a Redstone, the name was changed to Jupiter C—short for "Jupiter Composite Reentry Test Vehicle." It was similar to the Juno I but with important differences. The C was designed specifically to loft a nosecone above the atmosphere and allow it to reenter at high speed.

Jupiter A floodlit Jupiter C is prepared for launch from Cape Canaveral. *U.S. Army*

Both Juno I and Jupiter C used a Redstone first stage with stretched tanks to hold more fuel. But the Jupiter C booster burned LOX and alcohol, like the production version of the Redstone missile. The satellite-carrying Juno I first stage was modified to burn higher-energy Hydyne fuel. Both the C and Juno I used clustered Baby Sergeant solid motors on their second and third stages. However, since the C was never meant to launch satellites, it had no fourth-stage motor. The Jupiter C first stage, with a liftoff thrust of 348,000 N, burned for 150 seconds before falling away. This was followed by a 6-second firing from the second stage, separation and a 2-second pause, and finally a 6-second burst from the third stage. After this, the payload flew on its own, in a ballistic arch, toward the ocean. The intense heat of reentry enabled valuable scientific data to be gathered on flight dynamics and nosecone performance.

The first Jupiter C was launched from Cape Canaveral on September 20, 1956, and surpassed expectations. This was the only C to carry an inert fourth-stage ballasted with sand in case ABMA accidentally launched a satellite before Project Vanguard. The fourth stage soared 1,097 km high and 5,390 km downrange, setting new Cape records—a bittersweet success for ABMA. The Jupiter C had shown it was perfectly capable of launching a small satellite, yet bureaucracy forbade it to make the attempt.

Although 12 Jupiter Cs had been authorized for construction, the research program was canceled after just three launches in August 1957 by ABMA's chief, Maj. Gen. John Medaris. However, Medaris ordered that all remaining C booster hardware be mothballed, his intent being to keep the rocket components ready if the green light came to attempt a satellite launch. With Jupiter C performance data in hand, the Army continued to seek a go-ahead to launch instrumented satellites in support of the International Geophysical Year, but without success. Then came the shocking news that the Soviets had placed Sputnik 1 in orbit. By coincidence, Secretary of Defense Neil McElroy was visiting the Redstone Arsenal on that very day in October 1957. ABMA top brass tried to persuade McElroy on the spot that the Army deserved a chance to launch an American reply. Medaris secretly told Wernher von Braun, ABMA's technical director, to start preparing Jupiter C hardware for action, and the Army-sponsored JPL (Jet Propulsion Laboratory) was asked to evaluate satellite designs. McElroy did convince President Eisenhower to look favorably on the Army's request, but it was almost two months after Sputnik 1's triumph when ABMA finally got approval to schedule satellite launch dates. It was to be a major turning point in American space history. The Navy's Project Vanguard failed in its first attempt to launch a small tracking satellite on December 6, 1957, when the Vanguard rocket blew up on its pad. Then the Navy's second attempt, scheduled for January 24, 1958, was delayed until early February. Suddenly, ABMA's first shot at orbit, originally slated for five days after the Navy's second attempt, was next in line. But there would be one final twist before America entered space. In November 1957, William Pickering, head of JPL, suggested to Medaris that the satellite-bearing version of the Jupiter C be renamed. The recommendation was accepted by ABMA, and the Juno family of rockets was born.

K

Ka-band

See **frequency bands**.

Kagoshima Space Center

A Japanese launch site on the east coast of the Ohsumi peninsula at 31.3° N, 131° E. It is dedicated to **ISAS** (Institute of Space and Astronautical Science) missions.

Kakehashi

See **COMETS (Communications and Broadcasting Experimental Test Satellite)**.

Kaliningrad

The site of Russia's mission control center (TsUP). The Kaliningrad facility has eight control rooms, including a main one with 27 workstations, each able to accommodate two flight controllers. TsUP is integrated with a global network of mission-control centers, which includes those in the United States, France, and China.

Kaplan, Joseph (1902–1991)

A Hungarian-born American physicist who was prominent in efforts to launch the first American satellite. Kaplan came to the United States from his native Tapolcza in 1910, trained as a physicist at Johns Hopkins University, and worked on the faculty of the University of California, Berkeley, from 1928 until his retirement in 1970. He directed the university's Institute of Geophysics–later renamed the Institute of Geophysics and Planetary Physics–from the time of its creation in 1944. From 1953 to 1963, Kaplan served as the chair of the U.S. National Committee for the **International Geophysical Year**.[165]

Kaputsin Yar

A Russian launch complex, and also the country's oldest missile test site, located at 48.5° N, 45.8° E. Although used quite often for launches of smaller **Cosmos** satellites during the 1960s, the number of launches from this site fell dramatically during the 1970s and 1980s to about one orbital launch per year. On April 28, 1999, the **ABRIXAS** satellite was successfully launched from here (although the spacecraft's battery failed after only a few days in orbit).

K-band

See **frequency bands**.

Keldysh, Msitslav Vsevolodovich (1911–1978)

The chief theoretician of Soviet astronautics in the 1960s. Keldysh studied physics and mathematics at Moscow University, obtaining his Ph.D. in 1938. He worked for many years in a variety of positions at the Central Institute of Aerohydrodynamics, Moscow University, and the Steklov Mathematical Institute. From 1960 to 1961 he served as vice president of the Soviet Academy of Sciences, and thereafter as its president until 1975.

Kelly, Thomas J. (1930–2002)

An American engineer with Grumman Aircraft (now Northrop Grumman) who led the team that designed and built the **Apollo** Lunar Module (LM). Kelly helped develop the **lunar-orbit rendezvous** concept used by Apollo, then spearheaded Grumman's effort at Bethpage, New York, to realize the vehicle that would land a dozen astronauts on the Moon and, in the case of Apollo 13, serve as a lifeboat for the safe return of the stricken crew. A native of Brooklyn, Kelley earned a B.S. in mechanical engineering from Cornell University (1951) and a M.S. in the same field from Columbia University (1956). After graduating from Cornell, he joined Grumman as a propulsion engineer and later did the same job at Wright-Patterson Air Force Base, Ohio, when he was called up for military service. He worked as a space propulsion engineer for **Lockheed Martin**'s missiles and space division in 1958-1959, then returned to Grumman, where he stayed until his retirement in 1992. His experiences in building the LM are recounted in his book, *Moon Lander: How We Developed the Apollo Lunar Module* (2001).[166]

Kennedy, John F. (1916–1963)

U.S. president from 1961 to 1963. While on the presidential campaign trail, Kennedy attacked incumbent **Eisenhower**'s record in international relations, taking a Cold Warrior position on a supposed "missile gap" (which turned out to be false) wherein the United States supposedly lagged far behind the Soviet Union in intercontinental ballistic missile (ICBM) technology. On May 25, 1961, President Kennedy announced to the nation the goal of sending an American to the Moon before the end of the decade. The human spaceflight imperative was a direct outgrowth of it; Projects **Mercury** (in its latter stages), **Gemini**, and **Apollo** were designed to execute it.

Kennedy Space Center (KSC)

America's primary civilian space launch facility, operated by NASA and located at **Cape Canaveral**. All Space Shuttle launches and many unmanned launches take place from here. Selection of the site was influenced by the fact that the Missile Firing Laboratory had been launching missiles from Cape Canaveral since 1953. Originally called the Launch Operations Center, the facility was renamed the John F. Kennedy Space Center in 1963.

Kepler

A NASA Discovery mission, scheduled for launch in 2006, to detect and characterize Earth-sized planets by photometry (light-intensity measurements). Kepler's objectives over a four-year lifetime are to determine: (1) the frequency of Earth-like and larger planets in and near the habitable zone of a wide variety of spectral types of stars; (2) the distribution of diameter and orbital size of Earth-like planets; (3) the distribution of diameter, mass, density, albedo (reflectivity), and orbital size of giant inner planets; (4) the frequency of planets orbiting multiple star systems; and (5) the properties of stars that have planetary systems. Kepler's main instrument will be a one-m-aperture photometer with a 12° field of view, about equal to the area covered by a hand at arm's length; it will continuously and simultaneously monitor the light from 90,000 main-sequence stars in a star field in Cygnus. Planets will be discovered and characterized by the tiny periodic light dips they cause as they pass in front of (transit) their parent stars as seen from Earth. Detection of two transits will be taken as evidence of a candidate planet, with a third and subsequent transits providing confirmation.

Kepler, Johannes (1571–1630)

A German mathematician and astronomer who assisted the great observer Tycho Brahe and later used Brahe's observations to deduce three key laws of planetary motion, now known as **Kepler's laws**. Kepler lived when telescopes were first being used to look at the Moon and planets, and was one of the few vocal supporters of Galileo's claim that the planets orbited the Sun instead of the other way around. He was also the author of a lunar tale called *Somnium* (The Dream),[167] published posthumously in 1634. The hero, Duracotus, an Icelander and a self-proclaimed student of Brahe, is conveyed to the Moon by winged Selenites who trafficked with his mother, a notorious witch—an arrangement not unfamiliar to Kepler, since his own mother was tried for witchcraft.

Keplerian trajectory

An elliptical orbit followed by an object in accordance with the first of **Kepler's laws** of celestial motion.

Johannes Kepler *George Philip & Son*

Kepler's laws

Three laws of planetary motion, published by Johannes **Kepler** and based on accurate observations by Tycho Brahe; they were subsequently shown by Isaac **Newton** to be a direct result of his theory of gravitation and laws of motion. They are: (1) A planet orbits the Sun in an **ellipse** with the Sun at one focus; (2) A line drawn from the Sun to a planet (the radius vector) sweeps out equal areas in equal amounts of time; and (3) The square of the **period** of a planet's orbit varies as the cube of the planet's **semimajor axis**, the constant of proportionality being the same for all planets. The first law corrects the simpler model of Copernicus, which assumed circles. The second law expresses the way a planet speeds up when approaching the Sun and slows down when drawing away. The third law gives the exact relation by which planets move faster on orbits that are closer to the Sun.

kerosene

A commonly used rocket fuel, especially in a mixture known as **RP-1**.

Kerwin, Joseph Peter (1932–)

An American astronaut who, as science pilot aboard the first **Skylab** mission in 1973, became not only the first medical doctor to practice in orbit but also the first space repairman. He received a B.A. in philosophy from the College of the Holy Cross in 1953 and an M.D. from Northwestern University Medical School in 1957. Kerwin completed his internship at the District of Columbia

General Hospital, then joined the U.S. Navy and attended the Navy School of Aviation Medicine at Pensacola, Florida. He became a naval flight surgeon in 1958 and earned his wings in 1962. NASA selected Kerwin in its first group of six scientist-astronauts in June 1965. Following his mission aboard Skylab, Kerwin became director of Space and Life Sciences at the Johnson Space Center. He then resigned from NASA and from the Navy as a captain, and from 1987 to 1996 held management positions with Lockheed Martin Missiles and Space Company. Since 1996 he has headed Krug Life Sciences, which has a major research contract with NASA.

Kettering Bug

Along with the British **A.T.** (Aerial Target), one the earliest **guided missiles**. The development of the Kettering Bug, under the direction of Charles Kettering, began at the Delco and Sperry companies in 1917. It was an unpiloted biplane bomber made of wood, weighing just 270 kg (including a 135-kg bomb as payload), and was powered by a 40-hp Ford engine. Engineers employed an ingenious method of guiding the Kettering Bug to its target. Once wind speed, wind direction, and target distance had been determined, the number of revolutions the engine needed to take the missile to its target was calculated. A cam was then set to drop automatically into position when the right number of engine revolutions had occurred. The Kettering Bug took off using a four-wheel carriage that ran along a portable track. Once airborne, it was controlled by a small gyroscope, its altitude measured by an aneroid barometer. When the engine had completed the prescribed number of revs, the cam dropped into position, causing bolts that fastened the wings to the fuselage to be pulled in. The wings then detached, and the bomb-carrying fuselage simply fell onto its target. The Kettering Bug was successfully demonstrated in 1918 before Army Air Corps observers in Dayton, Ohio. However, World War I hostilities ended before the missile could be put into production.

Key Hole (KH)

The name used by the intelligence community for various series of U.S. military reconnaissance satellites. Missions in the **Corona** series, for example, were designated KH-1 to KH-4B. Argon missions were known as KH-5 and Lanyard as KH-6. Several launches with Key Hole designations, including KH-11, KH-12, and KH-13, have taken place since 1992. KH-11 photos of Osama bin Laden's training facilities at the Zhawar Kili Base Camp in Afghanistan were used in preparing for the cruise missile attack on this target in 1998. The ninth and final KH-11 satellite was launched in 1998, after which the KH-12 program, involving satellites about the size of the Hubble Space Telescope, began.

Kibalchich, Nikolai Ivanovitch (1854–1881)

A Russian medical student, journalist, and revolutionary who may have been the first person in history to propose using rocket power as a means of transport in space. Kibalchich had attempted several times to kill Czar Alexander II, who, ironically, had tried to introduce reforms in Russia during his reign. Along with his accomplices, Kibalchich finally succeeded in assassinating the czar on March 13, 1881, and was himself put to death on April 3, at the age of 27. While in prison awaiting execution, Kibalchich wrote a remarkable paper illustrating the principle of space propulsion. In it, he describes a means of propelling a platform by igniting gunpowder cartridges in a rocket chamber. Changing the direction of the rocket's axis, he realized, would alter the vehicle's flight path. In his notes, he explains: "I am writing this project in prison, a few days before my death. I believe in the practicability of my idea and this faith supports me in my desperate plight."

He was a close contemporary of two other rocket pioneers, his compatriot Konstantin **Tsiolkovsky** and the German Hermann **Ganswindt**. The great American rocketeer Robert **Goddard** was born a year after Kibalchich's execution.

kick stage

A propulsive stage used to provide an additional velocity increment required to put a spacecraft on a given trajectory. Two particular forms are the **apogee kick motor** and the **perigee kick motor**.

Kiku

See **ETS (Engineering Test Satellite)**.

kinetic energy

The energy that an object possesses by virtue of its motion. It is given by the expression $\frac{1}{2}\,mv^2$, where m is the object's mass and v its velocity.

Kiruna

See **Esrange**.

Kittinger, Joseph W. (1928–)

A U.S. Air Force test pilot who took part in extreme high-altitude balloon experiments in the years immediately preceding manned spaceflight. He flew the first mission in Project **Manhigh** in June 1957, reaching 29,500 m in a closed gondola, then bettered this in 1960 by jumping out of an open gondola at 31,300 m (102,800 feet) and plummeting 26 km (the longest free fall in history, during which he broke the sound barrier) before his main chute opened. At one point he radioed back, "There is a hostile sky above me. Man may live in space, but he will never conquer it."

Kiwi

A series of studies carried out by the Atomic Energy Commission aimed at developing nuclear reactors useful in high-thrust rocket engines.

Kodiak Launch Complex

A commercial spaceport at Narrow Cape on Kodiak Island, about 400 km south of Anchorage, Alaska, and 40 km southwest of the city of Kodiak. Kodiak Island is one of the best locations in the world for polar launch operations, providing a wide launch azimuth and an unobstructed downrange flight path. Development of the site began in January 1998 by the Alaska Aerospace Development Corporation, a public company formed in 1991 to foster an aerospace industry for the 49th state. Following three Air Force suborbital launches, the inaugural orbital launch from Kodiak of an **Athena** I carrying four small satellites—PICOSat, PCSat, Sapphire, and STARSHINE

3—was delayed two weeks because of the September 11, 2001, terrorist attacks on New York City and Washington, D.C., and then because of a massive solar flare. It finally took place September 29, 2001.

Komarov, Vladimir Mikhaylovich (1927–1967)

A Soviet cosmonaut and the first human being to die on a space mission. Born in Moscow and a graduate of four Soviet Air Force colleges, Komarov was selected as a military pilot cosmonaut in March 1960. After joining the **Vostok** training group in 1961, he was assigned to be backup pilot of Vostok 4 in 1962 and support for Vostok 5. On October 12, 1964, he served as commander of **Voskhod** 1 along with crewmates Boris Yegerov and Konstantin **Feoktistov**. Because all the Vostok missions ended with the cosmonaut ejecting at about 4 km and landing separately from the spacecraft, this crew also became the first to touch down on land inside their spacecraft. On April 24, 1967, having been launched as the pilot of **Soyuz** 1, a solo mission, he became the first person to die during a space mission after the lines of his spacecraft's parachute became tangled during descent. His ashes are buried in the Kremlin Wall.

Kodiak Launch Complex   The Launch Service Structure at Kodiak Launch Complex on Kodiak Island. *NASA*

Korabl-Sputnik

Early Soviet spacecraft that served as test vehicles for the **Vostok** manned flights; "Korabl-Sputnik" means "space-ship-satellite." They carried into orbit a variety of animals, including dogs, rats, and mice. Korabl-Sputnik (K-S) 1 to 5 were known in the West as Sputnik 4, 5, 6, 9, and 10. (Sputnik 7 and 8 were launched as Venus probes, the latter also known as **Venera** 1; see **Venus unmanned missions**.) All were successfully placed in orbit, but the first and third burned up during reentry—K-S 1 ended up in a higher orbit than intended and reentered more than two years later, while K-S 3 reentered on time but too steeply and was lost.

Although a number of dogs traveled aboard these Vostok precursors, the first dog in space was **Laika**, who rode aboard **Sputnik** 2 but died in orbit. The successful recovery of the Korabl canines proved that sizeable mammals could not only endure space travel but could survive the rigors of reentry, thus helping pave the way for manned spaceflight.

On K-S 4 and 5, the dog passengers were accompanied by a lifelike mannequin nicknamed Ivan Ivanovich. Ivan was dressed in the same SK-1 pressure suit that Gagarin would use, and looked eerily lifelike. In fact, technicians were so concerned that anyone finding him might take him for a real cosmonaut that they wrote "model" on his forehead. As in the case of the human Vostok pilots who would follow, Ivan was shot out of his capsule by ejection seat after reentering the atmosphere. A parachute then lifted him free of his seat to a soft landing. On his second flight, Ivan came down near the city of Izevsk in the Ural Mountains during a heavy snowstorm. The heroic dummy has remained in his spacesuit ever since. (See table, "Korabl-Sputnik Missions.")

Korolev, Sergei Pavlovich (1907–1966)

A Ukrainian-born rocket designer and engineer who masterminded the Soviet Union's early success in space. As a youngster he was inspired by aviation and trained in aeronautical engineering at the Kiev Polytechnic Institute. Upon moving to Moscow, he came under the influence and guidance of the aircraft designer Andrei Tupolev and cofounded the Moscow rocketry organization **GIRD**. Like the **Verein für Raumschiffahrt** in Germany and Robert **Goddard** in the United States, GIRD was, by the early 1930s, testing liquid-fueled rockets of increasing size. Seeing the potential of these devices, the Russian military seized control of GIRD and replaced it with RNII (Reaction Propulsion Scientific Research Institute). RNII developed a series of rocket-propelled missiles and gliders during the 1930s, culminating in Korolev's **RP-318**, Russia's first rocket-propelled aircraft. But before the aircraft could make a powered flight, Korolev was denounced by colleagues to the NKVD (forerunner of the KGB), arrested, beaten, and, along with other aerospace engineers, thrown into the Soviet prison system during the peak of Stalin's purges in 1937 and 1938. Korolev spent months in transit on the Trans-Siberian railway and on a prison vessel at Magadan, followed by a year in the Kolyma gold mines, the most dreaded part of the Gulag. However, Stalin soon recognized the importance of aeronautical engineers in preparing for the impending war with Hitler, and set up a system of sharashkas (prison-factories) to exploit the incarcerated talent. Loyalty to the Soviet Union was a hallmark of these gifted innovators: Korolev never believed that Stalin was behind his arrests, even though he wrote numerous letters to the Soviet leader protesting his innocence without getting a reply.

Korabl-Sputnik Missions

Korabl-Sputnik	Launch Date	Orbits	Notes
1	May 15, 1960	Many	Placed in orbit but planned reentry failed. Finally burned up after 844 days in space.
2	Aug. 19, 1960	17	Successful recovery of two dogs, Belka (Squirrel) and Strelka (Little Arrow), plus 12 mice, 2 rats, and fruit flies.
3	Dec. 1, 1960	17	Burned up during too-steep reentry. Onboard were 2 dogs, Pchelka (Bee) and Mushka (Little Fly), plus mice, insects, and plants.
4	Mar. 9, 1961	1	Successful recovery of dog Chernushka (Blackie), mice, and mannequin.
5	Mar. 25, 1961	1	Successful recovery of dog Zvezdochka (Little Star) and mannequin.

In the end, Korolev was saved by the intervention of Tupolev, himself a prisoner, who requested his services in the TsKB-39 sharashka. Later, Korolev was moved to another sharashka in Kazan, where he led design projects to build jet engines and rocket thrusters. His rehabilitation was complete when he was released and sent to Germany to gather information on the V-2 (see **"V" weapons**), collecting hardware and German expertise to reestablish Soviet rocket and missile technology. After the war and throughout the 1950s, Korolev concentrated on devising Russian alternatives to the V-2 and establishing a powerful Soviet rocket-production industry (see **"R" series of Russian missiles**). Trials produced the multi-stage R-7 with a range of 6,400 km, providing the Soviet Union with an intercontinental ballistic missile (ICBM) capable of reaching the United States. To speed development of the R-7, Korolev's other projects were spun off to a new design bureau in Dnepropetrovsk headed by his onetime assistant, Mikhail **Yangel**. This was the first of several design bureaus, some later competing with Korolev's, that would spring up once Korolev had perfected a new technology. Such immense strategic importance had Korolev's rocket and missile program acquired that it was controlled at a high level in the Soviet government by the secret Committee Number 2. In September 1953, Korolev proposed the development of an artificial satellite to this committee, arguing that the R-7–launched flight of **Sputnik** 1 would serve as a powerful public demonstration of the Soviet Union's ICBM capability. A year later, he put forward even more ambitious plans, for a "two- to three-ton scientific satellite," a "recoverable satellite," a "satellite with a long orbital stay for one to two people," and an "orbital station with regular Earth ferry communication." All of these were subsequently realized: Sputnik 3 flew in 1958, followed by the first spy satellite **Zenit** in 1962; the cosmonauts in **Vostok** in 1963 and **Voskhod** in 1964 broke long-endurance records; and the first space station **Salyut** was flown in 1971.

The Soviet lunar program depended heavily on the high technical performance of Korolev's rocket systems and the industrial infrastructure that he built up, as well as on his political influence, drive, and determination. At first, all went well. Plans to explore the Moon, eventually using astronauts, were presented in 1957. Successful flyby, landing, and lunar orbital flights were accomplished in quick succession during 1959. But it was clear that these missions were achieved at the limits of the technology available with the R-7 launcher. Sending people to the Moon demanded a much bigger launch vehicle and advances in electronics and guidance systems, and when the Moon race became official policy after President Kennedy's declaration in 1961, the Soviet military-industrial complex failed to keep up. Korolev

concentrated his resources on the **N-1** rocket, using a cluster of 30 R-7–type engines. An alternative from the military sector called UR-500K, which used storable propellants, emerged as a competitor. Bureaucratic intervention and personality clashes led to indecision and both projects were supported, but at inadequate levels. The Soviet heavy-lift launchers could not compete with NASA's Saturn and Apollo programs, and when Korolev died suddenly following surgery in 1966, the Soviet Moon program faltered. The N-1 failed its test firings and was canceled in 1976 (though the UR-500K survived to become the **Proton**).

Korolev's legacy is the town named for him and **Energiya Rocket & Space Corporation** (Energia RCS)– the modern Russian business organization that evolved from Korolev's design bureau–which built **Mir** and is now a partner with NASA in the production of the **International Space Station**. Korolev himself was classified as top-secret throughout his career, and his name became publicly known only after his death.[145]

Kourou
See **Guiana Space Centre**.

Kraft, Christopher Colombus, Jr. (1924–)
A NASA flight director involved with many of the early American manned spaceflights. He received a B.S. in aeronautical engineering from Virginia Polytechnic University in 1944 and joined the **Langley** Aeronautical Laboratory of **NACA** (National Advisory Committee for Aeronautics) the following year. In 1958, while still at Langley, he became a member of the Space Task Group developing Project **Mercury** and moved with the Group to Houston in 1962. He was flight director for all of the Mercury and many of the **Gemini** missions, and he directed the design of Mission Control at the Manned Spacecraft Center (MSC), redesignated the **Johnson Space Center** in 1973. Kraft moved up to become director of flight operations at MSC through the entire Apollo program, then director of Johnson Space Center until his retirement in 1982. Since then he has remained active as an aerospace consultant and published an autobiography, *Flight: My Life in Mission Control* (2001).[173]

Kranz, Eugene "Gene" F. (1933–)
The NASA flight director for the **Apollo** 11 lunar landing and head of the Tiger Team for the successful return of Apollo 13; he was memorably portrayed, wearing his trademark white waistcoat, in the Tom Hanks film of the ill-fated third moonshot. After receiving a B.S. in aeronautical engineering from Parks College of Saint Louis University (1954), Kranz was commissioned in the U.S. Air Force and flew high-performance jet fighters. In 1960,

Christopher Kraft   Kraft standing (far right) with (left to right) Terry J. Hart, NASA deputy administrator, Hans Mark, NASA administrator, and James M. Beggs, Johnson Space Center Director. President Ronald Reagan is seated next to astronaut Daniel C. Brandestein, serving as CAPCOM. Directly above Reagan in the background is the JSC flight operations director, Gene Kranz. *NASA*

he joined the NASA Space Task Group at **Langley** and was assigned as assistant flight director for Project **Mercury**. He served as flight director for all the **Gemini** missions, was appointed division chief for flight control in 1968, and continued his duties as flight director for the Apollo and **Skylab** programs before being assigned as deputy director of flight operations with responsibility for spaceflight planning, training and mission operations, aircraft operations, and flight crew operations. In 1983, Kranz was promoted to director of mission operations with added responsibility for the design, development, maintenance, and operations of all related mission facilities, as well as the preparation of the Space Shuttle flight software. Since his retirement from NASA in 1994, he has been involved with consulting, motivational speaking to youth groups, and lecturing on manned spaceflight. He also serves as a flight engineer on a B-17 Flying Fortress that performs at air shows throughout the United States. His autobiography, *Failure Is Not an Option: Mission Control from Mercury to Apollo 13 and Beyond*, was published in 2000.[174]

Krieger, Robert L. (1916–1990)

An American aerospace engineer and administrator. Krieger began his career with NACA (National Advisory Committee for Aeronautics) in 1936 as a laboratory apprentice at the **Langley** Aeronautical Laboratory. Having left NACA temporarily for college, he earned a B.S. in mechanical engineering at Georgia Tech in 1943 before returning to Langley. In 1945, he joined the group that set up the Pilotless Aircraft Research Station at **Wallops**

Island under Robert R. **Gilruth**, and in 1948, he was appointed director of Wallops Island–a facility that performed high-speed aerodynamic tests on instrumented models. In 1958, Wallops became an independent field center of NASA. Subsequently, Krieger led the first successful test flight of the Mercury capsule and oversaw the launch from Wallops of thousands of test vehicles, including 19 satellites. He retired as director in 1981.

Kristall

See **Mir**.

Krunichev State Research and Production Space Center

A major Russian aerospace enterprise, based in Moscow, that develops and builds Russian components for the **International Space Station** as well as a number of launch vehicles including the **Proton**, **Rockot**, and **Angara**. It was formed following a presidential decree in 1993 from the Khrunichev Machine-building Plant (originally set up in 1916 to make cars) and the Salyut Design Bureau, which had been responsible for building all elements of the **Salyut** and **Mir** space stations.

Ku-band

See **frequency bands**.

Kuiper Airborne Observatory (KAO)

A 0.915-m reflecting telescope mounted in a Lockheed C141 Starlifter jet transport aircraft and flown at altitudes of up to 13.7 km. Named after the Dutch-American astronomer Gerard Peter Kuiper (1905–1973), it was based at the **Ames Research Center** and operated as a national facility by NASA from 1975 to 1995. Among its discoveries was the ring system of Uranus. The KAO has since been replaced by **SOFIA** (Stratospheric Observatory for Infrared Astronomy).

Kummersdorf

A German Army artillery range, 25 km south of Berlin, in the vicinity of which the German Army Ordnance Department tested a variety of solid- and liquid-fueled rockets, built by Wernher **von Braun** and his team, in the first half of the 1930s. By the late 1930s, it had become clear the site was too small and congested for the kinds of tests that were needed. The staff at Kummersdorf already numbered around 80, and more engineers and laborers were needed. A new site was sought on the coast from which rockets could safely be launched, tracked, and/or recovered during testing, and which was also secure and secret. The Luftwaffe, too, had decided to invest funds in rocket research, and soon a joint operation with the German army resulted in the selection of the small fishing village of **Peenemünde** on the Baltic coast.

Kupon

A Soviet satellite originally developed for use with the third generation GKKRS (Global Space Command and Communications System)–a network that also included the **Potok** and **Geizer** satellites. When the project was canceled following the breakup of the Soviet Union, Kupon became instead the first communications satellite for the Russian banking system. Owned by the Russian Federation Central Bank, it relayed financial data for the Bankir network.

Launch
    Date: November 12, 1997
    Vehicle: Proton
    Site: Baikonur
    Orbit: geosynchronous

Kurchatov, Igor (1902–1960)

A Soviet nuclear scientist who was in charge of all nuclear research (aimed primarily at developing an atomic bomb as quickly as possible) in the Soviet Union from 1943 until his death. The Atomic Energy Institute in Moscow, which he founded, is named after him, as was element 104–kurchatovium–now known officially as rutherfordium.

Kurzweg, Hermann H. (1908–)

A German-born aerodynamicist with the V-2 (see **"V" weapons**) project and later with NASA. Kurzweg earned a Ph.D. from the University of Leipzig in 1933 and during World War II was chief of the research division and deputy director of the aerodynamic laboratories at **Peenemünde**. There he carried out aerodynamic research on the V-2 and the antiaircraft rocket Wasserfall, and he participated in the design of the supersonic wind tunnels. In 1946, he came to the United States and worked for the Naval Ordnance Laboratory at White Oak, Maryland, doing aerodynamics and aeroballistics research. He became associate technical director of the lab in 1956. In September 1960, he joined NASA headquarters as assistant director for aerodynamics and flight mechanics in the office of advanced research programs. The following year, he became director of research in the office of advanced research and technology, and in 1970 he was made chief scientist and chairman of the research council in the same office. He retired in 1974.

Kvant

See **Mir**.

Kwajalein Missile Range

An American military range and launch site for missiles and, occasionally (**Pegasus**-launched), small satellites, located on the Pacific atoll of Kwajalein in the Marshall Islands at 10° N, 168° E.

Kwangmyongsong

A North Korean test satellite that appears to have failed to reach orbit following its launch from Musudan on August 31, 1998. According to a North Korean statement:

> The satellite is running along the oval orbit 218.82 km in the nearest distance from the Earth and 6,978.2 km in the farthest distance. It will contribute to promoting scientific research for peaceful use of outer space. The satellite is now transmitting the melody of the immortal revolutionary hymns "Song of General Kim Il Sung" and "Song of General Kim Jong Il." The rocket and satellite which our scientists and technicians correctly put into orbit at one launch are a fruition of our wisdom and technology 100 percent. The successful launch of the first artificial satellite in the DPRK greatly encourages the Korean people in the efforts to build a powerful socialist state under the wise leadership of General Secretary Kim Jong Il.

This was followed on September 14 by the release of a photograph of the satellite and the claim that it had completed its 100th orbit. However, no foreign observer ever detected the satellite visually or by radar, or picked up its radio signals. It seems that the third stage either failed and fell into the Pacific, or misfired and put the satellite into a low orbit that quickly decayed.

Kyokko

A Japanese satellite, launched by **ISAS** (Institute of Space and Astronautical Science), that investigated and sent back images of auroras from orbit. Kyokko, whose name means "aura," was known as **Exos**-A before launch.

Launch
    Date: February 4, 1978
    Vehicle: M-3H
    Site: Kagoshima
Orbit: 630 × 3,970 km × 65°
Mass at launch: 126 kg

L

L series (Japanese launch vehicles)

Japan's first family of orbital launch vehicles. The Lambda, or L series, rockets were effectively scaled-up versions of the solid-propellant Kappa sounding rockets. An L-4S placed Japan's first artificial satellite in orbit in 1970. See **Japanese launch vehicles**.

L-1/L-3

See **Russian manned lunar programs**.

L5

Abbreviation for the fifth **Lagrangian point**.

L5 colony

See **O'Neill-type colony**.

L5 Society

An organization formed in 1975 to promote the **space colony** concepts of Gerard K. **O'Neill**. It merged with the **National Space Institute** in 1987 to form the **National Space Society**. The inspiration for the L5 Society came from O'Neill's first published paper on the subject, "The Colonization of Space."[223] This influenced a number of people who later became leaders of the society, including Keith Henson, a young entrepreneurial engineer, and his wife, Carolyn, both of Tucson, Arizona. The Hensons corresponded with O'Neill and were invited to present a paper on "Closed Ecosystems of High Agricultural Yield" at the 1975 Princeton Conference on Space Manufacturing Facilities, which was organized by O'Neill. A sign-up sheet at the conference eventually made its way to the Hensons, who also obtained O'Neill's mailing list. The Hensons incorporated the L5 Society in August 1975 and sent its first newsletter to those two lists. The society was founded partly because of Arizona congressman Morris Udall, who at the time was a serious candidate for U.S. president. Carolyn Henson arranged a meeting between O'Neill and Udall, from which Udall emerged enthusiastic about the idea. The first issue of the *L5 News,* published in September 1975, included a letter of support from Udall and the rallying cry that "our clearly stated long range goal will be to disband the Society in a mass meeting at L5."

Public excitement over the L5 scenario peaked in about 1977, the year in which NASA conducted its third consecutive summer study on Space Settlements and Industrialization Using Nonterrestrial Materials. In this study, the physicist and L5 director J. Peter Vajk and others developed the most detailed scenario yet for the production of solar power satellites (SPSs) from lunar materials. The scenario called for a space manufacturing facility that would house 3,000 workers in a rotating facility constructed from refurbished Space Shuttle External Tanks. The study identified how many launches of the Shuttle and a Shuttle-derived heavy-lift vehicle would be required, and it concluded that the project could begin in 1985 and have three SPSs online by 1992. Unfortunately, two assumptions were made that later proved overly optimistic: the Shuttle would significantly reduce the cost of space launch, and it would fly 60 times per year. However, the scenario did serve as a significant proof of concept. Also in 1977, two major books came out on the subject, O'Neill's *The High Frontier* and T. A. Heppenheimer's *Colonies in Space,* bringing in a new wave of members.

The L5 Society opposed the United Nations Moon Treaty in 1979 and 1980 on the grounds that the "common heritage" provision of the treaty would stifle the development of nonterrestrial resources, which were crucial to the construction of an O'Neill-type colony. The society hired the Washington lobbyist and lawyer Leigh Ratiner to train a number of L5 activists in the art of winning political support. This proved successful; the Senate Foreign Relations Committee was persuaded to oppose the treaty, which consequently was never signed by the United States. However, the L5 Society fared less well in its battle over solar power satellite funding. The Department of Energy had spent about $25 million on SPS research from 1977 to 1980, but the Carter administration eliminated the $5.5 million for SPS that was originally in the budget for fiscal year 1981. L5 director Mark Hopkins, who later formed **Spacecause** and **Spacepac**, lobbied intensively but unsuccessfully to get Congress to restore funding for SPS. With the loss of this program died the dream of realizing an L5 colony anytime in the foreseeable future.

LACE (Low-power Atmospheric Compensation Experiment)

Part of a dual payload with **RME** (Relay Mirror Experiment) carrying laser defense experiments. LACE, also

known as Losat-L, was built by the Naval Research Laboratory as a target for ground lasers, to investigate atmospheric distortion and compensation methods. Low-power lasers were beamed from the Air Force Maui Optical Station and picked up by onboard infrared and phased detectors. A laser then locked onto a reflector mounted on a 46-m boom to acquire the satellite and the sensor array returned data on the laser coverage, allowing the laser's adaptive optics to be adjusted to compensate for atmospheric distortion. The payload included ABE, the Army Background Experiment, an instrument that monitored background levels of neutron radiation in order to be able to discriminate between warheads and decoys, and the Ultraviolet Plume Instrument (UPI) to track rocket plumes.

Launch
   Date: February 14, 1990
   Vehicle: Delta 6925
   Site: Cape Canaveral
   Orbit: 463 × 480 km × 43.1°
   Size: 2.4 × 1.4 m
   Mass: 1,430 kg

Lacrosse

American all-weather reconnaissance satellites. Unlike previous spy satellites, Lacrosse uses side-looking radar to peer through clouds to form images of the target area. Equipped with a very large radar antenna powered by solar arrays almost 50 m across, Lacrosse may be able to resolve detail on the ground as small as one m—good enough to identify and track major military units such as tanks or missile transporter vehicles. However, this high resolution would come at the expense of broad coverage, and would be achievable over an area of only a few tens of square kilometers. Thus, Lacrosse probably uses a variety of radar scanning modes, some providing high-resolution images of small areas, and other modes offering lower-resolution images of areas covering several hundred square kilometers. Lacrosse 1 was launched on December 2, 1988, by the Space Shuttle, and was followed by Lacrosse 2 on March 8, 1991, and Lacrosse 3 more than six years later, both launched by Titan IVs from Vandenberg Air Force Base.

Lageos (Laser Geodynamics Satellite)

Solid, spherical, 60-cm-diameter passive satellites that provide reference points for laser ranging experiments. Lageos carries an array of 426 prisms that reflect ground-based laser beams back to their source. By measuring the time between the transmission of the beam and the reception of the reflected signal, stations can measure the distance between themselves and the satellite to an accuracy of 1 to 3 cm. Long-term data sets can be used to monitor the motion of Earth's tectonic plates, measure Earth's gravitational field, measure the wobble in Earth's rotational axis, and better determine the length of an Earth day. Lageos 1 was developed by NASA and placed into a high inclination orbit to permit viewing by ground stations around the world. Lageos 2 was a joint program between NASA and ASI (the Italian Space Agency), which built the satellite using Lageos 1 specifications and materials provided by NASA. Lageos 2's orbit was chosen to provide more coverage of seismically active areas, such as the Mediterranean Sea and California, and to help scientists understand irregularities noted in the motion of Lageos 1. Lageos 1 contains a message plaque addressed to humans and other beings of the far future with maps of Earth from three different eras—268 million years in the past, the present day, and 8 million years in the future (the satellite's estimated decay date). Both satellites are spherical bodies with an aluminum shell wrapped around a brass core. The design was a compromise between numerous factors, including the need to be as heavy as possible to minimize the effects of nongravitational forces, to be light enough to be placed in a high orbit, to accommodate as many retroreflectors as possible, and to minimize surface area to minimize the effects of solar pressure. The materials were chosen to reduce the effects of Earth's magnetic field on the satellite's orbit. (See table, "Lageos Missions.")

Lagrangian orbit

The orbit of an object located at one of the **Lagrangian points**.

Lagrangian point

One of five equilibrium points at which a spacecraft or some other small object can remain in the same relative position in the orbital plane of two massive bodies, such as the Earth and the Sun, or the Earth and the Moon. Lagrangian points are named after the Italian-born French mathematician Joseph Louis de Lagrange (1736–1813),

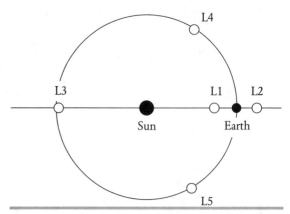

Lagrangian point   The five Lagrangian points in the Sun-Earth system.

who first demonstrated their existence mathematically. Two of these points, known as L4 and L5, are said to be stable because an object placed at one of them will, if slightly displaced, return to the point rather than move farther away. The L5 point of the Earth-Moon system, located 60° behind the Moon in its orbit (as seen from Earth), has been proposed as an ideal site for a **space colony**. However, to date, all practical spaceflight applications have focused on *unstable* Lagrangian points—L1, L2, and L3—so called because the slightest disturbance to any object located at one of them will cause the object to drift away permanently unless it makes a correction (a station-keeping maneuver) using an onboard propulsion system. The L1 and L2 points have proven particularly useful and are being used increasingly by a variety of scientific satellites. The L1 point of the Sun-Earth system—about 1.5 million km directly Sunward of Earth—affords a good spot for monitoring the solar wind, which reaches it about one hour before reaching Earth. Beginning in 1978, **ISEE-3** carried out solar wind observations from the L1 region for several years. More recently, **Wind, SOHO, ACE,** and **Genesis** have studied the Sun and its particle emissions from the same celestial perch. The preferred position is actually some distance to the side of L1, for if the spacecraft is right on the Sun-Earth line, the antennas that track it from Earth are also aimed at the Sun, which is an intense source of interfering radio waves. At the

Lageos Missions

| Spacecraft | Launch | | | Orbit | Mass (kg) |
	Date	Vehicle	Site		
Lageos 1	May 4, 1976	Delta 2914	Vandenberg	5,837 × 5,945 km × 109.8°	411
Lageos 2	Oct. 22, 1992	Shuttle STS-52	Cape Canaveral	5,615 × 5,952 km × 52.6°	400

and the Moon, which all lie effectively behind the vehicle. Among the missions heading to L2 over the next decade are NASA's large infrared observatory, **NGST**, and the European Space Agency's **Herschel, Planck, Eddington,** and **GAIA**. Unstable Lagrangian points also play a central role in a relatively new form of spacecraft maneuvering known as **chaotic control**.

Laika

The first animal to travel into orbit. Laika (Barker), or Muttnik, as the American press nicknamed her, was a female mongrel (mostly Siberian husky) stray, rounded up from the streets of Moscow, who flew aboard **Sputnik** 2. No provision was made for her safe return, and after about a week the dog's air supply ran out and she died. Forty years later the Russians unveiled a memorial to Laika–a plaque at the Moscow research center where she was trained.

Landsat

A continuing series of NASA Earth resources satellites, data from which are made freely available for public use. The Landsat program was originally known as the Earth Resources Technology (ERT) program, and Landsat 1 as ERTS-1; the name was changed by NASA in 1975. (See table, "Landsat Spacecraft.")

Langley Research Center

NASA's oldest primary field installation, named after the aviation pioneer Samuel Pierpont Langley (1834–1906). Its mission is to carry out broadscale research on vehicle configurations, materials, and structures for space, and aeronautics. Established in 1917 and formerly known as Langley Aeronautical Laboratory, it is located at Langley Field, Hampton, Virginia. During World War II, it was involved in developing military aircraft and then in programs that led to the first supersonic plane, the Bell X-1. Staff from Langley formed the Space Task Group, which was responsible for the Mercury Project before this program was transferred to the Manned Spacecraft Center at

Laika A mock-up of Laika aboard Sputnik 2 at the RKK Energia Museum. *Joachim Becker*

same distance from Earth on the opposite (nightward) side from the Sun is L2–a location scheduled to become increasingly visited by space observatories. It is an attractive site because it allows a spacecraft to make observations of the cosmos without interference from the Sun, the Earth,

Landsat Spacecraft

Launch site: Vandenberg Air Force Base				
	Launch			
Spacecraft	**Date**	**Vehicle**	**Orbit**	**Mass (kg)**
Landsat 1	Jul. 23, 1972	Delta 100	895 × 908 km × 99°	816
Landsat 2	Jan. 22, 1975	Delta 2914	898 × 912 km × 99°	953
Landsat 3	Mar. 5, 1978	Delta 2914	895 × 915 km × 99°	960
Landsat 4	Jul. 16, 1982	Delta 3925	693 × 705 km × 98°	1,942
Landsat 5	Mar. 1, 1984	Delta 3925	695 × 703 km × 98°	1,938
Landsat 6	Oct. 5, 1993	Titan II	Failed to reach orbit	1,740
Landsat 7	Apr. 15, 1999	Delta 7925	668 × 698 km × 98°	1,969

Houston. Langley was also involved in investigating the optimum shape for Gemini and Apollo spacecraft so that they could survive reentry. Among its other projects have been the **Echo** balloons, the **Pegasus** micrometeroid detection satellites, **Lunar Orbiter**, and developing the Mars **Viking** mission.

laser propulsion

Methods of propelling spacecraft using the energy of laser beams, all of which remain conceptual or in the early stages of experimentation. There are two main types of laser propulsion, depending on whether the laser is off-board or onboard the spacecraft. Off-board techniques have been proposed to boost lightweight vehicles either from the ground to orbit, or on interplanetary or interstellar missions. These techniques include laser-powered launching to orbit and laser light sails. Onboard methods involve the use of lasers as part of a **nuclear propulsion** system.

Off-board laser propulsion

This is part of a larger class of propulsive methods known as **beamed-energy propulsion**. Its great advantage is that it removes the need for the spacecraft to carry its own source of energy and onboard propulsion system. The propulsive energy comes instead from a fixed, high-power laser beam that is directed onto the spacecraft by a tracking and focusing system.

Laser-powered launching. Powerful lasers developed as part of the U.S. SDI (Strategic Defense Initiative) program promise to have a peaceful use—to launch lightweight spacecraft into low Earth orbit. Tests have already been conducted by Leik Myrabo of the Rensselaer Polytechnic Institute and other scientists from the U.S. Air Force and NASA, using a 10-kW infrared pulsing laser at the White Sands Missile Range and an acorn-shaped craft

laser propulsion A cold-flow propulsion test using lasers at Marshall Space Flight Center. *NASA*

with a diameter of 12.2 cm and a mass of 50 g. The base of the craft is sculpted to focus the beam from the laser onto a propellant. In tests so far, this propellant has been air, which is heated by the beam to a temperature of 10,000° to 30,000°C, expands violently, and pushes the craft upward. A height of 71 m was achieved in an October 2000 trial. To orbit a 1-kg spacecraft will demand a much more powerful, 1-MW laser and a supply of onboard propellant, such as hydrogen, to take over at altitudes where the air gets too thin.[163]

Laser light sails. These would be interplanetary or interstellar spacecraft powered by one or more extremely powerful, orbiting lasers. For details, see **space sail**.

Onboard laser propulsion

Lasers located onboard a spacecraft could be used in a variety of propulsion systems. In the laser-fusion concept, for example, pellets of deuterium-tritium fuel are bombarded by symmetrically positioned lasers for a few billionths of a second until the nuclei fuse, releasing a sudden burst of energy. A rapid series of such explosions could provide the basis for a **nuclear pulse rocket**. Alternatively, in one form of the **interstellar ramjet**, lasers are used to ionize interstellar gas ahead of the spacecraft so that the gas can be drawn in by a magnetic field and used as propellant.

launch

The initial motion of a space vehicle from its stationary, prelaunch position to dynamic flight. Also, the moment when the vehicle is no longer connected to or supported by the launch pad structure.

launch control center

A centralized control point for all phases of prelaunch and launch operations. The handover of control to a mission control center occurs at the moment of separation of the space vehicle from all hard ground connections.

Launch Escape System (LES)

See **Apollo**.

launch pad

(1) In a restricted sense, the load-bearing base or platform from which a rocket vehicle is launched. This is sometimes called the *launch pedestal*. (2) More generally, the area from which a rocket vehicle is fired, including all the necessary support facilities, such as the service tower, safety equipment, and cooling water and flame deflectors. Often called simply a *pad*.

launch support and hold-down arm

Equipment that provides direct support for a portion of the dead weight of the space vehicle and retains the vehicle on the launch pad during the period of thrust buildup.

launch umbilical tower
A frame platform upon which a steel towerlike structure, used to support and service the umbilical arms, is located. The tower also supports and houses equipment needed to perform certain service and checkout functions on the space vehicle prior to launch.

launch vehicle
The part of a space vehicle that supplies the propulsion and guidance to reach the prescribed velocity, position, and attitude required for injection into the desired trajectory. Today's launch vehicles consist of two, three, or more propulsive stages, and can be classified in a number of ways. One of these is according to launch capacity, as shown in the table ("Launch Vehicle Types"). Further classification is possible into expendable, or nonreusable, launch vehicles and nonexpendable launch vehicles.

launch window
The period during which a spacecraft must be launched to achieve its given mission trajectory.

launcher deflector
A device constructed of steel or reinforced concrete, or a combination of both, that deflects the booster engine flame in controlled directions.

L-band
See **frequency bands**.

LDEF (Long Duration Exposure Facility)
A package of 57 experiments placed in Earth orbit by the Space Shuttle to study the effects of exposure to the environment of space. The LDEF was supposed to have been recovered after about one year. However, delays in the Shuttle program meant that the package was not brought back until January 1990, just a few weeks before it would have reentered the atmosphere and been destroyed. LDEF was designed to provide long-term data on the space environment, including micrometeoroid and radiation bombardment, and its effects on materials and satellite systems (including power, propulsion, and optics), as well as the survivability of microorganisms in space. The mission's experiments included the participation of more than 200 principal investigators from 33 private companies, 21 universities, 7 NASA centers, 9 Department of Defense laboratories, and 8 foreign countries. Its planned retrieval was temporarily postponed in March 1985, and then indefinitely postponed following the *Challenger* disaster in 1986. Finally, it was brought back to Earth by STS-32 in January 1990. Many of the experiments benefited from their extended stay in space; in particular, it was found that many of the microorganisms onboard had survived, shielded beneath a layer or two of dead cells.

Shuttle deployment
    Date: April 8, 1984
    Mission: STS-41C
Orbit: 344 × 348 km × 28.5°
Size: 9.1 × 4.3 m
Total mass: 3,625 kg

Le Prieur, Yves Paul Gaston (1885–1963)
A French naval lieutenant who invented a small solid-fueled rocket designed to be fired from French or British biplanes against German captive observation balloons. La Prieur rockets were used in France in World War I.

Leasat (Leased Satellite)
A series of satellites, launched in the 1980s and developed as a commercial venture, to provide dedicated communications services to the American military. The program stemmed from Congressional reviews in 1976 and 1977 that advised increased use of leased commercial facilities. Owned by Hughes Communications, the satellites are designed to provide global UHF communications to military air, sea, and ground forces. The system's primary user was the U.S. Navy, with some support also provided to the Air Force and ground mobile forces. The Leasats, along with the Navy's FLTSATCOMs (Fleet Satellite Communications satellites), have now been

Launch Vehicle Types

Launch Vehicle (LV)	Capacity (kg)		Examples
	LEO	**GTO**	
HLLV (heavy-lift)	>10,000	>5,000	Ariane 5, Titan V, Space Shuttle, Zenit
LLV (large)	5,000–10,000	2,000–5,000	Ariane 4, Atlas IIAS, CZ-2E, H-2, GSLV
MLV (medium)	2,000–5,000	1,000–2,000	Delta 7925, CZ-3, PSLV, Molniya, Tsyklon
SLV (small)	<2,000	<1,000	Athena I/II, ASLV, Pegasus XL, Start, Taurus

LDEF Having spent almost six years in space, LDEF is removed from the payload bay of the Space Shuttle *Columbia* after its return to Kennedy Space Center in 1990. *NASA*

largely superseded by the **UFO** (UHF Follow-On) satellites.

Lebedev, Valentin V. (1942–)

A veteran Soviet cosmonaut who published a candid diary of his long stay aboard the **Salyut**-7 space station. After graduating from the Moscow Aviation Institute, Lebedev worked on the design and development of new spacecraft systems. He made his first spaceflight in December 1973 as flight engineer aboard Soyuz-13. In 1980, he missed out on the Soyuz-35 flight to Salyut-7 after injuring his leg but was assigned to a later mission to the space station that lasted from May to December 1982. His *Diary of a Cosmonaut: 211 Days in Space* is an intimate account of his experiences during this time and useful material for planners of future long-duration manned missions.

He comments on his problems in getting along with crewmate Anatoly Berezovoy: "July 11: Today was difficult. I don't think we understand what is going on with us. We silently pass each other, feeling offended. We have to find some way to make things better." He admits how hard it is to be away from his family: "October 3: Lusia [his wife] told me yesterday, 'Valia I miss you so much.' I said, 'We miss each other so much that this separation should be enough for the rest of our lives.' " He bemoans his space rations: "October 11: All night I dreamed of a bowl of steaming borscht with two scoops of sour cream." He speaks often of strained relations between ground controllers and cosmonauts: "September 12: They always want to know what our mood is and look inside of us. . . . [C]onstant questions such as, 'How do you feel?' 'What are you doing now?' or 'How did you sleep?' are very irritating. Future programs for space settlement and long-term flights should pay attention to . . . social-psychological problems, such as communications between people on Earth and those working in space." He becomes weary of the tedium: "October 13: Five months of flight. I cannot believe that we have flown for so long. We don't feel time anymore. It's getting more difficult now. I begin to count the days . . . I think our fatigue grows because our interest in work is fading. I don't even want to look out the porthole anymore." And he offers a frank account of his return to Earth: "December 10: It was dark outside. The air smelled fresh. The snow fell lightly on my face. But all I felt was sick to my stomach . . . I asked for a napkin and threw up into it. After I threw up a few more times, I felt better." There are some lighter moments, too, among them the tale of the constipated French visitor to the station, the sunbathing sessions at the porthole, the laborious bathing procedure, and the admission that extended weightlessness left the cosmonauts much "furrier."

LEO (low Earth orbit)

Definitions vary. According to some, LEO includes orbits having **apogees** (high points) and **perigees** (low points) between about 100 km and 1,500 km. Others extend that range up to 2,000 or 3,000 km. In some cases, the distinction between LEO and **MEO** (medium Earth orbit) is dropped, and LEO is considered to be any orbit below geosynchronous altitude. The majority of all satellites, as well as the Space Shuttle and International Space Station, operate from LEO, so that this region of space is getting very crowded. According to the U.S. Space Command, which keeps track of the various items in Earth orbit, there are now more than 8,000 objects larger than a softball in LEO.

Leonov, Alexei Archipovich (1934–)

A Soviet cosmonaut who, on March 18, 1965, during the flight of **Voskod** 2, became the first person to go on a spacewalk. He also served as the Soviet capsule commander during the **Apollo-Soyuz Test Project** (ASTP). Leonov graduated with honors from Chuguyev Higher Air Force School in 1957, then joined the Soviet Air Force as a fighter pilot before being selected as one of the original group of 20 cosmonauts on March 7, 1960. Following the Voskhod 2 mission, he attended the Zhukovsky Air Force Engineering Academy, graduating in 1968. For various reasons, his return to space was much delayed. He was assigned to the secret manned Soviet lunar project, but this was canceled. Later he was put on the flight crew of Soyuz 11, but a few days before launch, another crew member became ill, and the backup crew was sent instead. All three backup crew members died due to a malfunction during reentry. As a result, the flight that Leonov had been reassigned to, Soyuz 12a, was canceled. His next scheduled flight, aboard Soyuz 12b, was also canceled because of the in-orbit failure of Cosmos 557. Finally, Leonov returned to space as commander of Soyuz 19 on the ASTP. While in orbit, he made sketches, including one of the American astronaut Thomas **Stafford**, using a set of colored pencils attached to his wrist with a makeshift bracelet. Leonov was subsequently promoted to major-general, and served as commander of the cosmonaut team (March 1976 to January 1982) and as deputy director of the **Gagarin Cosmonaut Training Center**, until his retirement in 1991.

LES (Lincoln Experimental Satellite)

Spacecraft that were designed and built in the 1960s and 1970s by the Massachusetts Institute of Technology Lincoln Laboratory as part of the Lab's Air Force–sponsored

LES Spacecraft

Launch site: Cape Canaveral

Spacecraft	Launch Date	Vehicle	Orbit	Mass (kg)
LES 1	Feb. 11, 1965	Titan IIIA	2,783 × 2,809 km × 32.1°	31
LES 2	May 6, 1965	Titan IIIA	2,771 × 14,810 km × 32.2°	37
LES 3	Dec. 21, 1965	Titan IIIC	267 × 4,829 km × 26.5°	16
LES 4	Dec. 21, 1965	Titan IIIC	189 × 33,632 km × 26.6°	52
LES 5	Jul. 1, 1967	Titan IIIC	33,188 × 33,618 km × 11.6°	194
LES 6	Sep. 26, 1968	Titan IIIC	35,771 × 35,847 km × 13.0°	163
LES 8	Mar. 15, 1976	Titan IIIC	35,687 × 35,886 km × 17.0°	454
LES 9	Mar. 15, 1976	Titan IIIC	35,745 × 35,825 km × 17.0°	454

program in space communications. Their goal was to test realistically, in orbit, new devices and techniques developed for possible use in satellite communication systems. LES-5 was designed to aid in the development of a tactical satellite communication system for the Department of Defense and was used for the first-ever satellite communications among Army, Navy, and Air Force units. (See table, "LES Spacecraft.")

leveled thrust
Thrust maintained at a relatively constant level by a rocket engine equipped with a programmer or engine-control unit.

Lewis and Clark
Twin Earth observation satellites, named for the explorers Meriwether Lewis (1774–1809) and William Clark (1770–1838), and selected as part of NASA's SSTI (Small Spacecraft Technology Initiative). Only Lewis was actually built and launched—on August 23, 1997. However, once in orbit it spun out of control and reentered about a month later.

Lewis Research Center
See **Glenn Research Center**.

Ley, Willy (1906–1969)
A highly effective popularizer of spaceflight, first in Germany and then in the United States, to which he emigrated in 1935 after Hitler's accession to power. Ley helped to found the German rocketry society **Verein für Raumschiffahrt** in 1927, around the time that his first two books were published: *Die Fahrt ins Weltall* (Journey into Space, 1926) and *Die Möglichkeit der Weltraumfahrt* (The Possibility of Interplanetary Travel, 1928). The latter was the inspiration behind the film (and book) *The Woman in the Moon*. In the United States, he quickly earned a reputation as a visionary of the coming Space Age through well-researched articles. He also wrote several books that dealt with the dream of spaceflight, including *Rockets: The Future of Travel Beyond the Stratosphere*, first published in 1944. One of the earliest books on rocketry for the general public, it became a reference source for future science fiction and reality writing. A revised edition appeared in 1947, entitled *Rockets and Space Travel,* and another in 1952, entitled *Rockets, Missiles, and Space Travel.*[192] Perhaps his best-known book, however, was *The Conquest of Space* (1949),[191] outstandingly illustrated by Chesley **Bonestell**.

LH2
See **liquid hydrogen**.

lidar (light detection and ranging)
A technique that works by beaming pulses of laser light through the atmosphere and detecting the light reflected by dust and other small particles in the air called aerosols. The time between the pulse and the echo determines the distance, and the shift in color of the light determines the velocity of particles along the line of sight. True wind speed and direction, even in clear air, can be found in this way. Long used for ground-based observations of the atmosphere, lidar is now being carried by a new generation of Earth observing satellites.

Life Finder (LF)
A large successor to the **Terrestrial Planet Finder**, Life Finder would be capable of detecting the spectroscopic signs of life on nearby extrasolar planets. LF is identified in NASA's Office of Space Science Strategic Plan as a potential mission beyond 2007 and would be an important component of the **Origins Program**.

life support
The subsystems aboard a manned spacecraft or space station that maintain a livable environment within a pressurized crew compartment. They provide oxygen, drinking water, waste processing, temperature control, and carbon dioxide removal.

lift
The lifting force on a flying object (in particular, a wing or a whole aircraft), due to its motion relative to the surrounding air. Lift is one of the four forces acting on an airplane, the others being **weight, thrust**, and **drag**.

lift-drag ratio
The ratio of the lifting force to the **drag** force for either an isolated wing or a complete aircraft. This ratio is used in determining the rate of descent of a space vehicle in the atmosphere.

lifting body
A wingless aircraft that derives **lift** from the shape of its hull. Lifting bodies are used, for example, as models and prototypes for **space plane** designs.[286]

liftoff
(1) The point at which the upward **thrust** of a spacecraft exceeds the force due to local gravity and the spacecraft begins to leave the ground. (2) The rising of a spacecraft from its launch pad at the start of its ascent into space.

light
Electromagnetic **radiation** that is visible to the human eye. Light can also be considered to be a stream of tiny particles known as **photon**s.

light barrier

According to Einstein's **special theory of relativity**, the **speed of light** *(c)* represents an insurmountable barrier to any object that has real mass (leaving only the elusive **tachyon** possibly exempt). No ordinary material thing can be accelerated from sublight speeds up to the speed of light or beyond, the theory says, for two reasons. First, pumping more **kinetic energy** into an object that is already moving at high speed gives rise to the **relativistic effect** of mass increase rather than a substantial increase in speed. This strange phenomenon becomes so pronounced that at speeds sufficiently close to the speed of light, an object's relativistic mass would become so great that for it to approach still nearer to *c* would take more energy than is available in the entire universe. Second, faster-than-light (FTL) speeds would lead to violations of the fundamental principle of special relativity, which is that all **inertial reference frames** are equivalent—that is, all measurements made by observers at rest or moving at constant velocity with respect to one another are equally valid. In particular, FTL communication would enable simultaneity tests to be carried out on the readings of separated clocks, which would reveal a preferred reference frame in the universe—a result in conflict with the special theory.

Not all is lost, however, for there remains the **general theory of relativity**. This does not rule out faster-than-light travel or communication, but only requires that the local restrictions of special relativity apply. In other words, although the speed of light is still upheld as a local speed limit, the broader considerations of general relativity suggest ways around this statute. One example is the expansion of the universe itself. As the universe expands, new space is created between any two separated objects. Consequently, although the objects may be at rest with respect to their local environment and the cosmic background radiation, the distance between them may grow at a rate greater than the speed of light. Other possibilities, more directly relevant to interstellar travel, include **wormhole** transportation and the **Alcubierre Warp Drive**.

light flash phenomenon

A visual effect experienced when a person is exposed to high-energy particles; it appears similar to but brighter than a shooting star in the night sky. Although the exact mechanism that causes the phenomenon is unknown, it is believed that visual flashes result from direct ionizing energy loss as the particle traverses the cells of the eye.

light-time

The time taken for light or a radio signal to travel a certain distance at light speed.

light-year

The distance traveled by light, moving at 300,000 km/s (186,282 mi./s), in one year. It equals about 9.40 trillion km. The nearest star, Proxima Centauri, lies at a distance of 4.28 light-years, or just over 40 trillion km (25 trillion miles).

limit load

The maximum load calculated to be experienced by the structure under specified conditions of operation.

limit pressure

The maximum operating pressure, or operating pressure including the effect of system environment, such as vehicle acceleration and so forth. For hydraulic and pneumatic equipment, limit pressure will exclude the effect of surge.

Lindbergh, Charles Augustus (1902–1974)

An early aviator who gained fame as the first pilot to fly solo across the Atlantic Ocean in 1927. Such was his public stature following this flight that he became an important voice on behalf of aerospace activities until his death. He served on a variety of national and international boards and committees, including the central committee of **NACA** (National Advisory Committee for Aeronautics) in the United States. He became an expatriate living in Europe following the kidnapping and murder of his two-year-old son in 1932. There, during the rise of fascism, Lindbergh assisted American aviation authorities by providing them with information about European technological developments. After 1936, he was especially important in warning the United States of the rise of Nazi air power, and he helped with the war effort in the 1940s by serving as a consultant to aviation companies and the government. Subsequently, he lived in Connecticut and then Hawaii.[250]

Charles Lindbergh Lindbergh (front seat) with Fred E. Weick (rear seat), head of the Propeller Research Tunnel section (1925–1929), Langley, and Tom Hamilton. *NASA*

LiPS Missions

Launch site: Vandenberg Air Force Base

Mission	Launch Date	Vehicle	Notes
LiPS 1	Dec. 9, 1980	Atlas F	Launch vehicle failed on takeoff.
LiPS 2	Feb. 8, 1983	Atlas H	Carried UHF single-channel transponder.
LiPS 3	May 15, 1987	Atlas H	Data returned from over 140 solar-cell experiments.

LiPS (Living Plume Shield)

Plume shields ejected from the upper stage of **NOSS** (Naval Ocean Surveillance Satellite) primary payloads; they were used as subsatellites for conducting technology demonstrations. The plume shield, which deflects the exhaust of the upper-stage motor away from the primary payload, is customarily jettisoned after use to become space debris. However, on the LiPS experiments, the anterior surfaces of the shield, which are unaffected by the plume, were fitted with a variety of small experiments. (See table, "LiPS Missions.")

liquid hydrogen (H₂)

Hydrogen in its liquid state, used as a cryogenic rocket **fuel**; hydrogen gas turns to liquid under standard atmospheric pressure at −263°C. When oxidized by **liquid oxygen**, liquid hydrogen delivers about 40% more thrust per unit mass than other liquid fuels, such as **kerosene**. Commonly referred to in rocketry as LH2.[273]

liquid oxygen (O₂)

Oxygen in its liquid state, used as the **oxidizer** in many liquid fuel engines. Oxygen gas turns to liquid under standard atmospheric pressure at −183°C. Commonly referred to in rocketry as LOX.

liquid propellant

A liquid ingredient used in the **combustion chamber** of a rocket engine. Liquid propellants may be either **bipropellants**, which consist of a liquid **fuel** (such as kerosene or liquid hydrogen) and a liquid **oxidizer** (such as liquid oxygen or nitrogen tetroxide), or **monopropellants** (such as hydrazine). Liquid propellants, such as liquid oxygen and liquid hydrogen, that must be kept at a low temperature in order to remain liquid are known as **cryogenic propellants**. See **liquid-propellant rocket engine**.

liquid-air cycle engine

An advanced engine cycle that uses liquid hydrogen fuel to condense air entering an inlet. Liquid oxygen is separated from the condensed air and pumped into the **combustion chamber**, where it is burned with the hydrogen to produce **thrust**.

liquid-fueled rocket

A rocket that derives its thrust from one or more **liquid-propellant rocket engines**.

liquid-propellant rocket engine

A rocket propulsion system in which a **liquid fuel** (such as kerosene or liquid hydrogen) is combined in a **thrust chamber** with a liquid **oxidizer** (such as liquid oxygen or fuming nitric acid). Very efficient and controllable, such engines are used extensively in spaceflight. Unlike **solid-propellant rocket motors**, they can be shut off by remote command, simply by closing off their fuel line. In some cases, the thrust can also be varied over a certain range.

LISA (Laser Interferometer Space Antenna)

An array of three NASA spacecraft flying 5 million km apart in the shape of an equilateral triangle. The center of the triangle will be in the ecliptic plane one astronomical unit from the Sun and 20° behind Earth. The LISA mission, which is being designed at JPL (Jet Propulsion Laboratory), is intended to observe gravitational waves from galactic and extragalactic sources, including the supermassive black holes found in the centers of many galaxies. Gravitational waves are one of the fundamental building blocks of our theoretical picture of the universe, but although there is strong indirect evidence for their existence, they have not yet been directly detected. The LISA spacecraft, flying in formation, will be sensitive to the distortion of space caused by passing gravitational waves. Each spacecraft will contain two free-floating "proof masses." The proof masses will define optical paths 5 million km long, with a 60° angle between them. Lasers in each spacecraft will be used to measure changes in the optical path lengths with a precision of 20 pm (trillionths of a meter). If approved, the project will begin development in 2005 with a planned launch in 2008.

lithium hydroxide
A compound used to remove carbon dioxide from spacecraft cabins.

Little Joe
An American surface-to-air guided missile developed at the end of World War II in response to Japanese kamikaze attacks. A cluster of Little Joes was used for preliminary testing of the **Mercury** capsule in 1959–1960 and, later, of the **Apollo** Launch Escape System and Command and Service Module structure.

little LEO
Low Earth orbits in the lower part of the LEO range, typically with perigees and apogees up to a few hundred kilometers, although definitions vary.

load factor
The factor by which the steady-state loads on a structure are multiplied to obtain the equivalent static effect of **dynamic load**s.

loading
A term that can refer to gravitational loading, or the application of force to a bone or a part of the body.

Little Joe Launch of the first Little Joe to carry a Mercury capsule, on October 4, 1959, at Wallops Island. *NASA*

Under normal circumstances on Earth, gravity "loads" the bones and muscles with the force required to support the body.

local time
Time adjusted for location around Earth or other planets in time zones.

local vertical
The direction in which the force of gravity acts at a particular point.

Lockheed Martin
A major U.S. aerospace company formed from the 1995 merger of Lockheed and Martin Marietta. The company traces its roots to 1909, when aviation pioneer Glenn L. Martin (1886–1955) organized a modest airplane construction business. The Martin company later developed the **Vanguard** rocket, which launched several early U.S. satellites and the **Titan** I missile–the forerunner of the Titan family of space launch vehicles. Martin merged with the American Marietta Corporation in 1961 to form Martin-Marietta, which was chosen to develop the Titan II (used to launch **Gemini**) and the Titan III, the **Viking** Mars landers/orbiters, the propulsion system for **Voyager**, the **Space Shuttle** External Tank, and the **Manned Maneuvering Unit**. In 1994, Martin-Marietta took over the Space Systems Division of General Dynamics, the manufacturer of the **Atlas** and **Centaur** families of launch vehicles, before itself becoming part of Lockheed Martin as the Lockheed Martin Missiles and Space Systems Company a year later.

Long March (Chang Zeng, CZ)
See article, pages 240–241.

longitudinal axis
The fore-and-aft line through the center of gravity of a craft.

Longshot, Project
A design for an unmanned probe to Alpha Centauri developed by researchers from NASA and the U.S. Navy in 1988. Longshot would be assembled at the International Space Station, and be propelled on a 100-year flight to the nearest star by a pulsed fusion engine of the type proposed in the **Daedalus** design. The 6.4-ton spacecraft would carry a 300-kW fusion reactor to power instruments and engine startup, and use a 250-kW laser to transmit data to Earth.[18]

Long March (Chang Zeng, CZ)

A series of Chinese launch vehicles, the most recent of which support China's current unmanned and nascent manned space programs. Long March rockets are descendants of ballistic missiles developed by China in the 1960s following the breakdown of Sino-Soviet relations in June 1959. They fall into three generations.

First Generation

CZ-1

China's first orbital launch vehicle, derived from the DF-3 (Dong Fang, whose name means "east wind") missile, which was designed by Hsue-shen **Tsien** but never built. A 29.5-m-long three-stage rocket, the first two stages of which used liquid propellant and the third solid propellant, the CZ-1 launched the first two Chinese satellites in 1970 and 1971 after a failure in November 1969.

Second Generation

FB-1

Little is known of this launch vehicle except that like the CZ-2, it was developed from the two-stage, liquid-propellant DF-5 intercontinental ballistic missile (ICBM); FB stands for Feng Bao, which means "storm." It made four orbital flights between 1975 and 1981.

CZ-2A

Believed to be very similar to FB-1; it was launched only once, in November 1974.

CZ-2B

The B designation was reserved for FB-1.

CZ-2C

A successful LEO (low Earth orbit) rocket with a payload of about 2,800 kg, derived from the DF-5 ICBM, which formed the basis for future Long March launch vehicles. On October 6, 1992, on its thirteenth flight, the CZ-2C successfully launched the Swedish **Freja** satellite as a copassenger. On April 28, 1993, the **Chinese Great Wall Industrial Corporation** and Moto-

rola signed a contract for multiple launch of **Iridium** satellites using the CZ-2C/SD, a 2C variant with a newly developed Smart Dispenser and improved second-stage tanks and engines. The first deployment launches for the Iridium program (each carrying two satellites) went ahead successfully in 1997–1998.

Third Generation

CZ-3

A three-stage launch vehicle designed to place satellites into geostationary transfer orbit (GTO) or sun-synchronous orbit. The majority of the technology and flight hardware used in the CZ-3, including the first and second stages, was proven by the CZ-2C. The new third stage was equipped with a liquid oxygen/liquid hydrogen cryogenic engine. The CZ-3's GTO payload was about 1,500 kg.

CZ-3A

Similar to the CZ-3 but with a more powerful third stage, raising GTO payload capacity to about 2,700 kg, and a more capable attitude control system. By June 1998, the CZ-3A had made three consecutive successful launches and was offered to international customers.

CZ-3B

A powerful variant of the Long March launch vehicle, with a payload capacity of about 5,000 kg. It was based on the CZ-3A but had enlarged propellant tanks, larger fairing, and four boosters strapped onto the core stage.

CZ-3C

A launch vehicle that combined the CZ-3B core with two boosters from the CZ-2E. The standard CZ-3C fairing is 9.56 m long and 4.0 m in diameter.

CZ-4

A three-stage vehicle that uses storable propellants and is intended for launching satellites into polar or sun-synchronous orbits. It differs from the CZ-3 mainly in its third stage, which features thin-wall tankage and two gimbaled engines. On September 7, 1988, the CZ-4A made its inaugural flight, successfully launching China's first experimental meteorological satellite. The CZ-4B, introduced in 1999, is an improved model with enhanced third stage and fairing. With a length of 44.1 m and a first-stage thrust of 300 tons, it can launch a payload of 4,680 kg into LEO or 1,650 kg into a sun-synchronous orbit.

CZ-2D

A two-stage launch vehicle with storable propellants, suitable for launching a variety of LEO satellites. Developed and manufactured by the Shanghai Academy of Space Flight Technology, the CZ-2D had a typical payload capability of 3,500 kg in a 200-km circular orbit. Its first stage was identical to that of the CZ-4. The second stage was essentially the same as that of the CZ-4, except for an improved vehicle equipment bay.

CZ-2E

The largest Chinese launch vehicle, using four liquid rocket booster strap-ons to reach LEO payload capabilities approaching those of the Proton, Titan, or Ariane. The CZ-2E had a maximum payload capability of 9,500 kg. With a perigee kick motor, the CZ-2E could put 3,500 kg into GTO.

CZ-2F

A man-rated version of CZ-2E. Modifications probably are related to improved redundancy of systems, strengthened upper stage to handle large 921-1 spacecraft fairing, and launch escape tower.

CZ-2EA

A planned upgrade of CZ-2E with enlarged boosters the size of the CZ-2C first stage, probably intended for the eventual launch of Chinese space station modules.

Lorentz force turning

An innovative technique proposed for steering interstellar spacecraft. It would involve unreeling several extremely long **space tether**s and giving them a high electric charge. The Lorentz force (a force experienced by a charge moving in a magnetic field) between these tethers and the galactic magnetic field would result in the spacecraft undergoing a slow, constant course change that over time would be enough to switch the trajectory from one star to another.

Losat-X (Low Altitude Satellite Experiment)

A test flight of U.S. Department of Defense sensors as part of the Strategic Defense Initiative. Losat-X was launched alongside **Navstar GPS**-11. Its main mission was remote sensing using a multispectral imaging package and a wide field-of-view star camera. Designed for a three-month lifetime, the satellite accomplished its primary objectives in the first 24 hours. The secondary mission objectives were not met, however, due to a failure in the command receiver nine days after launch.

Launch
   Date: July 4, 1991
   Vehicle: Delta 7925
   Site: Cape Canaveral
   Orbit: 405 km × 417 km × 40.0°
   Size: 1.2 × 0.9 × 0.3 m
   Mass: 75 kg

Lousma, Jack Robert (1936–)

An American astronaut who served as pilot for **Skylab** 3 and backup docking module pilot of the American flight crew for the **Apollo-Soyuz Test Project**. On his second space mission, Lousma was commander of the third orbital test flight of Space Shuttle *Columbia* (STS-3), launched on March 22, 1982. He received a B.S. in aeronautical engineering from the University of Michigan in 1959, the year in which he also joined the Marine Corps, and earned an M.S. in aeronautical engineering from the U.S. Naval Postgraduate School in 1965. Lousma left NASA in 1983 and later became chairman of AeroSport, in Ann Arbor, Michigan.

Lovelace, William Randolph, II (1907–1965)

An accomplished physician, chief of surgery at the Mayo Clinic, and a leading figure in **aerospace medicine** during the early years of the American space program. Lovelace experimented on problems of high altitude escape. In World War II he survived deliberately parachuting from a B-17 at over 12,000 m to validate the use of a bail-out oxygen bottle during a high-altitude descent; the force of his chute opening was so violent that it knocked him unconscious and stripped the gloves from his hands. In 1958, Lovelace was appointed chairman of NASA's Special Advisory Committee on Life Science at NASA headquarters by the agency's first administrator, T. Keith **Glennan**, and went on to play a central role in selecting the **Mercury Seven** astronauts.

Lovell, James Arthur, Jr. (1928–)

An American astronaut who became the first human to go into space four times, flying aboard **Gemini** 7, Gemini 12, **Apollo** 8, and Apollo 13, and the only person ever to go to the Moon twice and not land. Lovell attended the University of Wisconsin for two years, then entered the Naval Academy at Annapolis, graduating with a B.S. in 1952. Following graduation, Lovell received flight training and served at a number of navy bases until entering Naval Test Pilot School in 1958. One of nine NASA astronauts chosen in September 1962, he was first assigned as backup pilot for Gemini 4. In January 1966, after returning from Gemini 7, he was named backup commander for Gemini 10, a job generally thought to be a dead end. Under the NASA rotation system, Lovell could have expected to command Gemini 13, but the program ended with flight number 12. His continued involvement with Gemini would keep him from being named to the early Apollo flights. But on February 28, 1966, Gemini 9 astronauts Elliott See and Charles Bassett were killed in a plane crash. In the subsequent shuffle of crew assignments, Lovell and his pilot Buzz **Aldrin** moved from the Gemini 10 backup to Gemini 9, and were later assigned to Gemini 12. A year after Gemini 12, in the wake of the tragic Apollo 1 fire, Lovell was assigned with Neil **Armstrong** and Aldrin to the backup crew for what eventually became Apollo 8. In July 1968, Lovell was promoted to prime crew, replacing Michael **Collins**. This sequence of events made it possible for Lovell to become one of the first humans to fly around the Moon, but cost him his participation in the first lunar landing. A similar mix of good and bad luck placed Lovell on Apollo 13. As backup commander of Apollo 11, he was in line to command Apollo 14. But an attempt by flight crew chief "Deke" **Slayton** to name Mercury astronaut Alan **Shepard**, recently returned to flight status, to Apollo 13 was blocked by NASA management, who thought Shepard needed more training. In August 1969, Lovell was asked if he and his crew could be ready in time to fly Apollo 13, eight months later. Lovell said they could and got the job. This turned out to be a dubious privilege, because a major explosion aboard Apollo 13's Service Module nearly cost the crew their lives. Lovell, with colleagues Fred **Haise** and Jack **Swigert**, took his second trip around the Moon without setting foot on the surface, before heading directly home in the stricken craft. Following Apollo 13 and a leave to attend Harvard, Lovell was named deputy director for science and applications at the Johnson Space Center in May 1971. On March 1, 1973, he retired from the Navy as a captain and resigned from NASA to enter private business. Lovell is president of Lovell Communications and also serves as chairman of Mission HOME, a campaign to rekindle enthusiasm and support for space.

low Earth orbit

See **LEO**.

low-gain antenna (LGA)

A small antenna that provides low amplification of either transmitted or received radio signals. LGAs have wide antenna patterns and therefore do not require precise pointing.

LOX

See **liquid oxygen**.

LSAT

See **Olympus**.

Luch

An element of the second generation Soviet global command and control system, deployed in the first half of the 1980s. Luch ("beam") satellites, analogous to the American **TDRSS** (Tracking and Data Relay Satellite System), provided communications service to **Mir**, the **Buran** shuttle, **Soyuz**-TM spacecraft, military satellites, and the TsUPK ground control center. They also supported mobile fleet communications for the Soviet Navy. The modernized Luch-2 allows two high data rate channels to operate at once, enabling real-time TV transmissions from Mir. Each satellite is equipped with three transponders and has a nominal life of five years. By January 1999, five had been launched.

Lucian of Samosata (c. A.D. 120–180)

A Syrian-Greek writer, born in Samosata, Syria, who penned the first known tales of interplanetary travel. In the first, mischievously called *The True History,* his hero's ship is swept up by a great whirlwind and deposited on the Moon. In the second, *Icaro-menippus,* his space-bound adventurer follows in the footsteps, or rather the wing-flaps, of his part-namesake, Icarus, son of Daedalus. He tells his friends, "I took, you know, a very large eagle, and a vulture also, one of the strongest I could get, and cut off their wings." Like many other writers for 1,500 years to come, Lucian made no distinction between *aero*nautics and *astro*nautics, assuming that normal air-assisted flight and breathing are possible on voyages between worlds. To be fair, however, his books were mainly satirical and not serious attempts to describe practical spaceflight.[197]

Lucid, Shannon Matilda Wells (1943–)

An American astronaut and the first woman to go into space more than twice. Lucid received a B.S. in chemistry (1963), and an M.S. (1970) and a Ph.D. (1973) in biochemistry, from the University of Oklahoma. Selected by

NASA in January 1978, she became an astronaut in August 1979 and is qualified for assignment as a mission specialist on Space Shuttle flights. A veteran of five space missions, Lucid has logged 223 days in space—the most by any woman—including flights aboard STS-51G (1985), STS-34 (1989), STS-43 (1991), STS-58 (1993), and Mir (launching March 22, 1996, aboard STS-76, and returning September 26, 1996, aboard STS-79).

Luna

See article, pages 244–246.

Lunar Landing Research Vehicle (LLRV)

A flying contraption used by American astronauts as part of their training in how to control the **Apollo** Lunar Module (LM) during descents to the Moon's surface. It had a weight distribution and throttle reaction time

(continued on page 247)

Lunar Landing Research Vehicle   LLRV Number 1 in flight at Edwards Air Force Base in 1964. *NASA*

Luna

A series of Soviet Moon probes, including orbiters, landers, and sample-return craft, launched between 1959 and 1976. Lunas were the first human-made objects to reach escape velocity, crash into the Moon, photograph the Moon's farside, soft-land on the Moon, automatically return lunar surface material to Earth, and deploy a rover on the Moon's surface.

The success of the first three Lunas (known in the West as "Luniks"), was followed by a gap of three and a half years while the Soviets developed a more sophisticated strategy for lunar exploration. This involved placing a probe in a temporary parking orbit around Earth before firing a rocket to put the craft on a lunar trajectory–in principle, a more accurate method than direct ascent (that is, shooting straight at the Moon from the ground). However, Lunas 4 through 8 all failed, for various reasons, in their attempts to soft-land. Success came again with Luna 9, the first spacecraft to send back photos from the lunar surface. Lunas 10 to 12 and 14 were orbiters, designed in part to provide detailed photographic maps and collect other data that were essential to the Soviet manned lunar program.

Then came a sudden shift in emphasis. With the Moon Race lost to the Americans, the Soviets began launching much larger Lunas–three times more massive than the earlier craft–requiring the more powerful but (then) less reliable Proton rocket. Several of the new generation of Lunas (though not officially named as such) were left stranded in Earth orbit before Luna 15 was successfully placed on a lunar trajectory just two days ahead of Apollo 11. Its audacious mission, to upstage Apollo 11, ended when it crashed on July 21, just as Armstrong and Aldrin were preparing to leave the Moon. Subsequent heavy Lunas, however, were for the most part highly successful, returning several samples along with other valuable data and delivering the first automated rovers to explore another world. (See table, "Luna Missions.")

Luna 1

The first human-made object to reach escape velocity; it was supposed to hit the Moon, but a failure of the launch vehicle's control system caused it to miss by about 6,000 km. En route, the probe released a cloud of sodium gas (as did Luna 2), the glowing orange trail of which allowed astronomers to track the progress of the spacecraft visually. Luna 1, also known as Mechta ("dream"), measured the strength of the solar wind and showed that the Moon had no magnetic field.

Luna 2

The first probe to hit the Moon. On impact, east of the Sea of Serenity, it scattered a number of Soviet emblems and ribbons across the surface. About 30 minutes later, the final stage of Luna 2's booster rocket made its own fresh crater.

Luna 3

The first probe to return images of the lunar farside. Luna 3 was launched on a figure-eight trajectory, bringing it within 6,200 km of the Moon and around the farside, which was sunlit at the time. The 17 indistinct pictures received from the spacecraft showed the farside to be mountainous with two dark regions, which were subsequently named Mare Moscovrae ("Sea of Moscow") and Mare Desiderii ("Sea of Dreams").

Lunas 4 to 8

The first Lunas to be placed in low Earth parking orbit prior to lunar trajectory insertion. All attempted soft landings, and all failed because of faulty rocket firings in Earth orbit, in midcourse, or during descent to the lunar surface.

Luna 9

The first successful lunar soft-lander. The landing capsule was a 100-kg sphere, which was dropped roughly onto the Ocean of Storms by a cylindrical mother craft that carried the main braking rocket and fell to destruction after its work was done. Once the lander had rolled to a stop, four petal-like covers opened and four radio antennas extended, allowing Luna 9 to remain in contact with Earth for the next three days. Its TV camera sent back images that, when pieced together, provided a panoramic view of the surface, including views of nearby rocks and of the horizon 1.4 km away.

Lunas 10 to 12 and 14

Lunar orbiter missions that returned photographs and took measurements of infrared, X-ray, and gamma emission from the Moon (to determine chemical composition), on-orbit radiation conditions, micrometeorite collisions, and variations in the lunar gravitational field caused by mass concentrations, or "mascons." They typically functioned for several months.

Luna 13

A soft-lander, similar to Luna 9, that arrived on the Ocean of Storms on Christmas Eve, 1966. In addition to a camera that sent back photographic panoramas under different lighting conditions, Luna 13 was equipped with two spring-loaded arms. One of these was used to determine the density of the surface by measuring the effect of the landing capsule's impact on the soil—a piece of information crucial to the designers of a manned mission. The other arm probed the chemical composition of the surface.

Lunas 15, 16, 18, 20, 23, and 24

Automated lunar sample-return craft, three of which were successful. Luna 15 entered lunar orbit two days ahead of Apollo 11 and, on the day Apollo 11 began circling the Moon, lowered its own orbit to 9 by 203 km. At this point there was concern in the United States that the Russian probe would somehow interfere with the manned mission. However, assurances were quickly given by the Soviets that this would not be the case. On July 20, just hours before Apollo 11's scheduled landing, Luna 15 carried out another maneuver to put it in a 16-by-110-km orbit. The next day, while Armstrong and Aldrin were on the surface, the little probe made its last retro-rocket burn and began to descend to what was supposed to be a soft landing. Unfortunately, it made contact instead at 480 km/hr in the Sea of Crises. Almost twenty years would pass before the Soviets officially admitted that Luna 15 was a failed sample-return attempt. Whether it could have beaten Apollo 11 if all had gone well is unclear. Even if its landing attempt had succeeded, it would

not have returned to Earth until the day after Apollo 11 splashed down. On the other hand, Luna 15 did spend one day longer in lunar orbit than was typical of later sample missions. If the probe had made it down in three days instead of four, or if Apollo 11 had failed to return samples, the Soviets might just have pulled off an outrageous coup.

Luna 16 landed safely on the Moon on September 20, 1970, on the Sea of Fertility and deployed an extendable arm with a drilling rig to collect 100 g of soil and rock. After 26 hours 25 minutes on the surface, the ascent stage, with a hermetically sealed soil sample container, took off and returned to Russia on September 24. The lower stage remained on the lunar surface and continued sending back data on temperature and radiation.

Luna 18 used a new method of navigation in lunar orbit and for landing. However, after 54 lunar orbits, it failed as it descended toward the Sea of Fertility on September 11, 1971, and crashed into mountainous terrain. On February 21, 1972, Luna 20 soft-landed in the Apollonius highlands, just 120 km from where Luna 18 had come down, collected 30 g of samples, and returned to Earth four days later.

A similar scenario played out with Lunas 23 and 24. The former actually survived its landing on the Sea of Crises but was sufficiently battered that its sample-collecting apparatus was knocked out of action. However, its successor, the final Luna mission, touched down just a few hundred meters away and returned triumphantly with 170 g of Moon rock.

Lunas 17 and 21

Soft-landers carrying **Lunokhod** automated rovers. Luna 17 entered lunar orbit on November 15, 1970, and landed two days later on the Sea of Rains. Having taken pictures of its surroundings, Lunokhod 1 then rolled down a ramp and began exploring the surface. The rover would run during the lunar day, stopping occasionally to recharge its batteries via the solar panels. At night it would hibernate until the next sunrise, kept warm by the radioactive source. Although intended to operate through three lunar days (earth months) the rover actually operated for eleven, officially ending its mission on October 4, 1971, the anniversary of Sputnik 1. By then it had traveled 10.5 km, transmitted more than 20,000 TV pictures and 200 TV panoramas, and conducted more than 500 soil tests.

On January 15, 1973, Luna 21 touched down in LeMonnier crater on the Sea of Serenity with Lunokhod 2 aboard. Over an operating period of four months, this second rover covered 37 km, including hilly upland areas and rilles, sent back over 80,000 TV pictures and 86 panoramic images, and carried out a variety of experiments, including mechanical tests of the surface and laser ranging.

Lunas 19 and 22

Heavy lunar orbiters that continued mapping the Moon's surface and extended earlier studies of the lunar gravitational field and the location of mascons, the on-orbit radiation environment, the gamma-active lunar surface, micrometeoroids, and the solar wind. Luna 22's orbit was eventually adjusted so that its perilune was as low as 25 km.

Luna Missions

Launch site: Baikonur

Luna	Launch Date	Launch Vehicle	Mass (kg)	Notes
1	Jan. 2, 1959	Luna	361	Missed Moon by 6,000 km.
2	Sep. 12, 1959	Luna	387	Crashed on Moon Sep. 13.
3	Oct. 4, 1959	Luna	279	Photographed Moon's farside.
4	Apr. 2, 1963	Molniya	1,422	Fell back to Earth.
5	May 9, 1965	Molniya-M	1,474	Attempted soft-landing; crashed.
6	Jun. 8, 1965	Molniya-M	1,440	Attempted soft-landing; missed Moon.
7	Oct. 4, 1965	Molniya	1,504	Attempted soft-landing; crashed.
8	Dec. 3, 1965	Molniya	1,550	Attempted soft-landing; crashed.
9	Jan. 31, 1966	Molniya	1,580	Landed in Oceanus Procellarum Feb. 3.
10	Mar. 31, 1966	Molniya	1,597	Entered lunar orbit Apr. 3.
11	Aug. 24, 1966	Molniya	1,638	Entered lunar orbit Aug. 27.
12	Oct. 22, 1966	Molniya	1,620	Entered lunar orbit Oct. 25.
13	Dec. 21, 1966	Molniya	1,700	Landed in Oceanus Procellarum Dec. 24.
14	Apr. 7, 1968	Molniya	1,700	Entered lunar orbit Apr. 10.
15	Jul. 13, 1969	Proton	5,600	Attempted sample-return; crashed.
16	Sep. 12, 1970	Proton	5,600	Landed in Mare Fecunditatis Sep. 30; returned to Earth with 100-g sample Sep. 24.
17	Nov. 10, 1970	Proton	5,600	Landed in Mare Imbrium Nov. 17; carried Lunokhod 1 rover.
18	Sep. 2, 1971	Proton	5,600	Attempted sample-return; crashed Sep. 11.
19	Sep. 28, 1971	Proton	5,810	Entered lunar orbit Oct. 3.
20	Feb. 14, 1972	Proton	5,600	Landed in Mare Fecunditatis Feb. 21, returned to Earth with 30-g sample Feb. 25.
21	Jan. 8, 1973	Proton	5,567	Landed in Mare Serenitatis Jan. 15; carried Lunokhod 2 rover.
22	May 29, 1974	Proton	5,835	Entered lunar orbit Jun. 2.
23	Oct. 28, 1974	Proton	5,300	Landed in Mare Crisium Nov. 6; damaged drill prevented sample return.
24	Aug. 9, 1976	Proton	5,306	Landed in Mare Crisium Aug. 18; returned to Earth with 170-g sample Aug. 23.

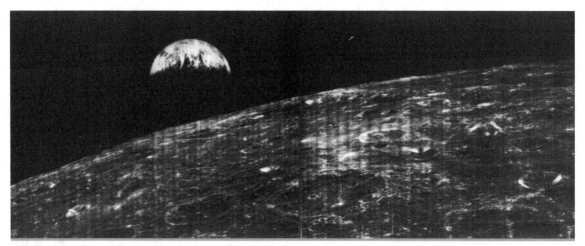

Lunar Orbiter   The world's first view of Earth taken from the Moon, transmitted by Lunar Orbiter 1 on August 23, 1966, at 16:35 GMT. *NASA*

Lunar Landing Research Vehicle (LLRV)

(continued from page 243)

similar to that of the real LM and a downward-facing turbofan engine to provide vertical thrust. The LLRV was based around the early VTOL "Flying Bedsteads," which were developed to study the potential of vertical takeoff and landing for jet aircraft. Neil **Armstrong** had a narrow escape when the LLRV he was flying went out of control and he was forced to eject.

lunar module (LM)

See **Apollo**.

Lunar Orbiter

A series of five highly maneuverable, Moon-orbiting NASA spacecraft, launched in 1966 and 1967. The Lunar Orbiters' primary mission was to obtain topographic data in the lunar equatorial region between 43° E and 56° W to help in the selection of suitable landing sites for the unmanned **Surveyor** and manned **Apollo** missions. With

this objective achieved by Lunar Orbiter 3, the remaining two flights were able to carry out further photography of lunar surface features for purely scientific purposes. Altogether, 99% of the Moon was photographed with a resolution of 60 m or better. The first three missions were dedicated to imaging 20 potential lunar landing sites that had been chosen from Earth-based observations. These were flown at low-altitude, low-inclination orbits. The fourth and fifth missions were flown in high-altitude polar orbits. Lunar Orbiter 4 photographed the entire nearside and 95% of the farside; Lunar Orbiter 5 completed the farside coverage and acquired medium- (20 m) and high- (2 m) resolution images of 36 preselected areas. The Lunar Orbiters also collected data on radiation and micrometeoroids in the circumlunar region. (See table, "Lunar Orbiter Missions.")

Launch
    Vehicle: Atlas-Agena D
Site: Cape Canaveral

Lunar Orbiter Missions

Spacecraft	Launch	Imaging	Mass (kg)
Lunar Orbiter 1	Aug. 10, 1966	Aug. 18–29, 1966	386
Lunar Orbiter 2	Nov. 6, 1966	Nov. 18–25, 1966	390
Lunar Orbiter 3	Feb. 5, 1967	Feb. 15–23, 1967	385
Lunar Orbiter 4	May 4, 1967	May 11–26, 1967	390
Lunar Orbiter 5	Aug. 1, 1967	Aug. 6–8, 1967	389

Lunar Prospector

The third mission in NASA's **Discovery Program** and the first to be competitively selected. Five days after its 1998 launch, Lunar Prospector entered a polar orbit around the Moon, 101 km high, at the start of a year-long primary mission. The spacecraft was equipped with five instruments, mounted on three 2.5-m long booms, including a neutron spectrometer—the first device of this type to be carried onboard an interplanetary probe. The spectrometer was designed to verify the existence of water ice at the lunar poles as first suggested by measurements made by **Clementine** in 1994. On March 5, 1998, it was announced that Lunar Prospector had found further evidence to backup Clementine's discovery. Lunar Prospector also provided a new map that shows the chemical composition and the magnetic and gravity fields of the Moon in unprecedented detail. The nominal one-year mission was followed by a two-year extended mission during which the orbit was lowered to an altitude of 50 km and then 10 km to obtain higher resolution measurements. On July 31, 1999, Lunar Prospector struck the Moon near the south pole in a controlled crash to look for further evidence of water ice, though none was found.

Launch
　Date: January 6, 1998
　Vehicle: Athena II
　Site: Cape Canaveral
Size: 1.4 × 1.2 m
Mass: 295 kg (fully fueled); 126 kg (dry)

Lunar Roving Vehicle (LRV)

See **Apollo**.

Lunar-A

A Japanese lunar probe, scheduled for launch in 2003 by **ISAS** (Institute of Space and Astronautical Science). It will be the first mission to study the Moon's internal state using **penetrators**. After entering lunar orbit, the spacecraft will deploy three 13-kg spear-shaped cases, 90 cm long and 13 cm in diameter. These will be individually released over a period of a month and impact the Moon at 250 to 300 m/s, burrowing 1 to 3 m into the surface. One penetrator will be targeted at the equatorial area of the nearside (in the region of the Apollo 12 and 14 landing sites), one at the equatorial farside, and one near the border between the nearside and the farside. The penetrators are equipped with seismometers and devices to measure heat flow, and they will transmit their data to the orbiter as it passes over each penetrator every 15 days. After deploying the penetrators, the orbiter will move to

a 200–300 km near-circular orbit and use its monochromatic camera to image features near the terminator with a resolution of up to 30 m.

lunar-orbit rendezvous (LOR)

A spacecraft maneuver involving docking and coupling, fueling, or transfer in a lunar **parking orbit**. The concept of lunar-orbit rendezvous was first discussed by the Russian rocket theoretician Yuri Kondratyuk in 1916 and later by the British scientist H. E. **Ross**.[284] As a means of reaching the Moon's surface, it has the advantage that the lunar landing vehicle can be designed specifically for this task and be of low mass, since all the equipment and supplies for the return to Earth can be left in lunar orbit. Against this, however, must be set the difficulty and danger of having to maneuver and dock two vehicles in lunar orbit. Despite this drawback, LOR was the technique chosen to convey astronauts to and from the Moon in the **Apollo** program.

Lundin, Bruce T. (1919–)

An American aerospace propulsion engineer. Lundin earned a B.S. in mechanical engineering from the University of California in 1942 and joined the Lewis Laboratory (see **Glenn Research Center**) in 1943. There he investigated heat transfer and methods to improve the performance of World War II aircraft engines. In 1946, he became chief of the jet propulsion research section, which conducted some of America's early research on turbojet engines. He became assistant director of Lewis in 1958 and directed much of the center's efforts in space propulsion and power generation. Lundin then advanced through the positions of associate director for development at Lewis (1961), managing the development and operation of the **Centaur** and **Agena** launch vehicles, and of deputy associate administrator for advanced research and technology at NASA headquarters (1968), before becoming acting associate administrator for advanced research and technology there (1969). Later that year, he was appointed director of the Lewis Research Center, a position he held until his retirement in 1977.

Lunik

The name by which the first three **Luna** spacecraft were known in the West.

Lunney, Glynn S. (1936–)

A longtime NASA official who trained as an aeronautical engineer and came to the Lewis Laboratory (see **Glenn Research Center**) around the time of NASA's creation in 1958 to join the Space Task Group developing Project

Mercury. He worked on the **Apollo** program in a number of positions, including manager of the Apollo Spacecraft Program, in 1973, and manager of the **Apollo-Soyuz Test Project** at the **Johnson Space Center**. Thereafter, he managed the development of the Space Shuttle and held several other NASA posts. Lunney retired from NASA in 1985 and became vice president and general manager of Houston Operations for Rockwell International's Space Systems Division.

Lunokhod

Two automated lunar rovers, launched by the former Soviet Union as part of the **Luna** program. The eight-wheeled vehicles carried TV cameras and instruments to analyze the Moon's soil and were guided over the lunar surface by remote control from Earth. Lunokhod 1 was delivered to the Sea of Rains aboard Luna 17 in November 1970 and had covered over 10 km by the end of its mission 11 months later. Lunokhod 2, an improved version with twice the top speed, landed aboard Luna 21 in January 1973 in the partially ruined crater Le Monnier on the edge of the Sea of Serenity and covered a total of 37 km in its five-month lifetime.

Lyndon B. Johnson Space Center

See **Johnson Space Center**.

M

M series (Japanese launch vehicles)

Rockets developed by **ISAS** (Institute of Space and Astronautical Science) from the earlier and smaller **L series** ("L" and "M" are short for "Lambda" and "Mu"). In order of debut after the L-4S, they are the M-4S, M-3C, M-3H, M-3S, M-3SII, and M-5. See **Japanese launch vehicles**. (See also the table "M Rockets.")

M-3

The three-stage M-3C was the second rocket in the M series, after the M-4S. With newly developed second and third stages, it improved the accuracy of orbital injection and launched the satellites **Tansei-2**, **Taiyo**, and **Hakucho**. The M-3H was derived by extending the first-stage motor casing to increase the propellant capacity and hence the amount of payload that could be carried. It launched Tansei-3, **Kyokko**, and **Jikiken**. The third generation M-3S brought further improvements in the accuracy of orbital insertion and launched Tansei-4, **Tenma**, **Hinori**, and **Ohzora**. A fourth generation began with M-3SII, which used the first stage of M-3S, and new upper stages to enhance its payload capability. The M-3SII-1 and -2, with an optional fourth kick stage, sent the first and second Japanese interplanetary probes, **Sakigake** and **Suisei**, on their way to an encounter with Halley's Comet. Seven M-3SII launches out of eight have been successful, including those of Sakigake, Suisei, **Ginga**, **Akebono**, **Hiten**, **Yohkok**, and **ASCA**.

M-4

Chronologically, the first member of the M series was the four-stage M-4S rocket. It launched the satellites **Tanpei**, **Shinsei**, and **Denpa**.

M-5

A new launch vehicle to meet demand for increased payload capability and support interplanetary missions anticipated in the late 1990s and beyond. It has more than twice the capability of the preceding M-3SII and enables missions to Mars, Venus, the asteroid belt, and the outer Solar System. The M-5 has the world's largest solid propellant satellite launch system, new lightweight materials and structures, flight control and guidance, and avionics.

M-2 (American lifting body)

A series of experimental **lifting bodies** tested by NASA in the pre–Space Shuttle era. From 1966 to 1975, following the cancellation of the **Dyna-Soar** project, NASA built and tested three different lifting body designs: the M-2, the HL-10, and the **X-24**. The resulting data on aerodynamic performance during reentry was crucial for the design of the Space Shuttle Orbiter. After 16 flights, the M2-F2 was involved in a crash on May 10, 1967, from which NASA test pilot Bruce Peterson was lucky to escape with his life. This dramatic accident has been replayed many times on television as the opening sequence to *The Six Million Dollar Man*. The vehicle was repaired in the wake of the crash, a center tail fin added to improve stability, and the modified craft, renamed the M2-F3, used to carry out a further 27 flights from 1967 to 1972.

M2P2 (Mini-Magnetospheric Plasma Propulsion)

A novel propulsion concept under study at the University of Washington, Seattle, with NASA funding. M2P2 would use the solar wind to accelerate a spacecraft by pushing on a miniature version of Earth's magnetosphere generated by the craft. The injection of plasma from the Sun into an artificially generated magnetic field would drag the magnetic field lines out and form a bubble some 30 to 60 km in diameter, depending on the strength of field that the spacecraft produced. An engine using this technology is estimated to be 10 to 20 times more efficient than the

M Rockets						
	M-4S	M3-C	M-3H	M-3S	M-3SII	M-5
Total length (m)	23.6	20.2	23.8	23.8	27.8	30.7
Diameter (m)	1.41	1.41	1.41	1.41	1.41	2.5
Total mass (tons)	43.6	41.6	48.7	48.7	61	139
Payload to LEO (kg)	180	195	300	300	770	1,800

M2 The M2-F2 lifting body returns to the Dryden Flight Research Center, Edwards, California, flanked by an F-104 chase plane. *NASA*

Space Shuttle Main Engine. With a bottle of just 3 kg of helium as plasma fuel, the magnetic bubble could be operated for three months—the size of the bubble growing and shrinking in response to changes in the solar wind. Calculations have shown that there is enough power in the solar wind to accelerate a 136-kg space probe to speeds of up to 80 km/s, or 6.9 million km/day. By contrast, the Space Shuttle travels at a mere 7.7 km/s, or 688,000 km/day. If launched by 2003, an M2P2 spacecraft could reach the heliopause (where the solar wind runs into the interstellar wind) by 2013—about six years ahead of **Voyager** 1, which was launched in 1977.

MABES (Magnetic Bearing Satellite)

A Japanese satellite, launched by **NASDA** (National Space Development Agency), that carried an experiment on the levitation of a magnetic bearing flywheel under zero-*g*. It was also known by its national name of Jindai ("cherry tree").

Launch
    Date: August 12, 1986
    Vehicle: H-1
    Site: Tanegashima
Orbit: 1,483 × 1,604 km × 50.0°

Mach number

A unit of speed, named after the Austrian physicist Ernst Mach (1838–1916), equal to the ratio of the speed of a moving object to the speed of sound in the surrounding medium under ambient conditions. Mach 1 under standard conditions at sea level is 1,218 km/hr (761 mph); it decreases with altitude.

MACSAT (Multiple Access Communications Satellite)

Two third-generation digital communications satellites launched simultaneously on May 9, 1990, to demonstrate tactical ultra-high frequency (UHF) voice, data, fax, and video store-and-forward capabilities for the U.S. military. The gravity gradient boom on one spacecraft failed to deploy; the other spacecraft was used during Operation Desert Storm for relaying messages to and from troops in the Gulf region.

MagConst (Magnetospheric Constellation)

A NASA mission that aims to put 100 nanosatellites, each with a mass of 10 kg, into Earth orbit in 2010 to observe how Earth's magnetosphere reacts to changes in solar radiation. The nanosats will be able to "think" for themselves and alter their configuration, without instructions from ground control, in reaction to events in space or to the failure of one or more components of the constellation. As a preparatory step, NASA plans to launch the **Nanosat Constellation Trailblazer** in 2003.

Magdeburg Project

A project headed by German scientists Rudolf Nebel and Herbert Schaefer, as part of which emerged in 1933 the first definite plan to build a manned rocket. A test rocket was launched on June 9, 1933, at Wolmirstedt near Magdeburg, but the rocket never cleared its 10-m launch tower. Several more tests followed with mixed results. On June 29, 1933, a rocket left the launch tower but flew horizontally at a low altitude for a distance of about 300 m. This rocket was recovered undamaged, was refashioned into a design more closely resembling the Repulsors built by **Verein für Raumschiffahrt**, and was eventually launched from Lindwerder Island in Tegeler Lake near Berlin. It climbed to an altitude of about 300 m before crashing 100 m from the launch tower. Additional test launches were conducted from a boat on Schwielow Lake through August 1933, at which time the Magdeburg Project was abandoned.

Magellan

A NASA Venus radar mapping mission, named after the Portuguese explorer Ferdinand Magellan (Fernão de Magalhaes, 1480–1521). Magellan surveyed almost the entire surface at high resolution and enabled a global gravity map to be compiled. Its extended mission ended on October 13, 1994, following an **aerobraking** experiment that (intentionally) caused entry into the Venusian atmosphere.

The primary objectives of the Magellan mission were to map the surface of Venus with a synthetic aperture radar (SAR) and to determine the topographic relief of the planet. At the completion of radar mapping, 98% of the surface was imaged at resolutions better than 100 m, and many areas were imaged multiple times. Other studies included measurements of surface altitude using radar altimetry and measurements of the planet's gravitational field using precision radio tracking. The mission was divided into "cycles," each cycle lasting 243 days (the time taken for Venus to rotate once under Magellan's orbit).

Shuttle deployment	
Date: May 5, 1989	
Mission: STS-30	
Length: 6.4 m	
Mass: 3,444 kg	

magnetic field

A field of force that is generated by electric currents and in which magnetic forces can be observed. See **electromagnetic field**.

magnetic field lines

Imaginary lines in space, used for visually representing the strength and direction of **magnetic field**s. At any point in space, the local field line points in the direction of the magnetic force that an isolated magnetic pole at that point would experience. In a **plasma**, magnetic field lines also guide the motion of **ion**s and **electron**s, and direct the flow of some electric currents.

magnetic levitation launch-assist

A method of getting spacecraft off the ground that uses magnets to accelerate a vehicle along a track. Just as powerful magnets lift and propel high-speed trains and roller coasters above a guideway, a magnetic levitation (maglev) launch-assist system would electromagnetically drive a space vehicle along a track. The spacecraft would be boosted to speeds of about 1,000 km/hr before switching to rocket engines for the climb into orbit. A 15-m track was built at the Marshall Space Flight Center in mid-1999 for testing and design analysis of maglev concepts for space propulsion. Scaled demonstrations of maglev technology will be conducted on a 120-m track also planned at Marshall.

magnetic sail

See **space sail**.

magnetometer

A device for measuring the strength and direction of the interplanetary and solar magnetic fields. Magnetometers typically detect the strength of magnetic fields in three planes.

Magellan The Magellan spacecraft with its attached Inertial Upper Stage booster in the Space Shuttle Orbiter *Atlantis* payload bay prior to launch. *NASA*

magnetoplasmadynamic (MPD) thruster

A form of **electromagnetic propulsion**. In MPD thrusters, a current along a conducting bar creates an azimuthal magnetic field that interacts with the current of an arc that runs from the point of the bar to a conducting wall. The resulting Lorentz force has two components: a radially inward force (pumping) that constricts the flow, and a force along the axis (blowing) that produces the directed thrust. Erosion at the point of contact between the current and the electrodes is a critical issue in MPD thruster design.

magnetosphere

The region of space in which the magnetic field of: (1) Earth or another planet dominates the **radiation pressure** of the **solar wind** to which it is exposed, or (2) the Sun or another star dominates the pressure of the surrounding **interstellar medium**.

Magnetosphere Probe

Three American suborbital missions, launched by Blue Scout Juniors from Cape Canaveral on March 13, 1964,

March 30, 1965, and May 12, 1965, the first of which failed.

Magnetospheric Multiscale (MMS)

A proposed NASA mission, involving five spacecraft in highly elliptical orbits, that would shed new light on the physics of Earth's magnetosphere. Scientists know that broad regions of the magnetosphere communicate by fundamental processes, of vastly different scale sizes operating in thin boundary layers, that interact strongly. MMS, a future Solar-Terrestrial Probe mission, would help our understanding of these processes by making simultaneous measurements at widely different points in the magnetosphere.

magnetotail

The portion of a planetary magnetosphere that is pushed in the direction of the **solar wind**.

Magnum

The code name for Space Shuttle–launched geostationary **ELINT** (electronic intelligence) satellites designed to replace the **Rhyolite**/Aquacade series. After the second Magnum launch in 1989, the code name was changed to Orion.

Magsat (Magnetic Field Satellite)

NASA spacecraft; the first designed specifically to investigate the near-Earth magnetic field. It also returned data on geologic structure and composition.

Launch
  Date: October 30, 1979
  Vehicle: Scout G
  Site: Vandenberg Air Force Base
  Orbit: 352 × 561 km × 96.8°
  Mass: 181 kg

main stage

Usually the lowest or first stage of a vehicle. In a single-stage rocket, it is the period when full thrust is attained; in a multistage rocket, the stage that develops the greatest amount of thrust; in a stage-and-a-half rocket, the sustainer engine.

major axis

The maximum diameter of an **ellipse**.

majority rule circuit

Critical electronic circuits of a spacecraft are triplicated for insurance against failure. A computer then chooses the two most correct signals coming from the three circuits; thus, if one circuit fails, its output will be ignored.

Makarov, Oleg G. (1933–)

A Soviet spacecraft designer and cosmonaut. Makarov joined the Soviet space program in 1964, helped design the **Soyuz** spacecraft, and was selected to be an engineer cosmonaut in November 1966. In 1967, he was assigned to the lunar training group with which he was involved until 1969, when he began to train as a **Salyut** crewman. He served as flight engineer on four Soviet missions between 1973 and 1980, including two visits to Salyut 6, which he helped design. In 1980, he was a backup for the Soyuz T2 crew. Makarov's first flight was Soyuz 12 in September 1973, the first Soviet manned mission in the wake of the Soyuz 11 tragedy, which killed three cosmonauts during their reentry. Following the accident, the Soyuz command module had been redesigned to allow two cosmonauts to wear pressure suits during launch and reentry. Makarov and commander Vasily Lazarev returned to Earth safely after only two days. On April 5, 1975, Makarov and Lazarev were launched aboard Soyuz 18-1 for a planned 60-day mission aboard Salyut 4. However, only minutes into the flight, separation problems occurred with the Soyuz booster. The Soyuz command module containing Makarov and Lazarev was separated from the booster and plunged back to Earth, eventually coming to rest on a Siberian mountainside near the Chinese border. The emergency reentry subjected the cosmonauts to 18g, twice the normal g-load, and may have caused injuries. Certainly Lazarev never flew in space again. Makarov did, twice, with better luck. He was aboard Soyuz 27 in January 1978, a weeklong flight during which he and commander Vladimir Dzhanibekov docked with the Salyut 6, swapping vehicles with the Soyuz 26 crew of Yuri Romanenko and Georgi Grechko, who were in the first month of a planned three-month mission. It was a rehearsal for future operations that permitted cosmonauts to remain aboard Salyut stations for missions lasting nine months. In November 1980, Makarov returned to Salyut 6 aboard Soyuz-T3. The three-man crew of Makarov, commander Leonid Kizim, and flight engineer Gennady Strekalov overhauled several systems inside Salyut 6 during their 13 days in space, allowing Salyut 6 to be occupied in early 1981 for another long-duration mission. In 1981 he earned his candidate of technical sciences degree (the equivalent of a Ph.D.) from the Moscow Bauman Higher Technical School (now Moscow Bauman State Technical University) and published a futuristic work, *The Sails of Stellar Brigantines*, written in collaboration with Grigory Nemetsky.

Malina, Frank J. (1912–1981)

An American pre–Space Age rocket engineer. As a Ph.D. student at the California Institute of Technology in the mid-1930s, Malina began a research program to design a high-altitude sounding rocket. Beginning in 1936, he and his colleagues started the static testing of rocket engines in the canyons above the Rose Bowl, with mixed results, but a series of tests eventually led to the development of the **WAC-Corporal** rocket during World War II. After the war, Malina worked with the United Nations and eventually retired to Paris to pursue a career as an artist.

Malyutka

A Soviet rocket point interceptor designed by Nikolai N. Polikarpov beginning in 1943. The small aircraft, powered by a Glushko engine burning nitric acid and kerosene, was designed to reach a speed of 845 km/hr on flights of 8- to 14-minute duration. Prototype construction was under way when Polikarpov died on July 30, 1944. He had Stalin's support but many enemies. The result was that his design bureau and projects were immediately canceled after his death.

maneuver, capture

A maneuver that carries out the transition from an open to a closed orbit.

maneuver, correction

A change of orbit during a spaceflight to a path closer to a preselected trajectory.

maneuver, escape

A maneuver carried out to escape from a celestial body.

maneuverability

The structural or aerodynamic quality of an air vehicle that determines the rate at which its attitude and direction of flight can be changed. It is commonly expressed in *g*s or as *g*-load.

Manhigh, Project

An experimental program begun in December 1955 to study the behavior of a balloon in an environment above 99% of Earth's atmosphere and to investigate cosmic rays and their effects on human beings. Three balloon flights to the edge of space were made during the program: Manhigh 1 to 29,500 m, by Captain Joseph **Kittinger** on June 2, 1957; Manhigh 2 to 30,900 m, by Major David **Simons** on August 19–20, 1957; and Manhigh 3 to 29,900 m by Lieutenant Clifton McClure on October 8, 1958. Including pilot and scientific equipment, the total mass of the Manhigh 2 gondola was 748 kg. At maximum altitude, the balloon expanded to a diameter of 60 m with a volume of over 85,000 cubic meters.[252]

Manned Maneuvering Unit (MMU)

A one-man propulsion backpack, used on Space Shuttle missions, that snaps onto the back of a spacesuit's portable life-support system. The MMU allows an astronaut to work outside without a tether up to 100 m away from the Orbiter and is designed to provide EVA support for as much as six hours at a stretch. The MMU weighs 140 kg and is propelled by nitrogen gas fed to 24 thruster jets. Two pressurized nitrogen tanks can be filled from the Orbiter's onboard supply. All systems on the MMU are dual-redundant—if one system fails, a second can completely take over. The MMU was first tested in space by Bruce **McCandless** in 1994.

Manned Orbiting Laboratory (MOL)

A U.S. Air Force two-man cylindrical space station, about 12.5 m long and 3 m in diameter, which would have been attached to a **Gemini** spacecraft for 30-day missions in orbit similar to those of the early **Salyut** space stations. It was intended to develop reconnaissance techniques and carry out other clandestine tasks. Approved by President Johnson in 1965, the MOL project went as far as having a mockup launched by a Titan IIIC on November 3, 1966, but was canceled in 1969. Much of the technology for MOL is believed to have been transferred to unmanned spy satellites.

Manned Maneuvering Unit   Bruce McCandless uses the MMU outside the Space Shuttle *Challenger* in 1984 for the first untethered spacewalk. *NASA*

MAP (Microwave Anisotropy Probe)

A MIDEX (Medium-class Explorer) mission developed by the Goddard Space Flight Center and Princeton University to study conditions in the early universe. It can detect anisotropy (directional variations) in the cosmic microwave background–the afterglow of the Big Bang–with a much higher resolution than that of the earlier **COBE** (Cosmic Background Explorer). As the microwave background ripples through the universe in waves of slightly different temperatures, it preserves a record, in expanded form, of the shape of the universe as it appeared about 13 billion years ago. MAP's advanced set of microwave radiometers will record temperature patterns across the full sky, providing the best indicator yet of the size and shape of the entire universe. The spacecraft's two-year mission is carried out from a **halo orbit** about the second Sun-Earth **Lagrangian point** 1.5 million km from Earth.

Launch
   Date: June 30, 2001
   Vehicle: Delta 7425
   Site: Cape Canaveral
Orbit: halo
Mass: 840 kg

MARECS

Two geostationary satellites–MARECS A (launched on December 20, 1981) and MARECS-B2 (launched on November 10, 1984)–that formed part of **Inmarsat's** global maritime communications satellite network. The program began as the experimental Maritime Orbital Test Satellite (MAROTS) in 1973, but subsequently evolved into an operational system resulting in a name change, a satellite redesign, and delayed development.

Mariner

See article, pages 257–259.

Mars

See article, pages 260–261.

Mars 96

An ambitious Russian Mars probe, launched on November 16, 1996, that fell back to Earth after a failed burn that should have taken it out of Earth orbit. The probe included an orbiter, two small autonomous landers, and two surface penetrators. Having achieved an initial 160 km-high circular orbit, Mars 96 was to have been boosted onto a Mars trajectory by the upper-stage engine still attached to the probe. Instead, a misfiring of the upper stage placed the probe on an orbit that caused it to reenter. No advance warning of the probe's imminent descent was given by the Russians despite the fact that Mars 96 was carrying 270 grams of plutonium-238 as an energy source. Its final whereabouts remain unknown, although parts of it are presumed to have fallen into the South Pacific and possibly regions of Bolivia and Chile.

Mars 1969 A and B

Two identical Soviet probes launched in the spring of 1969 and intended to orbit Mars. Both were lost following launch failures and not officially announced. Each spacecraft carried three television cameras designed to image the Martian surface, a radiometer, a water vapor detector, ultraviolet and infrared spectrometers, a radiation detector, a gamma spectrometer, a hydrogen/helium mass spectrometer, a solar plasma spectrometer, and a low-energy ion spectrometer. See **Mars, unmanned spacecraft**.

Mars 2005

See **Mars Reconnaissance Orbiter**.

Mars 2007

A long-range, long-duration NASA rover equipped to perform a variety of surface studies of Mars and to demonstrate the technology for accurate landing and hazard avoidance in difficult-to-reach sites. In the same year, CNES (the French space agency) plans to launch a remote sensing orbiter and four small **NetLanders**, and ASI (the Italian space agency) plans to launch a communications orbiter to link to the NetLanders and future missions. Also slated for possible launch in 2007 are one or more small **Scout** missions.

Mars Climate Orbiter

One of the spacecraft in the **Mars Surveyor '98** program. It was designed to study the weather, climate, and cycling of water and carbon dioxide on Mars from orbit but was lost following a navigational error. A failure review board set up by NASA found that the problem lay in some spacecraft commands having been issued in English units instead of their metric equivalents. As a result, the probe went into an orbit of less than half the intended altitude and burned up in the Martian atmosphere. This loss, together with that of the **Mars Polar Lander**, prompted NASA to revise its plans for the robotic exploration of Mars.

Launch
   Date: December 11, 1998
   Vehicle: Delta 7425
   Site: Cape Canaveral
Mass at launch: 629 kg

Mariner

An early series of NASA interplanetary probes developed and operated by JPL (Jet Propulsion Laboratory). The Mariners became the first probes to return significant data on the surface and atmospheric conditions of Venus, Mars, and Mercury. (See table, "The Mariner Series," on page 259.)

Mariner 1

The first American attempt to send a probe to Venus. Guidance instructions from the ground stopped reaching the rocket due to a problem with its antenna, so the onboard computer took control. However, a bug in the guidance software caused the rocket to veer off course, and it was destroyed by the range safety officer.

Mariner 2

Backup for Mariner 1, and the first probe to fly successfully by another planet. The data it sent back confirmed that Venus has a slow retrograde (backward) spin, a very high surface temperature, and a thick atmosphere made up mostly of carbon dioxide. Mariner 2 passed Venus at a distance of 34,773 km on December 14, 1962. The last transmission from the probe was received on January 3, 1963; it remains in solar orbit.[218]

Mariner 3

A failed Mars probe. Its launch fairing failed to separate, preventing a planned Mars flyby.

Mariner 4

A sister probe to Mariner 3 and the first spacecraft to photograph Mars at close range; it came within 9,846 km of the Martian surface on July 14, 1965. The 21 pictures it sent back showed a cratered terrain and an atmosphere much thinner than previously thought. Based on its findings, scientists concluded that Mars was probably a dead world, both geologically and biologically. Later missions, however, showed that the ancient region imaged by Mariner 4 was not typical of the planet as a whole. In 1967, Mariner 4 returned to the vicinity of Earth, and engineers were able to use the aging craft for a series of operational and telemetry tests to improve their knowledge of techniques needed for future interplanetary missions.

Mariner 5

A Venus flyby probe that came within 3,990 km of the planet's surface. Originally a backup for Mariner 4, Mariner 5 was refurbished and sent to Venus instead. Its main task was to find out more about Venus's atmosphere by using radio waves and measuring the brightness of the atmosphere in ultraviolet light. It also collected data on radiation and magnetic fields in interplanetary space.

Mariner 6

A Mars probe that returned 75 images of the Martian surface and flew by at a distance of 3,431 km. Disaster almost struck while it was still on the ground. Ten days before the scheduled launch, a faulty switch opened the main valves on the Atlas booster, releasing the pressure that supported the Atlas structure and causing the rocket to crumple. Two ground crewman who,

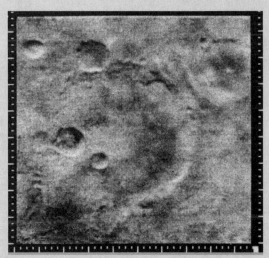

Mariner 4 One of the first close-up (yet grainy) pictures of Mars, sent back by Mariner 4. *NASA/JPL*

at risk to themselves, started pressurizing pumps and saved the 12-story structure from further collapse were awarded Exceptional Bravery Medals from NASA. Mariner 6 was moved to another Atlas-Centaur and launched on schedule. The pictures it sent back of Mars were less Moon-like than those of Mariner 4, and the south polar cap was identified as being composed mostly of carbon dioxide.

Mariner 7

A Mars flyby probe that was reprogrammed in the light of the Mariner 6 findings. It was instructed to go farther south than originally planned, take more near-encounter pictures, and collect more scientific data on the lighted side of Mars. In all, it returned 126 images and approached to within 3,430 km of the surface.

Mariner H (8)

An intended Mars flyby probe, which suffered a launch failure.

Mariner 9

The first spacecraft to orbit another planet. The Mariner Mars '71 mission was supposed to involve two spacecraft: Mariner 8, which was to map 70% of the Martian surface, and Mariner 9, which was to study changes in the Martian atmosphere and on the ground. The failure of Mariner 8 forced Mariner 9 to combine the mission objectives. On November 13, 1971, the probe arrived at Mars and, after a rocket burn lasting 15 minutes 23 seconds, entered orbit. Imaging was delayed by a dust storm, which quickly grew into one of the largest ever seen on the planet. Of the surface, nothing could be seen except the lofty summits of Olympus Mons and the three Tharsis volcanoes. The storm abated through November and December, and normal mapping operations began. The spacecraft gathered data on the atmospheric composition, density, pressure, and temperature and the surface composition, temperature, gravity, and topography. A total of 7,329 images were returned that covered the entire planet. After depleting its supply of attitude control gas, the spacecraft was turned off on October 27, 1972, and left in an orbit which should not decay for at least 50 years. Mariner 9 provided the first global map of the surface of Mars, including the first detailed views of the Martian volcanoes, Valles Marineris, and the polar caps, and of the satellites Phobos and Deimos.

Mariner 10

The first probe to send back close-up pictures of Mercury's surface. En route, it received a **gravity-assist** from Venus and returned images of the Venusian atmosphere in ultraviolet, revealing a previously unseen level of detail in the upper cloud deck. On March 29, 1974, it flew past Mercury at a distance of 704 km, sending back pictures of an intensely

Mariner 9 *NASA/JPL*

cratered, Moon-like surface and detecting a trace atmosphere of mostly helium. After the first flyby, Mariner 10 entered a solar orbit and went on to two further encounters with the innermost planet. On September 21, 1974, the second flyby, at an altitude of 47,000 km, afforded another opportunity to photo-graph the sunlit side of the planet and the south polar region. The third and final Mercury encounter on March 16, 1975, at an altitude of 327 km, yielded 300 photographs and magnetic field measurements. Mariner 10 was turned off on March 24, 1975, when its supply of attitude-control gas was depleted.

The Mariner Series

Launch site: Cape Canaveral

Mariner	Launch Date	Launch Vehicle	Mass (kg)	Notes
1	Jul. 22, 1962	Atlas-Agena B	200	Failed
2	Aug. 27, 1962	Atlas-Agena B	201	Venus flyby, Dec. 14, 1962
3	Nov. 5, 1964	Atlas-Agena D	260	Failed
4	Nov. 28, 1964	Atlas-Agena D	260	Mars flyby, Jul. 14, 1965
5	Jun. 14, 1967	Atlas-Agena D	244	Mars flyby, Oct. 19, 1967
6	Feb. 24, 1969	Atlas IIIC	412	Mars flyby, Jul. 31, 1969
7	Mar. 27, 1969	Atlas IIIC	412	Mars flyby, Aug. 5, 1969
H (8)	May 9, 1971	Atlas IIIC	996	Failed
9	May 30, 1971	Atlas IIIC	974	Mars orbit, Nov. 13, 1971
10	Nov. 3, 1973	Atlas IIID	526	Venus flyby, 3 Mercury flybys

Mars Exploration Rovers

Twin NASA missions to Mars carrying two identical rovers, each with a mass of about 130 kg, that will be landed by an airbag system similar to that used by **Mars Pathfinder**. The launches are scheduled for May-June 2003, and the landings for January 2004, at two widely separated sites on Mars that were selected because they appear to have been associated with liquid water in the past and may therefore have been favorable to life. Although the rovers will not search for organic remains directly, they will seek to determine the history of climate and water at their sites, which has a direct bearing on the issue of possible Martian biology.

Immediately after landing, the rovers will begin reconnaissance of their landing sites by taking 360° visible color and infrared image panoramas. Then they will drive off to begin their exploration. Using images and spectra taken daily from the rovers, mission scientists on Earth will command the vehicles toward rock and soil targets of particular interest and then evaluate their compositions and textures at microscopic scales. Initial targets will be close to the landing sites, but later targets are expected to be much farther afield. The rovers will be able to travel up to 100 m per Martian day (24 hours 37 minutes)—as far as the **Sojourner** rover did in its entire lifetime.

Each rover will carry the Athena scientific package (originally designed for a rover mission in 2001 but canceled in the wake of two mission failures in 1998), developed at Cornell University, which consists of a panoramic camera (Pancam), a rock abrasion tool (RAT) to expose fresh surfaces of rock, a miniature thermal infrared spectrometer (MiniTES), a microscopic camera, a Mossbauer spectrometer, and an alpha-proton-X-ray spectrometer (APXS). The rovers are each expected to function for at least 90 days.

Mars Express

An ESA (European Space Agency) mission scheduled for launch in May or June 2003 by a Soyuz-Fregat rocket and arrival at Mars in late December of the same year. It will be the first spacecraft to use radar to penetrate the surface of Mars and map the distribution of possible underground water deposits, and it will release the Beagle 2 lander. Mars Express will carry a remote observation payload

(continued on page 262)

Mars

An early series of Soviet spacecraft designed to fly by, orbit, and land on Mars. None of the probes was completely successful and most were almost total failures. The Soviet Union also launched several other spacecraft between 1960 and 1971 that were intended to explore Mars but, in the wake of their failure, did not acknowledge their true purpose (see **Mars, unmanned spacecraft**). These included **Marsnik** 1 and 2, Sputnik 22 and 24, Zond 2, Cosmos 419, and **Mars 1969 A and B**. Sputnik 22, an intended Mars flyby mission, was launched at the time of the Cuban missile crisis in 1962. When fragments of the exploded spacecraft showed up on the U.S. Ballistic Missile Early Warning System radar in Alaska, there was momentary alarm that a Soviet nuclear attack might be underway. Cosmos 419 had been intended to overtake Mariner 8, scheduled for launch two days earlier, and so become the first spacecraft to orbit Mars. In the event, Mariner 8 also failed, and the race to be first in Mars orbit was won by Mariner 9. (See table, "Mars Series.")

Mars 1

A flyby probe. On March 21, 1963, with the spacecraft 106 million km from home, communication was lost, perhaps due to a malfunction in the probe's orientation system. The closest approach to Mars took place on June 19, at a distance of about 193,000 km, after which the spacecraft entered solar orbit.

Mars 2

A Mars orbiter and descent probe. Mars 2 released its descent module—the first attempt to soft-land on the Red Planet—4.5 hours before entering orbit on November 27, 1971. However, the descent system malfunctioned and the lander crashed at 45° S, 302° W, delivering the Soviet Union coat of arms to the surface. Meanwhile, the orbiter engine performed a burn to put the spacecraft into a 1,380 × 24,940 km, 18-hour orbit about Mars with an inclination of 48.9°. Scientific instruments were generally turned on for about 30 minutes near periapsis (orbital low point), and data was sent back for several months.

Mars 3

Identical to Mars 2, except the Mars 3 orbiter also carried a French-built experiment called Spectrum 1, which measured solar radio waves in conjunction with Earth-based receivers to study the cause of solar outbursts. The descent module was released on December 2, 1971, about 4.5 hours before reaching Mars and, through a combination of aerobraking, parachutes, and retro-rockets, successfully soft-landed at 45° S, 158° W. However, 20 seconds later, its instruments stopped working for unknown reasons. The orbiter entered an elliptical, 11-day orbit about Mars from which it sent back data for several months.

Mars 4

An intended Mars orbiter. The probe's retro-rockets failed to fire to slow the craft into Mars orbit, and it flew by at a range of 2,200 km, returning just one swath of pictures and some radio occultation data.

Mars 5

A Mars probe intended to enter Martian orbit and comprehensively photograph the planet. The spacecraft reached Mars on February 12, 1974, and was inserted into a 1,760 × 32,586 km orbit. However, due to computer chip failures, the orbiter operated for only a few days, returning atmospheric data and images of a small portion of the Martian southern hemisphere.

Mars 6

A Mars probe consisting of a flyby bus and a descent module. Mars 6 reached Mars on March 12, 1974, and the descent module separated from the bus, opened its parachute, and began to fall through the Martian atmosphere. As the probe descended, it transmitted data for 150 seconds, representing the first data returned from the atmosphere of Mars. Unfortunately, the data were largely unreadable due to a flaw in a computer chip, and, shortly after the retro-rockets fired for landing, all contact was lost with the craft. Mars 6 landed at about 24° S, 25° W in the Margaritifer Sinus region.

Mars 7

A Mars probe consisting of a flyby bus and a descent module. Mars 7 reached Mars on March 9, 1974, but the landing probe separated prematurely and missed the planet by 1,300 km.

Mars Series

Launch site: Baikonur

Mars	Launch Date	Launch Vehicle	Mass (kg)	Notes
1	Nov. 1, 1962	Molniya	894	Radio contact lost en route
2	May 19, 1971	Proton	2,265	Mars orbit Nov. 27, 1971; lander crashed
3	May 28, 1971	Proton	2,270	Mars orbit Dec. 2, 1971; lander transmissions failed
4	Jul. 21, 1973	Proton	2,270	Intended orbiter; flew by on Feb. 10, 1974
5	Jul. 25, 1973	Proton	2,270	Mars orbit Feb. 12, 1974
6	Aug. 5, 1973	Proton	635	Mars flyby Mar. 12, 1974; lander crashed
7	Aug. 9, 1973	Proton	1,200	Mars flyby Mar. 9, 1974; lander missed planet

Mars Exploration Rover   An artist's rendering of one of the Mars 2003 Rovers as it begins its exploration of the Red Planet. *NASA/JPL*

Mars Express

(continued from page 259)

with some heritage from European instruments lost on the ill-fated **Mars 96**, as well as a communications package to support Mars lander missions from 2003 to 2007.

Beagle 2

The miniature lander to be carried aboard Mars Express. It is dedicated to looking for traces of life and conducting geochemical analyses, and is named after the ship in which Charles Darwin sailed to the Galapagos. Changes in the mission specification meant that the mass of Beagle 2 had to be trimmed from 90 kg to 60 kg, echoing economies that proved necessary on the original *Beagle*. (So cramped was that ship that a padre had to give up his berth so that Darwin could go.)

Once it has safely reached the surface by a combination of aeroshell, parachutes, and air bags, Beagle 2 will deploy a battery of ingenious, tiny instruments. These include a "mole" that can burrow along or through soil and collect samples for analysis by a mass spectrometer. The mole is placed on the surface by a robotic arm and can be pulled back into its storage tube aboard the lander by means of an attached cable. To expose fresh material from inside rocks, Beagle 2 also carries a modified dentist's drill. Samples will be subjected to "stepped combustion"—heated to successively higher temperatures—and then analyzed. An elevated ratio of carbon-12 to carbon-13 would be interesting, because living things preferentially use the lighter form of carbon. Beagle 2 will also

Mars Express The Beagle lander with its solar panels outspread and its instrument arm deployed.

sniff the atmosphere for tiny amounts of methane, which would suggest that there might be microbes alive today on Mars.

Mars Global Surveyor (MGS)

A NASA orbiter launched in November 1996 and designed to investigate, over the course of a full Martian year, the surface, atmosphere, and magnetic properties of Mars in unprecedented detail. It is the inaugural mission in a decade of planned intensive in situ research of the Red Planet and also a replacement mission to achieve most of the goals of the failed **Mars Observer**. It has been so successful that its operation has been extended until late 2004. During this bonus phase, MGS will help scout landing sites for the **Mars Exploration Rovers**.

MGS carries four main science instruments. The Mars Orbiter Camera provides daily wide-angle images of Mars similar to weather photos of Earth and narrow-angle images of objects as small as 1.5 m across. The Mars Orbiter Laser Altimeter bounces a laser beam off the surface to measure accurately the height of mountains and the depth of valleys. The Thermal Emission Spectrometer scans emitted heat to study both the atmosphere and the mineral composition of the surface. Finally, the Magnetometer and Electron Reflection experiment provides data on the magnetic state of the crustal rocks, which in turn sheds light on the early magnetic history of the planet.

Launch
   Date: November 7, 1996
   Vehicle: Delta 7925
   Site: Cape Canaveral

Mars, manned missions

For most of the twentieth century, humans dreamed of traveling to Mars and there has been no shortage of plans, proposals, and statements of intent to make the dream a reality. The first detailed study of a manned Mars mission appeared in Wernher **von Braun**'s book *The Mars Project*, published in 1952. Von Braun proposed flying 10 spacecraft, each with a crew of 70, in convoy to Mars. Fifty of the astronauts would land on the Red Planet and stay there for a year before returning to Earth. By 1956, in *The Exploration of Mars*, written in collaboration with Willy **Ley**, von Braun had scaled back the mission to a dozen crew members and two spacecraft: a winged glider (carrying a small ascent rocket) that would land on Mars, and a mother ship in orbit that would be used for the return home.

During the early years of the Apollo program, Mars was seen as the next obvious, and not too distant, step. A variety of studies appeared, including a proposal by Ernst

Stulinger, one of von Braun's old V-2 colleagues. Stulinger's idea was for a Mars-ship equipped with an ion drive. Those working on Project **Orion**, by contrast, believed that a nuclear-fission rocket was the way to go. Although at the time of the first Apollo landings Vice President Spiro Agnew spoke out in favor of a manned Mars mission, by the 1980s the cost of the Vietnam War and the general loss of interest after winning the Space Race left the idea with few political backers. Since then, there has been the occasional rallying cry from official circles, as when President Bush (senior) announced, on the 20-year anniversary of the Apollo 11 landing, that America would begin planning to send astronauts to Mars early in the twenty-first century.

Today, numerous robotic missions are planned to continue the exploration of the Red Planet and, in particular, look for signs of past or present life. These will culminate in sample-return missions sometime after 2010, and then, perhaps, NASA and other space agencies may begin once again to formulate definite plans to send people to Mars. In the meantime, preliminary work is being done in the form of the **Haughton-Mars Project** and further investigations into the effects of long-term habitation in space by individuals such as Edwin "Buzz" **Aldrin**, who has proposed the **orbital cycler**, and Robert Zubrin, and by interested groups such as the Mars Society. Also in May 2002, the Space Transportation Association called for linking the early opera-tions aboard the International Space Station to solving research issues associated with an early expedition to Mars by 2019.[322]

Mars Microprobe Mission
See **Mars Polar Lander**.

Mars Observer
A failed NASA mission to study the surface, atmosphere, interior, and magnetic field of Mars from orbit. The mission was designed to operate for one full Martian year (687 Earth days) to allow observations of the planet through its four seasons. However, communication with the spacecraft was lost on August 22, 1993, as it was preparing to enter orbit, and no significant data was returned. Later investigation pointed to an explosion in the propulsion system caused by propellant leaking through faulty valves.

Launch
    Date: September 25, 1992
    Vehicle: Titan 34D
    Site: Cape Canaveral
Mass (at launch): 2,573 kg

Mars Odyssey, 2001
A Mars orbiter that is the surviving part of the **Mars Surveyor 2001** project. Its main goals are to gather data to

Mars Odyssey, 2001   An artist's conception of the 2001 Mars Odyssey probe in orbit around Mars. *NASA/JPL*

help determine whether the environment on Mars was ever conducive to life, characterize the climate and geology of Mars, and study potential radiation hazards to future astronaut missions. It will also act as a communications relay for upcoming missions to Mars over a five-year period. After a seven-month cruise, the spacecraft reached Mars on October 24, 2001, transferred to a 25-hour elliptical capture orbit, and then used **aerobraking** over the next 76 days to achieve a two-hour, 400-km-high circular polar orbit. Mars Odyssey carries the Mars Radiation Environment Experiment (MARIE), to measure the near-space radiation environment (important to know about for future human missions); the Thermal Emission Imaging System (THEMIS), to map the mineralogy of the Martian surface using a high-resolution camera and a thermal infrared imaging spectrometer; and the Gamma-Ray Spectrometer (GRS), to map the elemental composition of the surface and determine the abundance of hydrogen in the shallow subsurface.

Mars Pathfinder The Sojourner rover on the Martian surface, having just left Mars Pathfinder's ramp. *NASA/JPL*

Launch
 Date: April 7, 2001
 Vehicle: Delta 7925
 Site: Cape Canaveral
 On-orbit dry mass: 376 kg

Mars Pathfinder (MPF)

A NASA Mars lander that deployed the first Martian rover—a miniature 10-kg vehicle called Sojourner. Pathfinder was the first successful Mars landing since **Viking** and the second mission in NASA's **Discovery Program**. Its main goals were to study the Martian atmosphere, surface meteorology, elemental composition of the rocks and soil, and the geology (or "areology") of the landing site. Pathfinder achieved safe planetfall in the lowland region known as Ares Vallis on July 4, 1997, using a novel technique that involved the inflation of a protective airbag and a subsequent hit, bounce, and roll landing like that of a beach ball thrown at the ground. After the probe came to rest, its airbag deflated and four metal petals opened to reveal the spacecraft's instruments and the Sojourner rover, which subsequently trundled down a ramp to begin its exploration. During almost three months of operation (nearly three times longer than its design lifetime), Pathfinder returned more than 16,000 images from the lander and 550 from the rover, carried out 15 chemical analyses of rocks, and gathered extensive data on winds and other aspects of the weather. Among the science highlights of the mission were these discoveries: the rock size distribution at the site is consistent with a flood-related deposit; the rock chemistry might be different from that of the SNC meteorites

(meteorites found on Earth and known to have come from Mars); and dust is the main absorber of solar radiation in the Martian atmosphere. The last transmission from the probe was on September 27, 1997. Pathfinder was operated by JPL (Jet Propulsion Laboratory) and later renamed the Sagan Memorial Station in honor of the late Carl **Sagan**.[243]

Launch
 Date: December 4, 1996
 Vehicle: Delta 7925
 Site: Cape Canaveral

Mars Polar Lander (MPL)

One of the spacecraft in NASA's **Mars Surveyor '98** program. MPL was intended to land on Mars near the south polar cap and, during its descent, deploy two surface penetrators known as the Mars Microprobe Mission. All contact with the spacecraft was lost at the point of separation of the lander and multiprobes, a failure subsequently blamed on shortcomings in project management and preflight testing. MPL was equipped with cameras, instruments (including a microphone to record the first sounds on the planet), and a robotic arm to sample and analyze the composition of the Martian soil. During its scheduled three-month operational period, it was intended to search for near-surface ice, possible surface records of cyclic climate change, and data bearing on the seasonal cycles of water vapor, carbon dioxide, and dust on Mars.

Mars Microprobe Mission

A mission, previously known as Deep Space 2, consisting of two basketball-sized penetrators that piggybacked aboard the Mars Polar Lander and were intended to detach as the main spacecraft began its descent. The penetrators, called Scott and Amundsen, were to have struck the Martian surface at about 200 m/s at the northernmost boundary of the area near the south polar ice cap referred to as the Polar Layered Terrain, about 200 km from the landing site of the main spacecraft. Each would have released a miniature two-piece science probe designed to punch into the soil to a depth of up to 2 m and search for subsurface ice.

Launch
    Date: January 3, 1999
    Vehicle: Delta 7925
    Site: Cape Canaveral
Length: 3.6 m
Total mass, less penetrators: 575 kg

Mars Reconnaissance Orbiter

A NASA orbiter, to be launched in August 2005, that will make high-resolution measurements of the surface from orbit, including images with resolution better than 1 m. The primary objectives of the mission will be to look for evidence of past or present water and to identify landing sites for future missions. The orbiter will also be used as a telecommunications link for future missions.

Mars Scout

A new class of robot explorer that will be able to reconnoiter Mars in any of a variety of ways—from orbit, from the planet's surface or subsurface, and by floating or flying at low altitude over the Martian terrain. NASA intends to use Mars Scouts to complement its core missions by focusing on new discoveries, particularly any revelations that might warrant rapid follow-up. The first Scout could be launched between late 2006 and mid-2007.

Mars Surveyor '98

A two-part mission consisting of the **Mars Climate Observer** and the **Mars Polar Lander**, designed to accomplish at lower cost the mission assigned to the failed **Mars Observer**. However, both of the Surveyor '98 probes also failed, due largely to human error, which prompted NASA to review and extensively revise its plans for the future exploration of Mars.

Mars, unmanned spacecraft

See **Mariner** 3, 4, and 6–9, **Mars, Mars 96, Mars 1969 A and B, Mars 2007, Mars Climate Orbiter, Mars Exploration Rovers, Mars Express, Mars Global Surveyor, Mars Observer, Mars Odyssey, Mars Pathfinder, Mars Polar Lander, Mars Reconnaisance Orbiter, Mars**

Scout, Marsnik, Netlander, Nozomi, Phobos, Viking, and **Zond** 2.[24, 126] (See also the table "Chronology of Mars Probes," on page 266.)

Marshall, George C. (1880–1959)

A career Army officer who served as General of the Army and U.S. army chief of staff during World War II. He became secretary of state (1947–1949), then secretary of defense (1950–1951), and he was the author of the European recovery program known to the world as the Marshall Plan; it played a crucial role in reconstructing a Europe ravaged by the war that Marshall had done so much to direct to a victorious end. In recognition of the effects of the Marshall Plan and his contributions to world peace, he received the Nobel Prize for Peace in 1953. It was fitting that a NASA center should be named after the only professional soldier to receive the prize, given NASA's charter to devote itself to the peaceful uses of outer space and yet to cooperate with the military services.[235, 278]

Marshall Space Flight Center (MSFC)

A major NASA field center located within the boundaries of the Army's **Redstone Arsenal**, in Huntsville, Alabama, which is responsible for developing new space launch vehicles and propulsion systems. Its programs focus on research, technology, design, development, and integration of space transportation and propulsion systems, including both reusable systems for Earth-to-orbit applications and vehicles for orbital transfer and deep space transportation. Marshall also carries out microgravity research and is the home of the **Neutral Buoyancy Tank**. The center was formed on July 1, 1960, by the transfer of buildings and staff from what was then the **Army Ballistic Missile Agency**. It subsequently played a central part in the realization of the Jupiter C, Centaur, Skylab, and the Space Shuttle programs, but it is best known for its development of the Saturn class rockets used to launch Apollo.

Marsnik

Two failed Soviet Mars probes launched on October 10 and 14, 1960. Marsnik 1 and 2 (also known as Mars 1960 A/B and Korabl 4/5) were the first attempted interplanetary spacecraft and were similar in design to **Venera** 1 with an on-orbit mass of 650 kg. They were intended to investigate interplanetary space between Earth and Mars, study Mars and return surface images from flyby trajectories, and study the effects of extended spaceflight on onboard instruments. Both probes were lost during launch when their third stages failed to ignite. At the time, Soviet premier Khruschev was on a visit to the

Chronology of Mars Probes

Spacecraft	Country	Launch Date	Notes
Marsnik 1	Soviet Union	Oct. 10, 1960	Attempted flyby, launch failure
Marsnik 2	Soviet Union	Oct. 14, 1960	Attempted flyby, launch failure
Sputnik 22	Soviet Union	Oct. 24, 1962	Attempted flyby
Mars 1	Soviet Union	Nov. 1, 1962	Flyby, contact lost
Sputnik 24	Soviet Union	Nov. 4, 1962	Attempted lander
Mariner 3	United States	Nov. 5, 1964	Attempted flyby
Mariner 4	United States	Nov. 28, 1964	Flyby
Zond 2	Soviet Union	Nov. 30, 1964	Flyby, contact lost
Mariner 6	United States	Feb. 24, 1969	Flyby
Mariner 7	United States	Mar. 27, 1969	Flyby
Mars 1969A	Soviet Union	Mar. 27, 1969	Attempted orbiter, launch failure
Mars 1969B	Soviet Union	Apr. 2, 1969	Attempted orbiter, launch failure
Mariner 8	United States	May 8, 1971	Attempted flyby, launch failure
Cosmos 419	Soviet Union	May 10, 1971	Attempted orbiter/lander
Mars 2	Soviet Union	May 19, 1971	Orbiter/attempted lander
Mars 3	Soviet Union	May 28, 1971	Orbiter/lander
Mariner 9	United States	May 30, 1971	Orbiter
Mars 4	Soviet Union	Jul. 21, 1973	Flyby, attempted orbiter
Mars 5	Soviet Union	Jul. 25, 1973	Orbiter
Mars 6	Soviet Union	Aug. 5, 1973	Lander, contact lost
Mars 7	Soviet Union	Aug. 9, 1973	Flyby, attempted lander
Viking 1	United States	Aug. 20, 1975	Orbiter and lander
Viking 2	United States	Sep. 9, 1975	Orbiter and lander
Phobos 1	Soviet Union	Jul. 7, 1988	Attempted orbit/Phobos landers
Phobos 2	Soviet Union	Jul. 12, 1988	Orbiter/attempted Phobos landers
Mars Observer	United States	Sep. 25, 1992	Attempted orbiter, contact lost
Mars Global Surveyor	United States	Nov. 7, 1996	Orbiter
Mars 96	Russia	Nov. 16, 1996	Attempted orbiter/landers
Mars Pathfinder	United States	Dec. 4, 1996	Lander and rover
Nozomi	Japan	Jul. 3, 1998	Orbiter
Mars Climate Observer	United States	Dec. 11, 1998	Attempted orbiter
Mars Polar Lander	United States	Jan. 3, 1999	Attempted lander and penetrators
Mars Odyssey, 2001	United States	Apr. 7, 2001	Orbiter

United States. Furious at the failures, he insisted that a third probe be hurriedly ready to dispatch before the 1960 launch window closed. On October 23, 1960, the rocket failed to lift off on time, and the Soviet commander, Marshall Nedelin, demanded that the vehicle be examined at once. Ignoring normal safety precautions, the technicians approached the fully fueled rocket, which suddenly exploded, killing Nedelin and almost the entire launch team. Although denied for years, the story eventually leaked to the West and became known as "The Nedelin Catastrophe."

Martin Marietta
See **Lockheed Martin**.

mass

In general, the amount of matter in a body. Mass can be defined more precisely in terms of how difficult it is to change a body's state of motion or how great is the body's gravitational effect on other objects. The first of these is called *inertial mass* and is given by the factor m in Newton's second law $F = ma$. The second is called *gravitational mass* and is the mass corresponding to an object's weight in a local gravitational field—the m in $F = mg$ for an object on or near the Earth. According to all experiments, the values for m arising from these two definitions are identical. Einstein's **mass-energy relationship** also shows that mass and energy are interchangeable.

mass driver

An electromagnetic cannon that would be able to accelerate payloads from the surface of a low-gravity world, such as the Moon or an asteroid, into space.[51]

mass ratio

The ratio of the initial mass of a spacecraft, or one of its stages, to the final mass after consumption of the **propellant**.

mass-energy relationship

Einstein's famous equation $E = mc^2$. This shows that mass and energy are really two sides of the same coin. It also gives the amount of energy E that results when a mass m is completely turned into energy, c being the speed of light. The fact that c squared is a huge number indicates that a vast amount of energy can come from even a tiny amount of matter. Mass-to-energy conversion is the basic principle at work in some advanced propulsion schemes such as the **nuclear pulse rocket** and **antimatter propulsion**.

MASTIF (Multiple Axis Space Test Inertia Facility)

A three-axis gimbal rig, built by the Lewis Research Center (now the **Glenn Research Center**), which simulated tumble-type maneuvers that might be encountered during a space mission. Three tubular aluminum cages could revolve separately or in combination to give roll, pitch, and yaw motions at speeds up to 30 rpm, greater than those expected in actual spaceflight. Nitrogen-gas jets attached to the three cages controlled the motion. At the center of the innermost cage, the pilot was strapped into a plastic seat similar to that in a Mercury capsule. His head, body, and legs were held in place, leaving only his arms free. The pilot actuated the jets by means of a right-hand control column. Communication was by radio, which was operated by a button atop the left-hand column. Complex tumbling motions were started by the operator at the control station, and control then switched to the pilot. By reading instruments mounted at eye level before him, the pilot interpreted his motions and made corrections accordingly. From February 15 to March 4, 1960, MASTIF provided training for all seven Project Mercury astronauts. Each experienced about five hours of "flight time." Later that year, a set of woman pilots also used the device as part of a broader assessment of abilities (see **Mercury Thirteen**). In addition, the rig was used to evaluate instrument control systems for spaceflight and to study the physiological effects of spinning, such as eye oscillation and motion sickness.

Matador

The first successful surface-to-surface, pilotless, tactical weapon developed for the U.S. Air Force. A highly mobile system designed to deliver a warhead on tactical missions in support of ground troops for a distance of up to 960 km, the Matador project began shortly after the end of World War II. Test firings began in 1949 in Alamogordo, New Mexico, and by January 1951 the Matador was in production.

Matagorda Island

Located 80 km northeast of Corpus Christi, Texas, at 28.5° N, 96.5° W, the site of a private launch pad used by Space Services to launch its **Conestoga** rocket in 1982.

Mattingly, Thomas K., II (1936–)

A veteran American astronaut involved in both the **Apollo** and **Space Shuttle** programs. Mattingly received a B.S. in aeronautical engineering from Auburn University in 1958 and subsequently flew carrier-based aircraft. Selected by NASA in April 1966, he served as a member of the astronaut support crews for the Apollo 8 and Apollo 11 missions, and he was also the astronaut representative in the development and testing of the Apollo spacesuit and backpack. Although designated as the Command Module pilot for Apollo 13, he was removed from flight status 72 hours before the scheduled launch because of exposure to German measles. Mattingly subsequently served as Command Module pilot of Apollo 16 and head of astronaut office support to the Space Shuttle program from January 1973 to March 1978. He was next assigned as technical assistant for flight test to the manager of the Orbital Flight Test Program. From December 1979 to April 1981, he headed the Astronaut Office ascent/entry group. He subsequently served as backup commander for STS-2 and STS-3, *Columbia*'s second and third orbital test flights. Mattingly was commander for the fourth and final orbital test flights of the Shuttle *Columbia* (STS-4) on June 27, 1982, and was also commander for the first Department of Defense (DoD)

MASTIF The MASTIF was engineered to simulate the tumbling and rolling motions of a space capsule and train the Mercury astronauts to control roll, pitch, and yaw by activating nitrogen jets. *NASA*

mission to be flown aboard Shuttle 51C. He was the head of the Astronaut Office DoD Support Group from June 1983 through May 1984.

Maul, Alfred (1864–1941)

A German engineer who, in 1906, successfully took aerial photographs of the ground by attaching cameras to a **black powder** rocket, thereby creating the first instrumented sounding rocket. His 1912-model rocket carried a 20- by 25-cm photographic plate stabilized by a gyroscope. This method of reconnaissance was discontinued, however, upon the advent of airplanes.

Max Q

Maximum dynamic pressure: the point during the powered flight phase of a rocket's ascent at which the acceleration stresses on the airframe are greatest.

MAXIM (Microarcsecond X-ray Imaging) Pathfinder

A spacecraft designed to pave the way for a full-scale MAXIM mission by demonstrating the feasibility in space of X-ray interferometry for astronomical applications. MAXIM Pathfinder would provide an imaging of celestial X-ray sources with a resolution of 100 microarcseconds, 5,000 times better than the **Chandra X-Ray Observatory**. It has been identified in NASA's Office of Space Science Strategic Plan as a potential mission beyond 2007 but remains at the stage of early concept definition.

McAuliffe, S. Christa (Corrigan) (1948–1986)

A high school teacher at Concord High School, New Hampshire, who was selected as the primary candidate for the NASA Teacher in Space Project, from more than 10,000 applicants, on July 19, 1985. She was assigned as a payload specialist on Space Shuttle mission STS 51-L, which was launched on January 28, 1986, and had planned to teach two lessons from orbit. However, she died along with the rest of the seven-member crew when the spacecraft exploded a little over a minute after liftoff (see *Challenger* **disaster**). Her backup for the mission, Barbara **Morgan**, is expected to make her inaugural flight in 2004.

McCandless, Bruce, II (1937–)

An American astronaut who became the first human to conduct an untethered extravehicular activity (EVA) in space. Selected by NASA as one of 19 new astronauts in April 1966, McCandless was appointed as a member of the astronaut support crew for the **Apollo** 14 mission and as backup pilot for **Skylab** 2. He was a co-investigator on the M-509 astronaut maneuvering unit experiment flown during the Skylab Program, collaborated on the development of the **Manned Maneuvering Unit** (MMU) used during Space Shuttle EVAs, and has been responsible for crew inputs to the development of hardware and procedures for the **Inertial Upper Stage**, the **Hubble Space Telescope**, the Solar Maximum Repair Mission, and the **International Space Station** program. McCandless was a mission specialist on the tenth Shuttle flight (STS 41-B), on February 3, 1984, during which the MMU and the Manipulator Foot Restraint were tested in space for the first time. McCandless made free flights on each of the two MMUs carried onboard and took part in other activities during two spectacular EVAs. The eight-day orbital flight of *Challenger* culminated in the first landing on the runway at the Kennedy Space Center on February 11, 1984. More recently, McCandless was a mission specialist on the crew of STS-31, launched on April 24, 1990. After leaving NASA, he joined the staff of Martin Marietta Astronautics Company, in Denver, Colorado.

McDermot, Murtagh (eighteenth century)

A pseudonymous Irish writer who, in his tale *A Trip to the Moon* (1728), gave one of the first descriptions of a **space cannon**. His protagonist is whisked to the Moon by a whirlwind after the manner of **Lucian**'s hero. There he meets the friendly Selenites, who help him build the means by which he can return. "We already know," says the lunar traveler, "the height of the Moon's atmosphere, and know how gunpowder will raise a ball of any weight to any height. Now I intend to place myself in the middle of ten wooden vessels, placed one within another, with the outermost strongly hooped with iron, to prevent its breaking. This I will place over 7,000 barrels of powder, which I know will raise me to the top of the [Moon's] atmosphere . . . but before I blow myself up, I'll provide myself with a large pair of wings, which I will fasten to my arms . . . by the help of which I will fly down to the earth." It would be another 137 years before Verne's characters used an even larger explosion to blast themselves on a lunar excursion.

McDivitt, James A. (1929–)

An American astronaut who served as command pilot of **Gemini** 4 and commander of **Apollo** 9. A career Air Force officer, retiring as a brigadier general, McDivitt was chosen as a NASA astronaut in the second group selected, in 1962. After his two spaceflights, he went on to manage the Apollo Spacecraft Program at the Johnson Space Center from September 1969 to August 1972. He then resigned from NASA and the Air Force and took on a variety of management positions in business, including senior vice president of Government and International Operations at Rockwell International.

McDonnell Douglas

An American aerospace company formed from the 1967 merger of McDonnell Aircraft Corporation and Douglas Aircraft. Both Douglas Aircraft, founded in 1928 by Donald Willie Douglas (1892–1981), and McDonnell Aircraft, founded in 1939 by James Smith McDonnell (1899–1980), made important contributions to the U.S. space program. Douglas built the F5D Skylancer prototype used in the development of the Boeing X-20 **Dyna-Soar**, developed the **Thor** missile (precursor of the **Delta**), and built the third stage of the **Saturn** V. Douglas was chosen to build the **Mercury** and **Gemini** spacecraft. Following the merger of the two companies, McDonnell Douglas was contracted to convert one of its Saturn V third stages into the **Skylab** space station. In 1997, McDonnell Douglas merged with **Boeing**.

MDS (Mission Demonstration Satellite)

A 450-kg Japanese satellite designed to evaluate the on-orbit performance of a variety of technologies, including testing the radiation resistance of some commercial semiconductor devices and new solar cells and the effectiveness in space of a new type of battery, a solid-state data recorder, and a parallel computer system. MDS-1 was launched by a **NASDA** (National Space Development Agency) H-2A rocket on February 4, 2002, into a geostationary transfer orbit (500 × 35,696 km × 28.5°).

MECO (main engine cutoff)

The time at which the main engines of a launch vehicle, or launch vehicle stage, are commanded to stop firing. MECO typically involves a sequence of events, including throttling back the engines before actual cutoff. For example, the MECO sequence for the Space Shuttle begins about 10 seconds before cutoff. About three seconds later, the Space Shuttle Main Engines (SSMEs) are commanded to begin throttling back at intervals of 10% thrust per second until they reach a thrust of 65% of rated power, called minimum power. Minimum power is maintained for just under seven seconds, then the SSMEs shut down.

Mentor

A fourth-generation **ELINT** (electronic intelligence) satellite launched for the CIA. See also **Mercury-ELINT**, **SB-WASS**, and **Trumpet**.

MEO (medium Earth orbit)

An orbit that is intermediate in altitude between that of **LEO** (low Earth orbit) and GSO (geostationary orbit) at 35,900 km.

Mercury, Project

See article, pages 271–275.

Mercury Seven

The original group of seven American astronauts selected on April 9, 1959, for the **Mercury** program. It included Scott **Carpenter**, Gordon **Cooper**, John **Glenn**, Virgil **Grissom**, Walter **Schirra**, Alan **Shepard**, and Donald "Deke" **Slayton**. All but Slayton actually flew a Mercury mission. The maximum age for candidate Mercury astronauts was set at 40, the maximum height at 5 ft. 11 in. (1.80 m), and the maximum weight at 180 lb. (81.6 kg). The first check of military test pilot records revealed that 508 test pilots met the basic astronaut requirements. With suggestions from commanding officers, this list was first cut to 110 and then, with the help of trainers and instructors who had brought these men up to flight

(continued on page 275)

Mercury Seven The Project Mercury astronauts: front row, left to right, Walter Schirra, Donald Slayton, John Glenn, and Scott Carpenter; back row, Alan Shepard, Virgil Grissom, and Gordon Cooper. *NASA*

Project Mercury

America's first manned space program, the object of which was to put a human being in orbit, test his ability to function in space, and return him safely to Earth. Project Mercury began on October 7, 1958– one year and three days after the launch of Sputnik 1– and included six manned flights between 1961 and 1963. It paved the way for the **Gemini** and **Apollo** programs.[43, 282]

History

Drawing on work done by its predecessor, NACA (National Advisory Committee for Aeronautics), NASA requested as early as 1959 proposals for a one-man capsule to be named Mercury and launched by either a **Redstone** or an **Atlas**. The development of

the spacecraft proceeded through a series of unmanned test flights involving different launch vehicles. Boiler-plate versions were lifted by solid-fueled **Little Joe**s, and later by Redstone and Atlas rockets, to test the capsule's structure and launch escape system. While some of these tests were successful, others blew up or veered off course. Next came flights involving real unmanned capsules, one of which (MR-1) failed moments after liftoff when the Redstone launcher simply settled back onto the launch pad. The launch escape system triggered, however, tricked into supposing an abort had happened during the ascent to orbit, and a few minutes later the capsule parachuted back to Earth within sight of the pad. It was collected and fitted to a new Redstone, which launched successfully a few weeks later. A major milestone was passed with

Mercury Test Flights

Test	Launch Date	Passenger	Notes
LJ-1	Aug. 21, 1959	—	Abort and escape test; failed
BJ-1	Sep. 9, 1959	—	Atlas-launched heat-shield test
LJ-6	Oct. 4, 1959	—	Capsule aerodynamics and integrity test
LJ-1A	Nov. 4, 1959	—	Abort and escape test
LJ-2	Dec. 4, 1959	Rhesus monkey Sam	Primate escape at high altitude
LJ-1B	Jan. 21, 1960	Rhesus monkey Miss Sam	Abort and escape test
BA-1	May 9, 1960	—	Pad escape system test
MA-1	Jul. 29, 1960	—	Qualification of spacecraft and Atlas; failed
LJ-5	Nov. 8, 1960	—	Qualification of Mercury spacecraft; failed
MR-1	Nov. 21, 1960	—	Qualification of spacecraft and Redstone; failed
MR-1A	Dec. 19, 1960	—	Qualification of systems for suborbital operation
MR-2	Jan. 31, 1961	Chimpanzee Ham	Primate suborbital and auto abort test
MA-2	Feb. 21, 1961	—	Qualification of Mercury-Atlas interfaces
MR BD	Mar. 24, 1961	—	Qualification of booster for manned operation
MA-3	Apr. 25, 1961	—	Test of spacecraft and Atlas in orbit; failed
LJ-5B	Apr. 28, 1961	—	Max Q escape and sequence
MA-4	Sep. 13, 1961	—	Test of spacecraft environmental control in orbit
MS-1	Nov. 1, 1961	—	Test of Mercury-Scout configuration; failed
MA-5	Nov. 29, 1961	Chimpanzee Enos	Primate test in orbit

Key

LJ = Little Joe, BJ = Big Joe, BA = Beach abort, MA = Mercury-Atlas, MR = Mercury-Redstone, MS = Mercury-Scout

the suborbital journey and safe return of the chimpanzee **Ham** in January 1961. A few months later, Alan **Shepard** became the second primate to fly a Mercury-Redstone to the edge of space. (See table, "Mercury Test Flights.")

Mercury Capsule

A bell-shaped capsule, 2.9 m tall, 1.88 m in diameter, built by McDonnell Aircraft Corporation for launch by either a Redstone or an Atlas booster (see **Mercury-Redstone** and **Mercury-Atlas**). Its basic design was proposed by Maxime **Faget** of NACA (National Advisory Committee for Aeronautics) Langley in December 1957. The pressurized cabin, with an internal volume about the same as that of a telephone booth, was made of titanium, while the capsule's outer shell consisted of a nickel alloy. Around the base, a fiberglass-reinforced laminated plastic heatshield was designed to ablate during reentry, then detach and drop about a meter to form the bottom of a pneumatic cushion to help soften the impact at splashdown. Attitude control in all three axes was achieved by 18 small thrusters linked to a controller operated by the astronaut's right hand. Three solid-fueled retro-rockets, held at the center of the heatshield by metal straps, were fired in quick succession to de-orbit and then were jettisoned. Above the cabin was a cylindrical section containing the main and reserve parachutes. Atop the whole capsule at launch was a latticework tower supporting a solid-rocket escape motor with three canted nozzles, which in an emergency could carry the spacecraft sufficiently clear of the booster for the capsule's parachute to be deployed. Inside the cabin was a couch, tailor-made for each astronaut, facing the control panel. Early capsules had two small round portholes, but following complaints from the astronauts about poor visibility, a larger rectangular window was installed on later versions. A retractable periscope was also provided. The capsule was filled with pure oxygen at about one-third atmospheric pressure, and the astronaut usually kept his helmet visor open. Only if the cabin pressure fell would he need to lower his visor and switch to his spacesuit's independent oxygen supply.

Manned Flights

See the table "Mercury Manned Flights."

Mercury MR-3

A suborbital mission that climbed high enough for MR-3 to be considered a true spaceflight and for Shepard to be considered the first American in space. After reaching a peak altitude of 187 km and a peak velocity of 8,335 km/hr, the *Freedom 7* capsule splashed down in the Atlantic Ocean 486 km downrange of the launch site. Shepard and his craft were recovered by helicopter within six minutes of splashdown and placed aboard the recovery vessel about five minutes later.

Because the planned flight was only 15 minutes long, no one gave much thought to the issue of personal waste disposal. But Shepard was strapped into his capsule some three hours before liftoff, and after a couple of hours on his back, he asked for "permission to relieve his bladder." After some debate, the engineers and medical team decided that this would be okay, presumably realizing that the alternative—postponing the launch while Shepard visited the bathroom—would not amount to a NASA publicity coup. Starting with Gus Grissom's flight in July 1961,

---

Mercury Manned Flights

Mission	Launch Date	Vehicle	Duration	Orbits	Pilot	Capsule
MR-3	May 5, 1961	Redstone	15 min 22 sec	Suborbital	Alan Shepard Jr.	*Freedom 7*
MR-4	Jul. 21, 1961	Redstone	15 min 37 sec	Suborbital	Virgil Grissom	*Liberty Bell 7*
MA-6	Feb. 20, 1962	Atlas	4 hr 55 min	3	John Glenn	*Friendship 7*
MA-7	May 24, 1962	Atlas	4 hr 56 min	3	M. Scott Carpenter	*Aurora 7*
MA-8	Oct. 3, 1962	Atlas	9 hr 13 min	6	Walter Schirra Jr.	*Sigma 7*
MA-9	May 15, 1963	Atlas	34 hr 19 min	22	L. Gordon Cooper	*Faith 7*

strap-on urine receptacles were provided for the astronauts' use.

During the flight, Shepard withstood a maximum 6*g* during ascent, about five minutes of weightlessness, and slightly under 12*g* during reentry. He successfully carried out all his assigned tasks, including manually guiding the capsule from the time it separated from the Redstone booster. This proved the important point that a human could competently handle a vehicle during both weightlessness and high-gravity situations. The *Freedom 7* capsule lacked a window, but Shepard was able to see out through a periscope—but only in black-and-white, because a gray filter had been mistakenly left on the lens.[130]

Mercury MR-4

Launch attempts on July 18 and 19, 1961, were scrubbed due to bad weather—the first scrubs in the history of American manned spaceflight. On his suborbital flight, Grissom became the second American in space, reached a maximum altitude of 190 km, and ended up 488 km downrange of the launch site. The mission goals were almost identical to those of Shepard's flight, but the *Liberty Bell 7* capsule had a window, easier-to-use hand controls, and explosive side hatch bolts that could be blown in an emergency. Unfortunately, after splashdown these bolts were unexpectedly blown, causing the capsule to start filling with water. Grissom made his way out but had to struggle to reach a sling lowered from a rescue helicopter because his spacesuit had become waterlogged. He was recovered without injury after being in the ocean for about four minutes, but an attempt to lift the water-laden *Liberty Bell 7* capsule by helicopter failed and the capsule sank—the only such loss following an American manned spaceflight. NASA ran a battery of tests and simulations to find out how Grissom might have blown the hatch and determined that it would have been nearly impossible for him to have done this accidentally. Instead, the loss of the capsule was blamed on an unknown failure of the hatch itself, although it became a standing joke that a crack that had been painted on the side of *Liberty Bell 7* (like the crack in the real bell) prior to the flight was the true cause of its sinking. Following the successes of missions MR-3 and MR-4, NASA decided that no more suborbital flights were needed before a manned orbital attempt, and it canceled the flights that had been designated MR-5 and MR-6.

Mercury MA-6

After a series of scrubs that delayed his launch for nearly a month, John Glenn became the first American to orbit the Earth. His *Friendship 7* capsule completed a 130,000-km flight, three times around the planet, with Glenn becoming the first American astronaut to view sunrise and sunset from space and to take pictures in orbit—using a 35-mm camera he bought in

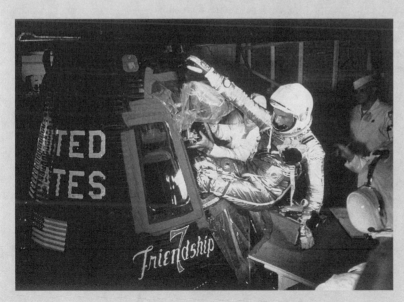

Mercury MA-6 John Glenn entering *Friendship 7* before becoming the first American in orbit. *NASA*

a drugstore. Glenn also had the distinction of being the first American astronaut to eat in space, consuming a small tube of applesauce, and then discovering what became known as the "Glenn Effect." Early in the flight, he noticed what looked like fireflies dancing outside his window. These were later identified as frost particles sparkling in the sunlight after being released from the spacecraft's attitude control jets. During the second and third orbits, Glenn took manual control of *Friendship 7* due to a failure of the automatic pilot caused by one of the control jets becoming clogged. Far more worrisome was a potential problem that mission managers became aware of as Glenn prepared to leave orbit. At first, they hid their concern from the astronaut. But when Glenn was asked to carry out unfamiliar instructions, he asked the reason and was told there was a possibility that the capsule's landing bag and heat-shield had come loose. The landing bag was designed to absorb the shock of water impact, and the heat-shield was essential to prevent the spacecraft from burning up during reentry. Glenn was instructed to delay jettisoning the capsule's retro-rocket package so that the straps holding it to the capsule might keep the potentially loose heat-shield in place until the last possible moment. Glenn later recalled that he saw the burning retro-rocket package pass by outside his window, causing him to think his spacecraft was on fire. Anxious moments passed for those on the ground during the unavoidable communications blackout as Glenn reentered, but Glenn's capsule returned safely and the problem was later traced to a faulty switch in the heat-shield circuitry that gave a false reading. *Friendship 7* splashed down in the Atlantic Ocean 267 km east of Grand Turk Island and remained in the water just 21 minutes before being picked up by helicopter. Glenn stayed inside his spacecraft until it was on the deck of the recovery vessel.[119]

Mercury MA-7

Scott Carpenter became the second American to orbit the Earth, flying three circuits like Glenn but with a much busier schedule. He was also the first American astronaut to eat an entire meal in space, squeezed, like Glenn's applesauce, out of tubes. A science experiment carried by *Aurora 7*, designed to see how fluids react in zero gravity, consisted of a small balloon to be inflated in orbit and remain tethered to the spacecraft. However, it failed to inflate properly and drifted lazily away from the capsule on its snaking line—essentially a failure. Carpenter's crowded flight plan also caused prob-

lems. Several times he was working so fast that he accidentally flipped switches that kicked in the spacecraft's manual thrusters, wasting valuable fuel. Then, toward the end of the mission, he became obsessed with the dancing lights—the "fireflies"—that Glenn had reported seeing. This caused him to miss the scheduled reentry burn by three seconds, which, together with a 25° yaw error at the time of firing, led to an overshoot of the intended recovery point by 400 km. Upon splashing down about 200 km northeast of Puerto Rico, Carpenter climbed out of his capsule and had to wait about three hours until the recovery vessel arrived.

Mercury MA-8

Although NASA was concerned on October 1, 1962, a couple of days before launch, that Tropical Storm Daisy might pose a threat, Wally Schirra took off aboard *Sigma 7* right on schedule. The capsule and flight plan had been modified to avoid problems that had cropped up on previous missions. Its reaction control system was to disarm the high-thrust jets during periods of manual maneuvering, and the mission had more scheduled "drift time" (periods of unmaneuvered flight), to save fuel. As it turned out, *Sigma 7* (named by Schirra to suggest engineering precision) was a model of efficient power and fuel utilization. Two high-frequency antennas were mounted onto the retro package to provide better communications between the capsule and the ground during the flight. Schirra operated an experimental handheld camera and took part in the first live television broadcast from an American manned spaceflight, the signal transmitted to North America and Western Europe via **Telstar**-1. Nine ablative-type material samples were included in an experiment package mounted onto the cylindrical neck of the capsule. Also, two radiation monitoring devices were carried inside the capsule, one on either side of the astronaut's couch. Shortly after Schirra's return, the Air Force announced that he would likely have been killed by radiation if his spacecraft had flown above an altitude of 640 km. Radiation monitoring devices on classified military satellites had confirmed this local zone of lethal radiation, the result of a high-altitude nuclear test carried out in July 1962. In fact, at the height at which Schirra actually flew (between 161 and 283 km), the radiation monitors inside the spacecraft showed that he had been exposed to much less radiation than predicted even under normal circumstances. Mission MA-8 proved that longer-duration spaceflights were feasible, and Schirra commented that both he and the spacecraft could have flown much longer than six orbits. Splashdown took place within the intended recovery zone, about 440 km northeast of Midway Island and just 7 km from the recovery vessel.

Mercury MA-9

Lying in his *Faith 7* capsule during countdown, Gordon Cooper was so relaxed that he even managed to nod off. He had another opportunity to sleep once he was in space, because this 22-orbit mission was the first in American manned spaceflight history to last more than a day. (Vostok 2, however, holds the record for the very first full-day manned mission.) During the flight, Cooper released a beacon sphere containing strobe lights—the first satellite to be deployed from a manned spacecraft—which he was able to see during his next orbit. He also spotted a 44,000-watt xenon lamp that had been set up as an experiment in visual observation in a town in South Africa, and was able to recognize cities, oil refineries, and even smoke from houses in Asia. Cooper tried twice without success to deploy an inflatable balloon. Between orbits 10 and 14

he slept for about eight hours, later reporting that he had anchored his thumbs to his helmet restraint strap to prevent his arms from floating freely—a potential hazard with so many switches within easy reach. On the 19th orbit a warning light came on indicating that the capsule had dipped to a dangerously low altitude. However, further tests showed that the capsule was still in its proper orbit and that most likely the warning system had failed because of a short-circuit caused by dampness in the electrical system. Concerned that there might be more such short-circuits in the automatic reentry system, mission managers instructed Cooper to reenter under manual control—the only such reentry of all four Mercury orbital flights. Cooper did a fine job bringing his capsule to a splashdown just 7 km from the prime recovery vessel. He was the last American astronaut to orbit Earth alone. NASA had considered one more Mercury flight, but the project officially ended on June 12, 1963, when NASA Administrator James Webb told the Senate Space Committee that no further Mercury missions were needed, and that NASA would press ahead with the Gemini and Apollo programs.

Mercury Seven

(continued from page 270)

status, to 69. All these remaining candidates were invited to apply, but only 32 volunteered. These volunteers went on to undergo rigorous physical examinations and tests; the number was reduced to 14. The selection of the final seven came on April 9, 1959. Shepard was the tallest Mercury astronaut at 5 ft. 11 in. (1.80 m), Grissom the shortest at 5 ft. 7 in. (1.70 m). Cooper was the youngest at 32, Glenn the oldest at 37. The first message received by the test pilots as to their impending call for astronaut service read: "You will soon receive orders to OP-05 in Washington in connection with a special project. Please do not discuss the matter with anyone or speculate on the purpose of the orders, as any prior identification of yourself with the project might prejudice that project."[41]

Mercury Thirteen

A group of 13 women who passed the physical tests given to candidate astronauts for the **Mercury** program, in some cases surpassing the performance of the **Mercury Seven**, but who were debarred from serving as astronauts. The group came about because aerospace physician Randolph **Lovelace** II, responsible for the

selection of the Mercury Seven, was curious to know whether female pilots could measure up to the same rigorous tests that the men had taken at his clinic in Albuquerque, New Mexico. Lovelace invited 26 women

Mercury Thirteen   Jerrie Cobb tests the Gimbal Rig at Lewis Research Center in April 1960. *NASA*

pilots to take the initial round of exams. The 13 who passed were Jerrie Cobb, Myrtle "K" Cagle, the twin sisters Jan and Marion Dietrich, Wally Funk, Jane Hart, Jean Hixson, Gene Nora Jessen, Irene Leverton, Sarah Gorelick Ratley, Bernice "B" Steadman, Jerri Sloan Truhill, and Rhea Allison Woltman, with Funk and Cobb being outstanding. Whereas the male candidates were tested in a sensory deprivation tank for only three hours, Funk floated for more than ten. She also endured a 5*g* centrifuge ride even though regulations prevented her from borrowing a regulation **anti-*g* suit**; she simply wore a full-length "merry widow" girdle borrowed from her mother instead. Lovelace concluded that Funk and Cobb, in particular, "would have made excellent astronauts," but their further testing and potential selection was stymied by bureaucracy and sexism. Cobb and Jane Hart lobbied for support, and thanks in part to Hart's husband, Senator Philip Hart of Michigan, a congressional hearing on official astronaut qualifications was set for July 1962. Cobb and Hart argued their own case, whereas NASA presented its views through agency administrators and the astronauts Scott Carpenter and John Glenn. Cobb began by saying, "We seek only a place in our nation's space future without discrimination." However, discrimination was evident from the outset when the committee chairman, the Republican Victor Anfuso of New York, replied in his opening remarks: "Miss Cobb, that was an excellent statement. I think that we can safely say at this time that the whole purpose of space exploration is to someday colonize these other planets, and I don't see how we can do that without women." Glenn did not help by arguing "that women are not in this field is a fact of our social order" and that women would need to be superior to be considered for astronaut selection: "If we could find any women that demonstrated they have better qualifications than men, we would welcome them with open arms." On Day 3 the hearings were canceled, and Congress ruled that future astronauts must come from the ranks of military jet test pilots. Since no women were allowed to train as test pilots until a decade later, the policy effectively slammed the door on the spaceflight ambitions of the Mercury Thirteen. However, the group was not forgotten. When, in July 1999, Eileen **Collins** became the first woman to command a Space Shuttle mission, she made sure that the Mercury Thirteen were included on the guest list for her launch.

Mercury-Atlas

An **Atlas** D missile adapted to carry a manned **Mercury** capsule. The engine thrusts were identical to those of the operational Atlas D, with only the flight profile altered to carry the Mercury capsule into orbit and certain safety features introduced to protect the astronauts.

Total length: 29.0 m
Core diameter: 3.1 m
Liftoff mass: 116,100 kg
Payload to 185-km low Earth orbit: 1,360 kg
Total thrust
    Boosters (2): 1,600,000 N each
    Sustainer: 280,000 N
    Verniers (2): 8,800 N each

Mercury-ELINT

Fourth-generation U.S. Air Force ELINT (electronic intelligence) satellites, introduced in the late 1990s and designed to pick up a wide spectrum of electromagnetic energy for use in broadband monitoring. Unlike most other ELINT satellites, which operate in geosynchronous orbits, Mercurys are placed in even higher orbits and move in complex, elliptical patterns. This not only gives them wider coverage but also allows them to take bearings on a given transmitter and thus accurately pinpoint its location.

Mercury-Redstone

A modified version of the **Redstone** used in NASA's initial effort to launch astronauts into space. In most respects, the Mercury-Redstone was similar to its missile relative. In fact, the vehicle was selected for the Mercury program because of its proven track record of safety and reliability. The Mercury-Redstone did incorporate added safety features as well as an upgraded Rocketdyne engine. About 800 engineering changes were made to the production version of the Redstone to qualify it as a manned space-launch vehicle. These included extending the fuel tank about 2 m to increase the burn time and thus achieve increased speed and altitude. Alan Shepard, the first American astronaut in space, was launched aboard a Mercury-Redstone from Cape Canaveral Launch Pad 5 on May 5, 1961, on mission MR-3. A nearly identical flight, designated MR-4, carried Virgil Grissom on July 21, 1961. Mercury-Redstone mission performance in support of suborbital manned flights MR-3 and MR-4 was so successful that two similar flights, which would have been designated MR-5 and MR-6, were canceled.

Length: 25.5 m
Diameter: 1.8 m
Thrust: 348,000 N

Mercury-Redstone A Mercury-Redstone launches *Freedom 7* and Alan Shepard on the first American manned suborbital spaceflight. *NASA*

Mercury-Scout

On May 5, 1961, the same day the first American astronaut was launched, NASA proposed using **Scout** rockets to evaluate the Mercury tracking and real-time computer network in preparation for manned orbital missions. The proposal was approved, and on June 13, 1961, the NASA Space Task Group issued detailed instrumentation requirements for the modified Scout, which became known as Mercury-Scout. Each Mercury-Scout was to carry a lightweight communications payload into orbit to allow a simulation of the tracking through the Mercury global network of an actual Mercury capsule. Blue Scout II number D-8 was modified for the first—and what would turn out to be the only—Mercury-Scout mission, MS-1. The launch was conducted by the U.S. Air Force, which had already launched other Blue Scout rockets from Cape Canaveral, on November 1, 1961. But 28 seconds after liftoff, the rocket veered off course and had to be destroyed by the range safety officer. Mercury program

managers decided against further Mercury-Scout missions, and by the end of 1961, Mercury-Atlas missions MA-4 and MA-5 had demonstrated that actual orbiting Mercury capsules could be successfully tracked.

mesosphere

A region of Earth's atmosphere between 50 and 90 km above the surface, where the temperature decreases with altitude from a maximum of 7°C at the stratopause to a minimum of −123°C at the mesopause.

MESSENGER (Mercury Surface, Space Environment, Geochemistry and Ranging)

A NASA **Discovery-class** probe to investigate Mercury from an orbit around the planet. Scheduled for launch in March 2004, MESSENGER will make two flybys of Venus and two of Mercury during a five-year voyage that will fine-tune its trajectory through **gravity-assist**s for Mercury-orbit insertion in 2009. Once in orbit, MESSENGER will have to contend with the intense heat in a region where the Sun is up to 11 times brighter than on Earth. However, sheltered behind a sunshield made of the same ceramic material that protects parts of the Space Shuttle, the probe's instruments will be able to operate at room temperature. Also, the spacecraft will pass only briefly over the hottest parts of the planet's surface, thereby limiting the instruments' exposure to reflected heat. During its nominal one-Earth-year period of operation in orbit, MESSENGER will analyze the planet with seven instruments, including X-ray, gamma-ray, infrared, and neutron spectrometers for exploring planetary composition, a magnetometer to learn more about the magnetic field, a laser altimeter for determining the height of surface features, and a high-resolution camera. MESSENGER will be designed, built, and operated by the Johns Hopkins University Applied Physics Laboratory for NASA. The only previous spacecraft to visit Mercury was **Mariner** 10 in 1974.

Meteor

Soviet weather satellites. The first Meteor system became operational in 1969 after several years of testing. Subsequently, about 30 Meteors were launched between 1969 and 1978 before the first of an upgraded version, Meteor 2, was orbited in July 1975. A further series, Meteor 3, debuted in October 1985, although Meteor 2 satellites continued to be launched in parallel. All of these spacecraft were placed in high-inclination orbits because a geostationary location—the usual norm for weather satellites—provides a poor view of the high latitudes at which much Russian territory lies. The Soviet Union did indicate that it would launch a geostationary weather satellite as early as 1978, but it was the mid-1990s before the promised spacecraft, **GOMS** (Geostationary Operational Meteorological Satellite), finally took off.

meteor deflection screen
See **Whipple shield**.

meteorological satellite
See **weather satellite**.

Meteosat
A series of European geostationary **weather satellites** developed in conjunction with **GARP** (Global Atmospheric Research Program). Although ESA (European Space Agency) originated and continues to control the Meteosat network, **Eumetsat** (European Meteorological Satellite Organisation), created between 1981 and 1986, has overall responsibility for the system. The main goal of the Meteosats is to supply 24-hour visible and infrared cloud cover and radiance data. Three preoperational craft were launched between 1977 and 1988 before the Meteosat Operational Program (MOP) began with the launch of Meteosat 4. Meteosat satellites closely resemble their American and Japanese counterparts. (See table, "Meteosat Series.")

Size: 3×2.1 m
On-orbit dry mass: 320 kg

MetOp (Meteorological Operational) satellites
Polar-orbiting meteorological satellites, three of which are planned as part of the **Eumetsat** Polar System (EPS) program. They will carry instruments provided by ESA (European Space Agency), Eumetsat (European Meteorological Satellite Organisation), NOAA (National Oceanic and Atmospheric Administration), and CNES (the French space agency), with the aim of improving upon the established service provided by NOAA satellites. MetOp 1 is scheduled for launch in 2005.

microgravity
The condition of near-weightlessness induced by **free fall** or unpowered spaceflight. It is characterized by the virtual absence of gravity-induced convection, hydrostatic pressure, and sedimentation.

MicroLab
A microsatellite that carries a NASA lightning mapper called the Optical Transient Detector. It also carries a weather sensor for the National Science Foundation, employing a GPS-MET receiver that monitors transmissions from any GPS (Global Positioning System) spacecraft near the horizon in order to infer values of temperature and humidity in its path. The spacecraft is now known as **OrbView**-1 and is operated commercially by ORBIMAGE.

Launch
    Date: April 3, 1995
    Vehicle: Pegasus
    Site: Vandenberg Air Force Base
Orbit: 733×747 km $\times 70.0°$
Mass: 76 kg

micrometeoroid protection
Methods used to protect spacecraft components from micrometeoroid impacts. They range from structural positioning to shield sensitive hardware, to placement of protective blankets on the spacecraft exterior. Interplanetary probes typically use tough blankets of Kevlar or other strong materials to absorb the energy of high-velocity dust particles.

Meteosat Series

| Spacecraft | Launch | | | GSO Location |
	Date	Vehicle	Site	
Meteosat 1	Nov. 23, 1977	Delta 2914	Cape Canaveral	0° E
Meteosat 2	Jun. 19, 1981	Ariane 1	Kourou	0° E
Meteosat 3	Jun. 15, 1988	Ariane 44LP	Kourou	0° E
Meteosat 4	Mar. 6, 1989	Ariane 44LP	Kourou	0° E
Meteosat 5	Jun. 15, 1988	Ariane 44LP	Kourou	4° E
Meteosat 6	Nov. 15, 1993	Ariane 44LP	Kourou	0° E
Meteosat 7	Sep. 2, 1997	Ariane 44LP	Kourou	10° W

microsat

A satellite with an on-orbit mass of 10 to 100 kg. See **satellite mass categories**.

MicroSat

A constellation of seven 22-kg microsatellites, sponsored by DARPA (Defence Advanced Research Projects Agency), and launched on July 16, 1991, by a **Pegasus** rocket into a 359×457 km $\times 82°$ orbit, rather than the higher intended orbit of 833×833 km. The satellites supported "bent-pipe" relay for voice, data, fax, and slow-speed video communications, as well as high-fidelity secure voice and encrypted data and limited store-and-forward communications, and were intended to provide data for ArcticSat, a system for communicating with nuclear submarines under the polar cap. However, this ability was compromised by the lower-than-planned orbit, which caused the satellites' footprints to be separated. All the spacecraft decayed in January 1992 after little operational use. The mission was also known as the Small Communications Satellite (SCS).

Microscale Coronal Features probe

A spacecraft designed to provide images and spectroscopic details of the Sun's corona at an unprecedented level of resolution. By showing activity in the corona that happens very quickly and at small spatial scales, this data will help distinguish among the various heating mechanisms that have been proposed. The Microscale Coronal Features probe has been identified in NASA's Office of Space Science Strategic Plan as a potential mission after 2007.

microwave plasma thruster

An experimental form of **electrothermal propulsion** that works by generating microwaves in a resonant, propellant-filled cavity, thereby inducing a plasma discharge through electromagnetic coupling. The microwaves sustain and heat the plasma as the working fluid, which is then thermodynamically expanded through a nozzle to create thrust.

microwave radiation

Electromagnetic radiation with wavelengths between 30 cm and 1 mm, corresponding to frequencies of 1 to 300 GHz. Microwaves appear in the electromagnetic spectrum between longer **radio waves** and **infrared** radiation.

MIDAS (Missile Defense Alarm System)

An obsolete and largely unsuccessful system of U.S. military early-warning satellites launched between 1960 and 1966; it preceded the **IMEWS** (Integrated Missile Early Warning Satellite) program. MIDAS spacecraft were designed to detect ballistic missile launches from low Earth orbit using infrared sensors. They represented one arm of the first American spaceborne reconnaissance (spy) system that also included **Corona** and **Samos**. The first MIDAS satellite, launched in February 1960, failed to reach orbit. MIDAS 2, launched in May 1960, did achieve orbit to become the first infrared reconnaissance satellite, but its telemetry system failed after two days. MIDAS 3, successfully launched in July 1961, also made it into orbit and was the heaviest American satellite up to that time. Altogether, there were 12 MIDAS launches, deploying four different types of increasingly sophisticated sensors, which paved the way for the development, launch, and use of IMEWS. Details of MIDAS became publicly available only on November 30, 1998, when the Air Force Space and Missile Systems Center declassified the information. (See table, "Midas Spacecraft," on page 280.)

midcourse

For lunar and planetary missions, the period between escape from the originating point and commitment to entry or orbit at the destination. A *midcourse maneuver* is a change to a spacecraft's flight path during midcourse to maintain the desired trajectory.

MIDEX (Medium-class Explorer)

Low-cost (typically not more than $140 million for development) NASA science missions. The first two to be launched were **IMAGE** (Imager for Magnetopause-to-Aurora Global Exploration) and **MAP** (Microwave Anisotropy Probe). The next spacecraft in the series, the **Swift Gamma Ray Burst Explorer**, is scheduled for launch in September 2003. NASA will select two from four MIDEX proposals—**ASCE** (Advanced Spectroscopic and Coronagraphic Explorer), **Astrobiology Explorer**, **NGSS** (Next Generation Sky Survey), and **THEMIS** (Time History of Events and Macroscale Interaction during Substorms)—for launch in 2007 and 2008.

Midori

See **ADEOS (Advanced Earth Observation Satellite)**.

Mightysat

Small, relatively inexpensive U.S. Air Force satellites intended to demonstrate new technologies in space. Mightysats are developed by the Air Force Research Laboratory and carry multiple experiments. The first was launched from the Space Shuttle via the Hitchhiker Ejection System, the second by a **Minotaur**. Mightysat 1's payload included experiments on composite materials, advanced solar cells, advanced electronics, and a shock device. Mightysat 2, also known as Sindri, carried a hyperspectral imager for Earth imaging and spectroscopy,

MIDAS Spacecraft

Mass: about 2,000 kg

| Spacecraft | Launch | | | Orbit |
	Date	Vehicle	Site	
MIDAS 1	Feb. 26, 1960	Atlas-Agena A	Cape Canaveral	Launch failure
MIDAS 2	May 24, 1960	Atlas-Agena A	Cape Canaveral	473 × 494 km × 33.0°
MIDAS 3	Jul. 12, 1961	Atlas-Agena B	Vandenberg	3,343 × 3,540 km × 91.1°
MIDAS 4	Oct. 21, 1961	Atlas-Agena B	Vandenberg	3,482 × 3,763 km × 95.9°
MIDAS 5	Apr. 9, 1962	Atlas-Agena B	Vandenberg	2,784 × 3,405 km × 86.7°
MIDAS 6	Dec. 17, 1962	Atlas-Agena B	Vandenberg	Launch failure
MIDAS 7	May 9, 1963	Atlas-Agena B	Vandenberg	3,607 × 3,676 km × 87.3°
MIDAS 8	Jun. 13, 1963	Atlas-Agena B	Vandenberg	Launch failure
MIDAS 9	Jul. 18, 1963	Atlas-Agena B	Vandenberg	3,672 × 3,725 km × 88.4°
I. MIDAS 1	Jun. 9, 1966	Atlas-Agena D	Vandenberg	154 × 3,678 km × 90.0°
I. MIDAS 2	Aug. 19, 1966	Atlas-Agena D	Vandenberg	3,658 × 3,708 km × 89.7°
I. MIDAS 3	Oct. 5, 1966	Atlas-Agena D	Vandenberg	3,656 × 3,721 km × 90.0°

as well as satellite technology experiments such as advanced solar arrays. (See table, "Mightysat Launches.")

Milstar (Military Strategic and Tactical Relay)

A series of advanced American military satellites designed to provide survivable, global jam-resistant communications for the command and control of strategic and tactical forces through all levels of conflict. Each Milstar satellite serves as a smart switchboard in space by directing traffic from ground terminal to ground terminal anywhere on Earth. Since the satellite actually processes the communications signal and can link with other Milstar satellites through crosslinks, the need for ground-controlled switching is much reduced. An operational Milstar constellation is supposed to consist of four satellites, positioned around the Earth in geosynchronous orbits, but the program has been plagued with fiscal and technical problems. The first and second Milstar 1 satellites were launched in 1994 and 1995. However, the first of a new generation of Milstar satellites was placed in a useless orbit because of a flight software error in its Cen-

taur upper stage during launch on April 30, 1999. With the satellite costing about $800 million, and the launcher a further $433 million, this is believed to be the most expensive unmanned loss in the history of Cape Canaveral launch operations. The second Milstar 2 was successfully launched on February 27, 2001, and the third on January 14, 2002. The final Milstar—the sixth overall—is scheduled for launch in 2003.

minisat

A satellite with an on-orbit mass of 100 to 1,000 kg. See **satellite mass categories**.

Minisat

A small, multipurpose satellite bus developed by **INTA**, the Spanish space agency. Minisat is intended to form the basis for several series of satellites, including ones for scientific applications (such as astronomy and microgravity), Earth observations, and communications experiments. Minisat-1—the only Minisat launch as of mid-2002—carried an extreme ultraviolet spectrograph, a

Mightysat Launches

| Spacecraft | Launch | | | Orbit | Mass (kg) |
	Date	Vehicle	Site		
Mightysat 1	Oct. 29, 1998	Shuttle STS-88	Cape Canaveral	381 × 395 km × 51.6°	320
Mightysat 2	Jul. 19, 2000	Minotaur	Vandenberg	547 × 581 km × 97.8°	130

gamma-ray burst detector, and an experiment on microacceleration in liquids.

Launch
> Date: April 21, 1997
> Vehicle: Pegasus XL
> Site: Gran Canaria
> Orbit: 561 × 580 km × 151.0°
> Mass: 209 kg

Minitrack

A tracking network originally established for **Vanguard** tracking and data acquisition during the **International Geophysical Year**. It is now the basic network for tracking small scientific Earth satellites.

Minotaur

A modified Minuteman intercontinental ballistic missile (ICBM) that uses the first two stages of the now-decommissioned Minuteman 2 ICBM and the upper two stages of the **Pegasus** XL booster. Minotaur can place up to 340 kg into a 740-km Sun-synchronous orbit—a payload about 50% greater than that of the Pegasus XL alone. Its first successful operational launch was on January 26, 2000, from Vandenberg Air Force Base, when it placed the **JAWSAT** payload adaptor, carrying several microsatellites, into orbit.

Mir

See article, pages 282–283.

Miranda

A small British satellite, also known as UK-X4, designed to test a new type of three-axis control system as an alternative to spin stabilization. It was built by Hawker Siddeley for the British Department of Trade and Industry and launched by NASA.

Launch
> Date: March 9, 1974
> Vehicle: Scout D
> Site: Vandenberg Air Force Base
> Orbit: 703 × 918 km × 97.9°
> Mass: 93 kg

Mishin, Vasily (1917–)

A Soviet rocket scientist and one of the first Russians to see Nazi Germany's V-2 (see **"V" weapons**) facilities at the end of World War II. He was subsequently a close collaborator of Sergei **Korolev** in the development of the first Soviet intercontinental ballistic missile (ICBM) and in the **Sputnik** and **Vostok** programs. After Korolev's death, Mishin became head of Korolev's OKB-1 design bureau and tried unsuccessfully to commit the Soviet space program to landing a man on the Moon.

missile

An unmanned aerial weapon of which there are two basic types, the **ballistic missile** and the **guided missile**.

mission control center

A room or building equipped with the means to monitor and control the progress of a spacecraft during all phases of its flight after launch.

Mission of Opportunity (MO)

An investigation that forms part of a non-NASA space mission of any size and has a total NASA cost of under $35 million. Missions of Opportunity are conducted on a no-exchange-of-funds basis with the organization sponsoring the mission. NASA solicits proposals for MOs with each Announcement of Opportunity (AO) issued for **UNEX** (University-class Explorer), **SMEX** (Small Explorer), and **MIDEX** (Medium-class Explorer) investigations. Examples include **HETE**-2, **TWINS**, and **CINDI**.

mission profile

A graphic or tabular presentation of the flight plan of a spacecraft showing all pertinent events scheduled to occur.

mission specialist

A Space Shuttle astronaut whose responsibilities may include conducting experiments in orbit, construction of the International Space Station, and control of the Orbiter's resources to a payload. Mission specialists are professional astronauts, employed by NASA or other space agencies, as distinct from **payload specialist**s.

Mississippi Test Facility

See **Stennis Space Center**.

Mitchell, Edgar Dean (1930–)

An American astronaut and a Navy captain who flew on **Apollo** 14 and became the sixth human to walk on the Moon. Mitchell obtained a B.S. in industrial management from the Carnegie Institute of Technology (now Carnegie-Mellon University), an M.S. from the U.S. Naval Postgraduate School, and a Ph.D. in aeronautics and astronautics from Massachusetts Institute of Technology. After retiring from the Navy in 1972, Mitchell founded the Institute of Noetic Sciences to sponsor research into the nature of consciousness as it relates to cosmology and causality. In 1984, he was a cofounder of the Association of Space Explorers, an international organization of those who have experienced space travel.

Mir

A large and long-lived Russian space station, the first segment of which was launched in February 1986. Bigger than its predecessors, the **Salyut** series, and composed of several modules, Mir ("peace") was designed to house more cosmonauts on longer stays than the Salyuts could support. The core of Mir was the "base block" living quarters, equipped with six docking ports to which visiting spacecraft and additional modules could be attached. Mir was gradually expanded by adding laboratory and equipment modules, rearranged for different missions, and upgraded without abandoning the original core unit. It was almost continuously occupied for 13 years–with just a four-month break in 1989–including the time during which the Soviet Union disintegrated. In 1995, Mir cosmonaut Valeri Polyakov set a new single-spaceflight endurance record of 439 days. The following modules were added to the base block: Kvant (in 1987, for astrophysics), Kvant 2 (in 1989, to provide more work space), Kristall (in 1990, for materials processing experiments and to provide a docking port for the Space Shuttle), Spektr (in 1995, for Earth and near-space observations), and Priroda (in 1996, to support microgravity research and remote sensing).

In 1993 and 1994, the heads of NASA and the Russian space agency, with government approval, signed historic agreements on cooperative ventures in space. The two agencies formed a partnership to develop the International Space Station and, in preparation for that project, to engage in a series of joint missions involving Mir and the Space Shuttle. The first docking mission of the Shuttle and Mir took place in 1995. Unlike the one-off **Apollo-Soyuz Test Project** of 1975, the Shuttle-Mir mission signaled an era of continuing cooperation between the United States and Russia in space.[39, 194]

After 15 years of service and more than 86,000 orbits, Mir returned to Earth. Three de-orbit burns brought it into the atmosphere, although the final engine burn was evidently more effective than planned, since Mir's plunge into the Pacific Ocean fell short of the target zone, treating people on a Fijian beach to an unexpected pyrotechnic display as glowing pieces of the space station streaked overhead.

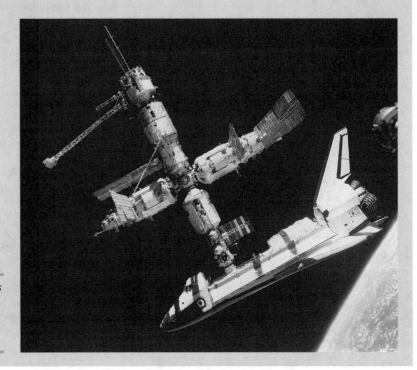

Mir The Space Shuttle *Atlantis* docked with Russia's Mir Space Station, photographed by the Mir-19 crew on July 4, 1995.
NASA

Mir Modules

See the table "Mir Modules."

Kvant (1987)

A habitable module attached to Mir's aft port for conducting research in astrophysics and supporting experiments in antiviral preparations and fractions. Kvant ("quantum") was divided into a pressurized laboratory compartment and a nonpressurized equipment compartment. The laboratory compartment was further divided into an instrumentation area and a living area, separated by an interior partition. A pressurized transfer chamber connected the Passive Docking Unit with the laboratory chamber.

Kvant 2 (1989)

Kvant 2 carried an airlock for spacewalks, solar arrays, and life-support equipment, and was based on the transport logistics spacecraft originally intended for the **Almaz** military space station program of the early 1970s. Its purpose was to provide biological research data, Earth observation data, and EVA capability. Kvant 2 added additional system capability to Mir. Kvant 2 included an additional life-support system, drinking water, oxygen provisions, motion control systems, and power distribution, as well as shower and washing facilities. Kvant 2 was divided into three pressurized compartments: instrumentation/cargo, science instrument, and airlock.

Kristall (1990)

One of Mir's science modules. Berthed opposite Kvant 2, Kristall ("crystal") carried two stowable solar arrays, science and technology equipment, and a docking port equipped with a special androgynous docking mechanism designed to receive heavy (up to about 100-ton) spacecraft equipped with the same kind of docking unit. Kristall's main aim was to develop biological and materials production technologies in microgravity. The androgynous unit was originally developed for the Russian Buran Shuttle program. *Atlantis* used the androgynous docking unit on Kristall during mission STS-71.

Spektr (1995)

A science module designed for Earth observations and berthed opposite Kvant 2. Spektr ("spectrum") was badly damaged on June 25, 1997, when Progress M-34, an unmanned supply vessel, crashed into it during tests of a new Progress guidance system. The module sustained a hole, lost pressure and electricity, and had to be shut down completely and sealed off from the remainder of the Mir complex. Its undamaged solar arrays were reconnected to the station's power system by exterior cables attached by two spacewalking cosmonauts on a later stay. The cosmonauts also installed a plate over the interior hatch to Spektr during a unique "inside spacewalk."

Priroda (1996)

A microgravity and remote sensing module that included equipment for American, French, and German experiments; its name means "nature." Soon after Priroda successfully reached orbit on April 23, 1996, a failure in its electrical supply system halved the amount of power available onboard. Since it had no solar panels, the module had only one attempt to dock with Mir before losing all its power and maneuverability. Given the fact that several previous modules had to abort their initial docking attempts, ground controllers were extremely nervous about the situation. Fortunately, the Priroda docking went flawlessly on April 26, 1996.

Mir Modules				
Name	Year	Length (m)	Diameter (m)	Mass (tons)
Core	1986	13.1	4.2	20.9
Kvant	1987	5.8	4.2	11
Kvant 2	1989	13.7	4.4	18.5
Kristall	1990	13.7	4.4	19.6
Spectr	1995	11.9	4.4	19.6
Priroda	1996	13.0	4.4	19.7

modulation

The process of modifying a radio frequency by shifting the phase, frequency, or amplitude in order to carry information.

module

A distinct and separable element of a spacecraft or space station.

Moffett Federal Airfield

A facility created in 1994, operated by NASA, and based on the old Naval Air Station Moffett Facility, near Mountain View, California.

Molniya (launch vehicle)

A **Soyuz** rocket with an additional upper stage used to launch Molniya satellites and, in the past, to send probes to the Moon and planets.

Molniya (satellite)

Russian communications satellites; "molniya" means "lightning." All move in highly elliptical, 12-hour orbits, with perigees (low points) of no more than a few hundred kilometers and apogees (high points) of up to 40,000 km. Such orbits, which have become known as **Molniya-type orbit**s, require less rocket power to achieve than a **geosynchronous orbit** and are better suited to communications with northern latitudes. The first flights of the Molniya-1 series, in 1964–1965, tested the satellite for use in military command and control. In 1968, a complete constellation of Molniya-1s became operational within the Soviet armed services. This constellation, like that of all subsequent Molniyas, consisted of four pairs of spacecraft with orbits inclined at 90° to one another, replenished as necessary. Flight trials of Molniya-2 took place from 1971 to 1974 and operational flights from 1974 to 1977. Molniya-2 was used in the Orbita television system and also for the military Korund communications system. The development of Molniya-2M, later renamed Molniya-3, began in 1972 and flight trials in November 1974. The Molniya-3 was used to create the Orbita communications system for northern regions, and later versions were incorporated into the Unified System of Satellite Communications (YeSSS). All Molniya satellites have masses of 1,500 to 1,800 kg and have been placed in orbit by Molniya 8K78 launch vehicles, mostly from Plesetsk but also from Baikonur.

Molniya-type orbit

A very elliptical 12-hour orbit, with a high apogee in the northern hemisphere and a relatively low perigee in the southern hemisphere. Soviet **Molniya**s were the first to be placed in such paths. Since satellites in Molniya-type orbits move very slowly at apogee, they appear to hover for hours at a time over northern latitudes, which enables them to relay for long periods in these locations. In addition, they can see two-thirds of the globe during a complete circuit—much more than a satellite in geosynchronous orbit. The disadvantage is that the sending/receiving dish on the ground must track the satellite, whereas for a geosynchronous satellite a fixed dish can be used. Various American military satellites, including those in the **SDS** (Satellite Data System) series, have been placed in Molniya-type orbits to spy on the Soviet Union and neighboring territories.

momentum

The product of the **mass** and **velocity** of a moving object; like velocity, it is a **vector**. The law of conservation of momentum states that when two or more objects interact, such as when a cannon fires a shell or a rocket shoots out a fast jet of hot gas, the total vector sum of their momenta is unchanged.

Momo

See **MOS (Marine Observation Satellite)**.

monopropellant

A single liquid, used in a **liquid-propellant rocket engine**, that serves as both **fuel** and **oxidizer**. A monopropellant decomposes into a hot gas when an appropriate **catalyst** is introduced. The most common example is **hydrazine**, often used in spacecraft attitude control jets.

Moon, manned missions

See **Apollo** and **Russian manned lunar programs**.

Moon, unmanned missions

See, in launch order: **Luna** (1959–1976), **Ranger** (1961–1965), **Zond** (1965–1970), **Surveyor** (1966–1968), **Lunar Orbiter** (1966–1967), **Hiten** (1990), **Clementine** (1994), **Lunar Prospector** (1998), and **Lunar-A** (2003).

Morelos

Mexican domestic communications satellites, named after the revolutionary Jose Maria Morelos (1765–1815). The first was launched in June 1985 by the Space Shuttle (STS-51G).

Morgan, Barbara R. (1951–)

NASA's first educator mission specialist. Morgan was selected as backup to Christa **McAuliffe** in the inaugural flight of the Teacher in Space Project in 1986, which ended in the *Challenger* disaster. She subsequently returned to her elementary school classroom in McCall, Idaho, but never gave up her dream of traveling in space.

In 1998, former NASA administrator Daniel Goldin announced that Morgan had been accepted as a full-time astronaut candidate; she then went through a year of training to become a flight-qualified, full-time astronaut. In April 2002, Goldin's successor, Sean O'Keefe, gave a speech at Syracuse University announcing a new initiative to recruit educator-astronauts and naming Morgan to an unspecified mission in 2004.

MORL (Manned Orbiting Research Laboratory)
A 1960s NASA concept, never built, for a cylindrical space station, 6.6 m in diameter (enabling it to fit on top of a Saturn IB launch vehicle), which would have a crew of six to nine and be powered by solar panels. The design, developed extensively by Douglas Aircraft, called for a crew quarters area, a laboratory area, and a hangar for receiving space freighters to transfer supplies.

MOS (Marine Observation Satellite)
Japan's first Earth resources satellites, also known by the national name Momo ("peach blossom"). MOS-1A and -1B, launched by **NASDA** (National Space Development Agency), monitored ocean currents and chlorophyll levels, sea surface temperature, atmospheric water vapor, precipitation, and land vegetation, and also acted as data relays for remote surface sensor platforms. (See table, "MOS Missions.")

Launch site: Tanegashima
Size: 2.4 × 1.5 m

MOST (Microvariability and Oscillations of Stars)
An astronomical microsatellite mission proposed by Dynacon Enterprises of Canada and sponsored by the **Canadian Space Agency**'s Small Payloads Program. Also participating in the mission are the University of Toronto's Institute for Aerospace Studies, and the physics and astronomy department of the University of British Columbia. With a mass of only 60 kg, MOST is the world's smallest space telescope; its 15-cm telescope and photometer are intended to measure tiny stellar pulsations and so, indirectly, the ages of stars. The spacecraft was scheduled for launch, in the last quarter of 2002, into a 780-km dawn-dusk sun-synchronous orbit, in order to

permit observations of single objects for up to eight weeks.

motor
A device that imparts motion through reaction.

motor chamber
In a liquid rocket engine, the combustion chamber. In a solid-fuel motor, the chamber that also contains the solid fuel.

MOUSE (Minimum Orbital Unmanned Satellite of the Earth)
An early design for an artificial satellite, proposed by Sam Fred **Singer** in 1953.

MSX (Midcourse Space Experiment)
A BMDO (Ballistic Missile Defense Organization) -sponsored mission designed to demonstrate various multispectral imaging technologies for identifying and tracking

Launch
  Date: April 24, 1996
  Vehicle: Delta 7925
  Site: Vandenberg
Orbit: 897 × 907 km × 99.4°
Mass: 2,700 kg

ballistic missiles during flight. It also studied the composition and dynamics of Earth's atmosphere by observing ozone, chlorofluorocarbons, carbon dioxide, and methane. Onboard instruments consisted of 11 optical sensors for making observations at infrared, visible, and ultraviolet wavelengths from 110 nm to 28 μm.

MTI (Multispectral Thermal Imager)
A satellite sponsored by the U.S. Department of Energy's Office of Nonproliferation and National Security to demonstrate advanced multispectral and thermal imaging, image processing, and associated technologies that could be used in future systems for detecting facilities that produce weapons of mass destruction. MTI carries a sophisticated telescope to collect day and night images of Earth in 15 spectral bands ranging from the visible to

MOS Missions

Spacecraft	Launch Date	Vehicle	Orbit	Mass (kg)
MOS-1A	Feb. 19, 1987	N-2	907 × 909 km × 99.1°	745
MOS-1B	Feb. 7, 1990	H-1	908 × 909 km × 99.1°	740

long-wave infrared. It also carries a High-energy X-ray Spectrometer (HXRS) sponsored by NOAA (National Oceanic and Atmospheric Administration) and the Czech Republic's Astronomical Institute of the Academy of Sciences. As a side benefit, HXRS gathers data needed to better understand powerful solar flares, which can endanger astronauts and damage space equipment.

Launch
   Date: March 12, 2000
   Vehicle: Taurus
   Site: Vandenberg
   Orbit: 574 × 609 km × 97.4°

Mu
See **M series (Japanese launch vehicles)**.

multiplexing
The simultaneous transmission of two or more signals within a single channel. Multiplexing is a **telemetry** technique used to circumvent the problem that not every instrument and sensor aboard a spacecraft can transmit its data at the same time.

multipropellant
A rocket **propellant** consisting of two or more substances fed separately to the **combustion chamber**.

Multi-Purpose Logistics Module
See **International Space Station**.

multistage ignition
An ignition system in a **ramjet** in which a portion of the fuel is ignited and the products then used to ignite the rest of the mixture.

multistage rocket
A launch system composed of successive **stages**, each capable of being jettisoned after its propellant is exhausted.

Munin
A Swedish **nanosatellite** meant to demonstrate the feasibility of using such tiny spacecraft—Munin's mass is a mere 5 kg—for on-orbit research; it is named after one of the god Odin's ravens. Munin collects data on the auroral activity in both the northern and southern hemispheres, enabling a global picture of the current state of activity to be placed online. Data acquired by the satellite then serve as input to the prediction of space weather. Munin images auroras with a miniature CCD (charge-coupled device) camera and measures high-energy particles with a solid-state detector. It was launched as a

secondary payload together with NASA's remote sensing satellite EO-1 and the Argentinean satellite SAC-C.

Launch
   Date: November 21, 2000
   Vehicle: Delta 7320
   Site: Vandenberg Air Force Base
   Orbit: 698 × 1,800 km × 95.4°

Muratori (fourteenth century)
An Italian historian credited with the first use of the word from which *rocket* is derived—*roccheta*—in 1379, when he described types of gunpowder-propelled fire arrows.

Muroc Army Air Field
See **Edwards Air Force Base**.

Murray, Bruce C. (1932–)
A professor of planetary science and geology at the California Institute of Technology and a faculty member since 1960, director of **JPL** (Jet Propulsion Laboratory) from 1976 to 1982, and current president of the **Planetary Society**, an organization he and Carl **Sagan** founded in 1979. Murray has been involved with planetary exploration since its inception in the 1960s, and he headed JPL during the **Viking** landings on Mars as well as the **Voyager** encounters with Jupiter and Saturn.

MUSES- (Mu Space Engineering Satellites)
The prelaunch designation of a series of missions implemented by Japan's **ISAS** (Institute of Space and Astronautical Science) to pioneer new satellite technologies, including lunar flyby, **aerobraking**, and large structure deployment. Members of the series include MUSES-A **(Hiten)**, MUSES-B **(HALCA)**, and **MUSES-C**.

MUSES-C
The first sample-return mission to an asteroid. This Japanese spacecraft, scheduled for launch by **ISAS** (Institute of Space and Astronautical Science) in late 2002, will use **solar-electric propulsion** to rendezvous with the asteroid 1998 SF36 in September 2005. It will then orbit its target for about five months using autonomous navigation and guidance while its optical navigation camera, **lidar**, laser range finder, and fan-beam sensors gather topographic and range information about the object's surface. After building a three-dimensional model of the asteroid during this mapping phase, the MUSES-C project team will decide on suitable sites for surface exploration. To capture samples, the spacecraft carries a horn that will be brought to the surface as the spacecraft makes a close approach. A small pyrotechnic charge will then fire a bullet into the surface, and fragments of the impact will be captured by

the horn and funneled into a sample container. The aim is to carry out several sample extractions, each taking only a second or so, from different locations. On each sampling attempt, the spacecraft will begin its approach at some distance from the asteroid, move in to collect the sample, and then back off to the same distance. In the final phase of the mission, MUSES-C will return to the vicinity of Earth and release a small reentry capsule containing the collected samples. Reentering at over 12 km/s directly from its interplanetary return trajectory, the capsule will use an advanced heat-shield as protection and finally deploy a parachute for soft-landing.

Musgrave, (Franklin) Story (1935–)

An American astronaut and a veteran of six Space Shuttle flights. Musgrave holds numerous degrees, including a B.S. in mathematics from Syracuse University (1958), an M.B.A. from the University of California, Los Angeles (1959), a B.A. in chemistry from Marietta College (1960), an M.D. from Columbia University (1964), an M.S. in physiology and biophysics from the University of Kentucky (1966), and an M.A. in literature from the University of Houston (1987). He joined the Marine Corps in 1953 and served as an aviation electrician and instrument technician and as an aircraft crew chief. He was selected by NASA as a scientist-astronaut in 1967, and he served as backup science-pilot for the first Skylab mission; CAPCOM (Capsule Communicator) for the second and third Skylab missions and several Space Shuttle missions; a mission specialist on Shuttle flights STS-6 (1983), STS-5F/Spacelab-2 (1985), STS-33 (1989), STS-44 (1991), and STS-80 (1996); and payload commander on STS-61 (1993). He left NASA in August 1997 and now works at Walt Disney Imagineering in the research and development division.

Musson

See GEO-IK.

MUSTARD (Multi-Unit Space Transport and Recovery Device)

A design for a reusable space vehicle proposed by the British Aircraft Corporation in 1964–1965. MUSTARD consisted of three fully reusable delta-shaped craft mated

Story Musgrave Musgrave, anchored on the end of the Space Shuttle's Remote Manipulator System, prepares to be lifted to the top of the Hubble Space Telescope to carry out servicing tasks in December 1993. *NASA*

together. Two of the components served as boosters for the third stage, which was orbital. These boosters could then be flown back either by pilots or by remote control. Fuel could be pumped from the boosters into the orbiter, so the orbiter could reach orbit with a full fuel load and therefore, if required, continue on to the Moon.

Musudan

Launch site at 40.5° N, 129.5° E, from which North Korea attempted to launch its first satellite, **Kwangmyongsong** 1, in 1998.

MX-324

An experimental American rocket-powered aircraft of World War II built by Northrop Corporation with an Aerojet engine. The MX-324 had a length of 4.3 m, a wingspan of 11.6 m, a top speed of 885 km/hr, and a maximum range of 1,600 km. Its maiden flight was on July 5, 1944, at Harper Dry Lake, California. The test pilot, Harry Crosby, lay prone in order to withstand the high acceleration. Only three of the planes were ever built: two were severely damaged during testing, and a third was dismantled.

Myojo

A NASDA (National Space Development Agency, of Japan) payload attached to the H-2 launch vehicle to monitor the rocket's performance, confirm the accuracy of the H-2 orbit injection, and collect data on the environment of the payload equipment; "Myojo" means "Venus." It was also known as the Vehicle Evaluation Payload (VEP).

Launch
  Date: February 3, 1994
  Vehicle: H-2
  Site: Tanegashima
Orbit: 449 × 36,261 km × 28.6°
Total mass: 2,391 kg

N

N series (Japanese launch vehicles)

Versions of the **Delta** rocket built under license in Japan using American and Japanese components and used to launch **NASDA** (National Space Development Agency) satellites. The N-1 launched seven satellites from 1975 to 1982 and could place up to 130 kg in geostationary orbit (GSO). The larger N-2 launched eight satellites from 1980 to 1986 and had a GSO payload capacity of 350 kg. The "N" prefix is taken from the first letter of "Nippon" (Japan). See **Japanese launch vehicles**.

N-1

See article, pages 290–291.

NACA (National Advisory Committee for Aeronautics)

A U.S. civilian agency for aviation research, chartered in 1915 and operational from 1917 to 1958, when it was absorbed into the newly formed **NASA**. NACA concentrated mainly on laboratory studies at its **Langley**, **Ames**, and **Lewis** centers, gradually shifting from aerodynamic research to military rocketry as the Cold War brought an increasing priority to missile development. Major NACA contributions to the military missile programs came from 1955 to 1957. Materials research led by Robert **Gilruth** at Langley confirmed **ablation** as a means of controlling the intense heat generated by warheads and other bodies reentering Earth's atmosphere; H. Julian Allen at Ames demonstrated the blunt-body shape as the most effective design for reentering bodies; and Alfred Eggers at Ames did significant work on the mechanics of ballistic reentry.[134]

Nadezhda

Soviet maritime navigation satellites, adapted from the **Tsikada** series, that can pinpoint distress signals from ships and aircraft. The original Nadezhda ("hope") satellites, launches of which began in 1982, carried the COSPAS/SARSat international space search and rescue system. In 1995, the Nadezhda-M series was introduced with the Kurs system for better location of air, sea, and ground craft in distress and for transmission of digital data to the Kurs Center.

NACA (National Advisory Committee for Aeronautics)
NACA's inaugural meeting in 1915. Seated left to right: William Durand, Stanford University; S. W. Stratton, director, Bureau of Standards; Brig. Gen. George P. Scriven, chief signal officer, War Dept.; C. F. Marvin, Chief, U.S. Weather Bureau; Michael I. Pupin, Columbia University. Standing: Holden C. Richardson, naval instructor; John F. Hayford, Northwestern University; Capt. Mark L. Bristol, director of Naval Aeronautics; Lt. Col. Samuel Reber, Signal Corps. Also present at the first meeting were Joseph S. Ames, Johns Hopkins University, and B. R. Newton, asst. secy. of the treasury. *NASA*

N-1

The Soviet counterpart to the **Saturn** V, intended as the launch vehicle for the Soviet Union's attempt at a manned landing on the Moon (see **Russian manned lunar programs**). The N-1 would have been the world's most powerful booster, but it was beset by technical problems and never made a successful flight. Work on it began in 1962 under the direction of Sergei **Korolev**.

On November 25, 1967, a mockup of the N-1 was installed on launch pad Number 1 at Baikonur for 17 days of electrical tests. The first flight-ready vehicle was erected on the same pad on May 7, 1968, but it had to be returned to the assembly building after cracks were found in the first-stage structure. After repairs, it was again set up on the launch pad in November 1968, only to be replaced with another mockup for further tests and launch team training. Finally, in mid-January 1969, N-1 Number 3L was moved to the pad ready for its first test flight. The test was a huge gamble since, in a desperate bid to get to the Moon first, the large cluster of first-stage engines had never been tested as a group. The payload was originally to have been an unmanned Lunar Orbiter Cabin (LOK) and Lunar Cabin (LK) designed to carry Soviet cosmonauts to a Moon landing. But as neither the LOK nor the LK were yet ready to fly, the payload was switched to a modified version of the circumlunar spacecraft L-1 known as the L-1S. After four weeks of on-pad preparation, the first N-1 was ready to go. Hours before the launch, the chief designer, Valery **Mishin**, who had taken over the project after Korolev's death in 1966, went to the launch pad while propellants were being loaded and christened the rocket by breaking a bottle of champagne against its hull. On February 21, 1969, the N-1 lifted off into clear skies—and trouble. Seconds into the flight, two of the 30 first-stage NK-33 engines shut down due to a problem with the KORD control system that guided the vehicle. The N-1 was designed to operate with as many as four first-stage engines out of action, so the flight continued with the other engines compensating for the loss. At T + 25 seconds the engines were throttled back as planned as the N-1 passed through the point of maximum dynamic pressure, and at T + 65 seconds they were throttled back up to full power. At this point the lack of adequate

testing of the main engines became apparent. The engines throttled up more quickly than expected, causing stronger than planned vibrations. A liquid oxygen (LOX) pipeline burst and started a fire in the engine compartment before KORD could shut down the affected engine. The surrounding engines and turbopumps quickly overheated and exploded. At T + 70 seconds, KORD finally shut down all engines, and the launch escape system was activated. The L-1S capsule was safely recovered about 35 km downrange, but the now powerless N-1 slammed into the Kazhak steppes and exploded some 48 km from its launch pad.

N-1 FACTS
Length: 105 m
Payload: 95 tons (LEO), 30 tons (Moon)
Number of engines
  First stage: 30
  Second stage: 8
  Third stage: 4
Propellant (all stages): RP-1 and LOX
First stage thrust: 44,100,000 N

In a last-ditch effort to win the Moon race, a second N-1 was modified and prepared for launch. Following the successful tests of the American Lunar Module in Earth orbit during Apollo 9 in March and the test in lunar orbit by Apollo 10 in May, only a failure of Apollo 11 and a successful N-1 test would leave the Soviets with any chance of being the first to land a man on the Moon. On July 3, the second N-1, Number 5L, took off from pad Number 1 with another L-1S as its primary payload. What followed was an even greater disaster than the first. At the very point of liftoff, debris in the LOX tank was sucked into one of the first-stage engines, causing its turbopump to explode. This led to a second engine failure, which in turn triggered a chain reaction, destroying several neighboring engines and damaging the KORD control system. The launch escape system was activated and carried the L-1S payload to safety. However, the fully fueled N-1 fell back onto its launch pad just 18 seconds after launch, destroying the rocket and pad Number 1, heavily damaging pad Number 2, and end-

N-1 A complete N-1 and its launch tower on the right, and another N-1 launch tower and a pair of taller towers, which provided protection from lightning strikes, on the left. The left-hand launch tower was destroyed during a launch failure on July 3, 1969. *Energia RSC*

ing any hope that the Soviets could reach the Moon ahead of the United States.

After two years of rocket modification and pad reconstruction, the N-1 was ready to launch again. However, there would be only two further flights, on June 27, 1971, and November 23, 1972. Both again ended in first-stage failures, at T + 51 seconds and T + 107 seconds, respectively. In early 1974, the N-1 project was abandoned and the six remaining rockets of the ten originally built dismantled.[185, 186, 317]

nadir

For spacecraft, the arbitrarily defined "down" direction; usually defined as the direction pointing toward the center of the body the spacecraft is orbiting. The opposite of nadir is **zenith**.

nanosat

A satellite with an on-orbit mass of 1 to 10 kg. See **satellite mass categories**.

Nanosat Constellation Trailblazer (NCT)

A NASA mission involving the launch of three identical small satellites, each weighing only 20 kg and the size of a hatbox, to test methods of operating a constellation of several spacecraft as a single system by autonomous decision-making and intersatellite communication. The spacecraft will maintain continuous contact with one another, sharing information and reconfiguring onboard instruments and systems to behave as a single unit. Throughout the mission, the three satellites will measure the effect of solar activity on Earth's magnetosphere. Scheduled for launch in 2003, the Trailblazer satellites are designed to pave the way for a much larger network of even smaller satellites that will make up **MagConst** in

2010. The Nanosat Constellation Trailblazer is the fifth mission in NASA's **New Millennium Program**.

NASA (National Aeronautics and Space Administration)

A civilian agency of the U.S. government, formally established on October 1, 1958, under the National Aeronautics and Space Act of 1958. It absorbed the former **NACA** (National Advisory Committee for Aeronautics), including its 8,000 employees and three major research laboratories–**Langley** Aeronautical Laboratory, **Ames** Aeronautical Laboratory, and **Lewis** Flight Propulsion Laboratory–and two small test facilities. Under its first administrator, T. Keith **Glennan**, NASA also incorporated several organizations involved in space exploration projects from other federal agencies to ensure that a viable scientific program of space exploration could be reasonably conducted over the long term. Glennan brought in part of the Naval Research Laboratory and created for its use the **Goddard Space Flight Center**. He also incorporated several disparate satellite programs, two lunar probes, and the important research effort to develop a million-pound-thrust, single-chamber rocket engine from the Air Force and the Department of Defense's Advanced Research Projects Agency. In December

1958, Glennan also acquired control of **JPL** (Jet Propulsion Laboratory), a contractor facility operated by the California Institute of Technology. In 1960, Glennan obtained the transfer to NASA of the **Army Ballistic Missile Agency**, which was then renamed the **Marshall Space Flight Center**. By mid-1960, Glennan had secured for NASA primacy in the federal government for the execution of all space activities except reconnaissance satellites, ballistic missiles, and a few other space-related projects, most of which were still in the study stage, that the DoD controlled.

The functions of the organization were conceived to plan, direct, and conduct all American aeronautical and space activities except those that are primarily military. NASA's administrator is a civilian appointed by the president, with the advice and consent of the Senate. The administration arranges for the scientific community to take part in planning scientific measurements and observations to be made through the use of aeronautical and space vehicles, and provides for the dissemination of the data that result. Under the guidance of the president, NASA helps develop programs of international cooperation in space.

With the advent of the Space Shuttle, NASA became more frequently involved in military activities despite its original intent as a civilian agency. Because of the long delay caused by the 1986 *Challenger* disaster, however, the military began expanding its own fleet of booster rockets. In 1996, NASA announced a $7-billion, six-year contract under which the agency would gradually turn over the routine operation of the Shuttle program to private industry.

In 1998, NASA established the Astrobiology Institute at its Ames Research Center to enhance research for new instruments and space probes to search for life in the Solar System and beyond. This Institute is a consortium of academic institutions, including the University of California at Los Angeles, Harvard University, and the University of Colorado, as well as the private sector and several NASA field centers.

NASA Institute for Advanced Concepts (NIAC)

A body run by USRA (Universities Space Research Association), under contract to NASA, to provide an independent forum for the external analysis and definition of space and aeronautics advanced concepts. Through a series of open solicitations, NIAC seeks proposals from the science and engineering community to develop advanced concepts in aeronautics and space. According to the terms of the contract, these advanced concepts should be "new," "revolutionary," "imaginative," and aimed at becoming aerospace missions over the next 10 to 40 years.

NASDA (National Space Development Agency)

One of two Japanese space agencies, the other being **ISAS** (Institute of Space and Astronautical Science). Founded in October 1969 by the Space Activity Commission, NASDA replaced NSDC (National Space Development Center). It is responsible for all of Japan's communications, weather, Earth observation, and technology demonstration satellites, together with the rockets used to launch them. NASDA satellites and spacecraft include: **ETS, OICETS, COMETS, DRTS, BS-, CS-, ECS, JAS**-1, **ALOS, TRMM, ADEOS, JERS**-1, **MOS**-1, **GMS, EGS, ISS, SELENE, MDS**-1, and **SFU** (see **Japanese satellite names**). Its launch vehicles include the **N-series**, the **H-series**, and the **J-1** (see **Japanese launch vehicles**).

National Aeronautics and Space Council

The Statutory Advisory Council to the president, consisting of the secretary of defense, the administrator of NASA, the chairman of the Atomic Energy Commission, the secretary of state, and other members selected by the president.

National Aerospace Plane (NASP)

See **X-30**.

National Association of Rocketry (NAR)

The largest sport rocket organization in the United States. Based in Altoona, Wisconsin, it promotes all aspects of flying model and high-power sport rockets.

National Reconnaissance Office (NRO)

A U.S. Department of Defense agency that designs, builds, and operates American reconnaissance (spy) satellites. These satellites are used by the CIA (Central Intelligence Agency) and other members of the American intelligence community. In recent years, NRO has declassified some of its earlier missions, including the 1960–1972 **Corona** photoreconnaissance operation. In December 1996, NRO also gave, for the first time, advance notice of the launch of one its spacecraft.

National Rocket Club

See **National Space Club**.

National Space Club (NSC)

A prospace lobbying group, open to general membership but composed mainly of professionals in the aerospace industry. Active in Washington, D.C., it acts as an information conduit and organizes monthly meetings and an annual "Outlook on Space" conference. It was formed as the National Rocket Club on October 4, 1957–the day Sputnik 1 was launched. See **National Space Institute**.

National Space Institute (NSI)

An organization founded by officers of the **National Space Club** to maintain public support for the U.S. space program; its first president was Wernher **von Braun**. The NSI was initially incorporated as the National Space Association in June 1974, but was renamed in April 1975. In April 1987, the NSI merged with the L5 Society to form the **National Space Society**.

National Space Science Data Center (NSSDC)

A department of NASA's **Goddard Space Flight Center**, founded in 1966, that is the repository for data from American space missions, including images from planetary probes.

National Space Society (NSS)

A prospace group formed in 1987 as a result of a merger between the **National Space Institute** and the **L5 Society**. Based in Washington, D.C., it has an extensive network of local chapters. The NSS supports a general agenda of space development and human space presence, sponsors a major annual space development conference, and publishes the bimonthly magazine *Ad Astra*. It is associated with the political lobbying organizations **Spacecause** and **Spacepac**.

NATO (satellites)

A series of military communications satellites, encrypted for use by NATO (North Atlantic Treaty Organization) and designed to link the capital cities of NATO countries. NATO-1 and -2 were launched in 1970 and 1971, respectively, and four of the much larger NATO-3 satellites between 1976 and 1984. Each NATO-3 could support hundreds of users and provide voice and facsimile services in UHF- (ultra-high frequency), X-, and C-bands. Two NATO-4 satellites, which operate in the same bands but have still more channels, were launched in 1991 and 1993.

Nauka

Scientific subsatellites carried aboard **Zenit** reconnaissance satellites. From the outset, Zenits carried small supplemental science packages—for example, for making measurements of meteoroids or cosmic rays. But the first Nauka ("science") autonomous subsatellite was flown in 1968 aboard Zenit number 80. The Nauka containers served a dual purpose. Mounted on the forward end of the Zenit's reentry sphere, they provided ballast during the primary spacecraft's mission. When the Zenit had completed its work, the Nauka would be released. Over 40 Naukas were installed on Zenit-2 and Zenit-2M satellites from 1968 to 1979 and used to study Earth's radiation belts, collect data on the Earth-space interface, and develop and test new instruments. Twenty-three autonomous Nauka subsatellites undertook research in geophysics, meteorology, and cosmic rays; others tested new systems for the Yantar reconnaissance satellite.

Navaho

One of three strategic surface-to-surface cruise missiles developed by the United States immediately after World War II, the others being the **Matador** and the **Snark**. Built by the Glenn L. Martin Company, the Navaho had a gross weight of 5,400 kg and the ability to carry a nuclear warhead. However, its limited range of only 1,050 km meant that in order to strike at the Soviet Union it had to be forward-based in areas such as Japan and Germany. The Navaho became operational in 1955 and in the early 1960s was replaced by the Mace missile, which had a longer range, greater speed, and superior accuracy. Although the Navaho program was canceled in 1957, it significantly influenced the development of large liquid-propellant rocket engine technology in the United States. From the Navaho were derived the engines for the **Redstone**, **Thor**, **Jupiter**, Atlas, **Titan** I, and **Saturn** I rockets. The Navaho program also led to fuel tank fabrication techniques, inertial and stellar navigation, and other technologies used in later vehicles. See **guided missiles, postwar development**.

navigational satellites

Spacecraft that are designed to assist in air, land, and oceanic navigation. They include satellites such as **GLONASS**, **GPS** (Global Positioning System), and **Transit**.

Navsat (Naval Navigation Satellite)

See **Transit**.

Navstar-GPS (Navigation Satellite Time and Ranging Global Positioning System)

See **GPS**.

NEAP (Near-Earth Asteroid Prospector)

A private venture to send a small spacecraft to an asteroid. NEAP, which would be the first commercial deep-space mission, is being developed by SpaceDev, a Colorado-based company, together with scientists from the University of California, San Diego, and other academic institutions. First conceived in 1997 as a 350-kg minisatellite to be launched on a Russian Eurokot, this low-cost mission has evolved to a 200-kg microsatellite to be launched as a secondary payload on an Ariane 5 in 2002–2005. Although plans have still to be finalized, the proposed target is the near-Earth asteroid Nereus.

NEAR-Shoemaker (Near-Earth Asteroid Rendezvous)

The first spacecraft to orbit and (though this was not originally planned) to land on an asteroid—Eros. Built and operated by the Applied Physics Laboratory at Johns Hopkins University, the probe was originally known simply as NEAR (Near-Earth Asteroid Rendezvous) but was renamed NEAR-Shoemaker in memory of the American geologist Eugene Shoemaker (1928–1997). It was the inaugural mission in NASA's **Discovery Program** and the first probe powered by solar cells to operate beyond the orbit of Mars.

On June 27, 1997, NEAR-Shoemaker flew by the asteroid 253 Mathilde at a distance of 1,200 km and found it to be composed of extremely dark material with many large impact craters, including one about 9 km deep. A deep-space maneuver in July 1997 brought the probe back around Earth on January 23, 1998, for a **gravity assist** that put the spacecraft on course for its rendezvous with the Manhattan-sized asteroid 433 Eros.

NEAR's instruments included a multispectral imager, a telescope with a CCD (charge-coupled device) array to determine the size, shape, and spin characteristics of the asteroid and to map its surface; an X-ray/gamma-ray spectrometer to determine the surface/near-surface elemental composition; a near-infrared spectrometer to map the mineralogical composition; a magnetometer to measure the magnetic field of Eros; and a laser altimeter to measure the distance between the spacecraft and the asteroid's surface.

NEAR-Shoemaker The location of the landing site of NEAR-Shoemaker on the asteroid Eros. *NASA*

Launch
  Date: February 17, 1996
  Vehicle: Delta 7925
  Site: Cape Canaveral
Size: 2.8 × 1.7 m
Mass: 818 kg (total), 55 kg (science payload)

NEAR entered an orbit of 323 × 370 km around Eros on February 14, 2000, then moved to gradually smaller orbits over the next year or so, returning a total of 160,000 images. During the final days of its mission, NEAR maneuvered to within 24 km and then, despite the fact that it had not been built as a lander, became the first spacecraft actually to touch down on an asteroid. NEAR-Shoemaker was designed only to orbit Eros. But with all of its objectives fulfilled, it was decided to try to bring the craft in for what mission controllers called a "controlled crash." In the final moments before it landed on February 12, 2001, NEAR-Shoemaker returned pictures showing surface details as small as a few tens of centimeters across. Finally, in one of the great moments of space exploration, the probe landed so smoothly that its radio beacon continued to send out a signal from its new home.

Nedelin Catastrophe

See **Marsnik**.

negative *g*

Acceleration that a subject experiences as acting from below (feet-to-head direction). Also known as "eyeballs up." See also **positive *g*** ("eyeballs down"), **prone *g*** ("eyeballs out"), and **supine *g*** ("eyeballs in").

Neptune Orbiter

An important component of NASA's future investigation of the outer Solar System, including Neptune's moon Triton, which may be an icy, organic-rich, captured Kuiper Belt object. The Neptune Orbiter is identified in NASA's Office of Space Science Strategic Plan as a potential mission beyond 2007 but remains in the early concept definition phase.

NERVA (Nuclear Engine for Rocket Vehicle Application)

A program conducted by NASA and the U.S. Atomic Energy Commission to develop a **nuclear propulsion** system for use on long-range manned missions. Canceled in 1969 because of high projected development costs, it was intended to result in large, heavy-lift launch vehicles capable of supplying a then-planned lunar base. NERVA was also to have played an integral part in a proposed manned mission to Mars, which was at one time scheduled for launch in November 1981 with a manned landing in August 1982.

NetLander

A European mission to Mars, led by CNES (the French space agency), that will focus on investigating the interior of the planet and the large-scale circulation of the atmosphere. NetLander consists of four landers that will be separated from their parent spacecraft and targeted to their locations on the Martian surface several days prior to the main craft's arrival in orbit. Once on the ground, the landers will form a linked network, each deploying a network science payload with instrumentation for studying the interior of Mars, the atmosphere, and the subsurface, as well as the ionospheric structure and geodesy. The mission is scheduled for launch by an Ariane 5 in 2007.

Neutral Buoyancy Simulator (NBS)

A 5.3-million-liter water tank at the **Marshall Space Flight Center**, first installed in the 1960s, that enables astronauts to train for extravehicular activities in conditions that simulate weightlessness. The astronaut is submerged fully suited with weights attached to maintain a stable position in the water. Floating at a predetermined depth, the experience is similar to that of floating in space. Professional divers with scuba gear are in attendance.

neutral point

A point in space at which an object experiences no overall gravitational force. A unique point would exist between any two *static* massive bodies. But when bodies are orbiting each other, there are five points, known as **Lagrangian point**s, at which the gravitational forces balance.

New Millennium Program (NMP)

A NASA program focused on the development and validation in space of advanced technologies that will be used routinely aboard unmanned spacecraft in the early part of the twenty-first century. Prototype scientific instruments, communications equipment, computer systems, and propulsion units will be tested during a series of space missions that will also return data of relevance to understanding how the planets formed and life evolved. The first five missions in the program are **Deep Space 1**, the **Mars Microprobe Mission** (Deep Space 2), **Starlight** (Space Technology 3), **Champollion** (Space Technology 4)—which was later canceled—and the **Nanosat Constellation Trailblazer** (Space Technology 5).

Neutral Buoyancy Simulator Astronauts training in 1985 with a space station mockup in the NBS. *NASA*

Newell, Homer E. (1915–1983)

An internationally recognized authority in atmospheric and space science. Newell earned his Ph.D. in mathematics at the University of Wisconsin in 1940 and served as a theoretical physicist and mathematician at the Naval Research Laboratory from 1944 to 1958. During part of that period, he was science program coordinator for Project **Vanguard** and acting superintendent of the atmosphere and astrophysics division. In 1958, he transferred to NASA to assume responsibility for planning and developing the new agency's space science program. He soon became deputy director of spaceflight programs. In 1961, he assumed directorship of the office of space sciences, and in 1963, he became associate administrator for space science and applications. Over the course of his career, he became an internationally known authority in the field of atmospheric and space sciences, as well as the author of numerous scientific articles and seven books, including *Beyond the Atmosphere: Early Years of Space Science*.[217] He retired from NASA in late 1973.

Newton, Isaac (1643–1727)

A great English scientist who organized our understanding of physical motion into three scientific laws, now known as **Newton's laws of motion**. Among other things, these laws explain the principle of rockets and how they are able to work in the vacuum of outer space. **Newton's law of universal gravitation** forms the basis of **celestial mechanics** and our understanding of the movement of satellites, both natural and artificial. See **Newton's orbital cannon**.

Newton's law of universal gravitation

Two bodies attract each other with equal and opposite forces; the magnitude of this force, F, is proportional to the product of the two masses, m_1 and m_2, and is also proportional to the inverse square of the distance, r_2 between the centers of mass of the two bodies. This leads to the equation:

$$F = \frac{Gm_1m_2}{r^2}$$

where G is the **gravitational constant**.

Newton's theory of gravity was superseded by Einstein's **general theory of relativity**.

Newton's laws of motion

Three laws that form the foundation of classical mechanics and describe how things move. The laws, which introduce the concepts of **force** and **mass**, are: (1) A body continues in its state of constant velocity (which may be zero) unless it is acted upon by an external force; (2) For an unbalanced force F acting on a body, the acceleration a produced is proportional to the force impressed, the constant of proportionality being the inertial mass m of the body; that is, $F = ma$; and (3) In a system where no external forces are present, every action force is always opposed by an equal and opposite reaction. Newton's laws (2) and (3) in Mach's formulation reduce to: "When two small bodies act on each other, they accelerate in opposite directions and the ratio of their accelerations is always the same."

Newton's orbital cannon

Isaac Newton discussed the use of a cannon to place an object in orbit in his *Principia Mathematica* (1687)–the book that defined classical physics and provided the theoretical basis for space travel and rocketry. Newton used the following thought experiment to explain the principle of Earth orbits. Imagine a mountain so high that its peak is above Earth's atmosphere; on top of this mountain is a cannon that fires horizontally. As more and more charge is used with each shot, the speed of the cannonball will be greater, and the projectile will fall to the ground farther and farther from the mountain. Finally, at a certain speed, the cannonball will not hit the ground at all but will fall toward the circular Earth just as fast as Earth curves away from it. In the absence of drag from the atmosphere, it will continue forever in Earth orbit.

NGSS (Next Generation Sky Survey)

A supercooled **infrared** space telescope designed to survey the entire sky with a sensitivity more than 1,000 times greater than that of previous missions. NGSS would result in the discovery of millions of new cosmic sources of infrared radiation, including many circumstellar disks, some of which are in the process of forming planetary systems. NGSS, which would be led by Edward L. Wright of the University of California, Los Angeles, is among four mission proposals chosen by NASA as candidates for two **MIDEX** (Medium-class Explorer) flights to be launched in 2007 and 2008.

NGST (Next Generation Space Telescope)

A large space telescope, to be known as the James Webb Space Telescope after NASA's second Administrator, that is scheduled for launch in 2010 and will be the successor to the **Hubble Space Telescope** (HST). It will have a primary mirror 6 m in diameter (2.5 times as large as HST's) and will operate at wavelengths between those at the red end of the visible spectrum and those at the middle of the infrared range. Located at the second **Lagrangian point**, 1.5 million km from Earth, so that it can point permanently away from the infrared glow of the Sun and Earth, it will not be serviceable by the Space Shuttle.

Nimbus

A series of seven NASA **weather satellite**s, launched between August 1964 and October 1978 into polar orbits, that photographed cloud cover and monitored air pollution. Nimbus helped develop the technology that would be used in later meteorological programs.

Nimiq

Canada's first **direct broadcast satellite**, launched by Telesat on May 20, 1999; "nimiq" is Inuit for "something that unifies." It has 32 Ku-band (see **frequency bands**) transponders and has been designed primarily to serve the Canadian market, with the capability of offering full North American coverage in the future.

NIMS (Navy Ionsopheric Monitoring System)

See **Transit**.

nitric acid (HNO₃)

A commonly used **oxidizer** in **liquid-propellant rocket engine**s between 1940 and 1965. It most often took the form of RFNA (red fuming nitric acid), containing 5% to 20% dissolved nitrogen dioxide. Compared to concentrated nitric acid (also known as white fuming nitric acid), RFNA is more energetic and more stable to store, but it produces poisonous red-brown fumes. Because nitric acid is normally highly corrosive, it can be stored and piped by only a few materials, such as stainless steel. However, the addition of a small concentration of fluoride ions inhibits the corrosive action and gives a form known as IRFNA (inhibited red fuming nitric acid).

nitrogen tetroxide (N₂O₄)

A yellow-brown liquid that is among the most common storable **oxidizer**s used by **liquid-propellant rocket engine**s today. For example, it is used with MMH (see **hydrazine**) in the Space Shuttle orbital maneuvering system. Although it can be stored indefinitely in sealed containers, its liquid temperature range is narrow and it is easily frozen or vaporized.

NNSS (Navy Navigational Satellite System)

See **Transit**.

NOAA (National Oceanic and Atmospheric Administration)

A U.S. government body that manages and operates environmental and meteorological satellites and provides data to users worldwide. Having absorbed ESSA (the Environmental Science Services Agency) in 1970, it took over responsibility for the **TIROS** program. All successful launches in this program now carry the NOAA designation.

non-coherent

A communications mode wherein a spacecraft generates its **downlink** frequency independent of any **uplink** frequency.

Noordung, Herman (1892–1929)

The pseudonym of Herman Potocnik, a relatively obscure officer in the Austrian Imperial Army who became an engineer and, encouraged by Hermann **Oberth**, published in 1928 a seminal book, *Das Problem der Befahrung des Weltraums* (The Problem of Space Travel: The Rocket Motor),[220, 221] which focuses largely on the engineering aspects of **space station**s. In it, Noordung deals with issues such as weightlessness, space communications, maintaining a habitable environment for the crew, and extravehicular activity. Noordung's proposed design consists of a doughnut-shaped structure for living quarters, a power-generating station attached to one end of the central hub, and an astronomical observation station. He was among the first to suggest a wheel-shaped design for a space station to produce **artificial gravity**; he also argued the scientific value of such a station in a synchronous orbit above Earth.

North American Aviation

A U.S. aircraft and aerospace company, founded in 1928, that played a central role in the early American space program. It was responsible for the development of the **X-15** rocket plane and, through its Rocketdyne Division, the propulsion system for the **Redstone** rocket, the **Apollo** Command and Service Modules, the second stage of the **Saturn** V, and the **F-1** and **J-2** rocket engines. Earlier, North American Aviation had provided some of the best-known combat planes of World War II and the postwar years, including the F-51 Mustang, the B-25 Mitchell bomber, and the F-86 Sabre. In 1966, North American Aviation merged with **Rockwell** Standard to form North American Rockwell, which, in 1976, was renamed Rockwell International. In 1996, Rockwell became part of **Boeing**.

nosecone

The cone-shaped leading end of a rocket vehicle, consisting of a chamber in which a satellite, instruments, organisms, or auxiliary equipment may be carried, or of an outer surface built to withstand high temperatures generated by aerodynamic heating.

NOSS (Naval Ocean Surveillance Satellite)

U.S. Navy ocean surveillance satellites. NOSS detected the location of naval vessels using radio interferometry, and consisted of a main spacecraft and several subsatellites linked by fine wires several hundred meters apart.

Eight of the spacecraft were placed in roughly circular orbits some 1,100 km high by **Atlas** F and H launch vehicles between 1976 and 1987.

Nova (rocket)

A proposed post-**Saturn** launch vehicle for space missions of 1970 and beyond. Larger even than the Saturn V, the Nova would have been capable of placing a manned spacecraft on a **direct ascent** to the Moon. See **Apollo**.

Nova (satellites)

A U.S. Navy satellite navigation system compatible with the **Transit** system, of which 15 spacecraft, together with three satellites of the **Transit Improvement Program**, had been placed in orbit by the time Nova 1 was launched. Nova was superseded by the **Navstar-GPS**. (See table, "Nova Satellites.")

Launch
  Vehicle: Scout G
  Site: Vandenberg Air Force Base

Nozomi

A Japanese Mars-orbiter, designed and launched by **ISAS** (Institute of Space and Aeronautical Science), to study the Martian upper atmosphere and its interaction with the solar wind and to develop technologies for use in future planetary missions. Its science instruments will measure the structure, composition, and dynamics of the ionosphere, aeronomy effects of the solar wind, the escape of atmospheric constituents, the intrinsic magnetic field, the penetration of the solar-wind magnetic field, the structure of the magnetosphere, and dust in the upper atmosphere and in orbit around Mars. It will also send back images of the surface.

Nozomi used several lunar and Earth **gravity-assists** to boost its energy for the cruise to Mars. A lunar swingby on September 24, 1998, was followed by another on December 18 to increase the apogee (high point) of its orbit. An Earth flyby on December 20 was coupled with a seven-minute burn of the probe's main engine to place Nozomi on an escape trajectory toward Mars. The scheduled arrival was October 11, 1999, but the Earth swingby

left the spacecraft with insufficient acceleration, and two course correction burns on December 21 used more propellant than planned, leaving the spacecraft short of fuel. Thus, a new plan was made in which Nozomi will remain in solar orbit for an additional four years and encounter Mars at a slower relative velocity in December 2003. Eventually, Nozomi will maneuver into a highly elliptical orbit about Mars with a periapsis (low point) of 150 km and an apoapsis (high point) of 15 Mars-radii. The periapsis portion of the orbit will allow in situ measurements of the thermosphere and lower exosphere, and remote sensing of the lower atmosphere and surface, while the more distant parts of the orbit will allow study of the ions and neutral gas escaping from Mars and their interactions with the solar wind. The nominal mission is planned for one Martian year (approximately two Earth years). An extended mission may allow a further operation of three to five years. Nozomi, whose name means "hope," was known before launch as **Planet**-B.

Launch
  Date: July 3, 1998
  Vehicle: M-5
  Site: Kagoshima
Size: 1.6 × 0.58 m
Mass: 540 kg (total at launch), 33 kg (science instruments), 282 kg (propellant)

nozzle

A specially shaped tube, or duct, connected to the **combustion chamber** of a rocket engine in which the gases produced in the chamber are accelerated to high velocities, thereby efficiently converting the pressure of the exhaust into **thrust**.

The nozzle must be carefully designed in order to change the high-pressure, low-velocity gas at the nozzle entrance (inside the combustion chamber) to a low-pressure, high-velocity flow at the exit. From the combustion chamber the hot gas is constricted at the throat—the nozzle's narrowest point—where it reaches Mach 1 (the local speed of sound). Then the nozzle expands along a carefully controlled contour, allowing the gas to gain speed and lose pressure. The larger the nozzle's cross sec-

Nova Satellites

Spacecraft	Launch Date	Orbit	Mass (kg)
Nova 1	May 25, 1981	1,163 × 1,183 km × 90.1°	170
Nova 2	Oct. 11, 1984	1,149 × 1,199 km × 89.9°	165
Nova 3	Jun. 16, 1988	1,150 × 1,198 km × 90.0°	174

nozzle The Space Shuttle main engine test firing in 1981. *NASA*

Like the combustion chamber, the nozzle throat gets very hot. Nozzle throats are often cooled by circulating fuel directly behind the pressure wall to cool it. The fuel passes through while it is still cool on its way to the combustion chamber. This has the added bonus of preheating the fuel before combustion, making more energy available from the combustion to provide thrust. Recently, new materials have become available, including ceramics and composites, that can withstand extremely high temperatures. These materials often slowly ablate under the extreme conditions in the nozzle throat of a high-performance rocket; however, the ablation rates may be tolerable in a rocket that is only fired once. In the case of a rocket designed for multiple firings, either in the same mission or in multiple missions, cooling rather than ablation is likely to be the method of choice.

nozzle area ratio
The ratio of a **nozzle**'s throat area to the exit area. The ideal nozzle area ratio allows the gases of combustion to exit the nozzle at the same pressure as the ambient pressure.

nozzle efficiency
The efficiency with which the **nozzle** converts the potential energy of a burned fuel into kinetic energy for **thrust**.

nozzle exit angle
The angle of divergence of a **nozzle**.

nuclear detection satellites
Spacecraft designed to detect nuclear explosions on the ground or in the atmosphere. Although primarily intended to monitor nuclear treaty compliance, these satellites have also been used to detect and observe galactic events such as supernovae. See **Vela** and **Advanced Vela**.

nuclear fission
See **fission, nuclear**.

nuclear fuel
Fissionable material of reasonably long life, used or usable in producing energy in a nuclear reactor.

nuclear fusion
See **fusion, nuclear**.

nuclear power for spacecraft
Nuclear power has essentially three applications to spaceflight: to provide a source of heat to keep equipment warm (see **radioisotope heater unit** [RHU]), to provide a source of electricity to power equipment (see **radioisotope thermoelectric generator** [RTG]), or to provide a means of propulsion either directly (see **nuclear**

tion at the exit, the higher the speed and the lower the pressure that the gas can achieve at the exit. For optimum thrust, the gas pressure at the nozzle exit should be exactly equal to the outside air pressure. In the vacuum of space the outside pressure is zero, so it is impossible for this optimum to be achieved. The bigger the exit area of the nozzle, the closer the thrust will be to optimum; however, there is a point at which gains in thrust are offset by the extra mass needed to make the nozzle wider. Even during the atmospheric portion of a rocket's flight it is impossible to achieve theoretical optimum performance, because the outside air pressure changes as the vehicle climbs. All designers can do is target the performance of a nozzle to some average outside pressure.

When a nozzle ends before the gas reaches the pressure of the outside air, it is called an under-expanded nozzle. In the under-expanded case, the rocket design is not getting all the thrust that it can from the engine. When a nozzle is too large and keeps trying to expand the gas flow, at some point the rocket plume will separate from the wall inside the nozzle. This is called an over-expanded nozzle. The performance from an over-expanded nozzle is worse than in the under-expanded case, because the nozzle's large exit area results in extra drag.

propulsion) or indirectly (see **nuclear-electric propulsion**). Research on nuclear power systems for prospective space applications began in the 1950s. Early American research on RTGs and space nuclear reactors was conducted under the auspices of the **SNAP** program. More recently, the **SP-100** program focused on the design of larger reactors for use in space. With one exception, the United States has flown nuclear material aboard spacecraft only to power RTGs and RHUs: **SNAPSHOT** was the only American space mission ever to carry a working nuclear reactor. By contrast, numerous Soviet **RORSAT** missions have been reactor-powered.[86]

nuclear propulsion
The use of energy released by a nuclear reaction to provide thrust *directly,* as distinct from **nuclear-electric propulsion**. A nuclear propulsion system derives its thrust from the products of nuclear **fission** or **fusion**,[169] and was first seriously studied by Stanislaw Ulam and Frederick de Hoffman in 1944 as a spinoff of their work on the Manhattan Project. During the quarter century following World War II, the Atomic Energy Commission (superseded by the Department of Energy in 1974) worked with various federal agencies on a series of nuclear engine projects, culminating in **NERVA**.

One way to achieve nuclear propulsion is to heat a working fluid by pumping it through a nuclear reactor and then let the fluid expand through a nozzle. Considering that nuclear fission fuel contains more than a million times as much energy per unit mass as chemical fuel does, this sounds promising. But the approach is limited by the temperature at which a reactor and key components of a rocket, such as a nozzle, can operate. The best working fluid to use is hydrogen, because it is the lightest substance and therefore, at any given temperature, consists of the fastest-moving particles. Chemical rockets cannot produce hydrogen as an exhaust, because hydrogen is not the sole product of any practical chemical reaction. With unlimited nuclear power, however, it is not necessary to react or burn anything; instead, hydrogen gas could simply be heated inside a nuclear reactor and then ejected as a high-speed exhaust. This was the idea of the NERVA project.[1]

Other concepts in nuclear propulsion have sought to circumvent the temperature limitation inherent in circulating the working fluid around a reactor by harnessing the power of runaway nuclear reactions. The most important and promising of these is the **nuclear pulse rocket**.

nuclear pulse rocket
A rocket propelled by a rapid and lengthy series of small atomic or nuclear explosions. A variety of schemes have been proposed since the 1940s. Project **Orion** was the longest study—involving actual model test flights—of the nuclear pulse concept based on **fission**. Project **Daedalus** provided a counterpart based on **fusion**.

Nuclear Rocket Development Station (NRDS)
A NASA-AEC (Atomic Energy Commission) facility concerned with performing research and development work on nuclear-powered rocket engines, such as the KIWI series, to be used from upper-stage space flight propulsion. The test site is located at Jackass Flats, which is approximately 95 km northeast of Las Vegas, Nevada.

nuclear-electric propulsion (NEP)
A form of **electric propulsion** in which the electrical energy used to accelerate the propellant comes from a nuclear power source, such as a space-based nuclear reactor.

nucleus
The center of an **atom**, around which orbits a cloud of **electrons**. It consists of **protons** and **neutrons** bound together by the strong force. Under the right conditions, nuclei may undergo nuclear **fission** or **fusion**, with the release of large amounts of energy.

NUSAT (Northern Utah Satellite)
An air traffic control radar calibration satellite built by Weber State University (WSU) and Utah State University (USU) students and staff at Ogden, Utah. It was deployed from a modified Get-Away Special canister on the Space Shuttle *Challenger.* NUSAT measured 48 cm in diameter and was an 18-sided cylinder. It orbited for 20 months until reentering on December 15, 1986, and it demonstrated that satellites could be built small, simple, and at low cost for special applications. With this satellite, WSU and USU shared the claim to be the first American university to place a satellite in space.

Shuttle deployment
    Date: April 29, 1985
    Mission: STS-51B
Orbit: 318 × 339 km × 57.0°
Mass: 54 kg

OAO (Orbiting Astronomical Observatory)

A series of large NASA astronomical satellites, launched from Cape Canaveral, of which two, OAO-2 and OAO-3, were highly successful, and two, OAO-1 and OAO-B, were failures. OAO-1 stopped working on April 10, 1966, two days after launch, when its primary battery overheated. Its replacement, known as OAO-B, was also lost, immediately after launch on November 30, 1970, when the Centaur nose shroud in which it was enclosed failed to open and the spacecraft tumbled back into the atmosphere.

OAO-2 made 22,560 observations over an operational lifetime of more than four years of objects both within the Solar System and beyond, and discovered a hydrogen cloud around Comet Tago-Sato-Kosaka and **ultraviolet** (UV) emissions from Uranus. Its instruments included seven UV telescopes and four large-aperture television cameras.

OAO-3 was renamed the Copernicus Observatory after launch in honor of the 500th anniversary of the birth of the Polish astronomer Nicolaus Copernicus (1473–1543). It involved a collaboration between NASA and the Science and Engineering Research Council of the United Kingdom. The main experiment onboard was the Princeton University UV telescope, but Copernicus also carried an X-ray astronomy experiment for use in the 0.5 to 10 keV range developed by the Mullard Space Science Laboratory of University College London. It operated until late 1980. (See table, "OAO Missions.")

Oberth, Hermann Julius (1894–1989)

A German space pioneer, born in the Transylanian town of Hermannstadt. Along with Konstantin **Tsiolkovsky** and Robert **Goddard**, he is considered to be one of the founding fathers of modern rocketry and astronautics. Oberth's interest was sparked at the age of 11 by Jules Verne's *From the Earth to the Moon*, a book that he later recalled he read "at least five or six times and, finally, knew by heart." By the age of 14, he had already envisioned a "recoil rocket" that could propel itself through a vacuum by expelling exhaust gases from a liquid fuel. He realized early on, too, that **staging** was the key to accelerating rockets to high speed. In 1912, Oberth enrolled in the University of Munich to study medicine and later served with a medical unit in World War I. During the war years, however, he realized that his future lay in a different direction, and upon returning to the University of Munich, he took up physics. In 1922, his doctoral thesis, in which he described a **liquid-propellant rocket**, was rejected as being too unorthodox. However, Oberth published it privately the following year as a 92-page pamphlet entitled *Die Rakete zu den Planetenräumen*[222] (The Rocket into Planetary Space), which set forth the basic principles of spaceflight. A much-expanded version, entitled *Ways to Spaceflight*, appeared in 1929 to international acclaim and served to inspire many subsequent spaceflight pioneers.

Unlike Robert Goddard, Oberth was not slow in seeking publicity for his work. In the same year his enlarged treatise went on sale, he became technical advisor for the film *Frau Im Mond* (*Woman in the Moon*), and was commissioned to build a rocket that would be sent up as a publicity stunt. Aided by the young Werner **von Braun**, Oberth managed to build and statically test a small rocket engine—a risky laboratory exercise in which he lost the sight in his left eye. But when it became clear that the rocket would not be ready to launch in time for the movie's release, the project was abandoned, and, shortly after, Oberth returned to teaching in Transylvania.

Following World War II, Oberth came to the United States to work again with his former student, von Braun, at the Army's Ballistic Missile Agency. However, three

OAO Missions

Spacecraft	Launch				
	Date	**Vehicle**	**Orbit**	**Mass (kg)**	
OAO-1	Apr. 8, 1966	Atlas-Agena D	783 × 793 km × 35.0°		1,774
OAO-2	Dec. 7, 1968	Atlas IIIC	749 × 758 km × 35.0°		2,012
OAO-B	Nov. 30, 1970	Atlas IIIC	Failed to reach orbit		2,121
OAO-3	Aug. 21, 1972	Atlas IIIC	713 × 724 km × 35.0°		2,204

Hermann Oberth   Oberth working on a demonstration rocket for the film *Frau in Mond* (Woman in the Moon). *NASA History Office*

years later, Oberth returned to Germany to continue to write books on rocketry and space travel.[240]

occultation

The period of time when the view to one celestial body is blocked by another body—for example, when a spacecraft in lunar orbit has its line of sight to Earth or to the Sun blocked by the Moon.

ODERACS (Orbital Debris Radar Calibration Sphere)

A series of microsatellites launched from the Space Shuttle to help calibrate ground-based radar used for keeping track of spacecraft debris in orbit. (See table, "ODERACS Missions.")

Ofeq

A series of Israeli indigenous satellites; "ofeq" (also transliterated as "ofek") is Hebrew for "horizon." Several Ofeqs have been launched, not all successfully, since the first in September 1988. They have conducted a variety of space science experiments and technology validation tests, but they have been used mainly for military reconnaissance. The launches are unusual in that they take place in a *westerly* direction. This is to avoid overflying Arab territories to the east, but it means that the launch vehicle must travel against the direction of Earth's spin and so work harder to achieve orbital velocity, thus reducing the payload that can be delivered. (See table, "Ofeq Missions," on page 303.)

Launch
   Vehicle: Shavit
   Site: Palmachim

OFO (Orbiting Frog Otolith)

A curiously named mission, undertaken by NASA's Office of Advanced Research and Technology, to study the effects of weightlessness on and the response to acceleration of that part of the inner ear which controls balance. The subjects of the experiment were two bullfrogs—animals that, surprisingly, have an inner ear structure similar to that of humans—which were placed in a water-filled centrifuge with microelectrodes surgically implanted in their vestibular nerves. The experimental package was designed for flight as part of the Apollo Applications Program; however, it was equipped for flight on an unmanned spacecraft.

Launch
   Date: November 9, 1970
   Vehicle: Scout B
   Site: Wallops Island
Orbit: 304 × 518 km × 37.4°
Mass: 133 kg

ODERACS Missions

| Spacecraft | Shuttle Deployment | | Orbit | Mass (kg) |
	Date	Mission		
ODERACS A to F	Feb. 3, 1994	STS-60	327–338 × 352–356 km × 57.0°	1–5
ODERACS 2 A to F	Feb. 4, 1995	STS-63	177–268 × 179–280 km × 51.6°	1–5

Ofeq Missions

Spacecraft	Launch Date	Orbit	Mass (kg)
Ofeq 1	Sep. 19, 1988	250 × 1,149 km × 143°	155
Ofeq 2	Apr. 3, 1990	250 × 1,149 km × 143°	160
Ofeq	Sep. 15, 1994	Failed; not acknowledged by Israel	
Ofeq 3	Apr. 5, 1995	366 × 694 km × 143°	189
Ofeq 4	Jan. 22, 1998	Failed; fell into the Mediterranean Sea	
Ofeq 5	May 28, 2002	Similar to that of Ofeq 3	n/a

OGO (Orbiting Geophysical Observatory)

A series of NASA spacecraft designed to study various geophysical and solar phenomena in Earth's magnetosphere and in interplanetary space. Their orbits were chosen, with low perigees (points of minimum altitude) and high apogees (points of maximum altitude), so that they could sample an enormous range of the near-Earth environment. Each OGO consisted of a main body, two solar panels each with solar-oriented experiments, two orbital plane experiment packages, and six boom experiment packages. (See table, "OGO Missions.")

Ohzora

A Japanese satellite, launched by **ISAS** (Institute of Space and Astronautical Science), that made optical observations of the stratosphere and the middle atmosphere, and in particular studied a geomagnetic anomaly over the southern Atlantic that had been discovered by **Taiyo** several years earlier (see **Van Allen Belts**). Ozhora, whose name means "sky," was known as **Exos**-C before launch.

Launch
  Date: February 14, 1984
  Vehicle: M-3S
  Site: Kagoshima
Orbit: 247 × 331 km × 75°
Mass at launch: 210 kg

OICETS (Optical Inter-orbit Communications Engineering Test Satellite)

A Japanese experimental communications satellite that was scheduled to be launched by **NASDA** (National Space Development Agency) in late 2002. OICETS will play a key role in the development of inter-satellite communications. The crux of its mission will be a conversation with **ESA**'s (European Space Agency's) **ARTEMIS** fixed-orbit communications satellite, using an array of pointing, targeting, and acquisition hardware. By proving the space-worthiness of these technologies, OICETS will help lay the foundation for a global communications network to support broadcast television, mobile communications, computer connections, and manned spacecraft.

OGO Missions

Spacecraft	Launch			Orbit	Mass (kg)
	Date	Vehicle	Site		
OGO 1	Sep. 5, 1964	Atlas-Agena B	Cape Canaveral	4,930 × 144,824 km × 40.7°	487
OGO 2	Oct. 14, 1965	Atlas-Agena D	Vandenberg	419 × 1,515 km × 87.4°	507
OGO 3	Jun. 9, 1966	Atlas-Agena B	Cape Canaveral	19,519 × 102,806 km × 77.6°	634
OGO 4	Jul. 28, 1967	Atlas-Agena D	Vandenberg	422 × 885 km × 86.0°	634
OGO 5	Mar. 4, 1968	Atlas-IIIA	Cape Canaveral	271 × 148,186 km × 54.0°	634
OGO 6	Jun. 5, 1969	Atlas-Agena D	Vandenberg	397 × 1,089 km × 82.0°	634

OKB-1

The classified Soviet name of Sergei **Korolev**'s design bureau in Kalinigrad, near Moscow, formed in 1950 within Department 3 of the Special Design Bureau of NII-88. It later evolved to become the **Energia Rocket & Space Corporation** (Energia RSC). Its rivals were **Yangel**'s OKB-586 and, later, **Chelomei**'s OKB-52.

OKB-52

The classified Soviet name of Vladimir **Chelomei**'s design bureau in Moscow. OKB-52, which became NPO Saliout, was mainly responsible for producing navy missiles. Among its greatest legacies are the **Proton** (UR-500) and **Rockot** (based on the SS-19 Stilleto intercontinental ballistic missile).

OKB-586

The classified Soviet name of Mikhai **Yangel**'s design bureau in Dnepropetrovsk in the Ukraine. It later became NPO Yuzhnoye.

Okean

A series of Soviet oceanographic and naval radar satellites, the first of which was placed in orbit in the 1980s. Okean-O, launched on July 17, 1999, marked the start of a new generation of larger Okean spacecraft that carry side-looking radar and a set of visible and infrared scanners and radiometers.

O'Keefe, Sean (1956–)

NASA's current administrator; he replaced the previous incumbent, Daniel **Goldin**, in December 2001. O'Keefe previously served as deputy director of the Office of Management and Budget, overseeing the management of the federal budget under the Bush (junior) administration and, before that, was the Louis A. Bantle Professor of Business and Government Policy at Syracuse University. Earlier still he had served as the secretary of the navy under George Bush (senior) and in a variety of other government posts. O'Keefe has been charged with sorting out NASA's accounting system and chronic overspending on the International Space Station. He earned a B.A. from Loyola University (1977) and a Master of Public Administration degree from the Maxwell School (1978).

Oko

Russian military satellites placed in **Molniya-type orbits**. They are equipped with an infrared telescope and are intended to provide early warning of missile attacks. "Oko" means "eye."

Olympus

An experimental communications satellite, originally known as LSAT and launched by ESA (European Space Agency). One of the largest comsats ever built, Olympus was primarily designed as a technology demonstrator. Its platform and payloads were built to test a range of space hardware and components, which would be necessary elements of future telecommunications satellites in Europe. As such, it tested new concepts in orbit, such as steerable spot beams, 20 to 30 GHz repeaters, high-speed switchable multiplexers, and very high power traveling wave tube amplifiers. A dust grain from the Perseid meteor stream was blamed for ending its mission. Although Olympus remained intact after the collision, it lost so much thruster fuel in trying to correct its attitude that it became unmanageable.

Launch
  Date: July 12, 1989
  Vehicle: Ariane 2
  Site: Kourou
Orbit: GSO at 19° W
Mass: 2,595 kg

---

omnidirectional antenna

A simple **antenna**, mounted on a spacecraft, that radiates energy equally in all directions.

O'Neill, Gerard Kitchen (1927–1992)

A particle physicist at Princeton University who worked out a strategy for the future expansion of the human race into space. He championed the idea of orbital settlements in several papers and in his book *The High Frontier*.[225] The **L5 Society** was formed to advocate and to develop his schemes, although O'Neill never served as officer in this organization. See **O'Neill-type space colony**.

O'Neill-type space colony

A large orbiting **space colony**, of the kind proposed by Princeton's Gerard K. **O'Neill**, consisting of an immense rotating aluminum cylinder, the inner wall of which would be inhabited. The structure would be built of material mined from the Moon or asteroids. O'Neill linked his ideas with Peter Glaser's Solar Power Satellite (SPS) concept. SPSs are large solar collectors in space that would beam energy for use on Earth or in space. O'Neill suggested that they, too, could be manufactured of extraterrestrial material and could provide an export valuable enough to make a colony economically self-sustaining. In 1973, George Hazelrigg, also of Princeton, suggested that the L4 and L5 **Lagrangian point**s might be ideal places for O'Neill's large habitats. (The idea of situating a large structure at one of these special orbital locations can be traced back to the 1961 novel *A Fall of*

O'Neill-type space colony An artist's rendering of the interior of an O'Neill-type space colony, showing settlements connected by a suspension bridge. *NASA*

Moondust[54] by Arthur C. Clarke.) L4 and L5 are points of gravitational equilibrium located on the Moon's orbit at equal distances from both Earth and the Moon. An object placed in orbit around L5 (or L4) will remain there indefinitely without having to expend fuel. The orbit around L5 has an average radius of about 140,000 km, which leaves room for a large number of space settlements even at this one location. O'Neill envisaged the construction of a series of colonies culminating in a structure 32 km long and 3.2 km in radius, and capable of permanently supporting hundreds of thousands or even millions of inhabitants. Normal Earth gravity would be achieved by rotating the colony at a rate of one revolution per 114 seconds. The interior of the cylinder would have three inhabited "valleys," each containing lakes, forests, towns, and so forth. Three large mirrors, capable of being opened and closed on a regular day/night basis, would shine sunlight into the valleys, and a large parabolic collector at one end of the cylinder would focus solar energy onto steam-driven generators to provide the colony's electricity needs.

1*g* spacecraft

A robot star probe could be structurally designed to withstand very high accelerations in order to quickly reach a cruising speed that is a substantial fraction of the speed of light, but the same would not be possible with manned interstellar craft. Human beings can tolerate 10*g* or more for a few seconds and around 3*g* (the peak rate of acceleration of the Space Shuttle) for longer periods, but such accelerations and decelerations would be out of the question for a journey lasting years. The optimum rate of acceleration for manned flights to the stars would be 1*g*, since this would allow the crew to live under normal Earth gravity conditions while still enabling the spacecraft to gain speed at a rate practicable for interstellar travel. Given such acceleration, it would be possible to reach the Orion Nebula (about 1,000 light-years away) in 30 years of shipboard time. As the 1*g* spacecraft drew closer and closer to the speed of light, **relativistic effects**, such as time dilation, would become increasingly apparent. Time would pass more slowly on the ship in comparison with time on Earth. For example, after a round-trip journey at 1*g* acceleration and deceleration lasting 10 years as measured by the crew, 24 years would have elapsed back home. Relativistic effects would also ensure that, as measured by stationary observers, a spacecraft could not continue to build speed at 1*g*, or 9.8 m/s^2, indefinitely. If it did, in just under one year it would break the light-barrier. According to the **special theory of relativity**, no object can be accelerated to the speed of light. Instead, as light-speed was approached, the relationship between space and time would alter in the spacecraft's frame of reference so that although the crew would continue to feel and register on their instruments a 1*g* acceleration, stationary observers would see the spacecraft simply drawing ever nearer—but never quite

Round-Trip Times Assuming an Acceleration of 1*g*

Time As Measured Onboard Ship (yr)	Time on Earth (yr)	Maximum Range (light-years)	Attainable Target
1	1	0.059	Oort Cloud
10	24	9.8	Sirius
20	270	137	Hyades
30	3,100	1,565	Orion Nebula
40	36,000	17,600	Globular cluster
50	420,000	209,000	Magellanic Clouds
60	5,000,000	2,480,000	Andromeda Galaxy

reaching–the ultimate speed limit of light. The table ("Round-Trip Times Assuming an Acceleration of 1g") shows some of the dramatic possibilities for lengthy excursions in a 1g spacecraft. These figures assume equal periods of acceleration and deceleration at 1g on both the outgoing and return legs of the journey.

ONERA (Office National d'Études et de Recherches Aérospatiale)

The French national aerospace research center. Since its creation in 1946 it has worked on all the major French and European aeronautical and space programs, including Mirage, Concorde, Airbus, and **Ariane**. It operates a number of major laboratories and wind tunnel facilities.

one-way light time (OWLT)

The time taken for an electromagnetic signal to travel one way between Earth and a spacecraft or another body in the Solar System.

OPAL (Orbiting PicoSat Launcher)

A small Stanford University satellite, launched from the **JAWSAT** payload adaptor on January 26, 2000, and which in turn deployed six even smaller **picosatellites**. Two of these picosatellites were tethered and built by the Aerospace Corporation for ARPA (Advanced Research Project Agency) research, three (Thelma, Louise, and JAK) were built by Santa Clara College, and one (Stensat) was provided by radio amateurs. OPAL had a mass of 23.1 kg and measured approximately 20 cm in each direction.

Operation Paperclip

See **Paperclip, Operation**.

Orbcomm

The first commercial venture to provide global data and messaging services; it became operational in 1998 and uses a constellation of 26 satellites in low Earth orbit. At full capacity, the Orbcomm system can handle up to 5 million messages from users utilizing small portable terminals to transmit and to receive messages directly to the satellites. The system is controlled from Orbcomm's headquarters in Dulles, Virginia.

orbit

The curved path an object follows under the gravitational influence of another body. A *closed orbit,* such as that followed by a satellite going around Earth, has the shape of a circle or an ellipse. A satellite in a circular orbit travels at a constant speed. The higher the altitude, however, the lower the speed relative to the surface of the Earth. Main-

taining an altitude of 35,800 km over the equator, a satellite is said to be in **geostationary orbit**. In an elliptical orbit, the speed varies and is greatest at **perigee** (minimum altitude) and least at **apogee** (maximum altitude). Elliptical orbits can lie in any plane that passes through Earth's center. A **polar orbit** lies in a plane that passes through the North and South Poles; in other words, it passes through Earth's axis of rotation. An equatorial orbit is one that lies in a plane passing through the equator. The angle between the orbital plane and the equatorial plane is called the **inclination** of the orbit. As long as the orbit of an object keeps it in the vacuum of space, the object will continue to orbit without propulsive power because there is no frictional force to slow it down. If part or all of the orbit passes through Earth's atmosphere, however, the body is slowed by aerodynamic friction with the air. This causes the orbit to decay gradually to lower and lower altitudes until the object fully reenters the atmosphere and burns up. An *open orbit* is one in which a spacecraft does not follow a closed circuit around a gravitating body but simply has its path bent into the shape of a **parabola** or a **hyperbola**.[210]

orbit decay

A gradual change in the orbit of a spacecraft caused by the aerodynamic **drag** of a planet's outer atmosphere and other forces. The rate of orbit decay rises as the spacecraft falls and encounters increasing atmospheric density, eventually resulting in **reentry**.

orbital curve

The trace of an orbit on a flattened map of Earth or another celestial body about which the spacecraft is orbiting. Each successive orbit is displaced by the amount of rotation of the body between each orbit.

orbital cycler

An economical method of travel within the Solar System proposed by Buzz **Aldrin**. It turns out that there are stable orbits that cross the orbits of both Earth and Mars. Aldrin has suggested placing a large permanent station in such an orbit. Travelers would embark on the station when it passed Earth and disembark as it passed Mars. The energy cost for the traveler is simply the cost to make the rendezvous at each end.

orbital elements

Six quantities that are used to describe the motion of an orbiting object, such as that of a satellite around the Earth. They are: (1) the **semi-major axis**, which defines the size of the orbit; (2) the **eccentricity**, which defines the shape of the orbit; (3) the **inclination**, which (in the case of an Earth satellite) defines the orientation of the orbit with respect to Earth's equator; (4) the argument of perigee, which defines where the **perigee** (low point) of the orbit is with respect to Earth's surface; (5) the right ascension of the ascending node, which defines the location of the ascending and descending points with respect to Earth's equatorial plane; and (6) the true/mean anomaly, which defines where the satellite is within the orbit with respect to perigee.

orbital energy
The sum of the potential and kinetic energies of an object in orbit.

orbital nodes
Points in an orbit where the orbit crosses a reference plane, such as the ecliptic (the plane of Earth's orbit) or the equatorial plane.

orbital period
The time taken for an object to go once around a closed orbit. It is related to the **semi-major axis** in a way that is defined by the third of **Kepler's laws**.

orbital plane
The plane defined by the motion of an object about a primary body. The position and velocity vectors lie within the orbital plane, whereas the **angular momentum** vector is at right angles to the orbital plane.

orbital velocity
The velocity required to establish and to maintain a satellite in orbit. The term refers to average velocity, since the velocity is greater at perigee (the point of minimum altitude) than at apogee (the point of maximum altitude).

orbiter
The portion of a spacecraft that orbits a celestial body, such as a planet.

Orbiter
The orbiting portion of the **Space Shuttle**.

Orbiter, Project
A joint U.S. Army/Office of Naval Research plan for launching satellites during the **International Geophysical Year**.

Orbiter Processing Facility (OPF)
A facility at **Kennedy Space Center** used for refurbishing the Space Shuttle Orbiter and for loading payloads.

OrbView
Satellites designed to provide high-resolution images of Earth from orbit, which ORBIMAGE (Orbiting Image Corporation), a company half-owned by Orbital Sciences, sells to civilian, government, and military customers. OrbView-1, launched in 1995, and originally called **MicroLab**, provides atmospheric imagery; OrbView-2, launched in 1997, and originally called **Seastar**, supplies images of both ocean and land. OrbView-4 carried a camera able to snap 1-m resolution black-and-white and 4-m resolution color images from a 470-km orbit. It also carried a hyperspectral imaging instrument for the Air Force Research Laboratory's Warfighter-1 program, which would have made it the first commercial satellite to produce hyperspectral imagery. The U.S. military is interested in this technique because it shows promise for detecting chemical or biological weapons, collecting bomb damage assessment for commanders, and finding soldiers and enemy vehicles hidden under foliage. However, OrbView-4 was lost minutes after a faulty launch by its Taurus booster on September 21, 2001. (Also destroyed on the same flight was NASA'S **QuikTOMS**.) OrbView-3, which carries the same camera as its ill-fated sibling, but not the hyperspectral imager, will be launched aboard a Pegasus rocket sometime after September 2002.

Ordway, Frederick Ira III (1927–)
A space scientist and a well-known author of visionary books on spaceflight. Ordway was in charge of space systems information at the **Marshall Space Flight Center** from 1960 to 1963 and before that performed a similar function for the **Army Ballistic Missile Agency**. For many years he was a professor at the University of Alabama's School of Graduate Studies and Research. However, his greatest contribution has been to the popularization of space travel through dozens of books that he has authored or coauthored. He was also a technical consultant to the film *2001: A Space Odyssey* and owns a large collection of original paintings depicting astronautical themes.[226] Ordway was educated at Harvard and completed several years of graduate study at the University of Paris and other universities in Europe.

ORFEUS (Orbiting and Retrievable Far and Extreme Ultraviolet Spectrometer)
See SPAS.

Origins program
A NASA initiative designed to explore, through a series of space projects over the first two decades of the twenty-

first century, the origin and evolution of life, from the Big Bang to the present day. Included in this investigation are the formation of chemical elements, galaxies, stars, and planets, the formation and development of life on Earth, and the quest for extrasolar planets and extraterrestrial life. Missions associated with the Origins program fall chronologically into four groups. Precursor missions include the **Hubble Space Telescope**, **FUSE** (Far Ultraviolet Spectroscopic Explorer), **WIRE** (Wide Field Infrared Explorer), **SOFIA** (Stratospheric Observatory for Far Infrared Astronomy), and **SIRTF** (Space Infrared Telescope Facility). Following these will be the first-generation missions employing either large, lightweight optics or collections of small telescopes working in harness to provide images equivalent to those obtainable with a single, much larger instrument. They include **SIM** (Space Interferometry Mission) and **NGST** (Next Generation Space Telescope). Prior to SIM, a preliminary mission known as **Starlight** will be launched to prove the technology of space-based interferometry. The first-generation missions will serve as technological stepping-stones to the second-generation missions, including the **Terrestrial Planet Finder**, and the third-generation missions, including the **Planet Imager** and **Life Finder**.

Orion, Project

See article, pages 309–312.

Orlets

Russian photo-reconnaissance satellites similar to the **Yantar** type but carrying a small amount of onboard propellant for maneuvering and also a film-return capsule mechanism with 22 separate capsules. These features extend the mission duration of an Orlets satellite to between six months and a year.

Ørsted

A Danish microsatellite designed to map Earth's magnetic field, measure the charged particle environment, and study auroral phenomena. Data from the mission complement those collected by **Magsat**. The spacecraft is named after the Danish physicist Hans Christian Ørsted (1777–1851), who discovered electromagnetism in 1820.

Launch
    Date: February 23, 1999
    Vehicle: Delta 7926
    Site: Cape Canaveral
Orbit: 644 × 857 km × 96.5°
Size: 34 × 45 × 68 cm
Mass: 62 kg

OSCAR (Orbiting Satellite for Communication by Amateur Radio)

A long-running series of small satellites built and operated by radio "hams" around the world. OSCARs have no fixed design but are lightweight (typically about 30 kg) and generally are launched free of charge, when space is available, as secondary payloads on rockets carrying much heavier, primary payloads. OSCAR 1 was launched by the American Department of Defense. In 1969, the Amateur Radio Satellite Corporation (AMSAT) was set up to coordinate work on the OSCAR series.

OSO (Orbiting Solar Observatory)

A series of stabilized orbiting platforms developed by the Goddard Space Flight Center for observing the Sun and
(continued on page 312)

OSO Missions

Launch site: Cape Canaveral

Spacecraft	Launch Date	Vehicle	Orbit	Mass (kg)
OSO 1	Mar. 7, 1962	Delta	522 × 553 km × 32.8°	208
OSO 2	Feb. 3, 1965	Delta C	294 × 306 km × 32.8°	247
OSO C	Aug. 25, 1965	Delta C	Launch failure	280
OSO 3	Mar. 8, 1967	Delta C	546 × 570 km × 32.8°	281
OSO 4	Oct. 15, 1967	Delta C	552 × 555 km × 32.9°	272
OSO 5	Jan. 22, 1969	Delta C	538 × 559 km × 33.0°	291
OSO 6	Aug. 9, 1969	Delta N	489 × 554 km × 32.9°	290
OSO 7	Sep. 29, 1971	Delta N	326 × 572 km × 33.1°	635
OSO 8	Jun. 21, 1975	Delta 1914	539 × 553 km × 32.9°	1,066

Project Orion

A project to explore the feasibility of building a **nuclear pulse rocket** powered by nuclear **fission**. It was carried out by the physicist Theodore Taylor and others over a seven-year period, beginning in 1958, with U.S. Air Force support. The propulsion system advocated for the Orion spacecraft was based on an idea first put forward by Stansilaw Ulam and Cornelius Everett in a classified paper in 1955. Ulam and Everett suggested releasing atomic bombs behind a spacecraft, followed by disks made of solid propellant. The bombs would explode, vaporizing the material of the disks and converting it into hot **plasma**. As this plasma rushed out in all directions, some of it would catch up with the spacecraft, impinge upon a pusher plate, and so drive the vehicle forward.

Project Orion originated at General Atomics in San Diego, a company (later a subsidiary of General Dynamics) founded by Frederick de Hoffman to develop commercial nuclear reactors. It was de Hoffman who persuaded Freeman **Dyson** to join Taylor in San Diego to work on Orion during the 1958–1959 academic year.

Ulam and Everett's idea was modified so that instead of propellant disks, the propellant and bomb were combined into a single pulse unit. Plastic was chosen as the propellant material, not only because of its effectiveness in absorbing the neutrons emitted by an atomic explosion but also because it breaks down into lightweight atoms, such as those of hydrogen and carbon, which move at high speed when hot. This approach, in tandem with the pusher plate concept, offered a unique propulsion system that could simultaneously produce high **thrust** and high **exhaust velocity**. The effective **specific impulse** could theoretically be as high as 10,000 to one million seconds. A series of abrupt jolts would be experienced by the pusher plate, jolts so powerful that, if these forces were not spread out in time, they would result in acceleration surges that were intolerable for a manned vehicle. Consequently, a shock-absorbing system was devised so that the impulse energy delivered to the plate could be stored and then gradually released to the vehicle as a whole.

Various mission profiles were considered, including an ambitious interstellar version. This called for a 40-million-ton spacecraft to be powered by the sequen-tial release of 10 million bombs, each designed to explode roughly 60 m to the vehicle's rear. In the shorter term, Orion was seen as a means of transporting large expeditions to the Moon, Mars, and Saturn.

Taylor and Dyson were convinced that chemical rockets, with their limited payloads and high cost, were the wrong approach to space travel. Orion, they argued, was simple, capacious, and above all affordable. Taylor originally proposed that the vehicle be launched from the ground, probably from the nuclear test site at Jackass Flats, Nevada. Sixteen stories high, shaped like the tip of a bullet, and with a pusher plate 41 m in diameter, the spacecraft would have used a launch pad surrounded by eight towers, each 76 m high. Remarkably, most of the takeoff mass of about 10,000 tons would have gone into orbit. The bomb units, ejected on takeoff at a rate of one per second, would each have yielded 0.1 kiloton; then, as the vehicle accelerated, the ejection rate would have slowed and the yield increased, until 20-kiloton bombs would have been exploding every 10 seconds.

It was a startling and revolutionary idea. At a time when the United States was struggling to put a single astronaut into orbit using a modified ballistic missile, Taylor and Dyson were hatching plans to send scores of people and enormous payloads on voyages of exploration throughout the Solar System. The original Orion design called for 2,000 pulse units—far more than the number needed to reach Earth **escape velocity**. In scale, Orion more closely resembled the giant spaceships of science fiction than the cramped capsules of Gagarin and Glenn. One hundred and fifty people could have lived aboard in relative comfort in a vehicle built without the need for close attention to weight-saving measures.

One of the major technical issues was the durability of the pusher plate, since the expanding bubble of plasma from each explosion would have a temperature of several tens of thousands of degrees, even at distances of 100 m or so from the center of detonation. For this reason, extensive tests were carried out on plate erosion using an explosive-driven helium plasma generator. The results showed that the plate would be exposed to extreme temperatures for only about one thousandth of a second during each explosion, and that the **ablation** would occur only within a thin surface layer. So brief was the duration of high temperatures that very little heat flowed into the plate, and the researchers concluded that active cooling was unnecessary and that either aluminum or steel would be durable enough to serve as plate material. The situation was similar to that in an automobile engine, in which the peak combustion temperatures far exceed the melting points of the cylinders and pistons. The engine remains intact because the period of peak temperature is short compared to the period of the combustion cycle.

Still, it was evident that some experimentation was needed, and so the Orion team built a series of models, called Put-Puts or Hot Rods, to test whether pusher plates made of aluminum could survive the momentary intense temperatures and pressures created by chemical explosives. Several models were destroyed, but a 100-m flight in November 1959, propelled by six charges, was successful and demonstrated that impulsive flight could be stable. These experiments also suggested that the plate should be thick in the middle and taper toward its edges for maximum strength with minimum weight.

There was no obvious technical flaw in the Orion scheme or any argument to suggest that it could not be implemented economically. Its huge weakness, however, was that it depended upon atomic explosions that would release potentially harmful radiation into the environment—a fact that would ultimately be its undoing.

Early on, Taylor and his team recognized that they would need substantial government funding. The Advanced Research Projects Agency was approached in April 1958 and, in July, agreed to sponsor the project at an initial level of $1 million per year. However, this funding was short-lived. The newly formed NASA was beginning to acquire all civil-oriented space projects run by the federal government, while the Air Force was assuming control over space projects with military applications. Orion was initially excluded from both camps because the Air Force felt it had no value as a weapon, and NASA had made a strategic decision in 1959 that the civilian space program would, in the near future at any rate, be nonnuclear. Most of NASA's rocket engineers were specialists in chemical propulsion and either did not understand or were openly opposed to nuclear flight. Moreover, NASA did not want to attract public criticism by being seen to favor atomic devices. Orion was ARPA's only space interest, and in 1959 it decided it

could no longer support the project on national security grounds.

Taylor then approached the Air Force, which, after much persuasion, agreed to support Orion, providing that some military use could be found for it. However, Secretary of Defense Robert McNamara was unconvinced that Orion could become a military asset, and his department consistently rejected any increase in funding, effectively limiting it to a feasibility study. For the project to take off literally, it was essential that NASA become involved, so Taylor and James Nance, a General Atomics employee and later director of the Orion project, made representation to **Marshall Space Flight Center** (MSFC). They put forward a new design that called for the Orion vehicle to be carried into orbit as a **Saturn** V upper stage, the core of the spacecraft being a 90,000-kg "propulsion module" with a pusher-plate diameter of 10 m (limited by the diameter of the Saturn). This smaller design would restrict the specific impulse to between 1,800 and 2,500 seconds—a figure that, though low by nuclear-pulse standards, still far exceeded those of other nuclear rocket designs. The proposed shock-absorbing system had two sections: a primary unit made up of toroidal pneumatic bags located directly behind the pusher plate, and a secondary unit of four telescoping shocks (like those on a car) connecting the pusher plate assembly to the rest of the spacecraft. At least two Saturn V launches would have been needed to put the components of the vehicle into orbit. One of the missions suggested for this so-called first-generation Orion was a 125-day round-trip to Mars, involving eight astronauts and about 100 tons of equipment and supplies. A great advantage of the nuclear-pulse method is that it offers so much energy that high-speed, low-fuel-economy routes become perfectly feasible.

Wernher **von Braun**, at MSFC, became a supporter of Orion, but his superiors at NASA were not so enthusiastic, and the Office of Manned Spaceflight was prepared only to fund another study. Serious concerns surrounded the safety of carrying hundreds of atomic bombs through Earth's atmosphere. And there was worse news to come for the project. With the signing of the nuclear test-ban treaty by the United States, Britain, and the Soviet Union in August 1963, Orion, as a military-funded program calling for the explosion of nuclear devices, became illegal under international law. The only way it could be saved was to be reborn as a peaceful scientific endeavor. The problem was that because Orion was classified, few people in the scientific and engineering community even knew it existed. Nance (now managing the project) therefore lobbied the Air Force to declassify at least the broad outline of the work. Eventually it agreed, and Nance published a brief description of the first-generation Saturn-launched vehicle in October 1964. The Air Force, however, also indicated that it would be unwilling to continue its support unless NASA also contributed significant funds. Cash-strapped by the demands of Apollo, NASA announced publicly in January 1965 that no money would be forthcoming. The Air Force then announced the termination of all funding, and Orion quietly died. Some $11 million had been spent over nearly seven years.

Overshadowed by the Moon race, Orion was forgotten by almost everybody except Dyson and Taylor.[84] Dyson reflected that "this is the first time in modern history that a major expansion of human technology has been suppressed for political reasons." In 1968 he wrote a paper[85] about nuclear pulse drives and even large starships that might be propelled in this way. But ultimately, the radiation hazard associated with the early ground-launch idea led him to become disillusioned with the idea. Even so, he argued that the most extensive flight program envisaged by Taylor and himself would have added no more than 1% to the atmospheric contamination then (c. 1960) being created by the weapons-testing of the major powers. A detailed account of the scheme is presented by Freeman Dyson's son, George, in his book *Project Orion*, published by Henry Holt in 2002.

Being based on fission fuel, the Orion concept is inherently "dirty" and probably no longer socially acceptable even if used only well away from planetary environments. A much better basis for a nuclear-pulse rocket is nuclear **fusion**—a possibility first explored in detail by the British Interplanetary Society in the **Daedalus** project.

Orion, Project   An artist's rendering of an interplanetary spacecraft propelled by pulsed nuclear fission.
NASA

OSO (Orbiting Solar Observatory)

(continued from page 308)

extrasolar sources at ultraviolet, X-ray, and gamma-ray wavelengths. OSO-1 was the first satellite to carry onboard tape recorders for data storage and instruments that could be accurately pointed. The two-section observatory was stabilized because the lower section, the "wheel," spun as a gyroscope at a near constant 30 rpm. The upper fan-shaped section, the "sail," was joined to the wheel by a connecting shaft and remained pointed at the Sun during the OSO daytime. Experiments in the wheel scanned the Sun every two seconds and those in the sail pointed continuously at the Sun. (See table, "OSO Missions," on page 308.)

Osumi

The first Japanese satellite. It comprised the fourth stage of a Lambda 4S launcher and transmitted in orbit for about 17 hours.

Launch
  Date: February 11, 1970
  Vehicle: Lambda 4S
  Site: Kagoshima
Mass: 24 kg
Orbit: 340 × 5,140 km × 31° orbit

OSO   An engineer checks two spectrometers on OSO 8.
Hughes Aircraft

overshoot boundary

The upper side of a **reentry corridor**, marking the region above which the atmospheric density is so low that the spacecraft cannot decelerate sufficiently and instead skips back into space.

OWL (Orbiting Wide-angle Light-collectors)

A proposed space mission to study **cosmic rays** of the highest energy. These particles are so rare that only a few have ever been seen; they may have as much energy as a fast-thrown baseball, and it is not even known whether they are protons, heavy nuclei, or photons. OWL will employ Earth as a particle detector by using two high-altitude spacecraft to obtain a binocular view of the light flashes produced by super-energetic cosmic rays interacting with the atmosphere. It has been identified in NASA's Office of Space Science Strategic Plan as a potential mission beyond 2007, but it remains in the very early concept definition phase.

oxidizer

A substance that supports the combustion reaction of a **fuel** or **propellant**.

oxidizer-to-fuel ratio

The ratio between the mass of **oxidizer** burned per mass of **fuel** burned (liquid engines only).

oxygen, regenerative

A spacecraft's oxygen supply that is recycled for repeated use after being cleaned of carbon dioxide. This is accomplished by human-made chemical devices or (in theory) by plant life onboard. The latter is suggested for interplanetary missions, which will be of long duration.

oxygen-hydrocarbon engine

A rocket engine that operates on a propellant of liquid oxygen as oxidizer and a hydrocarbon fuel, such as the propellant derivatives.

P

packetizing

A telecommunications technique used by modern spacecraft that has superseded the older method of **time-division multiplexing**. In packetizing, a burst or "packet" of data is transmitted from one onboard instrument or sensor, followed by a packet from another device, and so on, in a nonspecific order. Each burst carries an identification of the measurement it represents so that the ground data system can recognize it and handle it appropriately. The scheme adheres to the International Standards Organization's (ISO) Open Systems Interconnection (OSI) protocol suite, which recommends how different types of computers can intercommunicate. Being independent of distance, the ISO OSI holds for spacecraft light-hours away just as it does for local workstations.

pad abort

Halting the mission of a space vehicle still on the launch pad because of malfunction, change in plans, or other problems.

paddlewheel satellite

A satellite with large solar panels that are deployed to generate electrical energy. The panels resemble giant paddles alongside the satellite. The world's first paddlewheel satellite was Explorer 6, launched on August 7, 1959.

PAET (Planetary Atmospheric Entry Test)

A NASA suborbital experiment to study the feasibility of making mass spectrometer measurements from high-speed entry probes as they plunge into dense atmospheres.

Launch
    Date: June 20, 1971
    Vehicle: Scout B
    Site: Wallops Island

Pageos (Passive GEOS)

A 30-m diameter sphere made from 0.0125-mm Mylar externally coated with vapor-deposited aluminum. It was inflated in orbit to serve as a giant reflector of sunlight that could be photographed from the surface. In this way, over a five-year period, Pageos enabled the determination of the precise relative location of continents, islands, and other land masses. Pageos was the second NASA satellite (following GEOS 1) in the National Geodetic Satellites Program. The launch, orbit, separation, inflation, and operation went according to plan, with more than 40 ground stations taking part in the observation program. The orbit was generally considered too high for drag-density study, although some work was done in this area by the Smithsonian Astrophysical Observatory.

Launch
    Date: June 23, 1966
    Vehicle: TA Thor-Agena D
    Site: Vandenberg Air Force Base
Orbit: 2,953 × 5,207 km × 84.4°
Mass: 55 kg

Paine, Thomas O. (1921–1992)

NASA's third administrator, during whose term of office the first seven Apollo manned missions were flown, including the first two to the Moon's surface. Paine graduated from Brown University in 1942 with an A.B. degree in engineering, and in 1947, from Stanford University with a Ph.D. in metallurgy. During World War II, he served as a submarine officer in the Pacific. Subsequently, he held a variety of research and research-management positions before being appointed deputy administrator of NASA in 1968. Upon the retirement of James **Webb**, Paine was named acting administrator and nominated as administrator of NASA in March 1969, a position from which he resigned in September 1970, to return to General Electric. In 1985, the White House chose Paine as chair of a National Commission on Space to prepare a report on the future of space exploration. Since leaving NASA 15 years earlier, Paine had been a tireless spokesman for an expansive view of what should be done in space. The Paine Commission took almost a year to prepare its report, largely because it solicited public input in hearings throughout the United States. The Commission report, *Pioneering the Space Frontier,* published in May 1986, espoused "a pioneering mission for twenty-first-century America–to lead the exploration and development of the space frontier, advancing science, technology, and enterprise, and building institutions and systems that make accessible vast new resources and support human

Paine, Thomas O. Thomas Paine (center) being appointed NASA administrator by President Richard Nixon (left), with Vice-President Spiro Agnew standing alongside. *NASA*

settlements beyond Earth orbit, from the highlands of the Moon to the plains of Mars." The report also contained a "Declaration for Space" that included a rationale for exploring and settling the Solar System and outlined a long-range space program for the United States.

Palapa
Indonesia's national communications satellite system. Palapa 1 and 2, launched in 1976 and 1977, were among the first indigenous comsats. Stationed in geosynchronous orbits, Palapas provide voice circuits and television to Indonesia's more than 6,000 inhabited islands and its neighboring countries.

Palmachim
An Israeli air base and missile test site, located south of Tel Aviv at 31.9° N, 34.7° E, from which the **Shavit** satellite launch vehicle is deployed. Due-west launches over the Mediterranean are required to avoid overflying Arab countries, resulting in unique orbital inclinations (142°–144°) and direction.

Paperclip, Operation
The American scheme to detain top German scientists at the end of World War II and relocate them to the United States. These scientists included Wernher **von Braun** and more than 100 of his colleagues who had worked on the V-2 and other **"V" weapons**. Having been transferred to their new home at Fort Bliss, Texas, a large Army installation just north of El Paso, they were given the job of training military, industrial, and university personnel in the intricacies of rockets and guided missiles and helping refurbish, assemble, and launch a number of V-2s that had been shipped from Germany to the **White Sands Proving Ground** in New Mexico. President Harry Truman had given the go-ahead to Paperclip on condition that none of the detainees could be shown to have been members of the Nazi party or active supporters of the Hitler regime. However, it is now clear that many of the key figures in the roundup, including von Braun himself, Arthur **Rudolf**, and Hubertus **Strughold**, had been enthusiastic Nazis and, in some cases, had been aware of, or even involved in, atrocities inflicted on concentration camp detainees.

parabola
A curve, with an **eccentricity** of 1, obtained by slicing a cone with a plane parallel to one of the cone's sides. A parabola can be thought of as an **ellipse** with an infinitely long **major axis**. It is one of the **conic sections**.

parabolic flight
The following of a special parabolic trajectory by a suitably fitted aircraft in order to reproduce the conditions of **freefall** for a short period. Such flights are used as a routine part of astronaut training.

parabolic orbit
An orbit around a central mass that is followed by an object that, at any point on its path, has the minimum velocity needed to escape from the gravitating mass. See **conic sections**.

parabolic reentry
Reentry at speeds of less than 11,100 m/s. Above this speed, reentry is said to be hyperbolic.

parabolic velocity
In theory, if the velocity of an object moving in a circular orbit is multiplied by the square root of 2 (approximately 1.414), the orbit will become parabolic. For example, the Earth is moving in a near circular orbit about the Sun at a mean velocity of 29.8 km/s, so that its parabolic velocity would be 41.8 km/s. If the Earth's orbital velocity were to be increased to this value, our planet would be able to escape from the Solar System.

parallel operation of engines

The operation of two or more engines in a system to provide more thrust than from a single engine without having thrust misalignment or interference of one engine with another.

parallel staging

A practice similar to clustering, in which two or more stages mounted parallel to the main engine ignite simultaneously. The stages usually contain shorter-burning, higher-thrust motors than the main airframe, or sustainer, and drop off when their motors burn out.

PARASOL (Polarization and Anisotropy of Reflectances for Atmospheric Science coupled with Observations from a Lidar)

One of a constellation of satellites, which also includes **Aqua**, **CALIPSO**, and **Aura**, that will fly in similar orbits around the Earth studying various processes on the land and ocean, and in the atmosphere. PARASOL, a CNES (the French space agency) mission, is specifically designed to measure the direction and polarization of light reflected from areas of land observed by the lidar carried by CALIPSO. It is scheduled for launch in 2004.

Paris Gun

The supercannon with which the German army bombarded Paris from the woods of Crepy from March 1918 to the end of World War I. Also known as the Wilhelm Geschuetz (after Kaiser Wilhelm II) and Lange Max (Long Max), it is frequently confused with the Big Berthas, giant howitzers used by the Germans to smash the Belgian frontier fortresses, notably that at Liege in 1914. Although the famous Krupp-family artillery makers produced both guns, the resemblance ended there. The Paris Gun was a weapon like no other, capable of hurling a 94-kg shell to a range of 130 km and a maximum altitude of 40 km—the greatest height reached by a human-made object until the first successful flight of the V-2 (see **"V" weapons**) in October 1942. At the start of its 170-second trajectory, each shell from the Paris Gun reached a speed of 1,600 m/s (almost five times the **speed of sound**). The gun itself, which weighed 256 tons and was mounted on rails, had a 28-m-long, 210-mm-caliber rifled barrel with a 6-m-long smoothbore extension. After 65 shells had been fired, each of progressively larger caliber to allow for wear, the barrel was rebored to a caliber of 240-mm. The German goal of their great cannon was not to destroy the French capital—it was far too inaccurate a device for that (although it killed 256)—but to erode the morale of the Parisians. A similar giant gun with the same objective, the V-3, was built during World War II. Technologically, the Paris Gun can be seen as the first serious attempt to build a launch system with the capability of a **space cannon**.

parking orbit
An intermediate, waiting orbit adopted by a spacecraft between two phases of a mission.

Parus
A network of military satellites that, for a quarter of a century, has provided navigational information and store-dump radio communications for Russian/Soviet naval forces and ballistic missile submarines. Being the operational successor to Tsyklon, Parus is also known as Tsyklon-B. Flight trials of the new satellites began in 1974, and the system was accepted into military service in 1976. From 1974 to the present, there have been about 90 launches in the series, all by Cosmos-3M rockets from Plesetsk, into roughly circular orbits with an altitude of about 1,000 km and an inclination of 83°.

passive satellite
A satellite that simply reflects light or radio waves transmitted from one ground terminal to another without amplification or retransmission. An example is the **Echo** series of orbiting balloons.

Pathfinder
See **Mars Pathfinder**.

Patrick Air Force Base
A U.S. Air Force base located at the south end of **Cape Canaveral**, 19 km southeast of Cocoa, which is home to the 45th Space Wing of the **Air Force Space Command**. Patrick AFB operates the **Cape Canaveral Air Force Station**, from which many military and nonmilitary launches take place. The base is named after Major General Mason M. Patrick, chief of the Army Air Service (1921–1926) and chief of the Army Air Corps (1926–1927), and was established following the transfer of control of the Banana River Air Station, set up in 1940 to support Atlantic coastline defense, from the Navy to the Air Force.

Paulet, Pedro E.
A Peruvian chemical engineer reputed to have conducted experiments in Paris, beginning in 1895, on a rocket engine made of vanadium steel that burned a combination of nitrogen peroxide and gasoline. If true, this would credit Paulet as the designer of the first **liquid-fueled rocket**. The engine was described as weighing about 5 lb and using spark gap ignition of the fuels within a combustion chamber. It was said to have been capable of producing a 200-lb thrust at 300 sparks per minute. Paulet claimed that his rocket engine could burn continuously for as much as one hour without suffering any ill effects. However, news of what may have been a groundbreaking advance in rocketry did not surface until October 27, 1927, when a letter from Paulet appeared in an issue of the Peruvian publication *El Comercio,* in which he claimed legal ownership of his earlier rocket engine design. Recognizing that rocketry was beginning to boom in Europe, Paulet sought witnesses to help verify the work he said he had done years earlier. The letter was circulated across the world by a Russian named Alexander Scherschevsky in summary form. Had Paulet's work been authenticated, he would today be considered the undisputed father of liquid propellant rocketry. As it is, that title is more commonly attributed to Robert **Goddard**.

PAW ascent
"Powered all the way" ascent: one that follows the shortest distance between Earth and a point in Earth orbit. PAW ascents consume the most fuel and are not considered feasible for orbits higher than 230 km.

payload
The load carried by an aircraft or rocket over and above what is necessary for the operation of the vehicle during its flight.

Payload Assist Module (PAM)
A solid motor developed for the **Space Shuttle** program. Its purpose is to lift a satellite from the **parking orbit** into which it is placed by the Shuttle to a higher, designated orbit from which the spacecraft will actually operate. A special PAM, known as PAM-D, was adapted for use with **Delta** launch vehicles.

payload integration
The process of bringing together individual experiments, support equipment, and software into a single payload in which all interfaces are compatible and whose operation has been fully checked out.

payload ratio
The ratio of the mass of useful **payload** of a rocket-propelled vehicle to the total launch mass of the vehicle. See **rocket equation**.

payload specialist

A member of the crew aboard a Space Shuttle mission whose sole responsibility is the operation of the experiments of a payload. He or she is not necessarily a professional astronaut.

Peenemünde

Originally, a quiet, wooded region in Germany located on the Baltic Sea at the mouth of the river Peene on the island of Usedom. It came to the attention of the German army and the Luftwaffe in their search for a new rocket development site to replace **Kummersdorf**, having been suggested to Wernher **von Braun** by his mother. By the late 1930s, a massive construction project was under way involving new housing for engineers and scientists, a power plant, a liquid oxygen plant, a wind-tunnel facility,[310] barracks, a POW camp, a rocket production facility, a development works facility, and the Luftwaffe airfield. The site would eventually be home to over 2,000 scientists and 4,000 other personnel under the command of Walter **Dornberger**. To the northern end of Peenemünde, between the forest and the sandy foreshore, nine test stands for the firing of rockets were constructed, the largest and most infamous of them being Test Stand VII, from which the A-4/V-2 (see **"V" weapons**) would be launched.

Eventually, Allied intelligence became aware of the importance of Peenemünde, and on the night of August 17–18, 227 bombers from the RAF were ordered to attack the site. The objective of Operation "Hydra" was to kill as many German rocket scientists as possible and to destroy key targets such as the liquid oxygen plant, the power plant, the Experimental/Development Works, and the test stands. The Peenemünde West area was ignored because the Allies didn't know of the V-1 flying bomb development there. In the event, although the bombing left 732 people dead, no key people, except Walter **Thiel**, were killed, and much of the facility was undamaged. But the vulnerability of Peenemünde was now evident, and the remaining field trials of the V-2 were switched to Bliszna in Poland. Heinrich Himmler persuaded Hitler that the production and deployment of the V-2 should henceforth be handled by the SS, and a suitable site for mass production was found in the complex of tunnels beneath the Kohnstein Mountain near Nordhausen, in central Germany. This subterranean V-2 factory, in which POWs were forced to work under brutal conditions, became known as the Mittelwerke.

Research into rocketry and rocket weapons continued at Peenemünde but on a much-reduced scale. The autumn of 1944 saw testing of the Wasserfall antiaircraft missile, a winged version of the V-2 designed to carry explosives into the midst of Allied bomber formations where it would be detonated. At Peenemünde West, development continued of the Me 163 "Komet" rocket-powered fighter plane along with the Me 262 jet fighter.[216]

Pegasus (launch vehicle)

An air-launched space launch vehicle, developed by Orbital Sciences Corporation, that works in a similar way to an air-launched ballistic missile (ALBM). Like an ALBM, Pegasus is dropped from an aircraft and fires its rocket seconds later. But instead of completing a ballistic trajectory to carry a weapons payload to its target, Pegasus carries a payload into space. The original vehicle, now referred to as the Pegasus Standard, has been superseded by the Pegasus XL, which comes in three- or four-stage versions with two fourth-stage options. The Hydrazine Auxiliary Propulsion System (HAPS), burning hydrazine liquid fuel, provides precision orbital insertion capability for the payload. Alternatively, the fourth stage can be powered by a Thiokol Star 27 solid rocket motor, which

Pegasus XL Specifications

	Stages		
	First	**Second**	**Third**
Engine designation	Orion 50SXL	Orion 50XL	Orion 38
Length (m)	10.3	3.1	1.3
Diameter (m)	1.3	1.3	0.97
Fuel	HTPB	HTPB	HTPB
Thrust (N)	726,000	153,000	32,000
Payload (kg)			
LEO	450		
Earth escape	125		

Pegasus (launch vehicle) The Pegasus space booster attached to the wing pylon of NASA's B-52 launch aircraft. *NASA*

burns HTPB solid fuel and can boost the payload onto an Earth-escape trajectory. The Pegasus XL uses a standard payload fairing, 2.1 m long with a diameter of 1.1 m, which can accommodate one payload in a dedicated launch configuration or two payloads in a shared launch. In a typical mission profile without a fourth stage, the Pegasus XL is dropped over open water from a customized L-1011 aircraft flying at an altitude of 11,600 m. The drop is considered to be the actual launch, although the first stage isn't fired until five seconds later at an altitude of 11,470 m. The rocket has an inertial guidance system to pitch it upward after first stage ignition. The first stage burns out 77 seconds after launch and is jettisoned. The second stage kicks in 95 seconds after launch and burns out 73 seconds later, at which point the second and third stage combination begin an unpowered coast. During this coast period, the second stage stays attached to provide aerodynamic stability. Ultimately, the spent second stage is jettisoned and the third stage is fired 592 seconds after launch, burns out 65 seconds later, and is jettisoned from the payload, having successfully inserted it into orbit. (See table, "Pegasus XL Specifications.")

Pegasus (spacecraft)

A series of spacecraft that over a three-year period investigated the extent of the **micrometeroid** threat to the manned **Apollo** missions. Each was launched with a boilerplate Apollo capsule and, when in orbit, extended two large "wings," each measuring 4.3 by 14.6 m, mounted with detector panels. The fewer-than-expected number of impacting particles enabled about 450 kg of shielding to be trimmed from the operational Apollo vehicles. (See table, "Pegasus Missions.")

Launch:
  Vehicle: Saturn I
  Site: Cape Canaveral
Mass: approx. 10,500 kg

Pegasus Missions

Spacecraft	Launch Date	Orbit
Pegasus 1	Feb. 16, 1965	510 × 726 km × 31.7°
Pegasus 2	May 25, 1965	502 × 740 km × 31.7°
Pegasus 3	Jul. 30, 1965	441 × 449 km × 28.9°

penetration probe

A device that, by impact, tunneling, or melting, penetrates the surface of a planet, moon, asteroid, or comet, to investigate the subsurface environment. Data is typically transmitted to a mother craft for re-transmission to Earth. Among missions being designed or developed that will deploy penetration probes are the **Deep Impact** and the **Europa Ocean Explorer**. The first intended penetration probes, carried by the **Mars Polar Lander**, failed along with the rest of the mission.

periapsis

The point in an orbit closest to the body being orbited.

perigee

For a satellite in an elliptical Earth orbit, the point in the orbit that is nearest to Earth. Compare with **apogee** and **perihelion**.

perigee kick motor
A rocket motor fired at the **perigee** (low point) of an orbit to establish a more elliptical one. The new orbit will have the same perigee as the old, but a higher **apogee**. Compare with **apogee kick motor**.

perihelion
For an object in an elliptical solar orbit, the point on the orbit that is nearest to the Sun. Compare with **aphelion** and **perigee**.

perilune
The point on an elliptical lunar orbit that is nearest to the Moon.

perturbation
A disturbance to a planned trajectory or orbit, caused by atmospheric drag, gravitation, radiation pressure, or some other external influence.

Petrov, Boris N. (1913–1980)
A leading Soviet scientist whose later years were devoted to space exploration. Petrov was a senior academician in the Soviet Academy of Science and chair of the Inter-Cosmos Council, which promoted cooperation in space among eastern European nations during the height of the Cold War from 1966 to 1980.

PFS (Particles and Fields Subsatellite)
Subsatellites released from **Apollo** 15 and 16 into lunar orbit to collect data on particles and fields in the Moon's vicinity. (See table, "PFS Deployments.")

PFS Deployments

Spacecraft	Release Date	Mass (kg)
PFS, Apollo 15	Aug. 4, 1971	36
PFS, Apollo 16	Apr. 24, 1972	36

Phillips, Samuel C. (1921–1990)
A senior U.S. Air Force officer who led the **Apollo** program. Phillips trained as an electrical engineer at the University of Wyoming and also participated in the Civilian Pilot Training Program during World War II. Following his graduation in 1942, he entered the Army infantry but soon transferred to the air component. As a young pilot he served with distinction in the Eighth Air Force in England, earning two distinguished flying crosses, eight air medals, and the French croix de guerre. His growing interest in aeronautical research and development led him to become involved in the development of the B-52 bomber in the early 1950s and then to head the Minute-man intercontinental ballistic missile program in the latter part of the decade. In 1964, Phillips, by this time an Air Force general, was seconded to NASA to head the Apollo Moon landing program. He returned to the Air Force in the 1970s and commanded Air Force Systems Command up to his retirement in 1975.[295]

Phobos
See **Fobos**.

photometer
An instrument that measures the intensity of light from a source.

photon
A particle of light or, more precisely, a quantum of the electromagnetic field. It has zero **rest mass** and a spin of 1. Apart from its obvious role as the carrier of virtually all the information that we have so far been able to gain about phenomena beyond the Solar System (with the exception of cosmic rays and, possibly, gravitational waves), the photon is also the exchange particle of the electromagnetic force acting between any two charged particles. See **electromagnetic radiation**.

photon propulsion
A form of rocket propulsion, yet to be developed, in which the reaction is produced by **electromagnetic radiation**. Two specific forms of photon propulsion are the **solar sail** and **laser propulsion**.

photosynthetic gas exchanger
A device that utilizes plants and light energy to convert back into oxygen the carbon dioxide generated by man or animals through breathing oxygen.

photovoltaic cell
See **solar cell**.

physiological acceleration
The acceleration experienced by a human or animal test subject in an accelerating vehicle.

physiological factors
Factors that affect a crew's health and ability to function.

PICASSO-CENA (Pathfinder for Instruments for Cloud and Aerospace Spaceborne Observations—*Climatologie Etendue des Nuages et des Aerosols*)
An earlier name for the satellite now called **CALIPSO**.

Pickering, William Hayward (1910–)

The New Zealand–born director of **JPL** (Jet Propulsion Laboratory) for the first two decades of the Space Age. Pickering obtained his B.S. and M.S. in electrical engineering, then a Ph.D. in physics, from the California Institute of Technology before becoming a professor of electrical engineering there in 1946. In 1944, he organized the electronics efforts at JPL to support guided missile research and development, becoming project manager for the **Corporal**, the first operational missile to come out of JPL. He served as JPL's director (1954–1976), overseeing development of the first American satellite (**Explorer** 1), the first successful American circumlunar space probe (**Pioneer** 4), the **Mariner** flights to Venus and Mars in the early to mid-1960s, the **Ranger** photographic missions to the Moon in 1964 to 1965, and the **Surveyor** lunar landings of 1966 to 1967.

picket ship

An oceangoing ship used on a missile range to provide additional instrumentation for tracking or recovering missiles.

picosat

A satellite with an on-orbit mass of less than 1 kg (see **satellite mass categories**). Several experimental picosats have already been launched including **DARPA**'s Picosat-1A and -1B and Santa Clara University's Artemis Picosat—all released by **OPAL** (Orbiting Picosat Automatic Launcher) in January 2000.

Pierce, John Robinson (1910–2002)

A leading applied physicist who is commonly referred to as the father of the **communications satellite** for his work on it in 1954 (although the concept had first been suggested by Arthur C. **Clarke**). Pierce worked for 35 years as an engineer at Bell Telephone Laboratories, where he coined the term "transistor," and then at the California Institute of Technology and **JPL** (Jet Propulsion Laboratory). He urged NASA to build a satellite based on his design, and it was launched in 1960 as **Echo** 1. The project's success led to the construction and 1962 launch of the first commercial communications satellite, **Telstar** 1.

piggyback experiment

An experiment that rides along with the primary experiment on a space-available basis, without interfering with the mission of the primary experiment.

Pioneer

See article, pages 321–323.

pitch

The movement about an axis that is perpendicular to the vehicle's longitudinal axis and horizontal with respect to a primary body. See **roll** and **yaw**. *Pitch attitude* is the orientation of a spacecraft with respect to the pitch axis. The *pitching moment* is the rising and falling of a spacecraft's nose. When the nose rises, the pitching moment is positive; when the nose drops, the pitching moment is negative and is also called a diving moment.

William Pickering Left to right: Pickering, former JPL director, Theodore von Karman, JPL cofounder, and Frank J. Malina, cofounder and first director of JPL. *NASA*

Pioneer

A diverse series of NASA spacecraft designed for lunar and interplanetary exploration. The first few to be launched were either total or partial failures, although Pioneer 3 did send back data leading to the discovery of the outer of the **Van Allen Belts**. Pioneer 5 returned important data on solar flares, the solar wind, and galactic cosmic rays, and also established a record at the time of 36.2 million km for radio communication in space. Pioneers 6 through 9 went into elliptical orbit around the Sun and carried out further observations of the solar wind and interplanetary magnetic field. Pioneer 10 and Pioneer 11 were the first probes to explore the outer solar system, while the Pioneer Venus probes sent back valuable data from Earth's inner neighbor. (See table, "Pioneer Missions.")

Early Pioneers: A Failed Lunar Program

The Pioneer program began in 1958, when the Advanced Research Projects Agency of the U.S. Department of Defense authorized the launch of five spacecraft toward the Moon. The first was lost in a launch failure just 77 seconds after liftoff on August 17, 1958, and is referred to as Pioneer 0 or **Able** 1. The second, designated Pioneer 1, was the newly formed NASA's inaugural mission, but it failed to reach escape velocity and crashed into the South Pacific. Likewise, Pioneer 2 and 3 fell back to Earth. Pioneer 4 did at least manage to fly past the Moon, but its closest approach of 60,000 km was too remote for any lunar data to be returned. Four more probes followed, all of which failed, before the Pioneer lunar program was abandoned.

Pioneer 5 to 9: Success Further Afield

Beginning with Pioneer 5, the program shifted focus to interplanetary space—and immediately became more successful. Pioneer 5 was placed in a solar orbit, which ranged between Earth and Venus and provided the first experience of communicating with a spacecraft at distances of tens of millions of kilometers. It was followed by a series of spin-stabilized probes equipped with sensors to monitor the solar wind, magnetic fields, and cosmic rays. All vastly exceeded their design lifetimes of six months. Pioneer 9 finally

Pioneer Missions			
Launch site: Cape Canaveral			
	Launch		
Spacecraft	Date	Vehicle	Mass (kg)
Pioneer 1	Oct. 11, 1958	Thor-Able	38
Pioneer 2	Nov. 8, 1958	Thor-Able	39
Pioneer 3	Dec. 6, 1958	Juno 2	6
Pioneer 4	Mar. 4, 1959	Juno 2	6
Pioneer 5	Mar. 11, 1960	Thor-Able	43
Pioneer 6	Dec. 16, 1965	Delta E	43
Pioneer 7	Aug. 17, 1966	Delta E	63
Pioneer 8	Dec. 13, 1967	Delta E	63
Pioneer 9	Nov. 8, 1968	Delta E	63
Pioneer 10	Mar. 3, 1972	Atlas-Centaur	259
Pioneer 11	Apr. 6, 1973	Atlas-Centaur	259
Pioneer Venus 1	May 20, 1978	Atlas-Centaur	582
Pioneer Venus 2	Aug. 8, 1978	Atlas-Centaur	904

Early Pioneers Pioneer 3 being inspected by technicians before shipping to Cape Canaveral for launch in 1958. *NASA*

failed in 1983 after 15 years of service, and Pioneer 7 and 8 were last tracked successfully in the mid-1990s. Contact with Pioneer 6, the oldest operating spacecraft, was reestablished on December 8, 2000—35 years after launch.

Pioneer 10 and 11: Jupiter and Beyond

Twin probes that became the first to cross the asteroid belt and fly past Jupiter. Pioneer 11 used a Jupiter gravity-assist to redirect it to an encounter with Saturn. Both spacecraft, along with **Voyager** 1 and 2, are now leaving the solar system. Of this interstellar quartet, only Pioneer 10 is heading in the opposite direc-

tion to the Sun's motion through the galaxy. It continues to be tracked in an effort to learn more about the interaction between the **heliosphere** and the local **interstellar medium**. Pioneer 10's course is taking it generally toward Aldebaran (65 light-years away) in the constellation of Taurus, and a remote encounter about two million years from now. It was superseded as the most distant human-made object by Voyager 1 in mid-1998. The last communication from Pioneer 11 was received on November 30, 1995. With its power source exhausted, it can no longer operate any of its experiments or point its antenna toward Earth, but it continues its trek in the direction of the constellation of Aquila. (See table, "Pioneer 10/11 Facts.")

Pioneer Venus

After a six-month journey, Pioneer Venus 1 entered an elliptical orbit around Venus in December 1978 and began a lengthy reconnaissance of the planet. The spacecraft returned global maps of the Venusian clouds, atmosphere, and ionosphere, measurements of the interaction between the atmosphere and the solar wind, and radar maps of 93% of the planet's surface. In 1991, the Radar Mapper was reactivated to investigate previously inaccessible southern portions of the planet. In May 1992, Pioneer Venus began the final phase of its mission, in which the periapsis (orbital low point) was held at 150 to 250 km until the fuel ran out and atmospheric entry destroyed the spacecraft. Despite a planned primary mission duration of only eight months, the probe remained in operation until October 8, 1992.

Pioneer Venus 2 consisted of a bus that carried one large and three small atmospheric probes. The large

Pioneer 10/11 Facts

	Pioneer 10	Pioneer 11
Jupiter flyby		
Date	Dec. 3, 1973	Dec. 4, 1974
Closest approach	130,400 km	43,000 km
Saturn flyby		
Date	—	Sep. 1, 1979
Closest approach	—	21,000 km
Status on Jun. 1, 2002		
Distance	12.2 billion km	Mission ended
Speed	12.3 km/s	Mission ended

Pioneer 10   Pioneer 10's plaque, bearing an interstellar message and attached to the spacecraft's antenna support struts, can be seen to the right of center in this photo. *NASA*

probe was released on November 16, 1978, and the three small probes on November 20. All four entered the Venusian atmosphere on December 9, followed by the bus. The small probes were each targeted at a different part of the planet and were named accordingly. The North probe entered the atmosphere at about 60° latitude on the day side. The Night probe entered on the night side. The Day probe entered well into the day side, and it was the only one of the four probes that continued to send radio signals back after impact, for over an hour. With no heat-shield or parachute, the bus survived and made measurements only to about 110 km altitude before burning up. It afforded the only direct view of the upper atmosphere of Venus, as the probes did not begin making direct measurements until they had decelerated lower in the atmosphere.

pitchover

The programmed turn from the vertical that a rocket takes as it describes an arc and points in a direction other than vertical.

Planck

An ESA (European Space Agency) spacecraft designed to search for tiny irregularities in the cosmic microwave background over the whole sky with unprecedented sensitivity and angular resolution. Named after the quantum theorist Max Planck (1858–1947), it is expected to provide a wealth of information relevant to several key cosmological and astrophysical issues, such as theories of the early universe and the origin of cosmic structure. Known prior to its selection as COBRA/SAMBA, this will be the third of ESA's Medium-sized Missions in the Horizon 2000 Scientific Program. It is planned for launch in 2007, together with **Herschel**.

Planet-

The prelaunch designation of a series of planetary probes built and launched by the Japanese **ISAS** agency. This series includes Planet-A (**Suisei**) and Planet-B (**Nozomi**). See **ISAS missions**.

Planet Imager (PI)

A possible future NASA mission, part of the **Origins Program**, that would follow on from the **Terrestrial Planet Finder** (TPF) and produce images of Earth-like planets. To obtain a 25×25 pixel image of an extrasolar planet would call for an array of five TPF-class optical interferometers flying in formation. Each interferometer would consist of four 8-m telescopes to collect light and to null it before passing it to a single 8-m telescope, which would relay the light to a combining spacecraft. The five interferometers would be arranged in a parabola, creating a very long baseline of 6,000 km with the combining spacecraft at the focal point of the array. The Planet Imager is identified in NASA's Office of Space Science Strategic Plan as a potential mission beyond 2007 but remains in the very early concept definition phase.

planetary protection test

A test performed to assess that the potential for contamination of (1) Earth by returned spacecraft hardware, or (2) spacecraft destinations by spacecraft hardware is within acceptable limits. It typically consists of bacterial spore counts from swabs taken from hardware surfaces.

Planetary Society, The

A nonprofit, public organization, founded in 1980 by Carl **Sagan**, Bruce **Murray**, and Louis Friedman, and based in Pasadena, California, to encourage and support, through education, public information events, and special events, the exploration of the Solar System and the search for extraterrestrial life. With a membership of more than 100,000, from over 100 countries, the Planetary Society is the largest space advocacy group in the world. It publishes the bimonthly *Planetary Report* and provides financial support for SETI hardware development. It is also developing the **Cosmos 1** solar sail project.

plasma

A low-density gas in which the individual atoms are ionized (and therefore charged), even though the total number of positive and negative charges is equal, thus maintaining overall electrical neutrality. Plasma is sometimes referred to as the fourth state of matter. A partially ionized plasma, such as Earth's ionosphere, is one that also contains neutral atoms.

plasma detector

A device for measuring the density, composition, temperature, velocity, and three-dimensional distribution of **plasma**s that exist in interplanetary regions and within planetary magnetospheres.

plasma sheath

An envelope of ionized gas that surrounds a body moving through an atmosphere of hypersonic velocity.

plasma sheet

Low energy **plasma**, largely concentrated within a few planetary radii of the equatorial plane, distributed throughout the **magnetosphere**, throughout which concentrated electric currents flow.

plasma wave detector

A device for measuring the electrostatic and electromagnetic components of local plasma waves in three dimensions.

platform

A spacecraft, usually unmanned, that serves as a base for scientific experiments.

Plesetsk Cosmodrome

A Russian launch complex and missile test range, located at 62.8° N, 40.4° E, which has launched the most satellites since the beginning of the Space Age. Built in 1960, south of Archangel in northern Russia, its existence was not officially acknowledged by the Soviet Union until 1987, but it became publicly known in the West after the tracking of Cosmos 112 by a team of schoolboys at Kettering Grammar School in England. Prior to this, it had been the source of many UFO reports from people living below the site's launch trajectories. Western journalists allowed onto the site in 1989 were told of two fatal accidents that had happened there: on June 26, 1973, nine technicians were killed in a launch pad accident, and on March 18, 1980, 50 technicians were killed by an explosion while fueling a Soyuz booster.

Plesetsk's location makes it ideal for launching into polar or high-inclination orbits (63° to 83° inclination)— those typically favored by military reconnaissance and weather satellites. Between 1969 and 1993, it was the busiest spaceport in the world, accounting for more than a third of all orbital or planetary missions. It continues to be highly active today, especially for military launches and all **Molniya**-class communications satellites.

Plesetsk has traditionally supported four launch vehicle types: Cosmos-3M, Soyuz/Molniya, Tskylon-3, and Start. The Russian government is keen to shift more activity, especially unmanned launches, away from **Baikonur** (which first surpassed Plesetsk in number of launches in 1993), since its current agreement with Kazakhstan to use that facility expires in 2014. To this end, a program has been put in place to build the launch support infrastructure for the new **Angara** vehicle at the northernmost spaceport. The first launch of this new generation heavy-lift booster should take place from Plesetsk in 2003. Meanwhile, commercial space operations using the existing launch vehicles are set to be intensified at Plesetsk.

Plum Brook Research Station

A rocket propulsion test site, located at Sandusky, Ohio, and operated by the **Glenn Research Center**. It hosts the world's largest space environment simulation chamber, in which upper-stage engines can be tested under conditions like those encountered beyond Earth's atmosphere. Among Plum Brook's other facilities are a large hypersonic wind tunnel and a tank for experimenting with cryogenic fuels. NASA is currently decommissioning two experimental nuclear reactors at Plum Brook that were used in the 1960s to study the effects of radiation on materials used in spaceflight.

Pluto/Kuiper Belt mission

Also known as New Horizons, the first mission to the outermost planet, Pluto, and its only moon, Charon. The design of this spacecraft, whose future had seemed threatened on several occasions because of NASA budget cuts, is being conducted at the Johns Hopkins University Applied Physics Laboratory, by a team that also includes members from NASA and various other academic institutions. By mid-2002, the mission had successfully completed its first major product review. New Horizons is working toward a 2006 launch, arrival at Pluto and Charon in 2015, and exploration of various objects in the Kuiper Belt, beyond Pluto's orbit, up to 2026. Mission planners are anxious to intercept the ninth planet while it is still in the near-perihelion (closest to the Sun) part of its orbit; at greater distances from the Sun, Pluto's atmosphere may completely freeze and any surface activity, such as ice geysers, become less frequent.

PMG (Plasma Motor/Generator)

An experiment that involved lowering a 500-m-long **space tether** from a spent Delta upper stage into the lower part of the **thermosphere** where atmospheric gas exists in a **plasma**, or ionized state. The tether consisted of a conducting wire with hollow electrodes at either end. An electric current was produced in the tether in line with expectations, demonstrating the potential of this technique to generate power that could be used by satellites or space stations in low Earth orbit.

Launch
    Date: June 26, 1993
    Vehicle: Delta 7925
    Site: Cape Canaveral
Orbit: 185 × 890 km

Poe, Edgar Allan (1809–1849)

An American author and a pioneer of the mystery and science fiction genres. His *Unparalleled Adventure of One Hans Pfaal*[234] (1835) ranks among the first scientifically serious tales of spaceflight and had a powerful influence

Pluto/Kuiper Belt mission An artist's rendering of a probe flying through the Pluto-Charon system. *NASA/JPL*

on future writings on this subject. His description of Earth as seen from space is surprisingly accurate, and his sealed gondola-ship is reminiscent of stratospheric balloons of the 1930s. It is no coincidence that one of his university teachers was Joseph **Tucker**.

POES (Polar Operational Environmental Satellite)

NOAA (National Oceanic and Atmospheric Administration) satellites in polar orbits that collect global data on a daily basis for a variety of land, ocean, and atmospheric applications. Two satellites, a morning and an afternoon satellite, make up the POES constellation and provide global coverage four times daily. Each carries seven scientific instruments and two for search and rescue. Their measurements are used for weather analysis and forecasting, climate research and prediction, global sea surface temperature measurements, atmospheric soundings of temperature and humidity, ocean dynamics research, volcanic eruption monitoring, forest fire detection, global vegetation analysis, and many other applications. POES and the Department of Defense's **DMSP** (Defense Meteorological Satellite Program) are currently being merged to form the NPOESS program, which will be managed by an Integrated Program Office headed by NOAA.

pogo effect

Unstable, longitudinal (up and down) oscillations induced in a launch vehicle, mainly due to fuel **sloshing** and engine vibration.

Pogue, William R. (1930–)

An American astronaut who flew aboard **Skylab** and served on the support crews for the **Apollo** 7, 11, and 14 missions. His Air Force military career included a combat tour during the Korean conflict, two years as an aerobatic pilot with the Air Force's precision flying team, the Thunderbirds, and an exchange assignment with the Royal Air Force as a test pilot. In addition, he served as an assistant professor of mathematics at the U.S. Air Force Academy in Colorado Springs. In April 1966, Pogue was selected in the fifth group of NASA astronauts. Together with Gerald **Carr** and Edward **Gibson**, he flew on Skylab 4, the third and final manned visit to the orbiting laboratory, setting, along with Carr, a new record for the longest spacewalk of seven hours. In 1977, Pogue left NASA to pursue a new career as consultant to aerospace and energy companies. He worked with the Boeing Company in support of the Space Station Freedom Project and now works with them on the International Space Station Project, specializing in assembly extravehicular activity.

Polar

An Earth-observing satellite that, together with its sister spacecraft **Wind**, constitutes NASA's contribution to the International Solar Terrestrial Program (ISTP), an international effort to quantify the effects of solar energy on Earth's magnetic field. Polar is designed to measure the entry, energization, and transport of plasma into the magnetosphere. Data from Polar are correlated with data from ground-based scientific observatories and the other spacecraft in the ISTP program—Wind, **Geotail**, and **SOHO** (the first **Cluster** spacecraft would also have participated, but they were destroyed during launch)—to better understand the physical effects of solar activity on interplanetary space and Earth's space environment.

Launch
  Date: February 24, 1996
  Vehicle: Delta 7925
  Site: Vandenberg Air Force Base
Mass: 1,300 kg
Orbit: 5,554 × 50,423 km × 86.3°

Polar BEAR (Polar Beacon Experiment and Auroral Research)

An American military satellite designed to study communications interference caused by solar flares and increased auroral activity. The data from the mission complement data gathered by its predecessor, HILAT. The core vehicle was a **Transit** navigational satellite retrieved from the Smithsonian's National Air & Space Museum, where it had been on display for eight years. Polar BEAR's Auroral Imaging Remote Sensor imaged the aurora borealis, and its Beacon Experiment monitored ionospheric propagation over the poles.

Launch
  Date: November 14, 1986
  Vehicle: Scout G
  Site: Vandenberg Air Force Base
Orbit: 955 × 1,013 km × 89.6°
Size: 1.3 × 0.4 m
Mass: 125 kg

polar orbit

An orbit with an **inclination** of 90°, or very close to it. A spacecraft following such an orbit has access to virtually every point on Earth's (or some other planet's) surface, since the planet effectively rotates beneath it. This capability is especially useful for mapping or surveillance missions. An orbit at another inclination covers a smaller portion of the Earth, omitting areas around the poles. Placing a satellite into terrestrial polar orbit demands more energy and therefore more propellant than does achieving a direct orbit of low inclination. In the latter case, the launch normally takes place near the equator, where the rotational speed of the surface contributes a significant part of the final speed needed for orbit. Since a polar orbit is not able to take advantage of the free ride provided by Earth's rotation, the launch vehicle must provide all of the energy for attaining orbital speed.

polarimeter
An optical instrument that measures the direction and extent of the **polarization** of light reflected from its targets.

polarization
A state of electromagnetic radiation in which transverse vibrations take place in some regular manner–for example, all in one plane, a circle, an ellipse, or some other definite curve.

Polyakov, Valery Vladimirovich (1942–)
A Russian cosmonaut and physician who, in 1994–1995, set a new endurance record for manned spaceflight of 438 days (14½ months) aboard the Mir space station. A cardiac specialist, Polyakov is now deputy director of the Medico-Biological Institute.

Polyot
A name shared by a variety of Russian spacecraft, launch vehicles, and space organizations; "polyot" means "flight." (1) Polyot-1, launched on November 1, 1963, was the first satellite capable of maneuvering in space. (2) A design bureau (NPO Polyot). (3) A launch vehicle based on the Cosmos-3M.

posigrade motion
Orbital motion in the same direction as that which is normal to spatial bodies in a given system. In the case of Earth, this is anticlockwise looking down from the North Pole.

posigrade rocket
A small vernier rocket on a spacecraft used to control its attitude during spaceflight, the thrust of which is in the same direction as the movement of the spacecraft.

positive *g*
Acceleration experienced in the downward (head-to-feet) direction, expressed in units of gravity. Also known as "eyeballs down." See also **negative *g*** ("eyeballs up"), **prone *g*** ("eyeballs out"), and **supine *g*** ("eyeballs in").

positron
The positively charged **antiparticle** of the **electron**. When a positron and an electron meet, they destroy each other instantly and produce a pair of high-energy **gamma rays**.

potential energy
Stored energy or energy possessed by an object due to its position, for example, in a gravitational field.

Potok
An element of the second generation Soviet global command and control system. Potok satellites are integrated with the **Luch** system–the latter handling communications between spacecraft and ground stations, the former handling communications between fixed points and digital data from the Yantar-4KS1 electroptical reconnaissance satellite. About 10 Potoks have been placed in geostationary orbit since 1982 by Proton-K launch vehicles from Baikonur.

power
The rate at which energy is supplied. Power is measured in watts.

powered landing
The soft landing of a spacecraft on a celestial body achieved by firing one or more rockets in the direction opposite to that of the motion to act as brakes.

precessing orbit
See **walking orbit**.

precession
The change in the direction of the axis of rotation of a spinning body, or of the plane of the orbit of an orbiting body, when acted upon by an outside force.

prelaunch
The phase of operations that extends from the arrival of space vehicle elements at the launch site to the start of the launch countdown.

prelaunch test
A test of launch vehicle and ground equipment to determine readiness for liftoff. It may include a countdown and a flight readiness firing with all launch complex equipment operating, but it does not include the actual launch of the vehicle.

pressure feed system
The device in a liquid-propellant rocket engine that forces fuel into the **combustion chamber** under pressure.

pressure limit
The upper and lower limits of pressure within which a solid-fuel rocket motor delivers optimum performance.

pressure suit
See **spacesuit**.

pressurized cabin
A cabin in an aircraft or a spacecraft designed to provide adequate internal air pressure to allow normal respiratory and circulatory functions.

prevalve
A valve in the stage propellant feed systems that is used to keep the propellants from free-flowing out of the stage propellant tanks into the engine propellant feed system.

PRIME (Precision Recovery Including Maneuvering Entry)
The second part of the U.S. Air Force **START** project on **lifting bodies**, following **ASSET**. It involved the development and reentry testing of the SV-5D, also known as the X-23A, a 2-m-long, 1-m-wide subscale version of the lifting body configuration used in the **X-24A** manned aerodynamic test aircraft. Three missions were flown in 1966–1967, each beginning with a launch by an Atlas booster from Vandenberg Air Force Base. Outside Earth's atmosphere, the SV-5D maneuvered using nitrogen-gas thrusters, then reentered and descended to a controlled landing using an inertial guidance system and airplane-type flaps for pitch and roll control.

PRIME (Primordial Explorer)
A mission to study the birth of the first quasars, galaxies, and clusters of galaxies. PRIME is a 75-cm telescope with a large near-infrared camera that will capture images in four colors, surveying a quarter of the sky. PRIME is expected to establish the epoch during which the first galaxies and quasars formed in the early Universe, discover hundreds of Type Ia supernovae (the explosions of massive stars) for use in measuring the acceleration of the expanding Universe, and detect hundreds of small, cool stars known as brown dwarfs. PRIME was selected by NASA for study as a **SMEX** (Small Explorer).

Priroda
See **Mir**.

Private
A small, solid-propellant rocket developed for research purposes by the U.S. Army during World War II. The Private A was 2.4 m long, 0.85 m in diameter, and powered by an Aerojet solid-propellant sustainer engine, with liftoff thrust provided by four modified 11-cm barrage rockets attached by a steel casing. It had four guiding fins at the rear and a tapered nose. The rocket was launched from a rectangular steel boom employing four guide rails. A total of 24 Private A rockets were tested, and a maximum altitude of 18 km was achieved. It was followed by the Private F, designed to test different types of lifting surfaces for guided missiles. The Private F had a single guid-

PRIME (Precision Recovery Including Maneuvering Entry)
Three lifting bodies on the dry lake bed near Dryden Flight Research Center, from left to right: X-23A, M2-F3, and HL-10. *NASA*

ing fin and two horizontal lifting surfaces at the tail of the rocket, plus two wings at the forward section.

Prognoz

A series of Soviet research satellites meant to study processes of solar activity and their influence on the interplanetary environment and Earth's magnetosphere. Each spacecraft consists of a hermetically sealed cylindrical mainframe, 2 m in diameter, with a spherical base. Located inside the 0.925-m-high mainframe is the platform with instrumentation and equipment. Twelve Prognoz satellites were launched between 1972 and 1996. Prognoz 11 and 12 were also known as the **Interball** Tail Probe and Auroral Probe, respectively.

prograde

See **direct**.

Progress

An unmanned Russian supply vessel used to ferry materials to and from manned space stations. Progress was derived from the **Soyuz spacecraft**. The Soyuz descent module was replaced by a sealed unit that could carry about 1 ton of propellants for **Salyut**'s propulsion system, while the orbital module was transformed into a cargo bay and made accessible to astronauts by a docking system and a connecting hatch. In total, Progress had a mass of about 7,000 kg, including 1,000 kg of propellant for Salyut and 1,300 kg of cargo. It was first launched on January 20, 1978, to rendezvous with Salyut 6. With the arrival of **Mir**, a new version of the supply ship, known as Progress-M, was brought into service. This had a recoverable capsule that could be used to return material, such as samples of crystals grown under weightless conditions, to Earth. The latest version of Progress, called the Progress-M1, has a larger cargo capacity and has been developed for use with the **International Space Station**. Progress-M1 can carry up to 2,230 kg of cargo, of which a maximum of 1,950 kg can be propellant and a maximum of 1,800 kg can be equipment or supplies.

progressive burning

In **solid-propellant rocket motor**s, the burning of the fuel in such a way that the chamber pressure steadily rises

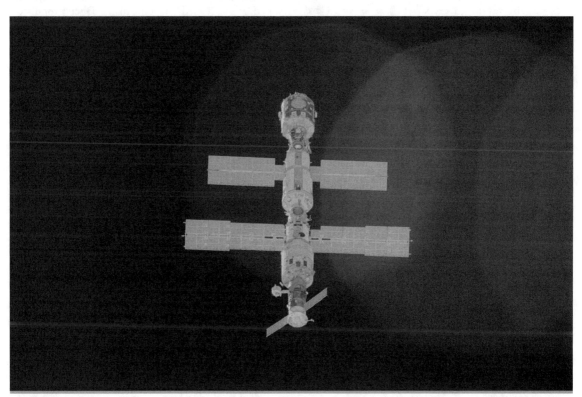

Progress   A Progress supply ship (at bottom) docked with the Zvezda module of the International Space Station in October 2000. *NASA*

throughout the burning time, thus delivering steadily increasing thrust.

Project Daedalus
See **Daedalus, Project.**

Project Longshot
See **Longshot, Project.**

Project Orbiter
See **Orbiter, Project**

Project Orion
See **Orion, Project.**

Project Rover
See **Rover, Project.**

Project Stargazer
See **Stargazer, Project.**

prone *g*
Acceleration experienced in a back-to-chest direction, expressed in units of gravity. Also known as "eyeballs out." See also **supine *g*** ("eyeballs in"), **positive *g*** ("eyeballs down"), and **negative *g*** ("eyeballs up").

propellant
A substance or substances used to furnish the exhaust material that, when allowed to escape from a rocket, provides **thrust**. Most rocket engines use chemical propellants, which can be classified as **liquid propellants**, **solid propellants**, or **gaseous propellants** depending on their physical state. Liquid propellants can be further subdivided into **bipropellants** and **monopropellants**. In an **ion propulsion** system, the propellant particles are first ionized and then accelerated to yield a high-speed exhaust.

propellant storage
Liquid rocket **propellants** need special storage and handling, the most appropriate methods depending on the type of **fuel** and **oxidizer** used. On most launch vehicles the fuel and oxidizer tanks are stacked vertically, with the fuel tank on top so that its greater density shifts the center of gravity forward and thereby increases the vehicle's stability. Typically, the space above the remaining propellant is pressurized by external gas lines at the top of each tank. This prevents a vacuum from developing and keeps the propellant flowing smoothly. The gas pressurization lines may use inert gas from a separate tank, or products from the engine itself. **Cryogenic propellants** have to be carefully insulated and, prior to launch, recirculated through an umbilical to an external cooler. Many tanks, and all cryogenic tanks, have a bleed valve to allow high-pressure gases to escape. Because cryogenic tanks aboard spacecraft cannot store propellants for long, they are usually limited to launch vehicles.

propellant stratification
The phenomenon of uneven temperature, density, and pressure distribution in a **propellant**. Propellant stratification increases tank pressure, which necessitates extra venting or releasing of gaseous propellant.

propellant utilization
The precise control over the ratio of **fuel** to **oxidizer** during operation of a liquid rocket.

propellant-mass ratio
The ratio of the effective **propellant** mass in the propulsion system to the gross mass.

propulsion
The means by which a body's motion can be changed. **Jet propulsion** works by ejecting matter from a vehicle. Two special cases of this are **rocket propulsion** and **duct propulsion**. In the former, thrust is produced by ejecting stored matter called the propellant. **Duct propulsion** involves **air-breathing engine**s that use mostly the surrounding medium together with some stored fuel.

propulsion system
The system that propels a space vehicle, comprising rocket motors (solid- or liquid-fueled), liquid-fuel tanks, pumps, ignition devices, combustion chambers, and so forth.

Prospector
An early proposed American unmanned lunar explorer able to move around the Moon's surface after landing.

Prospero
Britain's first and only satellite launched by an indigenous rocket. It tested equipment for future satellites and carried an experiment to measure the incidence of micrometeoroids. After 10 years in orbit, the satellite was still occasionally reactivated and telemetry received at the Lasham ground station. See **Britain in space.**

Launch
    Date: October 28, 1971
    Vehicle: Black Arrow
    Site: Woomera
Orbit: 531 × 1,403 km × 82.0°
Mass: 66 kg

quantum mechanics

A major branch of modern physics concerned with the way matter and energy behave on a very small scale.

quantum vacuum

"Empty space," when seen at the subatomic level, is far from empty; instead, it seethes with energy, the rapid appearance and disappearance of all sorts of elementary particles, and fluctuations in the fabric of **space-time** itself. It is the subject of the **Quantum Vacuum Forces Project** and the source of **zero point energy**, which is seen as a potential basis for radically new forms of propulsion in the far future.[256]

Quantum Vacuum Forces Project

A project, sponsored by the **Breakthrough Propulsion Physics Program**, to study ways of influencing the **energy density** of the **quantum vacuum** with a view to possible applications in spacecraft propulsion. It is being conducted by a team of scientists from various academic and corporate institutions in the United States and Mexico.

quasistationary orbit

An orbit that is almost, but not quite, a **geostationary orbit** (GSO). In a quasistationary orbit, a satellite does not hang still relative to Earth, but moves in a complex elliptical trajectory. As seen from the ground, the path of the satellite has the form of a closed intersecting loop, elongated along the horizon, and having angular dimensions of around 30° in azimuth and 5° to 6° in elevation. Quasistationary orbits offer a number of advantages for carrying out reconnaissance (spy) tasks, including a large monitoring area and the possibility of taking multiposition bearings on radio emitters to pinpoint their location. They were first proved out by the **Canyon** series and have

become a characteristic feature of American **SIGINT** (signals intelligence) satellites. In distinction to a GSO, which has an altitude of 35,800 km and an inclination of 0°, quasistationary orbits used by SIGINT satellites have a perigee (low point) of 30,000 to 33,000 km, an apogee (high point) of 39,000 to 42,000 km, and an inclination of 3° to 10°.

QuikScat (Quick Scatterometer)

A satellite, built under a NASA rapid delivery contract, that carries the **SeaWinds** scatterometer for remote sensing of ocean winds. QuikScat is among the first missions in NASA's EOS (Earth Observing System).

Launch
 Date: June 20, 1999
 Vehicle: Titan II
 Site: Vandenberg Air Force Base
Orbit: 804 × 806 km × 98.6°

QuikTOMS (Quick Total Ozone Mapping Spectrometer)

The fifth launch of NASA's **TOMS** instrument. TOMS-5 had been scheduled to fly aboard the Russian Meteor-3M(2) satellite, but this mission was terminated in April 1999. Because of the urgent need to continue daily mapping of the global distribution of atmospheric ozone, TOMS-5 was moved to a modified Microstar bus tailored for the QuikTOMS mission and procured through the Goddard Space Flight Center's Rapid Spacecraft Development Office. It took off on September 21, 2001, aboard a Taurus rocket as a secondary payload alongside the **OrbView**-4 satellite but ended up in the Indian Ocean minutes later, after the launch vehicle malfunctioned.

R

"R" series of Russian missiles
See article, pages 335–336.

radar
A detection and range-finding system based on the comparison of reference signals with radio signals reflected, or retransmitted, from the target.

RADARSAT
A Canadian satellite equipped with a powerful **synthetic aperture radar** (SAR)–the world's first civilian satellite SAR–that is an important source of environmental and resource information. RADARSAT's steerable, 15-m-wide SAR dish collects images of the ocean and land, with a resolution of 8 to 100 m, irrespective of weather conditions and the time of day. Its data is used by shipping companies in North America, Europe, and Asia and by government agencies that carry out ice reconnaissance and mapping, and is also a valuable tool for mapping Earth's structural features, such as faults, folds, and lineaments. These features provide clues to the distribution of ground water, mineral deposits, and oil and gas in the planet's crust. RADARSAT can facilitate the mapping and planning of land use and monitor disasters such as oil spills, floods, and earthquakes. It also provides the first routine surveillance of the entire Arctic–covering the Arctic daily and most of Canada every three days from a sun-synchronous orbit.

Launch
   Date: November 4, 1995
   Vehicle: Delta 7925
   Site: Vandenberg Air Force Base
Orbit: 790 × 792 km × 99°
Mass: 2,713

radiation
(1) **Electromagnetic radiation**. (2) A stream of particles, such as **proton**s, **electron**s, or atomic nuclei, as found, for example, in **cosmic rays**. Both forms of radiation exist commonly in space. In their high-energy forms, they pose a danger to both human space travelers and electronic devices aboard spacecraft (see **radiation protection in space**).

radiation belt
A layer of trapped charged particles that surrounds a body in space. See **Van Allen Belts**.

radiation pressure
The minute pressure exerted on a surface by **electromagnetic radiation** at right angles to the direction in which the incident radiation is traveling; it is crucial to the operation of a **space sail**. The existence of this pressure was first predicted by James Clerk Maxwell in 1899 and demonstrated experimentally by Peter Lebedev. In **quantum mechanics**, radiation pressure can be interpreted as the transfer of momentum from **photon**s as they strike a surface.

radiation protection in space
High-energy **radiation** poses a threat to astronauts, especially on long missions beyond the protection of Earth's **magnetosphere**. The greatest danger comes from two types of radiation: galactic cosmic rays (GCRs) and solar particle events (SPEs). These contain charged particles (mainly protons) that are trapped by Earth's magnetic field so that spacecraft in low Earth orbit, such as the Space Shuttle and International Space Station, are relatively well protected. But the danger of radiation exposure is very real, for example, in the case of a manned mission to Mars.

GCRs are unpredictable, come from all directions, and may have extremely high energy. However, they tend to be few and far between, consisting of an occasional handful of very-high-speed particles arriving from some random direction. The danger they pose is one of cumulative radiation exposure over the duration of a long mission. An SPE, on the other hand, is quite capable of killing an unprotected person in a single burst. It is associated with the most violent of solar flares and produces X-rays (which reach Earth in minutes), energetic particles (hours), and solar plasma (days). Though the X-rays are certainly not benign, it is the high-energy particles–mostly protons and alpha particles–that are the main concern.

Fortunately, because solar activity is continuously monitored by a number of satellites, a couple of hours' advance warning of potential SPEs is possible. Also, once the particle bombardment starts, it takes several hours to reach a peak before fading again. Not that this would have

(continued on page 336)

"R" series of Russian missiles

An early family of ballistic missiles developed in the postwar Soviet Union and based initially on German expertise and experience with the V-2 (see **"V" weapons**). From some of these missiles evolved launch vehicles that made Russia a space superpower and that are still in use today. The "R" stands for "rocket," which translates from the Russian "P" for "paketa." Other designations have been given to these vehicles in the West. See **Russian launch vehicles**.

R-1

A Soviet copy of the V-2. It was built at Sergei **Korolev**'s research institute in Podlipki, to German design specifications modified only slightly by Korolev's group, and powered by a V-2-based engine designed by Valentin **Glushko**. Following 30 test flights from Kapustin Yar in 1948–1949, the missile was put into military service in November 1950. Although the R-1 was perfected to serve as a mobile weapon, during the test flights scientists from the Physics Institute of the Academy of Sciences were able to loft instruments high into the atmosphere for pure research. A series of so-called geophysical rockets, the R-1A, was also derived from the R-1. The first in this series tested separable warheads that would be used on future missiles, including the R-2 (see below). But the last two of six R-1A launches were vertical scientific flights to sample the upper atmosphere using recoverable containers placed on the rocket's tail. As the R-1A reached its peak altitude of about 100 km, its engine was shut down and the containers jettisoned to land by parachute. Later variants of the R-1 were used to study cosmic rays, high-altitude winds, and other upper-atmosphere phenomena, and to fly recoverable biological payloads.

R-2

An enlarged version of the R-1 capable of doubling the R-1's range and carrying a warhead, which dispersed a radioactive liquid at altitude and resulted in a deadly rain over a wide area around the impact point. The ethanol used in the V-2 and R-1 was replaced by methanol in the R-2 to circumvent the problem of the launch troops drinking the rocket fuel. Several vari-

ants were developed, including the R-2A, which was used to extend the scientific work of the R-1 to a height of 200 km. Some equipment tested on the R-2A also found its way onto canine flights of **Sputnik** and **Vostok**. The first production rocket was rolled out in June 1953. In December 1957, an agreement was signed to license production of the R-2 to China. From this, China acquired the technological base for its future rocket programs. See **China in space**.

R-3

A long-range missile, authorized in 1947 at the same time as the R-1 and R-2 but far more ambitious in

"R" series of Russian missiles A Soyuz U—a direct descendant of the R7 missile—lifts off from Baikonur on May 3, 2000, carrying a Neman reconnaissance satellite. *Sergei Kazak*

scope. The R-3 would have been able to deliver a three-ton atomic bomb to any point in Europe from Soviet territory—a range of 3,000 km. Although it never left the drawing board, the R-3 had a lasting effect on Soviet rocketry. It challenged Russian designers with a new level of technical complexity and paved the way for a huge growth of the Soviet rocket industry, which was soon in full swing developing the R-7. The role of the R-3, as an intermediate-range ballistic missile (IRBM), was taken over by the R-5 and R-11. However, it was not until 1962, with the R-14, that the Soviet Union would put a 3,000-km-range IRBM into service.

R-7

The world's first intercontinental ballistic missile (ICBM), known in the West as the SS-6 or Sapwood. The R-7, designed by Korolev, formed the basis of a large family of space launch vehicles, which includes the **Sputnik, Vostok, Molniya, Voskhod,** and **Soyuz**—

the most frequently launched and the most reliable rockets in spaceflight history.

R-12 and R-14

IRBMs designed by Mikhail **Yangel** from which evolved the **Cosmos** family of space launch vehicles. The R-12 and R-14 were known to NATO as the SS-4 Sandal and SS-5 Skean, respectively. In October 1962, R-12s figured in the world's most dangerous nuclear standoff, following Khrushchev's decision to place them on Cuba. The Soviets backed down in the face of a U.S. naval blockade of the island.

R-36

An ICBM with a range of 12,000 km, designed by Yangel, which tilted the strategic balance in the 1960s and became known in the West as the SS-9 Scarp, or "city buster." It also formed the basis for the **Tsyklon** family of space launch vehicles.

radiation protection in space
(continued from page 334)

helped the Apollo astronauts once they were on their way to the Moon. The Apollo missions, being relatively brief, relied on the low probability of SPEs and had no extra shielding. By contrast, on a future Mars mission both a solar flare warning system and some form of radiation protection within the spacecraft will be an absolute necessity. The protection could take the form of a small shelter with radiation-resistant walls. However, this approach has limitations. For example, it is not effective against GCRs—in fact, unless the shelter is massive (in which case it places a heavy burden on the propulsion system), it is worse than no shielding at all, because the impact of a GCR nucleus on a light shield would be to spawn secondary radiation more hazardous than the original. Since GCR cumulative doses on missions lasting more than a year may exceed the recommended maximum allowable whole-body radiation dose, mission designers are considering an alternative to the simple shelter in the form of an active electromagnetic shield. This would work like a miniature version of Earth's magnetic field—by bending the trajectory of incoming charged particles away from the region to be protected.[28, 49, 104, 189]

radio astronomy satellites
See, in launch order: **RAE-1** (Jul. 1968), **Shinsei** (Sep. 1971), RAE-2 (Jun. 1973), **SWAS** (Dec. 1988), **COBE** (Nov. 1989), **HALCA** (Feb. 1997), **MAP** (Jun. 2001), **Herschel** (2007), **Planck** (2007), **ARISE**, **RadioAstron**, and **VSOP**.

radio waves
Electromagnetic radiation spanning a wide **frequency** range from about 3 kHz to about 300 GHz, corresponding to a **wavelength** range of 100 km to 0.1 cm.

radioactivity
A phenomenon displayed by some atomic nuclei in which they spontaneously emit radiation, at the same time shifting to a lower energy state or modifying the number of **protons** and neutrons they contain. The three types of radioactive emissions are alpha particles (helium nuclei), beta particles (high-speed electrons), and gamma rays (high-energy photons).

RadioAstron
A multinational mission to study radio galaxies and quasars with unprecedented angular resolution using an orbiting 10-m radio telescope. Through coordination with a global ground-based network, the telescope will provide information about the regions surrounding black holes, the distances to pulsars, and fundamental cosmological properties like the nature of hidden mass. Scientists from more than 20 nations are collaborating to build, plan, and support the RadioAstron mission.

radioisotope heater unit (RHU)

A device used to keep instruments warm aboard a spacecraft on a lunar or interplanetary mission. Each RHU provides about one watt of heat, derived from the decay of a few grams of a radioactive substance, such as plutonium 238-dioxide, contained in a platinum-rhodium alloy cladding. RHUs were used, for example, by the **Apollo** 11 to 17 missions (1969–1972), the Soviet **Luna** 17 (1970) and 21 (1973), and **Galileo**. **Cassini** and the Huygens Probe are equipped with 117 40-gram RHUs for temperature regulation.

radioisotope thermoelectric generator (RTG)

A power source often used for deep space missions because of its long life, steady output, and independence of solar illumination. RTGs employ banks of thermoelectric elements (typically silicon-germanium unicouples) to convert the heat generated by the decay of a radioisotope,

radioisotope thermoelectric generator Engineers examine the interface surface on the Cassini spacecraft before installing the third radioisotope thermoelectric generator (RTG). The other two RTGs are the long black objects projecting to the left and right near the base of Cassini in this photo. *NASA*

such as plutonium-238 (half-life of 87.7 years), into electricity. They are relatively expensive and heavy, and, because they produce nuclear and thermal radiation that can interfere with electronics and science instruments, are generally mounted some distance away from other equipment. On the positive side, they are reliable and durable and produce plenty of power. Each RTG used on recent NASA planetary spacecraft contains approximately 10.9 kg of plutonium dioxide fuel. **Galileo** was equipped with two RTGs, **Cassini** with three. RTGs have also been used to power **Pioneer** 10 and 11, **Voyager**, **Viking**, the **Apollo** surface experiments, and, more recently, **Ulysses**.

radiometer

An instrument that detects and measures the intensity of thermal radiation, especially **infrared** radiation. Satellites often carry radiometers to measure radiation from clouds, snow, ice, bodies of water, Earth's surface, and the Sun.

radiosonde

A balloon-borne instrument for the simultaneous measurement and transmission of meteorological data.

radius vector

The line joining an orbiting body to its center of motion at any instant, directed radially outward. For a circular orbit, the center of motion coincides with the center of the circle; for a parabolic or hyperbolic orbit, the center of motion is the **focus**, and for an elliptical orbit it is one of the two foci.

Radose

A series of American Air Force and Navy satellites, launched in 1963–1964, that carried out radiation dosimeter measurements.

Raduga

A series of Soviet communications satellites in geostationary orbit stationed at either 35° or 85° E; their name means "rainbow." Some Radugas are dedicated to military use; others support national and internal communications. Several dozen of these satellites have been launched since 1975. On July 5, 1999, the Proton launch vehicle carrying a military Raduga exploded after liftoff,

scattering debris near Karaganda. As a result of this accident, the Kazakh government suspended launches from Baikonur pending Russian agreement to pay back part of the rent it owed. Raduga satellites are known internationally as "Statsionar."

RAE (Radiation Astronomy Explorer)

NASA satellites that investigated low-frequency radio emissions from the Sun and planets as well as from galactic and extragalactic sources. RAE was a subprogram of the **Explorer** series. (See table, "RAE Missions.")

rain outage

The loss of signal at Ku- or Ka-Band frequencies (see **frequency bands**) due to absorption and increased sky-noise temperature caused by heavy rainfall.

RAIR (ram-augmented interstellar rocket)

See **interstellar ramjet**.

ramjet

The simplest possible **jet engine**, involving, as it does, no moving parts; it was invented in 1913 by Lorin in France. Air entering the ramjet is compressed solely by the forward movement of the vehicle. Also known as a "flying drainpipe," it consists essentially of a long duct into which fuel is fed at a controlled rate. Although straightforward in principle, the ramjet demands careful design if it is to work efficiently, and significant design variations are called for depending on the operating speed of the vehicle. In any case, a ramjet will function effectively only above a certain speed, so that below this speed some form of auxiliary propulsion system is needed. See **interstellar ramjet**.

Ramo, Simon (1913–)

An aerospace engineer, a leader of the scientific advisory teams for **Atlas**, **Thor**, and **Titan**, and a strong proponent of the use of electronics in spacecraft development. Ramo began his career with General Electric before moving to Hughes Aircraft, where he rose to become vice president of operations. Later, Ramo co-founded TRW (Thompson-Ramo-Wooldridge Corporation) and became scientific director for the American intercontinental guided missile

RAE Missions

Spacecraft	Launch			Orbit	Mass (kg)
	Date	Vehicle	Site		
Explorer 38	Jul. 4, 1968	Delta J	Vandenberg	5,835 × 5,861 km × 120.9°	190
Explorer 49	Jun. 10, 1973	Delta 1914	Cape Canaveral	Lunar orbit	328

program (1954–1958). When TRW merged with Fujitsu, Ramo was appointed vice-chairman and then chairman of the TRW-Fujitsu board. He also served as a member of the White House Energy Research and Development Advisory Council, the advisory committee on science and foreign affairs for the State Department, and chairman of the President's Commission of Science and Technology. Ramo earned a Ph.D. in electrical engineering from the California Institute of Technology (1960).

range safety officer

The person responsible for the safe launching of a spacecraft. The range safety officer has the authority to order the remote destruction of an unmanned rocket if the rocket shows signs of flying out of control.

Ranger

The first of three American programs of unmanned spacecraft intended to pave the way for the **Apollo** lunar landings. The Ranger probes were designed to return data en route to the Moon and then crash into the lunar surface, sending back images from an altitude of about 1,300 km up to the point of impact. Rangers 1 and 2 were test missions. Rangers 3, 4, and 5 carried a capsule, containing a seismometer, which was intended to be jettisoned and then decelerated by retro-rocket in order to make a rough but survivable landing. However, equipment failures dogged the first six missions, some of them blamed on heat-sterilization procedures meant to destroy microbes and thus avoid contaminating the Moon. Because of these failures, NASA announced that henceforth unmanned lunar landing spacecraft, including Rangers and Surveyors, would be assembled in clean rooms and treated with germ-killing substances to prevent contamination without damaging sensitive electronic components. The final three flights of the series, Ranger 7, 8, and 9, were a complete success, returning a total of more than

Ranger Ranger spacecraft. *NASA*

17,000 photos with a resolution of 0.25 to 1.5 m. Millions of Americans followed live TV coverage of the final descent of Ranger 9 into the crater Alphonsus. The Ranger program was followed by **Lunar Orbiter** and **Surveyor**. (See table, "Ranger Missions.")

Launch
    Vehicle: Atlas-Agena B
    Site: Cape Canaveral

Ranger Missions

Spacecraft	Launch	Lunar Impact	Impact Site	No. of Photos	Mass (kg)
Ranger 1	Aug. 23, 1961	—	—	—	306
Ranger 2	Nov. 18, 1961	—	—	—	304
Ranger 3	Jan. 26, 1962	—	—	—	327
Ranger 4	Apr. 23, 1962	Apr. 26, 1962	Far-side	—	328
Ranger 5	Oct. 18, 1962	—	—	—	340
Ranger 6	Jan. 30, 1964	Feb. 2, 1964	Sea of Tranquility	—	362
Ranger 7	Jul. 28, 1964	Jul. 31, 1964	Sea of Clouds	4,316	362
Ranger 8	Feb. 17, 1965	Feb. 20, 1965	Sea of Tranquility	7,137	366
Ranger 9	Mar. 21, 1965	Mar. 24, 1965	Crater Alphonsus	5,814	366

ranging

Techniques for determining the distance of a satellite or a spacecraft from a ground-based tracking system.

Rauschenbach, Boris Viktovich (1915–2001)

A Russian rocket engineer and Academician who headed the development of space vehicle control systems in the Soviet Union during the first 10 years of the Space Age, beginning with the launch of **Sputnik 1** in 1957. While a student in St. Petersburg in the 1920s, he carried out research at the famous Gas Dynamics Laboratory. In 1937, having graduated from the Institute of Aviation, he joined the **Moscow Scientific Rocket Research Institute**, where he began work with Sergei **Korolev**. During his career, he knew and spoke with Hermann **Oberth**, and in 1994, he published an important biography of the German rocket pioneer.[240]

RBM (Radiation Belt Mappers)

A mission that will result in a better understanding of the origin and dynamics of Earth's **radiation belt**s, and determine the temporal and spatial evolution of penetrating radiation during magnetic storms. RBM is a future mission in NASA's Living with a Star initiative.

reactant

Material used to provide energy from its **reaction** and also, in the case of a chemical rocket, a source of propulsion directly from its reaction products.

reaction control system (RCS)

A type of attitude control system that uses small, low-thrust **vernier engine**s to provide three-axis control of a spacecraft in the absence of aerodynamic forces. Most RCS systems employ a cold gas such as nitrogen, although some operate using **hydrazine**. Pointing accuracies of 0.1° to 0.5° are typically achieved.

reaction engine

An engine such as a rocket or a jet that propels the vehicle to which it is attached by expelling **reaction mass**.

reaction mass

Material expelled by a rocket to provide **thrust**.

reaction nozzle

The nozzle of an attitude control system.

Reaction Research Society (RRS)

The oldest continuously running amateur rocketry group in the United States.

reaction wheel

An electrically powered wheel aboard a spacecraft. Typically, three reaction wheels are mounted with their axes pointing in mutually perpendicular directions. To rotate the spacecraft in one direction, the appropriate reaction wheel is spun in the opposite direction. To rotate the vehicle back, the wheel is slowed down. The excess momentum that builds up in the system due to external torques must occasionally be removed via propulsive maneuvers.

reactionless drive

A hypothetical means of propulsion that does not depend on Newton's third law (action and reaction are always equal and opposite) and the expulsion of **reaction mass**. Some types of reactionless drive have been proposed as a means of achieving **faster-than-light travel**. In science fiction, such a system is often referred to as a "space drive" or "star drive."

Reagan, Ronald (1911–)

American president (1981–1989), during whose first term in office the maiden flight of the **Space Shuttle** took place. In 1984, Reagan mandated the construction of an orbital space station, known at the time as "Freedom" but now called the **International Space Station**. He declared that "America has always been greatest when we dared to be great. We can reach for greatness again. We can follow our dreams to distant stars, living and working in space for peaceful, economic, and scientific gain. Tonight I am directing NASA to develop a permanently manned space station and to do it within a decade."[57]

recombination

The process by which a positive and a negative **ion** join to form a neutral molecule or other neutral particle.

recovery

(1) The location and retrieval of the astronauts, scientific samples, data, and spacecraft at the termination of the mission. (2) The time and date when a spacecraft lands and human, animal, and plant organisms are reintroduced to Earth gravity; sometimes abbreviated to R+0, to indicate zero days after recovery.

Redstone

The first operational U.S. ballistic missile, and a rocket that played a key role in America's nascent space program. It was developed by Wernher **von Braun** and his German rocket team, together with hundreds of General Electric engineers, military personnel, and others, who staffed the newly formed Ordnance Guided Missile Center (OGMC) at the Redstone Arsenal in Huntsville, Alabama. Hardly had the OGMC been set up than the Korean War broke out, in June 1950. The Center was tasked with carrying out a feasibility study for a ballistic surface-to-surface missile with a range of 800 km (500 miles), a project that swiftly rose in priority as the conflict in Korea intensified and concerns grew about the Soviet ballistic missile threat. Rather than build a propulsion system from scratch, the OGMC designers decided to use a modified form of the engine developed by North American Aviation for the **Navaho**; this burned a mixture of liquid oxygen (LOX) and alcohol (75% ethanol and 25% water). As the program went on, the Army changed the desired range to 320 km (200 miles), with the result that the missile could now carry heavier payloads, including, if necessary, nuclear warheads (which, at the time, were relatively heavy). The Redstone (earlier called Ursa, and then Major, before being named, in April 1952, after its birthplace) had much in common with its ancestor, the V-2 (see **"V" weapons**), including its method of trajectory control, using vanes attached to four fixed fins at its base to deflect the engine's exhaust. Before launch, the rocket was supported on a circular stand fitted with a conical flame deflector. After launch, it was steered using an inertial guidance system based on air-breathing gyroscopes. The reentry vehicle separated from the rest of the missile at engine cutoff. Following its maiden launch, with moderate success, from Cape Canaveral, on August 20, 1953, the Redstone was fired 36 more times before being declared operational in June 1958. Chrysler was granted a production contract and the missile was deployed in West Germany. After just five years, it was rendered obsolete in the field by the Pershing. Yet by this time, the Redstone had secured its place in history—not as a weapon but as a space launch vehicle. As the first stage booster for the **Jupiter** C, later renamed the **Juno** I, it helped loft America's first satellite, **Explorer** 1. Just over three years later, modified and mated to a **Mercury** capsule (see **Mercury-Redstone**), it carried the first American astronaut, Alan **Shepard**, and the second, Virgil **Grissom**, on their 15-minute suborbital flights. It also remained at the heart of future designs for the Juno rocket family. The final generation, called the Juno V, was to have employed a first stage array of eight Redstones clustered around a Jupiter core. Although the Juno family was renamed **Saturn**, the Saturn I bore a striking resemblance to the proposed Juno V. Effectively, the mighty Saturns, which were instrumental in taking man to the Moon, were direct descendants of the modest Army missile that had first taken America into space.

Length: 21.0 m
Diameter: 1.8 m
Thrust: 335,000 newtons

---

reentry

The period of return to Earth when a spacecraft passes through the atmosphere before landing. During reentry, the spacecraft decelerates and is heated intensely due to aerodynamic friction. Radio communication may be blacked out for several minutes as a **plasma sheath**—an envelope of ionized air—surrounds the vehicle.

Reentry (test program)

A series of suborbital tests carried out by NASA in the 1960s to evaluate various types of heat ablative materials (see **ablation**) and atmospheric **reentry** technology, particularly in preparation for the **Apollo** program. The nosecones of the **Scout** rockets used in these tests were coated in heat-shield ablators and then caused to reenter the atmosphere at around 28,000 km/hr—the speed of reentry after a lunar excursion.

reentry corridor

A narrow corridor along which a spacecraft must travel during **reentry** in order to pass safely through the atmosphere and achieve a successful landing. The upper and lower margins of this corridor are known, respectively, as the **overshoot boundary** and the **undershoot boundary**.

Reentry Flight Demonstrator (RFD)

Two spacecraft launched on suborbital flights in 1963–1964 to test **reentry** effects on nuclear reactor mockups supplied by the U.S. Atomic Energy Commission (AEC). The tests were carried out at a time when NASA and AEC were engaged in the **NERVA** (Nuclear Engine for Rocket Vehicle Applications) program.

reentry thermal protection

Shielding that must be fitted to a spacecraft if it is to survive the intense heat generated during **reentry**, when the vehicle's high kinetic energy is transferred to the atmosphere. In order to use the upper atmosphere to slow down rapidly from a speed of many thousands of km/s, a reentering spacecraft must present foremost as large a surface area as possible—an orientation guaranteed to generate a huge amount of heat. One way to dissipate this thermal energy is with a heat-shield that works by **ablation**, that is, by parts of it melting or vaporizing and breaking off in order to carry the heat harmlessly away. This technique was used by reentering **Mercury**, **Gemini**, and **Apollo** spacecraft. Soviet manned capsules, which were spherical in shape and not oriented in any special way for reentry, simply had an all-over ablative covering.

The **Space Shuttle** is fitted with a far more complex reentry thermal protection system that works by dissipating heat rather than by ablation. This system includes ceramic tiles covering the underside and cockpit area, felt and ceramic blankets on the upper fuselage, and carbon-carbon composites along the wing's leading edges and nosecone. The Shuttle's thermal tiles come in three materials and two colors, are 2.5 to 7.5 cm thick, and cover most of the Orbiter's undercarriage and wings. All three materials are composites made mostly of silica fiber with various additives, heat-treated in a way similar to the firing of ceramics. They can be coated with either white or black glass. White tiles can be used in areas where the temperature does not exceed 650°C, black tiles where it does not exceed 1,260°C. Because the tiles are numbered and shaped individually to fit only at certain places, their replacement is both difficult and time-consuming. The leading edges of the Shuttle, which get hottest during reentry, are protected by panels made of reinforced carbon-carbon (RCC) that are fixed to the vessel's structural skeleton by floating joints to allow for differences in thermal expansion. The RCC panels can tolerate a maximum temperature of 1,630°C. At the other end of the spectrum of sophistication, the Chinese have successfully used a very low-tech approach for protecting their SKW series of recoverable satellites—panels of oak.

reentry vehicle

A space vehicle designed to return with its payload to Earth.

reentry window

The area at the limits of Earth's atmosphere through which a spacecraft, in a given trajectory, can pass to accomplish a successful reentry.

Rees, Eberhard F. M. (1908–)

One of the leading members of Wernher **von Braun**'s V-2 rocket team (see **"V" weapons**). A graduate of the Dresden Institute of Technology, Rees began his career in rocketry in 1940, when he became technical plant manager of the German rocket center at **Peenemünde**. In 1945, he came to the United States and worked with von Braun at Fort Bliss, Texas, moving to Huntsville in 1950, when the Army transferred its rocket activities to the Redstone Arsenal. From 1956 to 1960, he served as deputy

reentry thermal protection   Accompanied by former astronaut Michael McCulley, members of the STS-82 crew look at thermal protection system tiles under the Space Shuttle *Discovery* on the runway at the Shuttle Landing Facility shortly after the conclusion of a 10-day mission to service the Hubble Space Telescope. *NASA*

director of development operations at the **Army Ballistic Missile Agency**, and then as deputy to von Braun at the **Marshall Space Flight Center**. In 1970, he succeeded von Braun as Marshall SFC's director before retiring in 1973.

regenerative cooling
The cooling of part of an engine by the propellant. A regeneratively cooled engine is one in which the **fuel** or **oxidizer**, or both, is circulated around the **combustion chamber** or **nozzle**.

regimes of flight
Ranges of speed defined relative to the local **speed of sound**. They are **subsonic**, **transonic**, **supersonic**, and **hypersonic**.

regressive burning
In solid-propellant rocket motors, the burning of the fuel in such a way that the chamber pressure steadily decreases throughout the **burn time**, thus delivering steadily decreasing **thrust**.

relativistic effects
Several peculiar effects, predicted by Einstein's **special theory of relativity**, that come into play when objects— such as spacecraft—travel at speeds that are a substantial fraction of the **speed of light**. These effects include time dilation, mass increase, and length contraction. Time dilation refers to the slowing down of the passage of time at very high speeds, effectively making it possible to spend energy to buy time. A potentially important consequence follows for interstellar flight: simply by traveling fast enough, an astronaut can reach any destination within a specified amount of (shipboard) time. The table ("Shipboard Travel Times to Arcturus") shows the travel times, as measured by onboard clocks, to the star Arcturus, which is located 36 light-years away, for a spacecraft traveling at various speeds.

Shipboard Travel Times to Arcturus

Speed (fraction of light-speed)	0.1	0.5	0.9	0.99	0.999
Travel time (years)	35.8	31.2	15.7	5.1	1.6

While time dilation enables, in principle, a spacecraft to reach anywhere in the Galaxy (or the Universe) within a human lifetime simply by traveling fast enough, there is a serious downside. In exploiting this effect, astronauts would age less than everyone else, including their friends and family, who remained behind. For very long journeys at high fractions of the speed of light, the time dislocation would be so great that many generations, and even millennia, might pass on the home planet before the interstellar travelers returned. For example, an excursion from Earth to Rigel—900 light-years away—and back, at a constant 99.99% of light-speed, would take 1,800 years as measured on Earth but only about 28 years as experienced by those on the spacecraft.

The other major consequence for space travel at relativistic speed is the increase in mass of a spacecraft. This would make it more and more difficult to continue to accelerate the vehicle. The factor that determines the amount of mass increase and time dilation is called gamma (γ). For an object moving with speed v relative to an observer considered to be at rest (for example, on Earth), γ is given by

$$\gamma = 1 / \sqrt{(1 - v^2/c^2)}$$

where c is the speed of light.

The relativistic mass, m, of a body moving at velocity v is then

$$m = \gamma m_0 = m_0 / \sqrt{(1 - v^2/c^2)}$$

where m_0 is the rest mass. Note that when $v = 0$, this reduces to the non-relativistic result $m = m_0$. The impossibility of accelerating an object up to the speed of light is shown by the fact that when $v = c$, m becomes infinite.

Similarly, the relativistic time dilation is given by:

$$t = t_0 / \gamma = t_0 \sqrt{(1 - v^2/c^2)}$$

Relay
Two spacecraft, designed and built by RCA, to test intercontinental satellite communications technology. Relay was a communications satellite of the active repeater type, in which signals from one ground station were picked up and rebroadcast to another station by the satellite's internal equipment. Relay 1 remained operational for more than two years; Relay 2 was used in thousands of tests and experiments and in some 40 public demonstrations through September 1965. See **Telstar**. (See table, "Relay Missions.")

Launch
  Vehicle: Delta B
    Site: Cape Canaveral
Mass: 78 kg

Relay Missions

Spacecraft	Launch Date	Orbit
Relay 1	Dec. 13, 1962	1,319 × 7,440 km × 47.5°
Relay 2	Jan. 21, 1964	1,961 × 7,540 km × 46.4°

remaining mass
The mass of a spacecraft that reaches space after all its launch stages have been used and discarded.

Remek, Vladimir (1948–)
The first non-American/non-Soviet in space. Remek, a Czech, flew to the **Salyut** 6 space station as a guest cosmonaut on Soyuz 28 in 1978.

Remote Manipulator System (RMS)
Two large robot arms and associated equipment, one of which is attached to the **Space Shuttle** Orbiter for payload deployment and retrieval, the other to the **International Space Station** (ISS). The RMSs were designed, developed, tested, and built by the Canadian company Spar Aerospace, with some components from other Canadian firms, and so are also known as Canadarm (Shuttle version) and Canadarm 2 (ISS version).

The Shuttle's RMS is 15.3 m long and 38 cm in diameter and weighs 408 kg. It has six joints that correspond roughly to the joints of the human arm, with shoulder yaw and pitch joints, an elbow pitch joint, and wrist pitch, yaw, and roll joints. The end effector is the unit at the end of the wrist that actually grabs, or grapples, the payload. Two lightweight boom segments, called the upper and lower arms, connect the shoulder and elbow joints and the elbow and wrist joints, respectively. The Shuttle RMS can handle payloads with masses up to

29,500 kg. It can also retrieve, repair, and deploy satellites; provide a mobile extension ladder for spacewalking crew members; and be used as an inspection aid to allow flight crew members to view the orbiter's or payload's surfaces through a TV camera on the arm. One flight crew member operates the RMS from the aft flight deck control station, and a second flight crew member usually assists with television camera operations.

The 17.6-m-long RMS fitted to the ISS in April 2001 is crucial to the rest of the Space Station's assembly. It forms the main component of the Mobile Service System (MSS) for moving equipment and supplies around the ISS, supporting astronauts working in space, and servicing instruments and other payloads attached to the Station. The other parts of the MSS are the Special Purpose Dextrous Manipulator (SPDM), a small, highly advanced detachable two-armed robot that can be placed on the end of the RMS for doing detailed assembly and maintenance work, and the Mobile Remote Service Base System (MRSBS), a movable platform for the RMS and SPDM that slides along rails on the Station's main truss structure.

remote sensing
The observation of Earth from distant vantage points, usually by or from satellites or aircraft. Sensors, such as cameras, mounted on these platforms capture detailed pictures of Earth that reveal features not always apparent to the naked eye.

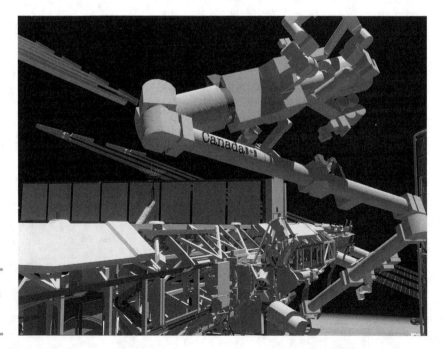

Remote Manipulator System
An artist's rendering of the RMS attached to the International Space Station. *NASA*

rendezvous

The meeting and joining together of two spacecraft in orbit at a prearranged place and time.[139]

resistojet

The simplest form of **electric propulsion**. A resistojet works by super-heating a propellant fluid, such as water or nitrous oxide, over an electrically heated element and allowing the resulting hot gas to escape through a converging-diverging nozzle. Thrust and specific impulse (a measure of the engine's efficiency) are limited by the material properties of the resistor. Resistojet thrusters, using a variety of propellants, are being tested for both low-Earth orbit and deep space missions. For example, the microsatellite UoSAT-12 carries 2.5 kg of nitrous oxide, sufficient for 14 hours running. A 60-minute firing of its resistojet would raise its 650-km orbit by 3 km.

rest mass

The mass of an object when at rest, relative to an observer.

Resurs

Zenit-derived satellites used for Earth resources and military studies as part of the Resurs ("resource") and Gektor-Priroda project. The first was launched in 1975.

retrofire

See **de-orbit burn**.

retrograde orbit

(1) A satellite orbit that goes east to west—the direction opposite to that of Earth's rotation. (2) In general, for any orbital or axial rotation in the Solar System, one that is clockwise as observed from above the Sun's north pole.

retro-rocket

A small rocket engine used to reduce the velocity of a spacecraft by providing **thrust** in the direction opposite to that of the motion. Also known as a **braking rocket**.

reusable launch vehicle (RLV)

A launch vehicle that can be refurbished and used for repeated flights.

revolution

The movement of one object around another. For artificial satellites, revolutions are figured relative to a given point on the planet's surface. Because Earth is spinning west to east, a spacecraft in a west-to-east orbit takes longer to complete one revolution than to complete one orbit relative to a fixed point in space. For example, the Space Shuttle may make 16 orbits of Earth in a day but only 15 revolutions.

REX (Radiation Experiment)

U.S. Air Force satellites designed to study scintillation effects of Earth's atmosphere on radio transmissions. REX 2 was also the first spacecraft to successfully employ GPS (Global Positioning System) navigation for full closed-loop attitude control. (See table, "REX Missions.")

Launch site: Vandenberg Air Force Base
Mass: 85 kg
Size: 0.6 × 0.8 m

RFNA (red fuming nitric acid)

See **nitric acid**.

RHESSI (Reuven Ramaty High Energy Solar Spectroscopic Imager)

An **SMEX** (Small Explorer) mission designed to explore the basic physics of particle acceleration and energy release in solar flares. It carries out simultaneous, high resolution imaging and spectroscopy of solar flares from 3 keV X-rays to 20 MeV gamma rays with high time resolution. "Reuven Ramaty" was added to the satellite's original name, HESSI, two months after takeoff, in memory of the physicist who was a driving force behind the mission and who died in 2001.

Launch
    Date: February 5, 2002
    Vehicle: Pegasus XL
    Site: Cape Canaveral
Orbit: 597 × 600 km × 38°
Mass: 293 kg

REX Missions

| Spacecraft | Launch | | Orbit |
	Date	Vehicle	
REX 1	Jun. 29, 1991	Scout G	766 × 869 km × 89.6°
REX 2	Mar. 9, 1996	Pegasus XL	799 × 835 km × 90.0°

Rhyolite

A series of U.S. **SIGINT** (signals intelligence) satellites placed in **quasistationary orbit**s in the 1970s to intercept, for example, telemetry in radio nets controlling the flight of Soviet bombers. Rhyolite satellites were also used to spy on communications during local conflicts in Vietnam, and also between India and Pakistan. When the project's code name was revealed during the espionage trial of Boyce and Lee—the "Falcon and the Snowman"—it was subsequently (from Rhyolite 3 on) changed to "Aquacade."

Ride, Sally Kristen (1951–)

The first American woman to fly in space. Ride was chosen as an astronaut in 1978 and served as a mission specialist for STS-7 (1983) and as a mission specialist on STS 41-G (1984). Following the *Challenger* **disaster** she was elected as a member of the presidential commission to investigate the accident and chaired a NASA task force (1986–1987) that prepared a report on the future of the civilian space program.[184] Ride resigned from NASA in 1987 to join the Center for International Security and Arms Control at Stanford University. In 1989, she assumed the directorship of the California Space Institute, part of the University of California at San Diego.

Riedel, Klaus (1901–1944)

The head of the test laboratory at Peenemünde, who first developed and perfected the handling of large rockets on the ground. Riedel was reportedly the true genius of the **Verein für Raumschiffahrt** team at Reinickendorf before World War II, although he had no formal training. He was killed in an automobile accident near the end of the war.

RKA

See **Russian Space Agency**.

RME (Relay Mirror Experiment)

Also known as USA 52 and Losat-R; RME was launched as a dual payload with **LACE** (Low-power Atmospheric Compensation Experiment). Both satellites carried defense experiments to help validate the technology needed for a space-based antimissile laser system. RME was designed to show stabilization, tracking, and pointing technologies working at the performance levels needed for Strategic Defense Initiative (SDI) missions. Its 61-cm mirror tested laser pointing technology by deflecting ground-based beams back to Earth. A problem occurred immediately after launch, when RME's attitude control malfunctioned, shutting off a reaction wheel.

Sally Ride Ride, mission specialist on STS-7, monitors control panels from the pilot's chair on the flight deck. Floating in front of her is a flight procedures notebook. *NASA*

The Maui Optical Observatory, on top of Mount Haleakala, established the first relay on June 26, 1990, with Kihei, Hawaii. RME's payload also included the Wideband Angular Vibration Experiment (WAVE), which measured low-level angular vibrations affecting the performance of acquisition, tracking, and pointing (ATP) systems.

Launch
  Date: February 14, 1990
  Vehicle: Delta 6925
  Site: Cape Canaveral
  Orbit: 261 × 281 km × 43°
  Mass: 1,040 kg

---

rocket

A projectile driven by reaction (jet) propulsion that carries its own **propellants**. A rocket is therefore independent of Earth's atmosphere in terms of both **thrust** and **oxidant**. See **rocket principle**.

rocket equation

The fundamental equation of rocketry. First derived by Konstantin **Tsiolkovsky** in 1895 for straight-line rocket motion with constant exhaust velocity, it is also valid for elliptical trajectories with only initial and final impulses. The rocket equation, which can be obtained from **Newton's laws of motion**, shows why high **effective exhaust velocity** has historically been a crucial factor in rocket design: the **payload ratio** depends strongly upon the effective exhaust velocity. In its simplest form the rocket equation can be written as

$$v = v_e \, ln(m_i \, / \, m_f)$$

where v is the maximum velocity of the rocket in gravity-free, drag-free flight, v_e is the effective exhaust velocity, ln is the natural logarithm, m_i is the initial or total rocket mass, and m_f is the final or empty rocket mass ($m_i \, / \, m_f$ is the payload ratio).

rocket principle

All rockets exploit Newton's third law of motion, namely, that action and reaction are equal and opposite. To appreciate the physics involved, imagine standing on a sledge on a frozen pond with a pile of 10 large pebbles on board, each with a mass of one kilogram. Suppose the total mass of you, the sledge, and the pebbles is 90 kg, and that there is absolutely no friction between the sledge and the ice. You pick up one of the pebbles and throw it from the back of the sledge at a speed of 10 m/s. According to Newton's third law, as the pebble shoots off in one direction, the sledge and its payload (you and the remaining pebbles) must move in the opposite direction.

Because the sledge plus payload now has a mass of 89 kg, the speed it acquires will be 1/89th that of the ejected pebble, or 11.24 cm/s. In real life, the sledge would immediately begin to slow down and eventually come to a halt because of friction between the sledge's runners and the ice. But since we have assumed zero friction, the sledge continues to move at whatever speed it has acquired. You now hurl another pebble overboard exactly as before. The sledge plus payload gains another increment of speed, but this time a marginally greater one than before, because its overall mass has dropped by one kilogram. The speed increase is 1/88 times 10 m/s, or 11.36 cm/s, so that the total speed is 22.60 cm/s. After all 10 pebbles have been thrown overboard, the speed of you and the sledge has climbed to 118.47 cm/s. This is the final speed, since nothing remains to be jettisoned.

Ignoring the effects of friction may seem unfair. But, on the contrary, it makes our example closer to the case of a real rocket, since friction-free conditions prevail in space. In fact, it is the absence of friction that dictates that rockets are the *only* means of space propulsion available, at least for the foreseeable future. A common mistake is to assume that rockets work by pushing against something, just as we move by pushing our feet against the ground, making use of the frictional force between two surfaces. But in space, there is nothing to push against.

Our example of the sledge and pebbles also points out one of the biggest problems in spaceflight. This is the need for a rocket to carry its own **reaction mass**–the mass it has to expel in order to gain speed. In the case of a jet aircraft, for example, which also works by the principle of action and reaction, new reaction mass is continually obtained from the surrounding air. A jet engine takes in air as it moves along (aided, especially at low speeds, by fast-spinning turbine blades), heats and compresses this air by using it to burn propellant, and then allows the resulting exhaust gases to expand and to escape as high-speed reaction mass at the rear. A rocket cannot do this. It must, with the interesting exception of the **interstellar ramjet**, be completely self-contained and have all of its reaction mass onboard from the outset. Unlike with an aircraft, this reaction mass also doubles as the **propellant**–the mixture of **fuel** and **oxidant** that is burned in the **combustion chamber** of the rocket to release energy. The resulting hot exhaust gases are then allowed to escape from a nozzle and, in accordance with Newton's third law, provide a forward **thrust**.

rocket propulsion

A type of reaction propulsion in which the propulsive force is generated by accelerating and discharging matter contained in the vehicle (as distinct from jet propulsion).[26]

rocket sled

A sled that runs on a rail or rails and is accelerated to high velocities by a rocket engine. Such devices were used, prior to the first manned space flights, for determining the *g* tolerance of subjects and in developing crash-survival techniques. See **Stapp, John**.

rocketry, history

See, in chronological order: **Archytas** (c. 428–c. 350 B.C.), **Hero of Alexandria** (first century A.D.), **Albertus Magnus** (1193–1280), Roger **Bacon** (c. 1214–c. 1292), **Muratori** (fourteenth century), Jean **Froissart** (c. 1337–1410), Giovanni da **Fontana** (1395–1455), Vannoccio **Biringuccio** (1480–1537), Conrad **Haas** (c. 1509–1579), Johan **Schmidlap** (sixteenth century), **Cyrano de Bergerac** (1618–1655), **Tipu Sultan** (c. 1750–1799), William **Congreve** (1772–1828), Nikolai **Kibalchich** (1854–1881), Hermann **Ganswindt** (1856–1934), Konstantin **Tsiolkovsky** (1857–1935), Alfred **Maul** (1828–1907), Y. P. G. **Le Prieur** (1885–1963), Pedro **Paulet**, Walter **Hohmann** (1880–1945), Guido **von Pirquet** (1880–1966), Robert **Esnault-Pelterie** (1881–1957), Robert **Goddard** (1882–1945), Franz **von Hoefft** (1882–1954), Hermann **Oberth** (1894–1989), Fridrikh **Tsander** (1887–1933), Louis **Damblanc** (1889–1969), Max **Valier** (1893–1930), Friedrich **Schmiedl** (1902–1994), James **Wyld** (1913–1953), Hsue-Shen **Tsien** (1909–), Eugen **Sänger** (1905–1964), and Wernher **von Braun** (1912–1977). Also see: **Verein für Raumschiffahrt, "A series of German rockets", "V" weapons, guided missiles, postwar development, NACA, Space Shuttle**.

Rockoon

A small sounding rocket launched from a high-altitude balloon. The first Rockoon was sent up in 1952 from the icebreaker *Eastwind* off Greenland by a research group headed by James **Van Allen**. It could carry a 9-kg payload to an altitude of 95 to 110 km. Rockoons were used in great numbers by Office of Naval Research and University of Iowa research groups from 1953 to 1955 and during the International Geophysical Year 1957 to 1958, from ships in the sea between Boston and Thule, Greenland.

Rockot

A three-stage launch vehicle based on the RS-18 (SS-19, or "Stiletto," as it was known to NATO) intercontinental ballistic missile—the most powerful missile in the Russian arsenal. Rockot is supplied and operated by Eurockot Launch Services GmbH of Bremen, Germany, a joint venture of Astrium GmbH and Krunichev State Research and Space Production Center. Eurockot markets existing RS-18 two-stage ICBMs with the restartable Breeze third stage for placing commercial spacecraft into low Earth orbit. Rockot successfully launched its first commercial payload, two Intersputnik communications satellites, in February 2001, and launched **GRACE** (Gravity Recovery and Climate Experiment) on March 17, 2002. Launches take place from a purpose-built pad at **Plesetsk**.

Length: 29 m
Diameter: 2.5 m
Launch mass: 107 tons
Payload to LEO: 1,900 kg

Rockwell

A major U.S. aerospace company that was taken over by **Boeing** in 1996. It had evolved in 1966 from the merger of **North American Aviation** and Rockwell Standard to form North American Rockwell, which changed its name to Rockwell International in 1973. Rockwell is best known for having developed and built the **Space Shuttle** Orbiter.

Roentgen Satellite

See **ROSAT**.

Rohini

Indian scientific satellites and developmental flights of the **SLV-3** launch vehicle. The Rohinis (Sanskrit for "Aldebaran") carried cameras for remote-sensing and radio beacons for accurate orbit and attitude determination. (See table, "Rohini Missions.")

Launch
Vehicle: SLV
Site: Sriharikota

Rohini Missions

Spacecraft	Launch Date	Orbit	Mass (kg)
Rohini 1A	Aug. 10, 1979	Failed to reach orbit	35
Rohini 1	Jul. 18, 1980	307 × 921 km × 44.7°	35
Rohini 2	May 31, 1981	Partial failure; orbit too low	38
Rohini 3	Apr. 17, 1983	326 × 485 km × 46.6°	41

roll

The motion of a spacecraft about its longitudinal, or nose-tail, axis.

Romb

Subsatellites, released by Soviet **Taifun** spacecraft, to enable the calibration of air and space defense radars. Their name means "rhombus."

Roosa, Stuart A. (1933–1994)

An American astronaut who orbited the moon as **Apollo 14** Command Module (CM) pilot in 1971 while crewmates Alan Shepard and Edgar Mitchell explored the surface. Roosa received a B.S. in aeronautical engineering from the University of Colorado, under the U.S. Air Force Institute of Technology Program. He served on active duty in the Air Force from 1953 to 1976, retiring as a colonel. Roosa was one of 19 astronauts selected by NASA in April 1966 and made the one spaceflight, during which he orbited the Moon alone for 35 hours. Roosa retired from NASA and the Air Force in 1976 and subsequently served in managerial positions with several companies.

RORSAT (Radar Ocean Reconnaissance Satellite)

Soviet naval reconnaissance satellites often launched to coincide with major NATO and American Navy maneuvers. A characteristic feature of RORSATs was their large radar antennas used to bounce signals off the ocean in order to locate ships. The strength of the signals required for this to work effectively, combined with the relatively backward state of electronics technology in the Eastern bloc, forced Soviet designers to adopt a radical solution—the use of a small nuclear reactor to power the RORSAT radar. Although prototype RORSATs in the mid-1960s flew with only chemical batteries, their operational counterparts, beginning in the 1970s, carried reactors. This introduced a serious risk of contamination, because all the RORSATs reentered after a few weeks or months in orbit. To counter the problem, each RORSAT consisted of three major components: the payload and propulsion section, the reactor, and a disposal stage used to lift the reactor into a higher orbit, with an altitude of 900 to 1,000 km, at the end of the mission. Thus, while the main spacecraft decayed, the reactor continued to circle the Earth. Unfortunately, the lifetime of an object at this altitude is about 600 years, whereas uranium-235 and -238 have a half-life of more than *one billion* years. This means there is at present, in 1,000 km × 65° orbits, about 940 kg of highly enriched uranium and a further 15 tons of other radioactive material including tens of thousands of droplets, 0.6 to 2 cm in diameter, which are the remains of the liquid sodium-potassium used to cool the RORSAT

reactors. There have also been accidents—the worst of them on January 24, 1978, when a RORSAT malfunctioned and crashed within Canada's Northwest Territory, showering radioactive debris onto the Great Slave Lake and surrounding region.

Size: 1.3 m (diameter) × 10 m (length)
Satellite mass
  Total: 3,800 kg
  Reactor + disposal stage: 1,250 kg
Reactor
  Fuel: highly enriched (90%) uranium-235
  Fuel mass: 31 kg
  No. of elements: 7
  Casing: beryllium

ROSAT (Roentgen Satellite)

German-American-British X-ray and ultraviolet astronomy satellite, named for Wilhelm Roentgen (1845–1923), who discovered X-rays; it operated for almost nine years. The first six months of the mission were dedicated to an all-sky survey in X-rays (0.1 to 2.5 keV) and ultraviolet (62 to 206 eV) using an imaging telescope with a sensitivity about 1,000 times greater than that achievable with the instruments on **Uhuru**. During the subsequent pointed phase of its mission, ROSAT made deep observations of a wide variety of objects. Its operation ended on February 12, 1999.

Launch
  Date: June 1, 1990
  Vehicle: Delta 6925
  Site: Cape Canaveral
Orbit: 539 × 554 km × 53.0°
Mass: 2,426 kg

Rosen, Milton W. (1915–)

A key figure in the development of the **Viking** rocket and **Vanguard** programs. Rosen trained as an electrical engineer and joined the staff of the Naval Research Laboratory in 1940, where he worked on guidance systems for missiles during World War II. From 1947 to 1955, he was in charge of Viking rocket development and subsequently was technical director of Project Vanguard, until he joined NASA in October 1958 as director of launch vehicles and propulsion in the Office of Manned Space Flight. In 1963, he became senior scientist in NASA's Office of the Deputy Associate Administrator for Defense Affairs and was later appointed deputy associate administrator for space science (engineering). In 1974, Rosen retired from NASA to become executive secretary of the National Academy of Science's Space Science Board.[248]

Rosetta

ESA (European Space Agency) and French mission, scheduled for launch in January 2003, to rendezvous with comet Wirtanen. En route, flybys of two asteroids—Otawara in 2006 and Siwa in 2008—are also planned. In 2011, Rosetta will encounter comet Wirtanen, go into orbit around it, and start remote sensing investigations. Following the selection of a landing site, the ROLAND lander, carrying the surface science package, will be released in August 2012. From Wirtanen's nucleus, the lander will transmit data to the main spacecraft, which will relay it to Earth. Rosetta is ESA's third Horizon 2000 Cornerstone mission and grew out of plans for the more ambitious Comet Nucleus Sample Return (CNSR) mission, envisaged in the 1980s as a collaboration with NASA.

Ross, H(arry) E. (twentieth century)

One of the leading figures in the **British Interplanetary Society** (BIS), from the time of its inception in 1933. Ross wrote a 1939 article in the BIS's *Journal* that outlined a method of accomplishing a lunar mission. The effort leading to the article had begun in London in February 1937, when the BIS formed a technical committee to conduct feasibility studies. In January 1949, Ross published a paper[249] in which he suggested that the most efficient strategy for a manned lunar expedition would be to leave the propellants for the return trip to Earth in lunar orbit—exactly the method eventually used by **Apollo**. He concluded that this method, now known as **lunar-orbit rendezvous** (LOR) would, compared with a direct flight to the Moon's surface from Earth, reduce the Earth-launch weight by a factor of 2.6. In his paper, he credited Hermann **Oberth**, Guido **von Pirquet**, Hermann **Noordung**, Walter **Hohmann**, Konstantin **Tsiolkovsky**, and Fridrikh **Tsander** for having earlier discussed ideas pertinent to the LOR concept.

rotation

Circular movement about an axis. The *rotational period* is the time taken for one complete turn through 360°.

Rosetta The Rosetta Orbiter swoops over the Lander soon after touchdown on the nucleus of Comet Wirtanen. *European Space Agency*

Rothrock, Addison M. (1903–1971)

A rocket propulsion specialist at the dawn of the Space Age. Rothrock graduated from Pennsylvania State University in 1925 with a B.S. in physics and joined the staff of the **Langley** Aeronautical Laboratory the next year. He worked in the areas of fuel combustion and fuel rating, contributing more than 40 papers and reports, and rising to the position of chief of the fuel injection research laboratory. In 1942, he moved to **Lewis** Laboratory, where he was chief of the fuels and lubricants division and then chief of research for the entire laboratory. In 1947, he became assistant director for research at **NACA** (National Advisory Committee for Aeronautics) Headquarters. With the foundation of NASA, he assumed the duties of assistant director of research (power plants). Two months later, he became the scientist for propulsion in, and then the associate director of, the office of program planning and evaluation. Rothrock retired from NASA in 1963 and taught for five years at George Washington University.

Roton

An American design for a piloted commercial space vehicle intended to provide routine access to orbit for a two-person crew and cargo. The Roton would be a reusable, single-stage-to-orbit (SSTO), vertical takeoff and landing (VTOL) vehicle, 19.5 m high and 6.7 m in diameter, able to carry up to 3,200 kg to and from a 300 km-high orbit. It would take off vertically powered by NASA's **Fastrac** engine burning liquid oxygen and jet fuel, then circularize its orbit following main engine cutoff with orbital maneuvering engines. Once its payload was deployed and any return cargo captured, the Roton would reenter and descend using a four-blade, nose-mounted rotor. During the high-speed phases of flight, the base of the vehicle would create most of the drag while the rotor remained windmilling behind, stabilizing the vehicle until it reached subsonic speed. Then the rotor would be spun up and the blades enter a helicopter-style autorotation flight mode, enabling the pilot to glide the craft to a precision landing. The manufacturer, Rotary Rocket Company, expected a market to develop whereby satellite operators and insurance companies sent rescue missions to repair or retrieve damaged or outdated satellites. But although the Roton Air Test Vehicle began flight tests in 1999, technical problems and NASA's decision not to select Roton for its **X-33** project stalled further development.

round-trip light time (RLT)

The time taken for an electromagnetic signal to travel from Earth to a spacecraft or another body in the Solar System and back again. It is roughly equal to twice the **one-way light time**.

Rover, Project

A nuclear rocket program begun by the United States in the 1950s with a view to a manned Mars mission. It involved the AEC (Atomic Energy Commission) and NASA developing the **KIWI**, Phoebus, Peewee, and Nuclear Furnace series of reactors at the Nevada Test Site using high-temperature nuclear fuels and long-life fuel elements to understand the basics of nuclear rocket technology. Rover also included the **NERVA** (Nuclear Engine for Rocket Vehicle Application) program. In 1969, the United States abandoned its immediate plans for the human exploration of Mars, and Project Rover was canceled in 1972.

RP-1 (rocket propellant 1)

A special grade of kerosene suitable for rocket engines. It has been used with liquid oxygen in the **Atlas**, **Thor**, **Delta**, **Titan** I, and **Saturn** rocket engines. RP-1's formulation stemmed from a program begun in 1953 by Rocketdyne to improve the engines being developed for the **Navaho** and Atlas missiles. Prior to that, a large number of petroleum-based rocket propellants had been used. Robert **Goddard** had begun with gasoline, and other experimental engines had been powered by kerosene, diesel oil, paint thinner, and jet-fuel kerosene. The wide variation in physical properties among fuels of the same class led to the identification of narrow-range petroleum fractions, embodied in 1954 in the standard United States kerosene rocket fuel RP-1, covered by Military Specification MIL-R-25576. In Russia, similar specifications were developed for kerosene under the specifications T-1 and RG-1. RP-1 is a kerosene fraction, obtained from crude oil with a high napthene content, that is subjected to further treatment, including acid washing and sulfur dioxide extraction.

RP-318

The Soviet Union's first rocket-powered aircraft. Built in 1936 by Sergei **Korolev** as an adaptation of his SK-9 glider, it was originally to have been used to flight test an early rocket engine designed by **Glushko**. When both Korolev and Glushko were arrested and sent to the Gulag in 1938, the development of the RP-318 was continued by others, culminating in the first powered flight on February 28, 1940. Test pilot V. P. Fedorov was towed to 2,600 m and cast off at 80 km/hr before firing the rocket engine and accelerating the aircraft to 140 m/s and an altitude of 2,900 m. In all, the RP-318 flew nine times before World War II ended the work.

Length: 7.4 m
Mass: 700 kg
Propellants: nitric acid/kerosene
Thrust: 1,370 newtons

Rudolph, Arthur L. (1906–1996)

A German-born rocket scientist who helped develop the American program that took men to the Moon in 1969. One of 118 top German rocket experts secretly brought to the United States after World War II, Rudolph became project manager of the **Saturn** V rocket, which powered the Apollo missions. In 1984, nearly a decade after he retired, the Department of Justice accused Rudolph of participating in a slave-labor program when he was operations director of a V-2 rocket factory during the war (see **"V" weapons**). Though he contested the charges, Rudolph relinquished his American citizenship and returned to Germany. While the Justice Department did not pursue his case, he was refused a U.S. visa in 1989 to attend a twentieth-anniversary celebration of the first Moon landing.

rumble

A form of combustion instability, especially in a liquid-propellant rocket engine, characterized by a low-pitched, low-frequency rumbling noise.

Russia in space

See the following biographical entries: Vladimir **Chelomei**, Yuri **Gagarin**, Valentin **Glushko**, Msitslav **Keldysh**, Nikolai **Kibalchich**, Sergei **Korolev**, Vassily **Mishin**, Boris **Rauschenbach**, Nikolai **Rynin**, Mikhail **Tikhonravov**, Konstantin **Tsiolkovsky**, Mikhail **Yangel**. Also see: **Apollo-Soyuz Test Project, Cosmos (spacecraft), Energia Rocket & Space Corporation, Glavcosmos, International Space Station, Luna, Mars, Mir, Molniya (satellite), Progress, "R" series of Russian missiles, Russian launch vehicles, Russian Space Agency (RKA), Russian-manned Moon programs, Salyut, Soyuz (spacecraft), Space Race, Sputnik (satellites), Vega, Venera, Voshkod, Vostok, Zenit,** and **Zond.**[149]

Russian launch sites

The three main Russian launch centers, or cosmodromes, are **Baikonur Cosmodrome, Kaputsin Yar,** and **Plesetsk Cosmodrome.** One of the problems with the location of these is that spent Russian rocket stages, unlike their American counterparts, tend to fall on land rather than into the ocean. The arctic tundra around Pletesk is littered with debris, which the locals refer to as "metal from heaven." Residents of Dzhezkazgan, which lies close to the flight path of rockets from Baikonur, have complained that toxic fuel from crashed rocket stages has contaminated the soil and made it impossible to graze lifestock. For this reason, the latest generation of Russian rockets, including the Angara and Zenit launch vehicles, burn a more environmentally friendly mixture of propellants.

Russian launch vehicles

Early Russian/Soviet launch vehicles were based on ballistic missiles (see **"R" series of Russian missiles**), and various naming schemes have sprung up to identify particular rockets. These include the official Russian "R" designations and various Western names (assigned by the U.S. Department of Defense, NATO, and others) for the original military missiles, and further names for the derived space rockets, including manufacturers' codes and names derived from the major types of satellite launched. For example, the rocket that placed the first satellite in orbit is known as Sputnik (after the satellite), 8K71PS (the manufacturer's index), R-7 (the ballistic missile from which it was derived), SL-1 (the American Department of Defense designation for the missile), and SS-66 and Sapwood (the NATO code number and name for the missile). More recently, the Russians have given specific names to their rockets. See **Angara, Buran, Cosmos (launch vehicle), Dnepr, Energia, Molniya (launch vehicle), N-1, Proton, Rockot, Shtil, Soyuz (launch vehicle), Sputnik (launch vehicle), Start, Strela, Tsyklon, Voshkod (launch vehicle), Vostok (launch vehicle),** and **Zenit.**

Russian manned lunar programs

See article, pages 353–354.

Russian Space Agency (RKA)

An agency formed after the breakup of the former Soviet Union and the dissolution of the Soviet space program. RKA (Rosaviacosmos) uses the technology and launch sites that belonged to the former Soviet space program and has centralized control of Russia's civilian space program, including all manned and unmanned nonmilitary spaceflights. The prime contractor used by the RKA is **Energia Rocket & Space Corporation,** which owns and operates the Mission Control Center in **Kaliningrad.** The military counterpart of RKA is the Military Space Forces (VKS), which controls the **Plesetsk Cosmodrome.** RKA and VKS share control of the **Baikonur Cosmodrome,** where the RKA reimburses the VKS for the wages of many of the flight controllers during civilian launches. RKA and VKS also share control of the **Gagarin Cosmonaut Training Center.**

Russian Space Research Institute (IKI)

The organization within the Russian Academy of Sciences mainly responsible for the long-term planning and development of space research programs of which a considerable part is performed within the framework of international space research cooperation. IKI are the Russian initials for Space Research Institute.

Russian manned lunar programs

For more than a decade, beginning in 1959, the Soviet Union made a concerted effort to be the first to send humans around the Moon and the first to achieve a manned landing. But whereas the **Apollo** program unfolded in a blaze of publicity, details of the Eastern aspect of the Moon race only began to emerge with the advent of Perestroika and the subsequent breakup of the Soviet Union in the early 1990s. Essentially, the Soviet Moon effort was three-pronged. The L-1 program was aimed at a manned circumlunar loop without a landing and involved the use of unmanned **Zond** craft to flight-validate the hardware. The L-3 program was designed to put a cosmonaut on the Moon's surface. Finally, the **Luna** series consisted of a variety of automated flyby, orbiter, hard- and soft-lander, sample-return, and rover vehicles. Only 20 of about 60 Soviet launches of all types of lunar craft from 1959 to 1976 were successful.

L-1

Sergei **Korolev**'s design bureau began work in 1965 on a manned spacecraft called L-1, which was intended to carry two cosmonauts on a single loop around the Moon. From the outside, the L-1 looked like the three-part **Soyuz** spacecraft, but it lacked its spherical orbital module. Other major differences were less obvious, including a modified propulsion system, a beefed-up heat-shield, and long-range communication systems. Because of repeated equipment failures, the L-1 never flew with a crew. However, unmanned L-1 spacecraft traveled to the Moon five times from 1968 through 1970 as Zonds 4 to 8. These missions tested the spacecraft and the maneuvers necessary for a manned mission.

L-3

Korolev also began designing spacecraft for a manned lunar landing mission, and some hardware was built under Vasily **Mishin**'s direction. This program, known as L-3, included an orbiter and a lander. The prototype lunar lander was successfully tested in Earth orbit, without a crew, three times in 1970–1971 under the name "Cosmos." The Soviet lunar lander, known as the Lunar Cabin (LK), was half the size and one-third the mass of the Apollo Lunar Module and was intended to carry one cosmonaut to the Moon's surface while the Lunar Orbiter Cabin (LOK) remained in lunar orbit with the second crew member. The program depended on the development of a super-rocket known as the **N-1**. When the N-1 was ready to test launch, the LK and LOK were still being built, so a modified L-1 spacecraft known as the L-1S was used as the primary payload. The N-1 was supposed to place the L-1S and a dummy LK on a trajectory toward the Moon. Once there, the L-1S alone would enter lunar orbit to take high-resolution photos of proposed landing sites and then return to Earth with the exposed film. For the mission, an Orientation Engine Module (DOK) would be attached to the front of the L-1S to slow it enough to place it into lunar orbit, after which the DOK would be jettisoned. The L-1S would then use its own propulsion system, located in the Instrument Module, to accelerate out of lunar orbit for the return to Earth. The spacecraft would perform a double-skip reentry as in previous L-1/Zond flights as a test of the nearly identical LOK Descent Module. However, four successive failures of the N-1's huge 30-engine first stage left the L-3 program in tatters.

Realizing that the Moon race was lost, the Soviets attempted to beat Apollo 11 at the last moment with an automated sample return, but Luna 15 crash-landed just as Armstrong and Aldrin were on their way back from the surface. Subsequently, the Soviets switched their main goal in manned spaceflight to establishing a permanent presence in Earth orbit, adapting their Moon-era hardware to launch a number of **Salyut** space stations, and using Soyuz spacecraft to ferry crews and supplies for missions of increasing duration. Their lunar ambitions were confined to large robotic sample-return, rover, and orbiter missions in the form of Lunas 16 through 24.[185, 186] (See table, "Soviet Launches Related to Manned Lunar Programs.")

Soviet Launches Related to Manned Lunar Programs

Name		Launch		
Official	Development	Date	Vehicle	Notes
Cosmos 133	7K-OK #2	Nov. 28, 1966	Soyuz	Destroyed on reentry
—	7K-OK #1	Dec. 14, 1966	Soyuz	Destroyed on launch pad
Cosmos 140	7K-OK #3	Feb. 7, 1967	Soyuz	Lost pressure during descent; fell in Aral Sea
Cosmos 146	7K-L1 #2P	Mar. 10, 1967	Proton	Tested Block D
Cosmos 154	7K-L1 #3P	Apr. 8, 1967	Proton	Second firing of Block D; failure
Soyuz 1	7K-OK #4	Apr. 23, 1967	Soyuz	Planned docking. Komarov in crash landing
—	7K-L1 #4	Sep. 28, 1967	Proton	First stage failure
Cosmos 186	7K-OK #6	Oct. 27, 1967	Soyuz	Docked with Cosmos 188
Cosmos 188	7K-OK #5	Oct. 30, 1967	Soyuz	Docked with Cosmos 186
—	7K-L1 #5	Nov. 22, 1967	Proton	Second stage failure
Zond 4	7K-L1 #6	Mar. 2, 1968	Proton	Reentry craft destroyed during reentry
Cosmos 212	7K-OK #8	Apr. 14, 1968	Soyuz	Docked with Cosmos 213
Cosmos 213	7K-OK #7	Apr. 15, 1968	Soyuz	Docked with Cosmos 212
—	7K-L1 #7	Apr. 23, 1968	Proton	Escape system self-initiated
—	7K-L1 #8	Jul. 14, 1968	Proton	On-pad explosion killed one person
Cosmos 238	7K-OK #9	Aug. 28, 1968	Soyuz	Test flight
Zond 5	7K-L1 #9	Sep. 15, 1968	Proton	Flew around Moon; returned and recovered
Soyuz 2	7K-OK	Oct. 25, 1968	Soyuz	Rendezvous with Soyuz 3
Soyuz 3	7K-OK	Oct. 28, 1968	Soyuz	Attempted docking with Soyuz 2 failed
Zond 6	7K-L1 #12	Nov. 10, 1968	Proton	Flew around Moon; crashed upon return
Soyuz 4	7K-OK	Jan. 14, 1969	Soyuz	Docked with Soyuz 5
Soyuz 5	7K-OK	Jan. 15, 1969	Soyuz	Docked with Soyuz 4; crew transfer
—	7K-L1 #13	Jan. 20, 1969	Proton	Second stage failed
—	7K-L1S	Feb. 21, 1969	N-1	First N-1 launch; failure at T + 69 seconds
—	7K-L1S	Jul. 3, 1969	N-1	Second N-1 launch; failed immediately
Zond 7	7K-L1 #11	Aug. 8, 1969	Proton	Flew around Moon; successfully recovered
Soyuz 6	7K-OK	Oct. 11, 1969	Soyuz	Joint mission with Soyuz 7 and 8
Soyuz 7	7K-OK	Oct. 12, 1969	Soyuz	Attempted docking with Soyuz 8 failed
Soyuz 8	7K-OK	Oct. 13, 1969	Soyuz	Attempted docking with Soyuz 7 failed
—	7K-L1 Ye	Nov. 28, 1969	Proton	Test of N-1 upper stage; failed on launch
Soyuz 9	7K-OK	Jun. 1, 1970	Soyuz	Longest human spaceflight to date
Zond 8	7K-L1 #14	Oct. 20, 1970	Proton	Flew around Moon; successfully recovered
—	T2K	Nov. 24, 1970	Soyuz	Lunar lander test in Earth orbit
Cosmos 382	7K-L1 Ye	Dec. 2, 1970	Proton	Successful test of N-1 upper stage in orbit
—	T2K	Feb. 26, 1971	Soyuz	Lunar lander test in Earth orbit
—	N1-L3	Jun. 27, 1971	N-1	Third N-1 launch; failed at T + 51 seconds
—	T2K	Aug. 12, 1971	Soyuz	Lunar lander test in Earth orbit
—	7K-LOK	Nov. 23, 1972	N-1	Fourth N-1 launch; failed at T + 107 seconds

moderate spectral resolution. It was named for the X-ray astronomy pioneer Bruno Rossi (1905–1993). Changes in X-ray brightness lasting from microseconds to months are monitored across a broad spectral range of 2 to 250 keV. RXTE was designed for a minimum operational lifetime of two years, with a goal of five years.

Launch
   Date: December 30, 1995
   Vehicle: Delta 7920
   Site: Cape Canaveral
Orbit (circular): 409 km × 29°
Mass: 3,200 kg

Rynin, Nikolai Alexsevitch (1887–1942)

A Russian author whose *Mezhplanetnye Soobschniya*[253] (Interplanetary Communications), a nine-volume encyclopedia of space travel, was published from 1928 to 1932. The first seven volumes appeared before a single book on interplanetary flight had been printed in the United States or Britain. Rynin also regularly sent out reports on Russian activities in rocketry to the West.

Ryusei

A ballistic capsule for testing materials and acquiring data related to atmospheric reentry for use in design of the Japanese **HOPE** space plane; "ryusei" means "meteor."

Launch
   Date: February 3, 1994
   Vehicle: H-2
   Site: Tanegashima
Orbit: 450 × 451 km × 30.5°
Mass: 865 kg

Russian launch vehicles   A Proton-K blasts off from Baikonur in February 2000 carrying the Garuda-1 satellite. *Sergei Kazak*

RXTE (Rossi X-ray Timing Explorer)

A NASA X-ray astronomy satellite designed to study variability in the energy output of X-ray sources with

S

Sagan, Carl Edward (1934–1996)

A professor of astronomy at Cornell University, a famous popular science writer and broadcaster, and an avid proponent of space exploration. Sagan contributed to the science investigations of many of the pioneering planetary spacecraft, including **Mariner** 9, **Viking**, **Pioneer**s 10 and 11, and **Voyager**s 1 and 2. He was a devotee and an advocate of broad international cooperation in space exploration, and one of the founders (in 1980) and president (until his death) of the **Planetary Society**. He expressed the opinion that carrying life from Earth to other planets is a duty of mankind, and that the conquest and colonization by mankind of other planets and extraterrestrial space are essential to our survival. "All civilizations," he wrote, "become either spacefaring or extinct." One of Sagan's chief interests was in the possibility of life in the Solar System and beyond. He helped conceive the contents of a message plaque attached to Pioneer 10 and a phonograph record carried by each of the Voyagers.[254]

Sagan Memorial Station

The name by which the defunct **Mars Pathfinder** is now known, in memory of the late Carl **Sagan**.

Sagdeyev, Roald Z. (1932–)

One of the leading figures in Soviet space science from the 1960s to the 1980s. Sagdeyev was involved in virtually every Soviet lunar and planetary probe in this period, including the highly successful **Venera** and **Vega** missions. He also advised Soviet leader Mikhail Gorbachev on space and arms control at the 1986 Geneva, 1987 Washington, and 1988 Moscow summits. In the late 1980s, Sagdeyev left the Soviet Union and settled in the United States, where he headed the East-West Science and Technology Center at the University of Maryland, College Park.[256]

SAGE (Stratospheric Aerosol and Gas Experiment)

A spaceborne instrument that collects data on how changes in the amount of solar radiation falling on Earth's upper atmosphere affect the concentrations of aerosol and ozone at different levels. SAGE I flew aboard **AEM**-2 (Applications Explorer Mission 2) in 1979, and SAGE II flew aboard **ERBS** (Earth Radiation Budget Satellite) in 1984. The first component of SAGE III, to monitor ozone depletion over the Arctic, is carried by the Russian **Meteor**-3M satellite, which was launched on December 10, 2001. Further SAGE III instrumentation

Sagan, Carl Carl Sagan stands next to a model of the Viking Lander in Death Valley, California. *NASA/JPL*

will be deployed aboard the International Space Station in 2005.

AEM-2/SAGE I
Launch
  Date: February 18, 1979
  Vehicle: Scout D
  Site: Wallops Island
Orbit: 456 × 506 km × 55.0°
Mass: 147 kg

Sakigake

A Japanese prototype interplanetary spacecraft launched by **ISAS** (Institute of Space and Astronautical Science) to encounter Halley's Comet. It was identical to **Suisei** except for its payload, which consisted of three instruments to measure plasma wave spectra, solar wind ions, and interplanetary magnetic fields. Like Suisei, it was designed to test basic technology for Japanese deep space missions, including communication and attitude control and determination, as well as to gather scientific data. Sakigake successfully flew within 7 million km of Halley's Comet on March 11, 1986, and swept past Earth on January 8, 1992, coming as close as 88,997 km–the first Japanese planetary flyby. However, controllers lost contact with the probe in November 1995, cutting short its extended mission, which would have taken it past comet Honda-Mrkos-Pajdusakova in 1996 and Giacobini-Zinner in 1998. "Sakigake" means "pioneer."

Launch
  Date: January 7, 1985
  Vehicle: M-3S
  Site: Kagoshima
Mass: 141 kg

Sakura
See **CS- (Communications Satellite)**.

Salyut
See article, pages 358–360.

Samos

A series of American military spacecraft launched in the 1960s to carry out global television surveillance from polar orbits. Samos (originally known as Sentry) satellites were one component of the first generation of U.S. Air Force orbital surveillance spacecraft, the other two components being **MIDAS** and **Corona.** The first satellites in the Samos series radioed back images from a frame readout camera as they passed over North America; later ones carried recoverable panoramic cameras. None of the satellites worked well, in contrast to the successful **Corona** project, and Samos was quietly wound up without any significant achievements. Some reporters have incorrectly said that "Samos" was an acronym of Satellite and Missile Observation System. In fact, the name came from the link with MIDAS–Samos being the island upon which King Midas lived.

SAMPEX (Solar Anomalous and Magnetospheric Particle Explorer)

A satellite designed to study the energy, composition, and charge states of four classes of charged particles that come from beyond the Earth: galactic cosmic rays (from supernova explosions in our galaxy), anomalous cosmic rays (from interstellar gas surrounding our Solar System), solar energetic particles (from explosions in the Sun's atmosphere), and magnetospheric electrons (particles from the solar wind trapped by Earth's magnetic field). SAMPEX was the first of NASA's Small Explorer (SMEX) missions.

Launch
  Date: July 3, 1992
  Vehicle: Scout G
  Site: Vandenberg Air Force Base
Orbit: 506 × 670 km × 81.7°
Size: 1.5 × 0.9 m
Mass: 158 kg

---

sample-return

An unmanned mission designed to collect material from another celestial body and bring it safely back to Earth. The first successful probe of this kind was **Luna** 16 in 1970. Over the next decade or so, a number of sample return missions are scheduled to harvest dust and/or rock from interplanetary space (see **Stardust** and **Genesis**), comets, asteroids, Mars, and other bodies in the Solar System. This has led to increased concern over the possibility of **contamination**.

San Marco (launch site)

Italy's launch site, consisting of two platforms in the Indian Ocean off the coast of Kenya (2.9° S, 40.3° E). The site's near-equatorial location makes it ideal for placing satellites into low-inclination orbits.

San Marco (satellites)

Italian satellites launched from San Marco and designed to carry out measurements of atmospheric density, temperature, and composition. San Marco D/L involved a collaboration with the United States and France. (See table, "San Marco Missions," on page 360.)

Salyut

An early series of Soviet **space station**s, of which seven were launched over a period of a decade beginning in 1971 with Salyut 1, the world's first space station. The Salyuts were intended to make human presence in space routine and continuous. As well as doing scientific research and spacecraft maintenance, cosmonauts tested equipment that would make space stations more habitable. In total, the Salyut ("salute") program involved 32 missions and cosmonauts from a variety of countries. The Soviet Union's guest cosmonaut program began in March 1978, when Soyuz 28 carried the Czech pilot Vladimir Remek to Salyut 6, and led to several firsts, including the first black person in space–Arnaldo Tamayo Mendez of Cuba. In 1986, Salyut 7, the last in the Salyut series, was replaced by **Mir**.

History

In 1969, the Soviet space program was in crisis. While American astronauts had reached the Moon, the Soviet Union's own effort to launch a Moon rocket resulted in two disastrous explosions that put the program years behind schedule. Many engineers working under Sergei **Korolev** sought a new direction. At the time, their competitors within the Soviet space industry, led by Vladimir **Chelomei**, had begun developing an ambitious military space station called Almaz. When this fell from government favor, Korolev's engineers proposed combining Chelomei's nascent hardware with a propulsion unit, solar arrays, and other equipment from the **Soyuz** spacecraft, to form the basis of a purely scientific orbiting outpost. It was suggested that this could be launched within a year of approval and before NASA's **Skylab**. In February 1970, the Soviet government officially endorsed the program, which was codenamed DOS 7-K. At the start of 1971, the world's first space station, DOS 1, was ready for launch.

Salyut 1

On April 19, 1971, DOS 1 was successfully placed in orbit. Shortly before launch, the name Zarya ("sunrise") was painted on the side of the station, but the mission staff were told to change this because a Chi-

nese spacecraft had already been given that name. No new name was put on the station, but the official Soviet press christened it Salyut 1. The first crew, Vladimir Shatalov, Alexei Eliseev, and Nikolai Rukavishnikov, took off aboard Soyuz 10 on April 23, 1971. In orbit, they docked with Salyut, but the Soyuz docking mechanism was damaged in the process, preventing the crew from entering the station. Fortunately, the Salyut docking port remained intact. The next crew, Georgy Dobrovoslky, Vladimir Volkov, and Victor Patsaev, successfully entered the station on June 6, 1971, and spent a record-breaking 23 days in orbit. However, disaster struck Soyuz 11 upon reentry when a pressure valve opened in the descent module, allowing the air to escape and killing the crew. Salyut 1 was abandoned on October 11, 1971, but several successor stations over the next 15 years helped pave the way for Mir.

Salyut 2

By the end of 1972, a team led by Chelomei had developed the first scaled-down Almaz military space station. However, in order to keep the true nature of Almaz secret, it was called Salyut 2. Legend has it that an embittered Chelomei had "Salyut" painted on the section that connected to the launch vehicle but was discarded after the craft reached orbit. Following a successful launch on April 3, 1973, the station quickly ran into trouble: its flight control system failed and there was a massive loss of pressure, rendering the station uninhabitable. A government investigation into the accident blamed the propulsion system, but Western radar provided a vital clue to what had probably gone wrong. Debris in the area of the spacecraft suggested that the **Proton** rocket's upper stage had exploded in orbit. Almost certainly the station had been punctured by a fragment from the resulting cloud of shrapnel.

DOS 3

Despite the failure of Salyut 2, the Soviet space station campaign continued on May 11, 1973, with the launch of DOS 3, just three days before Skylab went into orbit. DOS 3 featured a number of improvements, including custom-built solar arrays, which, unlike panels borrowed from Soyuz on the previous stations, could track the Sun and so supply more power to the spacecraft. Yet DOS 3 would never be inhabited. After a flawless launch, errors in the flight control system, which occurred while out of the range of ground control, caused the station to fire its orbit-correction engines until all its fuel was exhausted. Since the spacecraft was already in orbit and had been registered by Western radar, the Russians disguised the launch as Cosmos 557 and quietly allowed it to reenter and burn up a week later.

Salyut 3

The second Almaz, under the cover name "Salyut 3," was successfully launched on June 26, 1974, and its inaugural crew, Pavel Popovich and Yuri Artyukhin, docked with the station on July 3 for a stay lasting a couple of weeks. The next crew headed up on August 26, 1974, and immediately almost met with disaster. A failure of the Soyuz 15 rendezvous system caused the spacecraft to approach the station at a catastrophic 72 km/hr. Fortunately, the spacecraft was also about 40 m off target, and the crew were able to abort the mission and return safely home.

Salyut 4

On December 26, 1974, Sergei Korolev tried again with a nonmilitary space station by launching Salyut 4 (DOS-4), essentially a sibling of DOS-3. This time the mission went without a hitch. On January 11, 1975, the Soyuz 17 crew docked with the station and took up residence for about a month. The next visitors, Vasili Lazarev and Oleg Makarov, blasted off for Salyut 4 on April 5, 1975, but a faulty separation of the launch vehicle's second and third stages left the spacecraft spinning wildly. Luckily, the crew was able to make an emergency landing in the Altai mountains, but only after suffering nightmarish decelerations of up to 21g. On May 25, 1975, a new crew was sent to Salyut 4 for a stay of 63 days. In November 1975, an unmanned Soyuz 20 docked with Salyut 4 automatically and stayed docked for three months, demonstrating the future potential of such supply missions. The successful Salyut 4 was deorbited on February 3, 1977, bringing the highest civilian honor, "Hero of the Socialist Labor," to the chief designer of the spacecraft, Yuri Semenov, and one of the assembly technicians, V. Morozov (despite official objections that Morozov was not a member of the Communist Party).

Salyut 5

The third and last military space station was launched on June 22, 1976. Two crews visited Salyut 5, in July–August 1976 and February 1977. In between, another crew was launched aboard Soyuz 23 but never docked with the station due to a failure of the rendezvous system. The crew landed in the half-frozen Lake Tengiz, and was brought to safety only after a long and dangerous rescue effort.

Salyut 6

On September 29, 1977, Salyut 6 successfully reached orbit. Although it resembled its Salyut and Almaz predecessors, the spacecraft marked a revolution in space station technology. First, it sported a second docking port in the rear, which allowed two spacecraft to dock to the station. Furthermore, the rear docking port enabled an unmanned, cargo version of the Soyuz spacecraft, known as **Progress**, to refuel the station's propellant tanks. The Progress ship could also carry food and supplies to the station, enabling crews to stay much longer, and use its engines to boost the station to a higher orbit. These upgrades had an immediate effect on space station operations. From 1977 until 1982, Salyut 6 was visited by five long-duration crews as well as 11 shorter-term crews, including cosmonauts from Warsaw Pact countries. The first long-duration crew on Salyut 6 broke a record set onboard Skylab, staying 96 days in orbit. The longest flight onboard Salyut 6 lasted 185 days. The fourth Salyut 6 expedition deployed a 10-m radio-telescope delivered by a Progress ship. After Salyut 6 manned operations were discontinued in 1981, a heavy unmanned spacecraft called TKS, developed using hardware left from the canceled Almaz program, was docked to the station as a hardware test. This type of architecture would be used on the Mir spacecraft years later.

Salyut 7

When the sixth spacecraft in the pre-Mir space station program was deorbited on July 29, 1982, the Soviets already had a new station in place—Salyut 7, which had been launched three months earlier. The design of this latest station's solar panels allowed for the installation of additional sections to increase its power supply. Soviet designers also hoped eventually to equip Salyut 7 with complex electrically driven wheels, known as gyrodines, to allow the station to orient itself without propellant. Designers expected the gyrodines to be sent to Salyut 7 aboard a special module that would also carry a permanent astronomy payload called Kvant 1. Delays in development, however, kept Kvant 1 on the ground until Mir was launched. However, Salyut 7 crews further pushed the limits of human spaceflight. By 1986, four long-duration crews and five shorter-term crews had lived aboard the station. As is often the case in space exploration, some of the most valuable lessons of Salyut 7 came from its failures. One day in September 1983, fuel started spilling from a propellant line, apparently because of a meteor strike. The crew working aboard Salyut 7 in 1984 performed surgical work outside the station, isolating the damaged portion of the line and installing a bypass section. On February 11, 1985, as Salyut 7 was flying unmanned, a ground controller accidentally cut off communications with the station, leaving it out of control and disabled. All that could be done was to track the station's position with defense radar. A rescue crew, including Vladimir Djanibekov and Valeri Savinukh, was launched on June 6, 1985. For the first time, a crew manually docked to a totally disabled space station. When the cosmonauts entered the station, they found their future home with no lights, heat, or power. Large icicles hung from the life support system pipes, and all water aboard the station had frozen. As the cosmonauts' own limited supply of water and food was running out, it took an intensive effort to rehabilitate the facility. By the beginning of 1986, Salyut 7 was back in working order. Meanwhile, at Baikonur, preparations were under way for the launch of a new space station—Mir.

Sander, Friedrich Wilhelm

A German entrepreneur and manufacturer of black powder rockets who rose to fame in the 1920s and 1930s through his involvement, with Max **Valier** and Fritz **von Opel**, in the development of a variety of rocket-powered cars, sleds, gliders, and airplanes.

Sänger, Eugen (1905–1964)

An Austrian engineer, born in Pressnitz, Bohemia, who carried out pioneering work on the design of **space plane**s and other novel forms of space transport. As a 13-year-old, his thoughts were turned toward space travel by reading Kurd Lasswitz's science fiction novel *On Two Planets* (1897). In choosing a career he was again influenced by a space classic—Hermann **Oberth**'s *The Rocket into Interplanetary Space*. Reading this in the fall of 1923 encouraged him to switch from the course in civil engineering that he had just begun at the Technical University of Graz to a course in aeronautics. As an assistant at the Technical University in Vienna (1930–1935), he continued his systematic mathematical investigations of rocket engines. In contrast to Oberth and others, Sänger was convinced that the best means to reach space was through a combination of rocket and aircraft technology. He thus closely examined the idea of a space plane, made engine calculations and carried out investigations into the most suitable propellants, and finally set up a laboratory, in which he conducted experiments on forced-circulation-cooled liquid propellant rocket engines. The results of his work appeared in *The Technology of Rocket Flight*[257]—the first scientific study of space planes. From 1936 to 1945, Sänger directed rocket research for the

San Marco Missions

Spacecraft	Launch Date	Vehicle	Site	Orbit	Mass (kg)
San Marco 1	Dec. 15, 1964	Scout X-4	Wallops Island	200 × 842 km × 37.8°	254
San Marco 2	Apr. 26, 1967	Scout B	San Marco	219 × 741 km × 2.9°	129
San Marco 3	Apr. 24, 1971	Scout B	San Marco	222 × 707 km × 3.2°	164
San Marco 4	Feb. 18, 1974	Scout D	San Marco	270 × 875 km × 2.9°	164
San Marco D/L	Mar. 25, 1988	Scout G	San Marco	268 × 625 km × 3.0°	236

German Luftwaffe. Among other things, he constructed **ramjet** engines, which he tested on a Dornier 217 in April 1942. At the same time, his ideas on **hypersonic** aircraft began to take shape, and he proposed a winged rocket bomber that would skip in and out of the atmosphere to increase its range. After World War II, Sänger's work served as the basis for the development of the **X-15** and, ultimately, the **Space Shuttle.** From 1946 to 1954, Sänger served as a consultant engineer in France, where he worked mainly on developing rocket and large ramjet engines.[258] In 1954, he returned to Germany to take up a post at Stuttgart Technical University. Subsequently, he consulted for a number of German aerospace companies, assisting, for example, the Junkers Works (1961–1964) in their studies on the development of a European spacecraft. Work by Sänger also laid the foundations for the Sänger Project, a concept for a two-stage, reusable space plane for inexpensively transporting crews and payloads to and from orbit. In 1953, he suggested a design for a **photon** rocket, propelled by gamma rays produced by the annihilation of electrons with positrons.

Sarabhai, Vikram Ambalal (1919–1971)

A leading pioneer of India's space program. Born into a wealthy business family, Sarabhai broke with tradition and chose to enter science. He earned a Ph.D. in physics in England and studied cosmic rays with the eminent physicist C. V. Raman at the Indian Institute of Science in Bangalore. In 1948, he founded the Physical Research Laboratory in Ahmedabad, which later served as the nucleus of India's space program. Sarabhai expanded the **Indian Space Research Organisation**, oversaw India's first satellite **Aryabhatta** in 1975, and initiated the Satellite Instructional Television experiment (SITE). When SITE was put into action in 1975–1976, after Sarabhai's death, it brought education to 5 million people in 2,400 Indian villages.

SARSAT (Search and Rescue Satellites)

A system based on American, Soviet, and French satellites. The Soviet part of the system was known as COSPAS.

SAS (Small Astronomy Satellite)

A series of NASA spacecraft placed in orbit in the 1970s to observe celestial X-ray and gamma-ray sources. (See table, "SAS Missions.")

SAS-1

The first Earth-orbiting mission dedicated to X-ray astronomy. Launched from the **San Marco** platform off the coast of Kenya on the seventh anniversary of Kenyan independence, SAS-1 was renamed "Uhuru," which is Swahili for "freedom." It carried out observations in the 2 to 20 keV energy range, discovering 200 X-ray sources and finding evidence of the first known black hole. It stopped operating on April 5, 1979. Also known as Explorer 42.

SAS-2

A gamma-ray astronomy satellite sensitive in the energy range 20 MeV to 1 GeV. On June 8, 1973, a failure of the spacecraft's low-voltage power supply ended its mission. Also known as Explorer 48.

SAS-3

An X-ray astronomy satellite with three major scientific objectives: to fix the location of bright X-ray sources to an accuracy of 15 arcseconds, to study selected sources over the energy range of 0.1 to 55 keV, and to search the sky continuously for X-ray novae, flares, and other transient (short-lived) phenomena. It stopped operating in 1979. Also known as Explorer 53.

Satcom

A series of commercial communications and cable-TV satellites started by RCA Americom in 1975 and continued by GE when it took over RCA.

satellite

An object that revolves around a larger body. Thousands of artificial satellites have been placed in Earth orbit for a great variety of scientific and technological purposes, to support communication and navigation, and as military tools.

SAS Missions

Launch site: San Marco				
	Launch			
Spacecraft	**Date**	**Vehicle**	**Orbit**	**Mass (kg)**
SAS-1 (Uhuru)	Dec. 12, 1970	Scout B	$521 \times 570 \times 3°$	143
SAS-2	Nov. 15, 1972	Scout D	$526 \times 526 \times 1°$	185
SAS-3	May 7, 1975	Scout F	$498 \times 507 \times 3°$	195

satellite constellation

A set of satellites arranged in orbit to fulfill a common purpose. In the case of GPS (Global Positioning System), for example, the full operational constellation is composed of six orbital planes, each containing four satellites.

satellite mass categories

Small satellites are classified according to their mass, as shown in the table ("Satellite Mass Categories").

Satellite Mass Categories

Category	Mass Range (kg)
Picosat	<1
Nanosat	1–10
Microsat	10–100
Minisat	100–1,000

Saturn

See article, pages 363–366.

Saturn Ring Observer

A spacecraft that could perform detailed investigations of complex dynamic processes in Saturn's rings. The Saturn Ring Observer is identified in NASA's Office of Space Science Strategic Plan as a potential mission beyond 2007 but has yet to be clearly defined.

Savitskaya, Svetlana (1948–)

A Soviet cosmonaut; the second woman to orbit the Earth and the first to go on a space walk. She spent a total of over 19 days in space. The daughter of a World War II flying ace, who was also deputy commander of the Soviet Air Defenses and a two-time Hero of the Soviet Union, Savitskaya was refused entry to pilot school at age 16 but continued parachuting and by her 17th birthday had made 450 jumps. The following year, she entered the Moscow Aviation Institute (MAI), and in 1970 she won the world aerobatics competition in Hullavington as a member of the Soviet National Aerobatics Team. Following her graduation from the MAI in 1972, Savitskaya was accepted at a test pilot school, where she set world records in turboprop and supersonic aircraft, including the female airspeed record of 2,683 km/hr in a MiG-21. On July 30, 1980, she was selected to become a cosmonaut (the 53rd) and, on August 19, 1982, became the second woman in space when she and her fellow crew members, Leonid **Popov** and Alexander Serebrov, flew Soyuz T-7 to dock with the **Salyut** 7 space station. During her second spaceflight, aboard Soyuz T-12 on July 25, 1984, Savitskaya and fellow cosmonaut Vladimir Dzhanibekov carried out a three-and-a-half-hour spacewalk to conduct welding experiments on Salyut 7. Subsequently, she was appointed commander of an all-female crew to Salyut 7 for International Women's Day, but the mission was canceled because of problems with the space station and the limited availability of Soyuz T spacecraft. In 1987, Savitskaya was made deputy to the chief designer at Energia, and in 1989 became a member of the Soviet parliament. She retired as a cosmonaut on October 27, 1993.

S-band

See **frequency bands**.

SB-WASS (Wide Area Surveillance Satellite)

A series of **ELINT** (electronic intelligence) satellites developed by the U.S. Navy. Each SB-WASS consists of a primary infrared-scanning satellite and a triplet of associated drone craft. Triangulation, together with the known time lag between the satellites, enables the exact location of a target to be determined. Weapons targeting data are then sent from the SB-WASS to air and ground controllers by laser communications.

scan platform

An articulated, powered appendage to a spacecraft bus that points in commanded directions, allowing optical or other observations to be taken independently of the spacecraft's orientation.

SCATHA (Spacecraft Charging At High Altitudes)

A U.S. Air Force satellite designed to collect data on the electrical charging of spacecraft—an effect caused by repeated passage through the magnetosphere and known to have contributed to several on-orbit satellite failures. Specific goals of the mission were to obtain environmental and engineering data to allow the creation of design criteria, materials, techniques, tests, and analytical methods to control charging of spacecraft surfaces, and to collect scientific data about plasma wave interactions, substorms, and the energetic ring.

Launch
    Date: January 30, 1979
    Vehicle: Delta 2914
    Site: Cape Canaveral
Orbit: 28,018 × 42,860 km × 10.2°
Mass on-orbit: 360 kg

Schirra, Walter ("Wally") Marty, Jr. (1935–)

The only American astronaut to fly in all three **Mercury**, **Gemini**, and **Apollo** programs. Schirra earned a B.S. from the U.S. Naval Academy in 1945 and later flew combat missions during the Korean War. He was involved in the development of the Sidewinder missile at the Navy Ordnance Station at China Lake, California.
(continued on page 366)

Saturn

A family of large American liquid-fueled rockets that solved the problem of getting manned spacecraft to the Moon. Three models were developed—the Saturn I, the Saturn IB, and the Saturn V. In all, 32 Saturns were launched with no failures.

History

The Saturn story began in 1957 when Wernher **von Braun** and his team at the **Army Ballistic Missile Agency** (ABMA) came up with a design for a heavy-lift rocket with a first stage composed of eight **Redstone** missile stages clustered around one **Jupiter** missile central core. Each of the clustered stages was to have a Rocketdyne engine adapted from the **Thor** missile program. In August 1958, the Advanced Research Projects Agency (ARPA) authorized ABMA to develop a rocket along those lines. Redstone-based Juno I and Jupiter-based Juno II rockets had already been successfully developed, and Juno III and Juno IV were just concept vehicles that never left the drawing board, so the new project was called Juno V. Shortly after, though, ARPA approved von Braun's suggestion that the Juno V be renamed Saturn since the rocket was an evolutionary step beyond the Jupiter-based Juno II and Saturn is the next planet beyond Jupiter. On October 21, 1959, the ABMA Development Operations Division, including the Saturn program and von Braun, was transferred to NASA—a logical move, since the Army had no use for such a vehicle, whereas the newly formed NASA had embarked on an ambitious manned spaceflight program aimed at a Moon landing. A government committee recommended that NASA consider three options, known as Saturn A, B, and C. The last of these was eventually selected, and five versions of it, Saturn C-1 through C-5, ranging from the least powerful to the most powerful, were proposed. Of these, only C-1 and C-5 were chosen for development. The Saturn C-1 was a two-stage vehicle that closely followed the design of the Juno V, while the three-stage C-5 was the giant vehicle that would take **Apollo** astronauts to the Moon.

Saturn I

Flight tests of the Saturn C-1 began in 1961 and involved two configurations, called Block I and Block II. In the former, a live first stage was flown with a dummy second stage made up of an inert Jupiter shell ballasted with water. The first stage, known as the S-1, was made up of eight Redstone stages clustered around a single Jupiter stage. However, unlike the Juno V plans from which this design came, the Redstones served only as fuel tanks. Eight Rocketdyne **H-1** engines were employed in the first stage—four in a tight square at the center of the rocket's base and four at the corners. Four of the Redstones and the Jupiter stage carried liquid oxygen (LOX); the remaining Redstones carried RP-1 (kerosene mixture).

The first four Saturn C-1s were launched in the Block I configuration to test flight dynamics and first stage reliability. The Saturn C-1 was renamed Saturn I in February 1963, prior to the vehicle's fourth test flight. (The "C" was also dropped from the other Saturn rockets, which became Saturn IB and Saturn V.) From the fifth Saturn I test launch on, the Block II configuration was used, featuring an operational first and second stage. The Saturn I second stage, called S-IV, was powered by a Pratt and Whitney engine fueled by LOX and liquid hydrogen. In the Block II configuration, the Saturn I employed lengthened fuel tanks, improved H-1 engines with greater thrust, and eight stabilization fins on the first stage base. (See table, "Saturn I and IB Statistics.")

Saturn IB

An improved version of the Saturn I, introduced in 1966 to meet the growing demands of the Apollo program. The Saturn IB was a marriage of an uprated Saturn I first stage, called the S-IB, to a new second stage, known as the S-IVB. Using a more advanced engine design, Rocketdyne cut the weight of the first stage H-1 engine array while improving overall thrust. The second stage—used as the third stage on the Saturn V—was powered by a single Rocketdyne **J-2** engine fed with LOX and liquid hydrogen. Designed specifically for manned flight, the S-IVB carried three solid-propellant ullage motors and two auxiliary propulsion modules on its aft skirt. The ullage motors supplied positive acceleration between cutoff and separation of the first stage and ignition of the second stage, while the auxiliary engines were for on-orbit maneuvering.

A 6.5-m ring fixed to the top of the S-IVB contained an Instrument Unit (IU) that handled all electronic commands for control and guidance during ascent. Saturn IBs used for manned flights carried a space capsule and a launch escape system. Located at the top of the rocket, these components were attached to the IU by an adapter. The fourth scheduled launch of an Apollo Saturn IB, with a vehicle and mission designation of AS-204, met with tragedy. AS-204 was planned as the first manned Apollo flight. Launch of the 14-day mission had been set for February 1967. During a countdown rehearsal at Launch Complex 34 on January 27, 1967, Virgil Grissom, Ed White, and Roger Chaffee were killed when their capsule caught fire. In honor of the men, NASA later changed the designation of mission AS-204 to Apollo 1. The Saturn IB that would have launched Apollo 1 carried instead an unmanned Lunar Module into low Earth orbit on January 22, 1968. After the Apollo lunar program ended, Saturn IBs were used to ferry astronauts to **Skylab**. The Skylab Saturn IB was nearly identical to its Apollo sibling but featured uprated H-1 engines. The last Saturn IB flight supported the Apollo-Soyuz Test Project. Two unused Saturn IBs were later released by NASA for display purposes. (See table, "Saturn I and IB Statistics.")

Saturn IB The Saturn IB launch vehicle carrying the Skylab 4 crew lifts off on the final mission to the orbiting space station. *NASA*

Saturn V

The largest member of the Saturn family, developed at the **Marshall Space Flight Center** under the direction of Wernher von Braun. The three-stage Saturn V was taller than a 36-story building and the largest, most powerful rocket ever successfully launched (see **N-1**); fifteen of them were built. Among the facilities specially constructed to accommodate this huge rocket were the **Kennedy Space Center**, including the **Vehicle Assembly Building**, the Saturn V **crawler/transporter**, and launch pads 39A and 39B.

The Saturn V was flight-tested twice without a crew. The first manned Saturn V sent Apollo 8 into lunar orbit in December 1968. After two more missions to test the Lunar Module, a Saturn V sent Apollo 11 on its way to the first manned landing on the Moon.

In a typical Saturn V Apollo flight, the five F-1 first

Saturn I and IB Statistics

	Saturn I		Saturn IB
	Block I	Block II	
Length (m)	50	58	68.3
Diameter (m)	6.6	6.6	6.6
Payload to LEO (kg)	–	17,200	18,600
Thrust (N)			
1st stage	5,900,000	6,760,000	7,300,000
2nd stage	–	400,000	1,000,000

stage engines were ignited 6 seconds before liftoff. The center F-1 was shut down 135 seconds after launch and the outer four F-1s 15 seconds later. One second following cutoff of the four outer F-1s, the first stage separated. Simultaneously, eight retro-rockets were fired briefly to slow the first stage and to prevent it from bumping into the second stage. Following separation, the spent first stage fell into the Atlantic about 640 km downrange. One second after first-stage separation, eight solid-fueled motors mounted

Saturn V The Apollo 11 Saturn V lifts off with astronauts Neil Armstrong, Michael Collins, and Edwin Aldrin aboard. *NASA*

on the first/second stage adapter ring were fired for 4 seconds. As well as maintaining the positive motion of the rocket, this forced the second-stage fuel to the bottom of its tanks in order to feed the engines—a so-called ullage maneuver—and was the cue for the five J-2 second-stage engines to ignite. Thirty seconds later, the first/second-stage adapter ring fell away, and six seconds after that, the escape tower was jettisoned. The second-stage engines burned for 365 seconds before the next separation took place. Four solid-fueled retro-rockets on the second stage fired to keep the second and third stages from colliding. Then the second stage began its drop into the Atlantic about 4,000 km from the launch site. At this point, the Saturn V was traveling about 25,300 km/hr at an altitude of 185 km. Two solid-fueled motors on the third-stage aft skirt were fired briefly to settle the fuel, and, simultaneously, the S-IVB third-stage J-2 engine fired up for a burn of 142 seconds. This initial S-IVB burn carried Apollo into a 190-km orbit at a speed of 28,200 km/hr. During two or three checkout orbits, the S-IVB attitude control motors could be fired in sequence to make any necessary on-orbit corrections. Following these checkout orbits, the ullage motors were fired for 77 seconds to settle the fuel and provide forward spacecraft momentum. Then the third-stage J-2 engine reignited for 345 seconds to achieve a speed of 40,000 km/hr and to place Apollo on a lunar trajectory. The third stage separated from the CSM/LM combination, and the third-stage ullage motors fired for 280 seconds to move the S-IVB clear of Apollo. Finally, the J-2 fired for the last time, using up its remaining fuel and, depending on the mission profile, sending the spent third stage toward either deep space or a collision with the Moon. These intentional lunar impacts were to enable seismometers placed on the lunar surface by previous Apollo missions to detect the resulting "moonquakes" and to tell scientists about the Moon's internal structure.

Although a two-stage version of the Saturn V was used to place Skylab in orbit, the rocket was effectively retired at the end of the Apollo program. During its development, the Saturn V was expected to become a workhorse booster of a planned Apollo Applications space science program that would follow the lunar landing missions, but this program never materialized. In December 1976, NASA released components of two remaining vehicles along with test articles, which eventually enabled the completion of three Saturn V displays.

Overall length: 110.8 m (Apollo), 101.7 m (Skylab)
Payload: 129,300 kg (LEO), 48,500 kg (Moon)
First stage (S-IC)
    Size: 42 m (length); 10 m (diameter); 19.2 m (fin-span)
    Propellants: RP-1 and LOX
    Total thrust: 33,360,000 N (Apollo); 34,300,000 N (Skylab)
Second stage (S-II)
    Size: 24.8 m (length); 10 m (diameter)
    Propellants: liquid hydrogen and LOX
    Total thrust: 5,170,000 N (Apollo); 5,000,000 N (Skylab)
Third stage (S-IVB)
    Size: 17.9 m (length); 6.6 m (diameter)
    Thrust: 1,030,000 N

Schirra, Walter ("Wally") Marty, Jr. (1935–)

(continued from page 362)

Flying an F3D night fighter, Schirra was the first to fire a Sidewinder at a drone target—with almost disastrous results. The missile went out of control and started to loop around to chase the plane; Schirra's response was to make an even faster loop to stay on its tail. He flew the Mercury capsule Sigma 7 to become the third American in space, commanded Gemini 7, which made the first space rendezvous, and commanded Apollo 7.[262]

Schmidlap, Johann (sixteenth century)

A German fireworks maker and perhaps the first, in 1591, to experiment with **staging**—a technique for lifting fireworks to higher altitudes. A large skyrocket (first stage) carried a smaller skyrocket (second stage). When the large rocket burned out, the smaller one continued to a higher altitude before showering the sky with glowing cinders. Although Schmidlap appears to have been the first to fly staged rockets, priority for the idea may go to Conrad **Haas**.

Schmiedl, Friedrich (1902–1994)

An Austrian civil engineer who, beginning in 1918, carried out numerous experiments with solid-fueled rockets. In February 1931, he began a postal service using remote-controlled rockets that landed by parachute to carry mail between neighboring towns, primarily Schöckel and

Walter Schirra   Schirra in Mercury pressure suit with a model of the Mercury capsule behind him. *NASA*

Radegund, and Schöckel and Kumberg. So successful was Schmiedl's innovative method of delivery that another Austrian inventor, Gerhard Zucker, proposed a service that would carry rocket mail across the English Channel. Unfortunately for Zucker, all of his long-range solid-fueled rocket mail prototypes exploded at launch.

Schmitt, Harrison Hagan (1935–)

A Lunar Module pilot on **Apollo** 17 and the only geologist to walk on the Moon. Schmitt received a B.S. from the California Institute of Technology (1957) and a Ph.D. in geology from Harvard (1964). He had assignments with the Norwegian Geological Survey on the west coast of Norway and the U.S. Geological Survey in New Mexico and Montana. Schmitt was involved in photographic and telescopic mapping of the Moon with the Geological Survey's Astrogeology Center at Flagstaff, Arizona, when NASA selected him in June 1965 in its first group of scientist-astronauts. Unlike all previous astronauts, Schmitt was not a pilot and so had to attend a yearlong course at Williams Air Force Base, Arizona, before receiving his Air Force jet pilot wings and later his Navy helicopter wings. While training for his own Moon mission, Schmitt provided Apollo flight crews with detailed instruction in lunar navigation, geology, and feature recognition. He also helped integrate scientific activities into Apollo missions and analyze returned lunar soil samples. Schmitt was originally assigned to Apollo 18

but, when this flight was canceled, was moved up to Apollo 17 so that he could bring his geological expertise to bear on what would be the final Apollo journey. Stepping onto the mountain-ringed valley named Taurus-Littrow, Schmitt announced: "It's a good geologist's paradise if I've ever seen one!" He resigned from NASA in 1975 to run for the U.S. Senate in his home state of New Mexico. Elected on November 2, 1976, he served one six-year term and, in his last two years, was chairman of the Subcommittee on Science, Technology, and Space. He is currently a business and technical consultant.

Schneider, William C.

A senior NASA official who served as Gemini Mission Director for 7 of the 10 manned Gemini missions, NASA Apollo Mission Director and Apollo Program Deputy Director for Missions (1967–1968), Skylab Program Director (1968–1974), Deputy Associate Administrator for Space Transportation (1974–1978), and Associate Administrator for Space Tracking and Data Systems (1978–1980). He received a Ph.D. in engineering from Catholic University and joined NASA in June 1963.

Schriever, Bernard A. (1910–)

An aerospace engineer and administrator who figured prominently in American missile development. Schriever earned a B.S. in architectural engineering from Texas A&M in 1931 and was commissioned in the Army Air

Corps Reserve in 1933 after completing pilot training. He earned an M.A. in aeronautical engineering from Stanford in 1942 and then flew 63 combat missions in the Pacific Theater during World War II. In 1954, he became commander of the Western Development Division (soon renamed the Air Force Ballistic Missile Division) and from 1959 to 1966 was commander of its parent organization, the Air Research and Development Command (renamed Air Force Systems Command in 1961). As such, he oversaw the development of the **Atlas**, **Thor**, and **Titan** missiles, introducing a systems approach whereby the various components of the Atlas and succeeding missiles underwent simultaneous design and testing. Schriever also instigated the practice of concurrency, which allowed the components of missiles to enter production while still in the test phase, thereby speeding up development. He retired as a general in 1966.[215]

Schweikart, Russell ("Rusty") L. (1935–)

An American astronaut who served as the Lunar Module (LM) pilot on **Apollo** 9, the mission during which the LM was tested for the first time in space. Schweikart attended the Massachusetts Institute of Technology (MIT), earning a B.S. in aeronautical engineering and an M.S. in aeronautics and astronautics. Following graduation, he served as a pilot in the U.S. Air Force and Air National Guard (1956–1963). During part of this period he worked as a research scientist in the Experimental Astronomy Laboratory at MIT. Schweikart joined NASA in October 1963 as one of 14 selected in the third group of astronauts. Later he moved to NASA Headquarters in Washington as director of user affairs in the Office of Applications, responsible for transferring NASA technology to the outside world. He then held several technology-related positions with the California state government, including assistant to the governor for science and technology and, in 1979, as chairman of the California Energy Commission. Schweikart is currently president of Aloha Networks.

SCORE (Signal Corps Orbiting Relay Experiment)

An American satellite, nicknamed "Chatterbox," that relayed the first voice communications from space. Its payload included an audio tape machine that broadcast messages, including a 58-word prerecorded Christmas message from President Eisenhower, for 13 days.

Launch
    Date: December 18, 1958
    Vehicle: Atlas B
    Site: Cape Canaveral
    Orbit: 185 × 1,484 km × 32.3°
    Mass: 70 kg

Scott, David R. (1932–)

An American astronaut who was the pilot on **Gemini** 8, Command Module pilot on **Apollo** 9, and commander of Apollo 15. Scott received a B.S. from the U.S. Military Academy in 1954 and an M.S. in aeronautics and astronautics from the Massachusetts Institute of Technology in 1962. He entered the Air Force and graduated from the Experimental Test Pilot School and Aerospace Research Pilot School at Edwards Air Force Base before being selected by NASA as an astronaut in 1963. Later, he held administrative posts with NASA, including director of the Dryden Flight Research Center. Scott retired from the Air Force in 1975 as a colonel and is currently president of Scott Science and Technology.[265]

Scout

A small rocket, 21.9 m high and 1 m in diameter, able to launch lightweight satellites or perform high-altitude research at relatively low cost. Conceived by **NACA** (National Advisory Committee on Aeronautics) in 1958, the Scout program was taken over by the **Langley** Research Center when NASA was created in October 1958. In less than a year, Scout emerged as a four-stage vehicle that could serve as a sounding rocket or small satellite launcher. NASA decided that all four stages would be solid-fueled, citing the relative simplicity and reliability of previously demonstrated solid-fuel technology. The Scout first stage, called Algor, was adapted from the Polaris missile program. Controlled by the moveable outer tips of four stabilizer fins in conjunction with four exhaust deflector vanes, it burned for 40 seconds and could produce a thrust of 510,000 newtons (N). The second stage also had its roots in a military program. Called Castor, it originated with the Army **Sergeant** rocket program. It burned for 39 seconds, was stabilized by hydrogen peroxide jets, and could produce a thrust of 225,000 N. The third stage, Antares, was an upgraded version of the Vanguard Altair third stage. It burned for 39 seconds, was stabilized by hydrogen peroxide jets, and could produce 61,000 N of thrust. An actual Vanguard Altair third stage was incorporated as the Scout's fourth stage. This burned for 38 seconds, was spin-stabilized, and could produce a thrust of 13,700 N. Both the third and fourth stages were encased in a glass-fiber shield that included the payload shroud and a device to spin-stabilize the fourth stage. The Scout was able to carry a 23-kg payload on a ballistic trajectory to an altitude of 13,700 km or a 68-kg payload into low Earth orbit. The original NASA Scout was modified for Air Force applications under the designations Blue Scout I, Blue Scout II, and Blue Scout Junior. In addition, a single Blue Scout II rocket was modified by NASA for use in the Mercury program and became known as the **Mercury-Scout**.

Scout A Scout lifts off. *NASA*

scramjet

A supersonic combustion **ramjet** engine that operates–in fact, can *only* operate–at hypersonic (greater than Mach 5) speeds. Like its comparatively slower ramjet counterpart, the scramjet has engines with a simple mechanical design and no moving parts. However, scramjet combustion occurs at supersonic air speeds in the engine. Rather than using a rotating compressor like a turbojet engine, the forward velocity and vehicle aerodynamic design compress air into the engine. There, fuel, usually hydrogen, is injected, and the expanding hot gases from combustion accelerate the exhaust air and create thrust. Experimental scramjets include NASA's **X-43A,** the University of Queensland's (Australia) Hyshot, and a DARPA (Defense Advanced Research Projects Agency) scale model that can be fired from a special gun.[316] The latter achieved the first-ever free flight of a scramjet in September 2001, when it was gun-launched at Mach 7, then used its scramjet engine to travel 80 m in just over 30 milliseconds.

screaming

A form of combustion instability, especially in a liquid-propellant rocket engine, accompanied by a high-pitched noise.

scrub

To cancel a scheduled rocket firing.

SCS (Small Communications Satellite)

See **MicroSat.**

SDIO (Strategic Defense Initiative Organization)

A U.S. Department of Defense body formed in April 1984 to develop a space-based missile defense system. SDIO has been involved with a number of space missions, including **Clementine** and various Shuttle-launched payloads.

SDO (Solar Dynamics Observatory)

The first mission in NASA's Living with a Star program. SDO will investigate such problems as how the solar interior varies through a solar cycle, how this variation manifests itself in the structure of the Sun's corona and heliosphere, and the origin and effect of sunspots and solar active magnetic regions. The earliest expected launch date is 2010.

SDS (Satellite Data System)

United States Air Force satellites, in **Molniya-type orbit**s, used to relay images from optical and digital reconnaissance satellites. They also provide communications for Air Force units out of range of geostationary communications satellites, via an AFSATCOM (Air Force Satellite Communications System) transponder, and nuclear blast detection. Six first-generation SDS satellites were launched from 1976 to 1987. SDS-1 had a mass of about 630 kg, a cylindrical body about 4 m long and 3 m in diameter, and a main transmitting antenna over 3 m in diameter. SDS-2 satellites are much larger, with a mass of about 3,000 kg, and are equipped with several big antennas and the HERITAGE infrared missile launch detection sensors. Based on the **TDRS**-3 (Tracking and Data Relay Satellite 3) design, they are built to fit inside the Space Shuttle's payload bay. The first three were launched by the Shuttle from 1989 to 1992, and a fourth by a Titan IV in 1996.

Sea Launch

An international consortium, formed in April 1995, that places commercial satellites in orbit using **Zenit** 3SL rockets from a converted oil platform in the Pacific Ocean near the equator–the ideal location for launching into **geostationary orbit.** Sea Launch is a joint venture of the **Boeing Commercial Space Company, Energia Rocket & Space Corporation** of Moscow, Kvaerner Maritime of Oslo, and KB Yuzhnoye/PO Yuzhmash of Dnepropetrovsk, Ukraine. Kvaerner provides the Command Ship, KB Yuzhnoye provides the first two stages of the launch vehicle, NPO Energia provides the third stage, and Boeing provides the payload fairing and systems integration.

Seamans, Robert C., Jr. (1918–)

An aerospace engineer and administrator who served as deputy administrator of NASA, president of the National Academy of Engineering, and most recently as dean of the Massachusetts Institute of Technology's (MIT's) School of Engineering. Seamans earned a B.S. at Harvard (1939), an M.S. in aeronautics at MIT (1942), and an Sc.D. in instrumentation at MIT (1951). From 1955 to 1958, he worked at RCA, first as manager of the Airborne Systems Laboratory and later as chief engineer of the Missile Electronics and Controls Division. From 1948 to 1958, he also served on technical committees of **NACA** (National Advisory Committee for Aeronautics), NASA's predecessor. He served as a consultant to the Scientific Advisory Board of the U.S. Air Force from 1957 to 1959, as a member of the board from 1959 to 1962, and as an associate advisor from 1962 to 1967. He was a national delegate to the Advisory Group for Aerospace Research and Development (NATO) from 1966 to 1969. In 1960, Seamans joined NASA as associate administrator. In 1965, he became deputy administrator, and also served as acting administrator. During his years at NASA he worked closely with the Department of Defense in research and engineering programs and served as co-chairman of the Astronautics Coordinating Board.

Seasat

A NASA/JPL (Jet Propulsion Laboratory) satellite that carried out the first remote sensing of Earth's oceans using synthetic aperture radar (SAR). The mission ended on October 10, 1978, due to a failure of the vehicle's electric power system. Although only about 42 hours of real-time data was received, the mission proved the feasibility of using microwave sensors to monitor ocean conditions and laid the groundwork for future SAR missions. As well as the SAR, Seasat's payload included: a scatterometer to measure wind speed and direction; a multichannel microwave radiometer to measure surface wind speed, ocean surface temperature, atmospheric water vapor content, rain rate, and ice coverage; and a visible and infrared radiometer to identify cloud, land, and water features and to provide ocean thermal images.

Launch
   Date: June 27, 1978
   Vehicle: Atlas F
   Site: Vandenberg Air Force Base
Orbit: 761 × 765 km × 108.0°
Size: 21.0 × 1.5 m including antenna
Mass: 2,300 kg

Seastar/SeaWiFS

An Earth observation satellite that has a single instrument: the Sea-viewing Wide Field-of-view Sensor (SeaWiFS), designed to monitor the color of the world's oceans. Various colors indicate the presence of different types and quantities of marine phytoplankton, which play a role in the exchange of critical elements and gases between the atmosphere and oceans. Seastar monitors subtle changes in the ocean's color to assess changes in marine phytoplankton levels and provides data to better understand how these changes affect the global environmental and the oceans' role in the carbon cycle and other biogeochemical cycles. The satellite was built and launched and is operated by Orbital Sciences Corporation (OSC), who sell the data collected to NASA. NASA then retains all rights to data for research purposes, while OSC retains all rights for commercial and operational purposes. The mission, now renamed **OrbView**-2, is a follow-on to the Coastal Zone Color Scanner (CZCS) and the first spacecraft in NASA's **EOS** (Earth Observing System).

Launch
   Date: August 1, 1997
   Vehicle: Pegasus-XL
   Site: Vandenberg Air Force Base
Orbit: 707 × 708 km × 98.2°

Seawinds

A specialized microwave radar, known as a scatterometer, that measures near-surface wind velocity (both speed and direction) under all weather and cloud conditions over Earth's oceans. Built by JPL (Jet Propulsion Laboratory), it was carried aboard **QuikScat** in 1999 and is scheduled to fly on the Japanese **ADEOS**-2 satellite.

SECOR (Sequential Collation of Range)

Small U.S. Army geodetic satellites launched in the 1960s to determine the precise location of points on the Earth's surface (notably islands in the Pacific). Each SECOR satellite was linked to four ground stations—three at geographical points where the coordinates had been accurately surveyed and a fourth at the location whose coordinates were to be pinpointed. Radio waves were sent from the ground stations to the satellite and returned by a transponder. The position of the satellite at any time was fixed by the measured ranges from the three known stations. Using these precisely established positions as a base, ranges from the satellite to the unknown station were used to compute the position of the unknown station. SECOR allowed continents and islands to be brought within the same geodetic global grid. Experiments with SECOR led

SERT Missions

Total mass: 1,404 kg				
	Launch			
Spacecraft	**Date**	**Vehicle**	**Site**	**Orbit**
SERT 1	Jul. 20, 1964	Scout X-4	Wallops Island	30-min suborbital flight
SERT 2	Feb. 4, 1970	LT Thor-Agena D	Vandenberg	1,037 × 1,044 km × 99.3°

Service Module

See **Apollo**.

service tower

The vertical structure that provides access to various levels of a spacecraft to prepare it for launch. The service tower also supports the spacecraft until its own engines and guidance system take over.

Serviss, Garrett (Putnam) (1851–1929)

An American journalist and a writer of popular astronomy fiction and nonfiction, whose *Edison's Conquest of Mars*[267] introduces the first-ever scenes of fleets of interplanetary spacecraft. An experienced astronomer, Serviss was able to inject a dose of realism into his narratives that was missing from the space stories of many of his contemporaries.

Sfera

Soviet military geodetic satellites. Although Sfera ("sphere") data was applied to purely scientific problems such as measuring continental drift, it was used mostly by the Red Army General Staff in military topographical research. This work, in turn, allowed improved accuracy for long-range weapons. Flight tests for Sfera took place from 1968 to 1972 and operational launches from 1973 to 1980, using Cosmos-3M launch vehicles from Plesetsk.

SFU (Space Flyer Unit)

A Japanese satellite carrying materials, science, astronomy, and biological experiments. Launched by **NASDA** (National Space Development Agency), it was retrieved

Launch
   Date: March 18, 1995
   Vehicle: H-2
   Site: Tanegashima
Orbit: 471 × 483 km × 28.5°
Mass: 4,000 kg

SERT (Space Electric Rocket Test) Raymond J. Rulis, SERT-1 Program Manager, examining SERT-1 after its arrival at Lewis Research Center for preflight testing. A Lewis-built ion engine is in front of Rulis, and a Hughes-built engine is on the other side. *NASA*

by the Space Shuttle (STS-72) about 10 months later on January 20, 1996.

Shah-Nama

An epic poem by the Persian poet Firdausi (A.D. 1010) in which the hero king Kai-Kaus achieves flight with the help of four eagles, reminiscent of the method used by **Lucian**'s Icaro-Menippus.

SHARP (Super High Altitude Research Project)

A light-gas gun developed at Lawrence Livermore Laboratory, California, and funded by the Strategic Defense ("Star Wars") Initiative as a possible antimissile defense weapon. It consisted of an 82-m-long, 36-cm-caliber pump-tube and a 47-m-long, 10-cm-caliber gun barrel, in an L-shaped configuration. This arrangement was chosen to circumvent one of the main technical problems with a **space cannon**, namely that the projectile cannot outrun the gas molecules, which push it along the gun barrel. The speed of these molecules, at any temperature, can be ¹culated from the laws of physics and is higher the smaller the mass of the gas molecule. The lightest and best gas for the job is hydrogen, but hydrogen is not produced as a product of any explosive mixture. The solution in SHARP was to use a gun with two connected barrels, an auxiliary barrel and a main one in which the payload is accelerated. The two barrels were perpendicular to each other and separated by a partition that shattered when the pressure on it became too high. Rather than a shell, the auxiliary barrel carried a heavy piston, and its volume, between piston and partition, was filled with hydrogen. When the explosive mixture, consisting of methane and air, on the other side of the piston was detonated, the piston raced down the auxiliary barrel, compressing and heating the hydrogen ahead of it. At a certain point, the pressure of the hot hydrogen broke the partition, allowing the hydrogen to flow into the main barrel and propel the payload.

SHARP began operation in December 1992 and demonstrated velocities of 3 km/s (8 to 9 times the speed of sound) with 5 kg projectiles fired horizontally. Impressive as this capability sounds, it falls well short of what is needed to fire projectiles into space (at least 24 times the speed of sound for a low-altitude circular orbit), even if the barrel were pointed upward. Moreover, the $1 billion needed to fund space launch tests never materialized. In 1996, project leader John Hunter founded the Jules Verne Launcher (JVL) Company in an effort to develop the concept commercially. The company planned first to build a prototype Micro Launcher system that would fire 1.3-mm projectiles and demonstrate several new technologies, including the use of three pairs of supplemental gas injectors along the barrel, as used in the **Valier-**

SHARP (Super High Altitude Research Project) The SHARP light gas gun nearing completion in August 1992 at Lawrence Livermore National Laboratory. The gun's pump tube runs from top-right to middle-left where it meets the high pressure chamber. Originating in the high pressure chamber, the launch tube runs from middle-left to out-of-shot at the bottom-right corner. *Lawrence Livermore National Laboratory*

Oberth gun and the V-3 (see **"V" weapons**). The full-scale gun would have been bored into a mountain in Alaska for launches into high-inclination orbits, have a muzzle velocity of 7 km/s, and fire 5,000-kg projectiles, each 1.7 m in diameter and 9 m long. Following the burn of the rocket motor aboard the projectile, a net payload of 3,300 kg would have been placed into low Earth orbit. SHARP experience indicated that the space gun could have been fired up to once a day. Thus, a single gun could orbit over 1,000 tons a year into orbit at a cost per kilogram one-twentieth that of conventional rocket launchers. Payloads would be subjected to accelerations of about 1,000g during launch, so Hunter recruited specialists to design prototype hardened satellite systems. JVL was still operating in 1998, but no investors came forward to finance the multibillion dollar development cost.[266]

Shavit

An Israeli three-stage, solid-propellant launch vehicle developed by Israel Aircraft Industries from the Jericho-2 medium-range ballistic missile. Shavit ("comet") has successfully launched several **Ofeq** satellites; however, its launch capabilities are restricted by the need to avoid violating neighboring Arab countries' airspace. Launches take place from **Palmachim**.

Shepard, Alan Bartlett, Jr. (1923–2001)

One of the original **Mercury Seven** astronauts. Shepard was the first American in space, piloting **Mercury-Redstone 3** *(Freedom 7)* and serving as backup pilot for Mercury-Atlas 9. He was subsequently grounded due to an inner ear ailment until May 7, 1969, during which time he served as chief of the Astronaut Office. Upon returning to flight status, Shepard commanded **Apollo 14**, at age 47 becoming the oldest human to walk on the

Moon, and in June 1971 resumed duties in the Astronaut Office. He retired from NASA and the U.S. Navy in August 1974 with the rank of rear admiral, to become chairman of the Marathon Construction Company of Houston, Texas.[39, 258] Until his death he served on the boards of several companies and was president of Seven Fourteen Enterprises Inc. (named for his two flights, *Freedom 7* and Apollo 14), an umbrella company for several business concerns. He was also president emeritus of the Astronaut Scholarship Foundation, which raises scholarship money for science and engineering students. Shepard received a B.S. from the Naval Academy in 1944. After graduation, he served aboard the destroyer *Cogswell* in the Pacific and later entered flight training, receiving his wings in 1947. In 1950, he attended the Navy Test Pilot School at Patuxent, became a test pilot there, was assigned to a night fighter unit at Moffett Field, and then returned to Patuxent as a test pilot and an instructor. He later attended the Naval War College at Newport, Rhode Island and after graduation was assigned to the staff of the commander in chief, Atlantic Fleet, as aircraft readiness officer.[41, 270]

Alan Shepard   Shepard sealed inside the Mercury capsule undergoing a flight simulation test. *NASA*

Shesta, John

An American rocket engineer who designed the **Viking** rocket propulsion system. An early member of the American Rocket Society, Shesta was one of the founders of Reaction Motors, a company responsible for building many of the engines used in American space launch vehicles and missiles. As Milton **Rosen** wrote in *The Viking Rocket Story:* "Shesta was a conservative designer; he made the [Viking rocket] chamber overly large to ensure complete combustion and he chose an injector . . . that, while not the best known, was reasonably sure to work."

Shinsei

Japan's first scientific satellite. Shinsei ("new star"), launched by **ISAS**, carried out observations of high-frequency radio emissions from the sun, cosmic rays, and ionospheric plasma.

Launch
    Date: September 28, 1971
    Vehicle: M-4S
    Site: Kagoshima
Orbit: 870 × 1,870 km × 32°
Mass: 66 kg

shirtsleeve environment

An environment not requiring that a **pressure suit** be worn.

Shtil

A Soviet ballistic missile that, in 1998, carried the first submarine-launched satellites, the 8-kg Tubsat-N and the 3-kg Tubsat-N1, into low Earth orbit. A three-stage, liquid-fueled missile, the Shtil 1 uses its warhead fairing to hold the spacecraft. The Shtil 2 has a larger separating fairing that can hold bigger payloads.

shutdown

See **cutoff**.

SIGINT (signals intelligence) satellites

Reconnaissance (spy) satellites that intercept communications, radar, and other forms of electromagnetic transmissions. **ELINT** (electronic intelligence) and **COMINT** (communications intelligence) satellites are sub-categories. (See table, "Chronology of American SIGINT Satellites," on page 376.)

Silverstein, Abe (1908–2001)

A leading figure in twentieth-century aerospace engineering and director of the Lewis Research Center (now the **Glenn Research Center**) from 1961 to 1969. Silver-

Chronology of American SIGINT Satellites

	1st Generation	2nd Generation	3rd Generation	4th Generation	5th Generation
	1960s	1970s	1980s	1990s	2000+
GEO-USAF COMINT		Canyon	Chalet/Vortex	Mercury-ELINT	
GEO-CIA ELINT		Rhyolite/Aquacade	Magnum/Orion	Mentor	Intruder
HEO-USAF ELINT			Jumpseat	Trumpet	Prowler
LEO-USAF ELINT	Ferret	Subsats	Subsats		SB-WASS
LEO-Navy ELINT	GRAB	NOSS	NOSS	SB-WASS	

stein earned a B.S. in mechanical engineering (1929) and an M.E. (1934) from Rose Polytechnic Institute, and later received an honorary doctorate from Case Institute of Technology. He began his career with NASA's predecessor, **NACA** (the National Advisory Committee for Aeronautics), at its Langley Research Center in 1929. In 1943, he transferred to Lewis, where he carried out pioneering research on large-scale **ramjet** engines. After World War II, Silverstein conceived, designed, and built the first supersonic propulsion wind tunnel in the United States. The 10-ft by 10-ft Supersonic Wind Tunnel is still operational at Glenn. In 1958, Silverstein moved to NACA Headquarters in Washington, D.C., where he helped create and then direct efforts leading to the **Mercury** spaceflights. He later named and laid the groundwork for the **Apollo** missions. When he returned to Cleveland to become director of NASA Lewis, Silverstein was a driving force behind the creation of the **Centaur** launch vehicle, particularly the hydrogen-oxygen upper stage propulsion system. He retired from NASA in 1970 to take a position with Republic Steel. In 1997, he was awarded the Guggenheim Medal for his "technical contributions and visionary leadership in advancing technology of aircraft and propulsion performance, and foresight in establishing the Mercury and Gemini manned spaceflight activities."[73]

SIM (Space Interferometry Mission)
A major space-based observatory in NASA's **Origins Program** and the first spacecraft to carry an optical interferometer as its main instrument. SIM is designed specifically for the precise measurement of star positions. One of its prime goals will be to search for extrasolar planets as small as Earth in orbit around nearby stars. SIM will combine the light from two sets of four 30-cm-diameter telescopes arrayed across a 10-m boom to achieve a resolution approaching that of a 10-m-diameter mirror. This will allow it to perform extremely sensitive **astrometry** so that it will be able to detect very small wobbles in the movement of a star due to unseen companions. Objects of Earth mass could be inferred around a star up to 30 light-years away. Through a process known as synthesis imaging, SIM will also generate images of objects such as dust disks around stars and look for gaps or clearings in the debris that might indicate the presence of unseen worlds. Scheduled for launch in 2005, SIM will conduct its seven-year mission from a heliocentric, Earth-trailing orbit.

Simons, David G. (1922–)
A U.S. Air Force major who established a new human altitude record of 30,942 m (101,516 ft) when he flew aboard the second Project **Manhigh** mission on August 19–20, 1957. During his 32-hour flight, most of it at the edge of space, Simons recorded that the silence of the upper stratosphere is "like no earthly quiet" and that the color of the sky at this height is "deep indigo, intense, almost black." He was among the first to report the psychologically dangerous euphoria known as the **break-off phenomenon**.

Singer, Samuel Frederick (1924–)
A physicist at the University of Maryland who proposed a Minimum Orbital Unmanned Satellite of the Earth (MOUSE) at the fourth Congress of the International Astronautics Federation in Zurich, Switzerland, in the summer of 1953. It had been based upon two years of previous study conducted under the auspices of the **British Interplanetary Society**, which had built on the postwar research of the V-2 rocket. The Upper Atmosphere Rocket Research Panel at White Sands discussed Singer's plan in April 1954. In May, Singer presented his MOUSE proposal at the Hayden Planetarium's fourth Space Travel Symposium. MOUSE was the first satellite proposal widely discussed in nongovernmental engineering and scientific circles, although it was never adopted.

single-stage-to-orbit (SSTO)
A reusable single-stage rocket that can take off and land repeatedly and is able to boost payloads into orbit.

SIRTF (Space Infrared Telescope Facility)

The fourth and final element in NASA's **Great Observatories** Program; the other three are the **Hubble Space Telescope, Chandra X-Ray Observatory**, and **Compton Gamma-Ray Observatory**. SIRTF's cryogenically cooled optical system will enable scientists to peer through the dust clouds that surround many stars and will thus provide a tool for studying the birth of planets. The observatory will also help scientists explore the origin and evolution of objects ranging from the Milky Way to the outer reaches of the universe. SIRTF is scheduled for launch in January 2003.

Skylab

See article, pages 378–380.

Skylark

A British **sounding rocket** program that originated in 1955, when the Ministry of Supply announced that the Royal Aircraft Establishment and the Rocket Propulsion Establishment would develop a rocket in time for the **International Geophysical Year** in 1957. The first Skylark was designed for economy, with no guidance system and with launch towers built from spare parts of army bridges. Its maiden launch on February 13, 1957, from Woomera, Australia, was followed by two test flights and then, in November 1957, the first operational mission carrying scientific experiments. Over the years different versions evolved, including a two-stage Skylark in 1960. The rocket went international, being used by **ESRO** (European Space Research Organisation) from 1964 to 1972 and by Germany from 1970. Although the British Skylark program ended in 1979 after 266 launches and the rocket is no longer manufactured, enough of Skylark 7s remain in stock for ESA (European Space Agency) to continue launches of Britain's most successful rocket into the early part of the new century.

Skylon

A British design for a **space plane** intended to improve upon that of **HOTOL**. It stems from the work of engineers at Reaction Engines, a company formed in the early 1990s to continue work on HOTOL after the project was officially cancelled. Reaction Engines were not allowed to use the HOTOL RB454 engines, as they were still classified top secret. A new engine was designed called SABRE (Synergic Air Breathing Engine), which would use liquid hydrogen and air until Skylon reached Mach 5.5, then switch to an onboard supply of liquid oxygen for the final ascent to orbit. Skylon would be constructed from carbon fiber, with aluminum fuel tanks and a ceramic aeroshell to protect the craft from the heat of reentry. A payload bay measuring 12.3 by 4.6 m would allow the payload to use standard air transport containers. Skylon could take 12 tons of payload into orbit, or 9.5 tons to the International Space Station. Unlike other comparable projects, Skylon is intended to be run by commercial companies rather than government space agencies, resulting in a design similar to that of a normal aircraft with a vehicle turnaround of only two days between flights, rather than several weeks, as with the Space Shuttle. Operating on a commercial basis could also reduce the price of launches from $150 million for a 2- to 3-ton satellite to $10 million for all cargo. Eventually, costs could fall to allow a passenger seat to cost only $100,000, opening the way for regular space tourism. Reaction Engines envisage that by 2025 there could be several companies operating Skylons from specially constructed equatorial spaceports. In 1997, Skylon was considered by ESA (European Space Agency) for the FESTP (Future European Space Transportation Project). More recently, it has been suggested that Skylon could win the X-Prize, a sum of $10 million for the first team to send a passenger into space. Reaction engines have recently been attempting to put together a consortium of aerospace companies to fund the Skylon project.

Skynet

A British military satellite communications network developed with American assistance. With the launch of Skynet 1A in 1969, Britain became the third entity after the United States and Intelsat to orbit a geostationary satellite. Skynet 1B, launched in August 1970, was stranded in a transfer orbit when its apogee kick motor failed to fire. Then Skynet 1A failed prematurely, and it was not until 1974 that the first Skynet 2 was launched. Unfortunately, this too was stranded in a useless orbit. Skynet 2B, however, was successfully placed in an orbit over the Indian Ocean at 42° E and, remarkably, was still working 20 years later. Skynet 3 was never developed, and the next launch was the first of the Skynet 4 series in 1988. Skynet 4, with an on-orbit mass of 800 kg, has one experimental EHF (extremely high frequency) channel, two UHF (ultrahigh frequency) channels, and four X-band channels. The fifth and last Skynet 4, which replaced Skynet 4C, was launched by an Ariane on February 26, 1999. The first Skynet 5 is expected in 2005.

Skyrocket

See **Douglas Skyrocket**.

Slayton, Donald Kent "Deke" (1924–1993)

One of the original **Mercury Seven** astronauts selected in 1959, and the only one not to make a **Mercury** flight. He did, however, fly on the **Apollo-Soyuz Test Project** *(continued on page 381)*

Skylab

An experimental American **space station** adapted from **Apollo** hardware. Skylab consisted of the orbital workshop itself, an airlock module, a multiple docking adapter, and the Apollo telescope mount. In orbit, the station was 36 m long and, with a docked Apollo command and service module, had a mass of about 90.6 tons. The living volume was about the same as that of a small house. Because crews stayed for one, two, or three months, the orbital workshop was designed for habitability, with more amenities than previous spacecraft. Among the features especially appreciated by crews were a large window for viewing Earth, a galley and wardroom with a table for group meals, private sleeping quarters, and a shower, custom-designed for use in weightlessness.

Twin solar array wing panels were folded against the orbital workshop for launch, one on each side. When Skylab reached orbit, the arrays would extend, exposing solar cells to the Sun to produce 12 kW of power.

Skylab was equipped to observe Earth's natural resources and the environment, and activity on the Sun. Astronauts also studied the effects of long-term weightlessness on the human body and materials processing in microgravity, and performed experiments submitted by students for a "Classroom in Space." Because Skylab was a research laboratory, the composition of the crew differed from that of the Mercury, Gemini, and Apollo missions. Except for one scientist on the last Apollo mission, all previous crew members had been pilots. The Skylab crews included a number of scientist-astronauts.

Skylab 1

The unmanned mission to place Skylab in orbit. During launch, on May 14, 1975, the micrometeoroid shield accidentally deployed too soon, jamming one solar array wing and damaging the other so badly that both the wing and the shield were ripped away. Left with only one solar array wing and no micrometeoroid shield, which also served as a shade to keep the interior of the space station cool, Skylab could have been rendered useless. However, thanks to clever engineering and improvisation, the effect of the damage was minimized, and planned Skylab operations were able to go ahead despite some reduction in power.

Skylab 2

The first manned Skylab mission, devoted to starting up and checking out the space station and, most importantly, fixing the damage caused during the station's launch. With the temperature inside Skylab at a scorching 52°C, the astronauts' first priority was to set up a "parasol" shade as a makeshift replacement for the lost micrometeoroid shield. Once this had been deployed by reaching out of Skylab's main hatch, the temperature inside the station dropped to comfortable levels. On the third day of the mission (MD-3), the crew began turning on experiments. Then, on MD-14, Conrad and Kerwin went on a crucial three-hour spacewalk to extend Skylab's only remaining solar array wing, which was jammed by a strap of debris. Kerwin was able to cut through the debris using off-the-shelf barbed-wire snippers that NASA had bought for just $75. The success of this repair ensured that there would be enough power for the full 28-day mission and for the subsequent Skylab 3 and Skylab 4 missions. A couple of days before returning home, Conrad and Kerwin carried out another spacewalk to retrieve and replace film from the solar telescopes, repair a circuit breaker module, and do minor maintenance on experiment packages located outside the station. Upon completion of the mission, the Skylab 2 crew briefly set a new manned spaceflight endurance record, previously held by cosmonauts Dobrovolsky, Volkov, and Patsayev on **Salyut** 1 in June 1971.

Launch: May 25, 1973
Recovery: June 22, 1973
Mission duration: 28 days
Crew: Charles "Pete" **Conrad** Jr. (commander), Paul **Weitz** (pilot), Joseph **Kerwin** (science pilot)

Skylab 3

The second manned Skylab mission. Shortly after entering the space station, all three crew members fell victim to space-sickness, delaying the activation of onboard equipment. On Mission Day-5, an apparent failure of two of the four thruster quadrants on the Command and Service Module (CSM) was detected.

Skylab An overhead view of Skylab from the Skylab 4 Command and Service Module during its final flyaround before returning home. *NASA*

This not only threatened an early end to the mission but also put a question mark over whether the CSM would be able to return safely to Earth. Launch crews at Kennedy Space Center were put on a nonstop work schedule to prepare the Skylab 4 Saturn IB for a possible rescue operation. In the event, the mission was continued as planned and the CSM performed flawlessly during reentry. On August 6, 1973, Garriott and Lousma went on a 6-hour spacewalk to deploy a twin-pole thermal shield to replace the "parasol" installed by the Skylab 2 crew. Two further spacewalks followed for film retrieval and replacement, installation of a new rate gyro package, and routine maintenance.

Launch: July 28, 1973
Recovery: September 25, 1973
Mission duration: 59 days 11 hours
Crew: Alan **Bean** (commander), Jack **Lousma** (pilot), Owen **Garriott** (science pilot)

Skylab 4

The third and final manned visit to Skylab. Highlights of the record-setting 84-day mission included extensive photography and analysis of comet Kohoutek and the first complete photographic coverage of a solar flare from beginning to full size. A total of 1,563 hours of scientific experiments were performed, about twice that of Skylab 2 and Skylab 3 combined. The crew also completed four spacewalks, lasting a total of 22 hours 22 minutes and including the longest spacewalk up to that time of 7 hours 1 minute.

Launch: November 16, 1973
Recovery: February 8, 1974
Mission duration: 84 days 1 hour
Crew: Gerald **Carr** (commander), Edward **Gibson** (science pilot), William **Pogue** (pilot)

Epilogue: Skylab's Return

Before the Skylab 4 crew left, they boosted the space station to a slightly higher orbit, which varied from 430 to 455 km. Calculations indicated that this would enable Skylab to remain aloft for at least nine more years, giving NASA time to bring its Shuttle service online and mount a rescue mission. NASA originally planned to use the fifth Shuttle flight, slated for the second half of 1979, to dock with Skylab and lift it to a safe orbit so that it could be used again. But in late 1978, NOAA (National Oceanographic and Atmospheric Administration) warned that increased solar activity would result in a stronger solar wind that would likely push Skylab back into the atmosphere within a year. This, compounded with delays to the Shuttle program caused by engine development prob-

Skylab 4 Gerald Carr, commander of the Skylab 4 mission, jokingly demonstrates weight training in zero-gravity as he balances astronaut William Pogue, pilot, upside down on his finger. *NASA*

lems, meant that a rescue mission could not be launched in time. On December 15, 1978, NASA Administrator Robert Frosch informed President Carter that Skylab could not be saved, and that NASA would attempt to guide the space station to a controlled reentry as far away from populated areas as possible. At 3:45 A.M. EDT on July 11, 1979, controllers at Johnson Space Center commanded Skylab to tumble, hoping the space station would disintegrate upon reentry. However, Skylab did not break apart as much as expected and, 21 hours later, rained large chunks of debris near Perth, Australia.

Donald Slayton   Deke Slayton suits up for an altitude test of the Apollo Command Module in an altitude chamber of Kennedy Space Center's Manned Spacecraft Operations Building in preparation for the Apollo Soyuz Test Project. *NASA*

Slayton, Donald Kent "Deke" (1924–1993)

(continued from page 377)
(ASTP) in 1975. Slayton received a B.S. in aeronautical engineering from the University of Minnesota in 1949, joined the Air Force in 1942, and received his wings a year later. During World War II, he flew combat missions over Europe and Japan, then became an aeronautical engineer with Boeing in Seattle. He was recalled to active duty in the Minnesota Air National Guard in 1951 and remained in the Air Force, going on to attend the Air Force Test Pilot School at Edwards Air Force Base before becoming an experimental test pilot there. After being selected by NASA, Slayton was assigned to fly the second Mercury orbital mission but was grounded by an irregular heartbeat. He stayed with NASA to supervise the astronaut corps, first as chief of the Astronaut Office and then as director of flight crew operations. He eventually overcame his heart problem and was restored to flight status in 1972. Three years later, on July 17, 1975, Slayton made it into space aboard the ASTP after 16 years as an astronaut. For the next two years, Slayton was manager of the Space Shuttle approach and landing tests at Edwards. From 1977 until he retired from NASA in 1982, he was manager for orbital flight tests. Later, he became president of Space Services, a company that develops rockets for suborbital and orbital launch of small commercial space payloads.[272]

slingshot effect
See **gravity assist**.

sloshing
The back-and-forth splashing of a liquid fuel in its tank. Sloshing can lead to problems of stability and control in a launch vehicle.

slot
The longitudinal position at which a **communications satellite** in **geostationary orbit** is located.

SLV-3
An Indian four-stage, solid-propellant launch vehicle with a payload capacity of less than 50 kg into a low-Earth orbit with a mean altitude of 600 km at an inclination of 47°. Following an initial failure, the SLV-3 successfully orbited three **Rohini** satellites in 1980, 1981, and 1983. The SLV-3 formed the basis of the **ASLV** (Advanced Space Launch Vehicle).

SMART (Small Missions for Advanced Research in Technology)
An ESA (European Space Agency) program of small, relatively low-cost missions to test new technologies that will eventually be used on bigger missions. SMART-1, the first mission in this program, will become the first European spacecraft to visit the Moon. However, the Moon is not so much the focus of the mission as the venue around which SMART-1 will test a solar-electric propulsion system like that employed by **Deep Space 1**. If successful, the unit will be used by the Mercury probe **Bepi Colombo**. SMART-1 is scheduled for launch early in 2003.

SME (Solar Mesosphere Explorer)
A scientific satellite designed to investigate the processes that create and destroy ozone in Earth's upper atmosphere. The mission's specific goals were to examine the effects of changes in the solar ultraviolet flux on ozone densities in the mesosphere; the relationship between solar flux, ozone, and the temperature of the upper stratosphere and mesosphere; the relationship between ozone and water vapor; and the relationship between nitrogen dioxide and ozone. The mission was managed for NASA by JPL (Jet Propulsion Laboratory) and operated by the Laboratory for Atmospheric and Space Physics of the University of Colorado via the Goddard

Launch
   Date: October 6, 1981
   Vehicle: Delta 2914
   Site: Cape Canaveral
Orbit: 335 × 337 km × 97.6°
Size: 1.7 × 1.3 m
Mass: 437 kg

Space Flight Center. The scientific payload included an ultraviolet ozone spectrometer, a 1.27-micron spectrometer, a nitrogen dioxide spectrometer, a 4-channel infrared radiometer, a solar ultraviolet monitor, and a solar proton alarm detector. All instruments were turned off in December 1988, and contact was lost permanently on April 14, 1989, following a battery failure.

SMEX (Small Explorer)

A series of low-cost satellites launched by NASA for solar and astronomical studies. The first four spacecraft in the series were **SAMPEX** (Solar Anomalous and Magnetospheric Particle Explorer), **FAST** (Fast Auroral Snapshot Explorer), **TRACE** (Transition Region and Coronal Explorer), and **SWAS** (Submillimeter Wave Astronomy Satellite).

SMEX lite

A design, stemming from early experience with the **SMEX** program, for a new system architecture intended to provide ultra-low-cost small spacecraft whose performance exceeds that of the initial five SMEX missions. SMEX lite has been developed at the **Goddard Space Flight Center**.

SMM (Solar Maximum Mission)

A NASA spacecraft equipped to study solar flares and other high-energy solar phenomena. Launched during a peak of solar activity, SMM observed more than 12,000 flares and over 1,200 coronal mass ejections during its 10-year lifetime. It provided measurements of total solar radiative output, transition region magnetic field strengths, storage and release of flare energy, particle accelerations, and the formation of hot plasma. Observations from SMM were coordinated with in situ measurements of flare particle emissions made by **ISEE**-3 (International Sun-Earth Explorer 3). SMM was the first satellite to be retrieved, repaired, and redeployed in orbit: in 1984, the STS-41 Shuttle crew restored the spacecraft's malfunctioning attitude control system and replaced a failed electronics box. SMM collected data until November 24, 1989.

Launch
  Date: February 14, 1980
  Vehicle: Delta 3914
  Site: Cape Canaveral
Orbit: 405×408 km $\times 28.5°$
Mass: 2,315 kg
Length: 4.0 m

SMS (Synchronous Meteorological Satellite)

Weather satellites in geostationary orbit, managed by NASA for **NOAA** (National Oceanic and Atmospheric Administration). SMS-1 and -2 were development satellites leading up to the **GOES** (Geostationary Operational Environment Satellite) program; in fact, SMS 3 became operational as **GOES**-1. (See table, "SMS Series.")

Launch
  Vehicle: Delta 2914
  Site: Cape Canaveral

SNAP (Systems for Nuclear Auxiliary Power)

A program of experimental **radioisotope thermonuclear generator**s (RTGs) and space nuclear reactors flown during the 1960s. Odd-numbered SNAPs were RTG tests, and even-numbered SNAPs were reactor system tests. The United States flew only one complete nuclear reactor, aboard the **SNAPSHOT** mission.

SNAPSHOT

The first, and so far only, American flight of an operational nuclear space reactor. The **SNAP**-10A reactor produced 650 W of power from 1.3 kg of used uranium-235 fuel (embedded in uranium zirconium hydride) to run a small **ion propulsion** system.

Launch
  Date: April 3, 1965
  Vehicle: Atlas-Agena D
  Site: Vandenberg Air Force Base
Mass: 440 kg
Orbit: $1,270 \times 1,314$ km $\times 90°$

SMS Series

Spacecraft	Launch Date	Orbit	Mass (kg)
SMS 1	May 17, 1974	$36,216 \times 36,303$ km $\times 15.5°$	243
SMS 2	Feb. 6, 1975	$35,941 \times 36,060$ km $\times 12.0°$	627
SMS 3	See **GOES**-1		

Snark

An intercontinental subsonic cruise missile built for the U.S. Air Force by Northrop. Its first three test flights were from Cape Canaveral in 1956–1957.

SNOE (Student Nitric Oxide Explorer)

A small satellite built by a team at the University of Colorado to measure how the density of nitric oxide in the atmosphere varies with altitude. SNOE was the first satellite in the STEDI (Student Explorer Demonstration Initiative) program.

Launch
Date: February 26, 1998
Vehicle: Pegasus XL
Site: Vandenberg Air Force Base
Orbit: 529 × 581 km × 97.7°

Society for Space Travel

See **Verein für Raumschiffahrt.**

SOFIA (Stratospheric Observatory for Infrared Astronomy)

A Boeing 747-SP aircraft modified to accommodate a 2.5-m reflecting telescope. SOFIA, a joint project of NASA and **DLR** (the German center for aerospace research), will be the largest airborne telescope in the world, capable of observations that are impossible for even the largest and highest of ground-based telescopes. It is being developed and operated for NASA by a team of industry experts led by the **Universities Space Research Association,** will be based at **Moffett Federal Airfield,** and is expected to begin flying in 2004. SOFIA forms part of the **Origins Program.**

soft landing

Landing on a planetary body at a slow speed to avoid destruction of or damage to the landing vehicle.

SOHO (Solar and Heliospheric Observatory)

A joint ESA (European Space Agency) and NASA mission to investigate the dynamics of the Sun. SOHO was the first spacecraft to be placed in a **halo orbit**–in this

SOHO (Solar and Heliospheric Observatory) SOHO ready to be placed within its protective fairing before transfer to the launch pad. *NASA*

case, an elliptical orbit around the first **Lagrangian point** (L1), 1.5 million km ahead of Earth. In this orbit, it avoids solar eclipses by our planet. The data it collects will help astronomers understand how the solar corona is heated and how it expands into the solar wind, and will provide new insight on the Sun's structure and interior dynamics from the core to the photosphere. The spacecraft was built by ESA, tracking and data acquisition are shared by NASA and ESA, and mission operations are conducted by NASA.

Launch
  Date: December 2, 1995
  Vehicle: Atlas IIAS
  Site: Cape Canaveral
  Orbit: elliptical about L1 with semi-major axis of
    650,000 km
  Size: 3.6 × 3.6 m
  Mass: 1,850 kg

Sojourner

See **Mars Pathfinder**.

sol

The length of a Martian day. One sol equals 24 hours 37 minutes 22.6 seconds (39.6 minutes longer than an Earth day).

Solar-

The prelaunch designation of a series of satellites built by Japan's **ISAS** (Institute of Space and Astronautical Science) to study the Sun and its effect on Earth's environment. See **Yohkoh** (Solar-A) and **Solar-B**.

solar array

A panel that extends from a spacecraft, consisting of a large group of **solar cell**s cemented onto a substrate. Solar arrays are an important source of electric power for spacecraft operating no farther from the Sun than about the orbit of Mars.

solar cell

A crystalline wafer, also known as a photovoltaic cell, that converts sunlight directly into electricity without moving parts.

solar conjunction

The period of time during which the Sun is in or near the communications path between a spacecraft and Earth, thus disrupting the communications signals.

Solar Connections Initiative

A NASA program to better understand the interactions and coupling between the Sun's atmosphere and helio-sphere and Earth's magnetosphere and atmosphere. It involves a number of spacecraft, including **TIMED** (Thermosphere Ionosphere Mesosphere Energetic Dynamics satellite), **RHESSI** (Reuven Ramaty High Energy Solar Spectroscopic Imager), and the **Solar Probe**.

solar flare

A solar phenomenon that gives rise to intense ultraviolet and particle emission from the associated region of the Sun, affecting the structure of the ionosphere and interfering with communications.

Solar Polar Imager

A mission designed to improve understanding of the structure of the solar corona and to obtain a three-dimensional view of coronal mass ejections. The Solar Polar Imager would observe the Sun and inner heliosphere from an orbit about half the size of Earth's, perpendicular to the ecliptic. It is identified in NASA's Office of Space Science Strategic Plan as a potential mission beyond 2007 but remains in the early concept definition phase.

Solar Probe

A NASA spacecraft that will make the first-ever measurements within the atmosphere of a star–the Sun–to provide unambiguous answers to long-standing fundamental questions about how the corona is heated and how the solar wind is accelerated. The spacecraft, which will provide both imaging and in situ measurements, is targeted to pass within 3 solar radii of the Sun's surface. Solar Probe is tentatively scheduled for launch in 2007.

solar sail

See **space sail**.

Solar Terrestrial Probes

A NASA program to investigate the interaction between the Sun and the rest of the Solar System and, in particular, solar effects on the terrestrial environment. Six missions are currently identified within the program: **TIMED, Solar-B, STEREO, Triana, Magnetospheric Multiscale**, and **GEC**.

solar wind

A continuous stream of charged particles, mostly **protons** and **electrons**, that is ejected by the Sun and flows radially out at speeds of 200 to 850 km/s. The Sun's magnetic field is transported by the solar wind to become the interplanetary magnetic field, the lines of which are wound into spirals by the Sun's rotation. The solar wind confines Earth's magnetic field inside a cavity known as the **magnetosphere** and supplies energy to phenomena in the magnetosphere such as polar aurora and magnetic storms.

Solar-B

A Japan-led mission, with collaboration from the United States and Britain, to follow-on from the highly successful **Yohkoh** (Solar-A). It will look at the Sun in soft (longer wavelength) X-rays, as does Yohkoh, but will also provide very high-resolution images in visible light. Solar-B will carry a coordinated set of optical, extreme ultraviolet, and X-ray instruments to investigate the interaction between the Sun's magnetic field and its corona. The result will be an improved understanding of the mechanisms that underlie solar magnetic variability and how this variability affects the total solar output. **ISAS** (Institute of Space and Astronautical Science) will provide the spacecraft, the M-5 launch vehicle, and major elements of each of the scientific instruments. These include a 0.5-m solar optical telescope (SOT), an X-ray telescope (XRT), and an extreme ultraviolet imaging spectrometer (EIS). Britain's Particle Physics and Astronomical Research Council is responsible for the EIS. NASA will provide the focal plane package (FPP) for the SOT, as well as components of the XRT and EIS. Solar-B is scheduled for launch in the fall of 2005 into a sun-synchronous orbit that will keep the instruments in nearly continuous sunlight.

solar-electric propulsion (SEP)

A form of **electric propulsion** in which the electrical energy used to accelerate the propellant comes from a solar power source, such as arrays of **photovoltaic cell**s.

solid propellant

A rocket propellant in solid form, usually containing both fuel and oxidizer combined, or mixed and formed, into a monolithic (not powdered or granulated) grain.

Solid Rocket Boosters (SRBs)

See **Space Shuttle**.

solid-fueled rockets

Rockets that burn a solid mixture of fuel and oxidizer and that have no separation between combustion chamber and fuel reservoir. Gunpowder is such a mixture and was the earliest rocket fuel. They are somewhat less efficient than the best liquid-fueled rockets, but they are preferred for military use because they need no lengthy preparation and are easily stored in ready-to-fly condition. They are also used in auxiliary rockets that help heavily loaded liquid-fueled rockets, such as the Space Shuttle and Delta rockets, lift off and go through the first stage of their flight.

solid-propellant rocket motor

A rocket propulsion system in which the **propellant** is contained within the **combustion chamber** or case. The solid propellant charge is called the grain and contains all the chemical constituents for complete burning. Once ignited, it usually burns smoothly at a predetermined rate on all the exposed surfaces of the grain. The internal cavity grows as propellant is burned and consumed. The resulting hot gas flows through the supersonic nozzle to impart thrust. Once ignited, the combustion proceeds until all the propellant is used up. There are no feed systems or valves.

SOLO (Solar Orbiter)

A proposed ESA (European Space Agency) Sun-orbiting spacecraft that will carry optical instruments to view the Sun directly from close up and from high latitudes. SOLO will study fields and particles in the part of the heliosphere near to the Sun, the links between activity on the Sun's surface and the corona, and the Sun's polar regions. The spacecraft will tap much of the technology developed for **Bepi Colombo**, the Mercury Cornerstone mission. The orbital design follows the Mercury Orbiter trajectory design, and the mission will use **solar-electric propulsion** (SEP), powered by a set of large cruise solar arrays that are jettisoned after the last firing of the SEP thrusters. Using SEP together with multiple **gravity-assist** maneuvers, it will take Solar Orbiter only two years to reach a perihelion of 45 solar radii at an orbital period of 149 days.

Solrad (Solar Radiation program)

A series of missions, conceived by the U.S. Naval Research Laboratory in the late 1950s, to study the effects of solar emissions on the ionosphere and also to conceal the existence of classified intelligence satellites, known as **GRAB**, that were launched at the same time. The Solrad series was designed to provide continuous coverage of wavelength and intensity changes of solar radiation in the ultraviolet, soft (longer wavelength) X-ray, and hard (shorter wavelength) X-ray ranges. All missions up to Solrad 7B were GRAB copassengers. Solrad 8, 9, and 10 were also known as **Explorer** 30, 37 (or Solar Explorer B), and 44, respectively.

Solwind

A satellite designed to investigate how the solar wind interacts with Earth's ionosphere and magnetosphere. It was destroyed, while still functional, on September 13, 1985, as part of a U.S. Air Force antisatellite (ASAT) test.

Launch
    Date: February 24, 1979
    Vehicle: Atlas F
    Site: Vandenberg Air Force Base
Orbit: 310 × 317 km × 97.8°
Mass: 1,331 kg

sonic speed

The **speed of sound**, or the speed of a body traveling at Mach 1.

SORCE (Solar Radiation and Climate Experiment)

An Earth observation satellite that will carry four instruments to measure the total amount of solar radiation striking the atmosphere each day over a period of five years. Its data will allow the creation of a long-term climate record that will improve understanding of how changes in the Sun's intensity affect Earth's climate and the chemistry of the atmosphere. SORCE is part of NASA's **EOS** (Earth Observing System) and will be operated by the University of Colorado. It is scheduled for launch in December 2002.

sound barrier

Before 1947, it was believed that the **speed of sound** created a physical barrier for aircraft and pilots. As airplanes approach the speed of sound, a **shock wave** forms and the aircraft encounters sharply increased **drag**, violent shaking, loss of **lift**, and loss of control. In attempting to break the barrier, several planes went out of control and crashed, injuring many pilots and killing some. The barrier was eventually shown to be mythical, however, when Chuck **Yeager** surpassed the speed of sound in the **X-1**.

sounding rocket

A research rocket that sends equipment into the upper atmosphere on a suborbital trajectory, takes measurements, and returns to the ground.

Soviet

See "Russia"/"Russian" entries, pages 352–354.

Soyuz (launch vehicle)

The most frequently launched and the most reliable launch vehicle in the history of spaceflight. By the dawn of the twenty-first century, more than 1,600 Soyuz rockets of various kinds had been launched with an unparalleled success rate of 97.5% for production models. The two-stage Soyuz that currently transports cosmonauts to the International Space Station (ISS) is simply a more powerful variant of the original **Sputnik (launch vehicle)**, with a first stage derived directly from the R-7 ballistic missile (see **"R" series of Russian missiles**) and a two-engine second stage. It can lift up to 7,500 kg into low Earth orbit and, in addition to launching manned **Soyuz (spacecraft)**, has been used to launch a wide variety of scientific and military satellites. Soyuz is now marketed internationally by Starsem, a joint Russian-French consortium which, in addition to ISS cargo and manned versions of the rocket, offers the commercial four-stage

Soyuz and Soyuz/ST, each equipped with the Fregat upper stage for propelling payloads to all types of Earth orbit and escape trajectories.

Soyuz spacecraft

See article, pages 387–389.

SP-100

An ambitious American program to develop a nuclear reactor for spaceflight applications carried out from 1983 to 1994. The SP-100 program resulted in a realistic design for a system that could deliver 100 kilowatts of electric power using a reactor with 140 kg of uranium-235 (in uranium nitride). The modular construction of SP-100 would enable the reactor to be used either as a power source for a large **ion propulsion** system or as an electric power station for a manned lunar or planetary base.

space

The part of the universe lying outside of the limits of Earth's atmosphere. More generally, the volume in which all spatial bodies move.

space agencies

See: **ASI** (Italian space agency), **Canadian Space Agency**, **CNES** (French space agency), **DARA** (German space agency), **ESA** (European Space Agency), **INPE** (Brazilian space agency), **INTA** (Spanish space agency), **ISAS** (Japan), **NASA**, **NASDA** (Japan), and **Russian Space Agency** (RKA).

space ark

See **generation ship**.

space cannon

A gun so powerful that it can fire a projectile from Earth's surface into space. The idea of using a cannon to put objects into orbit was first suggested in the seventeenth century in Newton's *Principia Mathematica* (see **Newton's orbital cannon**). In 1865, in *From the Earth to the Moon*, Jules Verne envisioned a 274-m-long cannon, the *Columbiad*, sunk vertically into the ground not far from today's Cape Canaveral, that sent a three-man capsule to the Moon. Unfortunately, Verne was hopelessly optimistic about his astronauts' chances of survival. In attaining Earth escape velocity inside the barrel of a gun, the passengers would be subjected to lethal *g*-forces. Yet, cannons offer distinct advantages over rockets as a way of placing inert payloads in orbit. Rockets must lift, not only their own weight, but the weight of their fuel and oxidizer. Cannon "fuel," which is contained within the gun barrel, offers far more explosive power per unit cost

(continued on page 389)

Soyuz spacecraft

A series of manned Russian spacecraft, in use longer than any other; it is carried into orbit by the **Soyuz (launch vehicle)**. Designed during the Space Race era, Soyuz first carried a cosmonaut in April 1967. Since then the original Soyuz ("union") craft and its subsequent generations–the Soyuz T, TM, and TMA–have flown scores of manned missions. Although modifications have made the spacecraft more efficient and reliable, the basic structure remains the same.

A Soyuz spacecraft has three main components. The orbital module at the front is an egg-shaped section, 2.2 × 2.7 m, that provides space for the cosmonauts to work in once they are in space. The descent module in the middle is a bell-shaped section, about 2.2 m long and 2.2 m in diameter, in which the cosmonauts sit during launch and upon their return to Earth. The instrument module at the back is cylindrical and measures 2.7 m wide and 2.3 m long. As the spacecraft evolved, more efficient electronics and navigation systems were added, and the landing module interior was rearranged to make it more spacious. A version of Soyuz is being used as a crew rescue vehicle serving the **International Space Station**. (See table, "Versions of the Soyuz Spacecraft.")

History

The manned Soyuz spacecraft was originally conceived by Sergei **Korolev** in 1961 as a component of the "Soyuz complex," which also included unmanned booster modules and orbiting fuel tankers and was geared toward a manned mission to the Moon (see **Russian manned lunar programs**). When this plan was abandoned, only the crewed vessel remained: its new primary task was that of a space station ferry.

Soyuz 1 to 9

The first version of Soyuz could accommodate three cosmonauts in the cramped descent module, but only if spacesuits were not worn. Power was provided by a pair of solar arrays on either side of the instrument module that were folded during launch. Once in space the crew could enter the orbital module to conduct experiments. This module could also be depressurized to serve as an airlock for cosmonauts to perform spacewalks. Although the forward section of the orbital module did have provision for docking to another Soyuz, there was no connecting hatch, so that crew exchanges could be accomplished only by spacewalks. At the end of the flight, the crew returned to the descent module, which was equipped with a heatshield, separated from the other two modules, and reentered. The descent was slowed by a single large parachute and, just 2 m above the ground, by four small rocket motors on the capsule's base.

After several unmanned tests, the first Soyuz was launched on April 23, 1967, at 3:35 A.M. local time–the first night launch of a crewed vehicle–with pilot Vladimir **Komarov** on board. Once in orbit, it was to have served as a docking target for Soyuz 2 and its crew of three. After the docking, the two engineers were to have transferred to Soyuz 1 and returned home with Komarov. However, Soyuz 2 was never launched because of a problem with Soyuz 1 in orbit. After just over a day in space and several failed deorbit attempts, Soyuz 1 successfully reentered the atmosphere but then crashed to the ground after its parachute lines became tangled, killing Komarov. Although there had been a catalog of shortcomings with Soyuz before launch, Soviet premier Brezhnev

Versions of the Soyuz Spacecraft

Name	First Flown	Notes
Soyuz (1–9)	1967	Up to three cosmonauts without spacesuits
Soyuz (10–11)	1971	Addition of docking module
Soyuz (12–40)	1973	Limit of two spacesuited cosmonauts + equipment to Salyut
Soyuz T	1979	Longer-duration vehicle to carry three spacesuited cosmonauts to space station
Soyuz TM	1986	Further refinement to serve the Mir space station
Soyuz TMA	1997	"Lifeboat" for International Space Station

Soyuz spacecraft ESA astronaut Roberto Vittori (foreground) during training aboard the Soyuz simulator at Star City near Moscow. *European Space Agency*

wanted a spaceflight to commemorate the 50th anniversary of the Communist revolution. Cosmonauts had prepared a document listing 200 technical problems and given it to high-ranking officials. Yuri **Gagarin**, the first person in space and backup pilot for Soyuz 1, also tried unsuccessfully to convince Brezhnev to cancel the launch. Both Komarov and Gagarin, who were good friends, knew that the pilot of Soyuz 1 would be in great danger. A few weeks before launch, Komarov said, "If I don't make this flight, they'll send the backup pilot instead. That's (Yuri), and he'll die instead of me." Following the accident, Gagarin said, "... if I ever find out he (Brezhnev) knew about the situation and still let everything happen, then I know exactly what I'm going to do." Rumor has it that Gagarin did eventu-

ally catch up with Brezhnev—and threw a drink in his face.

A successful crew exchange first took place on a later mission. Three cosmonauts were launched on Soyuz 5 (Boris Volynov, Alexei Yeliseyev, and Yevgeni Khrunov), and one aboard Soyuz 4 (Vladmir Shatalov). For the trip back, Yeliseyev and Khrunov switched to Soyuz 4, thus becoming the first humans to return to Earth in a craft other than the one in which they left.

Soyuz 10 and 11

Three-man missions involving a Soyuz modified by the removal of a large fuel tank at the rear of the instrument module (not needed with the abandonment of the Moon plan) and the addition of a new docking system with a hatch to allow cosmonauts to transfer to a space station without a spacewalk. Soyuz 10 briefly docked with Salyut 1. Soyuz 11 docked with the station normally, but its crew was killed during reentry when a valve opened suddenly and allowed all the air in the descent module to escape. The cosmonauts, who wore no pressure suits, died apparently before they had realized what was happening.

Soyuz 12 to 40

Shocked by the catastrophe of Soyuz 11, Soviet engineers overhauled the Soyuz spacecraft to increase safety. Never again would cosmonauts go to and from orbit without pressure suits. To make room for the emergency life-support system needed for the suits, the standard Soyuz crew was reduced from three to two. There were no more fatalities in the program; however, Vasily Lazarev and Oleg Makarov, aboard Soyuz 18-1, had a lucky escape after half the explosive bolts holding the first and second stages of their Soyuz booster together failed to blow. Forced to abort the launch, they suffered a harsh reentry from 140 km up before their capsule hit a mountain near the Chinese border, rolled down the side, and left the crew to spend a day huddled around a fire waiting for rescue. The most famous of the Soyuz missions in this period was Soyuz 19, which took part in the **Apollo-Soyuz Test Project**.

Soyuz T

First flown in December 1979, the T was equipped with more compact electronics, which enabled the descent module to hold three spacesuited cosmonauts. It also carried solar panels–a feature of early Soyuz spacecraft that had been dispensed with after Soyuz 11 to save weight–enabling the vehicle to function independently of Salyut for up to four days.

Soyuz TM

A modernized version of the T with improved power supplies, new parachutes, and extra space for equipment. It can carry an extra 200 kg into orbit and return up to 150 kg to Earth–100 kg more than Soyuz T. First used with Mir, it remains in service ferrying cosmonauts, astronauts, and freight to and from the ISS.

Soyuz TMA

A version of the Soyuz spacecraft specially adapted to serve as a lifeboat for crew return to Earth in the event of an emergency aboard the ISS. See **ACRV**.

space cannon

(continued from page 386)

than rocket fuel. Cannon projectiles are accurate, thanks to the fixed geometry of the gun barrel, and are much simpler and cheaper than rockets. However, there are also serious drawbacks. The payload must be slender enough to fit into a gun barrel and sturdy enough to withstand the huge accelerations of launch, which can easily exceed 10,000g.

Long before the first test flights of the V-2 (see **"V" weapons**), the **Paris Gun** of World War I set impressive altitude and speed records for artificial objects. In the 1950s, as the rocket became established as the primary means of reaching space, the Canadian engineer Gerald **Bull** began a lifelong struggle to use guns for cheap access to the high atmosphere and Earth orbit. His **HARP** (High Altitude Research Project) in the 1960s showed that a suborbital cannon can be cost-effective for studying the upper atmosphere, between 50 km and 130 km, and has the potential to launch vast numbers of satellites each year in all kinds of weather. A further development of this concept was Lawrence Livermore Lab's **SHARP** (Super High Altitude Research Project).

Even if shot out of an extremely powerful cannon, a projectile would need to include a rocket in order to enter a stable orbit. This is for two reasons. First, reaching orbital velocity (with an extra margin for air resistance) is difficult using a cannon alone. Second, by Kepler's first law, any orbit is an ellipse with one focus at the Earth's center. If the payload is launched from a point A on the Earth's surface, its orbit necessarily would intersect the surface again at a symmetrically placed point B. An orbital adjustment is therefore essential.

Plans also exist for accelerating a payload by magnetic forces on a "rail gun" consisting of parallel conductors, into which a very large electric current is directed. The same problems apply here, plus the added one of storing and then suddenly releasing a great amount of electrical energy. This kind of technology might be appropriate for future use on the Moon but is at an even earlier stage than the space cannon. See **Valier-Oberth Moon gun**.

space colony

A large, self-contained, artificial environment in space, also known as a space habitat, which is the permanent home of an entire community. The first fictional account of a space colony appears in 1869, in Edward Everett **Hale**'s novel *The Brick Moon*.[140] Other early portrayals of the idea are to be found in novels by Jules Verne in 1878 and Kurd Lasswitz in 1897. In the 1920s, J. D. Bernal described spherical colonies that have come to be known as **Bernal spheres**.[25] The companion idea of mobile colonies, or **generation starships**, that could carry large numbers of people relatively slowly to other stars was envisioned in 1918 by Robert **Goddard**.[114] A vastly more ambitious scheme for completely encircling a star with artificial habitats was described by Freeman **Dyson**[83] in the 1960s. Indeed, the development of concepts about space colonies is deeply entwined with evolving notions about the colonization of other worlds, **terraforming**, and **space station**s. When does a space station become a space colony? In his 1952 novel, *Islands in the Sky*,[53] Arthur C. **Clarke** depicts structures that are somewhere between the two, while by 1961, in *A Fall of Moondust*,[54] he has moved on to even larger structures placed at the stable **Lagrangian point**s in Earth's orbit where they would remain fixed relative to both Earth and the Moon. In 1956, Darrell Romick advanced a yet more ambitious proposal for a rotating cylinder 1 km long and 300 m in diameter that would be home to 20,000 people. In 1963, Dandridge Cole suggested hollowing out an ellipsoidal asteroid about 30 km long, rotating it about the major axis to simulate gravity, reflecting sunlight inside with mirrors, and creating on the inner shell a pastoral setting as a permanent habitat for a colony. Related ideas, about the use of extraterrestrial resources to manufacture propellants and structure, go back to Goddard in 1920, and it became a common theme in science fiction, reappearing in technical literature after World War II. In 1950, Clarke noted the possibility of mining the Moon and of launching lunar material into space by an electromagnetic accelerator along a track on its surface.[51] Thus by the mid-1960s, the scene was set for a comprehensive proposal for building and sustaining large habitats at the Earth-Moon Lagrangian points—a proposal that took the form of the **O'Neill-type space colony**.

space drive
See **reactionless drive**.

space elevator
A continuous structure extending from a point on the equator to a point in geostationary orbit. The concept of the space elevator as a cheap and convenient way to access space was first put forward in 1960 by the Russian engineer Yuri **Artsutanov** (1929–).[13] It was later developed independently by John Isaacs[155] of the Scripps Institute of Oceanography, and Jerome Pearson[227] of Ames Research Center, but went largely unnoticed until 1979, when Arthur C. **Clarke** used it as the centerpiece for his novel *The Fountains of Paradise*.

A space elevator would consist of a cable some 47,000 km long, yet no more than a few cm wide. At the Earth end would be a base station, with all the facilities of a major international airport, above which would rise a launch structure tens of km tall. From this would extend the cable to a space station in geostationary orbit and, further out, a counterweight—perhaps a small asteroid—to which the end of the cable would be tethered. Passengers and cargo would travel up and down the cable at a cost estimated to be as low as $1.50 per kilogram. The idea is being actively explored at the Marshall Space Flight Center.

Magnetic levitation, or maglev, may provide the best means of propulsion for the space elevator. The greatest technological challenge is the cable itself, because the weight of the structure dangling from geostationary orbit would place extraordinary demands on the material used to make it. For a cable of practical dimensions, with a minimum diameter of 10 cm, NASA estimates that it would need to be made of something 30 times stronger than steel and 17 times stronger than Kevlar. One possibility is carbon in the form of so-called nanotubes: tiny, hollow cylinders made from sheets of hexagonally arranged carbon atoms. At present, nanotubes are extremely expensive and can be fabricated only in short lengths. However, it seems likely that production costs will fall dramatically in the future and that some way will be found to bind nanotubes into a composite material like fiberglass.

space equivalence
The upper reaches of Earth's atmosphere, where conditions for survival are almost the same as those required for survival in space.

Space Foundation
A public, nonprofit organization for promoting international education and understanding of space. The group, based in Colorado Springs, Colorado, hosts an annual conference for teachers and others interested in education. Other projects include developing lesson plans that use space to teach other basic skills, such as reading. It publishes *Spacewatch,* a monthly magazine of Foundation events and general space news.

Space Frontier Foundation
A California-based organization of space activists, scientists and engineers, media professionals, and entrepreneurs, that works to promote the large-scale, permanent settlement of space. The Foundation believes that free-market enterprise and cheap access to space are the keys to opening up the new frontier. Its board members have included Pete Conrad, physicist Freeman **Dyson**, George Friedman of the Space Studies Institute, John Lewis of the University of Arizona, and Herman Zimmerman, chief designer on *Star Trek*.

space gun
See **space cannon**.

space habitat
See **space colony**.

Space Launch Initiative (SLI)
A NASA plan, also known as the Second Generation Reusable Launch Vehicle (RLV) Program, the goal of which is to develop a successor to the **Space Shuttle** that will lower the cost of delivering payloads to low Earth orbit to less than $1,000 per pound (0.45 kg) and will also reduce the risk of crew loss to approximately 1 in 10,000 missions. The SLI is seen as the key to opening the space frontier for continued civil exploration and commercial development of space. It is the centerpiece of NASA's long-range Integrated Space Transportation Plan, which also includes near-term Space Shuttle safety upgrades and long-term research and development toward third and fourth generation RLVs.

space medicine
See article, pages 391–392.

space motion sickness (SMS)
A condition similar to ordinary travel sickness, produced by confusion of the human vestibular (balance) system

(continued on page 392)

space medicine

A branch of **aerospace medicine** concerned specifically with the physiological and psychological effects of spaceflight. Some of the potential hazards of space travel, such as acceleration and deceleration forces, the dependence on an artificial pressurized breathable atmosphere, and noise and vibration, are similar to those encountered in atmospheric flight and can be addressed in similar ways. However, space medicine must embrace a number of other issues that are unique to living and exploring beyond Earth's atmosphere.

The first information concerning the potential effects of space travel on humans was compiled in Germany in the 1940s under the direction of Hubertus **Strughold**. However, these seminal data are tainted by their purported link with Nazi atrocities. Both the United States and the Soviet Union conducted rocket tests with animals (see **animals in space**) beginning in 1948. In 1957, the Soviet Union put a dog, **Laika**, into Earth orbit, and, shortly after, the United States began sending primates on suborbital flights. These early experiments suggested that few biological threats were posed by short stays in space. This was confirmed when human spaceflight began on April 12, 1961, with the orbital flight of Yuri **Gagarin**.

Space motion sickness became a fairly regular, though not serious, side effect of lengthier missions. Of greater concern were the consequences of weightlessness that first became apparent in the 1970s and 1980s, when Soviet cosmonauts began spending months at a time in gravity-free environments aboard **Salyut** and then **Mir**. These included loss of bone matter (see **bone demineralization in space**) and loss of muscle strength. The atrophy of certain muscles, particularly those of the heart, was seen to be especially dangerous because of its effect on the functioning of the entire cardiovascular system. During extended spaceflight, the heart becomes smaller and pumps less blood with each beat. One way to try to counter this is by regular exercise on treadmills or bicycles. But some cardiovascular change appears inevitable. The blood itself is also affected, with a measurable decrease in the number of oxygen-carrying cells. To what extent these physiological changes are reversible is still not clear. The bones and

muscles of most space travelers have been observed to fully recover within weeks of their return. However, in 1997, serious effects on heart function were reported in some Russian cosmonauts who had served for unusually long periods in orbit.

The absence of gravity is particularly damaging to biological development. An early indication of this came from a seven-day Shuttle mission in 1985, involving 24 rats and 2 monkeys. Post-flight examination revealed not only the expected loss of bone and muscle strength but a decrease in the release of growth hormone as well. More recent findings point to a pervasive effect of gravity—or the lack of it—on cell metabolism, brain development, and DNA synthesis. A study of 18 pregnant mice launched into space carrying some 200 mouse fetuses at varying stages of development suggests that nerve cells, and possibly every cell in our bodies, may need gravity cues to grow and function properly. Profound changes were noted when the space fetuses were compared with carefully matched counterparts on Earth. Cell death, a normal aspect of development, slowed down in space, as did cell proliferation. Tiny structures that have to move about within the cells and that normally travel at high speed slowed to a crawl. Without gravity to guide the migration of nerve cells that form the outer layer of the cortex, the space-grown brains wound up smaller, and although they appeared to be normal otherwise, they turned out to have fewer nerve cells than normal mouse brains. Just what functional importance this would have in an adult animal awaits further study. But it appears as if the brain struggled to adapt as it developed, only to mask what could be deleterious changes. Women are already forbidden to head into orbit if they are pregnant. These new findings suggest that children born, or even conceived, in space might suffer permanent nervous-system damage unless exposed to Earth-like gravity at key points during their early development. At the very least, children who grew up under zero-gravity or low gravity (for example, on Mars) might have trouble walking on Earth, because their nervous systems would be permanently wired for a nonterrestrial environment.

Another concern on longer-duration spaceflights is radiation exposure. Short orbital flights result in exposures about equal to one medical X-ray. The crews on the longer Skylab flights sustained many times this dose. During deep-space exploration missions that may last 18 months or more, astronauts would receive radiation doses in excess of career maximums set by current medical standards unless preventive measures are taken. See **radiation protection in space**.

Long space missions will have not only physical consequences but also psychological effects arising from the close confinement of a few individuals with limited activity. No great problems have been encountered to date, perhaps because most astronauts are chosen for emotional stability and high motivation and because they are assigned enough tasks to keep them almost constantly busy. Even so, there have been some signs of strain—as revealed, for example, in the diary of cosmonaut Valentin **Lebedev**.[148]

space medicine On the middeck of Space Shuttle *Challenger,* Mission Specialist Guion Bluford, restrained by a harness and wearing a blood pressure cuff on his left arm, exercises on the treadmill. *NASA*

space motion sickness (SMS)

(continued from page 390)
and the absence of a gravitational vertical; it is also known as microgravity nausea. About one-third of Apollo crew members reported some level of SMS, rising to 60% on the longer-duration **Skylab** missions. During Space Shuttle flights the incidence of SMS in first-time astronauts has also been about 60%. Treatment with such medications as intramuscular promethazine (Phenergan) has been effective, and the nausea tends to wear off two to three days into the flight.

space passenger

A trip into orbit, it seems, can now be arranged for anyone who is fit, not pregnant—and has $20 million or so to spare. This is the amount that the first private space tourist, Denis **Tito**, paid to the Russian space authorities for his 10-day sojourn to the **International Space Sta-**

tion (ISS) in the spring of 2001. The South African businessman Mark Shuttleworth stumped up a similar amount for his stay at the high frontier a year later, and other millionaires, including 'N Sync's Lance Bass, are lining up for their chance to take a Soyuz ferry to the ISS. Although NASA is far from pleased at having the hugely expensive orbital facility turned into a space hotel, cash-strapped Russia is happy to have the spare berth in the three-seater Soyuz occupied by someone who effectively pays for the entire launch. In fact, Russia started the whole space tourism industry rolling a decade earlier by allowing journalist Tokohiro **Akiyama**'s employer to finance his flight to **Mir**.

space plane

An entirely reusable spacecraft equipped with engines that can operate effectively in both the atmosphere and in space. A variety of vehicles have been flown operationally or tested over the years that have some of the characteristics of space planes, including the **X-15**, lifting bodies such as the **M-2**, **PRIME**, and **X-24**, **Buran**, and, most notably, the **Space Shuttle**. Among the schemes that never left the drawing board are **Sänger**'s space plane, **Dyna-Soar**, **MUSTARD**, the X-30 (National Aerospace Plane), **Hermes**, **HOTOL**, and the **X-33** (VentureStar). Ongoing projects, in the early stage of development, include the **X-37/X-40A**, **HOPE**, **Skylon**, and the **X-43A** scramjet. The fully reusable space plane that can take off and land on an ordinary runway and that can economically transport substantial quantities of cargo and crew into low Earth orbit remains the Holy Grail of spaceflight research and development. It is the primary long-range goal of space agencies around the world, including NASA, with its **Space Launch Initiative**.

space probe

An unmanned spacecraft that undertakes a mission beyond Earth's orbit.

Space Race

What eventually became a race to land the first human beings on the Moon started in the late 1940s with American and Soviet strategists confronting the same challenge—how to strike at the heart of an enemy quickly in the event of war. Both nations began to look at means other than piloted aircraft to deliver bombs to distant targets. At first, they drew on German weapons technology from World War II. Technological improvements gradually transformed the V-1 into the modern long-range cruise missile and the V-2 (see **"V" weapons**) into the intercontinental ballistic missile (ICBM). But whereas the U.S. military gave relatively low priority to developing ICBMs in the early 1950s, the Soviet Union made huge strides in building large, long-range rockets

(see **guided missiles, postwar development**). The successful test of **Korolev**'s R-7 in August 1957 (see **"R" series of Russian missiles**) showed that the Soviets had the capability to place a satellite into orbit. Yet, the launch of **Sputnik** 1 in October 1957 still shook the world and gave the (mistaken) impression that the United States had fallen seriously behind the Soviets in key areas of science and technology. Subsequent U.S. launch failures (see **Viking** and **Vanguard**) did nothing to dispel that perception.

America's first success in space came in January 1958, when **Explorer** 1 was launched aboard an Army **Jupiter** C missile. But it was the Soviet Union that continued to break new ground and set records. From 1958 through 1961, six more Earth-orbiting Sputniks were successfully launched, all much larger than the first. The Soviets were the first to fly a probe past the Moon, then to hit it, and, in April 1961, to put a man in space. Yuri **Gargarin**'s flight took place a month before Alan **Shepard**'s suborbital flight and 10 months before John **Glenn** became the first American in orbit.

Immediately after Gagarin's flight, President **Kennedy** wanted to know what the United States could do in space to take the lead from the Soviets. Vice President Lyndon Johnson polled leaders in NASA, industry, and the military, and reported that "with a strong effort" the United States "could conceivably" beat the Soviets in sending a man around the Moon or landing a man on the Moon. As neither nation yet had a rocket powerful enough for such a mission, the race to the Moon was a contest in which America would not start at a disadvantage. On May 25, 1961, when Kennedy announced the goal of landing a man on the Moon before the end of the decade, the total time spent in space by an American was barely 15 minutes. Yet the Soviet Union's lead in rocketry was quickly overhauled, due in large part to the vision and genius of Wernher **von Braun**. To get to the Moon required the development of a super-rocket, and in this the United States succeeded, with the **Saturn** V, whereas the Soviets' giant **N-1** never rose above the atmosphere.

On July 20, 1969, as millions around the world watched on television, two Americans stepped onto another world for the first time. The United States landed men on the Moon and returned them safely, fulfilling Kennedy's vision and meeting the goal that inspired manned spaceflight during the 1960s. The Space Race had been won—but what next? In the United States, many hoped that **Apollo** would mark the dawn of an era in which humans moved out into space, to bases on the Moon and perhaps Mars. But support for manned missions to the Moon and beyond declined, and the focus for human activity in space shifted to near-Earth orbit.[36, 195, 201, 261] See **Space Shuttle** and **space station**.

space sail

A device that uses the pressure of sunlight, laser light, or some other form of radiated energy to propel a space vehicle in the same way that a sailing ship uses wind. It would consist of a large sheet of reflective material, such as Mylar film, and a framework of girders to keep it extended and to transmit its pressure to the spacecraft. Making the sheet reflective like a mirror is important. Sending most of the light back nearly doubles the pressure when the sail squarely faces the Sun. More important, the sail can be tilted at an angle to the Sun's rays and reflect them in some chosen direction, for example, in the direction opposite to the one in which the spacecraft is moving, which gradually increases the velocity of the spacecraft. When the sail faces the Sun squarely, the pressure of sunlight counteracts the Sun's pull and only enables the satellite to circle the Sun in a slightly larger orbit. On the other hand, placing it at an angle can put the spacecraft on an outward or inward spiral, which would be useful for interplanetary missions.

A solar sail can function only where sunlight is sufficiently intense. At Earth's distance from the Sun, the pressure of sunlight on a square kilometer of sail is so feeble that if sail and payload had a mass of 5,000 kg, the spacecraft would accelerate at a maximum of $0.0001g$ (1/10,000th the acceleration to gravity on Earth). Although this seems tiny, it would lead over a six-month period to a velocity of about 12 km/s.

Twenty years ago, NASA and ESA (European Space Agency) studied solar sail projects designed to rendezvous with Halley's Comet but decided against them because the technology available was insufficiently mature. Today, such an endeavor seems much more feasible, and several groups around the world have considered practical missions to solar sail from Earth to the Moon, or even to Mars. Efforts to build solar sails were made by amateur groups, including World Space Foundation, U3P Union for the Promotion of Photonic Propulsion, and Solar Sail Union of Japan, and semi-professional groups, such as the VSE European Solar Sail and the Russian Space Regatta Consortium, which succeeded in testing the deployment of a 20-m solar sail in orbit. The idea, first envisioned by Arthur C. **Clarke** in the 1960s, of racing solar sails between Earth and the Moon, was developed by U3P since 1981 and resulted in an official set of rules with the International Astronautical Federation. An experimental solar sail, **Cosmos 1**, was scheduled for launch in late 2002.

For more distant voyages, including interstellar flights, there is the possibility of laser light sails. Robert **Forward** and others have suggested propelling a star probe with a huge sail using tight beams of light from lasers that are powered by sunlight and in orbit around the Sun. As the craft approaches its target, part of the sail is cut free. The laser beams then push this section of sail ahead of the rest of the craft and are reflected back onto the parent portion of the sail from the opposite direction, slowing down the probe at its destination. Given sufficiently powerful

space sail A model of a solar sail on display at the K2001 trade fair. *European Space Agency*

lasers and a large enough sail, even a manned interstellar mission could be mounted on this basis.[96]

Recently, another variation on the space sail theme has been proposed: the magnetic sail, or magsail. One form of this would be a loop of superconducting wire reeled out from a spacecraft in which a current was made to flow. Once started, the current would continue cycling around the loop indefinitely, because a superconductor offers no resistance. The magnetic field generated by the current loop would interact with charged particles in the solar wind to impart momentum to the magsail and thus accelerate the spacecraft in the direction of the wind. At roughly the distance of Earth from the Sun, this would be sufficient over time for the magsail to reach speeds of several hundred km/s.[9, 320, 321]

Space Shuttle
See article, pages 396–401.

space-sickness
See **space motion sickness**.

space simulator
A device that simulates conditions in space and is used for testing equipment and training programs.

space station
A large orbiting structure in which humans can live and work for extended periods. The concept goes back at least as far as a tale called "The Brick Moon," written by Edward Everett **Hale** just after the American Civil War. Konstantin **Tsiolkovsky** tackled the idea more technically in a 1895 science fiction story and in 1903 expanded his description to include rotation for **artificial gravity**, the use of solar energy, and even a space greenhouse with a closed ecosystem. In 1923, Hermann **Oberth** coined the term "space station" to describe an orbiting outpost that would serve as the starting point for flights to the Moon and Mars. Five years later, Guido **von Pirquet** considered a system of three stations—one in a near orbit, one more distant, and a transit station in an intermediate elliptical orbit to link the other two—that he suggested might serve as refueling depots for deep space flights. The notion of a rotating wheel-shaped station was introduced in 1929 by Herman **Noordung** in his *Das Problem der Befahrung des Weltraums* (The Problem of Space Flight). He called his 30-m-diameter station "Wohnrad" (Living Wheel) and suggested that it be placed in geostationary orbit.

In the 1950s, Wernher **von Braun** worked with *Collier's* magazine (see *Collier's* **Space Program**) and Walt **Disney** Studios to produce articles and documentaries on spaceflight. In them, he described an updated version of Noor-

dung's wheel, enlarged to a diameter of 76 m and reached by reusable winged spacecraft. Von Braun saw the station as an Earth-observation post, a laboratory, and a springboard for lunar and Mars excursions. Later in the same decade, the dream slowly began to turn into reality. In 1959, a NASA committee recommended that a space station be established before a trip to the Moon, and the House of Representatives Space Committee declared a space station a logical follow-on to the **Mercury Project.** As it transpired, the Apollo lunar program preempted the goal of building an American space station in the early 1960s, although in 1969, the year Apollo 11 landed on the moon, NASA proposed a 100-person permanent space station to be completed by 1975. Known as Space Base, it was envisioned as a laboratory for scientific and industrial experiments and as a home port for nuclear-powered tugs designed to carry people and supplies to an outpost on the Moon. NASA realized that the cost of resupplying a space station using expendable rockets would quickly exceed the station's construction cost. A reusable spacecraft—the Space Shuttle—was the obvious solution. The Shuttle would ferry up the people and the supplies needed for a long stay in space, and ferry back down people and the industrial products and experiment samples they made on the station. But economic and political priorities shifted during the Apollo era, and, despite Apollo's success, NASA's annual budgets suffered dramatic cuts beginning in the mid-1960s. Because of this, plans for a permanent space station were deferred until after the Shuttle was flying.

Meanwhile, the Soviets, having lost the race to the Moon, focused instead on setting up a permanent human presence in orbit. Throughout the 1970s they launched a series of small, pioneering **Salyut** space stations and then, in 1986, began assembly of the multi-module **Mir**. NASA deferred post-Apollo station efforts to the 1980s, with the notable exception of **Skylab**. When the last Skylab crew headed home in February 1974, NASA proposed sending a Shuttle to boost the station—a converted Saturn V third stage—to a higher orbit, and even refurbishing it for further use. However, delays to the Shuttle program combined with greater-than-expected solar activity (which expanded Earth's atmosphere) hastened Skylab's fall from orbit.

In response to budgetary pressures and the magnitude of the task of building a large permanent space station, NASA began to explore the possibility of international cooperation. American and Soviet negotiators discussed joint Shuttle-Salyut missions as an outgrowth of the first manned American-Russian space effort, the **Apollo-Soyuz Test Project** in 1975. NASA offered the Shuttle for carrying crews and cargo to and from Salyut and in

(continued on page 401)

Space Shuttle

The core vehicle in NASA's space transportation system program and the only American craft used to carry humans into space since the **Apollo-Soyuz Test Project**. Each Space Shuttle consists of an Orbiter (of which four are in service), an External Tank (ET), two Solid Rocket Boosters (SRBs), and the Space Shuttle Main Engines (SSMEs). All these components are reusable except for the External Tank. Designed to operate on land, in the atmosphere, and in space, the Shuttle combines features of a rocket, an aircraft, and a glider. No other flying machine is launched, serves as a crew habitat and cargo carrier, maneuvers in orbit, then returns from space for an unpowered landing on a runway, and is ready to fly again within a few weeks or months. Its main engines and solid rocket motors are the first ever designed for use on multiple missions.

The Space Shuttle can take up to eight astronauts into low Earth orbit to carry out a wide variety of tasks, from satellite launching to construction of the **International Space Station**, on missions lasting up to two and a half weeks. It is used to support research in astronomy, biology, space medicine, and materials processing, and has delivered into orbit scientific, commercial, and military satellites and interplanetary probes. The Shuttle has also been used to carry **Spacelab** and to repair, refurbish, or recover satellites. Since the first launch on April 12, 1981–20 years to the day after Yuri Gagarin's historic flight–Shuttles have flown two to nine missions a year, except in 1986 and 1987, when flights were suspended following the *Challenger* **disaster**. Although its cost has proven much greater and its practical launch frequency much lower than originally anticipated, it represents a giant leap forward in manned spaceflight capability.

The Shuttle travels from the **Vehicle Assembly Building** at Kennedy Space Center (KSC), where its main components are put together, to its launch pad on a giant **crawler-transporter** vehicle–a trip that, at a maximum speed of 1.6 km/hr, takes about five hours. Launch complexes 39A and 39B were originally used for the Apollo missions and renovated for the Shuttle. A third intended Shuttle launch site built at Vandenberg Air Force Base, California, by renovating Titan III launch complex SLC-6 (nicknamed "Slick-6"), has never been employed.

History

Born in 1968 at the height of the Apollo program, the Shuttle was designed to provide NASA with an efficient, reusable method of carrying astronauts to and from a large, permanently manned space station (with a crew of 12 to 24), and a multipurpose satellite launch system with the potential to replace **Atlas-Centaur**, **Delta**, and **Titan** rockets. Originally slated to enter service by 1977, the Shuttle made its first spaceflight in 1981 following numerous design changes.

At the outset, NASA envisaged a two-stage Shuttle with a smaller manned winged vehicle (the Orbiter) riding piggyback on a larger manned winged vehicle (the Booster). These would be pad-launched from a vertical position. The Booster would carry the Orbiter to a height of about 80 km, at which point the Orbiter would separate and fire its own engines to reach orbit. The Booster–essentially a winged fuel tank–would immediately descend and land near the launch site, while the Orbiter would return at the end of its mission. As described by NASA in 1970, this two-stage Shuttle would be able to carry a 11,300-kg payload to a maximum 480-km circular orbit.

In a time of recession, it soon became clear that NASA could not afford a fleet of complicated two-stage Shuttles together with a space station. Furthermore, although military funding for the Shuttle was vital, the Air Force specified a payload capability nearly three times what NASA had in mind. These factors led to a dramatic redesign of both the Shuttle and the proposed space station. From a Skylab-like, single-structure station to be launched by a Saturn V and serviced by the Shuttle, NASA switched to a modular concept: the space station would be built over several years from separate, Shuttle-launched elements. This not only spread the financial outlay over a longer period, it meant that the more powerful Shuttle required would be able to carry heavy military payloads. The redesign would also allow NASA to secure private funding to carry commercial satellites aboard Shuttles while cutting costs by phasing out the Atlas, Delta, and Titan fleets. With an event like the *Challenger* disaster never anticipated, NASA put all of its eggs in one basket: the Shuttle would become the Agency's sole medium-to-heavy launcher into the

Space Shuttle *Atlantis* heads skyward from Launch Pad 39A on March 24, 1992, carrying a crew of seven and the Atmospheric Laboratory for Applications and Science-1. *NASA*

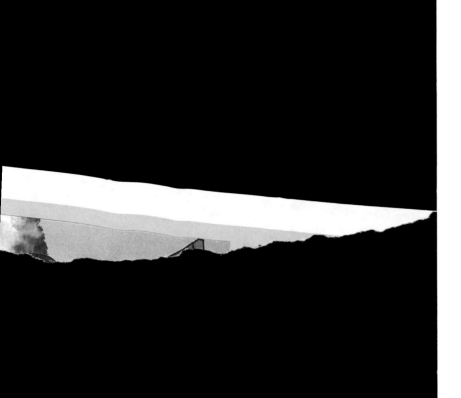

next century. To cut costs further, NASA opted for a more traditional, partly expendable system to carry the Orbiter into space, involving drop-away boosters and an expendable main engine fuel tank.

Rockwell began work on Orbiter *Enterprise*, officially known as Orbiter Vehicle-101 (OV-101), in June 1974. This was first rolled out of its hangar at Palmdale, California, on September 17, 1976, and subse-

quently used for flight testing at the Dryden Flight Research Center. Meanwhile Rockwell continued to develop Orbiters *Columbia* (OV-102), *Discovery* (OV-103), and *Atlantis* (OV-104). Although *Enterprise* was originally intended to be refitted as an operational Shuttle, NASA opted instead to upgrade *Challenger* (OV-099) from its status as a high-fidelity structural test article. Upon the loss of *Challenger* in January

1986, it was decided to build a fifth operational Shuttle, *Endeavor* (OV-105), as a replacement.

The initial concept of flying 25 to 60 missions per year never proved realistic, and by the mid-1980s, NASA was gearing toward a more modest launch rate of about 24 missions per year. Then the entire program ground to a halt when, on January 28, 1986, the 25th Shuttle mission ended tragically little more than a minute after takeoff. The explosion of *Challenger* made it stunningly clear that major changes in the entire Shuttle program were unavoidable. With the launch of *Discovery* on September 29, 1988, NASA entered a new era of Shuttle operations, adopting a more relaxed pace averaging about eight launches per year. Learning from its greatest failure, NASA was able to rebuild and maintain a Shuttle program that has since proven remarkably safe and reliable.

In April 1996, NASA began a four-phase plan to keep the existing Shuttle fleet flying through at least 2012 and also proposed the Space Shuttle upgrade program, which involves modifications and upgrades that might keep the fleet in action through 2030. Some of the modifications have already been implemented. Phase One of the plan called for improvements to the Shuttle to allow it to support the construction and maintenance of the International Space Station—the program's chief goal well into the first decade of the twenty-first century. Phase Two calls for operational and cost improvements in ground operations that will decrease Shuttle servicing and maintenance time in order to support an average of 15 launches per year. A new checkout and launch control system is being developed at KSC to facilitate this goal. Phase Three involves a number of modifications to orbiter onboard systems that will also lead to a drop in processing and maintenance time. More ambitious elements of this plan envisage the complete replacement of toxic fuels by nontoxic fuels in key orbiter systems. Phase Four calls for significant redesign of the Shuttle fleet and its basic configuration. An interesting proposal in this phase is the introduction of a liquid fly-back booster, which would return to the launch site and save servicing time.[128, 157, 158]

External Tank (ET)

The propellant tank for the Space Shuttle Main Engines (SSMEs). It is the largest and only non-reusable element of the Shuttle and provides a structural backbone for the entire system. Each ET contains a liquid oxygen (LOX) tank at the top and a liquid hydrogen (LH2) tank at the bottom, which are connected by an intertank, 6.9 m long by 8.4 m wide. The ET supplies LOX and LH2 to the Orbiter's three SSMEs through 43-cm-diameter feed lines: LOX at a rate of 72,300 kg (63,600 liters) per minute, LH2 at 12,100 kg (171,400 liters) per minute. At launch, the ET is attached to both the Orbiter and SRBs. It empties about 8.5 minutes after launch, at which time it is jettisoned, breaks up, and falls into a predetermined area of either the Pacific or the Indian Ocean.

The first six ETs delivered to NASA were called standard weight tanks (SWTs), each weighing 34,250 kg. SWTs were flown on missions STS-1 through STS-5 and STS-7. For STS-1 and STS-2, the ETs were painted white. Thereafter, hundreds of kilograms and thousands of dollars in preparation were saved by leaving ETs unpainted, so that all ETs flown from STS-3 on have sported an orange-brown color. In 1979, even before a Shuttle had completed a spaceflight, NASA directed that ETs be lightened so that heavier payloads could be flown. The resulting ET, called the Lightweight Tank (LWT), shaved 4,500 kg off the SWT by using new materials and design changes. The 29,700-kg LWT was first flown on STS-6, and then from STS-8 through STS-90. Starting with STS-91, a new ET called the super lightweight tank (SLWT) has been flown. Weighing 26,300 kg, it enables still heavier payloads to be carried in support of the International Space Station.

Length: 46.9 m
Diameter: 8.4 m
Mass (SLWT version): 757,000 kg (total); 730,000 (fuel)

LOX Tank
  Size: 16.6 × 8.4 m
  Capacity: 549,000 liters
  LOX mass: 626,000 kg

LH2 Tank
  Size: 29.5 × 8.4 m
  Capacity: 1.47 million liters
  LH2 mass: 104,000 kg

Orbiter

A wide-body, delta-winged airplane and space vehicle, built mainly of aluminum and covered with reusable surface insulation (see **reentry thermal protection**). It

from the rest of the Shuttle. About 225 seconds after separation, at an altitude of 4,800 m, the nose cap of each SRB is ejected and a pilot parachute is deployed. This serves to pull out the 16.5-m-diameter drogue parachute, which orients and stabilizes the descent of each SRB to a tail-first attitude ready for deployment of the main parachutes. At an altitude of 1,800 m, the three 41-m-diameter main parachutes open to slow each SRB to water impact at about 25 m/s. The SRBs splash down in the Atlantic about 225 km from the launch site. Retrieval ships locate each SRB by homing in on radio beacon signals transmitted from each booster and by flashing lights activated on each SRB. Once on location, recovery crews plug the SRB nozzles, empty the motors of water, and tow the rockets back to a receiving and processing site on Cape Canaveral. After inspection, the SRBs are disassembled, washed with fresh, deionized water to limit saltwater corrosion, and refurbished ready to fly again.

Length: 45.5 m
Diameter: 3.7 m
Mass (per SRB): 585,000 kg (with fuel); 85,000 kg (empty)
Thrust (per SRB): 14.7 million N

Space Shuttle Main Engines (SSMEs)

The most advanced liquid-fueled rocket engines ever built. Each SSME generates enough thrust to maintain the flight of 2.5 Boeing 747 jumbo jets and is designed for 7.5 hours of operation over an average lifespan of 55 starts. Each Orbiter has three SSMEs mounted on its aft fuselage in a triangular pattern. The Main Engines burn a combination of liquid oxygen and liquid hydrogen fed from the ET and employ a staged combustion cycle, in which the fuels are first partially burned at high pressure and low temperature, then burned completely at high pressure and high temperature. This allows the SSMEs to produce thrust more efficiently than other rocket engines.

SSME thrust can be varied from 65% to 109% of rated power at increments of 1%. Rated power is 100% thrust, or 1,670,000 N per each SSME at sea level. A thrust value of 104%, known as full power, is typically used as the Shuttle ascends, although, in an emergency, each SSME may be throttled up to 109% power. All three SSMEs receive identical and simultaneous throttle commands, which usually come from general purpose computers aboard the Orbiter. The firing of the three SSMEs begins at launch minus 6.6 seconds, at which time general-purpose computers aboard the Orbiter command a staggered start of each SSME. The first to fire is Main Engine number three (right), followed by number two (left) and number one (center) at intervals of 120 milliseconds. If all three SSMEs do not reach at least 90% thrust over the next 3 seconds, a main engine cutoff (MECO) command is issued automatically, followed by cutoff of all three SSMEs. If all SSMEs are performing normally, the Shuttle can be launched. The SSMEs achieve full power at launch, but they are throttled back at about launch plus 26 seconds to protect the Shuttle from aerodynamic stress and excessive heating. They are throttled up to full power again at about launch plus 60 seconds, and typically they continue to produce full power for about 8.5 minutes until shortly before the Shuttle enters orbit. During ascent, each SSME may be gimbaled plus or minus 10.5° in pitch and yaw to help steer the Shuttle. At about launch plus 7 minutes 40 seconds the SSMEs are throttled back to avoid subjecting the Shuttle and crew to gravitational forces over 3g. At about 10 seconds before main engine cutoff, a MECO sequence begins. About 3 seconds later, the SSMEs are commanded to begin throttling back at intervals of 10% thrust per second until they reach a thrust of 65% of rated power, called minimum power. This power is maintained for just under 7 seconds, then the SSMEs shut down.

Length: 4.3 m
Diameter: 2.3 m

space station

(continued from page 395)
return hoped to conduct long-term research onboard the Russian station until it could build its own. But these efforts ended with the collapse of East-West detente in 1979. Better progress was made on other fronts, and in

1973 Europe agreed to supply NASA with **Spacelab** modules—mini-labs that would ride in the Shuttle's payload bay. When the Shuttle flew for the first time in 1981, a permanently manned space station was once again touted as the next logical step for the United States in space. Budgetary, programmatic, and political

space station A model of an inflatable space station concept with a solar power system collector at Langley Research Center in 1961. *NASA*

pressures would all hinder this ambition, but by 1997 the orbital assembly of the **International Space Station** had begun.[40]

Space Studies Board

A committee, previously known as the Space Science Board, that was formed by the National Academy of Sciences in 1958, shortly after the launch of the first Earth satellites, to help establish U.S. goals in space science. It continues to provide external and independent scientific and programmatic guidance to NASA and other government agencies in the basic subdisciplines of space research.

Space Studies Institute (SSI)

A nonprofit research group founded by Gerard K. **O'Neill**, among the interests of which have been **mass-drivers**, lunar mining processes and simulants, composites from lunar materials, and solar power satellites. Its presidency was taken over by Freeman **Dyson** after O'Neill's death in 1992. SSI publishes a bimonthly newsletter, *SSI Update*, describing work-in-progress, and runs the biennial Princeton Conference on Space Manufacturing, in Princeton, New Jersey.

space survival

The question is often asked: For how long could an unprotected human survive in the vacuum of space? Fol-

lowing a sudden decompression to a vacuum, you could expect to remain conscious for, at most, 10 to 15 seconds. Quickly thereafter paralysis would set in, accompanied by the rapid formation of water vapor in the soft tissues, which would bloat your body to roughly twice its normal size (but not cause it to explode). Holding your breath in anticipation of decompression would be a bad move, since it would result in your lungs bursting almost instantly. Although your blood would not boil, your circulation would stop within a minute as the pressure in the venous system rose (again, due to the expansion of water vapor) until it surpassed that of the arterial system. Other health hazards of taking a spacewalk without a spacesuit, such as temperature extremes and exposure to intense ultraviolet, would not inconvenience you, as you would already be dead. Experiments to determine the biological effect of space exposure have, unfortunately, been carried out on animals, including chimpanzees. Data concerning the specific effect on humans have come from a number of accidents. One of these occurred at the Manned Spacecraft Center (now the Johnson Space Center) when a test subject was exposed to a near-vacuum after his spacesuit sprang a leak in a vacuum chamber. He remained conscious for about 14 seconds, and later, following repressurization after being in the near-vacuum for about 30 seconds, he reported that his last memory was of the water on his tongue starting to boil. Many deaths have followed rapid decompression, mostly involving crew and passengers of high-altitude planes, but also including those of the three cosmonauts aboard **Soyuz** 11 in June 1971. However, the only known decompression incident in American spaceflight happened during Space Shuttle mission STS-37 in April 1991. In the course of an EVA, the palm restraint in one of the astronaut's gloves came loose and punched an eighth-inch hole in the pressure bladder between thumb and forefinger. In the excitement of his spacewalk, the astronaut didn't realize what had happened and only later discovered a painful red mark on his hand. His skin had partly sealed the opening and he had then bled into space until his clotting blood had filled the rest of the gap. See **radiation protection in space** and **space medicine**.

Space Technology 3
See **Starlight**.

Space Technology 5
See **Nanosat Constellation Trailblazer**.

space tether
A chord, cable, or wire connection between a spacecraft and another object in orbit. The earliest tethers were those used as lifelines for astronauts carrying out spacewalks during the pioneering Soviet and American manned orbital missions. Much longer space tethers, however, provide a means of deploying probes to study Earth's outer atmosphere or generating electricity to power a spacecraft or space station. A number of such tethers have already been flown on missions such as **SEDS** (Small Expendable-tether Deployer System), **TSS** (Tether Satellite System), and **TiPS** (Tether Physics and Survivability experiment). The ability of a tether system to produce electric power was demonstrated by **PMG** (Plasma Motor/Generator). NASA's Marshall Space Flight Center now plans a more sophisticated version of PMG to show that an electrodynamic tether can serve as a propellant-free space propulsion system—a breakthrough that could lead to a revolution in space transportation.

Space Transportation System (STS)
The collective name for the **Space Shuttle** fleet, boosters and upper stages, launch and landing facilities, and training and control facilities.

Spacecause
A pro-space political lobbying organization, based in Washington, D.C., affiliated with the **National Space Society**. Active in supporting pro-space legislation, it arranges meetings with political leaders, interacts with legislative staff, and publishes a bimonthly newsletter, *Spacecause News*.

spacecraft system
The spacecraft and all equipment on the ground or in space that is associated with flight preparation and that is required during flight operation.

Spaceguard Foundation
An association aimed at the protection of Earth's environment against bombardment by comets and asteroids. The Foundation works to promote and coordinate

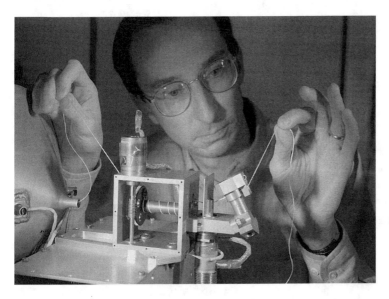

space tether   A scientist at the Marshall Space Flight Center inspects the non-conducting part of a tether as it exits a deployer similar to the system to be used in NASA's Propulsive Small Expendable Deployer (ProSEDS) experiment. ProSEDS technology draws power from the space environment around Earth and transfers it to a spacecraft. *NASA* activities for the discovery and characterization of near-Earth objects.

Spacehab

A small laboratory module, similar to the European **Spacelab** in concept, that is designed to be carried in the payload bay of the **Space Shuttle** to provide extra working space in which astronauts can carry out experiments. Its first, 16-day mission, aboard STS-107, was scheduled for the second half of 2002.

Spacelab

A modular, reusable scientific laboratory, built by ESA (European Space Agency) and carried into orbit on numerous missions between 1983 and 1998 by the Space Shuttle. Scientists from about a dozen nations used it to conduct research on a wide variety of subjects under conditions of microgravity. Spacelab consisted of an enclosed pressurized module, where the crew worked, and smaller unpressurized pallets. The latter carried telescopes and other instruments that required direct exposure to space or did not need a human operator. Some Spacelab missions were dedicated to a particular field, such as life sciences, materials science, astronomy, or Earth observation, while others were multidisciplinary. (See table, "Spacelab Missions.")

Spacepac

A political action committee, based in Washington, D.C., that is affiliated with the **National Space Society**. Spacepac carries out research on issues, policies, and candidates aimed at favorably influencing policy with

Spacelab Missions

Mission		Date
Space Shuttle	**Spacelab**	**Date**
STS-9	Spacelab-1	Nov. 28–Dec. 8, 1983
STS-51F	Spacelab-2	Apr. 29–May 6, 1985
STS-51B	Spacelab-3	Jul. 29–Aug. 6, 1985
STS-61A	Spacelab-D1	Oct. 30–Nov. 6, 1985
STS-35	Astro-1	Dec. 2–10, 1990
STS-40	Spacelab Life Sciences (SLS)-1	Jun. 5–14, 1991
STS-42	International Microgravity Laboratory (IML)-1	Jan. 22–30, 1992
STS-45	Atmospheric Lab. for Applications and Science (ATLAS)-1	Mar. 24–Apr. 2, 1992
STS-50	United States Microgravity Laboratory (USML)-1	Jun. 25–Jul. 29, 1992
STS-47	Spacelab-J	Sep. 12–20, 1992
STS-52	United States Microgravity Payload (USMP)-1	Oct. 22–Nov. 1, 1992
STS-56	ATLAS-2	Apr. 8–17, 1993
STS-55	Spacelab D-2	Apr. 26–May 6, 1993
STS-58	SLS-2	Oct. 18–Nov. 1, 1993
STS-62	USMP-2	Mar. 4–18, 1994
STS-65	IML-2	Jul. 8–23, 1994
STS-66	ATLAS-3	Nov. 3–14, 1994
STS-67	Astro-2	Mar. 2–18, 1995
STS-71	Spacelab-Mir	Jun. 27–Jul. 7, 1995
STS-73	USML-2	Oct. 20–Nov. 5, 1995
STS-75	TSS-1R/USMP-3	Feb. 22–Mar. 9, 1996
STS-78	Life and Microgravity Spacelab	Jun. 20–Jul. 7, 1996
STS-83	Microgravity Science Laboratory	Apr. 4–8, 1997
STS-87	USMP-4	Nov. 19–Dec. 5, 1997
STS-90	Neurolab	Apr. 17–May 30, 1998

Spacelab James Pawelczyk, payload specialist, runs an inventory of supplies in a stowage drawer on Neurolab during a simulation at the Johnson Space Center in May 1997. *NASA*

respect to space, and each year updates *The Space Activist's Handbook*. While Spacepac does not have a membership, it does have regional contacts to coordinate local activity. Operating primarily in the election process, it contributes funds and volunteers to pro-space candidates.

spacesuit
See article, pages 406–408.

space-time
The union of space and time into a four-dimensional whole.

spacesuit

An airtight fabric suit with flexible joints that enables a person to live and work in the harsh, airless environment of space (see **space survival**). A spacesuit maintains a pressure around the body to keep body fluids from boiling away, a comfortable temperature, and a supply of oxygen. The modern spacesuit is a development of the pressure suits worn by early high-altitude pilots.

Pre-Apollo Wardrobe

Pressure suits were suggested by the British physiologist J. B. S. Haldane as long ago as 1920, but they were first built in 1933 by the B. F. Goodrich company for the pioneer American aviator Wiley Post. By wearing a pressure suit, Post was able to fly his celebrated supercharged Lockheed Vega monoplane, *Winnie Mae*, in December 1934 to an altitude of 14,600 m. By the end of the decade, other nations had flown generally similar suits, and, in 1938, the Italian pilot Mario Pezzi reached an altitude of 17,080 m—a record that still stands for a piston-engine airplane.

The spacesuit worn by the **Mercury** astronauts was a modified version of the U.S. Navy high-altitude jet pressure suit. It had an inner layer of Neoprene-coated nylon fabric and an outer layer of aluminized nylon that gave it a distinctive silvery appearance. Simple fabric break lines sewn in to allow bending at the elbow and knee when the suit was pressurized tended not to work very well: as an arm or leg was bent, the suit joints folded in on themselves, reducing the suit's internal volume and increasing its pressure. Fortunately, the Mercury suits were worn "soft" or unpressurized and served only in case the spacecraft cabin lost pressure. Individually tailored to each astronaut, they needed, in Walter **Schirra**'s words, "more alterations than a bridal gown."

For the **Gemini** missions, which would involve astronauts intentionally depressurizing their cabins and going on spacewalks, mobility was a crucial issue. To address this, designers came up with a suit that consisted of a gas-tight, man-shaped pressure bladder, made of Neoprene-coated nylon, covered by a layer of fishnetlike fabric called Link-net, which was woven from Dacron and Teflon cords. This net layer served as a structural shell to prevent the bladder from bal-

looning when pressurized. Next came a layer of felt, seven layers of insulation to protect against temperature extremes, and an outer nylon cover. The suit was pressurized at one-quarter atmospheric pressure, and oxygen was piped in from the spacecraft's life-support system through an umbilical cord.

Suited for the Moon

The **Apollo** missions posed spacesuit designers with a new set of problems. Not only did the Moon explorers' outfits need to protect against sharp rocks and the heat of the lunar day, but they also had to be flexible enough to let astronauts stoop and bend to collect lunar samples, set up scientific equipment, and drive the lunar rover. Apollo spacesuit mobility was improved over earlier designs by using bellowslike molded rubber joints at the shoulders, elbows, hips, and knees. Further changes to the suit waist for Apollo 15 to 17 added flexibility, making it easier for crewmen to sit on the lunar rover. A Portable Life Support System (PLSS) backpack, connected to the suit by umbilicals at the waist, provided oxygen, suit pressurization, temperature and humidity control, and power for communications gear for moonwalks lasting up to 7 hours. A separate 30-minute emergency supply was carried in a small pack above the main PLSS.

The Gemini missions had taught that strenuous activity in space could soon cause an astronaut to overheat. So, from the skin out, the Apollo A7LB spacesuit began with a liquid-cooling garment, similar to a pair of long johns with a network of tubing sewn onto the fabric. Cool water, circulating through the tubing, transferred metabolic heat from the astronaut's body to the backpack and thence to space. Next came a comfort and donning improvement layer of lightweight nylon, followed by a pressure bladder, a nylon restraint layer to prevent ballooning, a lightweight thermal super-insulation of alternating layers of thin Kapton and glass-fiber cloth, several layers of Mylar and spacer material, and, finally, protective outer layers of Teflon-coated glass-fiber Beta cloth.

The fishbowl-like helmet was formed from high-strength polycarbonate and attached to the spacesuit by a pressure-sealing neck ring. Unlike Mercury and

spacesuit Daniel Barry, STS-96 mission specialist, wears a training version of the extravehicular mobility unit spacesuit during an underwater simulation of a spacewalk planned for an early Internal Space Station mission. *NASA*

Gemini helmets, which were closely fitted and moved with the crewman's head, the Apollo helmet was fixed and the head free to move within. While walking on the Moon, Apollo crewmen wore an outer, gold-coated visor to shield against ultraviolet radiation and help keep the head and face cool. Completing the Apollo astronaut's ensemble were lunar gloves and boots, both designed for the rigors of exploring, and the gloves for adjusting sensitive instruments. Modified Apollo suits were also used on the **Skylab** missions and the **Apollo-Soyuz Test Project.**

Shuttle and ISS Garb

During ascent to and descent from orbit, Shuttle astronauts wear special orange partial pressure suits with helmet, gloves, and boots in case of a loss of cabin pressure. Once in orbit, crew members inside the Shuttle enjoy shirtsleeve comfort. To work in the Shuttle's open cargo bay or perform other tasks outside the spacecraft, they don spacesuits known as extravehicular mobility units (EMUs), more durable and flexible than any previous suits. The EMU is modular, enabling it to be built up from a number of

parts depending on the particular task in hand. Also, the upper torso, lower torso, arms, and gloves are not individually tailored but made in a variety of sizes that can be put together in combinations to fit crew members of any size, man or woman. Each suit has supplies for a 6.5-hour spacewalk plus a 30-minute reserve and is pressurized to just under one third of atmospheric pressure. Before donning the suit, astronauts spend several hours breathing pure oxygen because the suit also uses 100% oxygen, whereas the habitable decks on the Shuttle use an Earth-normal 21% oxygen/79% nitrogen mixture at atmospheric pressure (reduced to 0.69 atmosphere before an EVA). This preparation is necessary to remove the nitrogen dissolved in body fluids to prevent its release as gas bubbles when pressure is reduced, a condition commonly called the bends.

The following parts of the EMU go on first: a urine-collection device, a liquid-cooled undergarment plumbed with 100 m of plastic tubing through which water circulates, an in-suit drink bag containing 600 grams of potable water, a communications system (known as the Snoopy Cap) with headphones and microphones, and a biomedical instrumentation package. Next, the astronaut pulls on the flexible lower torso assembly before rising into the stiff upper section that hangs on the wall of the airlock. The upper torso is a hard fiberglass structure that contains the primary life support system and the display control module. Connections between the two parts must be aligned to enable circulation of water and gas into the liquid cooling ventilation garment and return. Then the gloves are added and finally the extravehicular visor and helmet assembly.[171]

Russian Spacesuits

The pressure suit worn by **Vostok** cosmonauts was hidden under an orange coverall, and the Voskhod 1 crew flew without spacesuits at all. For his Voskhod 2 spacewalk in 1965, Alexi **Leonov** wore a special suit that drew supplies from a backpack, suggesting that this may have been a suit designed for use on the Moon. Four years later, when the crew of **Soyuz** 4 transferred to Soyuz 5, they wore a modified suit with no backpack but with air supplies attached to their legs. After the Soyuz 11 disaster, all Soviet cosmonauts wore pressure suits during launch, docking, and landing, but they began wearing the so-called Orlan spacesuit for EVAs. Versions of the Orlan suit have been used by cosmonauts on **Salyut** and **Mir** missions, and now for ISS spacewalks. It consists of flexible limbs attached to a one-piece rigid body/helmet unit that is entered through a hatch in the rear of the torso. The exterior of the hatch houses the life-support equipment.

spacewalk

An excursion outside a spacecraft or space station by an astronaut wearing only a spacesuit and, possibly, some sort of maneuvering device. A "standup spacewalk" is when the astronaut remains partially within the spacecraft; for example, in the case of a Gemini astronaut who stands up in his seat with one of the capsule hatches open while in orbit. Spacewalks and moonwalks are collectively known as extravehicular activity. The 100th EVA in history (both U.S. and Soviet, Earth Orbit and Lunar) occurred on September 15, 1992, when Anatoli Solovyov and Sergei Avdeyev completed their fourth EVA to install a 700-kg VDU on the Sofora Girder located on the external skin of the Kvant module of Mir.

Spadeadam Rocket Establishment

A facility in northern England, opened in the 1950s as a test site for the development of Britain's **Blue Streak** intermediate range ballistic missile. The first rocket firing took place in August 1959.

SPARCLE (Space Readiness Coherent Lidar Experiment)

Also known as **EO**-2, a flight intended to validate the technologies needed for a space-based **lidar** system used to measure tropospheric winds. It was to have been carried as a secondary Hitchhiker payload aboard the Space Shuttle, but it was canceled by NASA in 1999 due to cost overruns.

Spartan 201

A Shuttle-launched and -retrieved satellite for observing the Sun. Spartan 201's science payload consists of two telescopes: the ultraviolet coronal spectrometer and the white light coronagraph. To date it has been carried aboard five Shuttle missions—April 1993 (STS-56), September 1994 (STS-64), September 1995 (STS-69), November 1997 (STS-87), and October 1998 (STS-95). All but the fourth were successful.

SPAS (Shuttle Pallet Satellite)

A reusable free-flying vehicle built by Messerschmitt-Bolkow-Blohm that can be deployed and retrieved by the Space Shuttle's **Remote Manipulator System** (RMS). The original SPAS, with materials processing and SDI ("Star Wars")-related sensor payloads, was carried on three missions but not deployed on the second of these because of an electrical problem with the RMS.

ORFEUS (Orbiting and Retrievable Far and Extreme Ultraviolet Spectrometer) was a German astronomical payload that flew twice aboard SPAS. The main instrument was a 1-m telescope with extreme ultraviolet and far ultraviolet spectrometers of high spectral resolution. Also carried was the Princeton Interstellar Medium Absorption Profile Spectrograph, which studied the fine structure of UV absorption lines in stellar spectra caused by interstellar gas, and several non-astronomy payloads.

CRISTA, equipped for observation of the Earth's atmosphere, also flew on two SPAS missions. (See table, "SPAS Chronology.")

SPEAR (Spectroscopy of Plasma Evolution from Astrophysical Radiation)

An instrument to trace the energy flow in the gas between stars that would form the primary payload of the Korean KAISTSAT-4 mission, tentatively planned for

SPAS Chronology

| | Shuttle Deployment | | | | |
Spacecraft	Deployed	Retrieved	Mission	Orbit	Mass (kg)
SPAS-I	Jun. 18, 1983	Jun. 18, 1983	STS-7	295 × 300 km × 28.5°	
SPAS-1A	–	–	STS-11	Not deployed	
SPAS-II	Apr. 28, 1991	Apr. 29, 1991	STS-39	248 × 263 km × 57.0°	
ORFEUS-SPAS I	Sep. 13, 1993	Sep. 18, 1993	STS-51	301 × 331 km × 28.5°	3,202
CRISTA-SPAS I	Nov. 4, 1994	Nov. 12, 1994	STS-66	263 × 263 km × 57.0°	3,260
ORFEUS-SPAS II	Nov. 19, 1996	Dec. 4, 1996	STS-80	318 × 375 km × 28.5°	
CRISTA-SPAS II	Aug. 7, 1997	Aug. 16, 1997	STS-85	282 × 296 km × 57.0°	

launch in 2003. The SPEAR far-ultraviolet spectrograph would observe million-degree gas in the interstellar medium of our galaxy in order to identify the sources of the thermal energy and determine how this energy is transported through the gas. SPEAR was selected by NASA as a Small Explorer Mission of Opportunity.

special theory of relativity

The physical theory of space and time developed by Albert Einstein. It is based on two key ideas: that all the laws of physics are equally valid in all frames of reference moving at a uniform velocity, and that the speed of light from a uniformly moving source is always the same, regardless of how fast or slow the source or its observer is moving. Among the theory's consequences are the **relativistic effects** of mass increase and time dilation, and the principle of mass-energy equivalence.[284]

specific impulse

The ratio of a rocket engine's **thrust** to the weight of fuel burned in one second. Measured in seconds, specific impulse is an important gauge of the efficiency of a rocket propulsion system, similar to the idea of miles per gallon with cars. The higher its value, the better the performance of the rocket. See **total impulse**.

spectrometer

An instrument that splits the electromagnetic radiation it receives from a source into its component wavelengths by means of a diffraction grating, then measures the amplitudes of the individual wavelengths.

Spectrum-X-Gamma

A multinational high-energy astronomy satellite, expected to be launched in 2003. Its instruments include: JET-X, the Joint European Telescope, a British-Italian grazing-incidence X-ray telescope with CCD (charge-coupled device) detectors; SODART, a Danish-American-Russian telescope using grazing-incidence optics with proportional counters, solid-state detectors, a spectrometer, and an X-ray polarimeter; FUVITA, a Swiss-Russian instrument consisting of four telescopes operating in the far ultraviolet region; and MART, an Italian-Russian high-energy X-ray telescope.

speed of light (c)

In empty space, it equals 299,792 km/s (186,282 mi./s). The speed of light has the same value independent of the relative velocity between source and observer, an experimental fact that only makes sense if relative motion changes the relationship between space and time intervals to keep the distance covered by light per unit time the same for all observers. The fact that space and time are interchangeable to keep the speed of light constant implies that, in some sense, space and time must be the same, despite our habit of measuring space in meters and time in seconds. But if time and space are similar to the extent that they can be converted one into the other, then a quantity is needed to convert the units—something measured in meters per second that can be used to multiply seconds of time to get meters of space. This universal conversion factor is the speed of light. The reason that it is limited is due simply to the fact that a finite amount of space is equivalent to a finite amount of time.

The "speed of light" can also mean the speed at which light travels in a given medium. For example, light travels only two-thirds as quickly in glass as it does in a vacuum. If something, such as a subatomic particle, travels faster through a medium than light does, the result is a kind of electromagnetic shock wave known as Cerenkov radiation. However, there is no violation of the laws of physics, since the universal speed limit is how fast light travels in a vacuum.

speed of sound

The speed at which small disturbances travel through a medium. In the case of a gas, such as air, the speed of sound is independent of pressure but varies with the square root of temperature. Since temperature decreases with increasing altitude in the atmosphere, so too does the speed of sound; in air at 0°C, it is about 1,220 km/hr (760 mph), though it also varies slightly with humidity. The various regimes of flight are *subsonic* (well below the speed of sound), *transonic* (near sonic speed), *supersonic* (up to 5 times sonic speed), and *hypersonic* (above 5 times sonic speed). Compressibility effects start to become important in the transonic regime and very significant at supersonic speeds, when shock waves are present. In the hypersonic regime, the high energies involved have significant effects on the air itself. The important parameter in each of these situations is the **Mach number**–the ratio of the speed of the object to the local speed of sound.

Spektr

See **Mir**.

SPIDR (Spectroscopy and Photometry of the Intergalactic Medium's Diffuse Radiation)

A mission to map the web of hot gas that spans the universe. Half of the normal matter in the nearby universe is in filaments of hot gas, and SPIDR would observe it for the first time. SPIDR's data would answer fundamental questions concerning the formation and evolution of galaxies, clusters of galaxies, and other large structures in the universe. SPIDR has been selected by NASA for study as a Small Explorer mission.

spin-off

A commercial or other non-space benefit derived from space research. Some of the numerous devices and technologies developed originally for use in spaceflight or space science are listed in the table ("Some Space Spin-offs").

spin stabilization

A simple and effective method of keeping a spacecraft pointed in a certain direction. A spinning spacecraft resists perturbing forces in the same way that a spinning gyroscope or a top does. Also, because, in space, forces that slow the rate of spin are negligible, once a spacecraft is set spinning, the rate of rotation stays the same. With a spinner, there are inherent inefficiencies because only some of the solar cells are illuminated at any one instant and because most of the radio wave energy, radiating from the nondirectional antennas in all directions, is not directed at Earth. Thrusters are fired to make desired changes in the spin-stabilized attitude. They may require complicated systems to de-spin antennas or optical instruments that must be pointed at targets.

SPIRIT (Space Infrared Interferometric Telescope)

A highly sensitive orbiting observatory that would allow the far infrared background to be resolved almost completely into individual sources. SPIRIT is identified in NASA's Office of Space Science Strategic Plan as a poten-

Some Space Spin-offs

Device/Technique	Origin
Medical imaging	Signal-processing techniques to clarify images from spacecraft
Bar coding	NASA tracking system for spacecraft parts
Firefighter suits	Fire-resistant fabric for use in spacesuits
Smoke detector	Toxic vapor detectors aboard the Skylab space station
Sun tiger glasses	Protective lenses for welders working on spacecraft
Car design software	NASA software for use in spacecraft and airplane design
Cordless tools	Portable, self-contained drills to enable Apollo astronauts to obtain moon samples
Fisher space pen	Developed for use in space. Pressurized gas pushes the ink toward the ballpoint even in zero gravity, while special ink works at temperature extremes
Invisible braces	Translucent ceramic developed for use in spacecraft and aircraft
Joystick controller	Control stick for the Apollo Lunar Rover
Scratch-resistant lenses	Dual ion-beam bonding, in which a diamondlike coat is applied to plastic
Freeze-dried food	Meals for early astronauts

tial mission beyond 2007 but remains in the very early concept definition phase.

SPOT (Satellite Probatoire d'Observation de la Terre)

A civilian Earth observation program, sponsored by the French government with support from Belgium and Sweden. A single SPOT satellite provides complete coverage of the Earth every 26 days and produces images with a resolution of up to 10 m. Image products from SPOT are handled commercially by SPOT-Image Corp. (See table, "SPOT Missions.")

> Launch site: Kourou
> Length: 3.5 m
> Maximum diameter: 2.0 m

Sputnik (launch vehicle)

The world's first launch vehicle to reach Earth orbit. It was developed from the Soviet R-7 ballistic missile (see **"R" series of Russian missiles**) and comprised a central core of four RD-107 rockets, clustered around which were four tapered booster stages, each containing four RD-108 rockets. All the engines used liquid oxygen and kerosene. During ascent, the boosters were jettisoned, leaving the core to place the Sputnik satellite in orbit.

Sputnik (satellites)

A series of satellites launched by the Soviet Union at the dawn of the Space Age. "Sputnik" (satellite) was the abbreviated Western name for these spacecraft, known in Russia generically as "Iskusstvenniy Sputnik Zemli" (Artificial Earth Satellite). The spacecraft, known in the West as Sputnik 4, 5, 6, 9, and 10, were announced at the time in Russia as **Korabl-Sputnik** 1, 2, 3, 4, and 5. These were unmanned test launches of the Vostok space capsule, which would eventually carry Yuri **Gagarin** on his historic flight. Sputnik 7 and 8 were Venus probes. (See table, "Sputnik Series.")

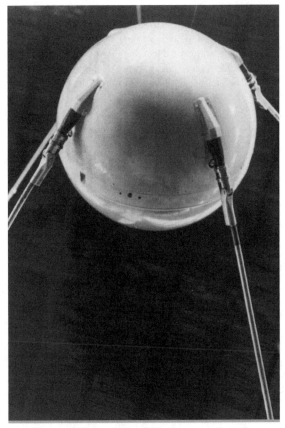

Sputnik 1   Sputnik 1, the world's first artificial satellite. *NASA*

Sputnik 1

The first spacecraft to be placed in orbit around the Earth. A basketball-sized (59-cm) aluminum sphere with four trailing spring-loaded antennae, it carried a small radio beacon that beeped at regular intervals and could verify, by telemetry, exact locations on the Earth's surface. It decayed three months after launch.[78, 183]

SPOT Missions

| Spacecraft | Launch | | Orbit | Mass (kg) |
	Date	Vehicle		
SPOT 1	Feb. 22, 1986	Ariane 1	821 × 823 km × 98.7°	1,830
SPOT 2	Jan. 22, 1990	Ariane 4	821 × 823 km × 98.7°	1,837
SPOT 3	Sep. 26, 1993	Ariane 4	821 × 823 km × 98.7°	1,907
SPOT 4	Mar. 24, 1998	Ariane 4	824 × 826 km × 98.7°	2,755
SPOT 5	May 4, 2002	Ariane 42P	821 × 823 km × 98.7°	3,085

Sputnik 2

The second artificial satellite and the first to carry a passenger—the dog **Laika.** Biological data were returned for approximately a week. However, as there was no provision for safe reentry, Laika was put to sleep after a week in space. The satellite itself remained in orbit 162 days.

Sputnik 3

The last in the series, although prior to a decision to be more cautious in the launch schedule, it may have been intended as the first. Sputnik 3 collected data on the Earth's ionosphere and magnetosphere.

Sputnik Series

Spacecraft	Launch Date	Orbit	Mass (kg)
Sputnik 1	Oct. 4, 1957	228 × 947 km	83.6
Sputnik 2	Nov. 3, 1957	225 × 1,671 km	508
Sputnik 3	May 15, 1958	230 × 1,880 km	1,327

Squanto Terror

The first known serious attempt by the U.S. Air Force to develop a practical antisatellite capability. The system involved **Thor**-launched spacecraft carrying nuclear explosives that could knock out enemy spacecraft by direct contact or explosive concussion. They were loosely based on an October 1962 Air Force/Atomic Energy Commission test, codenamed **Starfish.** The early Squanto terror tests were highly successful, with dummy warheads coming within a mile of their targets. By 1965, after only three tests, the system was declared operational. A total of 16 test launches, all from Johnson Island in the Pacific, were conducted by the time the tests were concluded in 1968. Meanwhile, a 1967 treaty was signed that banned nuclear weapons from space, and by 1975, the Johnson Island facility had been put out of service.

Sriharikota High Altitude Range Center (SHAR)

India's launch site located on Sriharikota Island (13.9° N, 80.3° E). Severe range safety restrictions on launch azimuth make it inefficient to launch into **polar orbit**s.

SROSS (Stretched Rohini Satellite Series)

Spacecraft developed by the Indian Space Research Organization (ISRO) as follow-ons to the successful **Rohini** satellite series. SROSS satellites were designed to carry small scientific and technology payloads, including astrophysics, remote sensing, and upper atmospheric monitoring experiments. They were launched aboard developmental flights of India's new **ASLV** (Advanced Space Launch Vehicle). Both SROSS 1 and SROSS 2 were lost due to launch failures. The third vehicle, SROSS C, was successfully placed in orbit, but one much lower than planned. The vehicle decayed on July 14, 1992, and, although it returned some scientific data, it was deemed only partially successful. SROSS C2 was the first unqualified success of the SROSS and ASLV programs. SROSS C and C2 carried a gamma-ray burst (GRB) experiment that monitored celestial gamma-ray bursts in the energy range 20 to 3,000 keV and a Retarded Potential Analyzer (RPA) experiment that measured temperature, density, and characteristics of electrons in the ionosphere. (See table, "SROSS Missions.")

Launch
   Vehicle: ASLV
   Site: Sriharikota
Size: 1.1 × 0.8 m

SRTM (Shuttle Radar Topography Mission)

An Earth-mapping mission that was conducted aboard the Space Shuttle in February 2000, supported jointly by NASA, NIMA (National Imagery and Mapping Agency), and international partners, as part of NASA's **Earth Probe** program.

SS (Scientific Satellite)

A Japanese satellite designation reserved for certain **ISAS** (Institute of Space and Astronautics Science) missions. For example, **Shinsei** was also known as SS-1, **Denpa** as SS-2, **Taiyo** as SS-3, and so on.

SROSS Missions

Spacecraft	Launch Date	Orbit	Mass (kg)
SROSS A	Mar. 24, 1987	Launch failure	150
SROSS B	Jul. 12, 1988	Launch failure	150
SROSS C	May 20, 1992	251 × 436 km × 46.0°	106
SROSS C2	May 4, 1994	437 × 938 km × 46.2°	113

SSF (subsatellite ferrets)

Classified U.S. Air Force subsatellites, released in association with other military launches, that carried out radio and radar signals intelligence gathering. There have been several types, dating back to the 1963 "Hitch Hiker" spacecraft, but details remain secret.

SSS (Small Scientific Satellite)

A NASA spacecraft launched from **San Marco** to measure aerodynamic heating and radiation damage during launch, and also to investigate particles and fields in the magnetosphere. Also known as **Explorer** 45.

stabilization of satellites

Satellite stabilization takes three possible forms: (1) **spin stabilization,** whereby the satellite is spun at 10 to 30 rpm; (2) **gravity gradient stabilization** using a large weight attached to the satellite by a length of line; and (3) **inertial stabilization** using heavy wheels rotating at high speed—typically three wheels, one for each axis, providing three-axis stabilization.

Stack, John (1906–1972)

An American aeronautical engineer who played an important role in the effort to achieve manned supersonic flight. Stack graduated from the Massachusetts Institute of Technology in 1928 and then joined the **Langley** Aeronautical Laboratory, becoming director of all the high-speed wind tunnels and high-velocity airflow research there in 1939. Three years later, he was named chief of the compressibility research division at Langley. In 1947, he was promoted to assistant chief of research, a title then changed to assistant director of the research center. Stack guided much of the research that paved the way for **transonic** aircraft, and in 1947 was awarded the Collier Trophy together with Charles **Yeager**, the pilot who first broke the sound barrier. From 1961 to 1962, Stack was director of aeronautical research at NASA Headquarters before leaving to become vice president for engineering at Republic Aircraft Corp (later part of Fairchild Industries), from which he retired in 1971.

Stacksat

An American military project that involved the simultaneous launch of three similar spacecraft: POGS (Polar Orbiting Geomagnetic Survey), TEX (Transceiver Experiment), and SCE (Selective Communications Experiment). POGS carried a magnetometer to map Earth's magnetic field. Data from the experiment was stored in an experimental solid state recorder and was used to help improve Earth navigation systems. TEX carried a variable power transmitter to study ionospheric effects on radio frequency (RF) transmissions and gather data to determine minimum spacecraft transmitter power levels for transmission to ground receivers. SCE carried a variable frequency transmitter to study ionospheric effects at various RF frequencies and was also designed to demonstrate message store and forward techniques. Six low-cost ground stations were designed, built, and located around the world to operate these spacecraft.

Launch
  Date: April 11, 1990
  Vehicle: Atlas E
  Site: Cape Canaveral
  Orbit (circular): 741 km × 90°
  Mass (each): 68 to 72.6 kg

Stafford, Thomas P. (1930–)

A veteran American astronaut and career Air Force officer, selected by NASA in 1962. Stafford served as backup

Thomas Stafford Stafford (standing) and Walter Schirra go through a suit-up exercise in preparation for their Gemini 6 flight in 1965. *NASA*

pilot for **Gemini** 3 and pilot for Gemini 6, became command pilot for Gemini 9 upon the death of a prime crew member, and was backup commander for **Apollo** 7, commander of Apollo 10, and commander of the **Apollo-Soyuz Test Project**. He resigned from NASA on November 1, 1975, to become commander of the Air Force Flight Test Center, Edwards Air Force Base. He was promoted to Air Force Deputy Chief of Staff for Research and Development in March 1978 and retired from the Air Force in November 1979. Subsequently, he held a number of senior management positions.

stage

A portion of a launch system that fires until its fuel supply is exhausted and then separates from the rest of the system.

staging

The jettisoning of self-contained propulsion units after consumption of their **propellant**s—a crucial technique for improving the **mass ratio** of space transport systems not using an **environmental engine**. Among the earliest pioneers of the idea appear to have been Conrad **Haas** and Johan **Schmidlap** in the sixteenth century. However, the first detailed theoretical analysis of staging was done by Konstantin **Tsiolkovsky**.

stand-up spacewalk

An **extravehicular activity** in which an astronaut stands up in the space capsule, with the hatch opened, so that part of his or her body extends outside the vehicle into space. Stand-up spacewalks were common in the early days of the space program when spacecraft were not equipped with airlocks.

stapp

A unit of exposure to acceleration or deceleration used in aerospace medicine and named after John **Stapp**. One stapp is the force exerted by one *g* acting on the human body for one second. Thus, an astronaut subjected to 3*g* for 12 seconds is said to have endured 36 stapps.

Stapp, John Paul (1910–1999)

See article, pages 416–417.

star drive

See **reactionless drive**.

Stardust

A NASA spacecraft that will pass within 160 km of comet Wild-2 (pronounced "vihlt") in December 2004 and return particles of dust from the comet's tail to Earth. Stardust will capture the samples of cometary material using a spongelike cushioning substance, known as **aerogel**, which is attached to panels on the probe. It will also send back pictures, take counts of the number of **comet** particles striking it, and analyze in real time the composition of substances in the comet's tail. Stardust will then bring back samples of cosmic dust, including those of recently discovered interstellar dust streaming into the Solar System from the direction of Sagittarius. Having been "soft-caught" and preserved in aerogel, the dust samples will be returned to Earth in a reentry capsule that will land in Utah in January 2006. Analysis of the material from Wild-2 and interstellar space, which will include pre–solar grains and condensates left over from the formation of the Solar System, is expected to yield important insights into the evolution of the Sun and planets and possibly the origin of life itself. During its seven-year mission, Stardust will complete three orbits of the Sun. On the first of these it passed close by the Earth for a **gravity-assist**, on the second it will encounter the comet, and on the third it will again pass by Earth and release its sample capsule, which will descend by parachute to the surface. Stardust is the fourth mission in NASA's **Discovery Program**, following **Mars Pathfinder**, **NEAR-Shoemaker**, and **Lunar Prospector**, and involves an international collaboration between NASA and university and industry partners.

Launch
  Date: February 7, 1999
  Vehicle: Delta 7925
  Site: Cape Canaveral
Length of main bus: 1.7 m
Mass: 370 kg

Starfish

Two early U.S. Air Force and Atomic Energy Commission satellites designed to collect data on radiation

(continued on page 417)

Starfish Missions

Spacecraft	Launch Date	Orbit	Mass (kg)
Starfish 1	Oct. 26, 1962	197 × 5,458 km × 71.4°	1,100
Starfish 2	Sep. 2, 1965	Launch failure	1,150

Stardust   An artist's conception of the Stardust probe encountering comet Wild-2. *NASA*

John Paul Stapp (1910–1999)

An American pioneer of **aerospace medicine**, famous for his extreme rocket-sled experiments. Stapp was born in Bahia, Brazil, 1910, the son of Southern Baptist missionaries. At age 12, he moved back to the United States with his family and later started college in Texas with the idea of becoming a writer. During Christmas vacation 1928, he witnessed a tragedy that changed his life. While he was visiting relatives, his baby cousin crawled into a fireplace and was badly burned. For three days before the child died, Stapp helped nurse the child; afterward he determined to become a doctor. Fifteen years later, having earned degrees in zoology and biophysics, he entered the University of Minnesota medical school to pursue his dream. When he graduated, he interned at St. Mary's Hospital, Duluth, before enlisting in the Army Medical Corps during World War II.

In 1946, Stapp joined the aeromedical laboratory at Wright Field and was serving as flight surgeon to Chuck **Yeager** when he broke the sound barrier. Stapp became convinced that a significant pattern lay behind the way some airmen died and others survived seemingly equally violent crashes. To solve the mystery, he used a high-speed rocket sled at Morocco Air Base.

Stapp planned a series of tests on humans and set out to develop a harness to hold them safely to the rocket-powered sled, known as the "Gee Whiz." First, he used a dummy named "Oscar Eight-Ball" to perfect the harness. Finally, after 32 sled runs, he was ready to try it out on a human guinea pig—himself. He was strapped into the sled facing rearward, refusing anesthetic because he wanted to study his reactions first-hand. Accelerated almost instantly to 145 km/hr (90 mph), Stapp was then crushed against the seat back as the sled ground to an abrupt halt. He suffered only a few sore muscles. Within a year, Stapp had made sled runs at up to 240 km/hr (150 mph), stopping in as little as 5.8 m (19 ft). He experienced up to 35 times the force of gravity ($35g$) and proved that the human body could withstand such stress, although in the process he suffered headaches, concussions, a fractured rib and wrist, and a hemorrhaged retina.

When Stapp's commanding officer learned he'd been his own test subject, he ordered the experiments to stop, fearing he'd miss out on promotion if Stapp were killed. However, Stapp secretly continued the tests, using chimpanzees, and found that when strapped in correctly they survived forces many times those experienced in most plane crashes. From this, he concluded that crash survival doesn't depend on a body's ability to withstand the high forces involved, but rather on its ability to withstand the mangling effects of the vehicle. To back up this idea, Stapp again unofficially began tests on humans—putting himself first in the firing line. Over the next four years, he lost fillings, cracked more ribs, and broke his wrist again. Yet, despite these daredevil exploits, Stapp was known as a quiet, philosophical man who loved classical music. He refused to marry until his test days were over.

In 1949, Stapp was involved in the birth of Murphy's Law. Stapp's harness held 16 sensors to measure the g-force on different parts of the body. There were exactly two ways each sensor could be installed, and it fell upon a certain Captain Murphy to make the connections. Before a run in which Stapp was badly shaken up, Murphy managed to wire up each sensor the wrong way, with the result that when Stapp staggered off the rocket sled with bloodshot eyes and bleeding sores, all the sensors read zero. Known for his razor-sharp wit, Stapp quipped: "If there are two or more ways to do something and one of those results in a catastrophe, then someone will do it that way."

The advent of supersonic flight and the need to bail out at very high speed demanded more extreme experiments. Transferred to head the aeromedical field lab at Holloman Air Force Base, New Mexico, Stapp built a much faster sled, called "Sonic Wind No. 1." Again, he began his studies using dummies, but in March 1954, he put himself forward as the subject. In his first ride on the new sled, Stapp reached 677 km/hr (421 mph)—a new land speed record.

On December 10, he took the sled chair for his final and most extreme ride. His wrists were tied together in front of him, because flapping limbs would be torn away in the ferocious air stream. His major concern was that the rapid deceleration might blind him. Earlier he had "practiced dressing and undressing with the lights out so if I was blinded I wouldn't be helpless." At the end of the countdown, Stapp was shot to 1,002 km/hr (623 mph) in 5 seconds and back to rest in just over a second. Subjected to $40g$, he temporarily blacked out, and his eyeballs bulged from their sockets. He was rushed to the hospital, where his eyesight gradually returned, and a checkup revealed he'd suffered no major injury. An hour later, he was eating lunch.

Later, he told the American Rocket Society that experiments with the rocket-powered sled would help pioneer the way to an early fulfillment of human space flight. He subsequently helped run tests on human and animal subjects in the giant **Johnsville Centrifuge**—the nightmare machine in which the early Mercury astronauts trained. In 1958, he married Lillian Lanese, who had danced with the Ballet Russe de Monte Carlo.

Because of his expertise in safety at high speed, the Air Force loaned Stapp to the National Highway Traffic Safety Administration in 1967 as a medical scientist. Upon retirement with the rank of colonel in 1970, he became a professor in the University of California's Safety and Systems Management Center and, later, a consultant to the surgeon general and to NASA. As chairman of the Stapp Foundation, he led the annual Stapp Car Crash Conference, which brought together automotive engineers, trauma surgeons, and other experts to study how people died in car crashes. These conferences continue today.

John Stapp Stapp prepares for a high-*g* rocket-sled run.
U.S. Air Force/Edwards Air Force Base

Starfish

(continued from page 414)

resulting from nuclear explosions in the atmosphere and in space. The second in the series was lost when its launch vehicle was destroyed by the range safety officer. See **Squanto Terror**. (See table, "Starfish Missions," on page 414.)

Launch
    Vehicle: Thor-Agena D
    Site: Vandenberg Air Force Base

Stargazer, Project

An early balloon-borne project to carry out astronomical studies at very high altitude. It involved only one flight. On December 13–14, 1962, Joseph **Kittinger** and William White, an astronomer, flew a gondola suspended beneath an 85-m-diameter Mylar balloon to a height of 25,000 m over New Mexico. In addition to obtaining telescopic observations from above 95% of Earth's atmosphere, the flight provided valuable data for the development of pressure suits and associated life support systems during a 13-hour stay at the edge of space.

Starlight

A NASA mission being developed by JPL (Jet Propulsion Laboratory) as part of the **Origins program**. Starlight consists of two spacecraft that will fly in formation to test techniques and instrumentation needed to carry out multiple-spacecraft interferometry. Following their expected launch in 2005, the two initially conjoined spacecraft will be placed in a solar orbit matching Earth's but trailing behind at a distance of up to 17 million km. After a brief checkout, the spacecraft will separate—the "daughter" to be parked at a nearby position, while the "mother" makes test interferometer observations for three months using its self-contained optics. Then the two spacecraft will perform experiments of the autonomous formation flying system, at separation distances ranging from 40 to 600 m. Finally, the technology for formation-flying optical interferometry will be put through its paces by making observations of specific target stars using both spacecraft. Previously known as Space Technology 3 and the New Millennium Interferometer, Starlight will validate the technique of space-based optical interferomety that will be used by later missions, such as the **Terrestrial Planet Finder**, to search for extrasolar Earth-like worlds and any life that may inhabit them.

STARSHINE (Student Tracked Atmospheric Research Satellite for Heuristic International Networking Equipment)

An educational microsatellite, built and launched to encourage optical tracking and satellite observation by students. STARSHINE was a passive, polished 48-cm-diameter hollow aluminum sphere, manufactured by the U.S. Naval Academy and covered by 878 25-mm-diameter mirrors that were ground at many participating American and international schools. Deployed from a hitchhiker slot on the Shuttle, STARSHINE was visible from the ground with the naked eye, flashing every few seconds due to its spin. It decayed on February 18, 2000.

Shuttle Deployment
   Date: June 5, 1999
   Mission: STS-96
Mass: 38 kg
Orbit (circular): 360 km × 52°

starship

A large, crewed spacecraft capable of rapid transit between stars.

Start-1

A small, four-stage Russian launch vehicle developed from the Topol intercontinental launch vehicle under a conversion program in the early 1990s. The first Start-1 was successfully launched in 1993, delivering a non-commercial payload to orbit. Several more launches have since taken place.

Length: 22.9 m
Diameter: 1.8 m
Payload (to LEO): about 400 kg

START (Spacecraft Technology and Advanced Reentry Test)

A U.S. Air Force program to experiment with **lifting bodies** that began after the cancellation of the **Dyna-Soar** project. It began in 1960, using subscale models of the X-20 Dyna-Soar to test materials, and continued with the **ASSET** and **PRIME** suborbital tests of subscale lifting body designs and B-52 drop tests of the **X-24**A and X-24B lifting bodies into the 1970s.

static firing

The firing of a rocket motor, rocket engine, or an entire stage in a hold-down position to measure thrust and to carry out other tests.

static testing

The testing of a device in a stationary or held-down position as a means of testing and measuring its dynamic reactions.

stationary orbit

See **geosynchronous orbit**.

station-keeping

Minor maneuvers that a satellite in **geostationary orbit** (GSO) must make over its mission life to compensate for orbital perturbations. The main source of perturbation is the combined gravitational attractions of the Sun and Moon, which cause the orbital **inclination** to increase by nearly one degree per year. This is countered by a north-south station-keeping maneuver about once every two weeks so as to keep the satellite within 0.05° of the equatorial plane. The average annual velocity change (delta v) needed is about 50 m/s, which represents 95% of the total station-keeping propellant budget. Additionally, the **bulge of the Earth** causes a longitudinal drift, which is compensated by east-west station-keeping maneuvers about once a week, with an annual delta v of less than 2 m/s, to keep the satellite within 0.05° of its assigned longitude. Finally, solar **radiation pressure** caused by the transfer of momentum from the Sun's light and infrared radiation both flattens the orbit and disturbs the orientation of the satellite. The orbit is compensated by an eccentricity control maneuver that can sometimes be combined with east-west station-keeping, whereas the satellite's orientation is maintained by momentum wheels supplemented by magnetic torquers and thrusters.

Statsionar
See **Raduga**.

stay time
The average time spent by a gas molecule in a liquid-fuel rocket engine's **combustion chamber** before exiting from the **nozzle** and producing **thrust**.

STEDI (Student Explorer Demonstration Initiative)
A program, managed by USRA (Universities Space Research Association) for NASA, that is designed to demonstrate that a small scientific or technology-based satellite could be designed, fabricated, launched, and operated for around $10 million. It is part of NASA's "smaller, cheaper, faster" initiative for research spacecraft.

Stella
A French geodetic satellite, carrying 60 laser reflectors mounted on a dense sphere of uranium alloy, that is tracked to measure small perturbations in Earth's gravitational field. Reflected laser beams enable measurements, accurate to about 1 cm, of the geoid, oceanic and terrestrial tides, and tectonic movements. Stella's twin, Starlette, launched in 1975, is also still in use.

Launch
  Date: September 26, 1993
  Vehicle: Ariane 4
  Site: Kourou
  Orbit: 793 × 803 km × 98.7°
  Mass: 48 kg

Stennis Space Center (SSC)
One of 10 NASA field centers in the United States. Previously known as the Mississippi Test Facility, it was renamed in 1988 after U.S. Senator John C. Stennis. Located in Hancock County, Mississippi, about 80 km northeast of New Orleans, it is the primary center for testing and flight certifying rocket propulsion systems for the Space Shuttle and future generations of space vehicles. In the past it was the static test site for **Saturn-** and **Nova-**class launch vehicles. All Space Shuttle Main Engines must pass a series of test firings at Stennis Space Center prior to being installed in the back of the Orbiter. Stennis is also NASA's lead center for commercial remote sensing within the Mission to Planet Earth Enterprise.

STENTOR (Satellite de Télécommunications pour Expériences de Nouvelles Technologies en Orbite)
An experimental communications satellite developed by CNES (the French space agency) in association with French Telecom. STENTOR will carry out propagation and transmission experiments, especially at wavelengths that are shorter than those currently used for satellites communications. Launch was scheduled for the second half of 2002.

STEP (Satellite Test of the Equivalence Principle)
A satellite designed to test, in microgravity, a well-established physical law–the Equivalence Principle–that all falling objects accelerate at the same rate. Until now, this principle escaped deep scrutiny because experiments have always been limited by gravitational conditions at the Earth's surface. STEP, which involves a collaboration between NASA and ESA (European Space Agency), will test the principle to an accuracy 100,000 times greater than that achieved in terrestrial laboratories. It will also conduct experiments in quantum mechanics and gravity variations. STEP was selected by NASA for study as a **SMEX** (Small Explorer).

step principle
A design feature of rockets in which one stage is mounted directly onto another. When a lower stage is used up, it is ejected, and the next upper stage takes over. With each stage's ejection, the weight of the spacecraft decreases, so the next stage has less work to do.

step rocket
A multistage rocket.

STEREO (Solar-Terrestrial Relations Observatory)
A mission to understand the origin of coronal mass ejections–powerful eruptions on the Sun in which as much as 10 billion tons of the solar atmosphere can be blown into interplanetary space–and their consequences for Earth. It will consist of two spacecraft, one leading and the other lagging Earth in its orbit. These spacecraft will each carry instrumentation for solar imaging and for in-situ sampling of the solar wind. STEREO, the third of NASA's **Solar Terrestrial Probe** missions, is being designed and built at the Johns Hopkins University Applied Physics Laboratory and is scheduled for launch in 2005.

Stewart, Homer J. (1915–)
A prominent aerospace engineer involved with the U.S. space program. Stewart earned a B.S. in aeronautic engineering from the University of Minnesota in 1936, joined the faculty of the California Institute of Technology (Caltech) in 1938, and earned his doctorate from Caltech two years later. In 1939, he took part in pioneering rocket research with other Caltech engineers and scientists, including Frank Malina, in the foothills of Pasadena. Out of their efforts, **JPL** (Jet Propulsion

Stennis Space Center   The firing of a Space Shuttle Main Engine at Stennis Space Center's A-2 test stand. *NASA*

Laboratory) arose, and Stewart maintained his interest in rocketry at that institution. He chaired the committee (see **Stewart Committee**) that made recommendations about the early direction of the U.S. space program and was heavily involved in developing the first U.S. satellite, Explorer 1, in 1958. In that year, on leave from Caltech, he became director of NASA's program planning and evaluation office, returning to Caltech in 1960 to a variety of positions, including chief of the advanced studies office at JPL (1963–1967) and professor of aeronautics at Caltech itself.[132]

Stewart Committee

The ad hoc Advisory Group on Special Capabilities set up in 1955 at the request of Donald Quarles, Assistant Secretary of Defense for Research and Development. Chaired by Homer J. **Stewart**, its purpose was to examine several proposals for launch vehicles designed to place the first American satellite in orbit. Three options were consid-

ered, based on **Atlas**, **Redstone**, and **Viking** first-stage launchers. Stewart himself favored the Redstone proposal (Project Orbiter) put forward by **von Braun**'s team; however, he was overridden by his committee, which preferred the Navy's Viking plan. In the event, this proved to be a mistake and led to the Soviet Union taking an early lead in the Space Race. See **United States in space**.

STEX (Space Technology Experiments)

A National Reconnaissance Office satellite that, for the most part, successfully tested over two dozen advanced technology subsystems. Among STEX's main equipment were Hall Effect electric thrusters derived from Russian technology, experimental solar arrays and batteries, and the Advanced Tether Experiment (ATeX), which was a follow-on to the earlier **TiPS** (Tether Physics and Survivability) satellite. ATeX comprised two end-masses connected by a 6-km polyethelyne tether. The upper end mass was deployed first, while the lower end mass was supposed to remain attached to STEX. However, this experiment failed on January 16, 1989, when, with only 21 m of tether deployed, the tether was so far off vertical that automatic safety systems jettisoned the base to protect the remainder of the STEX satellite.

Launch
Date: October 3, 1988
Vehicle: Taurus
Site: Vandenberg Air Force Base
Orbit: 744 × 759 km × 85.0°

strap-on booster

A rocket motor that is mounted to the first stage of a launch vehicle to provide extra thrust at lift-off and during the first few minutes of ascent. Most strap-ons, as in the case of the Space Shuttle, Titan, Delta, Atlas, and Ariane, are solid-propellant motors. Some Russian launch vehicles have used liquid-propellant engines. Typically, strap-on boosters burn out while the first stage is still firing, separate, and fall back to Earth. For this reason, they are sometimes referred to as a half stage or zero stage.

strap-on booster An Ariane 4 with strap-on boosters launching the Inmarsat 3 communications satellite. *Lockheed Martin*

stratosphere

The layer of Earth's atmosphere immediately above the **troposphere**, extending to the **mesosphere**; that is, between altitudes of 10 to 15 km and 50 km.

Strela

(1) A Russian space launch vehicle converted from the RS-18 intercontinental ballistic missile (NATO classification: SS-19 Stiletto) and marketed by Space Development Corporation Strela ("arrow") is 26.7 m long with a takeoff mass of 104 tons and can carry a payload of up to 1,700 kg into low Earth orbit. With production of Cosmos-3M rockets halted in 1995, Russian agencies have few choices other than converted RS-18s and RSD-10Ms, like Strela, Start, and Rockot. Of Russia's 160 nuclear warhead–bearing RS-18s, 55 must be decommissioned by 2007 under the START 2 treaty. The RS-18 has logged 146 launches over the past 27 years, with 143 of them a complete success. Strela will be launched from **Svobodny**, Russia's newest spaceport, a converted facility that used to serve as the base for a military ballistic missile unit. (2) A long-running series of Russian military, store-dump communications satellites, the first of which was launched in 1964. The latest batch of six Strela-3 satellites was placed in highly inclined 1400-km-high orbits by a Tsyklon rocket on December 28, 2001.

stressed limits

The environmental limits to which the crew may be subjected for limited periods of time such as launch, reentry, and landing.

Strughold, Hubertus (1898–1987)

A German-born pioneer of **space medicine** and the author of over 180 papers in the field. Strughold was brought to the United States at the end of World War II as part of Operation **Paperclip** and subsequently played an important role in developing the pressure suits worn by early American astronauts. In 1949, Strughold was made director of the department of space medicine at the School of Aviation Medicine at Randolph Air Force Base, Texas (now the School of Aerospace Medicine at Brooks Air Force Base, Texas). Randolph's Aeromedical Library was named after him in 1977, but it was later renamed because documents from the Nuremberg War Crimes Tribunal linked Strughold to medical experiments in which inmates from Dachau concentration camp were tortured and killed.

STRV (Space Technology Research Vehicle)

Satellites designed to allow in-orbit evaluation of new technologies at relatively low cost. The first two, STRV-1A and -1B, were designed, built, and tested at the U.K.

Defence Research Agency (DRA) at Farnborough and launched together in 1994. Each of the 52-kg spacecraft carried 14 experiments, most of them associated with ongoing research programs within the DRA's Space Department. In addition, there is a major international collaborative aspect to the project. The Ballistic Missile Defense Organization (BMDO) Materials and Structures Program sponsored four experiments that were built at JPL (Jet Propulsion Laboratory) and flown aboard STRV-1B. ESA (European Space Agency) submitted experiments for STRV-1B.

stub fins

Short-span aerodynamic surfaces, used on some launch vehicles for control or stabilization purposes.

Stuhlinger, Ernst (1913–)

A physicist who played an important part in the development of rocket instrumentation, first in Germany during World War II and then in the United States. He earned his Ph.D. at the University of Tübingen in 1936 and continued research into cosmic rays and nuclear physics until 1941 as an assistant professor at the Berlin Institute of Technology. In 1943, while serving in the German army on the Russian front, he was assigned to Wernher **von Braun**'s rocket development team at **Peenemünde**, where he worked on guidance and control systems until 1945. After the war, he came to the United States as part of Operation **Paperclip** to continue work in rocketry, first in Fort Bliss, Texas, and White Sands, New Mexico, and, from 1950 on, at the Redstone Arsenal in Huntsville, Alabama, which became the **Marshall Space Flight Center** (MSFC). Stuhlinger was director of the space science laboratory at MSFC (1960–1968) and then its associate director for science (1968–1975), after which he retired and became an adjunct professor and a senior research scientist with the University of Alabama at Huntsville. His main areas of work included guidance and control, instrumentation for scientific investigations, electric space propulsion systems, and space project planning.[280]

subcarrier

A second signal piggybacked onto a main signal to carry additional information. In satellite television transmission, the video picture is transmitted over the main carrier; the corresponding audio is sent via an FM subcarrier. Some satellite transponders carry as many as four special audio or data subcarriers, whose signals may or may not be related to the main programming.

subluminal

Less than the **speed of light**.

submillimeter band
That part of the **electromagnetic spectrum** lying between the far **infrared** and the **microwave** region, corresponding to wavelengths between about 300 microns and 1 mm.

suborbital
A flight or trajectory that reaches to Earth's upper atmosphere or the edge of space but does not involve the completion of an orbit. The first American manned flight into space, by Alan **Shepard** in May 1961, was suborbital.

subsatellite
A portion of a satellite that has a mission objective of its own. Once in space, the subsatellite is ejected and assumes its own orbit.

subsonic
Less than the **speed of sound**.

Suisei
A Japanese probe, launched by **ISAS** (Institute of Space and Astronautical Science), that rendezvoused with Halley's Comet on March 8, 1986. It was identical to **Sakigake** apart from its payload: a CCD (charge-coupled device) ultraviolet (UV) imaging system and a solar wind instrument. The main goal of the mission was to take UV pictures of the hydrogen corona for about 30 days before and after Halley's descending crossing of the ecliptic plane. Measurements of the solar wind were taken over a much longer period. Suisei began UV observations in November 1985, producing up to six images per day. The spacecraft encountered Halley on the sunward side at 151,000 km during March 8, 1986, suffering only two dust impacts. During 1987, ISAS decided to guide Suisei to a November 1998 encounter with comet Giacobini-Zinner, but due to depletion of the hydrazine, this, as well as plans to fly within several million kilometers of comet Tempel-Tuttle in February 1998, were cancelled. Suisei ("comet") was known before launch as Planet-A.

Launch
  Date: August 18, 1985
  Vehicle: M-35
  Site: Kagoshima
  Orbit around Sun: 1.012 × 0.672 AU × 0.89°
  Mass: 141 kg

Sullivan, Kathryn D. (1951–)
The first American woman to walk in space. A veteran of three Space Shuttle flights, Sullivan was a mission specialist on STS-41G (October 1984), STS-31 (April 1990), and STS-45 (March to April 1992). She received a B.S. in earth sciences from the University of California, Santa Cruz (1973) and a Ph.D. in geology from Dalhousie University in Halifax, Nova Scotia (1978). Sullivan left NASA in August 1992 to become chief scientist at **NOAA** (National Oceanic and Atmospheric Administration) and is currently president and CEO of Center of Science & Industry, Columbus, Ohio.

sun-synchronous orbit
An Earth orbit in which a satellite remains in constant relation to the Sun, passing close to both poles and crossing the meridians at an angle. The orbit, at an altitude of about 860 km, takes about 102 minutes and carries the satellite over a different swathe of territory at each pass, so every point on the surface is overflown every 12 hours, at the same local times each day. Another advantage of a sun-synchronous orbit is that a spacecraft's solar arrays are in almost continuous sunlight, enabling it to rely primarily on solar rather than battery power. Also known as a dawn-dusk orbit.

superluminal
Greater than the **speed of light**.

supersonic
Greater than the **speed of sound**.

supine *g*
Acceleration experienced in the chest-to-back direction, expressed in units of gravity. Also known as "eyeballs in." Also see **prone** *g* ("eyeballs out"), **negative** *g* ("eyeballs up"), and **positive** *g* ("eyeballs down").

surface gravity
The rate at which a freely falling body is accelerated by gravity close to the surface of a planet or other body.

SURFSAT (Summer Undergraduate Research Fellowship Satellite)
A small satellite built by undergraduate students and JPL (Jet Propulsion Laboratory) to support experiments by NASA's Deep Space Network. The satellite is designed to mimic signals from planetary spacecraft and radiates

Launch
  Date: November 4, 1995
  Vehicle: Delta 7925
  Site: Vandenberg Air Force Base
  Orbit: 934 × 1,494 km × 100.6°
  Size: 0.8 × 0.3 m
  Mass: 55 kg

milliwatt-level radio frequency signals in the X-, Ku-, and Ka-bands. These signals support research and development experiments supporting future implementation of Ka-band communications, tests of new 11-m ground stations built to support the Space Very Long Baseline Interferometry project, and training of ground station personnel. From conception through launch, the spacecraft cost about $3 million, including design, fabrication, test, and launch integration.

surveillance satellites

Military spacecraft that provide reconnaissance (spy) data in the form of high-resolution visible, infrared, and radar imagery.

Surveyor

A highly successful series of NASA/JPL (Jet Propulsion Laboratory) spacecraft that soft-landed on the Moon between 1966 and 1968 and, together with the **Ranger** and **Lunar Orbiter** programs, helped prepare for the **Apollo** manned landings. Once the early Surveyors had demonstrated an ability to make successful midcourse corrections and soft-landings (proving in the process that the lunar surface was not covered in a thick layer of dust, as some scientists had feared), the remaining Surveyors were used to evaluate potential Apollo landing sites. Seven spacecraft were launched, of which five arrived safely on the Moon and returned data. Surveyor 3 was the first of the series to carry a surface-sampling device with which the spacecraft excavated four trenches up to 18 cm deep. Eighteen months later, the crew of Apollo 12 landed nearby and recovered Surveyor 3's TV camera and other parts. Laboratory analysis showed, astonishingly, that terrestrial bacteria had remained alive in the camera's insulation during its time on the Moon. Surveyor 6 became the first spacecraft to (temporarily) lift off from the surface of another world. On November 17, 1967, its engines were fired for 2.5 seconds, enabling it to

rise 3.7 m above the ground. It was then commanded to move 2.4 m in a westerly direction and then touch down again. Following this maneuver, it continued its data-gathering mission, including the return of 30,027 pictures. (See table, "Surveyor Missions.")

Launch
    Vehicle: Atlas-Centaur IIIC (Surveyors 1 to 3, 6 to 7),
    Atlas-Agena (Surveyors 4 to 5)
    Site: Cape Canaveral

survival in space

See **space survival**.

suspended animation

A technique, popular in science fiction, and familiar through such films such as *2001: A Space Odyssey* and *Alien,* in which astronauts are placed in a state of deep hibernation for long-duration spaceflights. In the late 1960s to early 1970s, NASA ran a program to investigate "depressed metabolism" but abandoned it when it was clear that the technology was not available to make it feasible. Progress may be made by further observation of mammals, such as bears, that go into a deep slumber while still maintaining a sufficient level of metabolism to keep their kidneys from shutting down. More primitive creatures, including tardigrades, which display the extraordinary condition of cryptobiosis (where metabolism all but stops), may also have something to teach us. But whether techniques such as cryonics, which seeks to freeze individuals for later revival, will ever be usefully applied in space travel remains to be seen.

sustainer engine

A rocket engine that stays with a spacecraft during ascent after the booster has dropped off. It sustains or steadily increases the spacecraft's speed during ascent.

Surveyor Missions

| Spacecraft | Launch Date | Lunar Landing | | Mass (kg) |
		Date	Location	
Surveyor 1	May 30, 1966	Jun. 2, 1966	Ocean of Storms	269
Surveyor 2	Sep. 20, 1966	Sep. 22, 1966 (crashed)	Sinus Medii	292
Surveyor 3	Apr. 17, 1967	Apr. 20, 1967	Ocean of Storms	283
Surveyor 4	Jul. 14, 1967	Jul. 17, 1967 (lost contact)	Sinus Medii	283
Surveyor 5	Sep. 8, 1967	Sep. 11, 1967	Sea of Tranquility	279
Surveyor 6	Nov. 7, 1967	Nov. 10, 1967	Sinus Medii	280
Surveyor 7	Jan. 7, 1968	Jan. 10, 1968	Tycho North Rim	280

Surveyor Charles Conrad, Apollo 12 commander, examines the Surveyor 3 spacecraft. The Lunar Module *Intrepid* is in the right background. *NASA*

SUVO (Space Ultraviolet Optical Telescope)

A proposed follow-on mission to the **NGST** (Next Generation Space Telescope) that would enable the ultraviolet light from faint remote cosmic sources to be captured and analyzed in unprecedented detail. The information it provided would allow astronomers to trace galaxy evolution back to the initial era of star formation, to gain new insight into supermassive black holes, and to study metal element production in the present epoch. SUVO has been identified in NASA's Office of Space Science Strategic Plan as a potential mission beyond 2007 but remains in the early concept definition phase.

Svobodny Cosmodrome

A former Russian strategic missile base, 120 km north of Blagoveshensk, that has been converted to a launch site for the **Strela** and **Start-1** space launch vehicles. Further developments may eventually enable the launch of larger vehicles, including the **Angara** and **Proton**. Because of its location at a lower latitude than **Plesetsk Cosmodrome**, the only other spaceport on Russian soil, Svobodny is capable of launching vehicles into orbit with a payload 20% to 25% higher than this other base.

SWAS (Submillimeter Wave Astronomy Satellite)

A NASA satellite equipped with a 0.6-m telescope, for making observations in the 490 to 550 GHz submillimeter range, and an acousto-optical spectrometer. SWAS was used to study the cooling of molecular cloud cores–the sites of star formation in the Galaxy–by measuring spectral lines of molecular oxygen and water. It was the fourth **SMEX** (Small Explorer) spacecraft.

Launch
    Date: December 5, 1988
    Vehicle: Pegasus XL
    Site: Vandenberg Air Force Base
Orbit: 637 × 651 km × 69.9°
Mass: 288 kg

sweat cooling

A method of controlling the excessive heating of a reentering body. Surfaces subjected to excessive heating are made of porous material, through which liquid of high-heat capacity is forced. The evaporation of this coolant completes the sweat-cooling process.

Swift Gamma-Ray Burst Explorer

A NASA **MIDEX** (Medium-class Explorer) mission designed to detect and study the position, brightness, and physical properties of gamma-ray bursts—the most powerful energy blasts in the universe. Because the bursts are fleeting and unpredictable, Swift has been designed to detect and point, to collect images and measurements, and to send data back to Earth all within about a minute. During its three-year mission, scheduled to begin in September 2003, Swift is expected to record more than 1,000 gamma-ray bursts.

Swigert, John Leonard (Jack), Jr. (1931–1982)

An American astronaut who served as Command Module pilot on **Apollo** 13. Swigert received a B.S. in mechanical engineering from the University of Colorado in 1953 and an M.S. in aerospace science from Rensselaer Polytechnic Institute in 1965. Having served with the Air Force (1953–1956) and as a jet fighter pilot with the Connecticut Air National Guard (1960–1965), he became one of 19 astronauts selected by NASA in 1966. He served as a member of the astronaut support crews for Apollo 7 and Apollo 11, and was assigned to the Apollo 13 backup crew before replacing prime crewman Thomas **Mattingly** as command module pilot 24 hours prior to flight following Mattingly's exposure to German measles. From April 1973 to September 1977, Swigert served as executive director of the Committee on Science and Technology in the House of Representatives. In 1978, he ran unsuccessfully for the U.S. Senate. After several years in business, he ran for Congress and, in November 1982, won the new seat for Colorado's 6th congressional District. Swigert died of complications from cancer on December 27, 1982, a week before he was due to take office. He was the first circumlunar astronaut to die.

Syncom

A series of experimental **communications satellite**s, built by Hughes Aircraft, that demonstrated the feasibility of geosynchronous operation. A nitrogen tank explosion crippled Syncom-1 during its apogee burn, leaving Syncom 2 to become the first successful **geosynchronous satellite**. However, because its orbit was inclined to the equator, it did not remain absolutely fixed over the same spot but instead described a lazy figure-eight path north and south of the equator every day. Ground stations followed its movements in latitude, thus making it available 24 hours a day. The first **geostationary satellite** was Syncom 3, launched to provide live daily TV coverage of the 1964 Tokyo Olympics, which it did successfully. The two functioning Syncoms were eventually handed over to the Department of Defense to provide reliable transpacific communications; Syncom 2 was "walked" along the equator using its control thrusters, until it had joined its sister on the other side of the globe. Syncom was the descendant of **Relay** and **Telstar** and the immediate forerunner of more capable geostationary satellites such as **Intelsat**. (See table, "Syncom Series.")

Launch site: Cape Canaveral
Mass: 39 kg

synergic curve

A curve plotted for the ascent of a rocket or space vehicle that is calculated to maximize the vehicle's fuel economy and velocity.

synthetic aperture radar (SAR)

A high-resolution radar instrument capable of imaging surfaces covered by clouds and haze, and used for

Syncom Series

| Spacecraft | Launch | | Orbit | Mass (kg) |
	Date	Vehicle		
Syncom 1	Feb. 14, 1963	Delta B	Contact lost after orbital injection	39
Syncom 2	Jul. 26, 1963	Delta B	35,891 × 35,891 km × 32.7°	32
Syncom 3	Aug. 19, 1964	Delta D	35,784 × 35,792 km × 0.1°	39

ground-mapping. SAR images consist of a matrix in which lines of constant distance or range intersect with lines of constant Doppler shift.

Syromiatnikov, Vladimir S. (1934–)

The designer of one of the most successful pieces of space hardware built by the Soviet Union—the docking collar used to link two spacecraft together. It was successful in more than 200 dockings of Soviet/Russian missions, adapted for use in the **Apollo-Soyuz Test Project** in 1975, and adapted further for use aboard the International Space Station. Syromiatnikov was educated at Bauman Technical University in Moscow and went to work with RKK Energia of Kaliningrad upon his graduation in 1956.[303]

T

tachyon

A hypothetical particle that travels faster than the **speed of light**. The existence of tachyons is allowed in principle by Einstein's **special theory of relativity**; however, all attempts to detect them to date have been unsuccessful.[95, 241]

Taifun

Second-generation Soviet target and surveillance satellites used for testing air defense and space tracking systems. The Taifun-1 series consisted of nearly 40 spacecraft launched from the mid-1970s onward. Taifun-2 satellites, of which some 30 were launched between 1976 and 1995, differed from their predecessors in the type of equipment carried and also in the fact that they released up to 25 **Romb** subsatellites. All the Taifun spacecraft were placed into low Earth orbits of high inclination using Cosmos-3M launch vehicles from Plesetsk and Kapustin Yar.

Taiyo

A Japanese satellite designed to study how solar ultraviolet and X-rays affect Earth's **thermosphere**. Taiyo ("sun") is also known as Shinsei-3 and SRATS (Solar and Thermospheric Radiation Satellite).

Launch
  Date: February 24, 1975
  Vehicle: M-3C
  Site: Kagoshima
  Orbit: 255 × 3,135 km × 31°
Mass: 86 kg

Taiyuan Satellite Launch Center (TSLC)

A Chinese launch center, also known as Wuzhai, located in Shanxi province (37.5° N, 112.6° E). It is used to launch missions with highly inclined (polar) orbits.

takeoff mass

The mass of a launch vehicle, including all stages, fuel, and payload, at the time of takeoff.

Tanegashima Space Center

A Japanese launch site on the island of Tanegashima (30.2° N, 130.9° E), 980 km southwest of Tokyo. It is the largest such facility in Japan and is used for missions of **NASDA** (National Space Development Agency). Launches from here are normally restricted to two 90-day windows per year, due to safety range procedures set up as a result of pressure from local fishermen.

Tansei

A series of small Japanese satellites, launched from Kagoshima, designed to test the performance of new **ISAS** (Institute of Space and Astronautical Science) launch vehicles. (See table, "Tansei Series.")

Taurus

A four-stage launch vehicle that uses the same Orion solid motor combination as its smaller cousin, the **Pegasus**, stacked on top of a larger Castor 120 solid motor. First flown in 1994, the Taurus stands 27 m tall, weighs 69,000 to 101,000 kg at ignition, and is easily transported and launched. It was designed to extend Orbital Sciences' ability to launch small and Med-Lite satellites. Four variants of the Taurus launch vehicle exist. The smallest, known as the ARPA Taurus, uses a Peace Keeper first

Tansei Series

Spacecraft	Launch Date	Vehicle	Orbit	Mass (kg)
Tansei	Feb. 16, 1971	M-4S	990 × 1,110 km × 30°	63
Tansei-2	Feb. 16, 1974	M-3C	290 × 3,240 km × 31°	56
Tansei-3	Feb. 19, 1977	M-3H	790 × 3,810 km × 31°	129
Tansei-4	Feb. 17, 1980	M-3S	521 × 606 km × 39°	185

stage instead of a Castor 120 motor. The standard Taurus uses a Castor 120 first stage and a slightly larger Orion 50S-G second stage. The Taurus XL uses the Pegasus XL rocket motors and is considered a development stage launch vehicle. The largest Taurus variant, the Taurus XLS, is a study phase vehicle that adds two Castor IVB solid rocket boosters to the Taurus XL to increase payload capacity by 40% over the standard Taurus. For all Taurus configurations, satellite delivery to a geostationary transfer orbit can be achieved with the addition of a Star 37FM perigee kick motor. Five consecutive launch successes for Taurus, from 1994 through 2000, were followed by a failure on September 21, 2001, in which the **OrbView**-4 and **QuikTOMS** satellites were lost.

TDRSS (Tracking and Data Relay Satellite System)
A constellation of geosychronous communications satellites and ground support facilities for use by the Space Shuttle and other low-Earth-orbiting spacecraft. When first launched, the TDRS satellites were the largest, most sophisticated communications satellites ever built. The second vehicle in the series was lost in the *Challenger* **disaster** and later replaced by TDRS-7. TDRS-8 (H), -9 (I), and -J are higher performance replacements for the original satellites. (See table, "TDRS Series.")

Launch site: Cape Canaveral
Maximum diameter: 3.4 m (stowed), 21 × 13 m (solar
  panels and antennas deployed)
Mass (TDRS-8 through -J, on-orbit): 1,781 kg

Teledesic
A communications system designed to provide broadband and Internet access, videoconferencing, and high-quality voice and other digital data services through a constellation of 288 satellites in low Earth obit. The Teledesic network consists of terminals that interface between the satellite network and terrestrial end-users, network gateways, network operations and control systems that perform network management functions, and a space segment that provides the communication links and switching among terminals. Teledesic's space-based network uses fast-packet switching to provide seamless, global coverage. Each satellite is a node in the fast-packet-switch network and communicates through crosslinks to other satellites in the same and adjacent orbital planes. Communications are treated within the network as streams of short, fixed-length packets. Each packet carries the network address of the destination terminal, and each node independently selects the least-delay route to that destination. The Teledesic Network is planned to begin operations in 2003. Teledesic 1, the first satellite in the constellation, was launched on February 26, 1998, by a Pegasus XL from Vandenberg Air Force Base.

telemetry
Data received electronically from a spacecraft during flight. Telemetry informs ground control about the condition of the crew and of various critical parts and functions of the spacecraft.

Telstar
A global network of communications satellites operated by AT&T Skynet, later Loral Skynet. Telstar 1 was the first commercial communications satellite; owned by AT&T and flown by NASA, it relayed the first transatlantic television transmissions between Andover, Maine, and stations in Goonhilly, England, and Pleumeur-Bodou, France. Telstar 8 was scheduled for launch in the third

TDRS Series

| Spacecraft | Launch | | GSO Location |
	Date	Vehicle	
TDRS-1	Apr. 4, 1983	Shuttle STS-6	49° W
TDRS-3	Sep. 29, 1988	Shuttle STS-26	275° W
TDRS-4	Mar. 13, 1989	Shuttle STS-29	41° W
TDRS-5	Aug. 2, 1991	Shuttle STS-43	174° W
TDRS-6	Jan. 13, 1993	Shuttle STS-54	47° W
TDRS-7	Jul. 4, 1995	Shuttle STS-70	71° W
TDRS-8 (H)	Jun. 30, 2000	Atlas IIA	150° W
TDRS-9 (I)	Mar. 8, 2002	Atlas IIA	n/a*

*A problem with the satellite's propellant supply after launch left it unclear, as of mid-2002, whether TRDS-9 could be raised to its intended orbit.

Telstar Telstar 1, the first active-repeater communications satellite. *American Institute of Aeronautics and Astronautics*

quarter of 2002 to provide expanded C-band, Ku-band, and Ka-band coverage for North and South America. (See table, "Telstar Series.")

Tenma

A Japanese X-ray astronomy satellite, launched by **ISAS** (Institute of Space and Astronautical Science). It carried detectors developed at the **Goddard Space Flight Center** with a greater energy resolution (by a factor of two) than proportional counters and performed the first sensitive measurements of the iron spectral region for many cosmic sources. Tenma ("Pegasus") was known before launch as Astro-B. It stopped operating in October 1985.

Launch
    Date: February 20, 1983
    Vehicle: M-3S
    Site: Kagoshima
    Orbit: 497 × 503 km × 32°

Tereshkova, Valentina Vladimirovna (1937–)

A Soviet cosmonaut who became the first woman in space. Tereshkova was born in Maslennikovo, near Yaroslavl in western Russia. Her father was a tractor driver and her mother worked in a textile plant. Tereshkova began school at age 8, but she withdrew to work in the same factory as her mother at age 16. She continued her education through correspondence courses, during which time she parachuted as a hobby. When Tereshkova was selected for the Soviet space program in 1962, she became the first person to be recruited

Telstar Series

| Spacecraft | Launch | | Site | Orbit | Mass (kg) |
	Date	Vehicle			
Telstar 1	Jul. 10, 1962	Delta B	Cape Canaveral	945 × 5,643 km × 45°	77
Telstar 2	May 7, 1963	Delta B	Cape Canaveral	972 × 10,802 km × 43°	79
Telstar 3A	Jul. 28, 1983	Delta 3925	Cape Canaveral	GSO at 76° W	625
Telstar 3C	Aug. 30, 1984	Shuttle STS-41	Cape Canaveral	GSO at 125° W	625
Telstar 3D	Jun. 17, 1985	Shuttle STS-51	Cape Canaveral	GSO at 76° W	630
Telstar 401	Dec. 16, 1993	Atlas IIAS	Cape Canaveral	GSO at 97° W	3,375
Telstar 402	Sep. 9, 1994	Ariane 421	Kourou	Lost in orbit	3,485
Telstar 402R	Sep. 24, 1995	Ariane 421	Kourou	GSO at 89° W	3,410
Telstar 5	May 24, 1997	Proton 2KDM4 CK	Baikonur	GSO at 97° W	3,500
Telstar 6	Feb. 15, 1999	Proton 2KDM3 CK	Baikonur	GSO at 93° W	3,700

without experience as a test pilot, her selection being based instead on her parachuting skills. Tereshkova was assigned to be the pilot of **Vostok** 6 and was given the radio name "Chaika," Russian for "seagull." The Vostok craft lifted off from Baikonur Cosmodrome on June 16, 1963. It remained in space for nearly three days and orbited the Earth 48 times, once every 88 minutes. In a departure from earlier Soviet spaceflights, Tereshkova was allowed to operate the controls manually. The craft reentered Earth's atmosphere on June 19 and Tereshkova parachuted to the ground, in the manner typical of cosmonauts at this time. She landed, bruising her nose on impact, approximately 610 km northeast of Qaraghandy, Kazakhstan. It would be another 19 years before the next woman, Svetlana **Savitskaya**, flew in space. Tereshkova became a member of the Communist party and a representative of the Soviet government. In November 1963 she married fellow cosmonaut Andrian Nikolayev. The following year the couple had their first and only child, Elena—the first child whose parents had both traveled in space.

Valentina Tereshkova   *Joachim Becker*

terminal guidance

Guidance required in the final phase of a spacecraft's rendezvous maneuver.

terminal velocity

The hypothetical maximum speed that a body, under given conditions of weight and thrust, could attain along a specified straight flight path, if diving through an unlimited distance in air of uniform density.

Terra

An Earth observation satellite that carries multispectral imagers, a radiation budget instrument, a detector to measure carbon monoxide and methane pollution, and an instrument to study cloud-top and vegetation properties. Terra is part of NASA's **EOS** (Earth Observing System) and was formerly known as EOS AM-1.

Launch
    Date: December 18, 1999
    Vehicle: Atlas IIAS
    Site: Vandenberg Air Force Base
Orbit: 654 × 684 km × 98.2°
Mass: 4,854 kg

terraforming

The process of altering the environment of a planet to make it more clement and more suitable for human habitation. The possible future terraforming of Mars and Venus has been widely discussed. A major consideration before starting such a project would be its effects on any indigenous life. As a concept, terraforming goes back more than half a century. In 1948, the astrophysicist Fritz Zwicky, in expansive mood, suggested a reconstruction and reconfiguration of the entire universe, starting out by changing the positions of the planets, satellites, and asteroids of the Solar System with respect to the Sun. A more modest scheme to make Venus habitable by injecting colonies of algae to reduce atmospheric CO_2 concentration was proposed in 1961 by Carl **Sagan**. Hermann **Oberth** defined as the ultimate goal of terraforming: "To make available for life every place where life is possible. To make inhabitable all worlds as yet uninhabitable, and all life purposeful." An initial effort in this direction will most likely be directed at Mars.

Terrestrial Planet Finder (TPF)

A mission, currently under study, that would form an important future part of NASA's **Origins Program**. TPF would use an optical interferometer consisting of four 8-m telescopes, with a total surface area of 1,000 square meters, to identify Earth-sized planets around nearby

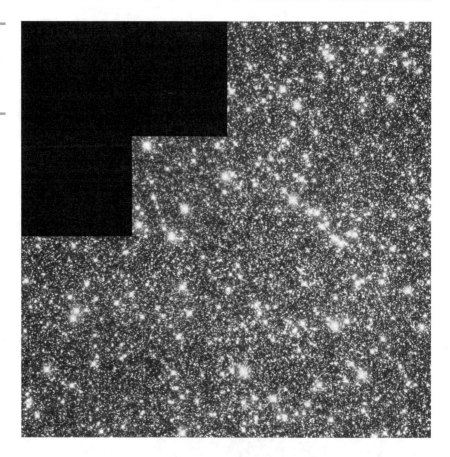

Terrestrial Planet Finder
According to one recent estimate, our galaxy may be home
to at least a billion Earth-like
planets. *NASA/Space Science
Institute*

stars and to analyze their spectra for the signatures of
terrestrial-type life. Its anticipated launch date is 2012.

terrestrial satellite

A satellite in Earth orbit.

TERRIERS (Tomographic Experiment using Radiative Recombinative Ionospheric EUV and Radio Sources)

A spacecraft designed to survey the upper atmosphere by
tomography, measuring ultraviolet light emissions to build
up a global three-dimensional map of electron density in
Earth's ionosphere. As secondary goals, TERRIERS was to
have examined related upper-atmospheric phenomena
and tested the utility of long-term solar irradiance measurements. It carried five imaging spectrometers, of which
four could operate in the night portion of the orbit; two
photometers; and a Gas Ionization Solar Spectral Monitor
(GISSMO). Riding piggyback at the base of the spacecraft
bus was a small payload, built by Cleveland Heights High

School. TERRIERS formed part of NASA's **STEDI** (Student Explorer Demonstration Initiative), a precursor program to the UnEX (University-class Explorer) series, and
was to be operated by the space physics group at Boston
University for ionosphere studies. However, although the
spacecraft was placed in the correct orbit, it failed to orient
its solar panel toward the Sun, and it ran out of battery
power by May 20, 1999, two days after its launch. All
attempts to revive the mission failed.

Launch
    Date: May 18, 1999
    Vehicle: Pegasus XL
    Site: Vandenberg Air Force Base
Mass: 120 kg
Orbit: 537 × 552 km × 97.8°

tethered satellite

See **space tether**.

Thagard, Norman E. (1943–)

An American astronaut and a veteran of five spaceflights (including the first by an American on Soyuz) who, in 1994–1995, set a new U.S. space endurance record of 115 days during his stay aboard **Mir** (the previous record of 84 days having been held by the crew of **Skylab** 4). Thagard received a B.S. (1965) and an M.S. (1966) in engineering science from Florida State University and a doctor of medicine degree from the University of Texas Southwestern Medical School in 1977 and was on active duty with the Marine Corps Reserve from 1966 to 1971, rising to the rank of captain. Selected as an astronaut candidate by NASA in 1978, he served as a mission specialist on Shuttle flights STS-7 (1983), STS-51B (1985), and STS-30 (1989), the payload commander on STS-42 (1992), and the cosmonaut/researcher on the Russian Mir 18 mission (1995).

THEMIS (Time History of Events and Macroscale Interaction during Substorms)

A proposed mission to study the onset of magnetic storms within the tail of Earth's **magnetosphere**. It would fly five microsatellite probes through different regions of the magnetosphere and observe the origin and evolution of storms. Led by Vassilis Angelopoulos of the University of California, Berkeley, THEMIS is one of four **MIDEX** (Medium-class Explorer) missions selected by NASA in April 2002 for further development, two of which will be selected for launch in 2007 and 2008.

thermal barrier

The speed at which frictional heat, generated by the rapid passage of an object through the atmosphere, exceeds endurance compatible with the function of the object.

thermal load

Stresses imposed upon a spacecraft or launch vehicle due to expansion or contraction (or both) of certain structural elements when exposed to a wide range of temperatures.

thermionic

Operating by means of electrically charged particles emitted by an incandescent material.

thermodynamics

The study of the relationship between heat and mechanical energy.

thermosphere

An upper part of Earth's atmosphere that includes the **ionosphere** and extends from an altitude of 85 km (above the mesopause), where the temperature is about −33°C, to the lower level of the **exosphere** at 500 km, where the temperature is about 1,500°C. The thermosphere is known to be very active with waves and vertical tides of thin air far above the highest clouds and storms of charged particles. However, it is still poorly understood and is difficult to explore. Weather balloons and research aircraft cannot reach it. Sounding rockets do travel through the upper atmosphere—however, they can at best take short snapshots of a specific region, and they do not provide global mapping. Furthermore, the Space Shuttle orbits in a region well above the lower thermosphere and passes through it only briefly during reentry. A promising alternative approach is to tether a probe from a larger orbiting platform and then lower the probe into the region of the thermosphere (see **space tether**). The feasibility of this technique has already been demonstrated by NASA's **SEDS** experiments.

Thiel, Walter (1910–1943)

A German engineer who directed the development of the V-2 (A-4) rocket motors (see **"V" weapons**). His designs resulted in a leap from engines that developed a few thousand kilograms of thrust to engines that were more than ten times as powerful. Thiel and his family were killed during the Allied attack on Peenemünde when their shelter took a direct hit from a falling bomb.

Thor

An intermediate-range ballistic missile (IRBM) that was adapted for use as a space launch vehicle and is the direct ancestor of the **Delta** rocket family.

Development of the Thor was authorized in November 1955 to give the U.S. Air Force an independent IRBM capability. At the time, the **Army Ballistic Missile Agency** (ABMA) was already developing the **Redstone** medium-range ballistic missile and the **Jupiter** IRBM. Since the Air Force was now free to compete with the Army in creating a similar weapon, an intense rivalry broke out between the two services. The Air Force requested proposals for a missile capable of carrying a nuclear warhead at least 2,400 km—the distance from England to Moscow. The Douglas Aircraft Company came up with a design that used the warhead and guidance system already being developed for the **Atlas** and the engine from the **Navaho**. The resulting missile could fit inside a Douglas C-124 Globemaster II for easy transportation to a launch site. The first Thor was ready to fly in August 1956, and the first operational Thors were deployed in England by the end of 1958.

Two-thirds of the Thor's body held the liquid propellants—liquid oxygen and RP-1—for the missile's single-stage engine, whose thrust capability was similar to that of both the Jupiter IRBM engine and each individual Atlas ICBM booster stage engine. Two vernier engines, similar to those on the Atlas, were attached at opposite sides of the base of the Thor and burned the same fuel as the main engine. Advances in ICBM technology led to the Thor's withdrawal from military service by 1965. But the Thor did not die: it remained critical to the evolution of the American space program, serving as the core booster for many satellite-carrying offspring, most notably the Delta.[228] (See table, "Data for Various Thor-Based Launch Vehicles," on page 435.)

Thor-Able

A three-stage rocket that used the Thor IRBM as a core booster. Introduced in 1958 and originally designed to support research of reentry vehicles, the Thor-Able was later modified to launch small satellites—the first Thor-based variant to do so and the first of what would become the Delta family of space launch vehicles. Its second and third stages were the same as those on the **Vanguard** rocket. The first three Thor-Able flights not only tested ablative materials short-listed for use on Atlas missile nosecones but also enabled pioneering biological research. Each of the nosecones in this series carried a live mouse, the monitoring of which showed that small animals, at least, could survive several minutes of both

Thor A Thrust Augmented Thor (TAT) Agena D stands on the launch pad. The TAT consisted of a central Thor with four strap-on boosters, which fell away in flight. *Douglas Aircraft/Boeing*

weightlessness and high *g*-forces on journeys that reached altitudes of 960 to 1,600 km. Unfortunately, none of these rodent astronauts was recovered. The first of the trio, named "Mouse-In-Able-1" (MIA-1) and nicknamed "Minnie Mouse," took off on April 23, 1958, and was lost when the Thor first stage exploded. Both the second and third mice passengers, called "Mouse-In-Able-2" (MIA-2) and "Wickie Mouse" (after Cape Canaveral journalist Mercer "Wickie" Livermore), launched on July 9, 1958, and on July 23, 1958, survived their flights, but recovery crews were not able to locate the nosecones. The 9,600-km range achieved on these two flights led to Douglas Aircraft proposing an ICBM version of the Thor called Thor-Intercontinental, or "Thoric." This proposal was rejected by the Air Force on the grounds that the Atlas and Titan ICBMs, with similar capabilities, were already well into development. However, Thor-Able research rockets were adapted instead to launch satellites, with a payload capacity of 140 kg into low Earth orbit (LEO).

Thor-Able I, II, III, and IV

Variants of the original Thor-Able used for a variety of different missions. The Thor-Able I had an added fourth stage, designed specifically to send early **Pioneer** spacecraft (unsuccessfully, as it turned out) toward the Moon. The solid-propellant fourth-stage motor remained attached to the payload. The Thor-Able II was used not for launching satellites but only for high-altitude reentry vehicle tests. Versions III and IV flew only one mission each, carrying the **Explorer** 6 and Pioneer 5 probes, respectively.

Thor-Able Star

A two-stage vehicle introduced in 1960 to launch military satellites. By using improved Thor-Able first and second stages, and by eliminating the weight of upper stages, Thor-Able Star was able to carry up to 450 kg into low Earth orbit. Moreover, the upgraded second-stage engine could be restarted to augment and adjust the orbit of the payload. Used until 1965, Thor-Able Star launched numerous **Transit** and **Solrad** satellites for the U.S. Navy.

Thor-Agena A, B, and D

Two-stage vehicles that used the Lockheed-built **Agena** A, B, or D as an upper stage. From 1959 to 1972, they launched numerous **Corona** spy satellites, along with others including **Echo**, **Nimbus**, **Alouette**, and **OGO**.

Thor-Delta

See **Delta**.

Thousand Astronomical Unit Probe (TAU)

A JPL (Jet Propulsion Laboratory) design for an **interstellar precursor mission**. TAU would use a **XIPS** (xenon-ion propulsion system), powered by a 150-kilowatt

Data for Various Thor-Based Launch Vehicles

	Thor ICBM	Thor-Able	Thor-Able Star	Thor-Agena D
Length (m)	22.0	26.9	29.0	31.0
Diameter (m)	2.4	2.4	2.4	2.4
Mass (kg)	49,300	51,600	53,000	56,500
Thrust (N)				
Stage 1	670,000	670,000	761,000	759,000
Stage 2	–	33,700	35,100	71,200
Stage 3	–	12,300	–	–
Propellants				
Stage 1	RP-1/LOX	RP-1/LOX	RP-1/LOX	RP-1/LOX
Stage 2	–	IRFNA/UDMH	IRFNA/UDMH	IRFNA/UDMH
Stage 3	–	solid	–	–
Payload (kg)				
LEO	–	140	450	1,200

nuclear reactor, to reach speeds of about 95 km/s (20 astronomical units per year), enabling it to travel 1,000 AU (0.016 light-years) within a 50-year mission time. Its primary science objective would be to measure directly the distance to stars throughout our galaxy using stellar parallax. Secondary science goals would include particles and fields measurements, a search for the **heliopause**, a search for the Oort Cloud (a postulated ring of ice-rock bodies, which, when perturbed by the passage of nearby stars, may fall into the inner solar system to become comets), tests of gravitational effects based on changes to the spacecraft's trajectory (which could be caused by a tenth planet or other dark companions in the solar system), and tests of relativity. TAU would be equipped with a 10-W laser communications system capable of transmitting 20 kilobits/s from interstellar space.

three-axis stabilization

A type of **stabilization** in which a spacecraft maintains a fixed attitude relative to its orbital track. This is achieved by nudging the spacecraft back and forth within a deadband of allowed attitude error, using small thrusters or reaction wheels. With a three-axis stabilized spacecraft, solar panels can be kept facing the Sun and a directional antenna can be kept pointed at Earth without having to be de-spun. On the other hand, rotation maneuvers may be needed to best utilize fields and particle instruments.

Three-Corner Constellation

A constellation of three nanosatellites due to be launched from the Space Shuttle in 2003. The mission is sponsored by the Air Force Research Laboratory, DARPA (Defense Advanced Research Projects Agency), and NASA, and supported by design teams at Arizona State University, the University of Colorado, and New Mexico State University. Among its goals are to experiment with stereo-imaging of the atmosphere and to test formation-flying techniques. The satellites will each have a mass of about 15 kg and be placed into a 380-km-high orbit inclined at 51°.

throat

In rocket and jet engines, the most constricted section of an exhaust **nozzle**.

throatable

A **nozzle** whose size and profile can be varied. A throatable nozzle can be especially useful in a solid-fuel rocket motor to maintain uniform thrust throughout the burn time of the fuel.

thrust

The forward force generated by a rocket. Thrust is produced by the expulsion of a reaction mass, such as the hot gas products of a chemical reaction.

In an optimum situation (see below), thrust equals the product of the mass expelled from the rocket in unit time (the propellant mass flow rate) and the **exhaust velocity** (the average actual velocity of the exhaust gases). If F is the thrust, m_p the propellant flow rate, and v_e the effective velocity, then

$$F = m_p v_e \qquad (1)$$

At first glance, it might seem that a constant thrust would lead to a constant acceleration, but this not the case. Even when the propellant flow rate and exhaust velocity are constant, so that the thrust is constant, a rocket will accelerate at an increasing rate because the rocket's overall mass decreases as propellant is used up. The total change in velocity of a rocket due to a specific thrust, acting in a straight line, is given by an important formula known as the **rocket equation**. In some situations, as of a rocket rising from Earth's surface, a large thrust acting over a relatively short period is essential. But in other situations, as of a probe on a deep space mission, the key factor is not so much the amount of thrust, which determines only the acceleration, but the final velocity. A high final velocity can be achieved by a propulsion system that produces a low thrust but expels material over long periods at a high velocity—for example, an ion engine.

Equation (1) holds true only when the pressure of the outgoing exhaust exactly equals the ambient (outside) pressure. If this is not the case, then an extra term comes into play, and the thrust is given by:

$$F = m_p v_e + (p_e - p_a) A_e \qquad (2)$$

where p_e is the exhaust pressure, p_a the ambient pressure, and A_e the area of the exit of the rocket **nozzle**. The first term in this equation is called the momentum thrust and the second the pressure thrust.

thrust chamber

The heart of all liquid propellant rocket engines. In its simplest form, the thrust chamber accepts propellant from the **injector**, burns it in the **combustion chamber**, accelerates the gaseous combustion products, and ejects them from the chamber to provide **thrust**.

thrust commit

The time, when all engines of a launch vehicle on the launch pad have been running for a designated period of time (typically about three seconds) and all other parameters are normal, that is the start of the final launch sequence.

thrust decay

The progressive decline of propulsive thrust, over some fraction of a second, after a rocket motor burns out or is cut off.

thrust equalizer

A safety device that prevents motion of a spacecraft if its solid-fuel rocket motor ignites accidentally. The device is usually a vent at the top of the thrust chamber that is left open until launch time. If the fuel ignites accidentally before launch, the gases of combustion will blow out from both the top and bottom of the motor, thus equal-izing thrust on both sides and preventing the spacecraft from launching prematurely.

thrust generator

A device that produces motive power. In an electric propulsion system, for example, it is composed of an electric power source and a device that expels a high velocity flow of the propellant.

thrust misalignment

Thrust directed accidentally in an undesired direction. Thrust misalignment can have serious consequences, especially during the initial stages of a spacecraft's ascent into orbit.

thrust vector control

Controlling the flight of a launch vehicle or spacecraft by controlling the direction of thrust.

thruster

A small rocket used by a spacecraft to control or change its attitude.

thrust-to-Earth-weight ratio

A quantity used to evaluate engine performance, obtained by dividing the thrust developed by the vehicle by the engine dry weight.

tidal force

A force that comes about because of the differences in gravitational pull on an object due to a large mass around which the object is moving. In the case of a space station in Earth orbit, parts of the station that are further away from the Earth are pulled less strongly, so that the **centrifugal force** of the orbit is not quite balanced by gravity, and there is a net *upward* tidal force. Similarly, for parts closer to the Earth, there is a *downward* tidal force. These opposing forces try to stretch the station along a line that passes through Earth's center. One effect is that tidal forces will make any elongated object tend toward an orbit with its long axis pointing to the Earth's center. Either the space station has to be designed to orbit in this way, or it must have an orientation correction system to counter the orientation drift that the tidal forces will produce. Another effect will be on objects within a space station. Tidal forces are one of the reasons it is impossible to have perfectly zero-gravity conditions in orbit. The fact that microgravity always exists has important consequences for some experiments and manufacturing processes in space.

Dangerous tidal effects would be most evident near highly condensed objects, such as black holes. Tidal forces are proportional to d/R^3, where d is the density of the gravitating mass and R is the distance from it. Using this formula, it is possible to calculate that an astronaut would be torn apart, head to toe, if he approached a 6-solar-mass black hole, feet first, closer than about 5,300 km.

Tikhonravov, Mikhail K. (1901–1974)
A chief theoretician of the early Soviet space program. Tikhonravov graduated from the Zhukovsky Air Force Academy in 1925 and subsequently became an expert in the field of aircraft engineering. In the early 1930s, as a member of GIRD, he worked with Sergei **Korolev** on the design of the first Russian liquid propellant rocket. Tikhonravov continued to work on engine design until Korolev's death in 1966 and played a key role in the development of **Sputnik** 3. From 1962 he was a professor at the Moscow Aviation Institute. Shortly before his death in 1974, Tikhonravov was awarded several of the Soviet Union's highest civilian honors, including the Orders of Lenin and Red Banner of Labor, the title "Hero of Socialist Labor," and the Lenin Prize.[288]

Tiling, Reinhold (1890–1933)
A German scientist who, in April 1931, launched four solid-fueled rockets at Osnabruck. One exploded 150 m above the ground, two rose to between 450 and 600 m, and one reached an altitude of 2,000 m and a maximum speed of 1,100 km/hr. Tiling later launched two more solid-fueled rockets, more advanced than the first four, from Wangerooge, one of the East Frisian Islands. Details of these tests are not certain, but one of the rockets is believed to have reached an altitude of about 10,000 m.

Timation
American military **navigation satellites**, launched as part of the **Navstar** program. (See table, "Timation Series.")

Launch site: Vandenberg Air Force Base
Mass: 700 kg

time dilation
See **relativistic effects**.

TIMED (Thermosphere Ionosphere Mesosphere Energetic Dynamics satellite)
The first science mission in NASA's **Solar-Terrestrial Probes Program**; it is managed for NASA by the Johns Hopkins University Applied Physics Laboratory. TIMED is designed to study the influences of the Sun and humans on Earth's **mesosphere** and lower **thermosphere**, at a height of 60 to 180 km—the least explored and understood region of our atmosphere.

Launch
 Date: December 7, 2001
 Vehicle: Delta 7920
 Site: Vandenberg Air Force Base
 Orbit (circular): 625 km × 74.1°

time-division multiplexing (TDM)
An older telecommunications technique in which a spacecraft's onboard computer samples one measurement at a time and transmits it. On Earth, the samples are demultiplexed, that is, assigned back to the measurements that they represent. In order to maintain synchronization between multiplexing and demultiplexing (mux and demux), the spacecraft introduces a known binary number many digits long, called the pseudo-noise code, at the beginning of every round of sampling (telemetry frame), which can be searched for by the ground data system. Once the pseudo-noise code is recognized, it is used as a starting point, and the measurements can be demuxed since the order of muxing is known. TDM has now been replaced by a newer method known as **packetizing**.

TiPS (Tether Physics and Survivability satellite)
An experiment, funded by the National Reconnaissance Office (NRO), that consisted of two end-masses (dubbed

Launch
 Date: May 12, 1996
 Vehicle: Titan IV
 Site: Vandenberg Air Force Base
 Orbit: 1,010 × 1,032 km × 63.4°

Timation Series

Spacecraft	Launch Date	Vehicle	Orbit
Timation 1	May 31, 1967	Thor-Agena D	894 × 900 km × 70.0°
Timation 2	Sep. 30, 1969	LT Thor-Agena D	898 × 925 km × 70.0°

"Ralph" and "Norton" and officially designated USA 123 and USA 124) connected by a 4-km **space tether**. After being jettisoned by their launch vehicle, the two masses (42 kg and 10 kg) moved apart as the tether was unwound from an **SEDS** box carried by the heavier mass. Data transmitted to the ground showed how the connected masses subsequently moved.

Tipu Sultan (c. 1750–1799)

A sultan of Mysore who successfully deployed rockets against the British army at Srirangapatana, India, in 1792 and 1799. Tipu Sultan was the first to use rockets in which the combustion powder was contained within a metal cylinder. The devices, which weighed about 2 kg, including 1 kg of gunpowder propellant, carried a sword blade as a warhead and were launched from bamboo tubes. Their effectiveness prompted the further development of military rockets by William **Congreve**.

TIROS (Television Infrared Observation System)

A long-running series of polar-orbiting meteorological satellites. It began with 10 experimental spacecraft, TIROS 1 to 10, launched between 1960 and 1965. These

TIROS (Television Infrared Observation System) TIROS 8 in simulated orbit. *NASA*

carried low-resolution television and infrared cameras, and were developed by the Goddard Space Flight Center and managed by ESSA (Environmental Science Services Administration). Then followed the **TOS** (TIROS Operational System), consisting of nine satellites with the ESSA designation. This gave way to the **ITOS** (Improved TIROS), beginning with TIROS-M and continuing with five satellites of the same design with the NOAA (National Oceanic and Atmospheric Administration) designation—NOAA having by this time absorbed ESSA and taken over management of the TIROS program. Further upgrades have come in the form of the TIROS-N, introduced in 1978, and the Advanced TIROS-N, introduced in 1984. (See table, "TIROS Series.")

Launch sites
    Cape Canaveral (TIROS 1–10, ESSA 1–2, 9)
    Vandenberg (ESSA 3–8, TIROS-M, TIROS-N,
      NOAA 1–16)

Titan

See article, pages 440–442.

Titan Explorer

A spacecraft that would conduct an in-depth analysis of the icy, organic-rich environment on Saturn's largest moon. Titan Explorer is a candidate mission in NASA's Outer Planet program and is identified in the Office of Space Science Strategic Plan.

Tito, Dennis (1941–)

The world's first space tourist. Millionaire businessman Tito blasted off aboard a Soyuz supply ship on April 28, 2001, in the company of two veteran cosmonauts, for a 10-day stay at the International Space Station (ISS), having paid Russian space chiefs about $20 million (£14 million) for the privilege. The 60-year-old American became the 415th person in space and the first as a private, paying traveler. Tito started his career in the 1960s as a space scientist with JPL (Jet Propulsion Laboratory), helping plan flight trajectories for several early Mars probes, before leaving to set up a finance company, through which he made his fortune. He first looked into a space vacation in 1991, on a trip to Moscow, and booked a berth aboard Mir—only to see it canceled when Russian space officials decided to de-orbit the station. Instead, Tito was offered a ride on a supply mission to the ISS. Tito's former employer, NASA, objected to the trip, citing safety concerns, but begrudgingly agreed to his visit, subject to his signing contracts relieving all national space agencies of responsibility in the event of a tragedy and saying that he would pay for any breakages he

(continued on page 442)

TIROS Series

Series	Spacecraft	Launch Date	Launch Vehicle	Orbit	Mass (kg)
TIROS	TIROS 1	Apr. 1, 1960	Thor-Able	656 × 696 km × 48°	120
	TIROS 2	Nov. 23, 1960	Delta	547 × 610 km × 49°	130
	TIROS 3	Jul. 12, 1961	Delta	723 × 790 km × 48°	129
	TIROS 4	Feb. 8, 1962	Delta	693 × 812 km × 48°	129
	TIROS 5	Jun. 19, 1962	Delta	580 × 880 km × 58°	129
	TIROS 6	Sep. 18, 1962	Delta	631 × 654 km × 58°	127
	TIROS 7	Jun. 19, 1963	Delta B	338 × 349 km × 58°	135
	TIROS 8	Dec. 21, 1963	Delta B	667 × 705 km × 59°	119
	TIROS 9	Jan. 22, 1965	Delta C	701 × 2,564 km × 96°	138
	TIROS 10	Jul. 2, 1965	Delta C	722 × 807 km × 99°	127
TOS	ESSA 1	Feb. 3, 1966	Delta C	684 × 806 km × 98°	138
	ESSA 2	Feb. 28, 1966	Delta E	1,352 × 1,412 km × 101°	132
	ESSA 3	Oct. 2, 1966	Delta E	1,348 × 1,483 km × 101°	145
	ESSA 4	Jan. 26, 1967	Delta E	1,323 × 1,437 km × 102°	132
	ESSA 5	Apr. 20, 1967	Delta E	1,352 × 1,419 km × 102°	145
	ESSA 6	Nov. 10, 1967	Delta E	1,406 × 1,482 km × 102°	132
	ESSA 7	Aug. 16, 1968	Delta N	1,428 × 1,471 km × 101°	145
	ESSA 8	Dec. 15, 1968	Delta N	1,411 × 1,461 km × 102°	132
	ESSA 9	Feb. 26, 1969	Delta E	1,422 × 1,503 km × 101°	145
ITOS	TIROS-M	Jan. 23, 1970	Delta N	1,431 × 1,477 km × 101°	309
	NOAA 1	Dec. 11, 1970	Delta N	1,421 × 1,470 km × 101°	306
	NOAA 2	Oct. 15, 1972	Delta 100	1,446 × 1,453 km × 102°	344
	NOAA 3	Nov. 6, 1973	Delta 100	1,499 × 1,508 km × 102°	345
	NOAA 4	Nov. 15, 1974	Delta 2914	1,442 × 1,457 km × 102°	340
	NOAA 5	Jul. 29, 1976	Delta 2914	1,504 × 1,519 km × 102°	340
TIROS-N	TIROS-N	Oct. 13, 1978	Atlas F	829 × 845 km × 99°	734
	NOAA 6	Jun. 27, 1979	Atlas F	785 × 800 km × 99°	723
	NOAA 7	Jun. 23, 1981	Atlas F	828 × 847 km × 99°	1,405
	NOAA 12*	May 14, 1991	Atlas E	805 × 824 km × 99°	1,416
Advanced TIROS-N	NOAA 8	Mar. 23, 1983	Atlas F	785 × 800 km × 99°	3,775
	NOAA 9	Dec. 12, 1984	Atlas E	833 × 855 km × 99°	1,712
	NOAA 10	Sep. 17, 1986	Atlas E	795 × 816 km × 99°	1,700
	NOAA 11	Sep. 24, 1988	Atlas E	838 × 854 km × 99°	1,712
	NOAA 13	Aug. 9, 1993	Atlas E	845 × 861 km × 99°	1,712
	NOAA 14	Dec. 30, 1994	Atlas E	847 × 861 km × 99°	1,712
	NOAA 15	May 13, 1998	Titan II	807 × 824 km × 99°	1,476
	NOAA 16	Sep. 21, 2000	Titan II	853 × 867 km × 99°	1,476

*N.B., NOAA 12 was launched out of sequence.

Titan

A large intercontinental ballistic missile (ICBM) that, like the **Atlas** and **Thor**, evolved into an important family of space launch vehicles that remain in use today. Its story begins with the U.S. Air Force seeking an ICBM that would surpass the **Atlas** in sophistication and delivery capacity.[228] (See table, "Titan Series," on page 442.)

Titan I

A silo-based missile, active from 1962 to 1966. A development contract for what would become the Titan ICBM went to the Martin Company in October 1955; subsequently, the missile was named after the Greek mythological father of Zeus. Unlike the one-and-a-half stage Atlas, Titan I used two stages, both equipped with Aerojet engines that burned liquid oxygen and RP-1 (kerosene mixture), and could deliver a four-megaton warhead over a distance of 12,900 km. Its all-inertial guidance system incorporated groundbreaking digital computer technology. The first Titan was test-launched on February 6, 1959, with a dummy second stage ballasted with water. In April 1962, the missile was declared operational and remained in active service for the next four years. It was also to have been used for suborbital tests of the X-20 Dyna-Soar; however, in the end, it was never modified for spaceflight—unlike its successor.

Titan II

The largest U.S. missile ever deployed and, in modified form, the launch vehicle for the **Gemini** program. In 1958, the Air Force gave Martin the go-ahead to develop an improved and a far more powerful version of the Titan I that would burn fuels that could remain in the missile's tanks for long periods, enabling the missile to be fired almost immediately to counter any Soviet threat. Aerojet-General first- and second-stage engines based on those of the Titan I were modified to burn **UDMH** (a type of hydrazine) and nitrogen tetroxide, substances that could be stored at room temperature for months. Furthermore, since they were hypergolic (self-igniting on coming into contact), the Titan II didn't need a complex igni-

tion system. Designed to be fired from within its underground silo (unlike its predecessor, which had to be raised to the surface), the Titan II could be readied for firing in under 60 seconds and could carry an 18-megaton warhead over a range of 15,000 km. The first successful launch took place in March 1962, and the missile was declared operational the following year. Yet, the Titan II was destined to serve not only as a weapon. In modified form, as the **Gemini-Titan II**, it became the launch vehicle for the two-man successor to Project Mercury. It was also eventually used to place satellites in orbit. When the Titan missile fleet was deactivated in the mid-1980s, the remaining Titan

Titan A Titan IV carrying the second Milstar satellite on the launch pad at Cape Canaveral. *Lockheed Martin Missiles & Space Co./Russ Underwood*

IIs were modified for unmanned spaceflight as the Titan II-B and II-G.[281]

Titan III

The first Titan specifically designed for launching satellites. It grew out of the Air Force's need for a heavy-lift vehicle, more powerful than the Atlas-Centaur, that could place large military payloads in orbit. This capability came initially from adding a third stage to the first two stages of the Titan II. In the Titan IIIA, which first flew successfully in December 1964, the third stage was the so-called Transtage, equipped with two Aerojet engines that burned aerozine 50 and nitrogen tetroxide and supplied a total thrust of 71,000 N. However, the IIIA was quickly replaced by the Titan IIIB, which used an **Agena** D third stage to launch classified Air Force satellites from the mid-1960s to the early 1970s, and the enormously powerful Titan IIIC, which was effectively a IIIA with two large solid-fueled rockets strapped onto the first stage. These strap-ons more than quadrupled the thrust available from Titan's first stage liquid-fueled engine and enabled payloads of well over a ton to be lofted into geostationary orbit or over 10 tons into low Earth orbit. The IIIC made its first successful test flight in June 1965 and went on to launch numerous members of the **IDCSP, DSCS, Vela**, and other series of military satellites. So effective did it prove that all succeeding members of the Titan family, and the Space Shuttle, have utilized solid rocket boosters, which are jettisoned when spent, to greatly increase their lifting capacity. The Titan IIID, introduced in 1971 and last launched in 1982, was a two-stage version of the IIIC with the Transtage eliminated in order to make room for bulky, low-Earth-orbiting spy satellites. The IIIE, which first flew in 1974, was a IIID fitted with a Centaur D upper stage, used for launching large NASA scientific payloads, including the **Viking, Voyager,** and **Helios** probes.

Titan 34 and Commercial Titan

In 1982, a stretched version of the Titan III was introduced, known as the Titan 34D. This was an evolution of the 34B, developed to carry the **Manned Orbiting Laboratory** (a project that was eventually cancelled) and flown on several satellite-launching missions between 1976 and 1983. The 34D had longer first and second stages than the III series and more powerful solid rocket strap-ons, and it could be fitted with either the Transtage or the new **Inertial Upper Stage** (IUS). The latter was a two-stage booster that could place military payloads of up to five tons in geostationary transfer orbit.

The so-called Commercial Titan, based on the 34D, was introduced in 1989 specifically to meet the needs of civilian clients, such as Intelsat. Capable of carrying either one payload or two separate payloads at the same time, it used upgraded first- and second-stage engines and enhanced solid boosters. Third stages available included the IUS, Centaur G-prime, Transfer Orbit Stage (TOS), Payload Assist Module (PAM), and Expendable Shuttle Compatible Orbit Transfer System (ESCOTS). The Commercial Titan's last flight was to carry the ill-fated **Mars Observer** in 1992.

Titan IV

Billed as "assured access to space" by the Air Force when it pulled out of the Shuttle program following the *Challenger* **disaster** in 1986, the Titan IV, with a payload capacity matching that of the Shuttle, is the most powerful unmanned launch vehicle in the U.S. fleet. It flies with two solid boosters and a two-stage liquid-propellant core, plus a wide-body Centaur upper stage, an Inertial Upper Stage, or no upper stage but one of several possible payload fairings. After 19 Titan IV launches from Cape Canaveral and 7 from Vandenberg Air Force Base, an improved version called the Titan IVB was introduced by its manufacturer, Lockheed Martin, in 1997. The IVB is similar to its predecessor (now named the IVA) but uses upgraded solid boosters to increase payload capacity by 25% and features a number of improvements to electronics, guidance, and vehicle interfaces. The first Titan IV B took off from Cape Canaveral Launch Complex 40 on February 23, 1997, carrying a Defense Department DSP satellite, and marked the first time a Cape-launched military Titan mission was declassified. In a change from previous policy, both the payload and the launch time were made public in advance. As well as various large reconnaissance satellites, the Titan IVB has launched the Cassini spacecraft to Saturn.

Titan Series

Titan Version	Length Max (m)	Diameter Max (m)	First-Stage Thrust (N)	Payload (kg) to LEO
I	31.0	3.1	670,000	–
II	36.0	3.1	1,920,000	3,100
IIIA	42.0	3.1	2,340,000	3,100
IIIB	45.0	3.1	2,340,000	3,300
IIIC	42.0	3.1	14,040,000	13,100
IIID	36.0	4.3	14,040,000	12,300
IIIE	48.0	4.3	14,040,000	15,400
34D	50.0	3.1	14,870,000	14,520
IVA	51.0	4.3	16,660,000	17,740
IVB	62.0	5.1	17,550,000	21,680

Tito, Dennis (1941–)

(continued from page 438)
caused. Tito may be the first to pay for a trip into space, but he is not the first civilian to make the journey. One of NASA's early attempts to put a non-professional in orbit ended in disaster when teacher Christa **McAuliffe** and her fellow crew members were killed in the *Challenger* disaster of 1986. Tito also follows confectionery scientist Helen Sharman, who in 1991 beat thousands to become Britain's first astronaut, and a Japanese journalist and a member of the Saudi royal family, who both went to Mir. See **space passengers**.

Titov, Gherman Stepanovich (1935–2000)

A Soviet cosmonaut who was the first person to spend an entire day in space, to sleep in space, and to experience space sickness. Titov served as backup to Yuri **Gagarin** on the first manned orbital mission, then flew aboard **Vostok** 2 on August 6, 1961, to become the fourth man in space that year and the youngest, at just 25 years of age—a record that still stands. During his 17 orbits of Earth, Titov was studied to discover if there were any effects of prolonged weightlessness on human beings. The nausea and irregular heartbeat he suffered during his flight concerned Soviet space engineers, who thought all space travelers might be similarly afflicted, although it later turned out that space sickness is an individual and temporary affliction. Titov also operated the spacecraft's controls manually, unlike Gagrin, whose capsule was guided automatically from Earth. Born in Verhnee Zhilino, Titov was in the Soviet Air Force when he was picked, on March 7, 1960, as one of the first 20 individuals for cos-

monaut training. After the Vostok 2 mission, he was assigned to a project known as the Spiral space plane, which was eventually canceled. He never flew in space again, but he became a top official in the Soviet military space forces and a became a member of Russian's lower house of parliament. Titov's accomplishments are honored through the naming of a lunar crater after him.[289]

Titov, Vladimir Georgievich (1947–)

A veteran Soviet cosmonaut of five spaceflights, including three **Soyuz** and two Space Shuttle missions. Titov was selected as a cosmonaut in 1976 and served as commander on Soyuz T-8 and Soyuz T-10 in 1983, and on Soyuz TM-4 in 1987. The first two of these missions nearly ended in disaster. On the final approach of Soyuz T-8 to the **Salyut** 7 space station, Titov realized that the capsule was coming in too fast, aborted the rendezvous, and returned to Earth. Then, just 90 seconds before the launch of Soyuz T-10, a valve in the propellant line failed to close, causing a fire to engulf the base of the launch vehicle. The automatic abort sequence failed because the wires involved had burned through and launch controllers manually aborted the mission. The Soyuz descent module was pulled clear by the launch escape system, seconds before the launch vehicle exploded, and, after being subjected to 15 to 17*g*, Titov and his companions landed uninjured about 4 km away. On his third trip into space, Titov stayed aboard **Mir** for 365 days 23 hours, setting a new endurance record and exceeding one year in space for the first time. In 1995, Titov was a mission specialist aboard Shuttle STS-63, which docked with Mir on the first flight of the new joint Russian-American pro-

Gherman Titov *Joachim Becker*

gram. He served again in this capacity in 1997 aboard STS-86, NASA's seventh mission to rendezvous and dock with the Mir.

Toftoy, Holger N. (1903–1967)
A career U.S. Army officer and an expert in ordnance, who was responsible for bringing the German rocket team under the leadership of Wernher von Braun to the United States in 1945 (see **Paperclip, Operation**). The idea for the **Bumper-WAC** was also his. Toftoy became commander of the Redstone Arsenal, Huntsville, in 1954 and worked closely with von Braun's teams in the development of the Redstone and Jupiter missiles. In the aftermath of Sputnik 1 in 1957, he persuaded the Department of Defense to allow the launch of America's first Earth-orbiting satellite aboard the Jupiter missile, with the result that Explorer 1 was placed in orbit on January 31, 1958. He also held a number of other positions in the Army, including head of the Rocket Research Branch of the Chief of Ordnance in Washington, D.C., and commander of the Aberdeen Proving Ground in Maryland. He retired from the Army in 1960 with the rank of major general.

TOMS (Total Ozone Mapping Spectrometer)
A NASA Earth-observing instrument that measures long-term changes in ozone concentrations to verify chemical models of the **stratosphere** and to help predict future climatic trends. The TOMS program is part of NASA's **ESSP** (Earth System Sciences Program). Various versions of TOMS have flown aboard **Nimbus**-7 (1978), a Soviet **Meteor**-3 (1991), TOMS-EP (1996), and **ADEOS**-1 (1996).

Originally intended for launch in 1994, TOMS-EP (Earth Probe) was delayed by failures of the first two Pegasus XL launch vehicles. As a result, TOMS-EP flew simultaneously with the ADEOS TOMS instrument (originally planned as TOMS-EP's successor). To prevent gathering redundant information, TOMS-EP was placed in an orbit lower than originally planned to obtain higher resolution measurements and data that was complementary to that gathered by ADEOS. When TOMS-EP began to show signs of premature aging, NASA ordered **QuikTOMS** as a gap-filler until ozone monitoring could be taken over by the EOS (Earth Observing System) **Aura** satellite in 2003. However, the fifth flight of TOMS ended in failure on September 21, 2001, when the Taurus rocket carrying the 162-kg Quik-TOMS (and also OrbView-4) broke up less than two minutes after liftoff.

TOMS-EP FACTS
Launch
   Date: July 2, 1996
   Vehicle: Pegasus XL
   Site: Vandenberg
Mass: 248 kg (total), 35 kg (TOMS alone)
Orbit: 490 × 510 km × 97.4°

TOPEX-Poseidon
A highly successful joint American-French satellite that, for almost a decade, has provided measurements of global sea levels accurate to within 4 cm. Its results have helped map ocean circulation patterns and improved our understanding of how the oceans interact with the atmosphere and our ability to predict the global climate. NASA/JPL (Jet Propulsion Laboratory) provided the satellite bus and five instruments, and is responsible for spacecraft operations; CNES (the French space agency) furnished two of the spacecraft's instruments and the launch vehicle. Much of the payload was devoted to measuring the exact height of the satellite above the ocean using a laser altimeter and correcting the altimetry data for effects such as the delay of radio pulses because of water vapor in the atmosphere. TOPEX-Poseidon has proven so successful that a follow-on mission, **Jason-1**,

has been launched to extend sea-surface height measurements into the next decade.

Launch
Date: August 10, 1992
Vehicle: Ariane 42P
Site: Kourou
Orbit: 1,330 × 1,342 km × 66.0°
Mass: 2,402 kg

TOS (TIROS Operational System)

The first operational U.S. polar-orbiting weather satellite system. It followed on from the experimental **TIROS** series, was managed by **ESSA** (Environmental Science Services Agency), and consisted of nine satellites launched between 1966 and 1969, and designated ESSA 1 to 9. It was succeeded by **ITOS** (Improved TIROS Operational System).

total impulse

The total **thrust** produced by a rocket engine during its entire **burn time**.

touchdown

The action or moment of landing a space vehicle, manned or unmanned, on the surface of a planet or another object.

Tournesol

A small French satellite that carried five experiments to study ultraviolet solar radiation and the distribution of stellar hydrogen. Tournesol ("sunflower") was also known as D2-A.

Launch
Date: April 15, 1971
Vehicle: Diamant B
Site: Kourou
Mass: 90 kg
Orbit: 456 × 703 km × 46°

TRACE (Transition Region and Coronal Explorer)

NASA's third **SMEX** (Small Explorer) spacecraft.

Launch
Date: April 2, 1998
Vehicle: Pegasus XL
Site: Vandenberg Air Force Base
Mass: 250 kg
Orbit: 602 × 652 km × 97.8°

TRACE carried a 30-cm extreme ultraviolet imaging telescope, with a field of view of 8.5 arcminutes and a resolution of 1 arcsecond, for studies of the Sun. It was placed in a sun-synchronous orbit to allow it to make continuous solar observations.

tracking

The process of following the movement of a satellite or rocket by radar or by homing in on signals transmitted by the spacecraft.

trajectory

In general, the path traced by any object, such as a rocket, that is moving as a result of externally applied forces.

transducer

A device that converts energy from one form to another for the purpose of the detection and measurement of information. Transducers are often used as sensors.

transfer

Any maneuver that changes a spacecraft's orbit.

transfer ellipse

The path followed by a body that is moving from one elliptical orbit to another. See **Hohmann orbit**.

Transit

The first and longest-running series of **navigation satellites**. Transits enabled nuclear submarines and surface vessels of the U.S. Navy to fix their position at sea, to within 150 m in the early days of the system and to within 25 m later. The first successful Transit, Transit 1B, was launched on April 13, 1960, to demonstrate the feasibility of using satellites as navigational aids. Four years later, the Navy put its first constellation of spin-stabilized Transits into operational service. Later Transits used gravity-gradient stabilization and were also known as Navy Navigation Satellites (NNS, or Navsat). Although Transit stopped being used for navigation on December 31, 1996—its role superseded by Navstar-**GPS** (Global Positioning System)—the satellites continued transmitting and became the Navy Ionospheric Monitoring System (NIMS).

A Transit receiver used the known characteristics of a satellite's orbit and measurements of the Doppler shift of the satellite's radio signal to establish an accurate position on Earth. An operational system consisted of six satellites (three in service plus three on-orbit spares) in 1,100-km polar orbits, three ground control stations, and receivers. The constellation eventually consisted of two types of spacecraft, the 50-kg Oscar, with an average operating lifetime of 12 years, and the more advanced, 160-kg Nova, with an average lifetime of nine years. The last Transit satellite launch was in August 1988. Day-to-day operations, including telemetry, tracking, and control, were conducted by the Naval Space Operations Center at Point Mugu, California, while the Applied Physics Laboratory at Johns Hopkins University devised and designed all aspects of the satellites.

translation

The motion of a spacecraft along its principal axis.

translational thrust

The **thrust** needed to propel a missile or space vehicle from one given position to another.

translunar

A term commonly used in referring to the phase of flight from Earth orbit to lunar orbit. Most reference books, however, describe translunar as referring to space *outside* the Moon's orbit around the Earth, while **cislunar** refers to space between the Earth and the Moon's orbit.

transonic

Flight in the range between the onset of compressibility effects (a Mach number of 0.7) and the establishment of fully **supersonic** flight conditions (a Mach number of 1.4).

transponder

A radio or radar system that is triggered by a received signal. Transponders are important components of communications satellites and consist of a receiver, a frequency converter, and a transmitter package. They have a typical output of 5 to 10 watts and operate over a 36 to 72 MHz bandwidth in the L-, C-, Ku-, or Ka-band, except for mobile satellite communications. Communications satellites typically have 12 to 24 onboard transponders, although the Intelsat 904, at the extreme end, has 98 (76 C-band and 22 Ku-band).

transporter/launcher

A transportable launcher that supports an integral umbilical tower and an erect space vehicle. It usually consists of the transporter (crawler) unit and the launcher (platform) vehicle.

transverse acceleration

The inertial force produced by an acceleration acting across the body, perpendicular to the long axis of the body, as in a chest-to-back direction.

trapped propellant

In a liquid-fuel rocket engine, the amount of fuel left in the tanks that cannot be used because of the suction limitations of the pumping systems.

Triana

An Earth observation satellite that is part of NASA's **Earth Probe** program. After reaching a **halo orbit** at the first **Lagrangian point** in the Earth-Sun system, Triana will transmit full-color images of the entire sunlit side of Earth once every 15 minutes. These images, which will be distributed continuously over the Internet, will highlight global vegetation structure, cloud and ozone cover, and atmospheric aerosol thickness, and provide a better understanding of how solar radiation affects Earth's climate. Named after Rodrigo de Triana, the first person to see the New World from Columbus's ship, Triana is being developed at the Scripps Institution of Oceanography, San Diego. No date has yet been set for the spacecraft's launch.

TRMM (Tropical Rainfall Measuring Mission)

A joint **NASDA** (Japan's National Space Development Agency) and NASA Goddard Space Flight Center mission dedicated to measuring tropical and subtropical rainfall, an important but poorly understood factor affecting global climate. TRMM, one of the first spacecraft in NASA's **EOS** (Earth Observing System), returns long-term data on rainfall and energy budget measurements, which will be used to better understand global climate changes and their mechanisms. The large spatial and temporal variations in tropical rain make it difficult to measure from Earth's surface, and TRMM provides measurement accuracies possible only from an orbiting platform. The satellite and four instruments are provided by the United States, while Japan has supplied one instrument and launch services. The scientific payload includes: Clouds and the Earth's Radiant Energy System (CERES), a passive broadband scanning radiometer with three spectral bands (visible through infared) that measures Earth's radiation budget and atmospheric radiation from the top of the atmosphere to the planet's surface; Lightning Imaging Sensor (LIS), an optical telescope and filter imaging system that investigates the distribution and variability of both intracloud and cloud-to-ground lightning; Precipitation Radar (PR), the first spaceborne

Launch
    Date: November 27, 1997
    Vehicle: H-2
    Site: Tanegashima
Orbit: 344×347 km $\times 35.0°$

TRMM (Tropical Rainfall Measuring Mission) The TRMM satellite in preparation for launch. *NASDA*

rain radar; TRMM Microwave Imager (TMI), a microwave radiometer that provides data related to rainfall rates over the oceans; and Visible Infrared Scanner (VIRS), a passive cross-track scanning radiometer that measures scene radiance in five spectral bands (visible through infrared).

troposphere

The lower layer of Earth's atmosphere, extending from the surface up to about 18,000 m at the equator and 9,000 m at the poles.

Truax, Robert C.

An American rocket engineer and a longtime advocate of small, low-cost launch vehicles. Truax began experimenting with rockets as early as 1932, then became involved with rocket programs as a Naval officer during World War II. He earned a B.S. in mechanical engineering from the Naval Academy (1939) and later a B.S. in aeronautical engineering from the Naval Postgraduate School and an

M.S. in nuclear engineering from Iowa State College. After the war, he worked on a number of ballistic missile and space launch vehicle programs including the Thor, Viking, and Polaris. In 1959, he retired from the Navy as a captain and joined Aerojet, where he headed the Advanced Development Division until leaving in 1966 to form his own company, Truax Engineering.

True of Date (TOD)

The most accurate **coordinate system** used to define a body's position relative to the center of the Earth. This coordinate system takes account of Earth's rotation, **UT** corrections, and small irregularities due to precession, nutation, and polar wander.

Truly, Richard H. (1937–)

An American astronaut and a senior NASA manager. Truly earned a B.S. in aeronautical engineering from Georgia Institute of Technology in 1959 and, that same year, received his commission in the U.S. Navy. Following flight school, he toured aboard U.S.S. *Intrepid* and *Enterprise*. From 1963 to 1965, he was a student and then an instructor at the Air Force Aerospace Research Pilot School, Edwards Air Force Base. In 1965, Truly became one of the first military astronauts selected to the Air Force's **Manned Orbiting Laboratory** program, transferring to NASA as an astronaut in August 1969. He served as CAPCOM (Capsule Communicator) for all three **Skylab** missions in 1973 and the **Apollo-Soyuz Test Project** in 1975, as pilot for one of the two-astronaut crews that flew the 747/Space Shuttle *Enterprise* approach and landing test flights in 1977, and as backup pilot for STS-1, the Shuttle's first orbital test. His first spaceflight came in November 1981, as pilot of *Columbia* (STS-2), and his second flight in August 1983 as commander of *Challenger* (STS-8), the first night launch and landing in the Shuttle program. Truly then temporarily left NASA to serve as the first commander of the Naval Space Command, Virginia, which was established October 1, 1983, but he came back as NASA's Associate Administrator for Space Flight on February 20, 1986. In this position, he led the rebuilding of the Shuttle program following the *Challenger* disaster. This was highlighted by NASA's celebrated return to flight on September 29, 1988, when *Discovery* lifted off from Kennedy Space Center on the first Shuttle mission in almost three years. After retiring from NASA, Admiral Truly returned to his alma mater as director of the Georgia Tech Research Institute.

Trumpet

Fourth-generation U.S. Air Force signals intelligence satellites equipped with a large deployable mesh antenna; they operate from **Molniya-type orbits** and are designed to monitor Soviet communications and missile tests. The first were launched in the mid-1990s.

Tsander, Fridrikh Arturovitch (1887–1933)

A gifted rocket scientist who designed the first Soviet liquid-propellant rocket, the GIRD-X, which reached a height of about 75 m when first flown in 1933, the year that Tsander died from typhoid fever. As a high school student in Riga, Latvia, Tsander had been exposed to the ideas of Konstantin **Tsiolkovsky** and became fanatical about spaceflight, especially about traveling to Mars. Apparently he even interested the Soviet leader Lenin in the subject at a meeting of inventors in Moscow in 1920. During a speech delivered at the Great Physics Auditorium at the Institute of Moscow on October 4, 1924, Tsander was asked why he wanted to go to Mars. He replied: "Because it has an atmosphere and ability to support life. Mars is also considered a red star and this is the emblem of our great Soviet Army." In 1931, Tsander became head of **GIRD** (the Moscow Group for the Study of Rocket Propulsion), and in 1932 he published *Problems of Flight by Means of Reactive Devices*.[291] Also active in rocket design at this time was Valentin **Glushko**.

Tselina

Russian **ELINT** (electronic intelligence) satellites, the first of which was launched on June 26, 1967; "tselina" means "untouched soil." There were two basic types. Tselina I was used mostly for tracking NATO shipping. Similar to the American **SB-WASS** system, it operated in a constellation that detected ELINT data and then compared readings from different locations over time to pinpoint the position of the source. It then transmitted weapons targeting data, which were relayed by communications satellites either to ground stations or directly to Russian ships. Tselina II was a more general-purpose system, similar to **Mercury-ELINT**.

Tsien, Hsue-Shen (1909–)

A leading Chinese rocket designer. Raised in Hang-zhou, a provincial capital in east China, Tsien was a precocious student who won a scholarship to study engineering at the Massachusetts Institute of Technology and then, in 1939, received a Ph.D. in aeronautics from the California Insti-

tute of Technology. At Caltech, Tsien was a protégé of Theodore **von Kármán** and a member of a group of students, known as the "Suicide Squad," whose rocketry experiments were considered so hazardous that they were banished to desert arroyos. Commissioned as a colonel in the U.S. Army Air Force and granted security clearance despite his Chinese citizenship, Tsien was a founding member of **JPL** (Jet Propulsion Laboratory). After World War II, he applied knowledge gained from the V-2 missile program (see **"V" weapons**) to the design of an intercontinental space plane. Tsien's work on this concept inspired the design of the **Dyna-soar** and, ultimately, that of the **Space Shuttle**. In 1950, at the start of the McCarthy era, Tsien was falsely accused of communist activities and for the next five years subjected to harassment and virtual house arrest before being deported to the People's Republic of China. Subsequently, he became the father of Chinese ICBM technology and of the **Long March** launch vehicle.[46]

Tsikada

The second-generation Soviet navigation satellite system; it arose from a collaboration between the Navy, the Academy of Sciences, and the Ministry of Shipping. The Tsikada system, which was accepted into military services in 1979, provides global navigation for both the Soviet Navy and commercial shipping. Each of the 20 Tsikada satellites, launched between 1976 and 1995, was placed in a roughly circular, 1,000-km-high orbit with an inclination of 83° by a Cosmos 11K65M from Plesetsk.

Tsiolkovsky, Konstantin Eduardovitch

See article, pages 448–449.

TSS (Tethered Satellite System)

A joint project of **ASI** (the Italian space agency) and NASA to deploy from the Space Shuttle a 20-km **space tether** connected to an electrically conductive 1.6-m-diameter satellite. During the first test of the Tethered Satellite System (TSS-1) in 1992, a fault with the reel mechanism allowed only 256 m of the tether to be deployed, although the satellite was recovered. On the second mission, TSS-1R in 1996, 19.6 km of tether was deployed before the tether suddenly broke and the satellite was lost. (See table, "TSS Missions.")

TSS Missions

| Spacecraft | Shuttle Deployment | | Orbit |
	Date	Mission	
TSS-1	Jul. 31, 1992	STS-46	299 × 306 km × 29°
TSS-1R	Feb. 22, 1996	STS-75	320 × 400 km × 29°

Konstantin Eduardovitch Tsiolkovsky
(1857–1935)

A Russian physicist and school teacher, regarded as the founder of modern rocket theory. Born in the small town of Izhevskoye almost exactly 100 years before his country placed the world's first artificial satellite in orbit, Tsiolkovsky developed the mathematics of rocketry and pioneered a number of ideas crucial to space travel, including that of multistage launch vehicles. He later recalled:

> For a long time I thought of the rocket as everybody else did—just as a means of diversion and of petty everyday uses. I do not remember exactly what prompted me to make calculations of its motions. Probably the first seeds of the idea were sown by that great fantastic author Jules **Verne**—he directed my thought along certain channels, then came a desire, and after that, the work of the mind.

At age nine, Tsiolkovsky went deaf following a bout of scarlet fever—an event that prevented him from attending school but led him to become an avid reader. An early interest in flight and model balloons was encouraged by his parents. His mother died when he was 13 and his father was poor, but he taught himself mathematics and went to technical college in Moscow. There he found an enlightened mentor named Nikolai Fyodorov (whose admirers were said to include Tolstoy, Dostoyevsky, and Leonid Pasternak, Boris's father). Fyodorov tutored the young Tsiolkovsky in the library daily for some three years, introducing him to books on mathematics and science and discoursing with him on the philosophical imperative leading humankind toward space exploration.

In 1878, Tsiolkovsky became a math teacher in Kaluga, two hours south of the capital. Although he carried out some experiments with steam engines, pumps, and fans in his home laboratory, his strength lay in theoretical work. "It was calculation that directed my thought and my imagination," he wrote. In 1898, he derived the basic formula that determines how rockets perform—the **rocket equation**. This formula was first published in 1903, a few months before the Wright brothers' historic manned flight. It

appeared, together with many other of Tsilokovsky's seminal ideas on spaceflight, in an article called "Investigating Space with Rocket Devices,"[293] in the Russian journal *Nauchnoye Obozreniye* (Science Review). Unfortunately, the same issue also ran a political revolutionary piece that led to its confiscation by the Tsarist authorities. Since none of Tsiolkovsky's subsequent writings were widely circulated at the time (he paid for their publication himself out of his meager teacher's wage), it was many years before news of his work spread to the West.

The 1903 article also discussed different combinations of rocket propellants and how they could be used to power a manned spacecraft:

> Visualize . . . an elongated metal chamber . . . designed to protect not only the various physical instruments but also a human pilot. . . . The chamber is partly occupied by a large store of substances which, on being mixed, immediately form an explosive mass. This mixture, on exploding in a controlled and fairly uniform manner at a chosen point, flows in the form of hot gases through tubes with flared ends, shaped like a cornucopia or a trumpet. These tubes are arranged lengthwise along the walls of the chamber. At the narrow end of the tube the explosives are mixed: this is where the dense, burning gases are obtained. After undergoing intensive rarefaction and cooling, the gases explode outward into space at a tremendous relative velocity at the other, flared end of the tube. Clearly, under definite conditions, such a projectile will ascend like a rocket . . .

One of the propellant combinations that Tsiolkovsky favored, used commonly today in launch vehicles, was liquid hydrogen and liquid oxygen because it produces a particularly high **exhaust velocity**. This factor, the rocket equation reveals, helps determine the maximum speed that a spacecraft of given mass can reach. There was the problem of converting hydrogen, especially, into liquid; yet, to begin with, Tsiolkovsky brushed this aside. He did note, however, that: "The hydrogen may be replaced by a liquid or condensed hydrocarbon; for example, acetylene or petroleum."

His rocket equation led him to another important realization:

> If a single-stage rocket is to attain cosmic velocity it must carry an immense store of fuel. Thus, to reach the first cosmic velocity [his term for the speed needed to enter Earth orbit], 8 km/s, the weight of fuel must exceed that of the whole rocket (payload included) by at least four times. . . . The stage principle, on the other hand, enables us either to obtain high cosmic velocities, or to employ comparatively small amounts of propellant components.

The concept of the multistage rocket had been known to firework makers for at least two centuries. But Tsiolkovsky was the first to analyze it in depth, and he concluded that it was the only feasible way of enabling a spacecraft to escape from Earth's gravity.

Having solved in principle many of the physical problems of spaceflight, Tsiolkovsky turned to the biological difficulties. He proposed immersing astronauts in water to reduce the effects of acceleration at takeoff, and cultivating plants onboard spacecraft to recycle oxygen and provide food. He wrote about spacesuits, zero-*g* showers, utilizing solar energy, and colonizing the solar system.

In Russia, his writings had a powerful influence on Valentin **Glushko**, Sergei **Korolev**, Igor Merkulov, Alexander Polyarny, and others, who would lay the foundations for the Soviet space program. In the West, his work was virtually unknown until the 1930s, by which time practical rocketry had developed independently through the efforts of men like Hermann

Konstantin Tsiolkovsky *NASA History Office*

Oberth in Germany and Robert **Goddard** in the United States.

After the Bolshevik revolution of 1917 and the creation of the Soviet Union, Tsiolkovsky was formally recognized for his accomplishments and, in 1921, received a lifetime pension from the state that allowed him to retire from teaching and devote himself fully to his studies. Appropriately, for a man who remained largely unrecognized during his life, his greatest monument is normally invisible to human eyes—the giant crater Tsiolkovsky, on the far side of the Moon.[292]

Tsyklon

A family of Russian launch vehicles derived from the **Yangel**-designed R-36 (Scarp) intercontinental ballistic missile and first introduced in 1966 (see **"R" series of Russian missiles**). The name Tsyklon ("cyclone") is relatively recent, appearing for the first time in 1986 when, in the wake of Gorbachev's Perestroika initiative, the Soviets began promoting their launchers for commercial use in the West. "Tsyklon" is the spelling widely accepted in Western technical literature, although the name that used to be painted on the side of the rocket is "Ciklon." (See table, "The Tsyklon Family.")

Yangel had proposed basing the first Tsyklon on the R-16, but this vehicle was never developed. Instead the Tsyklon-2 and -3 both evolved from the larger R-36. The Soviet government approved development of the Tsyklon-2 to orbit two types of large military satellites—the IS (Istrebitel Sputnik) **ASAT** (antisatellite) and the US (Upravlenniye Sputnik) naval intelligence satellites. This involved minimal changes to the R-36 because both the IS and US payloads had their own engines for insertion into final orbit. The first flight of the Tsyklon-2 was on October 27, 1967, from Baikonur. In the West, the rocket became known as F-1-r **FOBS** (Fractional Orbit Bombardment System).

Its eventual successor, the Tsyklon-3, was a medium-lift

The Tsyklon Family

Name				
Recent	Manufacturer Index	Western	Base Rocket	Payload
Tsyklon-1	11K64	—	R-16	—
Tsyklon-2A	11K67	SL-11, F-1-m	R-36	IS ASAT
Tsyklon-2	11K69	SL-11, F-1-r	R-36	US-A, -P, -PM
Tsyklon-3	11K68	SL-14, F-2	R-36	Meteor, Okean, Tselina-D

three-stage vehicle designed for multiple payload launches of **Strela**-class satellites and single payload launches of the **Tselina**-D ELINT (electronic intelligence) and Meteor weather satellites. By using an S5M third stage, the new rocket would have a low-Earth-orbit capacity 4 tons greater than that of the Tsyklon-2, combined with improved orbital injection accuracy. Later, the Tsyklon 3 was also used to launch AUOS scientific and **Okean**-O radar satellites.

With the disintegration of the Soviet Union in the early 1990s, the Ukrainian manufacturer of the Tsyklon ended up in a different country—an unacceptable situation for a rocket intended for launching national security payloads. Consequently, existing stocks of the vehicle were used, but no new ones were built. (See table, "Comparison of Tsyklon-2 and -3")

Tucker, George (1775–1861)

A science fiction writer (pseudonym: Joseph Atterley) and chairman of the faculty of the University of Virginia while Edgar Allan **Poe** was a student. In *A Voyage to the Moon with Some Account of the Manners and Customs, Science and Philosophy, of the People of Morosofia, and Other Lunarians*[14] (1827), he describes a trip to the Moon involving a spacecraft coated with the first antigravitic metal in literature, a forerunner of H. G. Wells's Cavorite. See **antigravity**.

tunneling

A phenomenon in **quantum mechanics** in which a jump can take place from one allowed energy state to another across a barrier of intermediate states that is forbidden by energy conservation in classical physics.

Tupolev, Andrei (1888–1972)

A Soviet aircraft designer. Since 1922, the design bureau that he founded and that now bears his name has built more than 100 different military and civilian aircraft, including the workhorse of the Aeroflot fleet, the Tu-154. In one of the more bizarre stories of the Great Purges, Stalin had the entire Tupolev design bureau arrested as "enemies of the people" in 1938. Nothing changed in their work, except that they were now all prisoners in the Gulag and that Tupolev's name was not mentioned in public until his release in 1943. Tupolev played an important role in the official rehabilitation of Sergei **Korolev**.

turbopump

A component of some high-thrust liquid-propellant rocket engines. The turbopump increases the flow of propellant to the **combustion chamber** to increase performance.

Comparison of Tskylon-2 and -3

	Tsyklon 2	Tsyklon 3
Number of stages	2	3
Length	35.5–39.7 m	39.3 m
Diameter	3 m	3 m
Liftoff mass	11.6 tons	12.1 tons
First launch	Oct. 27, 1967	Jun. 24, 1977

TWINS (Two Wide-angle Imaging Neutral-atom Spectrometer)

A mission consisting of two identical, widely separated satellites in high-altitude, high-inclination orbits, which will enable the three-dimensional visualization and resolution of large-scale structures and dynamics within the **magnetosphere**. A **MIDEX** (Medium-class Explorer), TWINS will extend the detailed study of energetic neutral atoms in the magnetosphere that began with the **IMAGE** mission in 2000. TWINS instrumentation is essentially the same as the MENA (Medium-Energy Neutral Atom imager) instrument on IMAGE. However, the capability of TWINS to carry out stereo imaging will greatly extend understanding of magnetospheric structure and processes. The TWINS satellites are scheduled for launch in early 2003 and early 2004 into orbits with apogees of 7.2 Earth-radii and inclinations of 63.4°.

two-body problem

The problem of describing the motion of two objects that are orbiting around each other. It is easily and accurately solved using Newton's law of universal gravitation. The solutions are **conic sections**. Far more complex is the three-body problem, for which precise solutions are available only in special cases, such as that of **Lagrangian points**.

two-stage-to-orbit (TSTO)

A reusable launch vehicle that uses an air-breathing first stage and a separate, or parallel, rocket-propelled second stage. Examples include **Sanger's** space plane and Boeing's Beta. Various two-stage-to-orbit schemes are being considered as a replacement for the Space Shuttle as part of NASA's **Space Launch Initiative**. See **single-stage-to-orbit**.

2001 Mars Odyssey
See **Mars Odyssey, 2001**.

Tyuratam
See **Baikonur Cosmodrome**.

UARS (Upper Atmosphere Research Satellite)

A NASA satellite designed to study the physical and chemical processes taking place in Earth's upper atmosphere, between 15 and 100 km. It was the first satellite dedicated to the science of the stratosphere, with special focus on the causes of ozone layer depletion. UARS complements and amplifies the measurements of total ozone made by **TOMS (Total Ozone Mapping Spectrometer)** onboard Nimbus-7 and Russian Meteor-3M satellites. Despite having a planned operational lifetime of only 18 months, it was still sending back useful data as of mid-2002.

Shuttle deployment
   Date: September 15, 1991
   Mission: STS-48
Orbit: 574 × 582 km × 57.0°
Size: 9.8 × 4.6 m
Mass: 6,795 kg

UDMH ([CH₃]₂NHH₂)

$UDMH\ ([CH_3]_2NHH_2)$

Unsymmetrical dimethylhydrazine; a liquid rocket fuel derived from **hydrazine**. UDMH is often used instead of or in mixtures with hydrazine because it improves stability, especially at higher temperatures.

UFO (UHF Follow-On) satellites

A new generation of U.S. Navy ultra-high-frequency communications satellites in geostationary orbits designed to replace the **FLTSATCOM (Fleet Satellite Communications)** and **Leasat** systems. They are bigger, more powerful, and have more channels than their predecessors but are compatible with the same terminals on the ground. UFO 4 to 9 carry an EHF (extremely high frequency) communications payload that enables the UHF (ultrahigh frequency) network to connect to the **Milstar** system. UFO 8 to 10 also carry a payload called the Global Broadcast Service, which can relay reconnaissance images and other data to a greater number of users at higher rates than any previous defense satellite. With the launch of the tenth satellite in November 1999, the UFO constellation was completed. It is controlled from the Naval Satellite Operations Center (NAVSOC) at Point Mugu, California.

Uhuru

See **SAS**-1 (Small Astronomy Satellite 1).

Ulinski, Franz Abdon (1890–1974)

A technical officer in the aviation corps of the Austro-Hungarian Empire army during World War I who, around 1919, became the first to propose an **ion rocket**. He suggested that solar energy could be used to disintegrate atoms, with the charged fragments expelled to power a spacecraft. A year later, he published his ideas in a journal of aeronautics in Vienna, but they were too far ahead of his time to be taken seriously.

ullage

In rocketry, the extra volume of gas above the liquid propellant held in sealed tanks. This space allows for thermal expansion of the propellant or the buildup of gases released by the propellant, just as ullage in wine bottles and casks helps prevent the vessels from bursting.

ullage rocket

Small rockets used to impart forward thrust to the vehicle or stage to shift the propellant to the rear of the tanks prior to firing the main engines.

Ultra Long-Duration Balloon (ULDB) Project

A project, managed by Goddard Space Flight Center, to fly balloons on missions lasting up to 100 days at the edge of space. The pumpkin-shaped ULDB is the largest single-cell, sealed balloon ever flown, with a fully inflated diameter of 58.5 m and a height of 35 m. It is designed to inflate gradually as it approaches its cruising altitude of 35 km—above 99% of Earth's atmosphere—and then circumnavigate the globe while taking measurements with its scientific payload. NASA sees the ULDB as offering a cheap alternative to certain kinds of low-Earth-orbiting satellite missions. Its first successful flight, on August 25, 2002, took it to a record altitude for a balloon of 49 km (161,000 ft.).

ultraviolet (UV)

That part of the **electromagnetic spectrum** lying between the regions of visible light and **X-rays,** with **wavelength**s of approximately 10 to 400 nanometers (nm). It is divided into near ultraviolet (300 to 400 nm),

Ultra Long-Duration Balloon (ULDB) Project   An artist's rendering of a ULD balloon flying at the edge of space. *NASA*

middle ultraviolet (200 to 300 nm), and far ultraviolet (10 to 200 nm).

ultraviolet astronomy satellites

See, in launch order: **OAO**-2 (Dec. 1968), OAO-3 (Aug. 1972), **ANS** (Aug. 1974), **IUE** (Jan. 1978), **Hubble Space Telescope** (Apr. 1990), **ROSAT** (Jun. 1990), **Astro** (Dec. 1990), **EUVE** (Jun. 1992), **ALEXIS** (Apr. 1993), **ORFEUS** (Sep. 1993), **IEH** (Sep. 1995), **FUSE** (Jun. 1999), **HETE** (Oct. 2000), **GALEX** (Oct. 2002), **CHIPS** (Dec. 2002), **Spectrum-X-Gamma** (2003), **SPEAR** (2003), and **SUVO** (2007+).

Ulysses

A joint NASA/ESA (European Space Agency) probe to study the Sun, the interplanetary medium, and the makeup of the **solar wind**; it provided the first opportunity

Shuttle deployment
    Date: October 6, 1990
    Mission: STS-41
Size: 3.0 × 2.0 m
Mass: 367 kg (total), 55 kg (science payload)
Orbit: 285 × 300 km × 28.4°

for measurements to be made over the Sun's poles. Originally named the International Solar Polar Mission, the spacecraft used a **gravity-assist** from Jupiter to leave the **ecliptic** plane. It completed passes over the Sun's southern pole in November 1994 and northern pole in October 1995, at which point its primary mission was completed. It then began an extended phase of observations.

umbilical connections
The electrical, hydraulic, and pneumatic connections between the ground support equipment and a launch vehicle.

umbilical cord
Often shortened to umbilical. (1) Any of the servicing electrical or fluid lines between the ground and an upright missile or space launch vehicle before launch. (2) The cord that, in early manned missions, attached a space-walking astronaut to his or her spacecraft.

umbilical swing arm
A metal arm that extends horizontally toward the space vehicle from the umbilical. It supports the service lines that link the space vehicle to the ground systems. The swing arm is part of the umbilical tower swing arm system and is supported by the tower and fastened to it by a hinged joint that contains a rotary hydraulic actuator.

umbilical tower
The vertical structure that supports the electrical serving and fluid lines running to a rocket in launch position.

Ume
See **ISS (Ionospheric Sounding Satellite)**.

UNEX (University-class Explorer)
The lowest-cost type of **Explorer** satellite. The UNEX Program is designed to provide frequent flight opportunities for highly focused and relatively inexpensive science missions whose total cost to NASA is limited to $13 million. The first two missions in the program were **CHIPS** (Cosmic Hot Interstellar Plasma Spectrometer) and **IMEX** (Inner Magnetosphere Explorer). The program is managed by the Goddard Space Flight Center.

undershoot boundary
The lower side of a **reentry corridor**, marking the region below which a spacecraft would suffer excessive heating due to atmospheric friction.

United Space Alliance
A Boeing/Lockheed Martin joint venture, with headquarters in Houston, to conduct the Space Flight Operations Contract for NASA. Since 1996, the United Space Alliance has been responsible for the day-to-day operation and management of the Space Shuttle fleet.

United States in space
See **Apollo**; **Explorer**; **Galileo**; **Gemini**; Robert **Goddard**; **guided missiles, postwar development**; **International Space Station**; **JPL**; **Lunar Orbiter**; **Mariner**; **Mercury, Project**; **NACA**; **NASA**; **Pioneer**; **Ranger**; **Skylab**; **Space Race**; **Surveyor**; **Vanguard**; **Voyager**; **White Sands Missile Range**; and **X planes**.

United States launch vehicles
See **Able, Aerobee, Agena, Athena, Atlas, Bumper WAC, Centaur, Conestoga, Corporal, Delta, Juno, Jupiter, Matador, Minotaur, Navaho, Pegasus (launch vehicle), Redstone, Saturn, Scout, Sergeant, Space Shuttle, Taurus, Thor, Titan, Viking (launch vehicle)**, and **WAC Corporal**.

Unity
See **International Space Station**.

unrestricted burn
In solid-fuel rocket motors, a burn achieved by boring a hole in the fuel along the motor's longitudinal axis so that the entire length of the fuel chamber burns simultaneously. This method of burning produces a great amount of thrust for a short time.

Uosat
Small satellites built by the University of Surrey, England, for a variety of purposes including amateur radio and educational experiments. They are launched as secondary payloads when room is available. Uosat 1 was also known as OSCAR 9 and was launched together with the **SME** (Solar Mesosphere Explorer). Uosat 5, which weighed just 50 kg and was launched on August 15, 1991, carried a CCD (charge-coupled device) camera able to relay pictures of the Earth to schools equipped with suitable receivers.

uplink
A signal sent to a spacecraft. See also **downlink**.

upper-air observation
A measurement of atmospheric conditions above the effective range of a surface weather observation. Also called sounding and upper-air sounding.

USA
Designation used by the United States since mid-1984 for military spacecraft. For example, Navstar 9 is also known as USA 1 and **LACE** as USA 51.

USRA (Universities Space Research Association)

A nonprofit corporation under the auspices of the National Academy of Sciences. Institutional membership has grown from 49 colleges and universities when USRA was founded 30 years ago to 90 in 2002. All member institutions have graduate programs in space sciences or aerospace engineering. Besides 77 member institutions in the United States, there are three member institutions in Canada, two in England, and two in Israel. USRA provides a mechanism through which universities can cooperate effectively with one another, with the government, and with other organizations to further space science and technology, and to promote education in these areas. Most of USRA's activities are funded by grants and contracts from NASA.

UT (Universal Time)

The same as GMT (Greenwich Mean Time) in England. Eastern Standard Time (EST) is five hours earlier than Universal Time.

"V" weapons

See article, pages 457–459.

vacuum

In the simplest sense, empty space. However, since a vacuum, either natural or artificial, is never completely empty, the term needs a modifier. Thus scientists speak of a **hard vacuum**, a **quantum vacuum**, and so forth. See also: **Casimir effect, energy, vacuum energy drive,** and **zero-point energy**.

vacuum energy drive

A hypothetical form of propulsion based on the discovery that a vacuum, far from being a pocket of nothingness, actually churns and seethes with unseen activity. This cosmic unrest is caused by quantum fluctuations, tiny ripples of energy–called zero-points–in the fabric of space and time. By interfering with these fluctuations, it may be possible to tap their energy. So far, only relatively crude demonstrations of the power of quantum fluctuations have been carried out. In one set of experiments, carried out by physicists led by the late Nobel prize–winner Hendrik Casimir, two metal plates were clamped and held together by zero-point forces. The crucial point is that the plates were brought together with a force that heated them up very slightly (see **Casimir effect**). While this isn't enough to run a starship, it showed that it is possible to tap the energy field of a vacuum and turn it into power. One proposal for creating a quantum fluctuation space-drive is based on the idea that these tiny energy ripples hold objects back as they fly through space; in other words, they are responsible for inertia. If this effect could be countered, rockets would need much less fuel to overcome their own inertia and would fly through space with far less effort.[58, 99]

Valier, Max (1893–1930)

An Austrian amateur rocketeer and space popularizer who advocated the use of rockets for spaceflight. Educated in engineering in Berlin, he experimented with rockets in the 1920s with the **Verein für Raumschiffahrt** (Society for Space Travel), of which Wernher **von Braun** and Hermann **Oberth** were prominent members. His non-technical book *Der Vorstoss in den Wltenraum* (The Advance into Space) spread Oberth's ideas to a wide audience. Valier was also interested in using rockets to propel ground vehicles and, together with Fritz **von Opel** and Friedrich **Sander**, built the world's first rocket-powered automobile in 1928. Two years later, aged only 31, Valier was killed when a small, steel-cased LOX/alcohol engine, designed for use in the Opel-Rak 7 rocket car, exploded during a test run in his laboratory.[91, 296] See **Valier-Oberth Moon gun**.

Valier-Oberth Moon gun

In the 1920s, members of the **Verein für Raumschiffahrt** amused themselves by redesigning Jules Verne's Moon gun (see *Columbiad*). In 1926, the rocket pioneers Max **Valier** and Hermann **Oberth** designed a gun that would correct Verne's technical mistakes and be capable of firing a projectile to the Moon. The projectile would be made of tungsten steel, practically solid, with a diameter of 1.2 m and a length of 7.2 m. Even using the latest gun propellants, a barrel length of 900 m would be necessary. To eliminate the compression of air in the barrel during acceleration, it was proposed that the barrel itself be evacuated to a near-vacuum, with a metal seal at the top of the barrel. Residual air would provide enough pressure to blast this seal aside before the shell exited the gun. To minimize drag losses in getting through the atmosphere, it was proposed to put the mouth of the gun above most of Earth's atmosphere: it would be drilled into a high mountain of at least 4,900 m altitude.

Van Allen, James Alfred (1914–)

An American space scientist who participated in 24 missions, including some of the early **Explorer**s and **Pioneer**s. His research focused on planetary **magnetosphere**s and the **solar wind**. He began high-altitude rocket research in 1945, initially used captured V-2s (see **"V" weapons**), and is best remembered for his discovery of the radiation belts that were subsequently named after him (see **Van Allen Belts**). Van Allen received a B.S. from Iowa Wesleyan College in 1935 and an M.S. (1936) and a Ph.D. (1939) from the California Institute of Technology. After a spell with the Department of Terrestrial Magnetism at the Carnegie Institution of Washington, where he studied photodisintegration, Van Allen moved in 1942 to the Applied Physics Laboratory at Johns Hopkins University, where he worked to develop a rugged vacuum tube. He also helped to develop proximity fuses for

(continued on page 459)

"V" weapons

A series of Vergeltungswaffen–"vengeance weapons"–designed and built in Nazi Germany for use against allied countries in World War II. The only ones to be significantly deployed were the V-1 and V-2. The latter was extraordinarily influential in the years leading up to the Space Age.

V-1

The infamous German flying bomb, known colloquially as the "doodlebug" or "buzz bomb," and officially as the Fieseler FZG-76; it was the world's first operational **cruise missile**. Carrying a 900-kg warhead, the V-1 was launched from an inclined ramp using a steam catapult and powered by a **pulse-jet** engine, firing at 40 times a second, that propelled it to speeds of over 580 km/hr and a range of over 320 km. Far from being the simple device it appeared to be, the V-1 was equipped with magnetic compass, autopilot, and range-setting and flight controls. Of the 10,492 flying bombs launched, 2,419 hit London. In July 1944, less than three weeks after the first V-1s struck England, American engineers began a remarkable effort to reverse-engineer their own version of the buzz bomb. A working copy of the V-1, built largely from damaged components salvaged from English crash sites, was fired at **Wright Field**. Further tests of the missile took place at **Muroc Army Air Base**, and contracts were let for the production of 2,000 weapons, designated JB-2 (Jet Bomb-2) and nicknamed the Loon. After World War II, the JB-2 played a significant role in the development of other surface-to-surface missile systems–in particular, the **Matador, Snark,** and **Navaho**.[162]

V-2

A 13-ton guided missile that carried a one-ton explosive charge. The liquid-fueled V-2 was the first large military rocket and the immediate ancestor of the launch vehicles used at the dawn of the Space Age. It was guided by an advanced gyroscopic system that sent radio signals to aerodynamic steering tabs on the fins and vanes in the exhaust. Its propellants–alcohol (a mixture of 75% ethanol and 25% water) and liquid oxygen–were delivered to the thrust chamber by two rotary pumps, driven by a steam turbine.

The V-2, or A-4 as it was initially known, was the ultimate development of the **"A" series of German rockets**. The first static tests of its engine took place in 1940, followed on October 3, 1942, by the first successful launch of the missile itself. Later that evening at dinner, Walter **Dornberger** said: "We have invaded space with our rocket and for the first time–mark this well–have used space as a bridge between two points on the Earth. . . . This third day of October 1942 is the first of a new era of transportation–that of space travel."

In the wake of an RAF raid on **Peenemünde** on August 17, 1943, V-2 production was taken away from Dornberger and moved to a massive network of underground tunnels near Nordhausen, an SS facility called the Mittelwerk. Disagreement arose over how to deploy the missile. Some wanted the rocket launched from hardened bunkers along the coast of France; others thought a mobile launching system would be best. In the end, both methods were tried. Massive concrete emplacements were started in France but never completed because of continual Allied bombing attacks. The Peenemünde engineers had developed a mobile transporter and erector for the V-2, known as the Meillerwagon. Along with rail transportation, it became the standard method of transportation and launch.

On Friday evening, September 8, 1944, the first V-2s were fired in earnest, one of them creating a huge explosion and crater in Staveley Road, west London. After the explosion came a double thunderclap caused by the sonic boom catching up with the fallen rocket. Over the next seven months, about 2,500 V-2s were launched, 517 of them striking London and hundreds more exploding in counties around the capital and Allied-held parts of France, Belgium, and Holland. By April 1945, the German rocket batteries were

Length: 14 m
Diameter (max.): 1.65 m
Launch mass: 13,000 kg
Thrust at liftoff: 245,000 N
Burn time: 65 sec
Maximum speed: 5,750 km/hr
Maximum altitude: 96 km
Range: 320 km

"V" weapons A captured V-2 is prepared for launch from White Sands Proving Ground, New Mexico. *U.S. Army/White Sands Missile Range*

in retreat before the advancing Allied armies. When it became obvious that Germany was facing collapse, von Braun agreed to surrender to the American army as part of Operation **Paperclip**. Soon, he and his group, together with hundreds of V-2s and rocket components taken from the Mittelwerk, were on their way to the United States.[81, 88, 162, 246]

V-3

A German super-cannon designed by Saar Roechling during World War II; also known as the Hochdruck-

pumpe (HDP), "Fleissiges Lieschen," and "Tausend Fussler." The 140-m V-3 was capable of sending a 140-kg shell over a 165-km range. Construction of a bunker for the cannons began in September 1943 at Mimoyecques, France. However, the site was damaged by Allied bombing before it could be put into operation, and it was finally occupied by the British at the end of August 1944. Two short-length (45-m long) V-3s were built at Antwerp and Luxembourg in support of the Ardennes offensive in December 1944. These were found to be unreliable and only a few shots were fired, without known effect.

Van Allen, James Alfred (1914–)

(continued from page 456)

weapons used in World War II, especially for torpedoes used by the U.S. Navy. By the fall of 1942, he had been commissioned as an officer in the Navy and sent to the Pacific to field test and complete operational requirements for the proximity fuses. After the war, Van Allen returned to civilian life and began working in high-altitude research, first for the Applied Physics Laboratory and, after 1950, at the University of Iowa. Van Allen's career took an important turn in 1955, when he and several other American scientists developed proposals for the launch of a scientific satellite as part of the research program conducted during the **International Geophysical Year** (IGY) of 1957–1958. After the success of the Soviet Union with Sputnik 1, Van Allen's Explorer spacecraft was approved for launch on a Redstone rocket. It flew on January 31, 1958, and returned enormously important scientific data about the radiation belts circling the Earth. Van Allen became a celebrity because of the success of that mission, and he went on to other important scientific projects in space. In various ways, he was involved in the first four Explorer probes, a number of the early Pioneers as well as Pioneer 10 and 11, several Mariner projects, and the **Orbiting Geophysical Observatory**. Van Allen retired from the University of Iowa in 1985 to become Carver Professor of Physics, Emeritus, after having served as the head of the Department of Physics and Astronomy from 1951.[219, 297]

Van Allen Belts

Two doughnut-shaped belts of high-energy charged particles trapped in Earth's magnetic field; they were discovered in 1958 by James **Van Allen**, based on measurements made by **Explorer 1**. The inner Van Allen Belt lies about 9,400 km (1.5 Earth radii) above the equator and con-

tains **protons** and **electrons** from both the **solar wind** and the Earth's **ionosphere**. The outer belt is about three times farther away and contains mainly electrons from the solar wind. Most manned missions take place in low Earth orbit, well below the inner Van Allen belt, and also avoid a region over the South Atlantic where a local weakness in Earth's magnetic field allows the lower belt to penetrate to a much lower altitude. The Apollo flight paths were chosen so as to only nick the (more energetic) inner belt; no special shielding was considered necessary to cope with the outer belt.

Vandenberg, Hoyt S. (1899–1954)

A career military aviator who served as chief of staff of the U.S. Air Force (1948 to 1953). He was educated at the Military Academy at West Point and entered the Army Air Corps after graduation, becoming a pilot and an air commander. After numerous command positions in World War II, most significantly as commander of Ninth Air Force, which provided fighter support in Europe during the invasion and march to Berlin, he returned to Washington and helped with the formation of the Department of Defense (DoD) in 1947. As Air Force chief of staff, he was a senior official in the DoD during the formative period of rocketry development and the work on intercontinental ballistic missiles.[203]

Vandenberg Air Force Base (VAFB)

The home of the **Western Space and Missile Center**, America's second major spaceport after Cape Canaveral. Vandenberg AFB lies about 240 km northwest of Los Angeles and is operated by the **Air Force Space Command**'s 30th Space Wing. The base was established in the late 1950s as a site from which ballistic missiles could be launched under simulated operational conditions. A **Thor** intermediate range ballistic missile was first test-fired from here in December 1958. Since then, Vandenberg has acquired pads and support facilities for launching other rockets, including versions of the Atlas, Delta, Titan, and Scout. Because of its location, with open water to the south, it is America's premier launch site for polar missions. It was also originally intended to serve as a launch center for polar Space Shuttle missions, and construction of Shuttle launch facilities at VAFB was begun. However, this was permanently halted after the *Challenger disaster* and the decision of the Department of Defense to use expendable launch vehicles for high-priority polar military payloads.

Vanguard
See article, pages 461–462.

VASIMR (Variable Specific Impulse Magnetoplasma Rocket)
A plasma engine that has been under development by a NASA astronaut for the past two decades. Astronaut Franklin Chang-Diaz began work on the rocket technology in 1979, when he was a graduate student at the Massachusetts Institute of Technology. Since late 1993, Chang-Diaz and colleagues have continued work on the engine at the Advanced Space Propulsion Laboratory at the Johnson Space Center. Unlike conventional rocket engines, which ignite a mix of fuel and oxidizer to generate thrust, VASIMR uses a series of magnetic fields to create and accelerate plasma, or high-temperature ionized gas. The process begins when neutral hydrogen gas is injected into the first of three magnetic cells. That cell ionizes the gas, stripping away the sole electron from each hydrogen atom. The gas then moves into the central magnetic cell, where radio waves, generated in a manner similar to a microwave oven, heat the gas to more than 50,000°C, turning it into a high-temperature plasma. The plasma is then injected into the last magnetic cell, a magnetic nozzle, which directs the plasma into an exhaust that provides thrust for the engine. A key advantage of the engine is that its specific impulse–a measure of the velocity of its exhaust–can be varied in flight to change the amount of thrust. The specific impulse of the engine can be turned down to provide additional thrust during key portions of a mission, then turned up during cruise phases to improve efficiency. If VASIMR is successfully developed, it could cut in half the time needed for travel to Mars. By firing continuously, accelerating during the first half of the flight, and then turning to deaccelerate the spacecraft for the second half, VASIMR could send a human spacecraft to Mars in just over three months. In addition, VASIMR would enable such a mission to abort to Earth if problems developed during the early phases, a

capability not available to conventional engines. A scale version of the VASIMR engine could fly in space as early as 2004 as part of the Radiation and Technology Demonstrator spacecraft.

VCL (Vegetation Canopy Lidar)
A spacecraft that will use a radarlike imaging technique called **lidar** (light detection and ranging) to map the height of the world's forests to an accuracy of one meter. VCL will shine five laser beams onto the forest's canopy. By reconstructing the reflected light, the satellite will produce an accurate map of both the heights of the trees and the topography of the underlying terrain. The data will be used to calculate the density and mass of the world's vegetation, enabling scientists to monitor the capacity of forests to absorb carbon dioxide. In addition, the satellite will provide the first accurate estimate of how much carbon is being released into Earth's atmosphere through the burning of tropical rainforests. VCL is a mission of NASA's **ESSP** (Earth System Science Pathfinder) program; a launch date has yet to be decided.

vector
A quantity having both magnitude and direction; examples include **velocity, acceleration**, and **force**. Vectors are added when, for instance, movement takes place in a frame of reference that is itself moving (for example, as when a swimmer tries to cross a flowing river). Vectors are added like arrows, end to end, so that in the case of two vectors, the sum is the vector from the tail of the first to the tip of the second.

vector thrust
A steering method in which one or more **combustion chambers** are mounted on **gimbals** so that the direction of the **thrust** force (thrust vector) may be tilted in relation to the **center of gravity** of the missile to produce turning.

Vega 1 and 2
Two identical Soviet spacecraft, each carrying a Venus lander and a Halley's Comet flyby probe. The spacecraft released balloons into the atmosphere of Venus and landers onto the surface before flying on to a rendezvous with Halley in March 1986. The twin Vegas passed 8,900 km and 8,000 km, respectively, from the comet's nucleus, returning photographs and analyzing ejected gas and dust. (See table, "Vega Missions," on page 463.)

Launch vehicle: Proton
Mass: 4,000 kg

Vanguard

An early U.S. Navy–led project intended to launch America's first Earth satellite. It began in 1955, at a time of intense interservice rivalry, having been selected ahead of two competitors based on the **Redstone** and **Atlas** launch vehicles. The fact that this decision was a mistake is now well documented. Multiple launch failures marred the project, and it was beaten to its goal by both the Soviet Union with **Sputnik 1** and the U.S. Army with **Explorer 1**. However, it eventually succeeded in placing three satellites in orbit. Shortly after the final launch, the Vanguard program was abandoned.

Vanguard 1 was the second U.S. satellite and the third spacecraft overall to achieve Earth orbit. Its two radio transmitters, which also functioned as temperature gauges, continued to transmit for seven years. Vanguard 1 remains the oldest human-made object in space, and is expected to orbit Earth for about 1,000 years. Vanguard 2 returned the first photo from space. (See table, "Vanguard Series.")

Vanguard Vanguard 1. *NASA*

Vanguard (Launch Vehicle)

A small three-stage rocket used in the Vanguard program. Only four of seven test launches (designated TV for "test vehicle") were successful, including TV-4, which placed Vanguard 1 in orbit, and only two of

Vanguard satellite chronology
Launch
    Vehicle: Vanguard
    Site: Cape Canaveral

Vanguard Series

Spacecraft	Launch Date	Orbit	Mass (kg)
Vanguard 1A*	Dec. 6, 1957	—	1.5
Vanguard 1B*	Feb. 5, 1958	—	1.5
Vanguard 1	Mar. 17, 1958	654 × 3,868 km × 34.2°	1.5
Vanguard 2A*	Apr. 29, 1958	—	1.5
Vanguard 2B*	May 28, 1958	—	
Vanguard 2C*	Jun. 26, 1958	—	
Vanguard 2D*	Sep. 26, 1958	—	
Vanguard 2	Feb. 17, 1959	557 × 3,049 km × 32.9°	10
Vanguard 3A*	Apr. 14, 1959	—	10
Vanguard 3B*	Jun. 22, 1959	—	10
Vanguard 3	Sep. 18, 1959	512 × 3,413 km × 33.4°	23

*Launch failures

seven operational launches (called SLV for "space launch vehicle") were successful—a failure rate unsurpassed by any other American launch vehicle. Adapted from the U.S. Navy's **Viking** and **Aerobee** research rockets, Vanguard was designed to fulfill the Eisenhower Administration's goal of placing a small scientific satellite in orbit during the **International Geophysical Year** (IGY). Its development stemmed from a 1955 study to determine which vehicle would be most suitable for launching the first American satellite. Among the proposals rejected were Project Orbiter, a Redstone-based plan put forward by the Army Ballistic Missile Agency (ABMA). At the height of the Cold War, Eisenhower was sensitive about allowing the military to launch the first American satellite. Although the Vanguard program was based on the Navy Viking, it was proposed under civilian management and so was selected ahead of its rivals. At least six Vanguard rockets had been scheduled for test launches prior to what would have been the first American attempt to loft a satellite. Since the IGY ran from July 1, 1957, through December 31, 1958, it was thought there would be plenty of time for the Vanguard to prove itself before attempting to reach orbit. However, technical problems quickly surfaced, and the first test launch was delayed until December 8, 1956. A second test took place on May 1, 1957, but before there could be a third, the Soviet Union shook the world with their launch of Sputnik 1 on October 4, 1957. To counter this unexpected Soviet coup, and that of Sputnik 2 in November 1957, a bold decision

was made to go for a satellite launch aboard the next available Vanguard (TV-3). However, TV-3 exploded on its launch pad before the eyes of the world on December 6, 1957. This public debacle cleared the way for ABMA to attempt, and successfully achieve, the launch of America's first satellite, Explorer 1, on January 31, 1958. Despite the highly publicized Vanguard failure, the program didn't immediately end. A second, but less spectacular, Vanguard failure occurred during a follow-up satellite launch attempt on February 5, 1958. But then came success. The next Vanguard rocket carried the Vanguard 1 satellite into orbit on March 17, 1958. The next four attempts to reach orbit failed, but Vanguard 2 was successfully launched on February 17, 1959. After two more failures, the last launch of a Vanguard rocket ended in triumph, placing Vanguard 3 in orbit on September 18, 1959. Though the program was then quietly wound up, Vanguard technology was successfully applied to other programs. The Vanguard's upper stages formed the basis for upper stage configurations used on Atlas-Able, Thor-Able, and Scout rockets, and even found their way into the Apollo program. A modified Vanguard upper stage served as the second stage of the Atlas-Antares, a rocket used to test various Apollo reentry vehicle designs. (See table, "Vanguard Launch Vehicle Specifications.")

Length: 21.9 m
Diameter: 1.1 m

Vanguard Launch Vehicle Specifications

	First Stage	Second Stage	Third Stage
Manufacturer	GE (Viking)	Aerojet	Altair
Fuel	LOX/kerosene	IWFNA/UDMH	Solid
Thrust (N)	124,000	33,000	13,800

Vega Missions

Spacecraft	Launch Date	Notes
Vega 1	Dec. 15, 1984	Ejected Venus lander Jun. 10, 1985; Halley flyby on Mar. 6, 1986
Vega 2	Dec. 21, 1984	Ejected Venus lander Jun. 14, 1985; Halley flyby on Mar. 9, 1986

Vehicle Assembly Building (VAB)

A large facility where the **Space Shuttle** Orbiter is joined to the Solid Rocket Boosters and the External Tank, and where, in the 1960s and 1970s, **Saturn** V components were assembled. The 52-story VAB occupies a ground area of 3.2 hectares (8 acres) and has an internal volume of 3,624,000 cubic meters. Designed to withstand winds of 200 km/hr, it has a foundation that rests on more than 4,200 steel pilings, each 40 cm in diameter, that goes to a depth of 49 m through bedrock. From the VAB, the Shuttle is moved to its launch pad by a **crawler/transporter**.

Vela

A series of spacecraft designed to monitor worldwide compliance with the 1963 nuclear test ban treaty ("vela" is Spanish for "watchman"). The Vela satellites, carrying X-ray, gamma-ray, and neutron detectors, were launched in pairs into high altitude orbits to detect possible nuclear explosions on Earth and in space, out to the distance of Venus or Mars. Interestingly, the gamma-ray detectors onboard the early Velas picked up the first signs of an important astrophysical phenomenon—gamma-ray bursters—but this information was only declassified about two decades after the military became aware of it. The project was directed by the Advanced Research Projects Agency of the Department of Defense; the U.S. Air Force Space and Missile Systems organization was responsible for the development of the spacecraft. The first three pairs of satellites were so successful, each operating for at least 5 years, that a planned acquisition of a fourth and fifth set of pairs was cancelled. Instead, TRW was awarded a further contract in March 1965, for an Advanced Vela spacecraft series (see **Vela, Advanced**). (See table, "Vela Series.")

Launch
　　Vehicle: Atlas IIIA-Agena D
　　Site: Cape Canaveral
　　Size: 1.4 × 1.4 m

Vela, Advanced

Like its predecessor, **Vela**, the Advanced Vela series of spacecraft was designed to monitor worldwide compliance with the 1963 nuclear test ban treaty; however, it could also detect atmospheric nuclear detonation. Additionally, the Advanced Vela monitored solar activity (providing radiation warnings for manned missions), terrestrial lightning activity, and celestial X-ray and gamma-ray radiation. All six spacecraft operated for more than 10 years—the last pair, Vela 11 and Vela 12, continuing to function well into the 1980s. Their nuclear detection role was taken over by IMEWS in the 1970s. Vela was touted as the longest continuously operating space system in 1985, when the Air Force shut down the last three craft. (See table, "Advanced Vela Series.")

Advanced Vela Series

Spacecraft	Launch Date
Vela 7	Apr. 28, 1967
Vela 8	Apr. 28, 1967
Vela 9	May 23, 1969
Vela 10	May 23, 1969
Vela 11	Apr. 8, 1970
Vela 12	Apr. 8, 1970

Vela, Rodolfo Neri (1952–)

The first Mexican in space. Vela, a teacher and a designer of satellites at the National Institute of Electrical Research in Mexico City, flew as a payload specialist aboard Space Shuttle mission STS-61B in November–December 1985 with responsibility for deploying the Mexican **Morelos** B satellite. As part of the food manifest he requested tortillas.

velocity

A **vector** quantity specifying both the **speed** and the direction of movement of an object in a given frame of reference.

Vela Series

Spacecraft	Launch Date	Orbit	Mass (kg)
Vela 1	Oct. 16, 1963	101,925 × 116,582 km × 37.8°	220
Vela 2	Oct. 16, 1963	101,081 × 116,582 km × 38.7°	220
Vela 3	Jul. 17, 1964	102,500 × 104,101 km × 39.1°	220
Vela 4	Jul. 17, 1964	92,103 × 114,000 km × 40.8°	220
Vela 5	Jul. 20, 1965	106,367 × 115,839 km × 35.2°	235
Vela 6	Jul. 20, 1965	101,715 × 121,281 km × 34.2°	235

Vehicle Assembly Building The Apollo 15 Saturn V leaving the Vehicle Assembly Building. *NASA*

Venera
See article, pages 467–468. See also **Vega 1 and 2.**

VentureStar
See **X-33.**

Venus, unmanned exploration
See **Venera** 1–16, **Zond** 1, **Mariner** 1, 2, 5, and 10, **Pioneer** Venus 1 and 2, and **Magellan.** (See table, "Chronology of Venus Exploration," on page 466.)

Venus Surface Sample Return
A potential mission beyond 2007 identified in NASA's Office of Space Science Strategic Plan. It remains in the early concept definition phase.

Verein für Raumschiffahrt (VfR)
Society for Space Travel. An association of German enthusiasts, formed in 1927, that carried out important development work on **liquid-fueled rockets.** The stimulus for the society was the publication of Hermann **Oberth's** 1923 book, *Die Rakete zu den Planetenräumen* (The Rocket into Interplanetary Space). Impressed by Oberth's mathematically sound theories that space travel was achievable, the founders of VfR set out to build the types of rockets he described. The group grew in size to about 500 members, produced its own journal, *Die Rakete* (The Rocket), and obtained permission to use an abandoned ammunition dump in Reinickendorf, a suburb of Berlin, as test site for its projects. The facility soon became known as the *Raketenflugplatz* (rocket airfield) and served as an early proving ground for several men who would go on to play a key role in the German army's World War II rocket program. One of the brightest young people involved at the Raketenflugplatz was 19-year-old Wernher **von Braun,** later chosen to head the army's rocket development program. By 1932, VfR had fallen on hard times and tried to secure funds from the army for further testing. After a demonstration launch failed to impress attending officers, society members knew their days at the Raketenflugplatz were numbered. Still, the army was impressed by von Braun, and he was invited to write his graduate thesis on rocket combustion at **Kummersdorf.** After Hitler came to power, Nazi Germany banned all rocket experimentation or rocket discussion outside of the German military. The rocket enthusiasts who had populated the Raketenflugplatz had to abandon their work or continue it in the military. Hauptmann Dornberger was now able to successfully recruit former members of the VfR, many of whom joined the army organization at Kummersdorf.

Verne, Jules (1828–1905)
A French novelist and playwright, considered one of the founding fathers of science fiction. He launched three travelers on a lunar journey in *From the Earth to the Moon*[299] (1865) and brought them back safely to Earth in the sequel *Around the Moon*[300] (1870). These tales formed the basis of one of the first science fiction films, *Le Voyage dans la Lune* (1902), produced by Georges Méliès.

vernier engine
A very small rocket engine used for fine adjustments of a spacecraft's velocity and trajectory.

VHO (very high orbit)
A category that includes all orbits with **perigees** at or above **geosynchronous orbit** (GEO) and **apogees** above GEO, yet remain in orbit around the Earth or Earth-Moon system. Such orbits are often highly elliptical, with apogees several hundred thousand kilometers in altitude.

Victory, John F. (1893–1975)
An influential administrator with **NACA** (National Advisory Committee for Aeronautics). Victory began work for the government in 1908 as a messenger for the patent office. After becoming the first employee of NACA in 1915, he became its secretary in 1921 and its executive secretary in 1948, in general charge of its administration. When NASA came into being, he served as a special assistant to T. Keith **Glennan** until his retirement at the end of July 1960. Over the years, he became known as "Mr. Aviation" to his friends, who ranged from Orville Wright to the builders of the fastest jet fighters. Although not an engineer or a technician, he helped NACA achieve working relationships with Congress, where he frequently testified, and with the military services, aerospace industry, and related groups engaged in government-sponsored research and development.

Viking (launch vehicle)
A U.S. Navy–developed rocket that formed the basis of the **Vanguard** launch vehicle. Viking started out as a **sounding rocket** program in 1946 led by Ernst Krause and Milton **Rosen** and drawing initially upon V-2 technology. From 1949 through 1957, 14 Vikings were built to test innovative features in control, structure, and propulsion, and to launch increasingly large instrument payloads. No two Vikings were identical. The Navy used the rockets to probe the region of the upper atmosphere that affects long-range radio communication and also to examine their potential as a tactical ballistic missile. When deciding which of several options, including the Atlas and the Jupiter C, to use for launching its first artificial satellite, the U.S. government chose the Navy Viking-based Vanguard.[248]

Chronology of Venus Exploration

Mission	Launch Date	Notes
Sputnik 7	Feb. 4, 1961	Attempted impact
Venera 1 (Sputnik 8)	Feb. 12, 1961	Flyby; contact lost
Mariner 1	Jul. 22, 1962	Attempted flyby; launch failure
Sputnik 19	Aug. 25, 1962	Attempted flyby
Mariner 2	Aug. 27, 1962	Flyby
Sputnik 20	Sep. 1, 1962	Attempted flyby
Sputnik 21	Sep. 12, 1962	Attempted flyby
Cosmos 21	Nov. 11, 1963	Possible attempted Venera test flight
Venera 1964A	Feb. 19, 1964	Attempted flyby; launch failure
Venera 1964B	Mar. 1, 1964	Attempted flyby; launch failure
Cosmos 27	Mar. 27, 1964	Attempted flyby
Zond 1	Apr. 2, 1964	Flyby; contact lost
Venera 2	Nov. 12, 1965	Flyby; contact lost
Venera 3	Nov. 16, 1965	Lander; contact lost
Cosmos 96	Nov. 23, 1965	Possible attempted lander
Venus 1965A	Nov. 23, 1965	Attempted flyby; launch failure
Venera 4	Jun. 12, 1967	Descent probe
Mariner 5	Jun. 14, 1967	Flyby
Cosmos 167	Jun. 17, 1967	Attempted descent probe
Venera 5	Jan. 5, 1969	Descent probe
Venera 6	Jan. 10, 1969	Descent probe
Venera 7	Aug. 17, 1970	Lander
Cosmos 359	Aug. 22, 1970	Attempted probe
Venera 8	Mar. 27, 1972	Lander
Cosmos 482	Mar. 31, 1972	Attempted probe
Mariner 10	Nov. 3, 1973	Venus/Mercury flybys
Venera 9	Jun. 8, 1975	Orbiter and lander
Venera 10	Jun. 14, 1975	Orbiter and lander
Pioneer Venus 1	May 20, 1978	Orbiter
Pioneer Venus 2	Aug. 8, 1978	Multiprobes
Venera 11	Sep. 9, 1978	Orbiter and lander
Venera 12	Sep. 14, 1978	Orbiter and lander
Venera 13	Oct. 30, 1981	Orbiter and lander
Venera 14	Nov. 4, 1981	Orbiter and lander
Venera 15	Jun. 2, 1983	Orbiter
Venera 16	Jun. 7, 1983	Orbiter
Vega 1	Dec. 15, 1984	Lander and balloon/comet Halley flyby
Vega 2	Dec. 21, 1984	Lander and balloon/comet Halley flyby
Magellan	May 4, 1989	Orbiter
Galileo	Oct. 18, 1989	Flyby; Jupiter orbiter and probe
Cassini	Oct. 15, 1997	Flyby; Saturn orbiter

Venera

A series of Soviet probes designed to fly by, orbit, and land on Venus. The Venera spacecraft, unlike the Soviet **Mars** probes, were tremendously successful and made the first soft landings on and sent back the first pictures from the surface of Venus. (See table, "Venera Missions.")

Venera 1

The first spacecraft to fly past Venus; however, all contact with the probe was lost just seven days after launch, when it was about 2 million km from Earth. After its remote Venusian rendezvous, Venera 1 went into orbit around the Sun. Note: the first successful Venus probe was **Mariner** 2.

Venera 2

Like its predecessor, a flyby mission that suffered a communications breakdown long before it arrived at the second planet. Note: the first spacecraft in the **Zond** series was actually the second Soviet probe launched toward Venus, and it too was lost.

Venera 3

The first attempted landing on Venus. The entry vehicle carried science instruments and medallions bearing the Soviet emblem. However, communication with the probe was lost during descent through the crushing Venusian atmosphere.

Venera 4

The first probe to send back information successfully during its parachute descent through the atmosphere of Venus.

Venera 5 and 6

Twin spacecraft similar to Venera 4 but of stronger design. Each deployed a 405-kg descent probe that sent back information about the atmosphere for about 50 minutes. Each also carried a medallion bearing the Soviet emblem and a bas relief of Lenin to the night side of Venus but failed to transmit from the surface.

Venera 7

The first probe to return data after landing on another planet. Following aerodynamic braking and deployment of its parachute system, Venera 7 extended its antenna and transmitted for 35 minutes during its descent and a further 23 minutes of very weak signals from the surface.

Venera 8

The second successful Venus lander. Venera 8 slowed from 41,696 km/hr to about 900 km/hr by aerobraking, then opened its 2.5-m-diameter parachute at an altitude of 60 km and transmitted data during its descent. A refrigeration system cooled the interior components and enabled signals to be sent back for 50 minutes after landing. The probe confirmed Venera 7's data on the high surface temperature and pressure, and also determined that the light level was suitable for surface photography, being similar to the illumination on an overcast day on Earth.

Venera 9 and 10

A pair of identical spacecraft, each consisting of an orbiter and a lander. After separation of the lander, the orbiter spacecraft entered orbit around Venus, studied the upper clouds and atmosphere, and served as a communications relay for the lander. Each lander was slowed down sequentially by protective hemispheric shells, three parachutes, a disk-shaped drag brake, and a compressible, metal, doughnut-shaped landing cushion. Each sent back data from the surface for 53 minutes and 65 minutes, respectively, from locations about 2,200 km apart, and they became the first probes to transmit black-and-white pictures from the Venusian surface. Full 360° shots were not returned, however, because on each probe one of two camera covers failed to come off, restricting the field of view to a half-circle.

Venera 11 and 12

A two-spacecraft mission, each craft consisting of a flight platform and a lander. After ejection of their landers, the flight platforms flew past Venus, serving as data relays for more than an hour and a half until they traveled out of range to continue their investigations of interplanetary space. The Venera-12 flyby bus successfully used its Soviet-French ultraviolet spectrometer to observe Comet Bradfield in February 1980. Both Venera 11 and 12 landers failed to return color TV views of the surface and to perform soil analysis experiments as planned. All of the camera protective covers failed to eject after landing, and the soil drilling experiments were damaged by exposure to the high Venusian atmospheric pressure. Results reported included evidence of lightning and thunder, and the discovery of carbon monoxide at low altitudes.

Venera 13 and 14

Identical orbiter/landers launched within the same week. The landing probes touched down 950 km apart to the northeast and east, respectively, of an elevated basaltic plain known as Phoebe Regio. Venera 13

became the first spacecraft to remotely analyze the Venusian surface. Its mechanical drilling arm obtained a sample, which was deposited in a hermetically sealed chamber and maintained at 30°C and a pressure of about .05 atmospheres. The makeup of the sample, as determined by the X-ray fluorescence spectrometer, appeared similar to that of oceanic basalts on Earth. Venera 14's attempt at surface analysis was foiled when, ironically, its drilling arm landed on a successfully ejected camera cover.

Venera 15 and 16

A two-spacecraft mission that used side-looking radar mappers to study the surface properties of Venus. The two probes entered nearly polar orbits around Venus a day apart with their orbital planes inclined about 4° apart. This made it possible to reimage an area if necessary. Over their eight months of operation, the spacecraft mapped the area from the north pole down to about 30° N latitude.

Venera Missions

Launch site: Baikonur

Spacecraft	Launch Date	Vehicle	Arrival Date	Closest Approach (km)	Mass (kg)
Venera 1	Feb. 12, 1961	Molniya	May 19, 1961	100,000	644
Venera 2	Nov. 12, 1965	Molniya-M	Feb. 27, 1966	24,000	962
Venera 3	Nov. 16, 1965	"	Mar. 1, 1966	Landed	958
Venera 4	Jun. 12, 1967	"	Oct. 18, 1967	Landed	1,104
Venera 5	Jan. 5, 1969	"	May 16, 1969	Landed	1,128
Venera 6	Jan. 10, 1969	"	May 17, 1969	Landed	1,128
Venera 7	Aug. 17, 1970	"	Dec. 15, 1970	Landed	1,180
Venera 8	Mar. 27, 1972	"	Jul. 22, 1972	Landed	1,180
Venera 9	Jun. 8, 1975	Proton	Oct. 21, 1975	Landed	4,936
Venera 10	Jun. 14, 1975	"	Oct. 25, 1975	Landed	5,033
Venera 11	Sep. 9, 1978	"	Dec. 21, 1978	Landed	4,715
Venera 12	Sep. 14, 1978	"	Dec. 25, 1978	Landed	4,715
Venera 13	Oct. 30, 1981	"	Mar. 1, 1982	Landed	4,500
Venera 14	Nov. 4, 1981	"	Mar. 5, 1982	Landed	4,000
Venera 15	Jun. 2, 1983	"	Oct. 10, 1983	Orbited	4,000
Venera 16	Jun. 7, 1983	"	Oct. 14, 1983	Orbited	4,000

Viking 1 and 2

Twin spacecraft that studied Mars both from orbit and on the surface, and carried out the first in situ experiments to look for life on another world. Each Viking consisted of an orbiter and a lander. After arrival in orbit around the planet on June 19 and August 7, 1976 (Viking 1 having departed second but arrived first), a detailed reconnaissance was carried out to identify suitable and safe landing sites, during which the United States celebrated its bicentennial. Subsequently, the Viking 1 lander touched down in the western part of Chryse Planitia (the Plains of Gold), followed by its sister craft 7,420 km to the northeast.

The orbiter was an enlarged version of the **Mariner 9** spacecraft and carried scientific instruments on a scan platform, the most important of which were two television cameras to photograph potential landing sites. The lander was a hexagonal box with four landing legs, each with a circular footpad. On top of the body were two TV cameras, a mast supporting meteorological experiments, two small nuclear power systems, a dish-shaped radio antenna, and a robot arm used to sample the surface. Inside the body was a package of three biology experiments.

During the flight to Mars, the lander, inside a protective shell, was attached to the orbiter by a truss structure. The shell included an outer biological shield to prevent terrestrial microbes from contaminating Mars and a clamlike aeroshell to protect the lander during entry into the Martian atmosphere. The automated landing sequence began when the lander was separated from the orbiter and fired thrusters on its aeroshell to start entry. Two hours later, the aeroshell plunged into the thin Martian air. A parachute opened at a height of 6,200 km and released at 1,200 m to enable the craft to touch down gently using its three braking thrusters. Minutes after landing, Viking 1 sent back the first pictures from the surface of Mars.

The results of the biological experiments were intriguing but inconclusive. Activity such as would be expected of microbes was found in some of the tests, but an instrument known as a gas chromatograph mass spectrometer, used to search for organic material, found none. The last data received from the *Viking 2* lander was on April 11, 1980, and contact with the other lander was lost on

Viking A technician checks the soil sampler of the Viking lander. *NASA*

November 11, 1982, after it had been accidentally sent a wrong command. (See table, "Viking Missions.")

Launch
    Vehicle: Titan IIIE
    Site: Cape Canaveral
Dry mass, orbiter + lander: 600 kg

Villafranca Satellite Tracking Station (VILSPA)

Part of ESA's (European Space Agency's) worldwide **ESOC** (European Space Operations Centre) Station Network (ESTRACK). VILSPA lies in the Guadarrama Valley, 30 km west of Madrid. Built in 1975, it supports many ESA and other international programs, including IUE, OTS 2, GOES 1, MARECS, Exosat, and ISO.

Viking Missions

Spacecraft	Launch Date	Landing Date	Location
Viking 1	Aug. 20, 1975	Jul. 20, 1976	Chryse Planitia 22.4° N, 47.5° W
Viking 2	Sep. 9, 1975	Sep. 3, 1976	Utopia Planitia 48° N, 226° W

visible light

Electromagnetic radiation that can be detected by the human eye. It extends from a wavelength of about 780 nm (red light) to one of 380 nm (violet light), and bridges the divide between **infrared** and **ultraviolet**.

VLS

An indigenous Brazilian four-stage, solid-propellant launch vehicle capable of placing satellites weighing 100 to 380 kg into equatorial circular orbits 200 to 1,200 km high. Configured as a missile, the VLS could fly 3,600 km with a 500-kg nuclear payload. The first flight of the 19.2-m-long VLS ended in failure on November 2, 1997, when it was destroyed 65 seconds into the flight following a strap-on booster problem. In December 1999, a second VLS had to be destroyed just three minutes into its flight when the rocket again veered off course. There have been no further launches.

Von Braun, Wernher Magnus Maximilian (1912–1977)

One of the most important rocket developers and champions of space exploration from the 1930s to the 1970s, and the son of a baron. Von Braun's enthusiasm for the possibilities of space travel was kindled early on by reading the fiction of Jules **Verne** and H. G. **Wells** and the technical writings of Hermann **Oberth**. It was Oberth's 1923 classic *Die Rakete zu den Planetenräumen* (By Rocket to Space) that prompted the young von Braun to master the calculus and trigonometry he needed to understand the physics of rocketry. At age 17, he became involved with the German rocket society, **Verein für Raumschiffarht** (VfR), and in November 1932 he signed a contract with the Reichswehr to conduct research leading to the development of rockets as military weapons. In this capacity, he worked for Captain (later Major General) Walter **Dornberger**–an association that would last for over a decade. In the same year, under an army grant, von Braun enrolled at the Friedrich-Wilhelm-Universität from which he graduated two years later with a Ph.D. in physics; his dissertation dealt with the theoretical and practical problems of liquid-propellant rocket engines.

Some of von Braun's colleagues from the VfR days joined him in developing rockets for the German army (see **"A" series of World War II German rockets**). By 1935, he and his team, now 80 strong, were regularly firing liquid-fueled engines at **Kummersdorf** with great success. Following the move to **Peenemünde**, von Braun found himself in charge of the A-4/V-2 (see **"V" weapons**) project. Less than a year after the first successful A-4 launch and following a British bombing raid on Peenemünde, mass production of the V-2 was switched to an underground factory in central Germany. Von Braun remained at Peenemünde to continue testing.

In mid-March 1944, von Braun was arrested by the Gestapo and imprisoned in Stettin. The alleged crime was that he had declared greater interest in developing the V-2 for space travel than for use as a weapon. Also, since von Braun was a pilot who regularly flew his government-provided airplane, it was suggested that he was planning to escape to the Allies with V-2 secrets. Only through the personal intervention of Munitions and Armaments Minister Albert Speer was von Braun released.

When, by the beginning of 1945, it became obvious to von Braun that Germany was on the verge of defeat, he began planning for the postwar era. Before the Allied capture of the V-2 rocket complex, von Braun engineered the surrender to the Americans of scores of his top rocket scientists, along with plans and test vehicles. As part of a military plan called Operation **Paperclip**, he and his rocket team were whisked away from defeated Germany and installed at Fort Bliss, Texas. There they worked on rockets for the U.S. Army, launching them at **White Sands Proving Ground**.

In 1950, von Braun's team moved to the Redstone Arsenal near Huntsville, Alabama, where they built the army's **Jupiter** ballistic missile. In 1960, his rocket development center transferred from the army to the newly established NASA and received a mandate to build the giant **Saturn** rockets. Von Braun was appointed director of the **Marshall Space Flight Center** and chief architect of the Saturn V. He also became one of the most prominent advocates of space exploration in the United States during the 1950s. In 1970, he was invited to move to Washington, D.C., to head NASA's strategic planning effort, but less than two years later, feeling that the U.S. government was no longer sufficiently committed to space exploration, he retired from the agency and joined Fairchild Industries of Germantown, Maryland.[232, 305]

Von Hoefft, Franz (1882–1954)

An Austrian rocket theorist who founded the first space-related society in Western Europe. Von Hoefft studied chemistry at the University of Technology, Vienna, the University of Göttingen, and Vienna University, graduating from the last in 1907. Subsequently, he worked as an engineer in Donawitz, a tester at the Austrian Patent Office, and a consultant. In 1926, he formed the Wissenschaftliche Gesellschaft für Höhen-forschung (Scientific Society for High Altitude Research) in Vienna and later wrote a series of articles titled "The Conquest of Space" for the **Verein für Raumschiffarht's** publication, *Die Rakete* (The Rocket), in which he laid out a remarkably visionary scheme for the exploration of the Solar System and beyond. The first step was the development of a

Wernher von Braun Von Braun stands in front of a Saturn IB launch vehicle at Kennedy Space Center in January 1968. *NASA*

liquid-fueled sounding rocket called the RH-I (RH for "Repulsion Hoefft"), which would be carried to a height of 5 to 10 km by balloon and then launched. Such rockets, he explained, could be used for delivering mail or photographic remote sensing of Earth. By stages, their capacity would be increased. RH-V, for example, would be able to fly around the Earth in elliptical orbits, yet take off and land on water like a plane. The ultimate development, the RH-VIII, would be launched from a space station and be able to reach other planets or even leave the Solar System.

Von Kármán, Theodore (1881–1963)

A Hungarian aerodynamicist who founded an Aeronautical Institute at Aachen before World War I and achieved a world-class reputation in aeronautics through the 1920s. In 1930, Robert Millikan and his associates at the California Institute of Technology lured von Kármán from Aachen to become the director of Caltech's Guggenheim Aeronautical Laboratory (GALCIT). There he trained a generation of engineers in theoretical aerodynamics and fluid dynamics. With its eminence in physics, physical chemistry, and astrophysics, as well as aeronautics, it proved to be an ideal site for the early development of U.S. ballistic rocketry. Von Kármán was the first chairman of the Advisory Group on Aeronautical Research and Development for NATO.[133]

Von Opel, Fritz (1899–1971)

A German automotive industrialist who took part, with Max **Valier** and Friedrich Wilhelm **Sander**, in experiments with rocket propulsion for automobiles and aircraft. The world's first rocket-propelled car, the Opel-Rak 1, was initially tested on March 15, 1928. Opel himself test-drove an improved version, the Opel-Rak 2, on May 23 of that year. On September 30, 1929, Opel piloted the second rocket airplane to fly, a Hatry glider fitted with 16 solid-fuel rockets.

Von Pirquet, Guido (1880–1966)

A talented Austrian engineer who carried out seminal studies of the most efficient way for spacecraft to travel to the planets. Von Pirquet, a member of a distinguished Austrian family (his brother Clemens was a world-renowned physician), studied mechanical engineering at the Universities of Technology in Vienna and Graz. His expertise in ballistics and thermodynamics soon won him recognition in European rocket circles. He was elected first secretary of the rocket society founded by Franz **von Hoefft** and made his most important contributions to rocketry through a 1928–1929 series of articles called "Travel Routes" in the **Verein für Raumschiffahrt**'s periodical, *Die Rakete* (The Rocket), and a book, *Die Möglichkeit der Weltraumfahrt* (The Possibility of Space Travel), edited by the young Willy **Ley**

in 1928. In these writings, he describes the most fuel-efficient trajectories for reaching the planets Venus, Mars, Jupiter, and Saturn. Through calculations of a rocket nozzle for a manned rocket to Mars, he shows that the rocket would be impossibly large if it had to fly directly from the Earth's surface–the nozzle area of the first stage being about 1,500 square meters. He concludes that a manned expedition to Mars could be achieved only by building the spacecraft at a space station in Earth orbit. His 1928 calculated trajectory for a space probe to Venus is identical to the one used by the first Soviet interplanetary spacecraft to Venus in 1961.[307]

Vortex

See **Chalet**.

Voskhod

A multiseater Soviet spacecraft hurriedly adapted from **Vostok** in order to upstage the two-man American **Gemini** program. Only two Voskhod ("sunrise") flights were

Voskhod The never-flown Voshkod 3 spacecraft on display in the RKK Energia museum. *Joachim Becker*

made. On the first, three seats were squeezed into a converted Vostok, giving their occupants no room to wear spacesuits–a life-threatening gamble in order to mount the first three-man mission. (A similar gamble during the early **Soyuz** program ended with the death of a crew after the cabin depressurized.) The Voskhod 1 cosmonauts, who included Konstantin **Feoktistov** and Boris Yegorov, the first scientist and the first physician in space, respectively, and Vladimir **Komarov**, spent a day in orbit five months before the first Gemini mission. Feoktistov earned his berth through having figured prominently in the design of Vostok and its transformation into Voskhod.

The two-man Voskhod 2 achieved another on-orbit spectacular when Alexei **Leonov** carried out the first spacewalk. To enable this, Soviet engineers had designed an airlock that could be inflated in orbit to form a tunnel for exiting and reentering without having to let the air out of the cabin. (By contrast, American capsules of this era had to be completely depressurized prior to EVAs.) However, during Leonov's 24-minute spacewalk, his suit ballooned up more than expected and became too large and rigid to fit back through the airlock. Leonov was compelled to vent some of the suit's air. Once he managed to get back inside, the main hatch refused to seal properly, causing the environmental control system to compensate by flooding the cabin with oxygen and creating a serious fire hazard in a craft only qualified for a sea-level nitrogen-oxygen mixture. More dangers were to follow. When the

time came for reentry, the primary retro-rockets failed, forcing a manually controlled retrofire one orbit later. Then the service module failed to separate completely, leading to wild gyrations of the joined reentry sphere and service module before connecting wires burned through. Vostok 2 finally landed, about 2,000 km off course, in heavy forest in the Ural mountains. The crew spent an uncomfortable night in the woods, surrounded by wolves, before being located.[142] (See table, "Voskhod Flights.")

Vostok

See article, pages 474–475.

Voyager

See article, pages 476–477.

VSOP (VLBI Space Observatory Program)

An international radio astronomy project led by **ISAS** (Institute of Space and Astronautical Science) and the National Astronomical Observatory of Japan. It involves correlating measurements by the **HALCA** (Highly Advanced Laboratory for Communications and Astronomy) satellite with those of various ground-based telescopes to form a telescope effectively 32,000 km across (three times larger than the Earth), with a maximum resolving power at radio wavelengths of about 90 microarcseconds (roughly 100 times the resolving power of the Hubble Space Telescope).

Voskhod Flights

Mission	Launch	Recovery	Crew
Voskhod 1	Oct. 12, 1964	Oct. 13, 1964	Feoktistov, Komarov, Yegorov
Voskhod 2	Mar. 18, 1965	Mar. 19, 1965	Belyayev, Leonov

Vostok

The first series of manned Russian spacecraft. Six Vostok ("East") missions, from 1961 through 1963, carried cosmonauts on successively longer flights, and each set a new first in spaceflight history. Vostok 1 was the first manned spacecraft to complete a full orbit, Vostok 2 the first to spend a full day in space. Vostoks 3 and 4 made up the first two-spacecraft mission. Vostok 5 was the first long-duration mission, and Vostok 6 was the first to carry a woman.[142]

Gagarin's historic flight in Vostok 1 was preceded by a number of unmanned missions to test the spaceworthiness of the Vostok capsule and the reentry and recovery method to be used. These test flights were known in the west as Sputnik 4, 5, 6, 9, and 10 but in the Soviet Union as **Korabl Sputnik** 1 to 5. (See table, "Vostok Flights.")

Vostok Spacecraft

A spherical cabin, 2.3 m in diameter, attached to a biconical instrument module. The cabin was occupied by a single cosmonaut sitting in an ejection seat that could be used if problems arose during launch and that was activated after reentry to carry the pilot free of the landing sphere. Also inside the cabin were three viewing portholes, film and television cameras, space-to-ground radio, a control panel, life-support equipment, food, and water. Two radio antennas protruded from the top of the capsule, and the entire sphere was coated with ablative material (see **ablation**) so that there was no need to stabilize it to any particular attitude during reentry. The instrument module, which was attached to the cabin by steel bands, contained a single, liquid-propellant retrorocket and smaller attitude control thrusters. Round bottles of nitrogen and oxygen were clustered around the instrument module close to where it joined the cabin.

Vostok Rocket

Essentially, the same rocket (a modified R-7 ballistic missile; see **"R" series of Russian missiles**) that had launched Sputnik 1, 2, and 3, but with an upper stage supported by a latticework arranged and powered by a single RD-7 engine. The combination could launch a payload of about 4,700 kg into low Earth orbit.

Vostok Missions

Vostok 1

Yuri Gagarin made history with his 108-minute, 181 × 327-km single-orbit flight around the world. Once in orbit, he reported that all was well and began describing the view through the windows. Gagarin had brought a small doll with him to serve as a gravity indicator: when the doll floated in midair he knew he was in zero-*g*. (On April 12, 1991, Musa Manarov, the man who had by then logged the most time in space

Vostok The Vostok 1 capsule used by Yuri Gagarin, on display in the RKK Energia museum. *Joachim Becker*

[541 days] carried the same doll back into orbit aboard Mir to mark the 30th anniversary of Gagarin's flight.) Gagarin had no control over his spacecraft; a "logical lock" blocked any actions he might make in panic because, at the time, little was known of how humans would react to conditions in space. In case of emergency, Gagarin had access to a sealed envelope in which the logical lock code was written. To use the controls he would have had to prove that he was capable of doing the simple task of reading the combination and punching three of nine buttons. However, in the event, this proved unnecessary and radio signals from the ground guided the spacecraft to a successful reentry. At a height of 8,000 m, Gagarin ejected from his capsule and parachuted to the ground, southeast of Moscow near the Volga river, some 1,600 km from where he took off. Official details of the flight were not released until May 30, when an application was issued to the International Aeronautical Federation (FAI) to make the flight a world record. Gagarin's midair departure from Vostok was kept a secret much longer because the FAI required the pilot to return in his craft in order for the record to be valid. It would be another month before Alan **Shepard** made his suborbital flight, and 10 months before John **Glenn** became the first American in orbit.

Vostok 2

The first manned spaceflight to last a whole day. The 36-year-old pilot, Gherman **Titov**, ate some food pastes on his third orbit and later took manual control and changed the spacecraft's attitude. About 10 hours into the mission, he tried to catch some sleep but became nauseous—the first of many space travelers to experience **space motion sickness**. However, Titov did eventually fall asleep for over seven hours before waking for a perfect reentry and landing, 25 hours 18 minutes after launch.

Vostok 3 and 4

The first manned double launch. Vostok 3 and 4 took off from the same launch pad a day apart and were placed in such accurate orbits that the spacecraft passed within 6.5 km of each other. No closer rendezvous than this was possible, however, because the Vostoks were not equipped for maneuvering. The joint flight continued, with the two cosmonauts, Nikoleyev and Popovitch, talking to each other and with ground control by radio. Finally, the spacecraft reentered almost simultaneously and landed just a few minutes apart.

Vostok 5 and 6

Another double launch, this time involving the first woman in space—26-year-old Valentina **Tereshkova**. She returned to Earth after almost three days in orbit, followed by Valery Bykovsky a few hours later at the conclusion of a five-day flight that has remained the longest mission by a single-seater spacecraft. (See table, "Vostok Flights.")

Vostok Flights

Mission	Launch	Recovery	Orbits	Pilot
Vostok 1	Apr. 12, 1961	Apr. 12, 1961	1	Yuri Gagarin
Vostok 2	Aug. 6, 1961	Aug. 7, 1961	17	Gherman Titov
Vostok 3	Aug. 11, 1962	Aug. 15, 1962	64	Adrian Nikolayev
Vostok 4	Aug. 12, 1962	Aug. 15, 1962	48	Pavel Popovich
Vostok 5	Jun. 14, 1963	Jun. 19, 1963	81	Valery Bykovsky
Vostok 6	Jun. 16, 1963	Jun. 19, 1963	48	Valentina Tereshkova

Voyager

Two identical spacecraft sent to Jupiter and Saturn to build on the knowledge acquired by **Pioneer** 10 and 11. Voyager 2 went on to fly past Uranus and Neptune. Both Voyagers, like the twin Pioneers, are now leaving the Solar System and are engaged in the Voyager Interstellar Mission.

The Voyager mission had its origins in the **Grand Tour** concept of the 1960s. When this ambitious project was cancelled, Voyager took shape in the form of two smaller, Mariner-type spacecraft. One major difference from earlier Mariners, however, was the substitution of **radioisotope thermoelectric generators** (RTGs) for solar panels because of the weakness of sunlight far beyond the orbit of Mars. Another difference was the size of the antenna dish needed for high-data-rate communications over such great distances—3.7 m in diameter.

The heart of each Voyager (a name selected in March 1977 after a competition) was an octagonal equipment section harboring most of the vital electronic systems. In the center of the octagon was a large spherical tank of hydrazine propellants for the 16 small thrusters used for attitude control and course corrections. Most of the science instruments were mounted on a 13-m-long fiberglass boom that projected from the side of the spacecraft opposite to the nuclear power sources. Several of the instruments were attached directly to the boom, but others, including two television cameras and the ultraviolet

and infrared spectrometers, were mounted on a movable scan platform so they could be pointed accurately at specific targets. To provide the extra thrust needed for a direct flight to Jupiter, each probe carried a 1,210-kg solid-propellant rocket motor with a thrust of 71,000 newtons, which was released after it had finished firing. (See table, "Voyager Missions.")

Voyager 1

Although launched later than its sister craft, Voyager 1 reached Jupiter first, having been placed on a faster trajectory. Among its discoveries as it passed the giant planet at a closer distance of 285,950 km were Jupiter's thin ring and active volcanoes on Io. The surfaces of the other large Jovian moons were also seen in sharp detail. Less than two years later, as Voyager 1 passed 123,900 km from Saturn, it returned more startling images of ring features, fast-moving clouds in the planet's atmosphere, the giant moon Titan, and several previously unknown small moons. The successful flyby of Titan was important because if Voyager 1 had failed, mission planners had decided that Voyager 2 would have to repeat the attempt but then miss out on flybys of Uranus and Neptune. After its Saturn rendezvous, Voyager 1 began its journey toward the edge of the Solar System, becoming the most remote human-made object when, in mid-1998, it surpassed the heliocentric distance of Pioneer 10. In November 2002, Voyager 1 was 12.9 billion km from the Sun, equivalent to a round-trip light-travel time of almost 24 hours. It is traveling at a speed of 17.2 km/s (38,550 mph) away from the Sun in the general direction of the Solar Apex (the direction of the Sun's motion relative to nearby stars), so that it will probably be the first of the four present star-bound craft to reach the **heliopause** and enter true interstellar space.

Launch
    Vehicle: Titan IIIE
    Site: Cape Canaveral
Mass at launch
    Total: 19,500 kg
    Science payload: 104 kg

Voyager Missions

| Spacecraft | Launch | Flybys | | | |
		Jupiter	Saturn	Uranus	Neptune
Voyager 1	Sep. 5, 1977	Mar. 5, 1979	Nov. 12, 1980	—	—
Voyager 2	Aug. 20, 1977	Jul. 9, 1979	Aug. 25, 1981	Jan. 24, 1986	Aug. 24, 1989

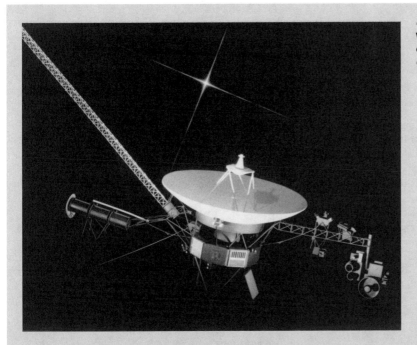

Voyager 2

The first spacecraft to encounter Uranus and Neptune. Having followed up on discoveries by its twin at Jupiter and Saturn, Voyager 2 used a gravity-assist from Saturn to send it on to the outermost gas giants. High points of these flybys included startling images of the moons Miranda (Uranus) and Triton (Neptune). In November 2002, Voyager 2 was 10.3 billion km from the Sun, equivalent to a round-trip light-travel time of just over 18 hours. Its departure speed from the Solar System is 15.7 km/s (35,180 mph). The future trajectory of Voyager 2 among the stars was determined by its final planetary encounter, with Neptune. An earlier planned route past Neptune would have resulted in the probe coming within 0.8 light-year of Sirius in just under 500,000 years from now—easily the closest and most interesting foreseeable stellar encounter of the four escaping probes. However, the Neptune flyby trajectory actually chosen (the "polar crown" trajectory) means that, like its sister craft, Voyager 2 will not come within a light-year of any star within the foreseeable future.

Voyager Interstellar Mission (VIM)

An extension of the Voyager primary mission approved in 1991, following the flyby of Neptune by Voyager 2. The objective of the VIM is to extend NASA's exploration of the Solar System beyond the neighborhood of the outer planets to the limits of the heliosphere, and possibly beyond. Data received from the two Voyagers (and Pioneer 10) into the first decade or two of the twenty-first century will be used to characterize the outer Solar System environment and search for the heliopause boundary, where the solar wind meets the interstellar medium. Penetration of the heliopause by either or both spacecraft, while still active, would allow the first-ever direct measurements to be made of the interstellar fields and particles.

W

WAC Corporal

A small liquid-propellant **sounding rocket** developed by the Jet Propulsion Laboratory for the U.S. Army. It used an attached solid-propellant booster to clear the launch tower. The first WAC Corporal was launched at the **White Sands Proving Ground** in October 1945 and reached an altitude of about 75 km. Improved versions of the rocket climbed to 100 km. Mated to the V-2 (see **"V" weapons**), the WAC Corporal served as the upper stage of the **Bumper-WAC**, the world's first two-stage rocket. The name Corporal derived from the fact that it came after the **Private** rocket program; WAC stood for "Without Any Control." Later versions of the **Corporal** missile were equipped with a guidance mechanism.

Length: 4.9 m
Mass: 300 kg
Thrust: 6,700 N
Propellants: aniline and nitric acid

Walker, Joseph Albert (1921–1966)

A chief research pilot at the Flight Research Center who set a world altitude record of 108 km (354,000 ft) in the X-15 rocket plane on August 22, 1963. He later tested the **Lunar Landing Research Vehicle** (LLRV). Walker was killed when his F-104 collided with an XB-70 bomber, with which he was flying in formation.

walking orbit

Also known as a *precessing orbit*, an orbit in which the **orbital plane** moves slowly with respect to fixed inertial space. It can be achieved intentionally by choosing the parameters of a spacecraft's orbit so that advantage is taken of some or all of the gravitational influences that give rise to **precession**. These factors include the deviation from perfect sphericity of a planet and the gravitational influence of other bodies such as the Sun or nearby moons.

Wallops Island Flight Facility

An old American launch site on Wallops Island on the east coast of Virginia, about 16 km from Chincoteague at 36.9° N, 75.5° W. It provides launch, tracking, and data acquisition capability for small launch vehicles and many of the sounding rockets used in NASA's programs. It also serves as an experimental station for advanced aerodynamic research programs, and provides for flight tests on instrumentation.

Walter, Hellmuth

A German rocket designer. The rocket engine he designed for the Heinkel He 52 aircraft (1937) was the first to use hydrogen peroxide.

warp drive

The principal means of interstellar propulsion in the *Star Trek* universe. According to the scriptwriters, it enables **faster-than-light travel** and uses matter-**antimatter** reactions inside a "warp core." One way of making it a reality may be the **Alcubierre Warp Drive**.

wavelength

The distance between two crests of a propagating wave. If wavelength is denoted by λ and frequency by f, the velocity v at which the wave advances is given by $v = f\lambda$.

weather satellites

Spacecraft that are used to monitor Earth's atmospheric conditions and to provide data to help understand and predict global weather patterns. The first was **TIROS**-1 in 1960. Over the past four decades, the United States, Russia, and other nations have launched weather satellites into both polar and geostationary orbits. See **NOAA**, **Meteor, SMS, GMS, Meteosat, GARP** (Global Atmospheric Research Program), **Eumetsat**, and **GOMS**.

Webb, James Edwin (1906–1992)

NASA's second administrator. Webb received an A.B. in education at the University of North Carolina in 1928 and studied law at George Washington University before being admitted to the bar of the District of Columbia in 1936. He enjoyed a long career in public service, rising to become undersecretary of state in 1949. At the end of the Truman administration in 1953, Webb took a post with the Kerr-McGee Oil Corporation, returning to Washington on February 14, 1961, to accept the position of administrator of NASA. Under his direction, the agency undertook President Kennedy's goal of landing an American on the Moon before the end of the decade. For seven years after Kennedy's May 25, 1961, lunar landing announcement, through October 1968, Webb politicked,

WAC Corporal   A WAC Corporal is prepared for launch at White Sands Proving Ground in the late 1940s. *U.S. Army/White Sands Missile Range*

cajoled, and maneuvered for NASA in the capital. By various means, Webb built a network of political liaisons that brought continued support for and resources to accomplish the Moon landing on the schedule Kennedy had announced. Webb was head of NASA when tragedy struck the Apollo program. On January 27, 1967, Apollo-Saturn (AS) 204 was on the launch pad at Kennedy Space Center, Florida, moving through simulation tests when a flash fire killed the three astronauts aboard–"Gus" Grissom, Edward White, and Roger Chaffee. Webb told the

media, "We've always known that something like this was going to happen soon or later. . . . who would have thought that the first tragedy would be on the ground?" As the nation mourned, Webb went to President Johnson and asked that NASA be allowed to handle the accident investigation and direct the recovery. He promised to be truthful in assessing blame and pledged to assign it to himself and NASA management as appropriate. The agency set out to discover the details of the tragedy, correct problems, and get back on schedule. Webb reported these findings to various congressional committees and took a grilling at nearly every meeting. While he was personally tarred with the disaster, the space agency's image and popular support were largely undamaged. He left NASA in October 1968, just as Apollo was nearing a successful completion. Subsequently he served on several advisory boards, including as a regent of the Smithsonian Institution.[178]

weight

The force with which a body such as a planet attracts a given **mass**. Weight is experienced as a reaction to this force against a solid surface. Although the mass of an object is independent of location, its weight varies depending on the strength of the local gravitational field. Weight is given by the product of mass *(m)* and the acceleration due to gravity *(g)*, that is: $W = mg$.

weight flow rate

In a liquid-fuel rocket motor, the rate at which the fuel flows into the **combustion chamber**; it is expressed in mass per unit of time.

weightlessness

The condition of zero-gravity (zero-*g*) or **microgravity** experienced by all spacecraft and their occupants when in orbit or traveling through space at constant speed. Weightlessness can be of great benefit in certain areas of research and manufacturing, and in large construction work in space. Large masses in orbit do not require support, and their movement is restricted only by inertia. Structures can be designed without provision for support against the forces of gravity—in free space there is no such thing as a static load. On the other hand, long-term exposure to weightlessness has adverse health effects, including muscular deconditioning and **bone demineralization**.

Weitz, Paul J. (1932–)

An American astronaut who served as the Command Module pilot on **Skylab** 2 in 1973 and commander of the sixth Space Shuttle flight in 1983. Weitz received a B.S. in aeronautical engineering from Pennsylvania State University in 1954 and an M.S. in aeronautical engineering from the U.S. Naval Postgraduate School in 1964. Subse-

weightlessness Astronaut Michael Gernhardt, weightless outside the Space Shuttle on mission STS-69 in 1995. Like all objects in orbit, Gernhardt is in continual free fall. *NASA*

quently, he served in various naval squadrons until NASA selected him as an astronaut in 1966. Weitz remained with NASA after his second spaceflight and became deputy director of the Johnson Space Center.

West Ford

A passive communications concept developed by the Massachusetts Institute of Technology Lincoln Laboratory for the Department of Defense in 1963. The reflector was to consist of a belt, 8 km wide and 25 km long, of 480 million hair-thin copper filaments, each about 2 cm long, in a 3,000-km-high orbit. Radio astronomers opposed the idea, believing it might affect their research. However, the copper cloud quickly dispersed, rendering it useless for communications and no threat to astronomy.

Western Space and Missile Center (WSMC)

An American launch site, located at **Vandenberg Air Force Base**, California, and operated by the 30th Wing of the **Air Force Space Command**. From WSMC, the Western Test Range extends westward over the Pacific Ocean and into the Indian Ocean, where it meets the Eastern Test Range. Most spacecraft launches from Vandenberg, however, take place not westward (which would be in opposition to the Earth's spin direction) but southward into **polar orbits** and include those of surveillance satellites, low-Earth-orbit weather satellites, and environmental and terrain monitoring satellites such as Landsat. Polar launches are particularly safe from WSMC because the next land mass south of the site is Antarctica.

wet emplacement

A launch emplacement that provides a deluge of water for cooling the **flame bucket**, launch vehicle engines, and other equipment before and during liftoff.

Wexler, Harry (1911–1962)

One of the first scientists to envision using satellites for meteorological purposes; he is remembered as the father of the **TIROS** weather satellite. Wexler received a Ph.D. in meteorology from the Massachusetts Institute of Technology in 1939 and worked for the U.S. Weather Bureau from 1934 until his death. From 1955 to 1958, he was also the chief scientist for the American expedition to Antarctica for the **International Geophysical Year**. In 1961, he was a lead negotiator for the United States in drafting plans for joint American-Soviet use of meteorological satellites.

Whipple, Fred L. (1906–)

An astrophysicist who did pioneering research on comets and, in the 1950s, helped expand public interest in the possibility of spaceflight through a series of symposia at the Hayden Planetarium in New York City and articles in *Collier's* magazine. Whipple established the first optical tracking system for artificial satellites. He earned a Ph.D. at the University of California, Berkeley, and subsequently served on the faculty of Harvard University. He was also heavily involved in planning for the **International Geophysical Year, 1957–1958**.[38]

Whipple shield

A thin shield, also known as a meteor deflection screen, that protects a spacecraft from damage due to collision with micrometeoroids. It is named after Fred **Whipple**, who first suggested it in 1946 and referred to it as a "meteor bumper." Whipple shields are based on the principle that small meteoroids explode when they strike a solid surface; therefore, if a spacecraft is protected by an outer skin about a tenth of the thickness of its main skin, an impinging body will be destroyed before it can cause any real damage.

White, Edward Higgins, II (1930–1967)

An American astronaut who carried out America's first spacewalk, on **Gemini** 4. Born in San Antonio, Texas, White received a B.S. from the U.S. Military Academy in 1952, an M.S. from the University of Michigan in 1959, and an honorary Ph.D. in astronautics from the University of Michigan in 1965. Following West Point, he undertook flight training in Florida and Texas, then spent 3½ years in Germany with an Air Force fighter squadron, flying F-86s and F-100s. In 1959, he attended the Air Force Test Pilot School at Edwards Air Force Base, and later was assigned to Wright-Patterson Air Force Base, Ohio, as an experimental test pilot with the Aeronautical System Division. NASA selected White, an Air Force lieutenant colonel, as an astronaut in 1962. He was pilot on the four-day Gemini 4 mission that began June 3, 1965, and was commanded by James **McDivitt**. During the first day, White stepped outside the spacecraft for a 21-minute spacewalk during which he maneuvered on the end of a 7.6-m lifeline by using a hand-held jet gun. During the remainder of the flight, McDivitt and White completed 12 scientific and medical experiments. White and fellow **Apollo** 1 astronauts Gus Grissom and Roger Chaffee died in a spacecraft fire during a launch pad test on January 27, 1967.

white room

The room in which astronauts prepare for a spaceflight before entering the spacecraft. The name is borrowed from a similar term used for clean rooms—free of dust and other contamination—in industries and hospitals.

Edward White   The Apollo 1 crew: Ed White (center), Virgil Grissom (left), and Roger Chaffee (right). *NASA*

White Sands Missile Range

A major facility that supports missile testing and development for the U.S. armed services, NASA and other government agencies, and private industry. Occupying almost 8,300 square kilometers in the Tularosa Basin of south-central New Mexico, it is the largest military installation in the country—larger than the states of Delaware and Rhode Island combined.

White Sands Proving Ground (its name changed in 1958) was established on July 9, 1945. A launch complex was quickly set up and used for the site's first "hot firing" of a Tiny Tim missile, on September 26, 1945. Soon this complex was the focal point for captured V-2 (see **"V" weapons**) launches and, in time, the developmental testing of such missiles as Nike, **Viking**, and **Corporal**.

NASA used the White Sands range for testing the Saturn's launch escape system and, later, the engines, components, and materials of the Space Shuttle Orbiter. White Sands also provides an alternate landing site for the Shuttle, and on March 30, 1982, the Orbiter *Columbia* touched down on the range's Northrup Strip after its third flight into space. Shuttle astronauts train over Northrup Strip (now named White Sands Space Harbor), practicing landings in a **Gulfstream** jet that simulates Orbiter glide characteristics, and on the ground terminal of NASA's **TDRSS** (Tracking and Data Relay Satellite System), also located at the range.

White Sands Proving Ground

The name of **White Sands Missile Range** from 1945 to 1958.

Whittle, Frank (1907–1996)

An English pioneer of the jet engine. Whittle was born in Earlsdon, at a time when powered flight was still in its infancy, his boyhood coinciding with the use of aircraft in World War I, the formation of the RAF in 1918, and Alcock and Brown's flight across the Atlantic in 1919. He grew up in an engineering environment, his father owning a general engineering business in Leamington Spa, and he was particularly fascinated by aircraft. After leaving school in 1923, he joined the RAF as an apprentice aircraft fitter and was later selected for pilot training at the RAF Staff College, Cranwell, where he was soon flying solo. The piston-engined planes of the 1920s were limited with respect to speed and altitude—as they would be for another two decades. But while at Cranwell, and aged only 21, Whittle began to consider the possibilities of jet propulsion. By 1930, he had designed and patented a jet aircraft engine. At first, many people refused to believe that his invention would work at all. The Air Ministry was approached, as were various industrial firms, but all were put off by the practical difficulties and high costs involved in making the engine: Britain in the 1930s was in the middle of a serious industrial depression, and money for new products was limited. With no one to back him, Whittle couldn't even afford to renew his patent. By 1935, Whittle had almost given up hope that his ideas would ever be developed. He was then at Cambridge, where the RAF had sent him to work for a degree in mechanical engineering. In May of that year, he was approached by two ex-RAF officers who suggested that they should cooperate to try to get work started on the jet engine. It proved to be a turning point. By 1936, some money had been obtained, a small company, Power Jets Ltd, was formed, and work on the first experimental engine started at the British Thomson Houston (now GEC) factory in Rugby. In April 1937, the first engine was tested. According to Whittle, it "made a noise like an air raid siren," causing onlookers to run for cover. Nevertheless, the demonstration was successful. The Air Ministry now began to take an interest and in 1939 gave Power Jets Ltd. an official contract for a flight engine. Subsequently, the Gloster Aircraft Company was asked to build an experimental aircraft. The result was the Gloster E.28/39, which, powered by the Whittle jet engine, took off from Cranwell on May 15, 1941, on an historic 17-minute flight. Toward the end of World War II, the now renamed Gloster Meteor entered service as the RAF's first jet fighter.

Wilkins, John (1614–1672)

A brother-in-law of Oliver Cromwell and an Oxford graduate who became Master of Trinity College, Cambridge, and eventually Bishop of Chester. As a young author he was one of several men, including Johannes **Kepler**, who speculated about the possibility of traveling to the Moon. His first book, *A Discourse tending to prove (tis probable) there may be another Habitable World in the Moon,* was published in 1638, the year Milton visited Galileo, then imprisoned by the Inquisition for promulgating the same ideas that Wilkins developed in his book. It was in the third edition of his *Discourse* that Wilkins spoke explicitly of space travel: "That tis possible for some of our Posterity to find a conveyance to this other World; and if there be Inhabitants there to have Commerce with them." He drew a parallel between the difficulty of crossing space and that of crossing the oceans. Ships must once have seemed strange, "And yet now, how easie a thing is this even to a timorous and cowardly nature, And questionless the Invention of some other means for our conveyance to the Moon cannot seem more incredible to us than did this at first to them." Wilkins recognized the problems of gravity and distance. He supposed that gravity became negligible above 20 miles; as for a means of transport, he suggested a flying chariot or bird-power—the latter a propulsion system also favored by Francis **Godwin**.

Williams, Walter C. (1919–1995)

A prominent engineer and administrator with the U.S. manned space program. Having earned a degree in aerospace engineering, Williams joined **NACA** (National Advisory Committee for Aeronautics) in 1940 and worked on improving the handling, maneuverability, and flight characteristics of World War II fighter planes. Following the war, he went to what became **Edwards Air Force Base** to set up flight tests for the **X-1**, including the first human supersonic flight by Charles **Yeager** in October 1947. Subsequently, he became the founding director of the organization that became **Dryden Flight Research Facility** and, in September 1959, assumed associate directorship of the new NASA space task group at Langley, created to carry out Project **Mercury**. He later became director of operations for this project, then associate director of the NASA Manned Spacecraft Center, renamed the **Johnson Space Center**. In 1963, Williams moved to NASA Headquarters as deputy associate administrator of the Office of Manned Space Flight. From 1964 to 1975, he was a vice president for Aerospace Corporation. Then from 1975 to 1982 he served as chief engineer of NASA, retiring in the latter year.

Wind

Together with its sister spacecraft **Polar**, NASA's contribution to the International Solar Terrestrial Program (ISTP), an international effort to quantify the effects of solar energy on Earth's magnetic field. From an L-1 **halo**

orbit, Wind provides continuous measurement of the solar wind, particularly charged particles and magnetic field data, and thus helps show how the solar wind affects magnetospheric and ionospheric behavior. It also carries the first Russian instrument to fly on an American spacecraft since cooperation in space between the two countries resumed in 1987.

Launch
    Date: November 1, 1994
    Vehicle: Delta 7925
    Site: Cape Canaveral
Size: 2.0 × 2.4 m
Mass: 1,195 kg
Orbit: 48,840 × 1,578,658 km × 19.6°

wind tunnel

A device used to investigate the aerodynamic properties of objects by passing a stream of velocity-controlled air over them. The largest wind tunnel in the world is the National Full-Scale Aerodynamics Complex at NASA

Ames. It is composed of a 40- by 80-foot section, more than 50 years old, and a newer 80- by 120-foot section that is large enough to test aircraft the size of a Boeing 737.

Winkler, Johannes (1887–1947)

A German engineer and the editor of the **Verein für Raumschiffahrt**'s journal *Die Rakete* who on February 21, 1931, supported in his research by Hugo A. Huckel, launched the first liquid-propellant rocket in Europe. The 60-cm-long, 30-cm-wide rocket, called the Huckel-Winkler 1, was powered by a combination of liquid oxygen and liquid methane. Launched near the city of Dessau, the Huckel-Winkler 1 reached an altitude of about 300 m. Unfortunately, its successor was not as impressive. The Huckel-Winkler 2 took off from a site near Pillau in East Prussia, on October 6, 1932, caught fire, and crashed, having reached an altitude of just 3 m.

WIRE (Wide-Field Infrared Explorer)

A **SMEX** (Small Explorer) satellite whose primary purpose was a four-month infrared survey of the universe,

wind tunnel   Looking down the throat of the 40- × 80-ft wind tunnel at NASA Ames Research Center. *NASA*

focusing specifically on starburst galaxies and luminous protogalaxies. WIRE experienced problems shortly after launch, and ground controllers attempted to recover the spacecraft. However, the entire supply of frozen hydrogen needed to cool its main scientific instrument was released into space, ending the spacecraft's primary mission. Operations have been redirected to use the onboard star tracker for long-term monitoring of bright stars in support of two separate science programs: astroseismology and planet-finding. The astroseismology program is intended to measure oscillations in nearby stars to probe their structure. The planet-finding program searches for stellar occultations by large planetary bodies as they pass through WIRE's line-of-sight to its target star.

Launch
    Date: March 4, 1999
    Vehicle: Pegasus XL
    Site: Vandenberg Air Force Base
Orbit: 540 × 590 km × 98°
Mass: 259 kg

Woman in the Moon

A German film *(Frau im Mond)* directed by Fritz Lang and released in 1929. Although it was never a box-office hit, being a silent movie when the new "talkies" were all the rage, it was technically accurate and visionary for its time. This was largely due to Hermann **Oberth**'s contribution as technical consultant. Indeed, *Frau im Mond*'s Moon rocket was virtually identical to the one Oberth had already described and illustrated in his *Die Rakete zu den Planetenraumen*. Oberth had also come to an agreement with Fritz Lang to build and fly a real rocket as a publicity stunt when the film opened. However, although Oberth and a young colleague by the name of Wernher **von Braun** got as far as testing their rocket engine in the lab, it was never destined to fly. One of the details of *Frau im Mond* would have a particularly lasting influence. As the Moon rocket nears the moment of launch, a loudspeaker announces (the words written on the screen): "Five . . . four . . . three . . . two . . . one . . . zero . . . FIRE!" Lang had invented the countdown—if only for dramatic effect—now so familiar in mission controls around the world.

women in space

See **Mercury 13**, astronauts/cosmonauts Eileen **Collins**, Shannon **Lucid**, Christa **McAullife**, Barbara **Morgan**, Sally **Ride**, Svetlana **Savitskaya**, Valentina **Tereshkova**, and Margaret **Cavendish**.

Woomera

An Australian launch complex at 31.1° S, 136.8° E. In the late 1940s, Britain, looking for a large remote area in which to test long-range weapons such as the planned "Super V-2," settled upon this region of the central Australian desert. Construction, and the building of a small town, began in 1947, the name "Woomera" being chosen because it is the Aboriginal word for a spear-throwing device. Woomera was used to test a variety of projects including the Blue Steel, Bloodhound, and Thunderbird missiles, and played an important role in the early British and European space programs. In 1957, the first **Skylark** was launched from here, as were the **Black Knight** reentry test rockets. Throughout the 1960s, Woomera was used by **ELDO** (European Launcher Development Organisation) to test the **Europa**-1 launch vehicle, but when this project ended, launches of Europa-2 were moved to the equatorial site of Korou in French Guinea. In 1967, Woomera became a spaceport when a **Redstone** rocket, purchased from the United States, placed Australia's first satellite in orbit. This was followed in 1971 by the first British satellite, **Prospero**, on a **Black Arrow**. By the 1970s, however, launches from Woomera had dwindled, and although the site was still used as part of the Anglo-Australian Joint Project until 1980, the town went into decline. Recently, though, the American company Kistler Aerospace has announced plans to use Woomera to launch satellites on reusable rockets.

Worden, Alfred Merrill (1932–)

An American astronaut, one of 19 selected by NASA in April 1966, who served on the support crew of **Apollo 9**, on the backup crew of Apollo 12, and as Command Module pilot of Apollo 15. Worden attended the Military Academy at West Point, graduating in 1955. From 1955 to 1961, he was a fighter pilot and armament officer with the 95th FIS, Andrews Air Force Base, Maryland. He became the chief of the Systems Studies Division at the Ames Research Center in 1972, a post he held until 1975. Worden worked as a consultant to the Northwood Institute and the state of Florida from 1975 to 1982, and later held executive positions with Maris-Worden Aerospace, Jet Electronics and Technology, and BF Goodrich Corporation before retiring in 1996.

working pressure

The maximum pressure to which a component in a rocket engine is subjected under steady-state conditions.

wormhole

See article, pages 486–487.

wormhole

A hypothetical "tunnel" connecting two different points in **space-time** in such a way that a trip through the wormhole could take much less time than a journey between the same starting and ending points in normal space. Speculation in recent years has centered on the possibility of exploiting this phenomenon for **faster-than-light (FTL) travel**. Wormholes arise as solutions to the equations of Einstein's **general theory of relativity**. In fact, they crop up so often and easily in this context that some theorists are encouraged to think that real counterparts may eventually be found or fabricated.

The idea of space-time tunnels first emerged from mathematical solutions of general relativity as they apply to **black holes**. Some of these solutions could be interpreted as two black holes connected by a "throat," known formally as an Einstein-Rosen bridge and now, more commonly, as a wormhole. The problem for would-be travelers along wormholes of this type is that they are inherently unstable—anything entering them, even a subatomic particle, would cause them immediately to collapse. However, all is not lost. In recent years, theorists have investigated several types of wormhole, both artificial and natural, not involving black hole mouths, that could possibly act as FTL conduits.[302]

Artificial Wormholes

Interest in so-called traversable wormholes gathered pace following the publication of a 1987 paper by Michael Morris, Kip Thorne, and Uri Yertsever (MTY) at the California Institute of Technology.[211] This paper stemmed from an inquiry to Thorne by Carl **Sagan**, who was mulling over a way of conveying the heroine in his novel *Contact* across interstellar distances at trans-light speed. Thorne gave the problem to his Ph.D. students, Morris and Yertsever, who realized that such a journey might be possible if a wormhole could be held open long enough for a spacecraft (or any other object) to pass through. MTY concluded that to keep a wormhole open would require matter with a negative **energy density** *and* a large negative pressure—larger in magnitude than the energy density. Such hypothetical matter is called **exotic matter**. Although the existence of exotic matter is speculative,

a way is known of producing negative energy density: the **Casimir effect**.

As a source for their wormhole, MTY turned to the **quantum vacuum**. "Empty space" at the smallest scale, it turns out, is not empty at all but seething with violent fluctuations in the very geometry of space-time. At this level of nature, ultra-small wormholes are believed to continuously wink into and out of existence. MTY suggested that a sufficiently advanced civilization could expand one of these tiny wormholes to macroscopic size by adding energy. Then the wormhole could be stabilized using the Casimir effect by placing two charged superconducting spheres in the wormhole mouths. Finally, the mouths could be transported to widely separated regions of space to provide a means of FTL communication and travel. For example, a mouth placed aboard a spaceship might be carried to some location many light-years away. Because this initial trip would be through normal space-time, it would have to take place at sub-light speeds. But during the trip and afterwards, instantaneous communication and transport through the wormhole would be possible. The ship could even be supplied with fuel and provisions through the mouth it was carrying. Also, thanks to relativistic **time-dilation**, the journey need not take long—even as measured by Earth-based observers. For example, if a fast starship carrying a wormhole mouth were to travel to Vega, 25 light-years away, at 99.995% of the speed of light (giving a time-dilation factor of 100), shipboard clocks would measure the journey as taking just three months. But the wormhole stretching from the ship to Earth directly links the space and time between both mouths—the one on the ship and the one left behind on (or near) Earth. Therefore, as measured by Earthbound clocks too, the trip would have taken only three months—three months to establish a more-or-less instantaneous transport and communications link between here and Vega.

Of course, the MTY scheme is not without technical difficulties, one of which is that the incredibly powerful forces needed to hold the wormhole mouths open might tear apart anything or anyone that tried to pass through. In an effort to design a more benign environment for travelers using a wormhole, Matt Visser of Washington University in St. Louis conceived an arrangement in which the space-time region of a wormhole mouth is flat (and thus force-free) but framed by "struts" of exotic matter that contain a region of very sharp curvature. Visser envisages a cubic design, with flat-space wormhole connections on the square sides and **cosmic strings** as the edges. Each cube-face may connect to the face of another wormhole-mouth cube, or the six cube faces may connect to six different cube faces in six separate locations.

Natural Wormholes

Given that our technology is not yet up to the task of building a wormhole subway, the question arises of whether they might already exist. One possibility is that advanced races elsewhere in the galaxy or beyond have already set up a network of wormholes that we could learn to use. Another is that wormholes might occur naturally. David Hochberg and Thomas Kephart of Vanderbilt University have discovered that in the earliest moments of the universe, gravity itself may have given rise to regions of negative energy in which natural, self-stabilizing wormholes may have formed.[151] Such wormholes, created in the Big Bang, might be around today, spanning small or vast distances in space.

WRESAT (Weapons Research Establishment Satellite)

Australia's first satellite. It was launched by a surplus U.S. Army **Redstone** and carried out measurements of the upper atmosphere and solar radiation.

Launch
  Date: November 29, 1967
  Vehicle: Redstone
  Site: Woomera
Mass: 45 kg
Orbit: 198 × 1,252 × 83°

Wright Air Development Division

A center located at Wright-Patterson Air Force Base, Ohio, for research, development, test and evaluation in aerodynamics, human factors, materials, electronics equipment, and aerospace sciences.

Wyld, James Hart (1913–1953)

An American engineer and one of the founders of the **American Rocket Society** who, in 1938, built the first rocket engine to use **regenerative cooling**. Some preliminary work with this technique had been done in 1993 by Harry W. Bull, who used it to cool the nozzle area only. The first tests of Wyld's 1-kg engine were carried out under the direction of the Rocket Society's Experimental Committee on December 10, 1938, with satisfactory results. In December 1941, Wyld helped found Reaction Motors, Inc., American's first commercial rocket company.

X

X planes

U.S. experimental aircraft designed to answer fundamental questions about the behavior of aircraft close to, at, or beyond the speed of sound, and as prototypes for advanced aerospace vehicles. Most of these planes have been flown from Muroc Air Field, later named **Edwards Air Force Base**.

X-1

The first of the X planes, based on the shape of a .22 caliber bullet, with revolutionary thin, straight wings. The rocket-powered Bell X-1 was carried to a height of about 12,200 m under the belly of a Boeing B-29 Superfortress before being released. Its prime mission was to break the sound barrier, a feat accomplished for the first time on October 14, 1947, when Charles "Chuck" **Yeager** accelerated his X-1 (christened by him "Glamorous Glennis," after his wife) to Mach 1.06 (1,130 km/hr). The X-1 program was the Air Force's first foray into experimental flight research and its first collaborative effort with **NACA**. Three X-1s were built, and they completed 157 flights between January 1946 and October 1951.[251]

X-2

A swept-wing, rocket plane designed to explore flight at high supersonic speeds and be the first aircraft to take man to the edge of space. The Bell X-2 was made of stainless steel and copper-nickel alloy and was powered by a two-chamber Curtiss-Wright XLR25 throttleable rocket engine. Following launch from a B-50 bomber, Bell test pilot Jean "Skip" Ziegler completed the first unpowered glide flight on June 27, 1952. This vehicle was subsequently lost in an explosion during a captive flight in 1953. Lt. Col. Frank "Pete" Everest completed the first powered flight in a second X-2 on November 18, 1955, and on his ninth and final flight in late July 1956 he established a new speed record of Mach 2.87. At high speeds, Everest reported that the X-2's controls were only marginally effective, and data from his flights combined with simulation and wind tunnel studies suggested that the aircraft would encounter very severe stability problems as it approached Mach 3. A pair of young test pilots, Captains Iven Kincheloe and Milburn "Mel" Apt, were assigned the job of further expanding the envelope, and on September 7, 1956, Kincheloe became the first pilot ever to climb above 100,000 ft (30,500 m), reaching a peak altitude of 38,500 m. Twenty days later, Apt made his first X-2 flight under instructions to follow the "optimum maximum energy flight path" and avoid any rapid maneuvers above Mach 2.7. Flying a precise profile, he became the first human to exceed Mach 3, accelerating to Mach 3.2 (3,369 km/hr) at 20,000 m. The flight had been flawless to this point, but shortly after reaching top speed, Apt attempted a banking turn while still above Mach 3. The X-2 tumbled violently out of control and Apt found himself struggling with the same problem of inertia coupling (the loss of stability at high speeds) that Yeager had experienced in the **X-1** nearly three years before. Unlike Yeager, however, Apt was unable to recover, and both he and the aircraft were lost. While the X-2 had supplied valuable research data on high-speed aerodynamic heat buildup and extreme high-altitude flight conditions, this tragic event ended the program before NACA could start detailed research with the plane. Finding answers to many of the riddles of high-Mach flight had to be postponed until the arrival, three years later, of the most ambitious of all rocket planes: the **X-15**.[92]

X-15

A rocket plane that set aircraft speed and altitude records that still stand today. First flown on June 8, 1959, the X-15 was used to provide data on thermal heating, control, and stability at extremely high speeds, and on atmospheric reentry. It was made primarily from titanium and stainless steel covered with Inconel X nickel, an alloy that can withstand temperatures up to 650°C. To sustain even higher temperatures, the X-15 was often covered with a pink ablative material (MA-25S) that could boil away, carrying the heat with it. The plane was dropped from a B-52 bomber at an altitude of 13,700 m and then ignited its Reaction Motors XLR99-RM-2 throttleable liquid propellant (liquid hydrogen and anhydrous ammonia) engine. The rear tail was movable and could be pivoted for control at altitudes where the air was sufficiently thick. At greater (nonatmospheric) heights, control was provided by 12 hydrogen peroxide jets—four in the wingtips and eight in the nose. The plane was piloted following a predetermined flight path, and came down on Rogers dry lakebed using unique landing gear. Just before landing, the lower half of the bottom tail section was jettisoned, and two landing skids deployed, together with a two-wheel conventional landing gear at the nose. Three X-15s were built and 199 missions flown, the last in November 1968.[244, 285] (See table, "Comparison of the X-1, X-2, and X-15.")

Comparison of the X-1, X-2, and X-15

	X-1	X-2	X-15
Length (m)	9.4	11.5	16.0
Wingspan (m)	8.5	9.8	6.7
Mass, fully fueled (kg)	5,560	11,300	15,400
Thrust (N)	26,700	66,700	254,000
Maximum speed (km/hr)	1,540 (Mach 1.45)	3,369 (Mach 3.20)	7,297 (Mach 6.72)
Maximum altitude (m)	21,916	38,500	107,960 (67.08 miles)

X-20
See **Dyna-Soar**.

X-23
See **PRIME**.

X-24
An early manned **lifting body** used by the U.S. Air Force and NASA to explore the low-speed flight characteristics of a subscale model, the X-23A, tested in the **PRIME** program. There were two configurations. The 7.5-m, 5,000-kg X-24A was carried under the wing of a B-52 bomber and released at an altitude of about 13,500 m. The X-24A could then either glide back to Earth or use its onboard rocket engine to accelerate to speeds of 1,600 km/s or climb up to 21,500 m to explore the characteristics of the design under various flight conditions. A total of 28 flights, including 10 glides, were flown from 1969 to 1971 before the design was changed to a delta-winged

X-15 Crew members secure the X-15 after a research flight, while the B-52 launch plane makes a low flyby. *NASA*

configuration and renamed the X-24B. In this form, a further 12 glides and 24 powered flights were made. The X-24 program was concluded in 1975.

X-33

An almost half-scale prototype of the VentureStar reusable single-stage-to-orbit launch vehicle proposed by Lockheed Martin. The X-33 program was begun by NASA in 1996 but ran into development problems with the lightweight composite materials from which the vehicle's propellant tanks were intended to be made. Budget constraints finally led NASA to abandon the X-33 and related X-34 programs in 2001.

X-37/X-40A

The X-40A is a 90%-scale model of the X-37 Space Maneuver Vehicle, intended to fly into space inside the Space Shuttle payload bay to deliver satellites and perform reconnaissance missions and other tasks. At the end of its mission, it can return to the Shuttle or land on its own. The X-40A has conducted several drop tests from a helicopter. The X-37 will be about 8.5 m long and should be capable of making at least 20 flights and landings.

X-38

A prototype of a proposed **ACRV** (Assured Crew Return Vehicle) for the International Space Station. The seven-seat X-38, with a design based on that of the **X-24** from the 1960s, was being developed at Johnson Space Center and Dryden Flight Research Center and had made several test flights. However, work on the program was stopped in 2001 following NASA budget cuts.

X-43A

A subscale prototype that forms part of NASA's Hyper-X program to develop a commercial hypersonic suborbital aircraft. The X-43A **scramjet** is designed to fly at speeds between Mach 7 and Mach 10, easily beating the world's previous fastest air-breathing plane, the Mach-3 SR 71 Blackbird. The X 43-A is launched aboard a Pegasus from a modified B-52 bomber. The first flight, on June 4, 2001, ended in failure when the Pegasus launch vehicle veered out of control and had to be destroyed just minutes before its passenger was due to fire its scramjet engine. NASA has other X-43As with which it will resume flights when the problem with the Pegasus has been resolved.

X-43A   An artist's conception of the X-43A in hypersonic flight. *NASA*

X-band
See **frequency bands**.

XEUS (X-ray Evolving Universe Spectroscopy mission)

A potential follow-on to the **XMM-Newton Observatory**. ESA's (European Space Agency's) XEUS would be a powerful telescope for X-ray astronomy: it would consist of two separate spacecraft, the mirror spacecraft and the detector spacecraft, capable of being periodically serviced and upgraded at the **International Space Station** (ISS). Its sensitivity would exceed that of XMM-Newton, enabling it to pick out the individual sources of the most distant X-ray emission. Following its launch by an **Ariane** 5, XEUS would begin observations from LEO (low Earth orbit), slightly higher than that of ISS. After a few years, the orbit of XEUS and the ISS would be synchronized and XEUS docked with ISS for refurbishment and a mirror upgrade. Four to six years later, following a second rendezvous, further mirror petals would be added to the mirror spacecraft and the entire detector spacecraft replaced with a new one, greatly increasing the observatory's capability. More such cycles would give XEUS an active life of more than 20 years.

Xichang Satellite Launch Center

A Chinese launch site in Sichuan province at 28.3° N, 102.0° E from which **geosynchronous satellites** are launched.

XIPS (xenon-ion propulsion system)

A type of **electron bombardment thruster** in which xenon, a heavy inert gas, serves as the propellant. XIPS ("zips") is regarded as one of the most effective forms of **ion propulsion** available. Investigations of xenon-based ion thrusters began in 1984 at Hughes Research Laboratories. These showed that xenon offered the highest thrust of any non-reactive gas; furthermore, being neither corrosive nor explosive, it posed no risk to spacecraft or to personnel loading the propellant tanks. PAS-5 became the first vehicle in space to be equipped with XIPS and was quickly followed by other satellites that use the system for station-keeping. The highly efficient ion system enables a reduction in propellant mass of up to 90% for a satellite designed for 12 to 15 years of operation. Less propellant translates to a reduced launch cost, a bigger payload, or a longer satellite lifetime, or any combination of these. A more powerful XIPS has been developed for use on deep-space missions. Flight-validated by **Deep Space 1**, this new system will be used with increasing frequency by interplanetary probes in the future.

XMM-Newton Observatory

An ESA (European Space Agency) orbiting X-ray observatory. XMM (X-ray Multi-Mirror)-Newton is the largest science satellite ever built in Europe and the most sensitive imaging X-ray observatory in the 250 eV to 12 keV range ever flown, exceeding the mirror area and energy range of **ROSAT, ASCA**, and even **Chandra**. It has three advanced X-ray telescopes, each containing 58 high-precision concentric mirrors nested to offer the largest possible collecting area. In addition, it carries five X-ray imaging cameras and spectrographs, and an optical monitoring telescope. The observatory moves in a highly elliptical orbit, traveling out to nearly one third of the distance to the Moon and enabling long, uninterrupted observations of faint X-ray sources. XMM-Newton is ESA's second Cornerstone mission.

Launch
> Date: December 10, 1999
> Vehicle: Ariane 5
> Site: Kourou
Size: 10×16 m
Mass at launch: 3.8 tons
Orbit: $7,000 \times 114,000$ km $\times 48°$

X-ray astronomy satellites

See, in launch order: **Uhuru** (Dec. 1970), **Copernicus Observatory** (Aug. 1972), **ANS** (Aug. 1974), **Ariel** 5 (Oct. 1974), **Taiyo** (Feb. 1975), **SAS**-3 (May 1975), **HEAO**-1 (Aug. 1977), HEAO-2 (Nov. 1978), **Hakucho** (Feb. 1979), Ariel 6 (Jun. 1979), HEAO-3 (Sep. 1979), **Hinotori** (Feb. 1981), **Tenma** (Feb. 1983), **Astron** (Mar. 1983), **Exosat** (May 1983), **Ginga** (Feb. 1987), **Granat** (Dec. 1989), **ROSAT** (Jun. 1990), **Gamma** (Jul. 1990), **BBXRT** (Dec. 1990), **ASCA** (Feb. 1993), **ALEXIS** (Apr. 1993), **BeppoSAX** (Dec. 1990), **DXS** (Jan. 1993), **RXTE** (Dec. 1995), **ARGOS** (Feb. 1999), **Chandra X-ray Observatory** (Jul. 1999), **XMM-Newton Observatory** (Dec. 1999), **HETE** (Oct. 2000), **Spectrum-X-Gamma** (2003), **High Resolution X-ray Spectroscopy Mission** (2007+), **EXIST** (2007+), **MAXIM** (2007+), **Constellation-X, Joule, XEUS**.

X-rays

Electromagnetic radiation of wavelength shorter than that of **ultraviolet** rays and longer than that of **gamma rays**. X-ray wavelengths range from one thousandth of a nanometer to 10 nm. Hard X-rays are at the high-energy end of the X-ray spectrum and have energies approximately in the range of 3 to 100 keV (0.4 to 0.0124 nm).

XTE (X-ray Timing Explorer)
See **RXTE** (Rossi X-ray Timing Explorer).

Y

Yangel, Mikhail K. (1911–1971)

One of the three most important designers of liquid-propellant rockets in the Soviet Union, the other two being Sergei **Korolev** and Vladimir **Chelomei**. Yangel began his career in the aviation industry and joined Korolev's OKB-1 enterprise almost by chance. As Korolev's associate, he set up a rocket propulsion center in Dnepropetrovsk in the Ukraine, which later formed the basis of his own OKB-586 bureau. At first, Yangel's facility served to mass-produce and further develop intercontinental ballistic missiles (ICBMs) that Korolev originated. However, Yangel quickly became a major competitor, not only as a developer of ICBMs, such as the R-12, R-14, and R-36 (see **"R" series of Russian missiles**), but also of several series of space launch vehicles, including the **Zenit**, **Cosmos (launch vehicle)**, and **Tsyklon**.

Yantar

Soviet photo-reconnaissance satellites. Scores of Yantars (their name means "amber") were launched in the 1970s and 1980s. However, the Russian photo-reconnaissance satellite program has been running at a low level in recent years, with only one launch in 1999 and three in 2000.

Currently three types of Yantar are flown. The close-look Yantar-4K2s, codenamed Kobalt, have three reentry vehicles: two small film-return capsules that can return data while the main satellite continues to operate in orbit, and a conical main descent module that is recovered with the camera system and remaining film at the end of the mission. Yantar-1KFTs, codenamed Kometa, carry out missions to update topographic and mapping data maintained by the Ministry of Defense. The operational lifetime of satellites like Kometa and Kobalt are limited by the amount of film they can carry. However this limitation is overcome with the Yantar-4KS1 Neman satellites. These return images digitally via radio link, either direct to Russian controllers or via data relay satellites in the **Potok** system that are in geosynchronous orbit. With Neman, the lifetime is limited by the amount of propellant carried for orbital maneuvering. A fourth type of photo-reconnaissance satellite currently used by Russia is the Orlets-2. Unlike the 6.6 to 7 ton Yantar-class satellites, which are launched by Soyuz-Us, the 12-ton Orlets requires the much larger Zenit-2.

Yardley, John F. (1925–2001)

A leading figure in the early days of human spaceflight and the Space Shuttle program. Yardley served as associate administrator for manned space flight, a post in which he led the **Apollo-Soyuz Test Project** and **Spacelab**, and was responsible for the development and acquisition of the **Space Shuttle**. In 1981, following the first successful Space Shuttle mission, he returned to private industry to serve as president of the former McDonnell Douglas Astronautics. He was awarded NASA's Public Service Medal for his outstanding contributions to the Mercury and Gemini Programs in 1963 and 1966.

Yavne

A classified Israeli launch complex located at or near Palmachim Air Force Base at 31.5° N, 34.5° E. Launches are limited to **retrograde orbit**s because of range safety restrictions.

yaw

The lateral, rotational, or oscillatory movement of a vehicle about its vertical axis. The amount of movement is measured in degrees.

Yeager, Charles ("Chuck") E. (1923–)

A U.S. Air Force test pilot who, on October 14, 1947, piloted the **X-1** on the first supersonic powered flight–an event he almost missed. The Sunday before his record-breaking mission, while horseback riding, Yeager hit a fence that had been closed across a road, and he cracked two ribs. Instead of informing the flight surgeon and risking being grounded, Yeager and a friend decided he could fly the plane but would have difficulty reaching over to lock the cockpit door. His friend cut off a piece of broomstick, which Yeager used to push the locking mechanism closed before taking off on his historic flight. Upon his return, only a few were able to congratulate him immediately on his achievement. The X-1 project was classified at the time, and his speed record was not made public until June 1948. In later years, Yeager served in several Air Force positions, retiring as a brigadier general.[318]

Yohkoh

A Japanese-led mission, launched by **ISAS** (Institute of Space and Astronautical Science) with collaboration from the United States and Britain, that in 2001 cele-

John Young   STS-1 crew members Commander John Young (left) and Pilot Robert Crippen (right). *NASA*

brated a decade of solar X-ray imaging. It is the first spacecraft to continuously observe the Sun in X-rays over an entire sunspot cycle and carries the longest-operating **CCD** (charge-coupled device) camera in space: the instrument has captured over six million images. According to the latest projections, Yohkoh will stay in orbit until the next solar maximum, around 2010. In the coming years, it will closely collaborate with **RHESSI** (Reuven Ramaty High Energy Solar Spectroscopic Imager), a NASA mission that will provide crucial calibration data for its high-resolution hard X-ray images. Yohkoh ("sunlight") was known before launch as Solar-A. See **Solar-B**.

Launch
  Date: August 30, 1991
  Vehicle: M-3S
  Site: Kagoshima
  Orbit: 516×754 km $\times 31.3°$
  Mass: 420 kg

Young, John Watts (1930–)

A veteran American astronaut who flew on **Gemini** 3 and 10, orbited the moon on **Apollo** 10, walked on the moon on Apollo 16, and commanded two **Space Shuttle** missions, STS-1 and STS-9. Born in San Francisco, Young received a B.S. in aeronautical engineering from Georgia

Institute of Technology in 1952 before joining the Navy. He served on a destroyer for a year, then took flight training and was assigned to Fighter Squadron 103 for four years, flying Cougars and Crusaders. After training at the Navy Test Pilot School in 1959, he was assigned to the Naval Air Test Center for three years, setting time-to-climb records in a Phantom jet. He was later maintenance officer of Phantom Fighter Squadron 103 and retired from the Navy as a captain in 1976. Young was selected as an astronaut in 1962 and flew with Gus Grissom on the first manned Gemini mission, Gemini 3, in 1965, and with Mike Collins on Gemini 10, in 1966. He was the Command Module pilot on Apollo 10, in 1969, and commander of Apollo 16, in 1972. Young's fifth flight was as commander of the first Space Shuttle mission, STS-1, on April 12, 1981. He was back in space aboard *Columbia* from November 28 to December 8, 1983, for the STS-9 mission. In 1973, Young was named chief of the Space Shuttle Branch of the Astronaut Office. The following year, he was selected chief of the astronaut office. He is currently the associate director (technical) at Johnson Space Center and remains an active astronaut.

Yuri

See **BS-(Broadcasting Satellite)**.

Yuri Gagarin Cosmonaut Training Center

See **Gagarin Cosmonaut Training Center**.

Z

Zarya
See **International Space Station**.

Zenit
A Russian launch vehicle, first introduced in 1985 but not generally announced until 1989. Currently, two- and three-stage Zenits are launched from Baikonur and three-stage Zenits from **Sea Launch**'s off-shore platform. Zenit ("zenith") traces its roots to work carried out by Mikhail **Yangel**'s design bureau in Dnepropetrovsk as early as December 1974. In March 1976, the Soviet government issued a decree to develop a two-stage vehicle, 11K77, which later became known as Zenit-2, principally for launching large spy satellites into polar low Earth orbits. The 11K77's first stage was also adopted for use as a strap-on booster for the Energia super-heavy launch vehicle. The most challenging part of the Zenit project was the development of the 804,000-kg thrust four-chamber RD-171 engine for the first stage. It took about 15 years and 900 firing tests for NPO Energomash in Moscow to perfect this ecologically clean, liquid oxygen (LOX)/RP-1 burning propulsion unit. Zenit's second stage is equipped with a single RD-120 engine that uses the same propellants. The two-stage Zenit-2 is able to deliver payloads up to 15 tons into low Earth orbit, making it currently the second most powerful Russian booster after the **Proton**. Topped by an NPO Energomash third stage, it becomes the Zenit 3SL, which is used to transfer satellites into high altitude orbits. The Zenit was designed to have a highly automated launch system and is able to launch in all types of weather. Officially, it can be assembled in 10 working days and made ready for launch just 21 hours after arrival on the pad.

zenith
The point directly overhead; for spacecraft this is arbitrarily defined as "up" and away from the primary body. The opposite of zenith is **nadir**.

zero gravity
See **weightlessness** and **microgravity**.

zero point energy (ZPE)
The energy left behind in a volume of space after all the matter and radiation has been removed. ZPE, also known as vacuum fluctuation energy, is predicted by **quantum mechanics** and gives rise to some unusual phenomena, such as the **Casimir effect**. It represents a vast unexploited potential: according to one estimate, there is enough ZPE in a volume the size of a coffee cup to boil away Earth's oceans. If ZPE can be tapped, it may be of future importance to space travel, a fact that has not gone unnoticed by the U.S. Air Force. A request for proposals by the Air Force Rocket Propulsion Laboratory in 1986 (AF86-77) read: "Bold, new non-conventional propulsion concepts are solicited.... The specific areas in which AFRPL is interested include ... esoteric energy sources for propulsion including the zero point quantum dynamic energy of vacuum space."[238]

zero stage
The common name given to a cluster of **strap-on boosters** that provides thrust additional to that of a sustainer engine, to a spacecraft during ascent into orbit. The zero stage helps the spacecraft carry a greater payload.

Zond
See article, pages 496–497.

Zvezda
See **International Space Station**.

Zenit A Zenit 3SL carrying the XM-1 satellite lifts off from the Sea Launch platform on May 8, 2001. *Boeing*

Zond

A series of Soviet probes ("zond" simply means "probe") that consisted of two entirely different sets of missions. Zonds 1 to 3 were 900-kg, **Venera**-class spacecraft sent to fly by Venus, Mars, and the Moon, respectively. Communication with the first two was lost en route. Zond 3 successfully sent back pictures of the Moon's farside that were much superior in quality to those returned by **Luna** 3, before going into orbit around the Sun. Zonds 4 to 8, by contrast, were much larger, 5-ton vehicles derived from the **Soyuz spacecraft** that formed an early stage of the Soviet Union's L-1 project to send humans on circumlunar flight (see **Russian manned Moon programs**).

Zond 4

Several flight tests of the L-1 had already failed because of problems with the new **Proton** launch vehicle. Then, in early March 1968, Zond 4 was placed in Earth parking orbit and successfully sent out to a lunar distance (though not in the direction of the Moon) to test the L-1 communication system at that range. During the six-day mission, cosmonauts in a bunker on Earth spoke to mission controllers through a relay transmitter onboard the spacecraft. Upon return from cislunar space, Zond 4 was to make a relatively low-g, double-skip entry and land in Soviet territory. However, the probe's astronavigation system failed, forcing the descent module to make a simple ballistic reentry. With the quickly falling probe heading for an Atlantic splashdown, far from any Soviet recovery ship, the spacecraft was deliberately blown up off the Bay of Biscay, France, to prevent it from falling into American hands.

Zond 5

The first spacecraft to loop around the Moon and return to Earth. Aboard was a small menagerie, including turtles, wine flies, meal worms, plants, seeds, and bacteria, to investigate the effects of radiation and other potential hazards in lunar space. On September 18, 1968, Zond 5 flew around the Moon, coming as close as 1,950 km. Upon its return, the astronavigation system again failed, but the wayward capsule splashed down in the Indian Ocean and was successfully recovered by a Soviet tracking ship.

Zond 6

By late 1968, the race for the Moon was reaching its climax. The CIA had informed NASA decision-makers of the Soviet Union's intent to fly a manned circumlunar mission in the near future. With the successful Earth-orbit mission of **Apollo** 7 in October 1968, the Command and Service Modules (CSM) were deemed ready for a lunar mission. However, with the ground testing of the Lunar Module (LM) months behind schedule, there was no way for the United States to mount a previously planned lunar orbit test of the LM to beat the Soviets if they went for an early attempt. Instead, it was decided that only the CSM would be launched on a lunar orbit mission in late December. For the Soviets to be first with their own manned mission, they needed the next L-1 test-flight to go without a hitch. In the event, Zond 6 did manage to do for the first time what all the L-1/Zond probes were supposed to do upon returning to Earth—dip into the atmosphere, skip off, and then enter a second time. The idea of this double reentry was to make it easier for humans returning from the Moon. Whereas Zond 4 experienced forces of $10g$ to $15g$, Zond 6 suffered only $4g$ to $7g$. Unfortunately, the capsule depressurized near the end of its flight, causing the altimeter to fail, which in turn led the parachute line to be jettisoned at a height of 5.3 km and the capsule to crash (just 15 km, as it turned out, from its launch point at Baikonur). What the Soviets publicly hailed a success was in fact a bitter blow to their lunar ambitions. As the December 7 lunar launch window approached, a heated debate took place over whether to put a man on board the next L-1 launch. The cosmonauts training for the mission wrote to the Politburo, saying they were ready to take the risk and arguing that success would be more likely with a human at the controls. Whether this persuaded officials to attempt a manned launch is not clear. What is certain is that after the Proton with its L-1 was rolled out to the pad, a series of problems meant that the launch window was missed. Meanwhile, on Christmas Eve 1968, the crew of Apollo 8 entered lunar orbit and the record books.

There was to be a chilling postscript. The following month, the delayed Proton was rolled out to its pad for with an unmanned L-1: there would be no risking a human passenger now that the race to circle the Moon first was over. The rocket took off and, as its second stage ignited, blew up. The launch escape system also failed, putting it beyond all doubt that had a cosmonaut been on board, he would have been killed.

Zond 7 and 8

Three weeks after Apollo 11 touched down on the lunar surface, Zond flights resumed. Zond 7 looped around the Moon, sent back the first color photographs of the Moon by a Soviet spacecraft, and executed a perfect entry and landing to become the first totally successful flight of the L-1 program. A little over a year later, Zond 8 became the last of its kind, ending its circumlunar foray with a single rather than the planned double reentry. (See table, "Zond Statistics.")

Zond Statistics

Zond	Launch Date	Vehicle	Mass (kg)	Notes
1	Apr. 2, 1964	Molniya	890	Venus flyby, contact lost en route
2	Nov. 30, 1964	Molniya	890	Mars flyby, contact lost en route
3	Jul. 18, 1965	Molniya	959	Lunar flyby, continued into solar orbit
4	Mar. 2, 1968	Proton	5,390	Circumlunar; destroyed during reentry
5	Sep. 15, 1968	Proton	5,390	Circumlunar; returned Sep. 21, 1968
6	Nov. 10, 1968	Proton	5,375	Circumlunar; returned Nov. 17, 1968
7	Aug. 7, 1969	Proton	5,975	Circumlunar; returned Aug. 14, 1969
8	Oct. 20, 1970	Proton	5,375	Circumlunar; returned Oct. 27, 1970

Acronyms and Abbreviations

Boldfaced terms indicate entry heading.

AAS **American Astronautical Society**

ABE **Astrobiology Explorer**

ABMA **Army Ballistic Missile Agency**

A.T. Aerial Target

ACCESS Advanced Cosmic-ray Composition Experiment on the Space Station

ACE Advanced Composition Explorer

ACRIMSAT Active Cavity Radiometer Irradiance Monitor Satellite

ACRV Assured Crew Return Vehicle

ACTS Advanced Communications Technology Satellite

ADE Air Density Explorer

ADEOS Advanced Earth Observation Satellite

AEM Applications Explorer Mission

AERCam Autonomous Extravehicular Robotic Camera

AEROS Advanced Earth Resources Observational Satellite

AFSATCOM Air Force Satellite Communications System

AFSPC **Air Force Space Command**

AIAA **American Institute of Aeronautics and Astronautics**

AIM Aeronomy of Ice in the Mesophere

AIRS Atmospheric Infrared Sounder

ALEXIS Array of Low Energy X-ray Imaging Sensors

ALOS Advanced Land Observing Satellite

ALSEP Apollo Lunar Science Experiment Package. See **Apollo**

AMPTE Active Magnetosphere Particle Tracer Explorer

AMR **Atlantic Missile Range**

AMS Alpha Magnetic Spectrometer

ANS Astronomische Nederlandse Satelliet

ARC **Ames Research Center**

ARGOS Advanced Research and Global Observation Satellite

ARISE Advanced Radio Interferometry between Space and Earth

ARTEMIS Advanced Relay Technology Mission

ARTV Advanced Reentry Test Vehicle

ASAT antisatellite

ASCA Advanced Satellite for Cosmology and Astrophysics

ASCE Advanced Solar Coronal Experiment

ASEB **Aeronautics and Space Engineering Board**

ASI Agenzia Spaziale Italiano

ASLV Advanced Satellite Launch Vehicle

ASSET Aerothermodynamic Elastic Structural Systems Environmental Tests

ASTP **Apollo-Soyuz Test Project**

ASTP **Advanced Space Transportation Program**

ATS Applications Technology Satellite

AU **astronomical unit**

AXAF Advanced X-ray Astrophysics Facility. See **Chandra X-ray Observatory**

BBXRT Broad-Band X-Ray Telescope

BE Beacon Explorer

BECO booster engine cutoff

BGRV **Boost Glide Reentry Vehicle**

BI-1 Bereznyak-Isayev 1

BIRD Bi-spectral Infrared Detection satellite

BIS **British Interplanetary Society**

BMDO **Ballistic Missile Defense Organization**

BNSC **British National Space Centre**

BPPP **Breakthrough Propulsion Physics Program**

BS- Broadcasting Satellite

CALIPSO Cloud-Aerosol Lidar and Infrared Pathfinder Observations

CATSAT Cooperative Astrophysics and Technology Satellite

CCD charge-coupled device

CERISE Characterisation de l'Environment Radio-electrique par un Instrument Spatial Embarque

CGRO **Compton Gamma Ray Observatory**

CGWIC **China Great Wall Industry Corporation**

CHAMP Challenging Minisatellite Payload

CHIPS Cosmic Hot Interstellar Plasma Spectrometer

CINDI Coupled Ion-Neutral Dynamics Investigations

CIRA Centro Italiano Richerche Aerospaziali

CM Command Module. See **Apollo**

CMBPOL Cosmic Microwave Background Polarization

CNES   Centre National d'Etudes Spatiales
CNSR   **Comet Nucleus Sample Return**
COBE   Cosmic Background Explorer
COMETS   Communications and Broadcasting Experimental Test Satellite
COMINT   communications intelligence
Comsat   Communications Satellite Corporation
CONTOUR   Comet Nucleus Tour
COROT   Convection and Rotation of Stars
COSPAR   Committee on Space Research
CRAF   Comet Rendezvous/Asteroid Flyby
CRRES   Combined Release and Radiation Effects Satellite
CRSP   Commercial Remote Sensing Program
CRV   Crew Return Vehicle. See **ACRV**
CS-   Communications Satellite
CSA   **Canadian Space Agency**
CSM   Command and Service Module. See **Apollo**
CSS   **Canadian Space Society**
CSS   **crew safety system**
CZ   Chang Zeng. See **Long March**
DARA   Deutsche Agentur für Raumfahrtangelegenheiten
DBS   **direct broadcast satellite**
DE   Dynamics Explorer
Demosat   Demonstration Satellite
DFH   Dong Fang Hong
DFRC   **Dryden Flight Research Center**
DFVLR   See **DLR**
DLR   Deutschen Zentrum für Luft und Raumfahrt
DME   Direct Measurement Explorer
DMSP   Defense Meteorological Satellite Program
DODGE   Department of Defense Gravity Experiment
DRTS   Data Relay Test Satellite
DS1   **Deep Space 1**
DS2   Deep Space 2. See **Mars Microprobe Mission**
DSCC   **Deep Space Communications Complex**
DSCS   Defense Satellite Communications System
DSN   **Deep Space Network**
DSPS   Defense Support Program Satellite. See **IMEWS**
DSS   **Deep Space Station**
DXS   Diffuse X-ray Spectrometer
ECS   European Space Agency Communications Satellite
ECS   Experimental Communications Satellite
EELV   Evolved Expandable Launch Vehicle
EGNOS   European Geostationary Navigation Overlay Service
EGS   Experimental Geodetic Satellite
ELDO   European Launcher Development Organisation
ELINT   electromagnetic intelligence

ELV   expendable launch vehicle
Envisat   Environmental Satellite
EO satellites   Earth Observing satellites
EORSAT   ELINT Ocean Reconnaissance Satellite
EOS   Earth Observing System
EOS Chem   Earth Observing System Chemistry. See **Aura**
EOS PM   Earth Observing System PM. See **Aqua**
ERBE   Earth Radiation Budget Experiment
ERBS   Earth Radiation Budget Satellite
ERPS   **Experimental Rocket Propulsion Society**
ERS   Earth Resources Satellite
ERS   Environmental Research Satellite
ERT   **Earth Received Time**
ERTS   Earth Resources Technology Satellites
ESA   European Space Agency
ESMC   **Eastern Space and Missile Center**
ESOC   European Space Operations Center
ESRIN   European Space Research Institute
ESRO   European Space Research Organisation
ESSA   Environmental Science Services Administration
ESSP   Earth System Science Pathfinder
ESTEC   European Space Research and Technology Centre
ET   External Tank. See **Space Shuttle**
ETS   Engineering Test Satellite
Eumetsat   European Meteorological Satellite Organisation
EURECA   European Retrievable Carrier
Eutelsat   European Telecommunications Satellite Organisation
EUVE   Extreme Ultraviolet Explorer
EVA   extravehicular activity. See **spacewalk**
EXIST   Energetic X-Ray Imaging Survey Telescope
ExPNS   **Exploration of Neighboring Planetary Systems**
EW   **early warning (satellite)**
FAIR   Filled-Aperture Infrared Telescope
FAST   Fast Auroral Snapshot Explorer
FIRST   Far Infrared and Submillimetre Space Telescope. See **Herschel**
FLTSATCOM   Fleet Satellite Communications
FOBS   Fractional Orbital Bombardment System
FORTE   Fast On-orbit Recording of Transient Events
FTL travel   **faster-than-light travel**
FUSE   Far Ultraviolet Spectroscopic Explorer
GAIA   Global Astrometric Interferometer for Astrophysics
GALEX   Galaxy Evolution Explorer
GARP   Global Atmospheric Research Project
GATV   Gemini Agena Target Vehicle. See **Gemini**
GCOM   Global Change Observing Mission
GEC   Global Electrodynamics Connections

GEO **geosynchronous orbit** or **geostationary orbit**
GEOS Geodetic Earth Orbiting Satellite
GFO Geosat Follow-On
GFZ-1 GeoForschungsZentrum
GIFTS Geostationary Imaging Fourier Transform Spectrometer
GIRD Gruppa Isutcheniya Reaktivnovo Dvisheniya
GLAST Gamma-ray Large Area Space Telescope
GLOMR Global Low Orbiting Message Relay
GLONASS Global Navigation Satellite System
GMS Geosynchronous Meteorological Satellite
GMT Greenwich Mean Time
GOES Geostationary Operational Environmental Satellite
GOMS Geostationary Operational Meteorological Satellite
GPS Global Positioning System
GRACE Gravity Recovery and Climate Experiment
GRC **Glenn Research Center**
GSFC **Goddard Space Flight Center**
GSLV Geosynchronous Satellite Launch Vehicle
GSO **geostationary orbit** or **geosynchronous orbit**
GTO **geosynchronous/geostationary transfer orbit**
HALCA Highly Advanced Laboratory for Communications and Astronomy
HARP High Altitude Research Project
HCMM Heat Capacity Mapping Mission
HEAO High Energy Astrophysical Observatory
HEO highly elliptical orbit or high Earth orbit
HEOS Highly Eccentric Orbiting Satellite
HESSI High Energy Solar Spectroscopic Imager. See **RHESSI**.
HETE High Energy Transient Experiment
HGA **high-gain antenna**
HLV **heavy-lift launch vehicle**
HNX Heavy Nuclei Explorer
HOPE H-2 Orbital Plane
HOTOL Horizontal Takeoff and Landing vehicle
HRF **Human Research Facility**
HST **Hubble Space Telescope**
IBSS Infrared Background Signature Survey
IAE **Inflatable Antenna Experiment**
IAF **International Astronautical Federation**
ICAN (rocket) Ion Compressed Antimatter Nuclear (rocket). See **antimatter propulsion**
ICE International Cometary Explorer
ICESAT Ice, Cloud, and Land Elevation Satellite
IDCSP Initial Defense Communications Satellite Program
IEH International Extreme Ultraviolet Hitchhiker
IEOS International Earth Observing System
IGY **International Geophysical Year**
IKI **Russian Space Research Institute**

ILS **International Launch Services**
IMAGE Imager for Magnetopause-to-Aurora Global Exploration
IMEWS Integrated Missile Early Warning Satellite
IMINT imagery intelligence satellite
IMP Interplanetary Monitoring Platform
INPE Instituto Nacional de Pesquisas Espacias
INTA Instituto Nacional de Tecnica Aeroespacial
INTEGRAL International Gamma-ray Astrophysics Laboratory
Intelsat International Telecommunications Satellite Organization
ION-F Ionospheric Observation Nanosatellite Formation
IQSY **International Quiet Sun Years**
IRAS Infrared Astronomy Satellite
IRBM Intermediate Range Ballistic Missile
IRFNA inhibited red fuming nitric acid. See **nitric acid**
IRIS Infrared Imaging Surveyor
IRS Indian Remote Sensing satellite
IRSI Infrared Space Interferometer. See **Darwin**
IRTS Infrared Telescope in Space
ISAS Institute of Space and Astronautical Sciences
ISM **interstellar medium**
ISRO **Indian Space Research Organisation**
ISEE International Sun-Earth Explorer
ISIS International Satellites for Ionospheric Studies
ISO Infrared Space Observatory
ISS **International Space Station**
ISS Ionospheric Sounding Satellite
ITOS Improved TIROS Operating System
IUE International Ultraviolet Explorer
IUS **inertial upper stage**
JAS Japanese Amateur Satellite
JERS Japanese Earth Resources Satellite
JMEX Jupiter Magnetospheric Explorer
JPL Jet Propulsion Laboratory
JSC **Johnson Space Center**
KAO **Kuiper Airborne Observatory**
KH **Key Hole**
KSC **Kennedy Space Center**
LACE Low-power Atmospheric Compensation Experiment
Lageos Laser Geodynamics Satellite
LDEF Long Duration Exposure Facility
Leasat Leased Satellite
LEO low Earth orbit
LES Launch Escape System. See **Apollo**
LES Lincoln Experimental Satellite
LF **Life Finder**
LGA **low-gain antenna**
LH2 **liquid hydrogen**

LiPS   Living Plume Shield
LISA   Laser Interferometer Space Antenna
LLRV   **Lunar Landing Research Vehicle**
LM   Lunar Module. See **Apollo**
LOR   **lunar-orbit rendezvous**
Losat-X   Low Altitude Satellite Experiment
LOX   **liquid oxygen**
LRV   Lunar Roving Vehicle. See **Apollo**
M2P2   Mini-Magnetospheric Plasma Propulsion
MABES   Magnetic Bearing Satellite
MACSAT   Multiple Access Communications Satellite
MagConst   Magnetospheric Constellation
Magsat   Magnetic Field Satellite
MAP   Microwave Anisotropy Probe
MASTIF   Multiple Axis Space Test Inertia Facility
MAXIM Pathfinder   Microarcsecond X-ray Imaging Pathfinder
MDS   Mission Demonstration test Satellite
MECO   main engine cutoff
MEO   medium Earth orbit
MESSENGER   Mercury Surface, Space Environment, Geochemistry and Ranging
MetOp satellites   Meteorological Operational satellites
MGS   **Mars Global Surveyor**
MIDAS   Missile Defense Alarm System
MIDEX   Medium-class Explorer
Milstar   Military Strategic and Tactical Relay
MMS   **Magnetospheric Multiscale**
MMU   **Manned Maneuvering Unit**
MO   **Mission of Opportunity**
MOL   **Manned Orbiting Laboratory**
MORL   Manned Orbiting Research Laboratory
MOS   Marine Observation Satellite
MOST   Microvariability and Oscillations of Stars
MOUSE   Minimum Orbital Unmanned Satellite of the Earth
MPF   **Mars Pathfinder**
MPL   **Mars Polar Lander**
MSFC   **Marshall Space Flight Center**
MSX   Midcourse Space Experiment
MTI   Multispectral Thermal Imager
MUSES-   Mu Space Engineering Satellites
MUSTARD   Multi-Unit Space Transport and Recovery Device
NACA   National Advisory Committee for Aeronautics
NAR   **National Association of Rocketry**
NASA   National Aeronautics and Space Administration
NASDA   National Space Development Agency
NASP   National Aerospace Plane. See **X-30**
Navsat   Naval Navigation Satellite
Navstar-GPS   Navigation Satellite Time and Ranging Global Positioning System. See **GPS**
NBS   **Neutral Buoyancy Simulator**

NCT   **Nanosat Constellation Trailblazer**
NEAP   Near-Earth Asteroid Prospector
NEAR-Shoemaker   Near Earth Asteroid Rendezvous
NEP   **nuclear-electric propulsion**
NERVA   Nuclear Engine for Rocket Vehicle Application
NGSS   Next Generation Sky Survey
NGST   Next Generation Space Telescope
NIAC   **NASA Institute for Advanced Concepts**
NIMS   Navy Ionsopheric Monitoring System. See **Transit**
NMP   **New Millennium Program**
NNSS   Navy Navigational Satellite System. See **Transit**
NOAA   National Oceanic and Atmospheric Administration
NOSS   Naval Ocean Surveillance Satellite
NRDS   **Nuclear Rocket Development Station**
NRO   **National Reconnaissance Office**
NSC   **National Space Club**
NSI   **National Space Institute**
NSS   **National Space Society**
NSSDC   **National Space Science Data Center**
NUSAT   Northern Utah Satellite
OAO   Orbiting Astronomical Observatory
ODERACS   Orbital Debris Radar Calibration Sphere
OFO   Orbiting Frog Otolith
OGO   Orbiting Geophysical Observatory
OICETS   Optical Inter-orbit Communications Engineering Test Satellite
ONERA   Office National d'Études et de Recherches Aérospatiale
OPAL   Orbiting PicoSat Launcher
OPF   **Orbiter Processing Facility**
ORFEUS   Orbiting and Retrievable Far and Extreme Ultraviolet Spectrometer. See **SPAS**
OSCAR   Orbiting Satellite for Communication by Amateur Radio
OSO   Orbiting Solar Observatory
OWL   Orbiting Wide-angle Light-collectors
OWLT   **one-way light time**
PAET   Planetary Atmospheric Entry Test
Pageos   Passive GEOS
PAM   **Payload Assist Module**
PARASOL   Polarization and Anisotropy of Reflectances for Atmospheric Science coupled with Observations from a Lidar
PFS   Particles and Fields Subsatellite
PI   **Planet Imager**
PICASSO-CENA   Pathfinder for Instruments for Cloud and Aerospace Spaceborne Observations–*Climatologie Etendue des Nuages et des Aerosols.* See **CALIPSO.**
PMG   Plasma Motor/Generator

POES   Polar Operational Environmental Satellite

Polar BEAR   Polar Beacon Experiment and Auroral Research

PPT   **pulsed-plasma thruster**

PRIME   Precision Recovery Including Maneuvering Entry

PRIME   Primordial Explorer

PSLV   Polar Satellite Launch Vehicle

QuikScat   Quick Scatterometer

QuikTOMS   Quick Total Ozone Mapping Spectrometer

RAE   Radiation Astronomy Explorer

RAIR   ram-augmented interstellar rocket. See **interstellar ramjet**

RBM   Radiation Belt Mappers

RCS   **reaction control system**

REX   Radiation Experiment

RFD   **Reentry Flight Demonstrator**

RFNA   red fuming nitric acid. See **nitric acid**

RHESSI   Reuven Ramaty High Energy Solar Spectroscopic Imager

RHU   **radioisotope heater unit**

RKA   **Russian Space Agency**

RLT   **round-trip light time**

RLV   **reusable launch vehicle**

RME   Relay Mirror Experiment

RMS   **Remote Manipulator System**

RORSAT   Radar Ocean Reconnaissance Satellite

ROSAT   Roentgen Satellite

RP-1   rocket propellant 1

RRS   **Reaction Research Society**

RTG   **radioisotope thermoelectric generator**

RXTE   Rossi X-ray Timing Explorer

SAGE   Stratospheric Aerosol and Gas Experiment

SAMPEX   Solar Anomalous and Magnetospheric Particle Explorer

SAR   **synthetic aperture radar**

SARSAT   Search and Rescue Satellites

SAS   Small Astronomy Satellite

SCATHA   Spacecraft Charging At High Altitudes

SCORE   Signal Corps Orbiting Relay Experiment

SCS   Small Communications Satellite. See **MicroSat**

SDIO   Strategic Defense Initiative Organization

SDO   Solar Dynamics Observatory

SDS   Satellite Data System

SECOR   Sequential Collation of Range

SECS   Special Experimental Communications System. See **GLOMR**

SEDS   Small Expendable-tether Deployer System

SEDS   Students for the Exploration and Development of Space

SELENE   Selenological and Engineering Explorer

SEP   **solar-electric propulsion**

SERT   Space Electric Rocket Test

SFU   Space Flyer Unit

SHAR   **Sriharikota Range Center**

SHARP   Super High Altitude Research Project

SIGINT satellites   signals intelligence satellites

SIM   Space Interferometry Mission

SIRTF   Space Infrared Telescope Facility

SLI   **Space Launch Initiative**

SM   Service Module. See **Apollo**

SMART   Small Missions for Advanced Research in Technology

SME   Solar Mesosphere Explorer

SMEX   Small Explorer

SMM   Solar Maximum Mission

SMS   **space motion sickness**

SMS   Synchronous Meteorological Satellite

SNAP   Systems for Nuclear Auxiliary Power

SNOE   Student Nitric Oxide Explorer

SOFIA   Stratospheric Observatory for Infrared Astronomy

SOHO   Solar and Heliospheric Observatory

SOLO   Solar Orbiter

Solrad   Solar Radiation program

SORCE   Solar Radiation and Climate Experiment

SPARCLE   Space Readiness Coherent Lidar Experiment

SPAS   Shuttle Pallet Satellite

SPEAR   Spectroscopy of Plasma Evolution from Astrophysical Radiation

SPIDR   Spectroscopy and Photometry of the Intergalactic Medium's Diffuse Radiation

SPIRIT   Space Infrared Interferometric Telescope

SPOT   Satellite Probatoire d'Observation de la Terre

SRBs   Solid Rocket Boosters. See **Space Shuttle**

SROSS   Stretched Rohini Satellite Series

SRTM   Shuttle Radar Topography Mission

SS   Scientific Satellite

SSC   **Stennis Space Center**

SSF   subsatellite ferrets

SSI   **Space Studies Institute**

SSME   Space Shuttle Main Engine. See **Space Shuttle**

SSS   Small Scientific Satellite

SSTO   **single-stage-to-orbit**

STARSHINE   Student Tracked Atmospheric Research Satellite for Heuristic International Networking Equipment

START   Spacecraft Technology and Advanced Reentry Test

STEDI   Student Explorer Demonstration Initiative

STENTOR   Satellite de Télécommunications pour Expériences de Nouvelles Technologies en Orbite

STEP   Satellite Test of the Equivalence Principle

STEREO   Solar-Terrestrial Relations Observatory

STEX Space Technology Experiments

STRV Space Technology Research Vehicle

STS **Space Transportation System**

SURFSAT Summer Undergraduate Research Fellowship Satellite

SUVO Space Ultraviolet Optical Telescope

SWAS Submillimeter Wave Astronomy Satellite

TAU **Thousand Astronomical Unit Probe**

TDA Target Docking Adapter. See **Gemini**

TDM **time-division multiplexing**

TDRSS Tracking and Data Relay Satellite System

TERRIERS Tomographic Experiment using Radiative Recombinative Ionospheric EUV and Radio Sources

THEMIS Time History of Events and Macroscale Interaction during Substorms

TIMED Thermosphere Ionosphere Mesosphere Energetic Dynamics satellite

TiPS Tether Physics and Survivability satellite

TIROS Television Infrared Observation System

TOD **True of Date**

TOMS Total Ozone Mapping Spectrometer

TOS TIROS Operational System

TPF **Terrestrial Planet Finder**

TRACE Transition Region and Coronal Explorer

TRMM Tropical Rainfall Measuring Mission

TSLC **Taiyuan Satellite Launch Center**

TSS Tethered Satellite System

TSTO **two-stage-to-orbit**

TWINS Two-wide Angle Imaging Neutral-atom Spectrometer

UARS Upper Atmosphere Research Satellite

UFO satellites UHF Follow-On satellites

ULDB Project **Ultra Long-Duration Balloon Project**

UNEX University-class Explorer

USRA Universities Space Research Association

UT Universal Time

UV **ultraviolet**

VAB **Vertical Assembly Building**

VAFB **Vandenberg Air Force Base**

VASIMR Variable Specific Impulse Magnetoplasma Rocket

VCL Vegetation Canopy Lidar

VHO very high orbit

VILSPA **Villafranca Satellite Tracking Station**

VIM Voyager Interstellar Mission. See **Voyager**

VSOP VLBI Space Observatory Program

WIRE Wide-Field Infrared Explorer

WRESAT Weapons Research Establishment Satellite

WSMC **Western Space and Missile Center**

XEUS X-ray Evolving Universe Spectroscopy mission

XIPS xenon-ion propulsion system

XMM-Newton X-ray Multi-Mirror

XTE X-ray Timing Explorer

ZPE **zero point energy**

References

1. Abbot, Alison. "Rubbia proposes a speedier voyage to Mars and back," news item about an idea by Carlo Rubbia, Nobel prize–winning physicist, a new concept of a NERVA-type nuclear rocket. *Nature* 397 (1999): 374.

2. Adams, James. *Bull's Eye: The Assassination and Life of Supergun Inventor Gerald Bull.* New York: Times Books, 1992.

3. Aharonian, F. A. *Very High Energy Cosmic Gamma Radiation: A Crucial Window on the Extreme Universe.* New York: World Scientific Pub. Co., 2001.

4. Alcubierre, M. "The Warp Drive: Hyper-fast travel within general relativity." *Classical and Quantum Gravity* 11 (1994): L73–77.

5. Aldrin, Edwin E., Jr. *Return to Earth.* New York: Random House, 1974.

6. Aldrin, Edwin E., Jr. *Men from Earth.* New York: Bantam Books, 1989.

7. Al-Khalili, J. S. *Black Holes, Wormholes and Time Machines.* London: Institute of Physics, 1999.

8. Allday, Jonathan. *Apollo in Perspective: Spaceflight Then and Now.* Bristol, England: Institute of Physics Publications, 2000.

9. Andrews, D. G., and R. M. Zubrin. "Use of Magnetic Sails for Advanced Exploration Missions." In *Vision-21: Space Travel for the Next Millennium,* edited by G. Landis. NASA CP-10059, April 1990.

10. Armstrong, Neil, et al. *First on the Moon: A Voyage with Neil Armstrong, Michael Collins and Edwin E. Aldrin, Jr.* Boston: Little, Brown, 1970.

11. Armstrong, Neil, et al. *The First Lunar Landing: 20th Anniversary/as Told by the Astronauts, Neil Armstrong, Edwin Aldrin, Michael Collins.* Washington, D.C.: National Aeronautics and Space Administration EP-73, 1989.

12. Arnold, Henry H. *Global Mission.* New York: Harper & Brothers, 1949.

13. Artsutanov, Yuri. "V Kosmos na Electrovoze" (Into Space on a Train). *Komsomolskaya Pravda,* July 31, 1960.

14. Atterley, Joseph [George Tucker]. *A Voyage to the Moon with some Account of the Manners and Customs, Science and Philosophy of the People of Morosofia and Other Lunarians.* New York: Elam Bliss, 1827.

15. Augustine, Norman R. *Augustine's Laws.* Washington, D.C.: American Institute for Aeronautics and Astronautics, 1984.

16. Baker, David. *Spaceflight and Rocketry Chronology.* New York: Facts on File, 1996.

17. Baker, David. *Inventions from Outer Space: Everyday Uses for NASA Technology.* New York: Random House, 2000.

18. Beals, K. A., M. Beaulieu, F. J. Dembia, J. Kerstiens, D. L. Kramer, J. R. West, and J. A. Zito. *Project Longshot: An Unmanned Probe to Alpha Centauri.* U.S. Naval Academy. NASA-CR-184718 (1988).

19. Bean, Alan, et al. *Apollo: An Eyewitness Account by Astronaut/Explorer/Artist/Moonwalker Alan Bean.* The Greenwich Workshop Press, 1998.

20. Beattie, Donald A. *Taking Science to the Moon: Lunar Experiments and the Apollo Program.* Baltimore: Johns Hopkins University Press, 2001.

21. Bennett, C. H., and S. J. Wiesner. "Communication via one- and two-particle operators." *Physical Review Letters* 69 (1992): 2881–84.

22. Benson, Charles D., and William B. Faherty. *Moon Launch!* (The Apollo History Series). Gainesville, Fla.: University Press of Florida, 2001.

23. Bergaust, Erik. *Reaching for the Stars.* New York: Doubleday, 1960.

24. Bergreen, Laurence. *Voyage to Mars: NASA's Search for Life Beyond Earth.* New York: Riverhead Books, 2000.

25. Bernal, J. D. *The World, the Flesh and the Devil.* London: Methuen, 1929.

26. Biblarz, Oscar, and George P. Sutton. *Rocket Propulsion Elements.* New York: John Wiley & Sons, 2001.

27. Bilstein, Roger E. *Flight in America: From the Wrights to the Astronauts.* Baltimore: Johns Hopkins University Press, 2001.

28. Birch, P. "Radiation Shields for Ships and Settlements." *Journal of the British Interplanetary Society* 35 (1982): 515–19.

29. Bond, Alan. "An Analysis of the Potential Performance of the Ram Augmented Interstellar Rocket." *Journal of the British Interplanetary Society* 27 (1974): 674–88.

30. Bond, A., A. R. Martin, R. A. Buckland, T. J. Grant, A. T. Lawton, et al. "Project Daedalus." *Journal of the British Interplanetary Society* 31 (Supplement, 1978).

31. Bondi, H. "Negative Mass in General Relativity." *Reviews of Modern Physics* 29 (1957): 423.

32. Borman, Frank with Robert J. Serling. *Countdown: An Autobiography.* New York: William Morrow, 1988.

33. Borrowman, Gerald L. "Walter R. Dornberger." *Spaceflight* 23 (1981): 118–19.

34. Broeck C. "A 'warp drive' with more reasonable total energy requirements." *Classical and Quantum Gravity* 16 (1999): 3973–79.

35. Bromberg, Joan Lisa. *NASA and the Space Industry.* Baltimore: Johns Hopkins University Press, 2000.

36. Bulkeley, Rip. *The Sputniks Crisis and Early United States Space Policy.* Bloomington: Indiana University Press, 1991.

37. Bull, G. V. "Development of Gun Launched Vertical Probes for Upper Atmosphere Studies." *Canadian Aeronautics and Space Journal* 10 (1964): 236.

38. Bullock, Raymond E. "Fred Lawrence Whipple." In *Notable Twentieth-Century Scientists,* edited by Emily J. McMurray. New York: Gale Research, 1995: 2167–70.

39. Burrough, Bryan. *Dragonfly: NASA and the Crisis Aboard Mir.* New York: Harper, 2000.

40. Caprara, Giovanni. *Living in Space: From Science Fiction to the International Space Station.* Toronto, Canada: Firefly Books, 2000.

41. Carpenter, M. S., et al. *We Seven—by the Astronauts Themselves.* New York: Simon and Schuster, 1962.

42. Casimir, H. G. B. "On the attraction between two perfectly conducting plates." *Proc. Kon. Ned. Akad. van Wetensch* B51 (7): 793–96 (1948).

43. Catchpole, John. *Project Mercury: NASA's First Manned Space Programme.* Chichester, England: Springer-Praxis, 2001.

44. Cernan, Eugene, and Don Davis. *The Last Man on the Moon: Astronaut Eugene Cernan and America's Race in Space.* New York: St. Martin's Press, 1999.

45. Chaiken, Andrew. *A Man on the Moon: The Voyages of the Apollo Astronauts.* New York: Viking, 1994.

46. Chang, Iris. *Thread of the Silkworm.* New York: Basic Books, 1995.

47. Chapman, J. L. *Atlas: The Story of a Missile.* New York: Harper, 1960.

48. "Chesley Bonestell," *Ad Astra,* July/August 1991, 9.

49. Clark, B. C., and L. W. Mason. "The Radiation Show-stopper to Mars Missions." *Case for Mars IV,* University of Colorado, June 4–8, 1990.

50. Clarke, Arthur C. "The Future of World Communications." *Wireless World,* October 1945.

51. Clarke, A. C. "Electromagnetic Launching as a Major Contributor to Space Flight." *Journal of the British Interplanetary Society* 9 (1950): 261–67.

52. Clarke, Arthur C. *The Exploration of Space.* New York: Harper, 1952.

53. Clarke, Arthur C. *Islands in the Sky.* Philadelphia: John C. Winston, 1952.

54. Clarke, Arthur C. *A Fall of Moondust.* New York: Harcourt, Brace & Co., 1961.

55. Clarke, Arthur C. *The Coming of the Space Age.* Des Moines, Iowa: Meredith, 1967.

56. Cleator, P. E. *Rockets Through Space.* New York: Simon & Schuster, 1936.

57. Coffey, Thomas M. *Hap: The Story of the U.S. Air Force and the Man Who Built It.* New York: Viking, 1982.

58. Cole, D. C., and H. E. Puthoff. "Extracting energy and heat from the vacuum." *Physical Review E* 48 (2): 1562–65 (1993).

59. Compton, William. *Where No Man Has Gone Before: A History of the Apollo Lunar Exploration Missions.* Washington, D.C., 1989: National Aeronautics and Space Administration Special Publication 4214.

60. Cooper, Gordon, and Bruce Henderson. *Leap of Faith: An Astronaut's Journey into the Unknown.* New York: Harper Collins, 2000.

61. Cooper, Henry, S. F. *Thirteen: The Apollo Flight That Failed.* Baltimore: Johns Hopkins University Press, 1995.

62. Copernicus, Nicolaus. *De Revolutionibus Orbium Coelestium.* New York: Johnson Reprint, 1965.

63. Cramer, J. G. "The Transactional Interpretation of Quantum Mechanics." *Reviews of Modern Physics* 58 (1986): 647.

64. Cramer, J. G. "An Overview of the Transactional Interpretation of Quantum Mechaincs." *International Journal of Theoretical Physics* 27 (1988): 227.

65. Crocco, Gaetano A. "One-Year Exploration-Trip Earth-Mars-Venus-Earth." *Rendiconti del VII Congresso Internanzionale Astronautico,* Associazione Italiana Razzi (1956): 227–52; paper presented at the Seventh Congress of the International Astronautical Federation, Rome, Italy, 1956.

66. Cromie, Robert. *A Plunge Through Space.* London: Frederick Warne & Co., 1891.

67. Crossfield, A. Scott, with Clay Blair, Jr. *Always Another Dawn.* Cleveland, Ohio: World Publishing, 1960.

68. Crouch, Tom D. *Aiming for the Stars: The Dreamers and Doers of the Space Age.* Washington, D.C.: Smithsonian Institution Press, 1999.

69. Cyrano de Bergerac. *Histoire comique des états et empires de la lune et du soleil.* Paris: Chez Charles le Sercy, 1656. See *Voyages to the Sun and Moon.* London: George Routledge & Sons, 1923, and *Other Worlds,* London: Oxford University Press, 1965.

70. "Cyrano de Bergerac, Savinien," in *The New Encyclopedia Britannica*. Chicago: Encyclopedia Britannica (1987 ed.): 3:829.

71. Damblanc, L. "Les Fusées Autopropulsives à Explosifs." *L'Aérophile* 43 (1935): 205–9, 241–47.

72. Damohn, Mark. *Back Down to Earth: The Development of Space Policy for NASA during the Jimmy Carter Administration*. San Jose, Calif.: Authors Choice Press, 2001.

73. Dawson, Virginia P. *Engines and Innovation: Lewis Laboratory and American Propulsion Technology*. Washington, D.C.: NASA SP-4306 (1991).

74. Day, Dwayne A. *Eye in the Sky: The Story of the Corona Spy Satellites*. Washington, D.C.: Smithsonian Institution Press, 1999.

75. Defoe, Daniel. *The Consolidator: Or Memoirs of Sundry Transactions from the World of the Moon*. London: Benj. Bragg at the Blue Ball, 1705.

76. Deser, S., R. Jackiw, and G. 't Hooft. "A critique of Gott. You can't construct his machine." *Physical Review Letters* 66 (1992): 267.

77. Dethloff, Henry C. *Suddenly Tomorrow Came: A History of the Johnson Space Center*. Washington, D.C.: NASA SP-4307 (1993).

78. Dickson, Paul. *Sputnik: The Shock of the Century*. New York: Walker & Co., 2001.

79. Divine, Robert A. *The Sputnik Challenge: Eisenhower's Response to the Soviet Satellite*. New York: Oxford University Press, 1993.

80. Doolittle, General James H. (Jimmy) with Carroll V. Glines. *I Could Never Be So Lucky Again: An Autobiography*. New York: Bantam Books, 1991.

81. Dornberger, Walter R. *V-2*. Translated by James Cleugh and Geoffrey Halliday. New York: Viking, 1958.

82. DuPre, Flint O. *Hap Arnold: Architect of American Air Power*. New York: Macmillan, 1972.

83. Dyson, F. J. "Search for Artificial Stellar Sources of Infrared Radiation." *Science* 131 (June 3, 1960).

84. Dyson, F. "Death of a Project." *Science* 149 (July 9, 1965): 141–44.

85. Dyson, F. "Interstellar Transport." *Physics Today*, October 1968, 41.

86. El-Genck, Mohamed S., ed. *A Critical Review of Space Nuclear Power and Propulsion*. New York: Springer-Verlag, 1994.

87. Ellis, Lee A. *Who's Who of NASA Astronauts*. New York: Americana Group Publishing, 2001.

88. Engelmann, Joachim. *V2: Dawn of the Rocket Age*. Schiffer Military History, Vol. 26. Atglen, Pa: Shiffer Publishing, 1990.

89. Esnault-Pelterie, Robert. *L'Astronautique*. Paris: A. Lahure, 1930.

90. Esnault-Pelterie, Robert. *L'Astronautique-Complément*. Paris: Société des Ingénieurs Civils de France, 1935.

91. Essers, I. *Max Valier: A Pioneer of Space Travel*. Washington, D.C.: NASA TT F-664 (1976).

92. Everest, F. K., Jr. *The Fastest Man Alive*. New York: Dutton, 1958.

93. Eyraud, Achille. *Voyage à Venus*. Paris: Michel Levy Frères, 1863.

94. Eyre, F. W. "The Development of Large Bore Gun Launched Rockets." *Canadian Aeronautics and Space Journal* 12 (1966): 143–49.

95. Feinberg, Gerald. "Particles That Go Faster Than Light." *Scientific American*, February 1970, 69–77.

96. Forward, R. L. "Roundtrip Interstellar Travel Using Laser-Pushed Lightsails." *Journal of Spacecraft* 21 (2): 187–95 (1984).

97. Forward, R. L. "Antiproton Annihilation Propulsion." *Journal of Propulsion* 1 (5): 370–74 (1985).

98. Forward, R. "Negative matter propulsion." *Journal of Propulsion* 6 (1): 28–37 (1990).

99. Forward, R. L. "Apparent Method for Extraction of Propulsion Energy from the Vacuum." Paper AIAA 98-3140, *34th AIAA/ASME/SAE/ASEE Joint Propulsion Conference*, Cleveland, Ohio (13–15 July 1998).

100. Fowler, Eugene. *One Small Step: Apollo 11 and the Legacy of the Space Age*. New York: Smithmark Publishing, 1999.

101. Freedman, S. J., and J. F. Clauser. "Experimental Test of Hidden Variable Theories." *Physical Review Letters* 28 (1972): 938.

102. Freeman, Marsha, et al. *How We Got to the Moon: The Story of the German Space Pioneers*. Washington, D.C.: Twenty First Century Science, 1994.

103. Freeman, Marsha. *Challenges of Human Space Exploration*. Chichester, England: Springer-Praxis, 2000.

104. French, F. W. "Solar Flare Radiation Protection Requirements for Passive and Active Shields." *Journal of Spacecraft* 7 (7): 794–800 (1970).

105. Fries, Sylvia D. "2001 to 1994: Political Environment and the Design of NASA's Space Station System." *Technology and Culture* 29 (1988): 568–93.

106. Furniss, Tim, and Alexa Stace. *The Atlas of Space Exploration*. New York: Gareth Stevens, 2000.

107. Furniss, Tim. *The History of Space Vehicles*. San Diego: Thunder Bay Press, 2001.

108. Gagarin, Yuri. *Road to the Stars*. Moscow: Foreign Languages Publishing, 1962.

109. Gavaghan, Helen. *Something New Under the Sun: Satellites and the Beginning of the Space Age*. New York: Copernicus, 1998.

110. Gedney, Richard T., Ronald Schertler, and Frank Gargione. *The Advanced Communications Technology Satel-

lite: An Insider's Account of the Emergence of Interactive Broadband Technology in Space. Mendham, N.J.: Scitech Publishing, 2000.

111. Gerom, Leonard, and Lennard Gerom. *Who's Who in Russia and the New States.* New York: St. Martin's Press, 1993.

112. Glaser, P. E. "Power from the Sun, It's Future." *Science* 162 (1968): 857–86.

113. Glines, Carroll V. *Jimmy Doolittle: Daredevil Aviator and Scientist.* New York: Macmillan, 1972.

114. Goddard, R. H. *The Ultimate Migration* (manuscript), Jan. 14, 1918, The Goddard Biblio Log, Friends of the Goddard Library, Nov. 11, 1972.

115. Goddard, Robert H. *Rockets.* New York: American Rocket Society, 1946. (Contains Goddard's two Smithsonian papers.)

116. Goddard, Robert H. *Rocket Development: Liquid-Fuel Rocket Research, 1929–1941.* Englewood Cliffs, N.J.: Prentice-Hall, 1948.

117. Godwin, Robert, comp. *Apollo 9: The NASA Mission Reports.* Burlington, Ontario: Apogee Books, 1999.

118. Godwin, Robert, comp. *Apollo 10: The NASA Mission Reports.* Burlington, Ontario: Apogee Books, 1999.

119. Godwin, Robert, comp. *Friendship 7: The NASA Mission Reports.* Burlington, Ontario: Apogee Books, 1999.

120. Godwin, Robert, comp. *Gemini 6: The NASA Mission Reports.* Burlington, Ontario: Apogee Books, 1999.

121. Godwin, Robert, comp. *Apollo 11: The NASA Mission Reports, Volume 2.* Burlington, Ontario: Apogee Books, 1999.

122. Godwin, Robert, comp. *Apollo 12: The NASA Mission Reports.* Burlington, Ontario: Apogee Books, 1999.

123. Godwin, Robert, comp. *Apollo 13: The NASA Mission Reports.* Burlington, Ontario: Apogee Books, 2000.

124. Godwin, Robert, comp. *Apollo 7: The NASA Mission Reports.* Burlington, Ontario: Apogee Books, 2000.

125. Godwin, Robert, comp. *Apollo 8: The NASA Mission Reports.* Burlington, Ontario: Apogee Books, 2000.

126. Godwin, Robert, comp. *Mars: The NASA Mission Reports.* Burlington, Ontario: Apogee Books, 2000.

127. Godwin, Robert, comp. *Apollo 14: The NASA Mission Reports.* Burlington, Ontario: Apogee Books, 2001.

128. Godwin, Robert, comp. *Space Shuttle STS 1–5: The NASA Mission Reports.* Burlington, Ontario: Apogee Books, 2001.

129. Godwin, Robert, comp. *Rocket and Space Corporation Energia: The Legacy of S. P. Korolev.* Burlington, Ontario: Apogee Books, 2001.

130. Godwin, Robert, and Steve Whitfield, comps. *Freedom 7: The NASA Mission Reports.* Burlington, Ontario: Apogee Books, 2001.

131. Gonsales, Domingo [Francis Godwin]. *The Man in the Moone: Or a Discourse of a Voyage Thither.* London: John Norton for Ioshua Kirton and Thomas Warren, 1638.

132. Goodstein, Judith R. *Millikan's School: A History of the California Institute of Technology.* New York: W.W. Norton, 1991.

133. Gorn, Michael H. *The Universal Man: Theodore von Kármán's Life in Aeronautics.* Washington, D.C.: Smithsonian Institution Press, 1992.

134. Gorn, Michael H. *Expanding the Envelope: Flight Research at NACA and NASA.* Lexington, Ky.: University Press of Kentucky, 2001.

135. Gott, J. R. *Astrophysical Journal* 288 (1985): 422.

136. Gott, J. R., III. "How pairs of cosmic strings can act as time machines." *Physical Review Letters* 66 (1991): 1126.

137. Greg, Percy. *Across the Zodiac: The Story of a Wrecked Record.* 2 vols. London: Trubner & Co., Ludgate Hill, 1880.

138. Grissom, Betty and Henry Still. *Starfall.* New York: Thomas Crowell, 1974.

139. Hacker, Barton C. "The Idea of Rendezvous: From Space Station to Orbital Operations in Space-Travel Thought, 1895–1951." *Technology and Culture* 15 (July 1974).

140. Hale, Edward Everett. "The Brick Moon." *Atlantic Monthly,* vol. XXIV (Oct., Nov., Dec. 1869).

141. Hall, R. Cargill. "The Eisenhower Administration and the Cold War: Framing American Astronautics to Serve National Security." *Prologue: Quarterly of the National Archives* 27 (1995): 59–72.

142. Hall, Rex, and David J. Shayler. *The Rocket Men: Vostok and Voskhod, the First Soviet Manned Spaceflights.* New York: Springer Verlag, 2001.

143. Hannah, E. C. "Meteoroid and Cosmic-Ray Protection." In *Space Manufacturing Facilities,* proceedings of the Princeton/AIAA/NASA Conference, edited by J. Grey. May 7–9, 1975, AIAA (1977): 151–57.

144. Hansen, James R. *Engineer in Charge: A History of the Langley Aeronautical Laboratory, 1917–1958.* Washington, D.C.: NASA SP-4305 (1987), 386–88.

145. Harford, James. *Korolev: How One Man Masterminded the Soviet Drive to Beat America to the Moon.* New York: John Wiley & Sons, 1997.

146. Harland, David. *Jupiter Odyssey: The Story of NASA's Galileo Mission.* New York: Springer Verlag, 2000.

147. Haroche, S., and J.-M. Raimond. "Cavity quantum electrodynamics." *Scientific American,* April 1993, 54–62.

148. Harrison, Albert A. *Spacefaring: The Human Dimension.* Berkeley: University of Calfornia Press, 2001.

149. Harvey, Brian. *Russia in Space: The Failed Frontier?* New York: Springer Verlag, 2001.

150. Heppenheimer, T. A. *A Brief History of Flight: From Balloons to Mach 3 and Beyond.* New York: John Wiley & Sons, 2001.

151. Hochberg, D., and T. W. Kephart. "Vacuum Squeezing." *Physics Letters B* 268 (1991): 377.

152. Holloway, Jean. *Edward Everett Hale: A Biography.* Austin, Tex.: University of Texas Press, 1956.

153. Hunley, J. D., ed. *The Birth of NASA: The Diary of T. Keith Glennan.* Washington, D.C.: U.S. Government Printing Office, NASA SP-4105 (1993).

154. Irwin, James B. *To Rule the Night.* Nashville, Tenn.: Broadman & Holman, 1982.

155. Isaacs, J., et al. "Satellite Elongation into a True Sky-Hook." *Science,* Vol. 151, February 11, 1966, 682–83; also Vol. 152, p. 800, and Vol. 158, p. 947.

156. Jackson, A. "Some Considerations on the Antimatter and Fusion Ram Augmented Interstellar Rocket." *Journal of the British Interplanetary Society* 33 (1980): 117–20.

157. Jenkins, Dennis R. *Space Shuttle: The History of the National Space Transportation System, the First 100 Missions,* 3rd ed. Cape Canaveral, Fla.: Dennis R. Jenkins, 2001.

158. Joels, Kerry Mark. *The Space Shuttle Operator's Manual.* New York: Ballantine, 1998.

159. Johnson, Clarence L. "Kelly" with Maggie Smith. *Kelly: More Than My Share of It All.* Washington, D.C.: Smithsonian Institution, 1985.

160. Johnson-Freese, Joan. *The Chinese Space Program: A Mystery Within a Maze.* Melbourne, Fla.: Krieger Publishing Co., 1998.

161. Jones, Eric M., and Ben R. Finney, eds. *Interstellar Migration and the Human Experience.* Berkeley: University of California Press, 1986.

162. Joubert de la Ferté, Sir Philip B. *Rocket.* New York: Philosophical Library, 1957.

163. Kantrowitz, A. "Propulsion to Orbit by Ground Based Lasers." *Aeronautics and Astronautics* 9 (3): 34–35 (May 1972).

164. Kaplan, Joseph. "The IGY Program." *Proceedings of the IRE* (June 1956): 741–43.

165. Kaplan, Joseph. "The Aeronomy Story: A Memoir." In *Essays on the History of Rocketry and Astronautics: Proceedings of the Third Through the Sixth History Symposia of the International Academy of Astronautics,* edited by R. Cargill Hall. Washington, D.C.: NASA Conference Publication 2014, 2 (1977): 423–27.

166. Kelly, Thomas J. *Moon Lander: How We Developed the Apollo Lunar Module.* Washington, D.C.: Smithsonian Institution Press, 2001.

167. Kepler, Johannes. See Lear, John. *Kepler's Dream, with Full Text and Notes of Somnium, Sive Astronomia Lunaris.* Berkeley: University of California Press, 1965.

168. Kluger, Jeffrey, et al. *Apollo 13: Anniversary Edition.* New York: Houghton Mifflin Co., 2000.

169. Kommash, Terry, ed. *Fusion Energy in Space Propulsion.* Reston, Va.: American Institute of Aeronautics, 1995.

170. Koppes, Clayton R. *JPL and the American Space Program: A History of the Jet Propulsion Laboratory.* New Haven, Conn.: Yale University Press, 1982.

171. Kozloski, Lillian D. *U.S. Space Gear: Outfitting the American Astronaut.* Washington, D.C.: Smithsonian Institution Press, 2000.

172. Kraemer, Robert S., and Roger D. Launius. *Beyond the Moon: The Golden Age of Planetary Exploration 1971–1978.* Washington, D.C.: Smithsonian Institution Press, 2000.

173. Kraft, Chris C., and James L. Schefter. *Flight: My Life in Mission Control.* New York: Dutton, 2001.

174. Kranz, Gene. *Failure Is Not an Option: Mission Control from Mercury to Apollo 13 and Beyond.* New York: Simon & Schuster, 2000.

175. Krige, John, and Arturo Russo. *A History of the European Space Agency, 1958–1987. Volume I: The Story of ESRO and ELDO, 1958–1973.* The Netherlands: ESTEC, ESA Publications Division, SP-1235 (2000).

176. Krige, John, Arturo Russo, and Lorenza Sebesta. *A History of the European Space Agency, 1958–1987. Volume II: The Story of ESA, 1973–1987.* The Netherlands: ESTEC, ESA Publications Division, SP-1235 (2000).

177. La Folie, Louis Guillaume de. *Le Philosophe Sans Prétention. . . .* Paris: Clusier, 1775.

178. Lambright, W. Henry. *Powering Apollo: James E. Webb of NASA.* Baltimore: Johns Hopkins University Press, 1995.

179. Lamoreaux, S. K. "Demonstration of the Casimir force in the 0.6 to 6mm range." *Physical Review Letters* 78 (1): 5–8 (1997).

180. Launius, Roger D. "A Western Mormon in Washington, D.C.: James C. Fletcher, NASA, and the Final Frontier." *Pacific Historical Review* 64 (1995): 217–41.

181. Launius, Roger D. "Jerome C. Hunsaker." In *Notable Twentieth-Century Scientists,* edited by Emily J. McMurray, et al. New York: Gale Research, 1995: 980–81.

182. Launius, Roger D. *Frontiers of Space Exploration.* Westport, Conn.: Greenwood Publishing Group, 1998.

183. Launius, Roger D., John M. Logsdon, and Robert W. Smith, eds. *Reconsidering Sputnik: Forty Years Since the Soviet Satellite.* Amsterdam, The Netherlands: Harwood Academic Publishers, 2000.

184. *Leadership and America's Future in Space.* Washington, D.C.: U.S. Government Printing Office, 1987.

185. Lebedev, D. A. "The N1-L3 Programme." *Spaceflight,* September 1992.

186. Lebedev, D. A. "The N1-L3 Programme." *Spaceflight,* February 1993.

187. Lehman, Milton. *Robert H. Goddard: A Pioneer of Space Research,* New York: Da Capo, 1988.

188. Leverington, David. *New Cosmic Horizons: Space Astronomy from the V-2 to the Hubble Space Telescope.* New York: Cambridge University Press, 2001.

189. Levy, R. H. "Radiation Shielding of Space Vehicles by Means of Superconducting Coils." *ARS Journal* 31 (11): 1568–70 (1961).

190. Lewis, R. A., K. Meyer, and T. Schmidt. "AIM-STAR: Antimatter Initiated Microfusion for Precursor Interstellar Missions. *AIAA,* 98-3404, July 1998.

191. Ley, Willy. *Conquest of Space.* New York: Viking, 1949.

192. Ley, Willy. *Rockets, Missiles, and Space Travel.* New York: Viking, 1961.

193. Lindsay, Hamish. *Tracking Apollo to the Moon.* New York: Springer Verlag, 2001.

194. Linenger, Jerry M. *Off the Planet: Surviving Five Perilous Months Aboard the Space Station Mir.* New York: McGraw-Hill Professional Publishing, 2000.

195. Logsdon, John M. *The Decision to Go to the Moon: Project Apollo and the National Interest.* Cambridge, Mass.: MIT Press, 1970.

196. Lowther, William. *Dr. Gerald Bull, Iraq, and the Supergun.* Toronto: McClelland-Bantam, 1991.

197. Lucian (of Samosata). *Trips to the Moon.* Translated by Thomas Francklin. London: Cassell & Co., 1887.

198. Maglich, Bogdan, ed. "Experiments with Bomb-Propelled Spaceship Models." *Adventures in Experimental Physics* β, World Science Education. (1972): 320.

199. Mallove, Eugene F., and Gregory L. Matloff. *The Starflight Handbook: A Pioneer's Guide to Interstellar Travel.* New York: John Wiley & Sons, 1989.

200. Martin, Richard E. *The Atlas and Centaur "Steel Balloon" Tanks: A Legacy of Karel Bossart.* San Diego: General Dynamics Corp., 1989.

201. McDougall, Walter A. *The Heavens and the Earth: A Political History of the Space Age.* New York: Basic Books, 1985.

202. McPhee, John. *The Curve of Binding Energy.* New York: Farrar Strauss Giroux, 1974.

203. Meilinger, Phillip S. *Hoyt S. Vandenberg: The Life of a General.* Bloomington: Indiana University Press, 1989.

204. Mellberg, William F. *Moon Missions: Mankind's First Voyages to Another World.* Vergennes, Vt.: Plymouth Press, 1997.

205. Miller, Ron, and Frederick C. Durant III. *The Art of Chesley Bonestell.* London: Paper Tiger, 2001.

206. Millis, M. "Challenge to Create the Space Drive." *AIAA Journal of Propulsion and Power* 13 (5): 577–82 (1997).

207. Millman, Peter M. "Big Gun on Barbados." *Sky and Telescope,* August 1966, 64.

208. Milonni, P. W., R. J. Cook, and M. E. Goggin. "Radiation pressure from the vacuum: Physical interpretation of the Casimir force." *Physical Review A* 38 (3): 1621–23 (1988).

209. Misner, C. W., K. S. Thorne, and J. A. Wheeler. *Gravitation.* New York: W. H. Freeman, 1973.

210. Montenbruck, Oliver and Eberhard Gill. *Satellite Orbits: Models, Methods, Applications.* New York: Springer Verlag, 2000.

211. Morris, M. S., K. S. Thorne, and U. Yurtsever. "Wormholes in Spacetime and Their Use for Interstellar Travel: A Tool for Teaching General Relativity." *American Journal of Physics* 56 (5): 395–412 (1988).

212. Muenger, Elizabeth A. *Searching the Horizon: A History of Ames Research Center, 1940–1976.* Washington, D.C.: NASA SP-4304, 1985, 12–14, 67–68, 131–132.

213. NASA Publication SP-428, *Space Resources and Space Settlement,* 1979.

214. Needell, Allan A. *Science, Cold War and the American State: Lloyd V. Berkner and the Balance of Professional Ideals.* Amsterdam, The Netherlands: Harwood Academic Publishers, 2000.

215. Neufeld, Jacob. "Bernard A. Schriever: Challenging the Unknown." *Makers of the United States Air Force.* Washington, D.C.: Office of Air Force History, 1986, 281–306.

216. Neufeld, Michael J. *The Rocket and the Reich: Peenemünde and the Coming of the Ballistic Missile Era.* Cambridge, Mass.: Harvard University Press, 1996.

217. Newell, Homer E. *Beyond the Atmosphere: Early Years of Space Science.* Washington, D.C.: NASA SP-4211 (1980).

218. Newlan, Irl. *First to Venus: The Story of Mariner II.* New York: McGraw-Hill, 1963.

219. Newton, David E. "James A. Van Allen." In *Notable Twentieth-Century Scientists,* edited by Emily J. McMurray et al. New York: Gale Research, 1995, 2070–72.

220. Noordung, H. *Das Problem der Befahrung des Weltraums* (The Problem of Space Flight). Berlin: Schmidt and Co., 1928.

221. Noordung, Hermann. *The Problem of Space Travel: The Rocket Motor.* Edited by Ernst Stuhlinger and J. D. Hunley with Jennifer Garland. Washington, D.C.: Government Printing Office, NASA SP-4026 (1995).

222. Oberth, Hermann. *Die Rakete zu den Planetenräumen* (The Rocket into Interplanetary Space). Munich: R. Oldenbourg, 1923.

223. O'Neill, G. K. "The Colonization of Space." *Physics Today* 27 (September 1974): 32–40.

224. O'Neill, G. K. "Space Colonies and Energy Supply to the Earth." *Science* 10 (5): 943–47 (December 1975).

225. O'Neill, Gerard Kitchen. *The High Frontier: Human Colonies in Space.* New York: William Morrow, 1977.

226. Ordway, Frederick I., III. *Visions of Spaceflight: Images from the Ordway Collection.* New York: Four Walls Eight Windows, 2001.

227. Pearson, Jerome. "The orbital tower: a spacecraft launcher using the Earth's rotational energy." *Acta Astronautica* 2 (9–10): 785–99 (September–October 1975).

228. Perry, Robert L. "The Atlas, Thor, Titan, and Minuteman." In *A History of Rocket Technology,* edited by Eugene M. Emme. Detroit, Mich.: Wayne State University Press, 1964, 143–55.

229. Petechuk, David. "Vannevar Bush." In *Notable Twentieth-Century Scientists,* edited by Emily J. McMurray et al. New York: Gale Research, 1995, 285–88.

230. Pfenning, M. J., and L. H. Ford. "The unphysical nature of 'warp drive'." *Classical and Quantum Gravity* 14 (1997): 1743–51.

231. Piszkiewicz, Dennis. *The Nazi Rocketeers: Dreams of Space and Crimes of War.* Westport, Conn.: Praeger, 1995.

232. Piszkiewicz, Dennis. *Wernher Von Braun: The Man Who Sold the Moon.* Westport, Conn.: Praeger, 1999.

233. Podkletnov, E., and R. Nieminen. "A Possibility of Gravitational Force Shielding by Bulk YBa2Cu307-x Superconductor." *Physica C* 203 (1992): 441–44.

234. Poe, Edgar Allan. "The Unparalled Adventure of One Hans Pfaall." In *Works of Edgar Allan Poe,* edited by J. H. Ingram. Edinburgh: A. & C. Black, 1875.

235. Pogue, Forrest C. *George C. Marshall.* New York: Viking, 1963–1966.

236. Powell, J. W. "The Flight of Able and Baker." *Journal of the British Interplanetary Society* 38 (1985): 94–96.

237. Price, R. "Negative mass can be positively amusing." *American Journal of Physics,* March 1993, 216–17.

238. Puthoff, H. E. "Source of vacuum electromagnetic zero-point energy." *Physical Review A* 40 (9): 4857–62 (1989); 44 (5): 3385–86 (1991).

239. Puthoff, H. E. "SETI, the velocity-of-light limitation, and the Alcubierre warp drive: An integrating overview." *Physics Essays* 9 (1): 156–58 (1996).

240. Rauschenbach, Boris V. *Hermann Oberth: The Father of Spaceflight.* Clarence, New York: West-Art, 1994.

241. Recami, E., ed. *Tachyons, Monopoles, and Related Topics.* Amsterdam: North Holland, 1978.

242. Redfield, Peter. *Space in the Tropics: From Convicts to Rockets in French Guiana.* Berkeley: University of California, 2000.

243. Reeves-Stevens, Judith, Garfield Reeves-Stevens, and Brian Muirhead. *Going to Mars: The Untold Story of Mars Pathfinder and NASA's Bold New Missions for the 21st Century.* New York: Pocket Books, 2000.

244. Regaskis, Richard. *X-15 Diary: The Story of America's First Space Ship.* Lincoln, Nebr.: iUniverse.com, 2000.

245. *Report of the Advisory Committee on the Future of the U.S. Space Program.* Washington, D.C.: Government Printing Office, 1990.

246. Reuter, Claus. *The V2, and the Russian and American Rocket Program.* New York: S. R. Research & Publishing, 2000.

247. Rich, H. T. "The Flying City." *Astounding Stories,* August 1930.

248. Rosen, Milton. *The Viking Rocket Story.* New York: Harper, 1955.

249. Ross, H. L. "Orbital Bases." *Journal of the British Interplanetary Society* 8 (1): 1949, 1–19.

250. Ross, Walter S. *The Last Hero: Charles A. Lindbergh.* New York: Harper & Row, 1967.

251. Rotundo, Louis, C., and Charles E. Yeager. *Into the Unknown: The X-1 Story.* Washington, D.C.: Smithsonian Institution Press, 1994.

252. Ryan, Craig. *The Pre-Astronauts: Manned Ballooning on the Threshold of Space.* Annapolis, Md.: United States Naval Institute, 1995.

253. Rynin, Nikolai Aleksevich. *Mezhplanetnye Soobshceniya* (Interplanetary Communications). Nine volumes. Various Leningrad publishers, 1928–1932.

254. Sagan, Carl. *Pale Blue Dot: A Vision of the Human Future in Space.* New York: Ballantine Books, 1997.

255. Sagdeyev, Roald Z. *The Making of a Soviet Scientist: My Adventures in Nuclear Fusion and Space from Stalin to Star Wars.* New York: John Wiley & Sons, 1995.

256. Sakharov, A. "Vacuum quantum fluctuations in curved space and the theory of gravitation." *Soviet Physics-Dokl.* 12 (11): 1040–41 (1968).

257. Sänger, Eugen. *Raketenflugtechnik.* Munich: R. Oldenbourg, 1933.

258. Sänger, Eugen, and Irene Bredt. *Rocket Drive for Long Range Bombers.* Whittier, Calif.: Robert Cornog, 1952.

259. Santarius, J. F. "Lunar Helium-3, Fusion Propulsion, and Space Development." In *Second Conference on Lunar Bases and Space Activities of the 21st Century* (Houston, Texas, April 5–7, 1988) (NASA Conf. Pub. 3166, 1, 75, 1992).

260. Scharnhorst, K. *Physics Letters* B236 (1990): 354.

261. Schefter, James. *The Race: The Complete True Story of How America Beat Russia to the Moon.* New York: Anchor Books, 2000.

262. Schirra, Walter M., and Richard Billings. *Schirra's Space*. Bluejacket Books. Annapolis, Md.: United States Naval Institute, 1995.

263. Schmidt, G., H. Gerrish, and J. J. Martin. "Antimatter Production for Near-term Propulsion Applications." 1999 Joint Propulsion Conference.

264. Schwinger, J. "Casimir light: The source." *Proceedings of the National Academy of Science* 90 (1993): 2105–6.

265. Scott, David, comp., and Robert Godwin. *Apollo 15: The NASA Mission Reports, Volume One*. Burlington, Ontario: Apogee Books, 2001.

266. Scott, William B. "SHARP Gun Accelerates Scramjets to Mach 9." *Aviation Week and Space Technology,* September 9, 1996, 63.

267. Serviss, Garrett P. *Edison's Conquest of Mars*. Los Angeles: Cacosa House, 1947.

268. Shayler, David J. *Disasters and Accidents in Manned Spaceflight*. Chichester, England: Springer-Praxis, 2000.

269. Shayler, David. *Gemini: Steps to the Moon*. New York: Springer Verlag, 2001.

270. Shepard, Alan, and Deke Slayton. *Moonshot: The Inside Story of America's Race to the Moon*. New York: Turner Publishing, 1994.

271. Shepherd, L. R. "Interstellar Flight." *Journal of the British Interplanetary Society* 11 (July 1952): 149–67.

272. Slayton, Donald K. *Deke! U.S. Manned Space: From Mercury to the Shuttle*. New York: Forge, 1995.

273. Sloop, John L. *Liquid Hydrogen as a Propulsion Fuel, 1945–1959*. Washington, D.C.: NASA, SP-4404 (1978), 173–77.

274. Smith, G. A., G. Gaidos, R. A. Lewis, K. Meyer, and T. Schmid. "Aimstar: Antimatter Initiated Microfusion for Precursor Interstellar Missions." *Acta Astronautica* 44 (1999): 183–86.

275. Smith, G. A., et al. "Antiproton Production and Trapping for Space Propulsion Applications." Unpublished paper. Read on-line at www.engr.psu.edu/antimatter/papers/anti_prod.pdf

276. Smith, Richard K. *The Hugh L. Dryden Papers, 1898–1965*. Baltimore: The Johns Hopkins University Library, 1974.

277. Stiernstedt, Jan. *Sweden in Space: Swedish Space Activities, 1959–1972*. The Netherlands: ESA Publications Division, ESA SP-1248, March 2001.

278. Stoler, Mark A. *George C. Marshall: Soldier-Statesman of the American Century*. Boston: Twayne, 1989.

279. Stone, R. W., Jr. "An Overview of Artificial Gravity." NASA Report SP-314 (1973).

280. Stuhlinger, Ernst, ed. *The Problems of Space Travel: The Rocket Motor (The NASA History Series)*. Washington, D.C.: NASA, 1995.

281. Stumpf, David K. *Titan II: A History of a Cold War Missile Program*. Fayetteville, Ark.: University of Arkansas Press, 2000.

282. Swenson, Jr., Lloyd S., James M. Grimwood, and Charles C. Alexander. *This New Ocean: A History of Project Mercury*. Washington, D.C.: NASA SP-4201 (1966).

283. Taubes, G. "The Art of the Orbit." *Science* 283 (1999): 620–22. A review of unusual orbits for solar-system exploration.

284. Taylor, Edwin F., and John Archibald Wheeler. *Spacetime Physics: Introduction to Special Relativity*. New York: W. H. Freeman, 1992.

285. Thompson, Milton O. and Neil Armstrong. *At the Edge of Space: The X-15 Flight Program*. Washington, D.C.: Smithsonian Institution Press, 1992.

286. Thompson, Milton O., and Curtis Peebles. *Flying Without Wings: NASA Lifting Bodies and the Birth of the Space Shuttle*. Washington, D.C.: Smithsonian Institution Press, 1999.

287. Thorne, Kip S. *Black Holes and Time Warps*. New York: Norton and Co., 1994.

288. Tikhonravov, M. K. *Raketnaya Tekhniya* (Rocket Technology). Moscow: ONTI, 1935.

289. Titov, G., and M. I. Caiden. *I Am Eagle!* Indianapolis: Bobbs-Merrill, 1962.

290. Tribble, Alan C. *The Space Environment*. Princeton, N.J.: Princeton University Press, 1995.

291. Tsandr, Fridrikh Arturovich. *Probleme Poleta Pri Pomoshchi Raketnykh Apparatov* (Problems of Flight by Jet Propulsion), 1932. English translation by Israel Program for Scientific Translations (Jerusalem), 1964.

292. Tsiolkovsky, K. E. *Dreams of Earth and Heaven, Nature and Man*. Moscow, 1895.

293. Tsiolkovsky, K. E. "The Rocket into Cosmic Space," Na-ootchnoye Obozreniye, Science Survey, Moscow, 1903.

294. Tucker, Wallace H., and Karen Tucker. *Revealing the Universe: The Making of the Chandra X-Ray Observatory*. New York: Harvard University Press, 2001.

295. Turner, Sarah H. "Sam Phillips: One Who Led Us to the Moon." *NASA Activities,* May/June 1990, 18–19.

296. Valier, Max. *Raketenfahrt*. Munich: R. Oldenbourg, 1930.

297. Van Allen, James A. *Origins of Magnetospheric Physics*. Washington, D.C.: Smithsonian Institution Press, 1983.

298. Vaughan, Diane. *The Challenger Launch Decision: Risky Technology, Culture, and Deviance at NASA*. Chicago: University of Chicago Press, 1997.

299. Verne, Jules. *De la terre à la lune*. Paris: J. Hetzel, 1866.

300. Verne, Jules. *Autour de la lune.* Paris: J. Hetzel, 1872. See *From the Earth to the Moon and Round the Moon.* London: Sampson, Low, Marston & Co., 1902.

301. Vilenkin, A. "Theory of Cosmic Strings." *Astrophysical Journal* 282 (1984): L51.

302. Visser, M. *Lorentzian Wormholes: From Einstein to Hawking.* New York: Springer Verlag, 1996.

303. "Vladimir S. Syromiatnikov, Russian Docking System Engineer." *Space News,* February 12–18, 1996, 22.

304. von Braun, W. "Crossing the Last Frontier." *Collier's,* March 22, 1952.

305. von Braun, Wernher. *The Mars Project.* Urbana, Ill.: University of Illinois Press, 1953; reprint ed., 1991.

306. von Braun, Wernher, and Frederick I. Ordway, III. *History of Rocketry and Space Travel.* New York: Thomas Crowell, 1966.

307. von Pirquet, G. Various articles, *Die Rakete,* vol. II, 1928.

308. Wachhorst, Wyn. *Dreams of Spaceflight: Essays on the Near Edge of Infinity.* New York: Basic Books, 2000.

309. Wald, Robert M. *General Relativity.* Chicago: University of Chicago Press, 1984.

310. Wegener, Peter P. *The Peenemünde Wind Tunnels: A Memoir.* New Haven, Conn.: Yale University Press, 1996.

311. Weinberg, Steven. *Gravitation and Cosmology: Principles and Applications of the General Theory of Relativity.* New York: John Wiley & Sons, 1972.

312. Wells, H. G. *The First Men in the Moon.* London: Newnes (1901).

313. Wilcox, D. "The Voyage that Lasted Six Hundred Years." *Amazing Stories,* October 1940.

314. Winter, Frank H. *The First Golden Age of Rocketry: Congreve and Hale Rockets of the Nineteenth Century.* Washington, D.C.: Smithsonian Institution Press, 1990.

315. Wolf, Mervin J. *Space Pioneers: The Illustrated History of the Jet Propulsion Laboratory and the Race to Space.* Santa Monica, Calif.: General Publishing Group (1999).

316. Wolkomir, Richard. "Shooting right for the stars with one gargantuan gas gun." *Smithsonian,* January 1996, 84.

317. Yasinsky, Alexander. "The N-1 Rocket Programme." *Spaceflight,* July 1993.

318. Yeager, Chuck. *Yeager.* New York: Bantam Books, 1982.

319. Zimmerman, Robert. *Genesis: The Story of Apollo 8: The First Manned Flight to Another World.* New York: Four Walls Eight Windows, 1998.

320. Zubrin, R. M., and D. G. Andrews. "Magnetic Sails and Interplanetary Travel." AIAA 89-2441, *AIAA/ASME 25th Joint Propulsion Conference,* Monterey, Calif., July 10–12, 1989.

321. Zubrin, Robert M. "The Magnetic Sail." *Analog,* May 1992.

322. Zubrin, Robert, and Richard Wagner. *The Case for Mars: The Plan to Settle the Red Planet and Why We Must.* New York: Simon & Schuster, 1997.

Web Sites

ACCESS (Advanced Cosmic-ray Composition Experiment on the Space Station)
http://lheawww.gsfc.nasa.gov/ACCESS/

ACE (Advanced Composition Explorer)
http://www.srl.caltech.edu/ACE/

ACRIMSAT (Active Cavity Radiometer Irradiance Monitor Satellite)
http://acrim.jpl.nasa.gov/

ADEOS (Advanced Earth Observing Satellite)
http://hdsn.eoc.nasda.go.jp/guide/satellite/satdata/adeos_e.html

AIM (Aeronomy of Ice in the Mesosphere)
http://www.sdl.usu.edu/programs/aim/index

AIRS (Atmospheric Infrared Sounder)
http://www-airs.jpl.nasa.gov/

ALEXIS (Array of Low Energy X-ray Imaging Sensors)
http://alexis-www.lanl.gov/

American Astronautical Society (AAS)
http://www.astronautical.org/

American Institute of Aeronautics and Astronautics (AIAA)
http://www.aiaa.org/

Ames Research Center
http://www.arc.nasa.gov/

AMS (Alpha Magnetic Spectrometer)
http://ams.cern.ch/AMS/ams_homepage.html

Andoya Rocket Range
http://www.rocketrange.no/

Angara
http://www.intertec.co.at/itc2/partners/KHRUNICHEV/angara.htm

Aqua
http://aqua.gsfc.nasa.gov/

ARGOS (Advanced Research and Global Observation Satellite)
http://www.te.plk.af.mil/stp/argos/argos.html

Ariane
http://www.esa.int/export/esaLA/

ARISE (Advanced Radio Interferometry between Space and Earth)
http://arise.jpl.nasa.gov/

ARTEMIS (Advanced Relay Technology Mission)
http://envisat.esa.int/support-docs/artemis/artemis.html

ASCE (Advanced Spectroscopic and Coronagraphic Explorer)
http://cfa-www.harvard.edu/asce/

ASI (Agenzia Spaziale Italiano)
http://www.asi.it/

Athena (launch vehicle)
http://www.ast.lmco.com/launch_athena.shtml

Atlas
http://www.ast.lmco.com/launch_atlas.shtml

Aura
http://eos-chem.gsfc.nasa.gov/

Australian Space Research Institute
http://www.asri.org.au/

Baikonur Cosmodrome
http://www.russianspaceweb.com/baikonur.html

Beagle 2
http:www.beagle2.com

BepiColombo
http://sci.esa.int/home/bepicolombo/index.cfm

BeppoSAX
http://bepposax.gsfc.nasa.gov/bepposax/

BIRD (Bi-spectral Infrared Detection satellite)
http://spacesensors.dlr.de/SE/bird/

Black Brant
http://www.bristol.ca/BlackBrant.html

Breakthrough Propulsion Physics Program (BPPP)
http://www.grc.nasa.gov/WWW/bpp/

British Interplanetary Society (BIS)
http://www.bis-spaceflight.com/

British National Space Center (BNSC)
http://www.bnsc.gov.uk/

CALIPSO
http://smsc.cnes.fr/CALIPSO/Satellite.html

Canadian Space Agency (CSA)
http://www.space.gc.ca/

Canadian Space Society (CSS)
http://css.ca/

Cassini/Huygens
http://www.jpl.nasa.gov/cassini/
http://sci.esa.int/huygens/

CATSat (Cooperative Astrophysics and Technology Satellite)
http://www.catsat.sr.unh.edu/

Celestis
http://www.celestis.com/

Center for Gravitational Biology Research
http://lifesci.arc.nasa.gov/CGBR/CGBR.html

CHAMP (Challenging Minisatellite Payload)
http://op.gfz-potsdam.de/champ/

Chandra X-ray Observatory
http://chandra.harvard.edu/

China Great Wall Industry Corporation (CGWIC)
http://www.cgwic.com/

CHIPS (Cosmic Hot Interstellar Plasma Spectrometer)
http://chips.ssl.berkeley.edu/chips.html

CloudSat
http://cloudsat.atmos.colostate.edu/

Cluster
http://sci.esa.int/cluster/

CNES (Centre National d'Etudes Spatiales)
http://www.cnes.fr/

Constellation-X
http://constellation.gsfc.nasa.gov/

CONTOUR (Comet Nucleus Tour)
http://www.contour2002.org/

Coriolis
http://www.te.plk.af.mil/vo/missions/coriolis/coriolis.html

COROT (Convection Rotation and Planetary Transit)
http://www.astrsp-mrs.fr/projets/corot/pagecorot.html

COSPAR (Committee on Space Research)
http://www.cosparhq.org/

Darwin
http://ast.star.rl.ac.uk/darwin/

Dawn
http://www-ssc.igpp.ucla.edu/dawn/

Deep Impact
http://www.ss.astro.umd.edu/deepimpact/

Deep Space 1
http://nmp.jpl.nasa.gov/ds1/

Deep Space Network
http://deepspace.jpl.nasa.gov/dsn/

Delta
http://www.boeing.com/defense-space/space/delta/

Discovery program
http://discovery.nasa.gov/

DMSP (Defense Meteorological Satellite Program)
http://www.ngdc.noaa.gov/dmsp/dmsp.html

Dryden Flight Research Center
http://www.dfrc.nasa.gov/

DSCS (Defense Satellite Communications System)
http://www.af.mil/news/factsheets/Defense_Satellite_Communicati.html

EchoStar
http://www.dishnetwork.com/content/aboutus/index.shtml

Edwards Air Force Base
http://www.edwards.af.mil/

Energia Rocket & Space Corporation (Energia RSC)
http://www.energia.ru/english/

Envisat
http://envisat.esa.int/

EO (Earth Observing) satellites
http://eo1.gsfc.nasa.gov/
http://nmp.jpl.nasa.gov/eo3/index.html

Equator-S
http://www.mpe-garching.mpg.de/EQS/eqs_home.html

ERS (Earth Resources Satellite)
http://earth.esa.int/ers/

ESA (European Space Agency)
http://www.esa.int/

ESOC (European Space Operations Center)
http://www.esoc.esa.de/pr/index.php3

ESRIN (European Space Research Institute)
http://www.esa.int/export/esaCP/GGGYA78RVDC_
index.0.html

ESTEC (European Space Research and Technology
Centre)
http://www.estec.esa.nl/

Eumetsat (European Meteorological Satellite
Organisation)
http://www.eumetsat.de/

Europa Orbiter
http://www.jpl.nasa.gov/europaorbiter/

Eurospace
http://www.eurospace.org/

Eutelsat (European Telecommunications Satellite
Organisation)
http://www.eutelsat.org/home/index.html

Experimental Rocket Propulsion Society (ERPS)
http://www.erps.org/

Explorer
http://explorers.gsfc.nasa.gov/

FAME (Full-Sky Astrometric Mapping Explorer)
http://www.usno.navy.mil/FAME/

Fastrac Engine
http://www.msfc.nasa.gov/NEWSROOM/background
/facts/fastrac.htm

FIRST (Far Infrared and Submillimetre Space Telescope)
See Herschel

FORTE (Fast On-orbit Recording of Transient Events)
http://nis-www.lanl.gov/nis-projects/forte/

FUSE (Far Ultraviolet Spectroscopic Explorer)
http://fuse.pha.jhu.edu/

Gagarin Cosmonaut Training Center
http://howe.iki.rssi.ru/GCTC/gctc_e.htm

GAIA (Global Astrometric Interferometer for
Astrophysics)
http://astro.estec.esa.nl/GAIA/

GALEX (Galaxy Evolution Explorer)
http://www.srl.caltech.edu/galex/

Galileo
http://www.jpl.nasa.gov/galileo/

GCOM (Global Change Observing Mission)
http://www.eorc.nasda.go.jp/EORC/Satellites/#gcom

Genesis
http://genesismission.jpl.nasa.gov/

Geotail
http://www-spof.gsfc.nasa.gov/istp/geotail/

GFO (Geosat Follow-On)
http://gfo.wff.nasa.gov/

GIFTS (Geostationary Imaging Fourier Transform
Spectrometer)
http://www.sdl.usu.edu/programs/gifts/

GLAST (Gamma-ray Large Area Space Telescope)
http://glast.gsfc.nasa.gov/

Glenn Research Center
http://www.grc.nasa.gov/

Globalstar
http://www.globalstar.com/

Goddard Institute for Space Studies
http://www.giss.nasa.gov/

Goddard Space Flight Center (GSFC)
http://www.gsfc.nasa.gov/

GPS (Global Positioning System)
http://www.colorado.edu/geography/gcraft/notes/gps
/gps_f.html

GRACE (Gravity Recovery and Climate Experiment)
http://www.csr.utexas.edu/grace/

Gravity Probe B
http://einstein.stanford.edu/

Guiana Space Centre
http://www.cnes.fr/cnes/etablissements/en
/centres_csg.htm

Haughton-Mars Project
http://www.arctic-mars.org/

Herschel Space Observatory
http://astro.estec.esa.int/First/

HETE (High Energy Transient Explorer)
http://space.mit.edu/HETE/

Hipparcos
http://astro.estec.esa.nl/Hipparcos/

HNX (Heavy Nuclei Explorer)
http://cosray2.wustl.edu/hnx/

Hubble Space Telescope (HST)
http://www.stsci.edu/hst/

Human Research Facility (HRF)
http://hrf.jsc.nasa.gov/

ICESat (Ice, Cloud, and Land Elevation Satellite)
http://icesat.gsfc.nasa.gov/

ICO
http://www.ico.com/

IEH (International Extreme Ultraviolet Hitchhiker)
http://sspp.gsfc.nasa.gov/hh/ieh3/index.html

Ikonos
http://www.spaceimaging.com/

IMAGE (Imager for Magnetopause-to-Aurora Global Exploration)
http://pluto.space.swri.edu/IMAGE/

IMEX (Inner Magnetosphere Explorer)
http://ham.space.umn.edu/spacephys/imex.html

Indian Space Research Organisation (ISRO)
http://www.isro.org/

Inmarsat
http://www.inmarsat.com/

INPE (Instituto Nacional de Pesquisas Espacias)
http://www.inpe.br/

INTEGRAL (International Gamma-ray Astrophysics Laboratory)
http://astro.esa.int/SA-general/Projects/Integral/integral.html

Intelsat (International Telecommunications Satellite Organization)
http://www.intelsat.int/

International Astronautical Federation (IAF)
http://www.iafastro.com/

International Launch Services (ILS)
http://www.ilslaunch.com/

International Space Science Institute
http://www.issi.unibe.ch/

International Space Station (ISS)
http://spaceflight.nasa.gov/station/index.html

Interstellar Probe
http://interstellar.jpl.nasa.gov/

Iridium
http://www.iridium.com/

IRIS (Infrared Imaging Surveyor)
http://www.ir.isas.ac.jp/ASTRO-F/index-e.html

ISAS (Institute of Space and Astronautical Sciences)
http://www.isas.ac.jp/e/index.html

ISO (Infrared Space Observatory)
http://isowww.estec.esa.nl/

Jason
http://topex-www.jpl.nasa.gov/mission/jason-1.html

Jodrell Bank
http://www.jb.man.ac.uk/

Johnson Space Center
http://www.jsc.nasa.gov/

JPL (Jet Propulsion Laboratory)
http://www.jpl.nasa.gov/

Kennedy Space Center
http://www.ksc.nasa.gov/

Kepler
http://www.kepler.arc.nasa.gov/

Kodiak Launch Complex
http://www.akaerospace.com/

Landsat
http://geo.arc.nasa.gov/sge/landsat/landsat.html

Langley Research Center
http://www.larc.nasa.gov/

Life Finder (LF)
http://origins.jpl.nasa.gov/missions/lf.html

LISA (Laser Interferometer Space Antenna)
http://lisa.jpl.nasa.gov/

Lunar Prospector
http://lunar.arc.nasa.gov/

MagConst (Magnetospheric Constellation)
http://stp.gsfc.nasa.gov/missions/mc/mc.htm

Magellan
http://www.jpl.nasa.gov/magellan/

MAP (Microwave Anisotropy Probe)
http://map.gsfc.nasa.gov/

Mars 2003 Rovers
http://mars.jpl.nasa.gov/missions/future/2003.html

Mars 2005+
http://mars.jpl.nasa.gov/missions/future/2005-plus.html

Mars Express
http://sci.esa.int/marsexpress/

Mars Global Surveyor (MGS)
http://mpfwww.jpl.nasa.gov/mgs/

Mars Odyssey, 2001
http://mpfwww.jpl.nasa.gov/odyssey/

Marshall Space Flight Center (MSFC)
http://www.msfc.nasa.gov/

MDS (Mission Demonstration test Satellite)
http://www.nasda.go.jp/sat/mds/index_e.html

MESSENGER (Mercury Surface, Space Environment, Geochemistry and Ranging)
http://messenger.jhuapl.edu/

MetOp (Meteorological Operational) satellites
http://earth.esa.int/METOP.html

MiniSat
http://laeff.esa.es/~trapero/EURD/minisat.html

Minotaur
http://www.orbital.com/LaunchVehicles/Minotaur/minotaur.htm

MOST (Microvariability and Oscillations of Stars)
http://www.astro.ubc.ca/MOST/

MTI (Multispectral Thermal Imager)
http://nis-www.lanl.gov/nis-projects/mti/

MUSES-C
http://www.isas.ac.jp/e/enterp/missions/muses-c/cont.html

NASA (National Aeronautics and Space Administration)
http://www.nasa.gov/

NASA Institute for Advanced Concepts (NIAC)
http://www.niac.usra.edu/

NASDA (National Space Development Agency)
http://www.nasda.go.jp/index_e.html

National Association of Rocketry (NAR)
http://www.nar.org/

National Space Science Data Center (NSSDC)
http://nssdc.gsfc.nasa.gov/

National Space Society (NSS)
http://www.nss.org

NEAR-Shoemaker (Near Earth Asteroid Rendezvous)
http://near.jhuapl.edu/

NetLander
http://discovery.nasa.gov/netlander.html

NGSS (Next Generation Sky Survey)
http://www.astro.ucla.edu/~wright/NGSS/

NGST (Next Generation Space Telescope)
http://www.stsci.edu/ngst/

Nozomi
http://www.isas.ac.jp/e/enterp/missions/nozomi/cont.html

ONERA (Office National d'Etudes et de Recherches Aérospatiale)
http://www.onera.fr/

OPAL (Orbiting PicoSat Launcher)
http://ssdl.stanford.edu/opal/

Orbcomm
http://www.orbcomm.com

OrbView
http://www.orbital.com/SciTechSatellites/OrbView-3/

Origins program
http://origins.jpl.nasa.gov

Pegasus (launch vehicle)
http://www.orbital.com/LaunchVehicles/Pegasus/pegasus.htm

Pioneer
http://spaceprojects.arc.nasa.gov/Space_Projects/pioneer/PNhome.html

Planck
http://astro.estec.esa.nl/Planck/

Planet Imager
http://origins.jpl.nasa.gov/missions/pi.html

Planetary Society
http://www.planetary.org

POES (Polar Operational Environmental Satellite)
http://poes.gsfc.nasa.gov/

Polar
http://www-istp.gsfc.nasa.gov/istp/polar/polar.html

QuikScat (Quick Scatterometer)
http://winds.jpl.nasa.gov/missions/quikscat/quikindex.html

RADARSAT
http://www.space.gc.ca/csa_sectors/earth_environment/radarsat/default.asp

Reaction Research Society (RRS)
http://www.rrs.org/

RHESSI (Reuven Ramaty High Energy Solar Spectroscopic Imager)
http://hesperia.gsfc.nasa.gov/hessi/

Rockot
http://www.eurockot.com/

Rosetta
http://www.esoc.esa.de/external/mso/roseta.html

Russian Aviation and Space Agency (RKA)
http://www.rosaviakosmos.ru/english/eindex.htm

RXTE (Rossi X-ray Timing Explorer)
http://heasarc.gsfc.nasa.gov/docs/xte/xte_1st.html

Seawinds
http://eospso.gsfc.nasa.gov/eos_homepage/Instruments/SeaWinds/

SEDS (Students for the Exploration and Development of Space)
http://seds.lpl.arizona.edu/seds/

SELENE (Selenological and Engineering Explorer)
http://www.isas.ac.jp/e/enterp/missions/selene/cont.html

SIM (Space Interferometry Mission)
http://sim.jpl.nasa.gov/

SIRTF (Space Infrared Telescope Facility)
http://sirtf.caltech.edu/

SMART (Small Missions for Advanced Research in Technology)
http://www.estec.esa.nl/spdwww/smart1/html/overview.html

SNOE (Student Nitric Oxide Explorer)
http://lasp.colorado.edu/snoe/

SOFIA (Stratospheric Observatory for Infrared Astronomy)
http://sofia.arc.nasa.gov/

SOHO (Solar and Heliospheric Observatory)
http://sohowww.nascom.nasa.gov/

Solar-B
http://science.nasa.gov/ssl/pad/solar/solar-b.stm

SORCE (Solar Radiation and Climate Experiment)
http://lasp.colorado.edu/sorce/

Space Foundation
http://www.spacefoundation.org

Space Frontier Foundation
http://www.space-frontier.org/

Space Shuttle
http://spaceflight.nasa.gov/shuttle/

Space Studies Board
http://www.nationalacademies.org/ssb/

Space Studies Institute (SSI)
http://www.ssi.org/

Spaceguard Foundation
http://spaceguard.ias.rm.cnr.it/SGF/

SPIDR (Spectroscopy and Photometry of the Intergalactic Medium's Diffuse Radiation)
http://www.bu.edu/spidr/indextoo.html

SPOT (Satellite Probatoire d'Observation de la Terre)
http://www.spot.com/

Stardust
http://stardust.jpl.nasa.gov/

StarLight
http://starlight.jpl.nasa.gov/

Stennis Space Center (SSC)
http://www.ssc.nasa.gov/

STEP (Satellite Test of the Equivalence Principle)
http://einstein.stanford.edu/STEP/

STEREO (Solar-Terrestrial Relations Observatory)
http://sd-www.jhuapl.edu/STEREO/

SWAS (Submillimeter Wave Astronomy Satellite)
http://cfa-www.harvard.edu/swas/

Swift Gamma-Ray Burst Explorer
http://swift.gsfc.nasa.gov/

Taurus
http://www.orbital.com/LaunchVehicles/Taurus/taurus.htm

TDRSS (Tracking and Data Relay Satellite System)
http://nmsp.gsfc.nasa.gov/tdrss/

Teledesic
http://www.teledesic.com

Terra
http://terra.nasa.gov/

Terrestrial Planet Finder (TPF)
http://planetquest.jpl.nasa.gov/TPF/tpf_index.html

Three-Corner Constellation
http://threecornersat.jpl.nasa.gov/

TIMED (Thermosphere Ionosphere Mesosphere Energetic Dynamics satellite)
http://www.timed.jhuapl.edu/

Titan
http://www.ast.lmco.com/launch_titan.shtml

TOMS (Total Ozone Mapping Spectrometer)
http://toms.gsfc.nasa.gov/

TOPEX-Poseidon
http://topex-www.jpl.nasa.gov/

TRACE (Transition Region and Coronal Explorer)
http://sunland.gsfc.nasa.gov/smex/trace/index.html

TRMM (Tropical Rainfall Measuring Mission)
http://trmm.gsfc.nasa.gov/

TWINS (Two Wide-Angle Imaging Neutral-atom Spectrometers)
http://nis-www.lanl.gov/nis-projects/twins/

UARS (Upper Atmosphere Research Satellite)
http://svs.gsfc.nasa.gov/stories/UARS/

Ultra Long-Duration Balloon (ULDB) Project
http://www.wff.nasa.gov/~code820/

Ulysses
http://ulysses.jpl.nasa.gov/

USRA (Universities Space Research Association)
http://www.usra.edu/

Vandenberg Air Force Base (VAFB)
http://www.vandenberg.af.mil/

VASIMR (Variable Specific Impulse Magnetoplasma Rocket)
http://www.nasatech.com/Briefs/Sep01/MSC23041.html

VCL (Vegetation Canopy Lidar)
http://essp.gsfc.nasa.gov/vcl/

Voyager
http://voyager.jpl.nasa.gov/

VSOP (VLBI Space Observatory Program)
http://www.vsop.isas.ac.jp/

Wallops Island Flight Facility
http://www.wff.nasa.gov/

White Sands Missile Range
http://www.wsmr.army.mil/

Wind
http://www-istp.gsfc.nasa.gov/istp/wind/

WIRE (Wide-Field Infrared Explorer)
http://sunland.gsfc.nasa.gov/smex/wire/

X-38
http://www.dfrc.nasa.gov/Projects/X38/

X-43A
http://www.orbital.com/LaunchVehicles/Hyper-X/index.html

XMM-Newton
http://xmm.vilspa.esa.es/

Category Index

Advanced propulsion systems and related concepts

Alcubierre Warp Drive
antigravity
antimatter propulsion
arcjet
beamed-energy propulsion
black hole
Breakthrough Propulsion Physics Program (BPPP)
Bussard interstellar ramjet. *See* interstellar ramjet
contact ion thruster
Cosmos 1
electric propulsion
electromagnetic propulsion
electron bombardment thruster
electrostatic propulsion
electrothermal propulsion
exotic matter
faster-than-light (FTL) travel
generation ship
Hall effect thruster
helicon thruster
ICAN (Ion Compressed Antimatter Nuclear) rocket. *See* antimatter propulsion
interstellar ark. *See* generation ship
interstellar ramjet
ion propulsion
Kiwi
laser propulsion
   Off-board Laser Propulsion
   Laser-powered launching
      Laser lightsails
   Onboard Laser Propulsion
light barrier
Lorentz force turning
M2P2 (Mini-Magnetospheric Plasma Propulsion)
magnetic levitation launch-assist
magnetic sail. *See* space sail
magnetoplasmadynamic (MPD) thruster
microwave plasma thruster
NERVA (Nuclear Engine for Rocket Vehicle Application)
nuclear power for spacecraft
nuclear propulsion

nuclear pulse rocket
Nuclear Rocket Development Station (NRDS)
nuclear-electric propulsion (NEP)
1*g* spacecraft
Orion, Project
photon propulsion
pulse detonation engine
pulsed-plasma thruster
Quantum Vacuum Forces Project
RAIR (ram-augmented interstellar rocket). *See* interstellar ramjet
reactionless drive
relativistic effects
resistojet
SERT (Space Electric Rocket Test)
solar-electric propulsion (SEP)
space ark. *See* generation ship
space drive. *See* reactionless drive
space sail
star drive. *See* reactionless drive
starship
subluminal
superluminal
tachyon
thermionic
time dilation. *See* relativistic effects
warp drive
vacuum energy drive
wormhole
   Artificial Wormholes
   Natural Wormholes
XIPS (xenon-ion propulsion system)
VASIMR (Variable Specific Impulse Magnetoplasma Rocket)
zero point energy (ZPE)

See also: **Interstellar and precursor missions**
Physics and astronomy

Aerodynamics and aeronautics

aeroballistics
aerobraking
aerocapture
aerodynamics
aeronautics
aerospace

aerothermodynamic border
air-breathing engine
airfoil
airframe
anacoustic zone
angle of attack
atmospheric braking. *See* aerobraking
atmospheric trajectory
buffeting
cavitation
center of pressure
coefficient of drag
cone stability
control surface
drag
duct
duct propulsion
dynamic behavior
dynamic load
dynamic pressure
dynamic response
dynamic stability
elevon
fin
flap
flight path
flight path angle
frontal area
glide
glide angle
glide path
heat barrier
hypersonic
jet steering
lift
lift-drag ratio
Mach number
maneuverability
propulsion
pulse-jet
ramjet
regimes of flight
scramjet
sonic speed
sound barrier
speed of sound
stub fins
subsonic